Engineering Digital Design

Second Edition

by

RICHARD F. TINDER
School of Electrical Engineering and Computer Science
Washington State University
Pullman, Washington

ACADEMIC PRESS
A Harcourt Science and Technology Company

SAN DIEGO / SAN FRANCISCO / NEW YORK / BOSTON / LONDON / SYDNEY / TOKYO

This book is printed on acid-free paper. ∞

Copyright © 2000 by Academic Press

All rights reserved.
No part of this publication may be reproduced or transmitted in any form or by any means, electronic or mechanical, including photocopy, recording, or any information storage and retrieval system, without permission in writing from the publisher.

Requests for permission to make copies of any part of the work should be mailed to the following address: Permissions Department, Harcourt, Inc., 6277 Sea Harbor Drive, Orlando, Florida, 32887-6777.

Explicit permission from Academic Press is not required to reproduce a maximum of two figures or tables from an Academic Press article in another scientific or research publication provided that the material has not been credited to another source and that full credit to the Academic Press article is given.

ACADEMIC PRESS
A Harcourt Science and Technology Company
525 B Street, Suite 1900, San Diego, CA 92101-4495, USA
http://www.academicpress.com

Academic Press
24–28 Oval Road, London NW1 7DX, UK
http://www.hbuk.co.uk/ap/

Library of Congress Catalog Number: 99-066780

ISBN: 0-12-691295-5

Printed in the United States of America
99 00 01 02 03 04 MV 9 8 7 6 5 4 3 2 1

This book is lovingly dedicated to my partner in life, Gloria

For the sake of persons of different types, scientific truth should be presented in different forms, and should be regarded as equally scientific, whether it appears in the robust form and the vivid coloring of a physical illustration, or in the tenuity and paleness of a symbolic expression.

James Clerk Maxwell
Address to the Mathematics and Physical Section,
British Association of Sciences, 1870

Contents

Preface		xix
1.	**Introductory Remarks and Glossary**	1
	1.1 What Is So Special about Digital Systems?	1
	1.2 The Year 2000 and Beyond?	3
	1.3 A Word of Warning	5
	1.4 Glossary of Terms, Expressions, and Abbreviations	5
2.	**Number Systems, Binary Arithmetic, and Codes**	31
	2.1 Introduction	31
	2.2 Positional and Polynomial Representations	32
	2.3 Unsigned Binary Number System	33
	2.4 Unsigned Binary Coded Decimal, Hexadecimal, and Octal	34
	2.4.1 The BCD Representation	34
	2.4.2 The Hexadecimal and Octal Systems	36
	2.5 Conversion between Number Systems	37
	2.5.1 Conversion of Integers	38
	2.5.2 Conversion of Fractions	40
	2.6 Signed Binary Numbers	43
	2.6.1 Signed-Magnitude Representation	44
	2.6.2 Radix Complement Representation	45
	2.6.3 Diminished Radix Complement Representation	48
	2.7 Excess (Offset) Representations	49
	2.8 Floating-Point Number Systems	49
	2.9 Binary Arithmetic	52
	2.9.1 Direct Addition and Subtraction of Binary Numbers	52
	2.9.2 Two's Complement Subtraction	53
	2.9.3 One's Complement Subtraction	54
	2.9.4 Binary Multiplication	55
	2.9.5 Binary Division	58
	2.9.6 BCD Addition and Subtraction	62
	2.9.7 Floating-Point Arithmetic	64
	2.9.8 Perspective on Arithmetic Codes	67
	2.10 Other Codes	68
	2.10.1 The Decimal Codes	68
	2.10.2 Error Detection Codes	69
	2.10.3 Unit Distance Codes	70
	2.10.4 Character Codes	70
	Further Reading	72
	Problems	72

3. Background for Digital Design — 79

- 3.1 Introduction — 79
- 3.2 Binary State Terminology and Mixed Logic Notation — 79
  - 3.2.1 Binary State Terminology — 79
- 3.3 Introduction to CMOS Terminology and Symbology — 82
- 3.4 Logic Level Conversion: The Inverter — 83
- 3.5 Transmission Gates and Tri-State Drivers — 84
- 3.6 AND and OR Operators and Their Mixed-Logic Circuit Symbology — 87
  - 3.6.1 Logic Circuit Symbology for AND and OR — 87
  - 3.6.2 NAND Gate Realization of Logic AND and OR — 88
  - 3.6.3 NOR Gate Realization of Logic AND and OR — 89
  - 3.6.4 NAND and NOR Gate Realization of Logic Level Conversion — 90
  - 3.6.5 The AND and OR Gates and Their Realization of Logic AND and OR — 92
  - 3.6.6 Summary of Logic Circuit Symbols for the AND and OR Functions and Logic Level Conversion — 94
- 3.7 Logic Level Incompatibility: Complementation — 95
- 3.8 Reading and Construction of Mixed-Logic Circuits — 97
- 3.9 XOR and EQV Operators and Their Mixed-Logic Circuit Symbology — 98
  - 3.9.1 The XOR and EQV Functions of the XOR Gate — 100
  - 3.9.2 The XOR and EQV Functions of the EQV Gate — 100
  - 3.9.3 Multiple Gate Realizations of the XOR and EQV Functions — 101
  - 3.9.4 The Effect of Active Low Inputs to the XOR and EQV Circuit Symbols — 102
  - 3.9.5 Summary of Conjugate Logic Circuit Symbols for XOR and EQV Gates — 103
  - 3.9.6 Controlled Logic Level Conversion — 103
  - 3.9.7 Construction and Waveform Analysis of Logic Circuits Containing XOR-Type Functions — 104
- 3.10 Laws of Boolean Algebra — 105
  - 3.10.1 NOT, AND, and OR Laws — 106
  - 3.10.2 The Concept of Duality — 107
  - 3.10.3 Associative, Commutative, Distributive, Absorptive, and Consensus Laws — 108
  - 3.10.4 DeMorgan's Laws — 110
- 3.11 Laws of XOR Algebra — 111
  - 3.11.1 Two Useful Corollaries — 114
  - 3.11.2 Summary of Useful Identities — 115
- 3.12 Worked Examples — 116
- Further Reading — 120
- Problems — 121

4. Logic Function Representation and Minimization — 131

- 4.1 Introduction — 131
- 4.2 SOP and POS Forms — 131
  - 4.2.1 The SOP Representation — 131
  - 4.2.2 The POS Representation — 134
- 4.3 Introduction to Logic Function Graphics — 137
  - 4.3.1 First-Order K-maps — 138
  - 4.3.2 Second-Order K-maps — 138
  - 4.3.3 Third-Order K-maps — 140
  - 4.3.4 Fourth-Order K-maps — 143

	4.4	Karnaugh Map Function Minimization	144
		4.4.1 Examples of Function Minimization	146
		4.4.2 Prime Implicants	148
		4.4.3 Incompletely Specified Functions: Don't Cares	150
	4.5	Multiple Output Optimization	152
	4.6	Entered Variable K-map Minimization	158
		4.6.1 Incompletely Specified Functions	162
	4.7	Function Reduction of Five or More Variables	165
	4.8	Minimization Algorithms and Application	169
		4.8.1 The Quine–McCluskey Algorithm	169
		4.8.2 Cube Representation and Function Reduction	173
		4.8.3 Qualitative Description of the Espresso Algorithm	173
	4.9	Factorization, Resubstitution, and Decomposition Methods	174
		4.9.1 Factorization	175
		4.9.2 Resubstitution Method	176
		4.9.3 Decomposition by Using Shannon's Expansion Theorem	177
	4.10	Design Area vs Performance	180
	4.11	Perspective on Logic Minimization and Optimization	181
	4.12	Worked EV K-map Examples	181
		Further Reading	188
		Problems	189
5.	**Function Minimization by Using K-map XOR Patterns and Reed–Muller Transformation Forms**		**197**
	5.1	Introduction	197
	5.2	XOR-Type Patterns and Extraction of Gate-Minimum Cover from EV K-maps	198
		5.2.1 Extraction Procedure and Examples	200
	5.3	Algebraic Verification of Optimal XOR Function Extraction from K-maps	204
	5.4	K-map Plotting and Entered Variable XOR Patterns	205
	5.5	The SOP-to-EXSOP Reed–Muller Transformation	207
	5.6	The POS-to-EQPOS Reed–Muller Transformation	208
	5.7	Examples of Minimum Function Extraction	209
	5.8	Heuristics for CRMT Minimization	217
	5.9	Incompletely Specified Functions	218
	5.10	Multiple Output Functions with Don't Cares	222
	5.11	K-map Subfunction Partitioning for Combined CRMT and Two-Level Minimization	225
	5.12	Perspective on the CRMT and CRMT/Two-Level Minimization Methods	229
		Further Reading	229
		Problems	230
6.	**Nonarithmetic Combinational Logic Devices**		**237**
	6.1	Introduction and Background	237
		6.1.1 The Building Blocks	237
		6.1.2 Classification of Chips	238
		6.1.3 Performance Characteristics and Other Practical Matters	238
		6.1.4 Part Numbering Systems	241
		6.1.5 Design Procedure	241

6.2	Multiplexers		242
	6.2.1	Multiplexer Design	242
	6.2.2	Combinational Logic Design with MUXs	245
6.3	Decoders/Demultiplexers		248
	6.3.1	Decoder Design	248
	6.3.2	Combinational Logic Design with Decoders	251
6.4	Encoders		254
6.5	Code Converters		257
	6.5.1	Procedure for Code Converter Design	257
	6.5.2	Examples of Code Converter Design	257
6.6	Magnitude Comparators		265
6.7	Parity Generators and Error Checking Systems		273
6.8	Combinational Shifters		275
6.9	Steering Logic and Tri-State Gate Applications		278
6.10	Introduction to VHDL Description of Combinational Primitives		279
	Further Reading		287
	Problems		288

7. Programmable Logic Devices — 295

7.1	Introduction		295
7.2	Read-Only Memories		295
	7.2.1	PROM Applications	299
7.3	Programmable Logic Arrays		301
	7.3.1	PLA Applications	302
7.4	Programmable Array Logic Devices		307
7.5	Mixed-Logic Inputs to and Outputs from ROMs, PLAs, and PAL Devices		310
7.6	Multiple PLD Schemes for Augmenting Input and Output Capability		312
7.7	Introduction to FPGAs and Other General-Purpose Devices		317
	7.7.1	AND–OR–Invert and OR–AND–Invert Building Blocks	317
	7.7.2	Actel Field Programmable Gate Arrays	319
	7.7.3	Xilinx FPGAs	321
	7.7.4	Other Classes of General-Purpose PLDs	328
7.8	CAD Help in Programming PLD Devices		328
	Further Reading		330
	Problems		331

8. Arithmetic Devices and Arithmetic Logic Units (ALUs) — 335

8.1	Introduction		335
8.2	Binary Adders		335
	8.2.1	The Half Adder	336
	8.2.2	The Full Adder	337
	8.2.3	Ripple-Carry Adders	338
8.3	Binary Subtractors		340
	8.3.1	Adder/Subtractors	342
	8.3.2	Sign-Bit Error Detection	343
8.4	The Carry Look-Ahead Adder		345
8.5	Multiple-Number Addition and the Carry-Save Adder		349
8.6	Multipliers		350
8.7	Parallel Dividers		353

	8.8	Arithmetic and Logic Units	357
		8.8.1 Dedicated ALU Design Featuring R-C and CLA Capability	358
		8.8.2 The MUX Approach to ALU Design	363
	8.9	Dual-Rail Systems and ALUs with Completion Signals	369
		8.9.1 Carry Look-Ahead Configuration	378
	8.10	VHDL Description of Arithmetic Devices	380
		Further Reading	383
		Problems	385
9.	**Propagation Delay and Timing Defects in Combinational Logic**	**391**
	9.1	Introduction	391
	9.2	Static Hazards in Two-Level Combinational Logic Circuits	392
	9.3	Detection and Elimination Hazards in Multilevel XOR-Type Functions	399
		9.3.1 XOP and EOS Functions	400
		9.3.2 Methods for the Detection and Elimination of Static Hazards in Complex Multilevel XOR-type Functions	403
		9.3.3 General Procedure for the Detection and Elimination of Static Hazards in Complex Multilevel XOR-Type Functions	408
		9.3.4 Detection of Dynamic Hazards in Complex Multilevel XOR-Type Functions	409
	9.4	Function Hazards	412
	9.5	Stuck-at Faults and the Effect of Hazard Cover on Fault Testability	412
		Further Reading	413
		Problems	415
10.	**Introduction to Synchronous State Machine Design and Analysis**	**419**
	10.1	Introduction	419
		10.1.1 A Sequence of Logic States	420
	10.2	Models for Sequential Machines	421
	10.3	The Fully Documented State Diagram: The Sum Rule	424
	10.4	The Basic Memory Cells	428
		10.4.1 The Set-Dominant Basic Cell	428
		10.4.2 The Reset-Dominant Basic Cell	431
		10.4.3 Combined Form of the Excitation Table	433
		10.4.4 Mixed-Rail Outputs of the Basic Cells	434
		10.4.5 Mixed-Rail Output Response of the Basic Cells	435
	10.5	Introduction to Flip-Flops	436
		10.5.1 Triggering Mechanisms	437
		10.5.2 Types of Flip-Flops	438
		10.5.3 Hierarchical Flow Chart and Model for Flip-Flop Design	438
	10.6	Procedure for FSM (Flip-Flop) Design and the Mapping Algorithm	440
	10.7	The D Flip-Flops: General	440
		10.7.1 The D-Latch	441
		10.7.2 The RET D Flip-Flop	444
		10.7.3 The Master–Slave D Flip-Flop	448
	10.8	Flip-Flop Conversion: The T, JK Flip-Flops and Miscellaneous Flip-Flops	450
		10.8.1 The T Flip-Flops and Their Design from D Flip-Flops	451
		10.8.2 The JK Flip-Flops and Their Design from D Flip-Flops	453
		10.8.3 Design of T and D Flip-Flops from JK Flip-Flops	455

	10.8.4	Review of Excitation Tables	457
	10.8.5	Design of Special-Purpose Flip-Flops and Latches	459
10.9		Latches and Flip-Flops with Serious Timing Problems: A Warning	461
10.10		Asynchronous Preset and Clear Overrides	463
10.11		Setup and Hold-Time Requirements of Flip-Flops	465
10.12		Design of Simple Synchronous State Machines with Edge-Triggered Flip-Flops: Map Conversion	466
	10.12.1	Design of a Three-Bit Binary Up/Down Counter: D-to-T K-map Conversion	466
	10.12.2	Design of a Sequence Recognizer: D-to-JK K-map Conversion	471
10.13		Analysis of Simple State Machines	476
10.14		VHDL Description of Simple State Machines	480
	10.14.1	The VHDL Behavioral Description of the RET D Flip-flop	480
	10.14.2	The VHDL Behavioral Description of a Simple FSM	481
		Further Reading	482
		Problems	483

11. Synchronous FSM Design Considerations and Applications   491

11.1		Introduction	491
11.2		Detection and Elimination of Output Race Glitches	491
	11.2.1	ORG Analysis Procedure Involving Two Race Paths	496
	11.2.2	Elimination of ORGs	496
11.3		Detection and Elimination of Static Hazards in the Output Logic	499
	11.3.1	Externally Initiated Static Hazards in the Output Logic	500
	11.3.2	Internally Initiated Static Hazards in the Output of Mealy and Moore FSMs	502
	11.3.3	Perspective on Static Hazards in the Output Logic of FSMs	509
11.4		Asynchronous Inputs: Rules and Caveats	510
	11.4.1	Rules Associated with Asynchronous Inputs	510
	11.4.2	Synchronizing the Input	511
	11.4.3	Stretching and Synchronizing the Input	512
	11.4.4	Metastability and the Synchronizer	514
11.5		Clock Skew	517
11.6		Clock Sources and Clock Signal Specifications	520
	11.6.1	Clock-Generating Circuitry	520
	11.6.2	Clock Signal Specifications	521
	11.6.3	Buffering and Gating the Clock	522
11.7		Initialization and Reset of the FSM: Sanity Circuits	522
	11.7.1	Sanity Circuits	523
11.8		Switch Debouncing Circuits	526
	11.8.1	The Single-Pole/Single-Throw Switch	526
	11.8.2	The Single-Pole/Double-Throw Switch	528
	11.8.3	The Rotary Selector Switch	529
11.9		Applications to the Design of More Complex State Machines	530
	11.9.1	Design Procedure	530
	11.9.2	Design Example: The One- to Three-Pulse Generator	532
11.10		Algorithmic State Machine Charts and State Tables	536
	11.10.1	ASM Charts	537
	11.10.2	State Tables and State Assignment Rules	539
11.11		Array Algebraic Approach to Logic Design	542

CONTENTS   xv

	11.12	State Minimization	547
		Further Reading	549
		Problems	551

12. Module and Bit-Slice Devices   561

	12.1	Introduction	561
	12.2	Registers	561
		12.2.1 The Storage (Holding) Register	562
		12.2.2 The Right Shift Register with Synchronous Parallel Load	562
		12.2.3 Universal Shift Registers with Synchronous Parallel Load	565
		12.2.4 Universal Shift Registers with Asynchronous Parallel Load	568
		12.2.5 Branching Action of a 4-Bit USR	570
	12.3	Synchronous Binary Counters	572
		12.3.1 Simple Divide-by-N Binary Counters	573
		12.3.2 Cascadable BCD Up-Counters	575
		12.3.3 Cascadable Up/Down Binary Counters with Asynchronous Parallel Load	579
		12.3.4 Binary Up/Down Counters with Synchronous Parallel Load and True Hold Capability	581
		12.3.5 One-Bit Modular Design of Parallel Loadable Up/Down Counters with True Hold	584
		12.3.6 Perspective on Parallel Loading of Counters and Registers: Asynchronous vs Synchronous	588
		12.3.7 Branching Action of a 4-Bit Parallel Loadable Up/Down Counter	589
	12.4	Shift-Register Counters	590
		12.4.1 Ring Counters	590
		12.4.2 Twisted Ring Counters	593
		12.4.3 Linear Feedback Shift Register Counters	594
	12.5	Asynchronous (Ripple) Counters	600
		Further Reading	605
		Problems	606

13. Alternative Synchronous FSM Architectures and Systems-Level Design   613

	13.1	Introduction	613
		13.1.1 Choice of Components to be Considered	613
	13.2	Architecture Centered around Nonregistered PLDs	614
		13.2.1 Design of the One- to Three-Pulse Generator by Using a PLA	615
		13.2.2 Design of the One- to Three-Pulse Generator by Using a PAL	617
		13.2.3 Design of the One- to Three-Pulse Generator by Using a ROM	618
		13.2.4 Design of a More Complex FSM by Using a ROM as the PLD	622
	13.3	State Machine Designs Centered around a Shift Register	626
	13.4	State Machine Designs Centered around a Parallel Loadable Up/Down Counter	632
	13.5	The One-Hot Design Method	636
		13.5.1 Use of ASMs in One-Hot Designs	640
		13.5.2 Application of the One-Hot Method to a Serial 2's Complementer	643
		13.5.3 One-Hot Design of a Parallel-to-Serial Adder/Subtractor Controller	645
		13.5.4 Perspective on the Use of the One-Hot Method: Logic Noise and Use of Registered PLDs	647

	13.6	System-Level Design: Controller, Data Path, and Functional Partition	649
		13.6.1 Design of a Parallel-to-Serial Adder/Subtractor Control System	651
		13.6.2 Design of a Stepping Motor Control System	655
		13.6.3 Perspective on System-Level Design in This Text	666
	13.7	Dealing with Unusually Large Controller and System-Level Designs	666
		Further Reading	668
		Problems	670
14.	**Asynchronous State Machine Design and Analysis: Basic Concepts**		**683**
	14.1	Introduction	683
		14.1.1 Features of Asynchronous FSMs	684
		14.1.2 Need for Asynchronous FSMs	685
	14.2	The Lumped Path Delay Models for Asynchronous FSMs	685
	14.3	Functional Relationships and the Stability Criteria	687
	14.4	The Excitation Table for the LPD Model	688
	14.5	State Diagrams, K-maps, and State Tables for Asynchronous FSMs	689
		14.5.1 The Fully Documented State Diagram	689
		14.5.2 Next-State and Output K-maps	690
		14.5.3 State Tables	691
	14.6	Design of the Basic Cells by Using the LPD Model	692
		14.6.1 The Set-Dominant Basic Cell	692
		14.6.2 The Reset-Dominant Basic Cell	694
	14.7	Design of the Rendezvous Modules by Using the Nested Cell Model	695
	14.8	Design of the RET D Flip-Flop by Using the LPD Model	698
	14.9	Design of the RET JK Flip-Flop by Flip-Flop Conversion	700
	14.10	Detection and Elimination of Timing Defects in Asynchronous FSMs	701
		14.10.1 Endless Cycles	702
		14.10.2 Races and Critical Races	703
		14.10.3 Static Hazards in the NS and Output Functions	705
		14.10.4 Essential Hazards in Asynchronous FSMs	711
		14.10.5 Perspective on Static Hazards and E-hazards in Asynchronous FSMs	718
	14.11	Initialization and Reset of Asynchronous FSMs	719
	14.12	Single-Transition-Time Machines and the Array Algebraic Approach	720
	14.13	Hazard-Free Design of Fundamental Mode State Machines by Using the Nested Cell Approach	730
	14.14	One-Hot Design of Asynchronous State Machines	734
	14.15	Perspective on State Code Assignments of Fundamental Mode FSMs	738
	14.16	Design of Fundamental Mode FSMs by Using PLDs	740
	14.17	Analysis of Fundamental Mode State Machines	741
		Further Reading	758
		Problems	759
15.	**The Pulse Mode Approach to Asynchronous FSM Design**		**773**
	15.1	Introduction	773
	15.2	Pulse Mode Models and System Requirements	773
		15.2.1 Choice of Memory Elements	774
	15.3	Other Characteristics of Pulse Mode FSMs	777
	15.4	Design Examples	779
	15.5	Analysis of Pulse Mode FSMs	788

	15.6	Perspective on the Pulse Mode Approach to FSM Design	795
		Further Reading	796
		Problems	797

16. Externally Asynchronous/Internally Clocked (Pausable) Systems and Programmable Asynchronous Sequencers — 805

	16.1	Introduction	805
	16.2	Externally Asynchronous/Internally Clocked Systems and Applications	806
		16.2.1 Static Logic DFLOP Design	807
		16.2.2 Domino Logic DFLOP Design	812
		16.2.3 Introduction to CMOS Dynamic Domino Logic	814
		16.2.4 EAIC System Design	816
		16.2.5 System Simulations and Real-Time Tests	817
		16.2.6 Variations on the Theme	820
		16.2.7 How EAIC FSMs Differ from Conventional Synchronous FSMs	821
		16.2.8 Perspective on EAIC Systems as an Alternative Approach to FSM Design	822
	16.3	Asynchronous Programmable Sequencers	823
		16.3.1 Microprogrammable Asynchronous Controller Modules and System Architecture	823
		16.3.2 Architecture and Operation of the MAC Module	824
		16.3.3 Design of the MAC Module	827
		16.3.4 MAC Module Design of a Simple FSM	830
		16.3.5 Cascading the MAC Module	832
		16.3.6 Programming the MAC Module	833
		16.3.7 Metastability and the MAC Module: The Final Issue	834
		16.3.8 Perspective on MAC Module FSM Design	834
	16.4	One-Hot Programmable Asynchronous Sequencers	835
		16.4.1 Architecture for One-Hot Asynchronous Programmable Sequencers	835
		16.4.2 Design of a Four-State Asynchronous One-Hot Sequencer	837
		16.4.3 Design and Operation of a Simple FSM by Using a Four-State One-Hot Sequencer	838
		16.4.4 Perspective on Programmable Sequencer Design and Application	839
	16.5	Epilogue to Chapter 16	842
		Further Reading	842
		Problems	844

A Other Transistor Logic Families — 849

A.1	Introduction to the Standard NMOS Logic Family	849
A.2	Introduction to the TTL Logic Family	850
A.3	Performance Characteristics of Important IC Logic Families	852
	Further Reading	852

B Computer-Aided Engineering Tools — 855

B.1	Function Minimization Tools	855
B.2	Schematic Capture, Simulation, and Timing Analysis Tools	855
	Further Reading	857

C	**IEEE Standard Symbols**		**859**
	C.1 Gates		859
	C.2 Combinational Logic Devices		859
	C.3 Flip-Flops, Registers, and Counters		860
	Further Reading		862

Index   **863**

Preface

TEXT OVERVIEW

This text emphasizes the successful engineering design of digital devices and machines from first principles. A special effort has been made *not* to "throw" logic circuits at the reader so that questions remain as to how the circuits came about or whether or not they will function correctly. An understanding of the intricacies of digital circuit design, particularly in the area of sequential machines, is given the highest priority — the emphasis is on error-free operation. From an engineering point of view, the design of a digital device or machine is of little or no value unless it performs the intended operation(s) correctly and reliably.

Both the basics and background fundamentals are presented in this text. But it goes well beyond the basics to provide significant intermediate-to-advanced coverage of digital design material, some of which is covered by no other text. In fact, this text attempts to provide *course coverage at both the first and second levels* — an ambitious undertaking. The aim is to provide the reader with the tools necessary for the successful design of relatively complex digital systems from first principles. In doing so, a firm foundation is laid for the use of CAD methods that are necessary to the design of large systems. In a related sense, VHDL behavioral and architectural descriptions of various machines, combinational and sequential, are provided at various points in the text for those instructors and students who wish to have or require a hardware description language in the study of digital design.

The text is divided into 16 relatively small chapters to provide maximum versatility in its use. These chapters range from introductory remarks to advanced topics in asynchronous systems. In these chapters an attempt is made to replace verbosity by illustration. Students generally do not like to read lengthy verbal developments and explanations when simple illustrations suffice. Well more than 600 figures and tables help to replace lengthy explanations. More than 1000 examples, exercises, and problems (worked and unworked, single and multiple part) are provided to enhance the learning process. They range in complexity from simple algebraic manipulations to multipart system-level designs, each carefully chosen with a specific purpose in mind. Annotated references appear at the end of each chapter, and an appendix at the end of the text provides the details of subjects thought to be peripheral to the main thrust of the text. Chapter 1 breaks with tradition in providing a complete glossary of terms, expressions, and abbreviations that serves as a conspicuous and useful source of information.

SUBJECT AREAS OF PARTICULAR STRENGTH IN THIS TEXT

Like others, this text has its subject areas of strengths — those that are uniquely presented in sufficient detail as to stand out as significant didactic and edifying contributions. This text breaks with tradition in providing unique coverage in several important areas. In addition to the traditional coverage, the following 20 subject areas are of particular strength in this text:

1. Thorough coverage of number systems, arithmetic methods and algorithms, and codes
2. Mixed logic notation and symbology used throughout the text
3. Emphasis on CMOS logic circuits
4. Unique treatment of conventional Boolean algebra and XOR algebra as these subjects relate to logic design
5. Entered variable mapping methods as applied throughout the text to combinational and sequential logic design
6. Applications of Reed–Muller transformation forms to function minimization
7. Nonarithmetic combinational logic devices such as comparators, shifters, and FPGAs
8. Arithmetic devices such as carry-save adders, multipliers, and dividers
9. Three uniquely different ALU designs, including an introduction to dual-rail systems and ALUs with completion signal and carry look-ahead capability
10. Detection and elimination methods for static hazards in two-level and multilevel (e.g., XOR-type) circuits including the use of binary decision diagrams (BDDs)
11. Design and analysis of flip-flops provided in a simple, well organized fashion
12. Detection and elimination of timing defects in synchronous sequential circuits
13. Input synchronization and debouncing, and FSM initialization and reset methods
14. Use of unique modular methods in the design of shift registers and counters
15. Complete coverage of ripple counters, ring counters and linear feedback shift register (LFSR and ALFSR) counters
16. Application of the array algebraic and one-hot approaches to synchronous FSM design
17. Detection and elimination of timing defects in asynchronous fundamental mode FSMs
18. Design and analysis of asynchronous FSMs including the nested cell approach, single transition time (STT) machines by using array algebra, and the one-hot code method
19. High speed externally asynchronous/internally clocked systems, including an introduction to dynamic domino logic applications
20. Programmable asynchronous sequencers

READERSHIP AND COURSE PREREQUISITES

No prior background in digital design is required to enter a *first* course of study by using this text. It is written to accommodate both the first- and second-level user. What is required is that the reader have sufficient maturity to grasp some of the more abstract concepts that are unavoidable in any digital design course of study. It has been the author's experience that digital design makes an excellent introduction to electrical and computer engineering because of the absolute and precise nature of the subjects—there are no approximation signs. This text is designed to make first reading by a user a rewarding experience. However, there is sufficient advanced material to satisfy the needs of the second level students and professionals in the field. A first-level understanding of the subject matter is necessary before entering a second-level course using this text.

SUGGESTED TEXT USAGE

Perhaps the best advice that can be given to instructors on the use of this text is to study the table of contents carefully and then decide what subject matter is essential to the course under consideration. Once this is done the subject area and order of presentation will usually become obvious. The following two course outlines are offered here as a starting point for instructors in making decisions on course subject usage:

The Semester System

[1] First-Level Course — Combinational Logic Design

Block I

    Introduction (Chapter 1)
    Number systems, binary arithmetic and codes (Sections 2.1 through 2.5 or choice)
    Binary state terminology, CMOS logic circuits, and mixed-logic symbology (Sections 3.1 through 3.7)
    Reading and construction of logic circuits (Section 3.8)
    XOR and EQV operators and mixed-logic symbology (Section 3.9)
    Laws of Boolean and XOR algebra (Sections 3.10 through 3.12)
    Review

EXAM #1

Block II

    Introduction; logic function representation (Sections 4.1 and 4.2)
    Karnaugh map (K-map) function representation and minimization, don't cares, and multioutput optimization (Sections 4.3 through 4.5)
    Entered variable mapping methods and function reduction of five or more variables (Sections 4.6, 4.7 and 4.12)
    Introduction to minimization algorithms (Section 4.8)
    Factorization and resubstitution methods (Subsections 4.9.1 and 4.9.2)
    Function minimization by using XOR K-map patterns (Sections 5.1 through 5.4)
    Review

EXAM #2

Block III

    Introduction to combinational logic design (Section 6.1)
    Multiplexers, decoders, priority encoders, and code converters (Sections 6.2 through 6.5; Section 2.10)
    Magnitude comparators, parity generators and shifters (Sections 6.6 through 6.8)
    Programmable logic devices — ROMs, PLAs and PALs (Sections 7.1 through 7.6)

Adders, subtractors, multipliers, and dividers (Section 2.6 and Subsections 2.9.1 through 2.9.5 or choice; Sections 8.1 through 8.7 or choice)

Arithmetic and logic units — ALUs (Section 8.8) — may be omitted if time-limited

Static hazards in combinational logic devices (Sections 9.1 and 9.2)

Review

EXAM #3 and/or FINAL

[2] Second-Level Course — State Machine Design and Analysis

Block IV

Introduction; models, the state diagram, and heuristic development of the basic memory cells (Sections 10.1 through 10.4)

Design and analysis of flip-flops, flip-flop conversion; timing problems; asynchronous overrides; setup and hold time requirements (Sections 10.5 through 10.11)

Design of simple synchronous finite state machines; K-map conversion; analysis of synchronous FSMs (Sections 10.12 and 10.13)

Review

EXAM #1

Block V

Introduction; detection and elimination of timing defects in synchronous state machines (Sections 11.1 through 11.3)

Synchronizing and stretching of asynchronous inputs; metastability; clock skew and clock sources (Sections 11.4 through 11.6)

Initialization and reset of FSMs, and debouncing circuits (Sections 11.7 and 11.8)

Applications to the design and analysis of more complex synchronous FSMs; ASM charts and state assignment rules; array algebraic approach to FSM design; state minimization (Sections 11.9 through 11.12)

Review

EXAM #2

Block VI

Introduction; design of shift registers and synchronous counters; synchronous vs asynchronous parallel loading (Sections 12.1 through 12.3)

Shift register counters and ripple counters; special purpose counters (Sections 12.4 through 12.5)

Alternative architecture — use of MUXs, decoders, PLDs, counters and shift registers; the one-hot design method (Sections 13.1 through 13.5)

The controller, data path, functional partition, and system-level design (Sections 13.6 and 13.7)

Introduction to asynchronous sequential machines — fundamental mode FSMs (Sections 14.1 through 14.9)

Pulse mode approach to asynchronous FSM design (Sections 15.1 through 15.6)
   Selected topics in Chapter 16
Review

EXAM #3 and/or FINAL

The choice of course content is subject to so many variables that no one course outline will suffice even within a single institution where several instructors may teach a given course. It is for this reason that the text is divided up into 16 relatively small chapters. This offers the instructor somewhat more flexibility in the choice of subject matter. For example, if it is desirable to offer a single (combined) semester course in digital design, it might be desirable to offer both combinational and sequential (synchronous FSM) logic design. Such a course might include the following subject areas taken from Blocks I through VI in sample course outlines [1] and [2]:

[3] Single (Combined) Semester Course in Digital Design

Binary state terminology, and mixed-logic symbology (Sections 3.1 through 3.7)
Reading and construction of logic circuits (Section 3.8)
XOR and EQV operators and mixed-logic symbology (Section 3.9)
Laws of Boolean and XOR algebra (Sections 3.10 through 3.12)
Review

EXAM #1

Logic function representation (Sections 4.1 and 4.2)
K-map function representation and minimization, don't cares and multioutput optimization (Sections 4.3 through 4.5)
Entered variable mapping methods and function reduction of five or more variables (Sections 4.6, 4.7 and 4.12)
Multiplexers, decoders, priority encoders, and code converters (Sections 6.2 through 6.5)
Comparators, parity generators, and shifters or choice (Sections 6.6 through 6.8)
Adders, subtractors, and multipliers (Sections 8.1 through 8.3; Section 8.6)
Static hazards in combinational logic devices (Sections 9.1 and 9.2)
Review

EXAM #2

Heuristic development of the basic memory cells (Sections 10.1 through 10.4)
Design and analysis of flip-flops, flip-flop conversion (Sections 10.5 through 10.8)
Asynchronous overrides; setup and hold time requirements; design and analysis of simple synchronous state machines (Sections 10.10 through 10.13)
Detection and elimination of timing defects in synchronous state machines (Sections 11.1 through 11.3)
Synchronizing of asynchronous inputs (Section 11.4)
Initialization and reset of FSMs; debouncing circuits (Sections 11.7 and 11.8)
Shift registers and counters (Sections 12.1 through 12.3)

Alternative architecture—use of MUXs, decoders, PLDs; the one-hot method (Sections 13.1 through 13.3, Section 13.5)

The controller, data path, and functional partition and system-level design (Sections 13.6 and 13.7)

Review

EXAM #3 and/or FINAL

Though the subject coverage for EXAM #3 in course sample outline [3] seems large in proportion to those required for EXAM #2, a close inspection will indicate that the number of sections are the same. The sections required for EXAM #1 number about half that of the other two.

The Quarter System

Not all courses at colleges and universities are operated on a semester basis. Some are operated on the quarter system. This requires that the course subject areas be divided up in some logical and effective manner, which may require that both combinational and sequential machines be covered within a given quarter course. As a guide to subject area planning on the quarter system when using this text, the following quarter system may be considered (refer to sample course outlines [1] and [2]):

First Quarter

Block I
    EXAM #1
Block II
    EXAM #2

Second Quarter

Block III
    EXAM #1
Block IV
    EXAM #2

Third Quarter

Block V
    EXAM #1
Block VI
    EXAM #2

Fourth Quarter (if applicable)

Chapters 14 and 15
    EXAM #1
Chapter 16
    PROJECT and/or EXAM #2

Certainly, there are an endless number of ways in which the subject areas can be divided up to meet the requirements of digital design courses that are offered on the basis of a semester, quarter, or trimester system. The presence of 16 relatively small chapters should make the decision process less complicated and lead to a meaningful and productive treatment of digital logic design.

INSTRUCTIONAL SUPPORT SOFTWARE AND MATERIALS

For the Student

Bundled with this text on CD-ROM are all end-of-chapter problems and two powerful tools: a logic minimization program called *BOOZER*, and an interactive logic simulator called *EXL-Sim2000*. BOOZER is capable of returning an optimal or near optimal result on single or multioutput logic systems of up to 16 functions of 16 variables, by using either entered variables or canonical data entry. EXL-Sim2000 operates in the Windows environment and accepts either positive or mixed-logic input and generates either positive or mixed-logic waveform results. EXL-Sim2000 incorporates the best features of the leading simulator programs available, yet it is unique in its operational simplicity and versatility. More complete descriptions of BOOZER and EXL-Sim2000 are provided in Appendix B of this text.

For the Instructor

An instructor's manual is placed on CD-ROM together with BOOZER and EXL-Sim2000. The manual contains the detailed solutions to all problems presented in the text and provides suggestions and caveats as to problem assignments and text usage. All figures (In PDF format) are included separately for selective use in creating transparencies or hard copies.

ERRORS

Any text of this size and complexity is bound to contain errors and omissions that have been overlooked throughout the extensive review and editing process. Identification of any error or omission would be greatly appreciated by the editors of Academic Press and by the author. Constructive comments regarding matters of clarity, organization and coverage of subject matter are also valued. Such information should be directed to the author via the editorial staff at Academic Press:
PECS Editor
Academic Press
200 Wheeler Road
Burlington, MA 01803
Tel 781-221-2212
Fax 781-221-1615

ACKNOWLEDGMENTS

Of the many people who have contributed to the completion of this project, certain individuals stand out as having played very significant roles. First, my sincere thanks go to the five reviewers of portions of this text: Professors Ward D. Getty of the University of Michigan, James C. Harden of Mississippi State University, John P. Robinson of the University of Iowa, Harpreet Singh of Wayne State University, and Murali Varanasi of the University of South Florida. Three other persons are here acknowledged for many helpful conversations and suggestions. These are professors Mark Manwaring, Jack Meador, and Mircea Dabacan, all of the School of EECS at Washington State University and friends and colleagues of the author. Special thanks is owed to Professor Manwaring, author of the logic minimizer called BOOZER, for permitting BOOZER to be bundled with this text on CD ROM. A debt of gratitude also goes to Professor Marek Perkowski of Portland State University for his help and suggestions regarding material on Reed–Muller transformation forms covered in Chapter 5. Finally, of great importance to this text is the work of Bob McCurdy, who, with only sketchy ideas from the author, is responsible for the student-friendly but powerful logic simulator, called EXL-Sim2000, that is bundled with this text on CD-ROM.

Four students are gratefully acknowledged for their work in proofing portions of the manuscript: Ryan O'Fallon, Becky Richardson, Rebecca Sheats, and Parag Upadhyaya. Finally, sincere thanks go to the hundreds of students that have over several years made many helpful suggestions and who have helped identify and eliminate many errors and omissions. Furthermore, it must be acknowledged that the students, more than anyone else, have played an essential role in shaping the pedagogical content of this text.

These acknowledgments would not be complete without recognizing the encouragement of and many helpful conversations with Joel Claypool, Executive Editor of Academic Press, a Division of Harcourt, Inc. Most importantly, the person to whom the author owes much more than just a statement of gratitude is his loving wife, his friend and confidant, Gloria.

Richard F. Tinder
Pullman, Washington

Engineering Digital Design

Second Edition

CHAPTER 1

Introductory Remarks and Glossary

1.1 WHAT IS SO SPECIAL ABOUT DIGITAL SYSTEMS?

No area of technology has had or is likely to continue to have more of a profound impact on our lives than digital system development. That's quite a statement, but its truth is obvious when one considers the many ways we have become dependent on "digitized" technology. To put this in perspective, let us review the various areas in which digital systems play an important role in our lives. As this is done, keep in mind that there is significant, if not necessary, overlap in the digital system technologies that make possible those areas we have come to take for granted: computing, information retrieval, communication, automatic control systems, entertainment, and instrumentation.

Computing: A computer, like the telephone and television, has become almost an essential part of every household. Word processing, information retrieval, communication, finance and business management, entertainment, art and graphics — these are but a few of the functions performed by our beloved computers. In the span of a little more than 10 years, computers in the home and in small businesses have advanced from what was termed microcomputers to the present computers with nearly mainframe capability. Home computers can now perform relatively sophisticated operations in the areas just mentioned. Of course, vastly improved computer speed and memory, together with powerful software development, are primarily responsible for the rapid rise in personal computer capabilities. In addition to the digital computer itself, there are other digital devices or peripherals that are normally part of a computer system. These include disk drives, CD-ROM drives, modems, CRT and LCD monitors, sound cards, scanners, and printers. Then there are the hand-held calculators that now have nearly microcomputer capability and are quite inexpensive. All of these things have been made possible because of the advances in digital system technology. But this is just the beginning.

Information Retrieval: The ability to consult one's favorite encyclopedia via CD-ROM or surf (browse) the World Wide Web (WWW) has become a very important part of computer use in the home, at school, and in business. The use of CD-ROMs also permits access to information in the specialized areas of literature, music, religion, health, geography, math, physical science, biology, and medicine, to name a few. But information retrieval is not limited to these functions. Network communication between computers and our ability to tap into huge university libraries are other important sources of information. Think of where businesses would be without access to data-base information that is critical to day-to-day operation. Local and national security operations depend heavily on data-base information stored on computers that are most likely part of a network. Yes, and then there is education. What an invaluable source of information the computer has become both in the classroom and in the home.

Communications: It would be hard to imagine what our world would be like without the ability to send facsimile (fax) communications or e-mail. These are digital transmission methods that were developed to a high degree of sophistication over a period of about 10 years. Of course, the modem, another digital device, has made this possible. Digital communication is hardly limited to fax and e-mail. One's home phone or cellular phone is likely to be digital, permitting a variety of features that were difficult if not impossible to provide by means of an analog transmission device. Scientific data, national security information, and international communications, all of which are collected and transmitted back to earth by satellite, are accomplished by digital transmission methods with accuracy not possible otherwise.

Automatic Control Systems: Digital automatic control systems have replaced the old analog methods in almost all areas of industry, the home, and transportation. Typical examples include rapid transit systems, integrated circuit fabrication systems, robot systems of all types in assembly-line production, space vehicle operations, a variety of automobile associated operations, guidance systems, home security systems, heating and air-conditioning systems, many home appliances, and a host of medical systems.

Entertainment: Who cannot help but be awed by the impressive computer generated graphics that have become commonplace in movies and in games produced on CDs. Movies such as Jurassic Park and the new Star Wars series will perhaps be remembered as having established a new era in the art of make-believe. The games that are available on the home computer include everything from chess and casino-type games to complex and challenging animated aircraft operations and adventure/fantasy games. Then add to these the high-quality sound that CDs and the Internet produce, and one has a full entertainment center as part of the personal computer. Of course, the incursion of digital systems into the world of entertainment extends well beyond movies and games. For example, one has only to listen to digitally recorded or remastered CDs (from the original analog recordings) to enjoy their clear, noise-free character. Also, don't forget the presence of electronic keyboard instruments ranging from synthesizers to Clavinovas and the like. Then for those who consider photography as entertainment, digital cameras and camcorders fall into this category. And the list goes on and on.

Instrumentation: A listing of the many ways in which digital system technology has affected our lives would not be complete without mentioning the myriad of measurement and sensing instruments that have become digitized. Well known examples of electronic laboratory testing equipment include digital voltmeters, ammeters, oscilloscopes, and waveform generators and analyzers. Then there are the sophisticated medical instruments that include MRI and CAT scan devices. Vital signs monitoring equipment, oximeters, IV pumps, patient controlled analgesia (PCA) pumps, digital ear thermometers, and telemetry equipment are typical examples of the many other ways the medical industry has made use of digital systems technology.

1.2 THE YEAR 2000 AND BEYOND?

If one considers what has happened in, say, the past 15 years, the path of future technological development in the field of digital systems would seem to be limited only by one's imagination. It is difficult to know where to begin and where to end the task of forecasting digital system development, but here are a few examples in an attempt to accomplish this:

Computer power will continue to increase as the industry moves to 0.20μ (and below) CMOS technology with speeds approaching the gigahertz range and with a demand for more efficient ways to sink the heat generated by up to a billion transistors per processor operated with supply voltages below 2 volts. There will be dramatic changes in the peripherals that are now viewed as part of the computer systems. For example, vacuum (CRT) monitors will eventually be replaced by picture-frame style LCD monitors, or by *micropanel displays* using either DLP (*Digital Light Processing*) or FED (*field emission display*) technologies. Digitized high-definition TV (HDTV) will eventually replace all conventional TV sets, and the World Wide Web (WWW) will be viewed on HDTV via special dedicated computers. In all, larger, sharper, brighter, and clearer computer and TV displays are to be expected, together with a fast-growing and impressive assortment of wireless hand-held and wrist-bound devices.

Expect that the mechanically operated magnetic storage systems (disk drives) of today will soon be replaced by a MR (*magneto-resistive*) technology that will increase the *areal storage density* (gigabits per square inch) by a factor of 100 to 200, or by OAWD (*optically assisted Winchester drive*) and MO (*magneto-optical*) technologies that are expected to increase the areal density even further. Eventually, a holographic storage technology or a *proximal probe* technology that uses a scanning tunneling microscopic technique may provide capabilities that will take mass storage to near its theoretical limit. Thus, expect storage systems to be much smaller with enormously increased storage capacity.

Expect that long-distance video conferencing via computer will become as commonplace as the telephone is today. Education will be a major beneficiary of the burgeoning digital age with schools (K–12, and universities and colleges both public and private) being piped into major university libraries and data banks, and with access to the ever-growing WWW. Look for the common film cameras of today to be replaced by digital cameras having megapixel resolution, audio capability, and with the capability to store a large number of pictures that can be reviewed on camera and later presented on screen by any computer. Expect that certain aspects of laser surgery will be microprocessor controlled and that X-ray imaging methods (e.g., mammography) and radiology generally will be digitally enhanced as a common practice. Also, health facilities and hospitals will be linked for immediate remote site consultation and for specialized robotics surgery.

Expect digital systems to become much more sophisticated and pervasive in our lives. Interconnectivity between "smart" electrically powered systems of all types in the home, automobile, and workplace could be linked to the web together with sophisticated fail-safe and backup systems to prevent large-scale malfunction and possible chaos. Such interconnected systems are expected to have a profound effect on all aspects of our lives — what and when we eat, our exercise habits, comfort and entertainment needs, shopping activities, medical requirements, routine business transactions, appointment schedules, and many others imaginable.

Optical recognition technology will improve dramatically in the fields of robotics, vehicular operation, and security systems. For example, expect that *iris and retinal pattern recognition* will eventually be used to limit access to certain protected systems and areas, and may even replace digital combination locks, IDs, and licenses for such purposes. Taxation, marketing, and purchasing methods will undergo dramatic changes as digital systems become commonplace in the world of government, commerce, and finance. Even the world of politics, as we now know it, will undergo dramatic change with the use of new and more efficient voting and voter sampling methods. Mass production line manufacturing methods by using robots and other digitally automated mechanical devices will continue to evolve at a rapid pace as dictated by domestic and world market forces. Expect that logic minimization tools and automated digital design tools will become more commonplace and sophisticated, permitting designers with little practical experience to design relatively complex systems.

Business networking will undergo dramatic improvements with the continued development of gigabit Ethernet links and high-speed switching technology. Home connectivity will see vast improvements in *satellite data service downloading* (up to 400 kbps), 56-kbps (and higher) modems that need high-quality digital connections between phones and destination, improved satellite data service with bidirectional data transmission, and DSL (digital subscriber line) cable modem systems.

Finally, there are some really exciting areas to watch. Look for speech recognition, speech synthesis, and handwriting and pattern recognition to dramatically change the manner in which we communicate with and make use of the computer both in business and in the home. Somewhere in the future the computer will be equipped with speech understanding capability that allows the computer to build ideas from a series of spoken words — perhaps like HAL 9000 in the film *2001: A Space Odyssey*. Built-in automatic learning capability may yet prove to be the most challenging undertaking facing computer designers of the future. Thus, expect to see diminished use of the computer keyboard with time as these technologies evolve into common usage.

Revolutionary computer breakthroughs may come with the development of radically different technologies. *Carbon nanotube technology*, for example, has the potential to propel computer speeds well into the gigahertz range together with greatly reduced power dissipation. The creation of carbon nanotube transistors could signal the dawn of a new revolution in chip development. Then there is the specter of the *quantum computer*, whose advent may lead to computing capabilities that are trillions of times faster than those of conventional supercomputers. All of this is expected to be only the beginning of a new millennium of invention limited only by imagination. Remember that radically different technological breakthroughs can appear at any time, even without warning, and can have a dramatic affect on our lives, hopefully for the better.

To accomplish all of the preceding, a new generation of people, technically oriented to cope with the rapidly changing digital systems technology, will result as it must. This new generation of people will have a dramatic impact on education, labor, politics, transportation, and communications, and will most certainly affect domestic and global economies. Thus, expect that more pressure and responsibility will be placed on universities to produce the quality training that can match up to this challenge, not just over a short period but also in the long term.

1.3 A WORD OF WARNING

Not yet mentioned are the changes that must take place in the universities and colleges to deal with this rapidly evolving technology. It is fair to say that computer aided design (CAD) or automated design of digital systems is on the upswing. Those who work in the areas of digital system design are familiar with such hardware description languages as VHDL or Verilog, and the means to "download" design data to program PLAs or FPGAs (field programmable gate arrays). It is possible to generate a high-level hardware description of a digital system and introduce that hardware description into circuit layout tools such as Mentor Graphics. The end result would be a transistor-level representation of a CMOS digital system that could be simulated by one of several simulation tools such as HSPICE and subsequently be sent to the foundry for chip creation. The problem with this approach to digital system design is that it bypasses the need to fully understand the intricacies of design that ensure proper and reliable system operation. As is well known, a successful HSPICE simulation does not necessarily ensure a successful design. In the hands of a skilled and experienced designer this approach may lead to success without complications. On the other hand, if care is not taken at the early stages of the design process and if the designer has only a limited knowledge of design fundamentals, the project may fail at one point or another. Thus, as the use of automated (CAD) designs become more attractive to those who lack design detail fundamentals, the chance for design error at the system, device, gate, or transistor level increases. *The word of warning*: Automated design should never be undertaken without a sufficient knowledge of the field and a thorough understanding of the digital system under consideration — *a little knowledge can be dangerous*! This text is written with this warning in mind. The trend toward increasing CAD use is not bad, but automated design methods must be used cautiously with sufficient background knowledge to carry out predictably successful designs. Computer automated design should be used to remove the tedium from the design process and, in many cases, make tractable certain designs that would otherwise not be possible. But CAD is *not* a replacement for the details and background fundamentals required for successful digital system design. It is the goal of this text to provide the reader with the necessary details and background fundamentals so as to permit a successful transition into the CAD domain.

1.4 GLOSSARY OF TERMS, EXPRESSIONS, AND ABBREVIATIONS

Upon entering any new field, there is always the problem of dealing with the "jargon" that is peculiar or unique to that field. Conspicuously absent in most texts on digital design is a glossary of terms, expressions, and abbreviations that are used — yes, and even overused — in presenting the subject matter. Readers of these texts are often left leafing through back pages and chapters in search of the meaning of a given term, expression or abbreviation. In breaking with tradition, this text provides an extensive *glossary*, and does so here at the beginning of the text where it can be used — not at the end of the text where it may go unnoticed. In doing this, Chapter 1 serves as a useful source of information.

ABEL: advanced Boolean expression language.
Accumulator: an adder/register combination used to store arithmetic results.
Activate: to assert or make active.

Activation level: the logic state of a signal designated to be active or inactive.
Activation level indicator: a symbol, (H) or (L), that is attached to a signal name to indicate positive logic or negative logic, respectively.
Active: a descriptor that denotes an action condition and that implies logic 1.
Active device: any device that provides current (or voltage) gain.
Active high (H): indicates a positive logic source or signal.
Active low (L): indicates a negative logic source.
Active state: the logic 1 state of a logic device.
Active transition point: the point in a voltage waveform where a digital device passes from the inactive state to the active state.
Addend: an operand to which the augend is added.
Adder: a digital device that adds two binary operands to give a sum and a carry.
Adder/subtractor: a combinational logic device that can perform either addition or subtraction.
Adjacent cell: a K-map cell whose coordinates differ from that of another cell by only one bit.
Adjacent pattern: an XOR pattern involving an uncomplemented function in one cell of a K-map and the same function complemented in an adjacent cell.
ALFSR: autonomous linear feedback shift register.
ALFSR counter: a counter, consisting of an ALFSR, that can sequence through a unique set of pseudo-random states that can be used for test vectors.
Algorithm: any special step-by-step procedure for accomplishing a task or solving a problem.
Alternative race path: one of two or more transit paths an FSM can take during a race condition.
ALU: arithmetic and logic unit.
Amplify: the ability of an active device to provide current or voltage gain.
Analog: refers to continuous signals such as voltages and current, in contrast to digital or discrete signals.
AND: an operator requiring that all inputs to an AND logic circuit symbol be active before the output of that symbol is active — also, Boolean product or intersection.
AND function: the function that derives from the definition of AND.
AND gate: a physical device that performs the electrical equivalent of the AND function.
AND laws: a set of Boolean identities based on the AND function.
AND-OR-Invert (AOI) gate: a physical device, usually consisting of two AND gates and one NOR gate, that performs the electrical equivalent of SOP with an active low output.
AND plane: the ANDing stage or matrix of a PLD such as a ROM, PLA, or PAL.
Antiphase: as used in clock-driven machines to mean complemented triggering of a device relative to a reference system, such as, an FET input device to an RET FSM.
Apolar input: an input, such as CK, that requires no activation level indicator to be associated with it.
Arbiter module: a device that is designed to control access to a protected system by arbitration of contending signals.
Arithmetic and logic unit (ALU): a physical device that performs either arithmetic or logic operations.

Arithmetic shifter: a combinational shifter that is capable of generating and preserving a sign bit.

Array algebra: the algebra of Boolean arrays and matrices associated with the automated design of synchronous and STT machines.

Array logic: any of a variety of logic devices, such as ROMs, PLAs or PALs, that are composed of an AND array and an OR array (see Programmable logic device or PLD).

ASIC: application-specific IC.

ASM: algorithmic state machine.

Assert: activate.

Assertion level: activation level.

Associative law: a law of Boolean algebra that states that the operational sequence as indicated by the location of parentheses in a p-term or s-term does not matter.

Associative pattern: an XOR pattern in a K-map that allows a term or variable in an XOR or EQV function to be looped out (associated) with the same term or variable in an adjacent cell provided that the XOR or EQV connective is preserved in the process.

Asynchronous: clock-independent or self-timed — having no fixed time relationship.

Asynchronous input: an input that can change at any time, particularly during the sampling interval of the enabling input.

Asynchronous override: an input such as preset or clear that, when activated, interrupts the normal operation of a flip-flop.

Asynchronous parallel load: the parallel loading of a register or counter by means of the asynchronous PR and CL overrides of the flip-flops.

Augend: an operand that is added to the addend in an addition operation.

Barrel shifter: a combinational shifter that only rotates word bits.

Base: radix. Also, one of three regions in a BJT.

Basic cell: a basic memory cell, composed of either cross-coupled NAND gates or cross-coupled NOR gates, used in the design of other asynchronous FSMs including flip-flops.

BCD: binary coded decimal.

BCH: binary coded hexadecimal.

BCO: binary coded octal.

BDD: binary decision diagram.

Bidirectional counter: a counter that can count up or down.

Binary: a number system of radix 2; having two values or states.

Binary code: a combination of bits that represent alphanumeric and arithmetic information.

Binary coded decimal (BCD): a 4-bit, 10-word decimal code that is weighted 8, 4, 2, 1 and that is used to represent decimal digits as binary numbers.

Binary coded hexadecimal (BCH): the hexadecimal number system used to represent bit patterns in binary.

Binary coded octal (BCO): the octal number system used to represent bit patterns in binary.

Binary decision diagram (BDD): a graphical representation of a set of binary-valued decisions, beginning with an input variable and proceeding down paths that end in either logic 1 or logic 0.

Binary word: a linear array of juxtaposed bits that represents a number or that conveys an item of information.

Bipolar junction transistor (BJT): an npn or pnp transistor.
Bipolar PROM: a PROM that uses diodes as fusible links.
BIST: built-in-self-test.
Bit: a binary digit.
Bit slice: partitioned into identical parts such that each part operates on one bit in a multibit word — part of a cascaded system of identical parts.
BJT: bipolar junction transistor.
BO: borrow-out.
Bond set: in the CRMT method, a disjoint set of bond variables.
Bond variable: one of two or more variables that form the axes of an EV K-map used in the CRMT method of function minimization.
Boolean algebra: the mathematics of logic attributed to the mathematician George Boole (1815–1864).
Boolean product: AND or intersection operation.
Boolean sum: OR or union operation.
BOOZER: Boolean ZEro-one Reduction — a multioutput logic minimizer that accepts entered variables.
Borrow-in: the borrow input to a subtractor.
Borrow-out: the borrow output from a subtractor.
Boundary: the separation of logic domains in a K-map.
Bounded pulse: a pulse with both lower and upper limits to its width.
Branching condition (BC): the input requirements that control a state-to-state transition in an FSM.
Branching path: a state-to-state transition path in a state diagram.
Buffer: a line driver.
Buffer state: a state (in a state diagram) whose only purpose is to remove a race condition.
Bus: a collection of signal lines that operate together to transmit a group of related signals.
Byte: a group of eight bits.
C: carry. Also, the collector terminal in a BJT.
CAD: computer-aided design.
CAE: computer-aided engineering.
Call module: a module designed to control access to a protected system by issuing a request for access to the system and then granting access after receiving acknowledgment of that request.
Canonical: made up of terms that are either all minterms or all maxterms.
Canonical truth table: a 1's and 0's truth table consisting exclusively of minterms or maxterms.
Capacitance, C: the constant of proportionality between total charge on a capacitor and the voltage across it, $Q = CV$, where C is given in farads (F) when charge Q is given in coulombs and V in volts.
Capacitor: a two-terminal energy storing element for which the current through it is determined by the time-rate of change of voltage across it.
Cardinality: the number of prime implements (p-term or s-term cover) representing a function.
Carry generate: a function that is used in a carry look-ahead (CLA) adder.
Carry-in: the carry input to a binary adder.
Carry look-ahead (CLA): same as look-ahead-carry.

1.4 GLOSSARY OF TERMS, EXPRESSIONS, AND ABBREVIATIONS

Carry-out: the carry output from an Adder.
Carry propagate: a function that is used in a CLA adder.
Carry save (CS): a fast addition method for three or more binary numbers where the carries are saved and added to the final sum.
Cascade: to combine identical devices in series such that any one device drives another; to bit-slice.
Cell: the intersection of all possible domains of a K-map.
Central processing unit (CPU): a processor that contains the necessary logic hardware to fetch and execute instructions.
CGP: carry generate/propagate.
CI: carry-in.
Circuit: a combination of elements (e.g., logic devices) that are connected together to perform a specific operation.
CK: clock.
CL or CLR: clear.
CLA: carry look-ahead.
CLB: configurable logic block. Also, a logic cell (LC).
Clear: an asynchronous input used in flip-flops, registers, counters and other sequential devices, that, when activated, forces the internal state of the device to logic 0.
Clock: a regular source of pulses that control the timing operations of a synchronous sequential machine.
Clock skew: a phenomenon that is generally associated with high frequency clock distribution problems in synchronous sequential systems.
C-module: an RMOD.
CMOS: complementary configured MOSFET in which both NMOS and PMOS are used.
CNT: mnemonic for count.
CO: carry-out.
Code: a system of binary words used to represent decimal or alphanumeric information.
Code converter: a device designed to convert one binary code to another.
Collapsed truth table: a truth table containing irrelevant inputs.
Collector: one of three regions in a BJT.
Combinational hazard: a hazard that is produced within a combinational logic circuit.
Combinational logic: a configuration of logic devices in which the outputs occur in direct, immediate response to the inputs without feedback.
Commutative law: the Boolean law that states that the order in which variables are represented in a p-term or s-term does not matter.
Comparator: a combinational logic device that compares the values of two binary numbers and issues one of three outputs indicative of their relative magnitudes.
Compatibility: a condition where the input to a logic device and the input requirement of the device are of the same activation level, that is, are in logic agreement.
Compiler: converts high-level language statements into typically a machine-coded or assembly language form.
Complement: the value obtained by logically inverting the state of a binary digit; the relationship between numbers that allows numerical subtraction to be performed by an addition operation.
Complementary metal oxide semiconductor (CMOS): a form of MOS that uses both p- and n-channel transistors (in pairs) to form logic gates.

Complementation: a condition that results from logic incompatibility; the mixed-logic equivalent of the NOT operation.

Composite output map: a K-map that contains entries representing multiple outputs.

Computer: a digital device that can be programmed to perform a variety of tasks (e.g., computations) at extremely high speed.

Concatenation: act of linking together or being linked together in a series.

Conditional branching: state-to-state transitions that depend on the input status of the FSM.

Conditional output: an output that depends on one or more external inputs.

Conjoint: as used in "mutually conjoint" to mean a set of s-terms whose ORed values taken two at a time are always logic 1. Thus, mutually conjoint terms never take logic 0 at the same time.

Conjugate gate forms: a pair of logic circuit symbols that derive from the same physical gate and that satisfy the DeMorgan relations.

Connective: a Boolean operator symbol (e.g., $+, \oplus, \cap$).

Consensus law: a law in Boolean algebra that allows simplification by removal of a redundant term.

Consensus term: the redundant term that appears in a function obeying the consensus law.

Controlled inverter: an XOR gate that is used in either the inverter or transfer mode.

Controller: that part of a digital system that controls the data path.

Conventional K-map: a K-map whose cell entries are exclusively 1's and 0's.

Counter: a sequential logic circuit designed to count through a particular sequence of states.

Counteracting delay: a delay placed on an external feedback path to eliminate an E-hazard or d-trio.

Count sequence: a repeating sequence of binary numbers that appears on the outputs of a counter.

Coupled term: one of two terms containing only one coupled variable.

Coupled variable: a variable that appears complemented in one term of an expression (SOP or POS) and that also appears uncomplemented in another term of the same expression.

Cover: a set of terms that covers all minterms or maxterms of a function.

CPU: central processing unit.

Creeping code: any code whose bit positions fill with 1's beginning at one end, and then fill with 0's beginning at the same end.

Critical race: a race condition in an asynchronous FSM that can result in transition to and stable residence in an erroneous state.

CRMT: Contracted Reed–Muller transformation.

Cross branching: multiple transition paths from one or more states in the state diagram (or state table) of a sequential machine whereby unit distance coding of states is not possible.

CU: control unit.

Current, I: the flow or transfer of charged matter (e.g., electrons) given in amperes (A).

Cutoff mode: the physical state of a BJT in which no significant collector current is permitted to flow.

Cycle: two or more successive and uninterrupted state-to-state transitions in an asynchronous sequential machine.

1.4 GLOSSARY OF TERMS, EXPRESSIONS, AND ABBREVIATIONS

Data bus: a parallel set of conductors which are capable of transmitting or receiving data between two parts of a system.

Data lockout: the property of a flip-flop that permits the data inputs to change immediately following a reset or set operation without affecting the flip-flop output.

Data lockout flip-flop: a one-bit memory device which has the combined properties of a master/slave flip-flop and an edge triggered flip-flop.

Data path: the part of a digital system that is controlled by the controller.

Data path unit: the group of logic devices that comprise the data path.

Data selector: a multiplexer.

Data-triggered: referring to flip-flops triggered by external inputs (no clock) as in the pulse mode.

DCL: digital combination lock.

Deactivate: to make inactive.

Deassert: deactivate.

Debounce: to remove the noise that is produced by a mechanical switch.

Debouncing circuit: a circuit that is used to debounce a switch.

Decade: a quantity of 10.

Decoder: a combinational logic device that will activate a particular minterm code output line determined by the binary code input. A demultiplexer.

Decrement: reduction of a value by some amount (usually by 1).

Delay: the time elapsing between related events in process.

Delay circuit: a circuit whose purpose it is to delay a signal for a specified period of time.

Delimiter: a character used to separate lexical elements and has a specific meaning in a given language. Examples are @, #, +, /, ', >.

DeMorgan relations: mixed logic expressions of DeMorgan's laws.

DeMorgan's laws: a property that states that the complement of the Boolean product of terms is equal to the Boolean sum of their complements; or that states that the complement of the Boolean sum of terms is the Boolean product of their complements.

Demultiplexer: a combinational logic device in which a single input is selectively steered to one of a number of output lines. A decoder.

Depletion mode: a normally ON NMOS that has a conducting n-type drain-to-source channel in the absence of a gate voltage but that looses its conducting state when the gate voltage reaches some negative value.

D flip-flop: a one-bit memory device whose output value is set to the D input value on the triggering edge of the clock signal.

D-flop module: a memory element that is used in an EAIC system and that has characteristics similar to that of a D flip-flop.

Diagonal pattern: an XOR pattern formed by identical EV subfunctions in any two diagonally located cells of a K-map whose coordinates differ by two bits.

Difference: the result of a subtraction operation.

Digit: a single symbol in a number system.

Digital: related to discrete quantities.

Digital combination lock: a sequence recognizer that can be used to unlock or lock something.

Digital engineering design: the design and analysis of digital devices.

Digital signal: a logic waveform composed of discrete logic levels (e.g., a binary digital signal).

Diode: a two-terminal passive device consisting of a *p–n* junction that permits significant current to flow only in one direction.
Diode-transistor logic: logic circuits consisting mainly of diodes and BJTs.
Direct address approach: an alternative approach to FSM design where PS feedback is direct to the NS logic.
Disjoint: as used in "mutually disjoint" to mean a set of p-terms whose ANDed values taken two at a time are always logic zero. Thus, mutually disjoint terms never take logic 1 at the same time.
Distributed path delays: a notation in which a path delay is assigned to each gate or inverter of a logic circuit.
Distributive law: The dual of the factoring law.
Divide-by-n counter: a binary counter of n states whose MSB output divides the clock input frequency by n.
Dividend: the quantity that is being divided by the divisor in a division operation.
Divider: a combinational logic device that performs the binary division operation.
Divisor: the quantity that is divided into the dividend.
DLP: digital light processing.
DMUX: demultiplexer (see decoder).
Domain: a range of logic influence or control.
Domain boundary: the vertical or horizontal line or edge of a K-map.
Don't care: a non-essential minterm or maxterm, denoted by the symbol ϕ, that can take either a logic 1 or logic 0 value. Also, a delimiter ϕ that, when attached to a variable or term, renders that variable or term nonessential to the parent function.
DPU: data path unit; also data processing unit.
Drain: one of three terminals of a MOSFET.
DRAM: dynamic RAM.
Driver: a one-input device whose output can drive substantially more inputs than a standard gate. A buffer.
DTL: diode-transistor logic.
D-trio: a type of essential hazard that causes a fundamental mode machine to transit to the correct state via an unauthorized path.
Duality: a property of Boolean algebra that results when the AND and OR operators (or XOR and EQV operators) are interchanged simultaneously with the interchange of 1's and 0's.
Dual-rail systems: as used in this text, a system of split signals in an ALU configuration that permits a completion signal to be issued at the end of each process, be it arithmetic or logic.
Dual relations: two Boolean expressions that can be derived one from the other by duality.
Duty cycle: in a periodic waveform, the percentage of time the waveform is active.
Dyad: a grouping of two logically adjacent minterms or maxterms.
Dynamic domino logic: buffered CMOS logic that requires complementary precharge and evaluate transistors for proper operation.
Dynamic hazard: multiple glitches that occur in the output from a multilevel circuit because of a change in an input for which there are three or more asymmetric paths (delay-wise) of that input to the output.
Dynamic RAM: a volatile RAM memory that requires periodic refreshing to sustain its memory.

1.4 GLOSSARY OF TERMS, EXPRESSIONS, AND ABBREVIATIONS

EAIC system: externally asynchronous/internally clocked system.
ECL: emitter-coupled logic.
Edge-triggered flip-flop: a flip-flop that is triggered on either the rising edge or falling edge of the clock waveform and that exhibits the data-lock-out feature.
EEPROM: electrically erasable PROM.
E-hazard: essential hazard.
EI: enable-in.
Electron: the majority carrier in an n-type conducting semiconductor.
Electronic switch: a voltage or current controlled switching device.
Emitter: one of three terminals of a BJT.
Emitter-coupled logic (ECL): a high-speed nonsaturating logic family.
EN: enable.
Enable: an input that is used to enable (or disable) a logic device, or that permits the device to operate normally.
Encoder: a digital device that converts digital signals into coded form.
Endless cycle: an oscillation that occurs in asynchronous FSMs.
Enhancement mode: a normally OFF NMOS that develops an n-channel drain-to-source conducting path (i.e., turns ON) with application of a sufficiently large positive gate voltage.
Entered variable (EV): a variable entered in a K-map.
EO: enable-out.
EPI: essential prime implicant.
EPLD: erasable PLD.
EPROM: erasable programmable read-only memory.
EQPOS: EQV-product-of-sums.
Equivalence: the output of a two-input logic gate that is active if, and only if, its inputs are logically equivalent (i.e., both active or both inactive).
EQV: equivalence.
EQV function: the function that derives from the definition of equivalence.
EQV gate: a physical device that performs the electrical equivalent of the EQV function.
EQV laws: a set of Boolean identities based on the EQV function.
Erasable programmable read-only memory (EPROM): a ROM that can be programmed many times.
Error catching: a serious problem in a JK master/slave flip-flop where a 1 or 0 is caught in the master cell when clock is active and is issued to the slave cell output when clock goes inactive.
Essential hazard: a disruptive sequential hazard that can occur as a result of an explicitly located delay in an asynchronous FSM that has at least three states and that is operated in the fundamental mode.
Essential prime implicant (EPI): a prime implicant that must be used to achieve minimum cover.
EU: execution unit.
EV: entered variable.
EV K-map: a K-map that contains EVs.
EV truth table: a truth table containing EVs.
Even parity: an even number of 1's (or 0's) in a binary word depending on how even parity is defined.

EVM: entered variable K-map.

Excess 3 BCD (XS3) code: BCD plus three.

Excitation table: a state transition table relating the branching paths to the branching condition values given in the state diagram for a flip-flop.

Exclusive OR: a two-variable function that is active if only one of the two variables is active.

EXOP: XOR-sum-of-products.

Expansion of states: opposite of merging of states.

Extender: a circuit or gate that is designed to be connected to a digital device to increase its fan-in capability — also called an expander.

Factoring law: the Boolean law that permits a variable to be factored out of two or more p-terms that contain the variable in an SOP or XOR expression.

Fall time: the period of time it takes a voltage signal to change from 90% to 10% of its high value.

Falling edge-triggered (FET): activation of a device on the falling edge of the triggering (sampling) variable.

False carry rejection: the feature in an ALU where all carry-outs are disabled for all nonarithmetic operations.

False data rejection (FDR): the feature of a code converter that indicates when unauthorized data has been issued to the converter.

Fan-in: the maximum number of inputs a gate may have.

Fan-out: the maximum number of equivalent gate inputs that a logic gate output can drive.

FDR: false data rejection.

FDS diagram: fully documented state diagram.

FED: field emission display.

Feedback path: a signal path of a PS variable from the memory output to the NS input.

FET: falling edge-triggered. Also, field effect transistor.

Fetch: that part of an instruction cycle in which the instruction is brought from the memory to the CPU.

FF: flip-flop.

Field programmable gate array (FPGA): a complex PLD that may contain a variety of primitive devices such as discrete gates, MUXs and flip-flops.

Field programmable logic array (FPLA): one-time user programmable PLA.

FIFO: first-in-first-out memory register.

Fill bit: the bit of a combinational shifter that receives the fill logic value in a shifting operation.

Finite state machine (FSM): a sequential machine that has a finite number of states in which it can reside.

Flag: a hardware or software "marker" used to indicate the status of a machine.

Flip-flop (FF): a one-bit memory element that exhibits sequential behavior controlled exclusively by a clock input.

Floating-gate NMOS: special NMOS used in erasable PROMs.

Floating point number (FPN) system: a binary number system expressed in two parts, as a fraction and exponential, and that is used in computers to arithmetically manipulate large numbers.

Flow chart: a chart that is made up of an interconnection of action and decision symbols for the purpose of representing the sequential nature of something.

Flow table: a tabular realization of a state diagram representing the sequential nature of an FSM.
Fly state: a state (in a state diagram) whose only purpose is to remove a race condition. A buffer state.
Forward bias: a voltage applied to a *p–n* junction diode in a direction as to cause the diode to conduct (turn ON).
FPGA: field programmable gate array.
FPLA: field programmable logic array.
FPLS: field programmable logic sequencer.
Free set: the variables of a function not used as the bond set in establishing the CRMT forms.
Frequency, f: the number of waveform cycles per unit time in Hz or s^{-1}.
Frequency division: the reduction of frequency by a factor of f/n usually by means of a binary counter, where n is the number of states in the counter.
FSM: finite state machine, either synchronous or asynchronous.
Full adder (FA): a combinational logic device that adds two binary bits to a carry-in bit and issues a SUM bit and a carry-out bit.
Full subtractor (FS): a combinational logic device that subtracts a subtrahend bit and a borrow-in bit from a minuend bit, and issues a difference bit and a borrow-out bit.
Fully documented state diagram: a state diagram that specifies all input branching conditions and output conditions in literal or mnemonic form, that satisfies the sum rule and mutually exclusive requirement, and that has been given a proper state code assignment.
Function: a Boolean expression representing a specific binary operation.
Functional partition: a diagram that gives the division of device responsibility in a digital system.
Function generator: a combinational logic device that generates logic functions (usually via a MUX).
Function hazard: a hazard that is produced when two or more coupled variables change in near proximity to each other.
Fundamental mode: the operational condition for an asynchronous FSM in which no input change is permitted to occur until the FSM has stabilized following any previous input change.
Fusible link: an element in a PLD memory bit location that can be "blown" to store a logic 1 or logic 0 depending on how the PLD is designed.
GAL: general array logic.
Gate: a physical device (circuit) that performs the electrical equivalent of a logic function. Also, one of three terminals of a MOSFET.
Gated basic cell: a basic cell that responds to its S and R input commands only on the triggering edge of a gate or clock signal.
Gate/input tally: the gate and input count associated with a given logic expression — the gate tally may or may not include inverters, but the input count must include both external and internal inputs.
Gate-minimum logic: logic requiring a minimum number of gates; may include XOR and EQV gates in addition to two-level logic.
Gate path delay: the interval of time required for the output of a gate to respond to an input signal change.
Glitch: an unwanted transient in an otherwise steady-state signal.

Go/No-Go configuration: a single input controlling the hold and exit conditions of a state in a state diagram.
Gray code: a reflective unit distance code.
Ground: a reference voltage level usually taken to be zero volts.
GS: group signal.
Half adder (HA): a combinational logic device that adds two binary bits and issues a sum bit and a carry-out bit.
Half subtractor: a combinational logic device that subtracts one binary bit from another and issues a difference bit and a borrow-out bit.
Hamming distance: as used in this text, the number of state variables required to change during a given state-to-state transition in an FSM.
Handshake interface: a configuration between two devices whereby the outputs of one device are the inputs to the other and vice versa.
Hang state: an isolated state in which an FSM can reside stably but which is not part of the authorized routine.
Hardware description language (HDL): a high-level programming language with specialized structures for modeling hardware.
Hazard: a glitch or unauthorized transition that is caused by an asymmetric path delay via an inverter, gate, or lead during a logic operation.
Hazard cover: the redundant cover that removes a static hazard.
HDL: hardware description language.
Heuristic: by empirical means or by discovery.
Hexadecimal (hex): a base 16 number system in which alphanumeric symbols are used to represent 4-bit binary numbers 0000 through 1111. (See Binary coded hexadecimal.)
Hold condition: branching from a given state back into itself or the input requirements necessary to effect such branching action.
Holding register: a PIPO (storage) register that is used to filter output signals.
Hold time: the interval of time immediately following the transition point during which the data inputs must remain logically stable to ensure that the intended transition of the FSM will be successfully completed.
Hole: the absence of a valence electron—the majority carrier in a p-type conducting semiconductor.
HV: high voltage.
Hybrid function: any function containing both SOP and POS terms.
IC: integrated circuit.
ICS: iterated carry-save.
Implicant: a term in a reduced or minimized expression.
Inactive: not active and implying logic 0.
Inactive state: the logic 0 state of a logic device.
Inactive transition point: the point in a voltage waveform where a digital device passes from the active state to the inactive state.
Incompatibility: a condition where the input to a logic device and the input requirement of that device are of opposite activation levels.
Incompletely specified function: a function that contains nonessential minterms or maxterms (see Don't care).
Increment: to increase usually by 1.

Indirect address approach: an alternative approach to FSM design where PS feedback to the NS logic is by way of a converter for the purpose of reducing MUX or PLD size.
Inertial delay element: a delay circuit based mainly on an R–C component.
Initialize: to drive a logic circuit into a beginning or reference state.
Input: a signal or line into a logic device that controls the operation of that device.
Input/state map: a K-map, with inputs as the axes and state identifiers as cell entries, that can be used to determine if the sum rule and the mutually exclusive requirement of any state in an FSM have been violated.
Integrated circuit (IC): an electronic circuit that is usually constructed entirely on a single small semiconductor chip called a monolith.
Intersection: AND operation.
Inversion: the inverting of a signal from HV to LV or vice versa.
Inverter: a physical device that performs inversion.
Involution: double complementation of a variable or function.
I/O: input/output.
IOB: I/O block.
Irredundant: not redundant, as applied to an absolute minimum Boolean expression.
Irrelevant input: an input whose presence in a function is nonessential.
Island: a K-map entry that must be looped out of a single cell.
Iterative: repeated many times to achieve a specific goal.
JEDEC: Joint Electron Device Engineering Council as it pertains to PLD programming format.
JK flip-flop: a type of flip-flop that can perform the set, reset, hold, and toggle operations.
Juxtapose: to place side by side.
Karnaugh map (K-map): graphical representation of a logic function named after M. Karnaugh (1953).
Keyword: a word specific to a given HDL.
Kirchhoff's current law: the algebraic sum of all currents into a circuit element or circuit section must be zero.
Kirchhoff's voltage law: the algebraic sum of all voltages around a closed loop must be zero.
K-map: Karnaugh map.
LAC: look-ahead-carry (see also CLA).
Large-scale integrated circuits (LSI): IC chips that contain 200 to thousands of gates.
Latch: a name given to certain types of memory elements as, for example, the D latch.
Latency: the time (usually in clock cycles) required to complete an operation in a sequential machine.
LCA: logic cell array.
LD: mnemonic for load.
Least significant bit (LSB): the bit (usually at the extreme right) of a binary word that has the lowest positional weight.
LED: light-emitting diode.
Level: a term used when specifying to the number of gate path delays of a logic function (from input to output) usually exclusive of inverters. See, for example, two-level logic.
Level triggered: rising edge triggered (RET) or falling edge triggered (FET).
Linear state machine: an FSM with a linear array of states.

Line driver: a device whose purpose it is to boost and sharpen a signal so as to avoid fan-out problems.
LFSR: linear feedback shift register.
LFSR counter: a counter, consisting of an LFSR, that can sequence through a unique set of pseudorandom states controlled by external inputs.
Logic: the computational capability of a digital device that is interpreted as either a logic 1 or logic 0.
Logic adjacency: two logic states whose state variables differ from each other by only one bit.
Logic cell: a configurable logic block (CLB).
Logic circuit: a digital circuit that performs the electrical equivalent of some logic function or process.
Logic diagram: a digital circuit schematic consisting of an interconnection of logic symbols.
Logic family: a particular technology such as TTL or CMOS that is used in the production of ICs.
Logic instability: the inability of a logic circuit to maintain a stable logic condition. Also, an oscillatory condition in an asynchronous FSM.
Logic level: logic status indicating either positive logic or negative logic.
Logic level conversion: the act of converting from positive logic to negative logic or vice versa.
Logic map: any of a variety of graphical representations of a logic function.
Logic noise: undesirable signal fluctuations produced within a logic circuit following input changes.
Logic state: a unique set of binary values that characterize the logic status of a machine at some point in time.
Logic waveform: a rectangular waveform between active and inactive states.
Look-ahead-carry (LAC): the feature of a "fast" adder that anticipates the need for a carry and then generates and propagates it more directly than does a parallel adder (see also carry look-ahead).
Loop-out: the action that identifies a prime implicant in a K-map.
Loop-out protocol: a minimization procedure whereby the largest 2 group of logically adjacent minterms or maxterms are looped out in the order of increasing n ($n = 0, 1, 2, 3, \ldots$).
LPD: lumped path delay.
LPDD: lumped path delay diagram.
LSB: least significant bit.
LSD: least significant digit.
LSI: large-scale integration.
Lumped path delay diagram (LPDD): a diagram that replaces discrete gates with other logic symbols for the purpose of comparing path delays from input to output.
Lumped path delay (LPD) model: a model, applicable to FSMs that operate in the fundamental mode, that is characterized by a lumped memory element for each state variable/feedback path.
LV: low voltage.
Magnitude comparator: comparator.

1.4 GLOSSARY OF TERMS, EXPRESSIONS, AND ABBREVIATIONS

Majority function: a function that becomes active when a majority of its variables become active.
Majority gate: a logic gate that yields a majority function.
Mantissa: the fraction part of a floating point number.
Map: usually a Karnaugh map.
Map compression: a reduction in the order of a K-map.
Map key: the order of K-map compression; hence, 2^{N-n}, where N is the number of variables in the function to be mapped and n is the order of the K-map to be used.
Mapping algorithm: In FSM design, the procedure to obtain the NS functions by ANDing the memory input logic value in the excitation table with the corresponding branching condition in the state diagram for the FSM to be designed, and entering the result in the appropriate cell of the NS K-map.
Master/slave (MS) flip-flop: a flip-flop characterized by a master (input) stage and a slave (output) stage that are triggered by clock antiphase to each other.
Mask: to prevent information from passing a certain point in a given process.
Mask programmed: refers to the bit patterns produced in a PLD chip at the foundry.
Maxterm: a POS term that contains all the variables of the function.
Maxterm code: a code in which complemented variables are assigned logic 1 and uncomplemented variables are assigned logic 0 — the opposite of minterm code.
Mealy machine: an FSM that conforms to the Mealy model.
Mealy model: the general model for a sequential machine where the output state depends on the input state as well as the present state.
Mealy output: a conditional output.
Medium-scale integrated circuits (MSI): IC chips that contain 20 to 200 gates according to one convention.
Memory: the ability of a digital device to store and retrieve binary words on command.
Memory element: a device for storing and retrieving one bit of information on command. In asynchronous FSM terminology, a fictitious lumped path delay.
Merge: the concatenation of buses to form a larger bus.
Merging of states: in a state diagram, the act of combining states to produce fewer states.
Metal-oxide-semiconductor: the material constitution of an important logic family (MOS) used in IC construction.
Metastability: an unresolved state of an FSM that resides between a Set and a Reset condition or that is logically unstable.
Metastable exit time: the time interval between entrance into and exit from the metastable state.
MEV: Map entered variable.
Minimization: the process of reducing a logic function to its simplest form.
Minimum cover: the optimally reduced representation of a logic expression.
Minterm: a term in an SOP expression where all variables of the expression are represented in either complemented or uncomplemented form.
Minterm code: a logic variable code in which complemented variables are assigned logic 0 while uncomplemented variables are assigned logic 1 — the opposite of maxterm code.
Minuend: the operand from which the subtrahend is subtracted in a subtraction operation.
Mixed logic: the combined use of the positive and negative logic systems.

Mixed-rail output: dual, logically equal outputs of a device (e.g., a flip-flop) where one output is issued active high while the other is issued active low, but the two are not issued simultaneously.

Mnemonic: a short single group of symbols (usually letters) that are used to convey a meaning.

Mnemonic state diagram: a fully documented state diagram.

Model: the means by which the major components and their interconnections are represented for a digital machine or system.

Module: a device that performs a specific function and that can be added to or removed from a system to alter the system's capability. A common example is a full adder.

Modulus-n counter: (see divide-by-n counter)

Monad: a minterm (or maxterm) that is not logically adjacent to any other minterm (or maxterm).

Moore machine: a sequential machine that conforms to the Moore model.

Moore model: a degenerate form of the Mealy (general) model in which the output state depends only on the present state.

Moore output: an unconditional output.

MOS: metal-oxide-semiconductor.

MOSFET: metal-oxide-semiconductor field effect transistor.

Most significant bit (MSB): the extreme left bit of a binary word that has the highest positional weight.

MSB: most significant bit.

MSD: most significant digit.

MSI: medium scale integration.

MTBF: mean time between failures.

Muller C module: a rendezvous module (RMOD).

Multilevel logic minimization: minimization involving more than two levels of path delay as, for example, that resulting from XOR-type patterns in K-maps.

Multiple-output minimization: optimization of more than one output expression from the same logic device.

Multiplex: to select or gate (on a time-shared basis) data from two or more sources onto a single line or transmission path.

Multiplexer: a device that multiplexes data.

Multiplicand: the number being multiplied by the multiplier.

Multiplier: a combinational logic device that will multiply two binary numbers. Also, the number being used to multiply the multiplicand.

Mutually exclusive requirement: a requirement in state diagram construction that forbids overlapping branching conditions (BCs) — i.e., it forbids the use of BCs shared between two or more branching paths.

MUX: multiplexer.

NAND-centered basic cell: cross-coupled NAND gates forming a basic cell.

NAND gate: a physical device that performs the electrical equivalent of the NOT AND function.

NAND/INV logic: combinational logic consisting exclusively of NAND gates and inverters.

Natural binary code: a code for which the bits are positioned in a binary word according to their positional weight in polynomial notation.

1.4 GLOSSARY OF TERMS, EXPRESSIONS, AND ABBREVIATIONS

Natural binary coded decimal: a 4-bit, 10-word code that is weighted 8, 4, 2, 1 and that is used to represent decimal numbers. Same as binary code.
NBCD: natural binary coded decimal. Same as binary coded decimal (BCD).
n-channel: an n-type conducting region in a p-type substrate.
Negative logic: a logic system in which high voltage (HV) corresponds to logic 0 and low voltage (LV) corresponds to logic 1. The opposite of positive logic.
Negative pulse: a 1–0–1 pulse.
Nested cell: a basic cell that is used as the memory in an asynchronous FSM design.
Nested machine: any asynchronous machine that serves as the memory in the design of a larger sequential machine. Any FSM that is embedded within another.
Next state (NS): a state that follows the present state in a sequence of states.
Next state forming logic: the logic hardware in a sequential machine whose purpose it is to generate the next state function input to the memory.
Next state function: the logic function that defines the next state of an FSM given the present state.
Next state map: a composite K-map where the entries for each cell are the next state functions for the present state represented by the coordinates of that cell (see flow table).
Next state variable: the variable representing the next state function.
Nibble: a group of four bits.
NMH: noise margin high — the lower voltage limit of logic 1 and the upper boundary of the uncertainty region.
NML: noise margin low — the upper voltage limit of logic 0 and the lower boundary of the uncertainty region.
NMOS: an n-channel MOSFET.
Noise immunity: the ability of a logic circuit to reject unwanted signals.
Noise margin: the maximum voltage fluctuation that can be tolerated in a digital signal without crossing the switching threshold of the switching device.
Non-restoring logic: logic that consists of passive switching devices such as diodes or transmission gates that cannot amplify but that dissipate power.
Nonvolatile: refers to memory devices that require no power supply to retain information in memory.
NOR-centered basic cell: cross-coupled NOR gates forming a basic cell.
NOR gate: a physical device that performs the electrical equivalent of the NOT OR function.
NOR/INV logic: combinational logic consisting exclusively of NOR gates and inverters.
NOT function: an operation that is the logic equivalent of complementation.
NOT laws: a set of Boolean identities based on the NOT function.
npn: refers to a BJT having a p-type semiconductor base and an n-type semiconductor collector and emitter.
NS: next state.
Octad: a grouping of eight logically adjacent minterms or maxterms.
Octal: a base 8 number system in which numbers 1 through 7 are used to represent 3-bit binary numbers 000 through 111. (See Binary coded octal.)
Odd parity: an odd number of 1's or 0's depending on how odd parity is defined.
Offset pattern: an XOR pattern in a K-map in which identical subfunctions are located in two nondiagonal cells that differ in cell coordinates by two bits.

Ohm's law: voltage is linearly proportional to current, $V = RI$, where R is the constant of proportionality called the resistance (in ohms).
One-hot code: a nonweighted code in which there exists only one 1 in each word of the code.
One-hot design method: use of the one-hot code for synchronous and asynchronous FSM design.
One-hot-plus-zero: one-hot code plus the all-zero state.
One's complement: a system of binary arithmetic in which a negative number is represented by complementing each bit of its positive equivalent.
Operand: a number or quantity that is to be operated on.
Operation table: a table that defines the functionality of a flip-flop or some other device.
Operator: a Boolean connective.
OPI: optional prime implicant.
Optional prime implicant (OPI): a prime implicant whose presence in a minimum function produces alternative minimum cover.
OR: an operator requiring that the output of an OR gate be active if one or more of its inputs are active.
OR-AND-Invert gate: a physical device, usually consisting of two OR gates and one NAND gate, that performs the electrical equivalent of POS with an active low output.
Order: refers to the number of variables on the axes of a K-map.
OR function: a function that derives from the definition of OR.
ORG: output race glitch.
OR gate: a physical device that performs the electrical equivalent of the OR function.
OR laws: a set of Boolean identities based on the OR function.
OR plane: the ORing stage of a PLD.
Outbranching: branching from a state exclusive of the hold branching condition.
Output: a concluding signal issued by a digital device.
Output forming logic: the logic hardware in a sequential machine whose purpose it is to generate the output signals.
Output holding register: a register, consisting of D flip-flops, that is used to filter out output logic noise.
Output race glitch (ORG): an internally initiated function hazard that is produced by a race condition in a sequential machine.
Overflow error: a false magnitude or sign that results from a left shift in a shifter when there are insufficient word bit positions at the spill end.
Packing density: the practical limit to which switches of the same logic family can be packed in an IC chip.
PAL: programmable array logic (registered trademark of Advanced Micro Devices, Inc.).
PALU: programmable arithmetic and logic unit.
Parallel adder: a cascaded array of full adders where the carry-out of a given full adder is the carry-in to the next most significant stage full adder.
Parallel load: the simultaneous loading of data inputs to devices such as registers and counters.
Parity: related to the existence of an even or odd number of 1's or 0's in a binary word.
Parity bit: a bit appended to a binary word to detect, create, or remove even or odd parity.
Parity detector: a combinational logic device that will detect an even (or odd) number of 1's (or 0's) in a binary word.

Parity generator: a combinational logic device that will append a logic 1 (or logic 0) to a binary word so as to generate an even (or odd) number of 1's (or 0's).

Passive device: any device that is incapable of producing voltage or current gain and, thus, only dissipates power.

Pass transistor switch: a MOS transistor switch that functions as a nonrestoring switching device and that does not invert a voltage signal. A transmission gate.

PCB: printed circuit board.

p-channel: a p-type conducting region in an n-type substrate.

PDF: portable document format.

PDP: power–delay product.

PE: priority encoder.

Period: the time in seconds (s) between repeating portions of a waveform; hence, the inverse of the frequency.

Physical truth table: an I/O specification table based on a physically measurable quantity such as voltage.

PI: prime implicant.

Pipeline: a processing scheme where each task is allocated to specific hardware (joined in a line) and to a specific time slot.

PIPO: parallel-in/parallel-out operation mode of a register.

PISO: parallel-in/serial-out operation mode of a register.

PLA: programmable logic array.

Planar format: a two-dimensional K-map array used to minimize functions of more than four variables.

PLD: programmable logic device.

PLS: programmable logic sequencer.

PMOS: a p-channel MOSFET.

p–n **junction diode:** (see Diode)

pnp: refers to a BJT having an n-type semiconductor base and a p-type semiconductor emitter and collector.

Polarized mnemonic: a contracted signal name onto which is attached an activation level indicator.

Port: an entry or exit element to an entity (e.g., the name given to an input signal in a VHDL declaration).

POS: product-of-sums.

POS hazard: a static 0-hazard.

Positional weighting: a system in which the weight of a bit in a binary word is determined by its polynomial representation.

Positive logic: the logic system in which HV corresponds to logic 1 and LV corresponds to logic 0.

Positive pulse: a 0–1–0 pulse.

Power, P: the product of voltage, V, and current, I, given in units of watts (W).

Power–delay product (PDP): the average power dissipated by a logic device multiplied by its propagation delay time.

PR or PRE: preset.

Present state (PS): the logic state of an FSM at a given instant.

Present state/next state (PS/NS) table: a table that is produced from the next state K-maps and that is used to construct a fully documented state diagram in an FSM analysis.

Preset: an asynchronous input that is used in flip-flops to set them to a logic 1 condition.
Prime implicant (PI): a group of adjacent minterms or maxterms that are sufficiently large that they cannot be combined with other groups in any way to produce terms of fewer variables.
Primitive: a discrete logic device such as a gate, MUX, or decoder.
Priority encoder: a logic device that generates a coded output based on a set of prioritized data inputs.
Product-of-sums (POS): the ANDing of ORed terms in a Boolean expression.
Programmable logic array (PLA): any PLD that can be programmed in both the AND and OR planes.
Programmable logic device (PLD): any two-level, combinational array logic device from the families of ROMs, PLAs, PALs or FPGAs, etc.
Programmable read-only memory (PROM): a once-only user-programmable ROM.
PROM: programmable read-only memory.
Propagation delay: in a logic device, the time interval of an output response to an input signal.
PS: present state.
PS/NS: present state/next state.
P-term: a Boolean product term–one consisting only of ANDed literals.
P-term table: a table that consists of p-terms, inputs, and outputs and that is used to program PLA-type devices.
Pull-down resistor: a resistor that causes a signal on a line to remain at low voltage.
Pull-up resistor: a resistor that causes a signal on a line to remain at high voltage.
Pulse: an abrupt change from one level to another followed by an opposite abrupt change.
Pulse mode: an operational condition for an asynchronous FSM where the inputs are required to be nonoverlapping pulse signals.
Pulse width: the active duration of a positive pulse or the inactive duration of a negative pulse.
Quad: a grouping of four logically adjacent minterms or maxterms.
Quadratic convergence: a process as in "fast division" whereby the error per iteration decreases according to the inverse square law.
Quotient: the result of a division operation.
R: reset.
Race condition: a condition in a sequential circuit where the transition from one state to another involves two or more alternative paths.
Race gate: the gate to which two or more input signals are in race contention.
Race path: any path that can be taken in a race condition.
Race state: any state through which an FSM may transit during a race condition.
Radix: the number of unique symbols in a number system — same as the base of a number system.
RAM: random access memory.
Random access memory (RAM): a read/write memory system in which all memory locations can be accessed directly independent of other memory locations.
R–C: resistance/capacitance or resistor/capacitor.
Read only memory (ROM): a PLD that can be mask programmed only in the OR plane.

Read/write memory (RWM): a memory array (e.g., RAM) that can be used to store and retrieve information at any time.
Redundant cover: nonessential and nonoptional cover in a function representation.
Redundant prime implicant: a prime implicant that yields redundant cover.
Reflective code: a code that has a reflection (mirror) plane midway through the code.
Register: a digital device, configured with flip-flops and other logic, that is capable of storing and shifting data on command.
Remainder: in division, the dividend minus the product of the divisor and the quotient.
Rendezvous module: an asynchronous state machine whose output becomes active when all external inputs become active and becomes inactive when all external inputs become inactive.
Reset: a logic 0 condition or an input to a logic device that sets it to a logic 0 condition.
Residue: the part of term that remains when the coupled variable is removed (see consensus term).
Resistance, R: the voltage drop across a conducting element divided by current through the element (in ohms).
Resistor-transistor logic: a logic family that consists of BJTs and resistors.
Restoring logic: logic consisting of switching devices such as BJTs and MOSFETs that can amplify.
RET: rising edge triggered.
Reverse bias: a voltage applied to a p–n junction diode in a direction that minimizes conduction across the junction.
Reverse saturation current: the current through a p–n junction diode under reverse bias.
Ring counter: a configuration of shift registers that generates a one-hot code output.
Ripple carry (R-C): the process by which a parallel adder transfers the carry from one full adder to another.
Ripple counter: a counter whose flip-flops are each triggered by the output of the next LSB flip-flop.
Rise time: he period of time it takes a voltage (or current) signal to change from 10% to 90% of its high value.
Rising edge triggered (RET): activation of a logic device on the rising edge of the triggering variable.
RMOD: rendezvous module.
ROM: read-only memory.
Round-off error: the amount by which a magnitude is diminished due to an underflow or spill-off in a shifter undergoing a right shift.
RPI: redundant prime implicant.
RTL: resistor–transistor logic.
Runt pulse: any pulse that barely reaches the switching threshold of a device into which it is introduced.
S: set. Also, the source terminal of a MOSFET.
Sampling interval: sum of the setup and hold times.
Sampling variable: the last variable to change in initiating a state-to-state transition in an FSM.
Sanity circuit: a circuit that is used to initialize an FSM into a particular state, usually a resistor/capacitor (R–C) type circuit.

Saturation mode: the physical state of a BJT in which collector current is permitted to flow.

Schmitt trigger: an electronic gate with hysteresis and high noise immunity that is used to "square up" pulses.

Selector module: a device whose function it is to steer one of two input signals to either one of two outputs depending on whether a specific input is active or inactive.

Self-correcting counter: a counter for which all states lead into the main count sequence or routine.

Sequence detector (recognizer): a sequential machine that is designed to recognize a particular sequence of input signals.

Sequential machine: any digital machine with feedback paths whose operation is a function of both its history and its present input data.

Set: a logic 1 condition or an input to a logic device that sets it to a logic 1 condition.

Setup time: the interval of time prior to the transition point during which all data inputs must remain stable at their proper logic level to ensure that the intended transition will be initiated.

S-hazard: a static hazard.

Shift register: a register that is capable of shifting operations.

Shift: the movement of binary words to the left or right in a shifter or shift register.

Shifter: a combinational logic device that will shift or rotate data asynchronously upon presentation.

Sign bit: a bit appended to a binary number (usually in the MSB position) for the purpose of indicating its sign.

Sign-complement arithmetic: 1's or 2's complement arithmetic.

Sign-magnitude representation: a means of identifying positive and negative binary numbers by a sign and magnitude.

Single transition time (STT): a state-to-state transition in an asynchronous FSM that occurs in the shortest possible time, that is, without passing through a race state.

SIPO: serial-in/parallel-out operation mode of a register.

SISO: serial-in/serial-out operation mode of a register.

Slice: that part of a circuit or device that can be cascaded to produce a larger circuit or device.

Small-scale integration: IC chips that, by one convention, contain up to 20 gates.

SOP: sum-of-products.

SOP hazard: a static 1-hazard.

Source: one of three terminals of a MOSFET. The origin of a digital signal.

Spill bit: the bit in a shifter or shift register that is spilled off (lost) in a shifting operation.

SPDT switch: single-pole/double-throw switch.

SPST switch: single-pole/single-throw switch.

Square wave: a rectangular waveform.

SRAM: static RAM.

SSI: small-scale integration.

Stability criteria: the requirements that determine if an asynchronous FSM, operated in the fundamental mode, is stable or unstable in a given state.

Stable state: any logic state of an asynchronous FSM that satisfies the stability criteria.

Stack format: a three-dimensional array of conventional fourth-order K-maps used for function minimization of more than four variables.

1.4 GLOSSARY OF TERMS, EXPRESSIONS, AND ABBREVIATIONS

State: a unique set of binary values that characterize the logic status of a machine at some point in time.

State adjacency set: any 2^n set of logically adjacent states of an FSM.

State code assignment: unique set of code words that are assigned to an FSM to characterize its logic status.

State diagram: the diagram or chart of an FSM that shows the state sequence, branching conditions, and output information necessary to describe its sequential behavior.

State machine: a finite state machine (FSM). A sequential machine.

State identifier: any symbol (e.g., alphabetical) that is used to represent or identify a state in a state diagram.

State table: tabular representation of a state diagram.

State transition table: (see excitation table).

State variable: any variable whose logic value contributes to the logic status of a machine at any point in time. Any bit in the state code assignment of a state diagram.

Static hazard: an unwanted glitch in an otherwise steady-state signal that is produced by an input change propagating along asymmetric path delays through inverters or gates.

Static-1 hazard: a glitch that occurs in an otherwise steady-state 1 output signal from SOP logic due to a change in an input for which there are two asymmetric paths (delay-wise) to the output.

Static-0 hazard: a glitch that occurs in an otherwise steady-state 0 output signal from POS logic due to a change in an input for which there are two asymmetric paths (delay-wise) to the output.

Static RAM: a nonvolatile form of RAM — does not need periodic refreshing to hold its information.

Steering logic: logic based primarily on transmission gate switches.

S-term: a Boolean sum term — one containing only ORed literals.

Stretcher: an input conditioning device that catches a short input signal and stretches it.

STT: single transition time.

Stuck-at fault: an input to a logic gate that is permanently stuck at logic 0 or logic 1 because of a shorted connection, an open connection, or a connection to either ground or a voltage supply.

Substrate: the supporting or foundation material in and on which a semiconductor device is constructed.

Subtractor: a digital device that subtracts one binary word from another to give a difference and borrow.

Subtrahend: the operand being subtracted from the minuend in a subtraction operation.

Sum-of-products (SOP): the ORing of ANDed terms in a Boolean expression.

Sum rule: a rule in state diagram construction that requires that all possible branching conditions be accounted for.

Switching speed: a device parameter that is related to its propagation delay time.

Synchronizer circuit: a logic circuit (usually a D flip-flop) that is used to synchronize an input with respect to a clock signal.

Synchronous machine: a sequential machine that is clock driven.

Synchronous parallel load: parallel loading of a register or counter via a clock signal to the flip-flops.

System level design: a design that includes controller and data path sections.

Tabular minimization: a minimization procedure that uses tables exclusively.

T flip-flop: a flip-flop that operates in either the toggle or hold mode.
TG: transmission gate.
Throughput: the time required to produce an output response due to an input change.
Time constant: the product of resistance and capacitance given in units of seconds (s) — a measure of the recovery time of an R–C circuit.
Timing diagram: a set of logic waveforms showing the time relationships between two or more logic signals.
Toggle: repeated but controlled transitions between any two states, as between the Set and Reset states.
Toggle module: a flip-flop that is configured to toggle only. Also, a divide-by-2 counter.
Transfer characteristic: for a transistor switch, a plot of current (I) vs voltage (V).
Trans-HI module: a transparent high (RET) D latch.
Trans-LO module: a transparent low (FET) D latch.
Transistor: a three-terminal switching device that exhibits current or voltage gain.
Transistor–transistor logic: a logic family in which bipolar junction transistors provide both logic decision and current gain.
Transition: in a digital machine, a change from one state (or level) to another.
Transmission gate: a pass transistor switch.
Transparent D latch: a two-state D flip-flop in which the output, Q, tracks the input, D, when clock is active if RET or when clock is inactive if FET.
Tree: combining of like gates, usually to overcome fan-in limitations.
Triggering threshold: the point beyond which a transition takes place.
Triggering variable: sampling (enabling) variable.
Tri-state bus: as used in this text, the wire-ORed output lines from a multiplexed scheme of PLDs having tri-state enables. Note: tri-state is a registered trademark of NSC.
Tri-state driver: an active logic device that operates in either a disconnect mode or an inverting (or noninverting) mode. Also, three-state driver. Note: tri-state is a registered trademark of NSC.
True hold: the condition whereby a device can sustain the same logic output values over any number of clock cycles independent of its input logic status.
Truth table: a table that provides an output value for each possible input condition to a combinational logic device.
TTL: transistor–transistor (BJT) logic.
Twisted ring counter: a configuration of shift registers that generates a creeping code output.
Two-level logic: logic consisting of only one ANDing and one ORing stage.
Two-phase clocking: two synchronized clock signals that have nonoverlapping active or nonoverlapping inactive waveforms.
Two's complement: one's complement plus one added to the LSB.
Unconditional branching: state-to-state transitions that take place independent of the input status of the FSM.
Unconditional output: an output of an FSM that does not depend on an input signal.
Union: OR operation.
Unit distance code: a code in which each state in the code is surrounded by logically adjacent states.
Universal flip-flop: a JK flip-flop.
Universal gate: a NAND or NOR gate.

Universal shift register: a shift register capable of performing PIPO, PISO, SIPO, and SISO operations in addition to being capable of performing the true hold condition.

Unstable state: any logic state in an asynchronous FSM that does not satisfy the stability criteria.

Unweighted code: a code that cannot be constructed by any mathematical weighting procedure.

USR: universal shift register.

UVEPROM: ultraviolet erasable PROM.

VEM: variable entered map.

Very large scale integrated circuits: IC chips that contain thousands to millions of gates.

VHDL: VHSIC hardware description language.

VHSIC: very high speed integrated circuit.

VLSI: very large scale integrated circuits.

Voltage, V: the potential difference between two points, in units of volts (V). Also, the work required to move a positive charge against an electric field.

Voltage waveform: a voltage waveform in which rise and fall times exist.

Weighted code: a binary code in which the bit positions are weighted with different mathematically determined values.

Wired logic: an arrangement of logic circuits in which the outputs are physically connected to form an "implied" AND or OR function.

WSI circuits: wafer-scale integrated circuits.

XNOR: (see Equivalence and EQV)

XOR: exclusive OR.

XOR function: the function that derives from the definition of exclusive OR.

XOR gate: a physical device that performs the electrical equivalent of the XOR function.

XOR laws: a set of Boolean identities that are based on the XOR function.

XOR pattern: any of four possible K-map patterns that result in XOR type functions.

XS3 code: BCD code plus three.

ZBI: zero-blanking input.

ZBO: zero-blanking output.

Zero banking: a feature of a BCD-to-seven-segment conversion that blanks out the seven-segment display if all inputs are zero.

CHAPTER 2

Number Systems, Binary Arithmetic, and Codes

2.1 INTRODUCTION

Number systems provide the basis for conveying and quantifying information. Weather data, stocks, pagination of books, weights and measures — these are just a few examples of the use of numbers that affect our daily lives. For this purpose we find the decimal (or Arabic) number system to be reliable and easy to use. This system evolved presumably because early humans were equipped with a crude type of calculator, their 10 fingers. But a number system that is appropriate for humans may be intractable for use by a machine such as a computer. Likewise, a number system appropriate for a machine may not be suitable for human use.

Before concentrating on those number systems that are useful in computers, it will be helpful to review those characteristics that are desirable in any number system. There are *four* important characteristics in all:

- Distinguishability of symbols
- Arithmetic operations capability
- Error control capability
- Tractability and speed

To one degree or another the decimal system of numbers satisfies these characteristics for hard-copy transfer of information between humans. Roman numerals and binary are examples of number systems that do not satisfy all four characteristics for human use. On the other hand, the binary number system is preferable for use in digital computers. The reason is simply put: current digital electronic machines recognize only two identifiable states, physically represented by a high voltage level and a low voltage level. These two physical states are logically interpreted as binary symbols 1 and 0.

A fifth desirable characteristic of a number system to be used in a computer should be that it have a minimum number of easily identifiable states. The binary number system satisfies this condition. However, the digital computer must still interface with humankind. This is done by converting the binary data to a decimal and character-based form that can be readily understood by humans. A minimum number of identifiable characters (say 1 and 0, or true and false) is not practical or desirable for direct human use. If this is difficult to understand, imagine trying to complete a tax form in binary or in any number system other than decimal. On the other hand, use of a computer for this purpose would not only be practical but, in many cases, highly desirable.

2.2 POSITIONAL AND POLYNOMIAL REPRESENTATIONS

The *positional form* of a number is a set of side-by-side (juxtaposed) digits given generally in *fixed-point* form as

$$N_r = (\underbrace{a_{n-1} \cdots a_2 a_1 a_0}_{Integer} \cdot \underbrace{a_{-1} a_{-2} a_{-3} \cdots a_{-m}}_{Fraction})_r \tag{2.1}$$

where MSD $\downarrow$ points to a_{n-1}, the Radix Point is between a_0 and a_{-1}, and LSD $\downarrow$ points to a_{-m}.

where the radix (or base), r, is the total number of digits in the number system, and a is a digit in the set defined for radix r. Here, the radix point separates n integer digits on the left from m fraction digits on the right. Notice that a_{n-1} is the most significant (highest order) digit called MSD, and that a_{-m} is the least significant (lowest order) digit denoted by LSD.

The *value* of the number in Eq. (2.1) is given in *polynomial form* by

$$N_r = \sum_{i=-m}^{n-1} a_i r^i = (a_{n-1} r^{n-1} + \cdots + a_2 r^2 + a_1 r^1 + a_0 r^0 + a_{-1} r^{-1}$$
$$+ a_{-2} r^{-2} + \cdots + a_{-m} r^{-m})_r, \tag{2.2}$$

where a_i is the digit in the ith position with *a weight r^i*.

Applications of Eqs. (2.1) and (2.2) follow directly. For the decimal system $r = 10$, indicating that there are 10 distinguishable characters recognized as decimal numerals $0, 1, 2, \ldots, r - 1 (= 9)$. Examples of the positional and polynomial representations for the decimal system are

$$N_{10} = (d_3 d_2 d_1 d_0 \cdot d_{-1} d_{-2} d_{-3})_{10}$$
$$= 3017.528$$

and

$$N_{10} = \sum_{i=-3}^{n-1} d_i 10^i$$
$$= 3 \times 10^3 + 0 \times 10^2 + 1 \times 10^1 + 7 \times 10^0 + 5 \times 10^{-1} + 2 \times 10^{-2} + 8 \times 10^{-3}$$
$$= 3000 + 10 + 7 + 0.5 + 0.02 + 0.008,$$

2.3 UNSIGNED BINARY NUMBER SYSTEM

where d_i is the decimal digit in the ith position. Exclusive of possible leading and trailing zeros, the MSD and LSD for this number are 3 and 8, respectively. This number could have been written in a form such as $N_{10} = 03017.52800$ without altering its value but implying greater accuracy of the fraction portion.

2.3 UNSIGNED BINARY NUMBER SYSTEM

Applying Eqs. (2.1) and (2.2) to the binary system requires that $r = 2$, indicating that there are two distinguishable characters, typically 0 and $(r - 1) = 1$, that are used. In positional representation these characters (numbers) are called *binary digits* or *bits*. Examples of the positional and polynomial notations for a binary number are

$$N_2 = (b_{n-1} \cdots b_3 b_2 b_1 b_0 \cdot b_{-1} b_{-2} b_{-3} \cdots b_{-m})_2$$
$$= 1\ 0\ 1\ 1\ 0\ 1\ .\ 1\ 0\ 1_2$$
$$\quad\uparrow \qquad\qquad\qquad\quad \uparrow$$
$$\text{MSB} \qquad\qquad\qquad \text{LSB}$$

and

$$N = \sum_{i=-m}^{n-1} b_i 2^i$$
$$= 1 \times 2^5 + 0 \times 2^4 + 1 \times 2^3 + 1 \times 2^2 + 0 \times 2^1$$
$$\quad + 1 \times 2^0 + 1 \times 2^{-1} + 0 \times 2^{-2} + 1 \times 2^{-3}$$
$$= 32 + 8 + 4 + 1 + 0.5 + 0.125$$
$$= 45.625_{10},$$

where $n = 6$ and $m = 3$, and b_i is the bit in the ith position. Thus, the bit positions are weighted,...16, 8, 4, 2, 1, 1/2, 1/4, 1/8,...for any number consisting of integer and fraction portions. Binary numbers, so represented, are sometimes referred to as *natural* binary. In positional representation, the bit on the extreme left and extreme right are called the MSB (most significant bit) and LSB (least significant bit), respectively. Notice that by obtaining the value of a binary number, a conversion from binary to decimal has been performed. The subject of radix (base) conversion will be dealt with more extensively in a later section.

For reference purposes, Table 2.1 provides the binary-to-decimal conversion for two-, three-, four-, five-, and six-bit binary. The six-bit binary column is only halfway completed for brevity.

In the natural binary system the number of bits in a unit of data is commonly assigned a name. Examples are:

4 data-bit unit — nibble (or half byte)
8 data-bit unit — byte
16 data-bit unit — two bytes (or half word)
32 data-bit unit — word (or four bytes)
64 data-bit unit — double-word

Table 2.1 Binary-to-decimal conversion

Two-Bit Binary	Decimal Value	Three-Bit Binary	Decimal Value	Four-Bit Binary	Decimal Value	Five-Bit Binary	Decimal Value	Six-Bit Binary	Decimal Value
00	0	000	0	0000	0	10000	16	100000	32
01	1	001	1	0001	1	10001	17	100001	33
10	2	010	2	0010	2	10010	18	100010	34
11	3	011	3	0011	3	10011	19	100011	35
		100	4	0100	4	10100	20	100100	36
		101	5	0101	5	10101	21	100101	37
		110	6	0110	6	10110	22	100110	38
		111	7	0111	7	10111	23	100111	39
				1000	8	11000	24	101000	40
				1001	9	11001	25	101001	41
				1010	10	11010	26	101010	42
				1011	11	11011	27	101011	43
				1100	12	11100	28	101100	44
				1101	13	11101	29	101101	45
				1110	14	11110	30	101110	46
				1111	15	11111	31	101111	47
								.	.
								.	.
								.	.

The word size for a computer is determined by the number of bits that can be manipulated and stored in registers. The foregoing list of names would be applicable to a 32-bit computer.

2.4 UNSIGNED BINARY CODED DECIMAL, HEXADECIMAL, AND OCTAL

Although the binary system of numbers is most appropriate for use in computers, this system has several disadvantages when used by humans who have become accustomed to the decimal system. For example, binary machine code is long, difficult to assimilate, and tedious to convert to decimal. But there exist simpler ways to represent binary numbers for conversion to decimal representation. Three examples, commonly used, are natural binary coded decimal (BCD), binary coded hexadecimal (BCH), and binary coded octal (BCO). These number systems are useful in applications where a digital device, such as a computer, must interface with humans. The BCD code representation is also useful in carrying out computer arithmetic.

2.4.1 The BCD Representation

The BCD system is an 8, 4, 2, 1 weighted code. This system uses patterns of four bits to represent each decimal position of a number and is converted to its decimal equivalent by

2.4 UNSIGNED BINARY CODED DECIMAL, HEXADECIMAL, AND OCTAL

Table 2.2 BCD bit patterns and decimal equivalent

BCD Bit Pattern	Decimal	BCD Bit Pattern	Decimal
0000	0	1000	8
0001	1	1001	9
0010	2	1010	NA
0011	3	1011	NA
0100	4	1100	NA
0101	5	1101	NA
0110	6	1110	NA
0111	7	1111	NA

NA = not applicable (code words not valid)

polynomials of the form

$$N_{10} = b_3 \times 2^3 + b_2 \times 2^2 + b_1 \times 2^1 + b_0 \times 2^0$$
$$= b_3 \times 8 + b_2 \times 4 + b_1 \times 2 + b_0 \times 1$$

for any $b_3b_2b_1b_0$ code integer. Thus, decimal 6 is represented as $(0 \times 8) + (1 \times 4) + (1 \times 2) + (0 \times 1)$ or 0110 in BCD code. As in binary, the bit positional weights of the BCD code are derived from integer powers of 2^n. Table 2.2 shows the BCD bit patterns for decimal integers 0 through 9.

Decimal numbers greater than nine or less than one can be represented by the BCD code if each digit is given in that code and if the results are combined. For example, the number 63.98 is represented by (or converted to) the BCD code

$$\begin{array}{cccc} 6 & 3 & . & 9 & 8 \end{array}$$
$$63.98_{10} = (0110\ 0011\ .\ 1001\ 1000)_{BCD}$$
$$= 1100011.10011_{BCD}$$

Here, the code weights are 80, 40, 20, 10; 8, 4, 2, 1; 0.8, 0.4, 0.2, 0.1; and 0.08, 0.04, 0.02, 0.01 for the tens, units, tenths, and hundredths digits, respectively, representing four decades. Notice that the leading and trailing 0's can be dropped. Pencil-and-paper conversion between binary and BCD requires conversion to decimal as an intermediate step. For example, to convert from BCD to binary requires that groups of four bits be selected in both directions from the radix point to form the decimal number. If necessary, leading and trailing zeros are added to the leftmost or rightmost ends to complete the groups of four bits as in the example above. Negative BCD numbers are coded by using 10's complement notation as discussed in a later section.

Another code that is used for number representation and manipulation is called Excess 3 BCD (or XS3 BCD or simply XS3). XS3 is an example of a *biased-weighted* code (a bias of 3). This code is formed by adding $0011_2 (= 3_{10})$ to the BCD bit patterns in Table 2.2.

Thus,

$$XS3 = BCD + 0011.$$

For example, the number 63.98_{10} is represented in XS3 code as $1001\,0110\,.\,1100\,1011_{XS3}$. To convert XS3 to BCD code, 0011 must be subtracted from XS3 code. In 4-bit quantities the XS3 code has the useful feature that when two numbers are added together in XS3 notation, a carry will result and yield the correct value any time a carry results in decimal (i.e., when 9 is exceeded). This feature is not shared by either binary or BCD addition.

2.4.2 The Hexadecimal and Octal Systems

The hexadecimal number system requires that $r = 16$ in Eqs. (2.1) and (2.2), indicating that there are 16 distinguishable characters in the system. By convention, the permissible hexadecimal digits are 0, 1, 2, 3, 4, 5, 6, 7, 8, 9, A, B, C, D, E, and F for decimals 0 through 15, respectively. Examples of the positional and polynomial representations for a hexadecimal number are

$$N_{16} = (h_{n-1} \cdots h_3 h_2 h_1 h_0 \cdot h_{-1} h_{-2} h_{-3} \cdots h_{-m})_{16}$$
$$= (AF3 \cdot C8)_{16}$$

with a decimal value of

$$N = \sum_{i=-m}^{n-1} h_i 16^i$$
$$= 10 \times 16^2 + 15 \times 16^1 + 3 \times 16^0 + 12 \times 16^{-1} + 8 \times 16^{-2}$$
$$= 2803.78125_{10}.$$

Here, it is seen that a hexadecimal number has been converted to decimal by using Eq. (2.2).

The octal number system requires that $r = 8$ in Eqs. (2.1) and (2.2), indicating that there are eight distinguishable characters in this system. The permissible octal digits are 0, 1, 2, 3, 4, 5, 6, and 7, as one might expect. Examples of the application of Eqs. (2.1) and (2.2) are

$$N_8 = (o_{n-1} \cdots o_2 o_1 o_0 \cdot o_{-1} o_{-2} o_{-3} \cdots o_{-m})_8$$
$$= 501.74_8,$$

with a decimal value of

$$N = \sum_{i=-m}^{n-1} o_i 8^i$$
$$= 5 \times 8^2 + 0 \times 8^1 + 1 \times 8^0 + 7 \times 8^{-1} + 4 \times 8^{-2}$$
$$= 321.9375_{10}.$$

2.5 CONVERSION BETWEEN NUMBER SYSTEMS

Table 2.3 The BCH and BCO number systems

Binary	BCH	BCO	Decimal	Binary	BCH	BCO	Decimal
0000	0	0	0	1010	A	12	10
0001	1	1	1	1011	B	13	11
0010	2	2	2	1100	C	14	12
0011	3	3	3	1101	D	15	13
0100	4	4	4	1110	E	16	14
0101	5	5	5	1111	F	17	15
0110	6	6	6	10000	10	20	16
0111	7	7	7	11011	1B	33	27
1000	8	10	8	110001	31	61	49
1001	9	11	9	1001110	4E	116	78

When the hexadecimal and octal number systems are used to represent bit patterns in binary, they are called binary coded hexadecimal (BCH) and binary coded octal (BCO), respectively. These two number systems are examples of *binary-derived radices*. Table 2.3 lists several selected examples showing the relationships between BCH, BCO, binary and decimal.

What emerges on close inspection of Table 2.3 is that each hexadecimal digit corresponds to four binary digits, and that each octal digit corresponds to three binary digits. The following example illustrate the relationships between these number systems:

$$10110111111.11011_2 = \begin{matrix} 5 & B & F & . & D & 8 \\ 0101 & 1011 & 1111 & . & 1101 & 1000 \end{matrix}$$

$$= 5BF.D8_{16}$$

$$= \begin{matrix} 2 & 6 & 7 & 7 & . & 6 & 6 \\ 010 & 110 & 111 & 111 & . & 110 & 110 \end{matrix}$$

$$= 2677.66_8$$

$$= 1471.84375_{10}.$$

To separate the binary digits into groups of four (for BCH) or groups of three (for BCO), counting must begin from the radix point and continue outward in both directions. Then, where needed, zeros are added to the leading and trailing ends of the binary representation to complete the MSDs and LSDs for the BCH and BCO forms.

2.5 CONVERSION BETWEEN NUMBER SYSTEMS

It is not the intent of this section to cover all methods for radix (base) conversion. Rather, the plan is to provide general approaches, separately applicable to the integer and fraction portions, followed by specific examples.

2.5.1 Conversion of Integers

Since the polynomial form of Eq. (2.2) is a geometrical progression, the integer portion can be represented in *nested radix* form. In source radix s, the nested representation is

$$N_s = (a_{n-1}s^{n-1} + a_{n-2}s^{n-2} + \cdots + a_1s^1 + a_0s^0)_s$$
$$= a_0 + s(a_1 + s(a_2 + \cdots + a_{n-1}))\cdots)_s$$
$$= a_0 + s\left(\sum_{i=1}^{n-1} a_i s^{i-1}\right) \quad (2.3)$$

for digits a_i having integer values from 0 to $n-1$. The nested radix form not only suggests a conversion process, but also forms the basis for computerized conversion.

Consider that the number in Eq. (2.3) is to be represented in nested radix r form,

$$N_r = b_0 + r(b_1 + r(b_2 + \cdots + b_{m-1}))\cdots)_r$$
$$= b_0 + r\left(\sum_{i=1}^{m-1} b_i r^{i-1}\right), \quad (2.4)$$

where, in general, $m \neq n$. Then, if the source number N_s is divided by r, the results are of the form

$$\frac{N_s}{r} = Q + \frac{R}{r}, \quad (2.5)$$

where Q is the integer quotient rearranged as $Q_0 = b_1 + r(b_2 + \cdots + b_{m-1})\cdots)$ and R is the remainder $R_0 = b_0$. A second division by r yields $Q_0/r = Q_1 + R_1/r$, where Q_1 is arranged as $Q_1 = b_2 + r(b_3 + \cdots + b_{m-1})\cdots)_r$ and $R_1 = b_1$. Thus, by repeated division of the integer result Q_i by r, the remainders yield $(b_0, b_1, b_2, \ldots, b_{m-1})_r$ in that order.

The conversion method just described, called the *radix divide method*, can be used to convert between any two integers of different radices. However, the requirement is:

> The arithmetic required by N_s/r must be carried out in source radix, s.

Except for source radices 10 and 2, this creates a problem for humans.

Table 2.4 provides the recommended procedures for integer conversion by noncomputer means. The radix divide method is suitable for use in computers only if they are programmed to carry out the arithmetic in different radices. Notice the partitioning required for conversion between binary and BCH and BCO integers.

The following two algorithms offer noncomputer methods for integer conversion:

> **Algorithm 2.1: $N_r \leftarrow N_s$ Positive Integer Conversion**
>
> Use Eq. (2.2) and the substitution method with base 10 arithmetic to convert N_s to N_{10}, then use the radix divide method and base 10 arithmetic to convert N_{10} to N_r.

2.5 CONVERSION BETWEEN NUMBER SYSTEMS

Table 2.4 Summary of recommended methods for integer conversion by noncomputer means

Integer Conversion	Conversion Method
$N_{10} \to N_r$	Radix division by radix r using Eq. (2.5)
$N_s \to N_{10}$	Eq. (2.2) or (2.3)
$(N_s)_{s \neq 10} \to (N_r)_{r \neq 10}$	$N_s \to N_{10}$ by Eq. (2.2) or (2.3)
	$N_{10} \to N_r$ radix division by r using Eq. (2.5)
Special Cases for Binary Forms	
$N_2 \to N_{10}$	Positional weighting
$N_2 \to N_{BCH}$	Partition N_2 into groups of four bits starting from radix point, then apply Table 2.3
$N_2 \to N_{BCO}$	Partition N_2 into groups of three bits starting from radix point, then apply Table 2.3
$N_{BCH} \to N_2$	Reverse of $N_2 \to N_{BCH}$
$N_{BCO} \to N_2$	Reverse of $N_2 \to N_{BCO}$
$N_{BCH} \to N_{BCO}$	$N_{BCH} \to N_2 \to N_{BCO}$
$N_{BCO} \to N_{BCH}$	$N_{BCO} \to N_2 \to N_{BCH}$
$N_{BCD} \to N_{XS3}$	Add $0011_2 (= 3_{10})$ to N_{BCD}
$N_{XS3} \to N_{BCD}$	Subtract $0011_2 (= 3_{10})$ from N_{XS3}

> **Algorithm 2.2:** $N_{2^k} \leftarrow N_{2^n}$ Positive Integer Conversion
>
> To convert numbers N_{2^n} to N_{2^k}, where n and k are integers, convert the groups of n digits in N_{2^n} to N_2, then reorganize the result in groups of k beginning with the LSB and proceeding toward the MSB.[1] Finally, replace each group of k, reading from the LSB, with the digit appropriate for number system N_{2^k}.

The integer conversion methods of Table 2.4 and Algorithms 2.1 and 2.2 are illustrated by the following simple examples:

EXAMPLE 2.1 $139_{10} \to N_2$

$$
\begin{array}{rcll}
N/r & & Q & R \\
139/2 & = & 69 & 1 \\
69/2 & = & 34 & 1 \\
34/2 & = & 17 & 0 \\
17/2 & = & 8 & 1 \\
8/2 & = & 4 & 0 \\
4/2 & = & 2 & 0 \\
2/2 & = & 1 & 0 \\
1/2 & = & 0 & 1 \quad 139_{10} = 10001011_2
\end{array}
$$

EXAMPLE 2.2 $10001011_2 \to N_{10}$

By positional weights $N_{10} = 128 + 8 + 2 + 1 = 139_{10}$.

---

[1] Note that leading 0's may be needed to complete the groups of k.

EXAMPLE 2.3 $139_{10} \rightarrow N_8$

$$
\begin{array}{rcll}
N/r & & Q & R \\
139/8 & = & 17 & 3 \\
17/8 & = & 2 & 1 \\
2/8 & = & 0 & 2 \quad 139_{10} = 213_8
\end{array}
$$

EXAMPLE 2.4 $10001011_2 \rightarrow N_{\text{BCO}}$

$$
\begin{array}{ccc}
2 & 1 & 3 \\
010 & 001 & 011 = 213_{\text{BCO}}
\end{array}
$$

EXAMPLE 2.5 $213_{\text{BCO}} \rightarrow N_{\text{BCH}}$

$$
\begin{array}{cccccc}
2 & 1 & 3 & & 8 & B \\
\end{array}
$$
$213_{\text{BCO}} = 010\ 001\ 011 = 10001011_2 = 1000\ 1011 = 8B_{16}$

EXAMPLE 2.6 $213_8 \rightarrow N_5$

$$213_8 = 2 \times 8^2 + 1 \times 8^1 + 3 \times 8^0 = 139_{10}$$

$$
\begin{array}{rcll}
N_r & & Q & R \\
139/5 & = & 27 & 4 \\
27/5 & = & 5 & 2 \\
5/5 & = & 1 & 0 \\
1/5 & = & 0 & 1 \quad 213_8 = 1024_5
\end{array}
$$

Check: $1 \times 5^3 + 0 \times 5^2 + 2 \times 5^1 + 4 \times 5^0 = 125 + 0 + 10 + 4 = 139_{10}$

2.5.2 Conversion of Fractions

By extracting the fraction portion from Eq. (2.2) one can write

$$
\begin{aligned}
\cdot N &= (a_{-1}s^{-1} + a_{-2}s^{-2} + \cdots + a_{-m}s^{-m})_s \\
&= s^{-1}(a_{-1} + s^{-1}(a_{-2} + \cdots + a_{-m})) \cdots)_s \\
&= s^{-1}\left(a_{-1} + \sum_{i=2}^{m} a_{-i}s^{-i+1}\right)_s
\end{aligned}
\tag{2.6}
$$

in source radix s. This is called the *nested inverse radix* form and provides the basis for computerized conversion.

If the fraction in Eq. (2.6) is represented in nested inverse radix r form, then

$$
\begin{aligned}
\cdot N &= (b_{-1}r^{-1} + b_{-2}r^{-2} + \cdots + b_{-p}s^{-p})_r \\
&= r^{-1}(b_{-1} + r^{-1}(b_{-2} + \cdots + b_{-p})) \cdots)_r \\
&= r^{-1}\left(b_{-1} + \sum_{i=2}^{p} b_{-i}r^{-i+1}\right)_r
\end{aligned}
\tag{2.7}
$$

for any fraction represented in radix r. Now, if source N_s is multiplied by r, the result is of the form

$$\cdot N_s \times r = I + F, \tag{2.8}$$

2.5 CONVERSION BETWEEN NUMBER SYSTEMS

where I is the product integer, $I_1 = b_{-1}$, and F_0 is the product fraction arranged as $F_1 = r^{-1}(b_{-2} + r^{-1}(b_{-3} + \cdots + b_{-p}))\cdots)_r$. By repeated multiplication by r of the remaining fractions F_i, the resulting integers yield $(b_{-1}, b_{-2}, b_{-3} \ldots, b_{-m})_r$ in that order.

The conversion just described is called the *radix multiply method* and is perfectly general for converting between fractions of different radices. However, as in the case of integer conversion, the requirement is that *the arithmetic required by $\cdot N_s \times r$ must be carried out in source radix, s*. For noncomputer use by humans, this procedure is usually limited to fraction conversions $N_{10} \rightarrow N_r$, where the source radix is 10 (decimal). Algorithm 2.3 gives the recommended methods for converting between fractions of different radices. The radix multiply method is well suited to computer use.

Algorithm 2.3: $\cdot N_r \leftarrow \cdot N_s$ Fraction Conversion

(1) Use Eq. (2.2) and the substitution method with base s arithmetic, or
(2) Use the radix multiply method of Eq. (2.8) with source radix s arithmetic.

In either case for noncomputer means, if the source radix is other than 2 or 10, convert the fraction as follows: $\cdot N_s \rightarrow \cdot N_{2 \text{ or } 10} \rightarrow \cdot N_r$ so that base 2 or 10 arithmetic can be applied.

Shown in Table 2.5 are the recommended methods given in some detail for fraction conversion by noncomputer means. Notice again the partitioning that is now required for conversion between binary fractions and those for BCH and BCO.

For any integer of source radix s, there exists an exact representation in radix r. This is not the case for a fraction whose conversion is a geometrical progression that never converges.

Table 2.5 Summary of recommended methods for fraction conversion by noncomputer means

Fraction Conversion	Conversion Method
$\cdot N_{10} \rightarrow \cdot N_r$	Radix multiplication by using Eq. (2.8)
$\cdot N_s \rightarrow \cdot N_{10}$	Eq. (2.2) or (2.6)
$\cdot N_s)_{s \neq 10} \rightarrow \cdot N_r)_{r \neq 10}$	$\cdot N_s \rightarrow \cdot N_{10}$ by Eq. (2.2) or (2.6)
	$\cdot N_{10} \rightarrow \cdot N_r$ radix multiplication by Eq. (2.5)
Special Cases for Binary Forms	
$\cdot N_2 \rightarrow \cdot N_{10}$	Positional weighting
$\cdot N_2 \rightarrow \cdot N_{BCH}$	Partition $\cdot N_2$ into groups of four bits starting from radix point, then apply Table 2.3
$\cdot N_2 \rightarrow \cdot N_{BCO}$	Partition $\cdot N_2$ into groups of three bits starting from radix point, then apply Table 2.3
$\cdot N_{BCH} \rightarrow \cdot N_2$	Reverse of $\cdot N_2 \rightarrow \cdot N_{BCH}$
$\cdot N_{BCO} \rightarrow \cdot N_2$	Reverse of $\cdot N_2 \rightarrow \cdot N_{BCO}$
$\cdot N_{BCH} \rightarrow \cdot N_{BCO}$	$\cdot N_{BCH} \rightarrow \cdot N_2 \rightarrow \cdot N_{BCO}$
$\cdot N_{BCO} \rightarrow \cdot N_{BCH}$	$\cdot N_{BCO} \rightarrow \cdot N_2 \rightarrow \cdot N_{BCH}$
$\cdot N_{BCD} \rightarrow \cdot N_{XS3}$	Add $0011_2 (= 3_{10})$ to N_{BCD}
$\cdot N_{XS3} \rightarrow \cdot N_{BCD}$	Subtract $0011_2 (= 3_{10})$ from N_{XS3}

Terminating a fraction conversion at n digits (to the right of the radix point) results in an error or uncertainty. This error is given by

$$\varepsilon = a_{-n}r^{-n} + a_{-(n+1)}r^{-(n+1)} + a_{-(n+2)}r^{-(n+2)} + \cdots$$

$$= r^{-n}\left[a_{-n} + \sum_{i=1}^{\infty} a_{-(n+i)}r^{-(n+i)}\right]_r,$$

where the quantity in brackets is less than $(a_{-n} + 1)$. Therefore, terminating a fraction conversion at n digits from the radix point results in an error with bounds

$$0 < \varepsilon \le r^{-n}(a_{-n} + 1). \tag{2.9}$$

Equation (2.9) is useful in deciding when to terminate a fraction conversion.

Often, it is desirable to terminate a fraction conversion at $n + 1$ digits and then round off to n from the radix point. A suitable method for rounding to n digits in radix r is:

Algorithm 2.4: Rounding Off to n Digits for Fraction Conversion in Radix r

Perform the fraction conversion to $(n - 1)$ digits from the radix point, then drop the $(n - 1)$ digit if $a_{-(n+1)} < r/2$; add $r^{-(n-1)}$ to the result if $a_{-(n-1)} \ge r/2$.

After rounding off to n digits, the maximum error becomes the difference between the rounded result and the smallest value possible. By using Eq. (2.9), this difference is

$$\varepsilon_{\max} = r^{-n}(a_{-n} + 1) - r^{-n}\left(a_{-n} + \frac{a_{-(n+1)}}{r}\right)$$

$$= r^{-n}\left(1 - \frac{a_{-(n+1)}}{r}\right).$$

Then, by rounding to n digits, there results an error with bounds

$$0 < \varepsilon \le r^{-n}\left(1 - \frac{a_{-(n+1)}}{r}\right). \tag{2.10}$$

If $a_{-(n+1)} < r/2$ and the $(n + 1)$ digit is dropped, the maximum error is r^{-n}. Note that for $N_s \to N_{10} \to N_r$ type conversions, the bounds of errors aggregate.

The fraction conversion methods given in Table 2.5 and Algorithms 2.3 and 2.4 are illustrated by the following examples:

EXAMPLE 2.7 $0.654_{10} \to N_2$ rounded to 8 bits:

$$\begin{array}{ccc}
N_s \times r & F & I \\
0.654 \times 2 & 0.308 & 1 \\
0.308 \times 2 & 0.616 & 0 \\
0.616 \times 2 & 0.232 & 1 \\
0.232 \times 2 & 0.464 & 0 \\
\end{array}$$

$$
\begin{array}{ll}
0.464 \times 2 & 0.928 \quad 0 \\
0.928 \times 2 & 0.856 \quad 1 \\
0.856 \times 2 & 0.712 \quad 1 \\
0.712 \times 2 & 0.424 \quad 1 \quad 0.654_{10} = 0.10100111_2 \\
0.424 \times 2 & 0.848 \quad 0 \quad \varepsilon_{\max} = 2^{-8}
\end{array}
$$

EXAMPLE 2.8 $0.654_{10} \to N_8$ terminated at 4 digits:

$$
\begin{array}{lll}
\cdot N_s \times r & F & I \\
0.654 \times 8 & 0.232 & 5 \\
0.232 \times 8 & 0.856 & 1 \quad 0.654_{10} = 0.5166_8 \\
0.856 \times 8 & 0.848 & 6 \quad \text{with error bounds} \\
0.848 \times 8 & 0.784 & 6 \quad 0 < \varepsilon \leq 7 \times 8^{-4} = 1.71 \times 10^{-3} \text{ by Eq. (2.9)}
\end{array}
$$

EXAMPLE 2.9 Let $0.5166_8 \to N_2$ be rounded to 8 bits and let $0.5166_8 \to N_{10}$ be rounded to 4 decimal places:

$$0.5166_8 = 5 \times 8^{-1} + 1 \times 8^{-2} + 6 \times 8^{-3} + 6 \times 8^{-4}$$

$$= 0.625000 + 0.015625 + 0.011718 + 0.001465$$

$$= 0.6538_{10} \text{ rounded to 4 decimal places; } \varepsilon_{10} \leq 10^{-4}$$

$$
\begin{array}{lll}
\cdot N_s \times r & F & I \\
0.6538 \times 2 & 0.3076 & 1 \\
0.3076 \times 2 & 0.6152 & 0 \\
0.6152 \times 2 & 0.2304 & 1 \\
0.2304 \times 2 & 0.4608 & 0 \\
0.4608 \times 2 & 0.9216 & 0 \\
0.9216 \times 2 & 0.8432 & 1 \\
0.8432 \times 2 & 0.6864 & 1 \\
0.6864 \times 2 & 0.3728 & 1 \quad 0.5166_8 = 0.10100111_2 \text{ (compare with Example 2.7)} \\
0.3728 \times 2 & 0.7457 & 0 \quad \varepsilon_{10} \leq 10^{-4} + 2^{-8} = 0.0040
\end{array}
$$

EXAMPLE 2.10 $0.10100111_2 \to N_{\text{BCH}}$

$$
\begin{array}{cc}
\cdot A & 7 \\
\end{array}
$$
$$0.10100111_2 = 0.1010 \; 0111 = 0.A7_{\text{BCH}}$$

2.6 SIGNED BINARY NUMBERS

To this point only unsigned numbers (assumed to be positive) have been considered. However, both positive and negative numbers must be used in computers. Several schemes have been devised for dealing with negative numbers in computers, but only four are commonly used:

- Signed-magnitude representation
- Radix complement representation

- Diminished radix complement representation
- Excess (offset) code representation

Of these, the radix 2 complement representation, called 2's complement, is the most widely used system in computers.

2.6.1 Signed-Magnitude Representation

A signed-magnitude number in radix r consists of a magnitude $|N|$ together with a symbol indicating its sign (positive or negative) as follows:

$$N_{rSM} = \left(\underbrace{a_{n-1}}_{Sign} \underbrace{\overbrace{a_{n-2} \cdots a_2 a_1 a_0}^{Integer} \cdot \overbrace{a_{-1} a_{-2} a_{-3} \cdots a_{-m}}^{Fraction}}_{Magnitude\ |N|} \right)_{rSM}, \qquad (2.11)$$

where the subscript rSM refers to signed-magnitude in radix r. Such a number lies in the decimal range of $-(r^{n-1} - 1)$ through $+(r^{n-1} - 1)$ for n integer digits in radix r. The fraction portion, if it exists, consists of m digits to the right of the radix point.

The most common examples of signed-magnitude numbers are those in the decimal and binary systems. The sign symbols for decimal (+ or −) are well known. In binary it is established practice to use the following convention:

0 *denotes a positive number*

1 *denotes a negative number.*

One of these (0 or 1) is placed at the MSB position of each SM number. Four examples in 8-bit binary are:

EXAMPLE 2.11

$$+45.5_{10} = 0\ \overbrace{101101.1}^{Magnitude}{}_{2SM}$$
$$\uparrow$$
$$Sign\ Bit$$

EXAMPLE 2.12

$$+0_{10} = 0\ 0000000_{2SM}$$

EXAMPLE 2.13

$$-123_{10} = 1\ \overbrace{1111011}^{Magnitude}{}_{2SM}$$
$$\uparrow$$
$$Sign\ Bit$$

2.6 SIGNED BINARY NUMBERS

EXAMPLE 2.14

$$-0_{10} = 1\ 0000000_{2SM}$$

Although the sign-magnitude system is used in computers, it has two drawbacks. There is no unique zero, as indicated by the previous examples, and addition and subtraction calculations require time-consuming decisions regarding operation and sign, for example, (-7) minus (-4). Even so, the sign-magnitude representation is commonly used in *floating-point* number systems as discussed in Section 2.8.

2.6.2 Radix Complement Representation

The *radix complement* N_{rC} of an n-digit number N_r is obtained by subtracting N_r from r^n, that is,

$$N_{rC} = r^n - N_r$$
$$= \bar{N}_r + 1_{LSD} \qquad (2.12)$$

where

$$\bar{N}_r \equiv \text{Digit complementation in radix } r$$

This operation is equivalent to that of replacing each digit a_i in N_r by $(r-1) - a_i$ and adding 1 to the LSD of the result as indicated by Algorithm 2.5. The digit complements $\bar{N}_r$ for three commonly used number systems are given in Table 2.6. Notice that the digit complement of a binary is formed simply by replacing the 1's with 0's and 0's with 1's required by $2^n - N_2 - 1 = \bar{N}_2$ as discussed in Subsection 2.6.3. The range of representable numbers is $-(r^{n-1})$ through $+(r^{n-1} - 1)$.

Application of Eq. (2.12) or Algorithm 2.5 to the binary and decimal number systems requires that for 2's complement representation $N_{2C} = \bar{N}_2 + 1_{LSB}$ and for 10's complement $N_{10C} = \bar{N}_{10} + 1_{LSD}$, where $\bar{N}_2$ and $\bar{N}_{10}$ are the binary and decimal digit complements given in Table 2.6.

Table 2.6 Digit complements for three commonly used number systems

Digit	Complement ($\bar{N}_r$)		
	Binary	Decimal	Hexadecimal
0	1	9	F
1	0	8	E
2		7	D
3		6	C
4		5	B
5		4	A
6		3	9
7		2	8
8		1	7

> **Algorithm 2.5:** $N_{rC} \leftarrow N_r$
>
> Replace each digit a_i in N_r by $(r-1) - a_i$ and then add 1 to the LSD of the resultant.

A simpler, "pencil-and-paper" method exists for the 2's complement of a number N_2 and is expressed by Algorithm 2.6:

> **Algorithm 2.6:** $N_{2C} \leftarrow N_2$
>
> For any binary number N_2 and beginning with the LSB, proceed toward the MSB until the *first* 1 bit has been reached. Retain that 1 bit and complement the remainder of the bits toward and including the MSB.

With reference to Table 2.6, Eq. (2.12), and Algorithm 2.5 or 2.6, the following examples of radix complement representation are provided:

EXAMPLE 2.15 The 10's complement of 47.83 is $\bar{N}_{10} + 1_{LSD} = 52.17$.

EXAMPLE 2.16 The 2's complement of 0101101.101 is $\bar{N}_2 + 1_{LSB} = 1010010.011$.

EXAMPLE 2.17 The 16's complement of A3D is $\bar{N}_{16} + 1_{LSD} = 5C2 + 1 = 5C3$.

The decimal value of Eq. (2.12) can be found from the polynomial expression as

$$N_{rC})_{10} = -(a_{n-1}r^{n-1}) + \sum_{i=-m}^{n-2} a_i r^i \tag{2.13}$$

for any n-digit number of radix r. In Eqs. (2.12) and (2.13) the MSD is taken to be the position of the sign symbol.

2's Complement Representation The radix complement for binary is the 2's complement (2C) representation. In 2's complement the MSB is the sign bit, 1 indicating a negative number and 0 a positive one. The decimal range of representation for n integer bits in 2's complement is from $-(2^{n-1})$ through $+(2^{n-1} - 1)$. From Eq. (2.12), the 2's complement is formed by

$$N_{2C} = 2^n - N_2 = \bar{N}_2 + 1_{LSB} \tag{2.14}$$

for any binary number N_2 of n integer bits. Or by Algorithm 2.5, the 2's complement of a binary number N_2 is obtained by replacing each bit a_i in N_2 by $(1 - a_i)$ and adding 1 to the LSB of the result. The simpler pencil-and-paper method, often used to generate 2's complement from a binary number N_r, results from application of Algorithm 2.6. In this case $\bar{N}_2$ is the bit complement of the number as given in Table 2.6. A few examples of 8-bit 2's complement numbers are shown in Table 2.7. Notice that application of Eq. (2.14) or Algorithm 2.6 changes the sign of the decimal value of a binary number (+ to − and vice versa), and that only one zero representation exists.

Application of Eq. (2.13) gives the decimal value of any 2's complement number, including those containing a radix point. For example, the pattern $N_{2C} = 11010010.011$ has

2.6 SIGNED BINARY NUMBERS

Table 2.7 Examples of eight-bit 2's and 1's complement representations (MSB = sign bit)

Decimal Value	2's Complement	1's Complement
−128	10000000	
−127	10000001	10000000
−31	11100001	11100000
−16	11110000	11101111
−15	11110001	11110000
−3	11111101	11111100
−0	00000000	11111111
+0	00000000	00000000
+3	00000011	00000011
+15	00001111	00001111
+16	00010000	00010000
+31	00011111	00011111
+127	01111111	01111111
+128		

a decimal value of

$$(N_{2C})_{10} = -1 \times 2^7 + 1 \times 2^6 + 1 \times 2^4 + 1 \times 2^1 + 1 \times 2^{-2} + 1 \times 2^{-3}$$
$$= -128 + 64 + 16 + 2 + 0.25 + 0.125$$
$$= -45.625_{10}.$$

But the same result could have easily been obtained by *negation* of N_{2C} followed by the use of positional weighting to obtain the decimal value. Negation is the reapplication of Eq. (2.12) or Algorithms 2.5 or 2.6 to any 2's complement number N_{2C} to obtain its *true value*. Thus, from the forgoing example the negation of N_{2C} is given by

$$N_{2C})_{2C} = 00101101.101$$
$$= 32 + 8 + 5 + 0.5 + 0.125$$
$$= 45.625_{10},$$

which is known to be a negative number, -45.625_{10}.

Negative BCD numbers are commonly represented in 10's complement notation with consideration of how BCD is formed from binary. As an example, $-59.24_{10} = 40.76_{10}$ is represented in BCD 10's complement (BCD,10C) by

$$-0101\ 1001.0010\ 0100)_{BCD} = 0100\ 0000.0111\ 0110)_{BCD,10C},$$

where application of Eq. (2.12), or Algorithm 2.5 or 2.6, has been applied in radix 10 followed by the BCD representation as in Subsections 2.4.1. Alternatively, the sign-magnitude (SM) representation of a negative BCD number simply requires the addition of a sign bit to the BCD magnitude according to Eq. (2.11). Thus,

$$-0101\ 1001.0010\ 0100)_{BCD} = (1\ 01011001.0010\ 0100)_{BCD,2SM}.$$

2.6.3 Diminished Radix Complement Representation

The *diminished radix complement* $N_{(r-1)C}$ of a number N_r having n digits is obtained by

$$N_{(r-1)C} = r^n - N_r - 1, \qquad (2.15)$$

where, according to Eq. (2.12), $N_{(r-1)C} + 1 = N_{rC}$. Therefore, it follows that

$$N_{(r-1)C} = \bar{N}_r.$$

This means the diminished radix complement of a number is the digits complement of that number as expressed by Algorithm 2.7. The range of representable n digit numbers in diminished radix complement is $-(r^{n-1} - 1)$ through $+(r^{n-1} - 1)$ for radix r.

Algorithm 2.7: $N_{(r-1)C} \leftarrow N_r$

(1) Replace each digit a_i of N_r by $r - 1 - a_i$ or
(2) Complement each digit by $\bar{N}_r$ as in Table 2.6.

In the binary and decimal number systems the diminished radix complement representations are the 1's complement and 9's complement, respectively. Thus, 1's complement is the binary digits complement given by $N_{1C} = \bar{N}_2$, while the 9's complement is the decimal digits complement expressed as $N_{9C} = \bar{N}_{10}$. Examples of eight-bit 1's complements are shown in Table 2.7 together with their corresponding 2's complement representation for comparison. Notice that in 1's complement there are two representations for zero, one for $+0$ and the other for -0. This fact limits the usefulness of the 1's complement representation for computer arithmetic.

Shown in Table 2.8 are examples of 10's and 9's complement representations in n digits numbering from 3 to 8. Notice that leading 0's are added to the number on the left to meet the n digit requirement.

Table 2.8 Examples of 10's and 9's complement representation

Number	n	10's Complement	9's Complement
0	5	[1]00000	99999
3	3	997	996
14.59	6	9985.41	9985.40
225	4	9775	9774
21.456	5	78.544	78.543
1827	8	99998173	99998172
4300.50	7	95699.50	95699.49
69.100	6	930.900	930.899

2.7 EXCESS (OFFSET) REPRESENTATIONS

Other systems for representing negative numbers use *excess* or *offset* (*biased*) codes. Here, a bias B is added to the true value N_r of the number to produce an excess number, N_{xs}, given by

$$N_{xs} = N_r + B. \tag{2.16}$$

When $B = r^{n-1}$ exceeds the usable bounds of negative numbers, N_{xs} remains positive. Perhaps the most common use of the excess representation is in floating-point number systems — the subject of the next section. The biased-weighted BCD code, XS3, was discussed in Subsection 2.4.1.

Two examples of excess 127 representation are given below.

EXAMPLE 2.18

$$\begin{array}{rll} -43_{10} & 11010101 & N_{2\text{'s Compl.}} \\ +127_{10} & 01111111 & B \\ \hline 84_{10} & 01010100 & N_{xs} = -43_{10} \text{ in excess 127 representation} \end{array}$$

EXAMPLE 2.19

$$\begin{array}{rll} 27_{10} & 00011011 & N_{2\text{'s Compl.}} \\ +127_{10} & 01111111 & B \\ \hline 154_{10} & 10011010 & N_{xs} = 27_{10} \text{ in excess 127 representation} \end{array}$$

The representable decimal range for an excess 2^{n-1} number system is -2^{n-1} through $+(2^{n-1} - 1)$ for an n-bit binary number. However, if $N_2 + B > 2^{n-1} - 1$, *overflow* occurs and 2^{n-1} must be subtracted from $(N_2 + B)$ to give the correct result in excess 2^{n-1} code.

2.8 FLOATING-POINT NUMBER SYSTEMS

In *fixed-point* representation [Eq. (2.1)], the radix point is assumed to lie immediately to the right of the integer field and at the left end of the fraction field. The fixed-point system is the most commonly used system for representing bounded orders of magnitude. For example, with 32 bits a binary number could represent decimal numbers with upper and lower bounds of the order of $\pm 10^{10}$ and $\pm 10^{-10}$. However, for greatly expanded bounds of representation, as in scientific notation, the *floating-point* representation is needed. This form of number representation is commonly used in computers.

A floating-point number (FPN) in radix r has the general form

$$\text{FPN})_r = M \times r^E, \tag{2.17}$$

where M is the *fraction* (or *mantissa*) and E is the *exponent*. Only fraction digits are used for the mantissa! Take, for example, Planck's constant $h = 6.625 \times 10^{-34}$ J s. This number can be represented many different ways in floating-point notation:

$$\text{Planck's constant, } h = 0.6625 \times 10^{-33}$$
$$= 0.06625 \times 10^{-32}$$
$$= 0.006625 \times 10^{-31}.$$

All three adhere to the form of Eq. (2.17) and are, therefore, legitimate floating-point numbers in radix 10. Thus, as the radix point *floats* to the left, the exponent is *scaled* accordingly. The first form for h is said to be *normalized* because the most significant digit (MSD) of M is nonzero, a means of standardizing the radix point position. Notice that the sign for M is positive while that for E is negative.

In computers the FPN is represented in binary where the normalized representation requires that the MSB for M always be 1. Thus, the range in M in decimal is

$$0.5 \leq M < 1.$$

Also, the fraction (mantissa) M is represented in sign-magnitude from. The normalized format for a 32-bit floating-point number in binary, which agrees with the IEEE standard [3], is shown in Fig. 2.1. Here, the sign bit (1 if negative or 0 if positive) is placed at bit position 0 to indicate the sign of the fraction. Notice that the radix point is assumed to lie between bit positions 8 and 9 to separate the E bit-field from the M bit-field.

Before two FPNs can be added or subtracted in a computer, the E fields must be compared and equalized, and the M fields adjusted. The decision-making process can be simplified if all exponents are converted to positive numbers by using the excess representation given by Eq. (2.16). For a q-digit number in radix r, the exponent in Eq. (2.17) becomes

$$E_{xs} = E_r + r^{q-1}, \qquad (2.18)$$

where E is the actual exponent augmented by a bias of $B = r^{q-1}$. The range in the actual exponent E_r is usually taken to be

$$-(r^{q-1} - 1) \leq E_r \leq +(r^{q-1} - 1).$$

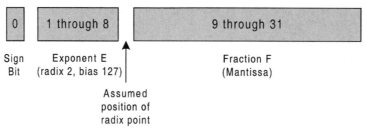

FIGURE 2.1
IEEE standard bit format for 32-bit normalized floating-point representation.

2.8 FLOATING-POINT NUMBER SYSTEMS

In the binary system, required for computer calculations, Eq. (2.18) becomes

$$E_{xs} = E_2 + 2^{q-1}, \tag{2.19}$$

with a range in actual exponent of $-(2^{q-1} - 1) \leq E_2 \leq +(2^{q-1} - 1)$. In 32-bit normalized floating-point form, the exponent in Eq. (2.19) is stored in either excess 127 or excess 128 code (depending on the FPN system used) while the mantissa is stored in sign-magnitude form.

There still remains the question of how the number zero is to be represented. If the M field is zero, then the exponent can be anything and the number will be zero. However, in computers the normalized FPN_2 limits M to $(0.5 \leq M < 1)$ since the MSB for M is always 1. The solution to this problem is to assume that the number is zero if the exponent bits are all zero regardless of the value of the mantissa. But this leads to a discontinuity in normalized FPN_2 representation at the low end.

The IEEE standard for normalized FPN_2 representation attempts to remove the problem just described. The IEEE system stores the exponent in excess $2^{q-1} - 1$ code and limits the decimal range of the actual exponent to

$$-(2^{q-1} - 2) \leq E_2 \leq +(2^{q-1} - 1).$$

For 32-bit FPN single precision representation, the exponent is stored in excess 127 code as indicated in Fig. 2.1. Thus, the allowable range of representable exponents is from

$$-126_{10} = 00000001_2 \text{ through } +127_{10} = 11111110_2.$$

This system reserves the use of all 0's or all 1's in the exponent for special conditions [3]. So that the M field magnitude can diminish linearly to zero when $E = -126$, the MSB $= 1$ for M is not specifically represented in the IEEE system, but is implied.

The following example attempts to illustrate the somewhat confusing aspects of the IEEE normalized representation:

EXAMPLE 2.20 The number 101101.11001_2 is to be represented in IEEE normalized FPN_2 notation:

$$101101.11001_2 = .10110111001 \times 2^6 \text{ Sign bit} = 0 \text{ (positive)}$$

where

$$E_{xs} = 6 + 127 = 133_{10} = 10000101_2$$

$$M = 0110111001\ldots00 \text{ (the MSB} = 1 \text{ is not shown)}.$$

Therefore, the IEEE normalized FPN is

$$FPN_2 = 0\ 10000101\ 0110111001\ldots 0.$$

Still, other forms of FPNs are in use. In addition to the IEEE system, there are the IBM, Cray, and DEC systems of representation, each with their own single- and double-precision forms. For example, the IEEE double-precision FPN system requires an exponent of 11 bits in excess 1023 code and a mantissa (in sign-magnitude) of 53 bits for a total 64 bits.

2.9 BINARY ARITHMETIC

There are many ways in which to manipulate data for the purpose of computation. It is not the intent of this section to discuss all these methods. Rather, the emphasis will be on the basic addition, subtraction, multiplication, and division methods for binary number manipulation as a foundation for the design of the arithmetic units featured in Chapter 8. The coverage will include the basic heuristics involved in fixed-point binary arithmetic together with simple examples. Advanced methods applicable to computer arithmetic operations are provided as needed for the device design. A limited treatment of floating-point binary arithmetic will be given in a later section.

2.9.1 Direct Addition and Subtraction of Binary Numbers

The addition of any two positive binary numbers is accomplished in a manner similar to that of two radix (base) 10 numbers. When the addition of two binary bits exceeds 01_2, a carry bit is added to the next MSB, and this process is continued until all bits of the addend and augend have been added together. As an example consider the addition of the following two 8-bit numbers:

EXAMPLE 2.21

$$
\begin{array}{rll}
 & 1\ 1\ 1\ 1 & \leftarrow \text{Carries} \\
59_{10} & 0\ 0\ 1\ 1\ 1\ 0\ 1\ 1_2 & = \text{Augend} \\
+122_{10} & +0\ 1\ 1\ 1\ 1\ 0\ 1\ 0_2 & = \text{Addend} \\ \hline
181_{10} & 1\ 0\ 1\ 1\ 0\ 1\ 0\ 1_2 & = \text{Sum}
\end{array}
$$

Notice that in binary addition the carry is rippled to the left in much the same manner as in base 10 addition. The binary numbers are easily converted to base 10 by the method of positional weight described in Section 2.3.

Algorithm 2.8: $A_2 + B_2$

(1) Set operands $A_2 = a_{n-1}a_{n-2}\cdots a_1 a_0$ and $B_2 = b_{n-1}b_{n-2}\cdots b_1 b_0$, and their sum $A_2 + B_2 = S_n S_{n-1}\cdots S_1 S_0 = S_2$.
(2) Set $i = 0$ and $S_2 = 0$.
(3) If $a_0 + b_0 < 10_2$, $S_0 = a_0 + b_0$ and a carry $C_1 = 0$ is generated for position $i+1 = 1$. If $a_0 + b_0 \geq 10_2$, then $S_0 = a_0 + b_0 - 10_2$ and a carry $C_{i+1} = 1$ is generated into position $i+1 = 1$.
(4) Continue steps (2) and (3) in the order $i = 1, 2, 3, \ldots, n-1$ with carries generated into position $i + 1$.
(5) The most significant sum bit is $S_n = C_n$, where C_n is the carry resulting from the addition of a_{n-1}, b_{n-1}, and C_{n-1}.

Direct subtraction of two binary numbers also parallels that for base 10 subtraction. Now however, when the subtrahend bit is 1 when the minuend bit is 0, a borrow is required from the next MSB. Thus, the borrowing process begins at the MSB and ends with the LSB — the opposite of the carry process for addition. Remember that a borrow of 1_2 from the next MSB, creates a 10_2 in the column being subtracted. The following 8-bit example illustrates the subtraction process in base 2:

EXAMPLE 2.22

$$
\begin{array}{r}
 & & & & 10 & 1 & & & \\
 & & 0 & \emptyset & 1\emptyset & 10 & 0 & 10 & \leftarrow \text{Borrows} \\
101_{10} & & 0 & 1 & 1 & \emptyset & \emptyset & 1 & \emptyset & 1_2 = \text{Minuend} \\
-58_{10} & -0 & 0 & 1 & 1 & 1 & 0 & 1 & 0_2 = \text{Subtrahend} \\
\hline
43_{10} & & 0 & 0 & 1 & 0 & 1 & 0 & 1 & 1 = \text{Difference}
\end{array}
$$

Here, the notation $\emptyset$ or 1 represents denial of the 0 or 1 when a borrow is indicated. Notice, as in the example just given, that the borrowing process may involve more than one level of borrowing as the process proceeds from right to left.

2.9.2 Two's Complement Subtraction

Computer calculations rarely involve direct subtraction of binary numbers. Much more commonly, the subtraction process is accomplished by 2's complement arithmetic — a considerable savings in hardware. Here, subtraction involves converting the subtrahend to 2's complement by using Eq. (2.14) in the form $\bar{N}_2 + 1$ and then adding the result directly to the minuend. For an n-bit operand subtraction, $n + 1$ bits are used where the MSB bit is designated the *sign bit*. Also, the *carry overflow* is discarded in 2's complement arithmetic. The following example illustrates the process for two four-bit numbers, A and B:

EXAMPLE 2.23

$$
\begin{array}{rl}
A & 0\,|\,1101 \\
-B & -0\,|\,0111 \\
\end{array}
\longrightarrow
\begin{array}{rl}
 & 0\,|\,1101 = +13_{10} \\
 & +1\,|\,1001 = -7_{10} \\
\hline
\boxed{1} & 0\,|\,0110 = +6_{10}
\end{array}
$$

Discard overflow — Sign bit positive

Further illustration continues by interchanging the minuend and subtrahend so as to yield a negative number:

EXAMPLE 2.24

$$
\begin{array}{rl}
A & 0\,|\,0111 \\
-B & -0\,|\,1101 \\
\end{array}
\longrightarrow
\begin{array}{rl}
 & 0\,|\,0111 = +7_{10} \\
 & +1\,|\,0011 = -13_{10} \\
\hline
\boxed{0} & 1\,|\,1010 = -6_{10}
\end{array}
$$

Discard overflow — Sign Bit negative

In this case, the difference $1 \vdots 1010$ is given in 2's complement. The answer in true form is the 2's complement (*negation*) of this result given by $1 \vdots 1010 \rightarrow 0 \vdots 0110_2$ or 6_{10} which is negative. Algorithm 2.9 summarizes the steps required for the 2's complement subtraction.

Algorithm 2.9: $\{A_2 + (-B_2) \rightarrow A_2 + B_{2C}\}$ or $\{-A_2 + (-B_2) \rightarrow A_{2C} + B_{2C}\}$

(1) Set n-bit operands $A = a_n a_{n-1} \cdots a_1 a_0$ and $B = b_n b_{n-1} \cdots b_1 b_0$, where the MSB a_n and b_n are reserved as sign bits, initially both 0.
(2) Generate B_{2C} by applying Eq. (2.14).
(3) Add operands $A_2 + B_{2C}$ according to Algorithm 2.8.
(4) If $|A_2| > |B_2|$, then the sum ($S_2 > 0$) is the true value with the sign bit 0. If $|A_2| < |B_2|$, then the sum ($S < 0$) is given in 2's complement with sign bit equal to 1.
(5) The true value of a 2's complement sum is obtained by negation, $S_{2C})_{2C}$.
(6) If A_2 and B_2 are both negative numbers, the sum ($S_{2C} < 0$) is obtained by steps (2), (3), and (5).

2.9.3 One's Complement Subtraction

Somewhat less common is the 1's complement subtraction of two binary numbers. In this case the subtrahend is converted to 1's complement by using Eq. (2.15) in the form $\bar{N}_2$. The result is then added to the minuend with the carry overflow *carried-end-around* and added to the LSB. Clearly, this is a more complex process than that for 2's complement. The following two examples illustrate the subtraction process:

EXAMPLE 2.25

$$
\begin{array}{rl}
A & 0 \vdots 1101 \\
-B & -0 \vdots 0111 \\
\end{array}
\longrightarrow
\begin{array}{rl}
& 0 \vdots 1101 = +13_{10} \\
& +1 \vdots 1000 = -7_{10} \\
\hline
\boxed{1}\ & 0 \vdots 0101 \\
& \downarrow 1 \\
\hline
& 0 \vdots 0110 = +6_{10}
\end{array}
$$

Sign Bit ↑

Again, the minuend and subtrahend are interchanged for comparison purposes, yielding a negative difference as follows:

EXAMPLE 2.26

$$
\begin{array}{rl}
A & 0 \vdots 0111 \\
-B & -0 \vdots 1101 \\
\end{array}
\longrightarrow
\begin{array}{rl}
& 0 \vdots 0111 = +7_{10} \\
& +1 \vdots 0010 = -13_{10} \\
\hline
\boxed{0}\ & 1 \vdots 1001 \\
& \downarrow 0 \\
\hline
& 1 \vdots 1001 = -6_{10}
\end{array}
$$

Sign Bit ↑

2.9 BINARY ARITHMETIC

In this latter case the result is given in 1's complement. The true value for the difference is obtained by negation and is $1\,\vdots\,1001 \to 0\,\vdots\,0110_2 = 6_{10}$, which is known to be negative.

2.9.4 Binary Multiplication

Like binary addition and subtraction, binary multiplication closely follows base 10 multiplication. Consider two n-bit binary integers $A_2 = (a_{n-1}\cdots a_2 a_1 a_0)$ and $B_2 = (b_{n-1}\cdots b_2 b_1 b_0)_2$. Their product $A_2 \times B_2$ is expressed in n-bits assuming that both numbers are expressible in $n/2$ bits excluding leading zeros. Under this assumption the product is

$$\mu P = A \times B = \left(\sum_{i=0}^{n-1} a_i \cdot 2^i\right) \cdot B$$
$$= a_{n-1} \times 2^{n-1} B + \cdots + a_1 2^1 B + a_0 B, \quad (2.20)$$

meaning that if $B = b_{n-1}\cdots b_2 b_1 b_0$, the product $2^i \times B$ is

$$2^i \times B = b_{n-1}\cdots b_2 b_1 b_0 \underbrace{00\cdots 0}_{i\ zeros}.$$

Thus, the product $A \times B$ is expressed as the sum of the partial products p_i in the form

$$P = \sum_{i=0}^{n-1} p_i = \sum_{i=0}^{n-1} a_i \,(b_{n-1}\cdots b_2 b_1 b_0 \underbrace{00\cdots 0}_{i\ zeros}). \quad (2.21)$$

Therefore it should seem clear that binary multiplication requires addition of all terms of the form $2^i \times B$ for all i for which $a_i = 1$. The following example illustrates this process.

EXAMPLE 2.27

$$\begin{array}{rll}
A & 00001111 & = \text{Multiplicand} \\
\times B & \times 00001011 & = \text{Multiplier} \\
\hline
 & 00001111 & 2^0 \times B \\
 & 000011110 & 2^1 \times B \\
 & 0000000000 & \\
 & 00001111000 & 2^3 \times B \\
 & 111011 & \text{Level 1 Carries} \\
 & 1 & \text{Level 2 Carries} \\
\hline
 & 000\ \underbrace{10100101}_{\text{8-bit representation}} & = \text{Product } P_2
\end{array}$$

Notice that the carry process may involve more than one level of carry as is true in this example. The following algorithm avoids multiple levels of carry by adding each product to P as it is formed.

Algorithm 2.10: $A_2 \times B_2$

(1) Set $n = 2k$, where k is the number of bits of the larger number exclusive of leading zeros.
(2) Set $A = a_{n-1} \cdots a_2 a_1 a_0$ and $B = b_{n-1} \cdots b_2 b_1 b_0$ for an n-bit multiplier and an n-bit multiplicand, respectively.
(3) Set $P = 0$ and $i = 0$.
(4) If $a_i = 1$, calculate $2^i \times B = (b_{n-1} \cdots b_1 b_0 \underbrace{00 \cdots 0}_{i \text{ zeros}})$ and add it to P.

(5) Increment i by 1.
(6) Repeat steps (3) and (4) for all $0 \leq i \leq (n-1)$ ending with a product P_2 of n bits or less.

If A_2^k and B_2^m represent operands expressible by a different number of bits, k and m, exclusive of leading zeros, then their product is $P_2^n = A_2^k \times B_2^m$ given in $n \leq (k+m)$ bits. For numbers containing both integers and fractions, k and m must each include all bits exclusive of leading integer zeros. For example, if $B_2^m = 1101.11$ ($m = 6$) and $A_2^k = 110.1$ ($k = 4$), their product P_2^n will be given in $n = 6 + 4 = 10$ bits. The following example illustrates the multiplication process for these two operands.

EXAMPLE 2.28

$$
\begin{array}{rl}
1101.11 & \text{Multiplicand } B_2^m \\
\times 110.1 & \text{Multiplier } A_2^k \\
\hline
110\,111 & 2^0 \times B \\
0000\,00\,0 & \\
11011\,100 & 2^2 \times B \\
110111\,00\,0 & 2^3 \times B \\
111011 & \text{Level 1 Carries} \\
1 & \text{Level 2 Carries} \\
\hline
1011001.01\,1 & \text{Product } P_2^n \\
\end{array}
$$

$\underbrace{}_{\text{10-bit representation}}$

2's Complement Multiplication To understand 2's complement multiplication it is helpful to introduce the concept of *modulo* 2^n (Mod 2^n) arithmetic. In Mod 2^n arithmetic multiplication is carried out in 2's complement numbers (if negative) ignoring the overflow beyond n bits. For example, $2^4 \times 1111$ (Mod 2^4) = $10000 = 2^4$ or generally, for number B of n bits,

$$2^n \times B \,(\text{Mod } 2^n) = 2^n.$$

Consider the n-bit integer operands $A_2 = a_{n-1} \cdots a_1 a_0$ and $B_2 = b_{n-1} \cdots b_1 b_0$. Then, if the product is $P = A \times (-B)$, there results, after converting B to 2's complement,

$$
\begin{aligned}
P_2 &= A_2 \times (B_{2C}) \\
&= A_2 \times (2^n - B) && \text{Mod } 2^n \\
&= A_2 \times 2^n - A_2 \times B_2 && \text{Mod } 2^n
\end{aligned}
$$

or
$$P_2 = 2^n - A_2 \times B_2 \qquad \text{Mod } 2^n \qquad (2.22)$$

2.9 BINARY ARITHMETIC

Thus, $P_2 = A_2 \times (B_{2C})$ generates the 2's complement of $A_2 \times B_2$ — the form in which a computer stores the results. The "true" value of P_2 can be obtained by negation, which is the 2's complement of the result, that is, $P_2)_{2C} = (2^n - A_2 \times B_2)_{2C}$.

Likewise, if both A_2 and B_2 are negative n-bit operands, then the product $(-A_2) \times (-B_2)$ becomes, after conversion to 2's complement by Eq. (2.14),

$$\begin{aligned} P_2 &= A_{2C} \times (B_{2C}) \\ &= (2^n - A_2) \times (2^n - B) & \text{Mod } 2^n \\ &= 2^n \times 2^n - 2^n A_2 - 2^n B_2 + A_2 \times B_2 & \text{Mod } 2 \\ &= 2^{2n} - 2^n - 2^n + A_2 \times B_2 & \text{Mod } 2^n \end{aligned}$$

or $\qquad P_2 = A_2 \times B_2,$ (2.23)

where $2^{2n} - 2^n - 2^n \,(\text{Mod } 2^n) = 0$. Thus, the product of two negative binary numbers in 2's complement notation is the positive product of the two numbers. In dealing with numbers whose bit representations are, say, $k > m$, excluding leading zeros for both, the 2's complement product P_{2C}^n must be given in $n = 2k$ bits. This count for k must include fraction bits, if present, but exclude both leading and trailing zeros.

The following example illustrates 2's complement multiplication of two numbers with fractions each of $k = m = 4$ bits and represented as $2k = 8$-bit operands:

EXAMPLE 2.29

$$\begin{array}{rl} -2.25 & -000010.01 \\ \times 6.5 & \times 0000110.1 \\ \hline -14.625 & \end{array} \longrightarrow \begin{array}{ll} 111101.11 & \textit{Multiplicand, } B_{2C} \\ \times 0000110.1 & \textit{Multiplier, } A_2 \\ \hline 11110111 & 2^0 \times B \\ 000000000 & \\ 1111011100 & 2^2 \times B \\ 11110111000 & 2^3 \times B \\ \hline 1010101100 & \text{Level 1 Carries} \\ 10101 & \text{Level 2 Carries} \\ \hline 110010001011 & \textit{Product, } P_{2C} \text{ Mod } 2^8 \end{array}$$

8-bit representation

The true value of the 8-bit representation is obtained by negation,

$$10001.011_{2C})_{2C} = 01110.101_2 = 14.625,$$

which, of course, is a negative number. This example illustrates what is called *Booth's algorithm* for fast signed multiplication, and is expressed as follows:

Algorithm 2.11: $A_2 \times B_{2C}$ or $A_{2C} \times B_{2C}$

(1) Set $n = 2k$, where $k(>m)$ is the larger of two numbers[2] counting both integer and fraction bits in k but excluding leading and trailing zeros.

---

[2]The two numbers are initially $|A_2^k|$ of k bits and $|B_2^m|$ of m bits or vice versa.

58 CHAPTER 2 / NUMBER SYSTEMS, BINARY ARITHMETIC, AND CODES

(2) Generate the 2's complement of the negative number(s) by applying Eq. (2.14).
(3) Carry out steps (3) through (6) of Algorithm 2.10 applied to operands A_2 and B_{2C} or A_{2C} and B_{2C}, represented as n-bit operands, to generate the product P_{2C}^n or P_2^n. Use Mod 2^n arithmetic where applicable.

2.9.5 Binary Division

The division operation is generally more complex than that for multiplication. This is so because the result is often not an integer, though the dividend and divisor may be. Consider that A_2 and B_2 are binary operands each of n bits and that

$$A \div B = \sum_{i=0}^{n-1} Q_i 2^i + R/B, \qquad (2.24)$$

where A is the *dividend*, B is the *divisor*, Q is the *quotient*, and R is the *remainder* such that $0 \leq R < B$. An integer quotient is expressed as the binary number $Q_{n-1} \cdots Q_1 Q_0$. From Eq. (2.24) there results the expression

$$A = B \cdot \sum_{i=0}^{n-1} Q_i 2^i + R, \qquad (2.25)$$

which forms the basis for a *restoring* type of binary division procedure:

Begin with $n-1$ for a k-bit divisor and a $(k+n)$-bit dividend to yield an n-bit quotient and a k-bit remainder. If

$$A - 2^{n-1}B = A - b_{n-1} \cdots b_1 b_0 \underbrace{00 \cdots 0}_{n-1 \text{ zeros}} \geq 0,$$

$Q_{n-1} = 1$ or otherwise $Q_{n-1} = 0$. If $Q_{n-1} = 1$, the remaining quotient bits $Q_{n-1} \cdots Q_1 Q_0$ are found beginning with $A' = A - 2^{n-1}B$. Then, if $A' - 2^{n-2}B \geq 0$, $Q_{n-2} = 1$ or otherwise $Q_{n-2} = 0$. Consequently, if $Q_{n-2} = 1$, the remaining quotients $Q_{n-3} \cdots Q_1 Q_0$ are found beginning with $A'' = A' - 2^{n-2}B$, etc. The procedure just described mimics the familiar pencil-and-paper division method.

As an example, consider the following division operation $A \div B$ with 5-bit operands:

EXAMPLE 2.30

$$
\begin{array}{r}
00000101 = Q \\
B = 0101\overline{)00011011} = A \\
\underline{-00010100} = 2^2 \cdot B \\
00000111 = A' = A - 2^2 B \\
\underline{-00000101} = 2^0 \cdot B \\
0010 = R = A' - 2^0 B
\end{array}
$$

2.9 BINARY ARITHMETIC

In this example, not all steps are shown, but the implied division process is

$$A - 2^4 B = A - 01010000 < 0, \quad Q_4 = 0$$
$$A - 2^3 B = A - 00101000 < 0, \quad Q_3 = 0$$
$$A - 2^2 B = A - 00010100 > 0,$$
$$= 00111 = A', \quad Q_2 = 1$$
$$A' - 2^1 B = A' - 00001010 > 0, \quad Q_1 = 0$$
$$A' - 2^0 B = A' - 00000101 > 0, \quad Q_0 = 1$$
$$= 00000010 = A'' = R,$$

where $A - 2^5 B < 0$, $A - 2^6 B < 0$, $A - 2^7 B$, etc., all yield quotients bits $Q = 0$. Notice that the subtractions can be carried out with 2's complement arithmetic according to Algorithm 2.9.

The following algorithm generalizes the binary division process as just presented:

Algorithm 2.12: $A_2 \div B_2$

(1) Set B to k-bits and A to $(k + n)$-bits.
(2) Set $i = n - 1$ and the remainder $= A$.
(3) Set $Q_i = 1$ if $R - 2^i B \geq 0$ and subtract $2^i B$ from A; otherwise set $Q_i = 0$ if $R - 2^i B < 0$.
(4) Repeat step (2) for $i = n - 2, n - 3, \ldots, 1, 0$ to generate quotient bits Q_{n-2}, $Q_{n-3}, \ldots, Q_1, Q_0$ ending with the final n-bit quotient $Q = Q_{n-1} \ldots Q_1 Q_0$.

Binary division involving numbers with fractions is handled in a manner similar to that for decimals. The bit position of the radix point measured from the LSB in the dividend is the same for the quotient. If a fraction exists in the divisor, the radix point of the quotient is that of the dividend minus that of the divisor all taken from the LSB.

Division involving negative numbers is most easily carried out as unsigned division with the result determined by the normal laws of algebra — that is, operands of like sign produce a positive quotient, while those of unlike sign produce a negative quotient. The remainder is given the same sign as the dividend. Signed division can be performed directly by using 2's complement, but the process requires many decision-making steps and, for this reason, is rarely used.

High-Speed Division by Direct Quadratic Convergence A great deal of effort has gone into making multiplication as fast and efficient as possible for use in high-speed computers. So it is only logical that use be made of this fact in generating suitable algorithms for high-speed division. Such methods, commonly used in modern computers, involve *iterative divide algorithms* and are *nonrestoring*. One such method features a system that operates on both the dividend D_D and the divisor D_S with equal multipliers so as to cause the divisor to converge quadratically on unity, thereby yielding the dividend as the quotient. This requires that at least the divisor D_S be represented as a fraction. Therefore, *in binary* the divisor must be represented in the decimal range $0.5 \leq operand < 1$. If both operands are so represented, the direct quadratic convergence method is ideally suited for use with the mantissas in normalized FPN notation described in Section 2.8.

For this approach the quotient is given by

$$Q = \frac{DIVIDEND}{DIVISOR} = \frac{D_D}{D_S}$$
$$= \frac{D_D \cdot k_0 \cdot k_1 \cdot k_2 \cdots}{D_S \cdot k_0 \cdot k_1 \cdot k_2 \cdots} \Rightarrow \frac{Q}{1}. \quad (2.26)$$

The process begins with $D_S = 1 - \alpha$, where $\alpha = 1 - D_S$. But since $D_S < 1$, it follows that $\alpha < 1$. Next choose

$$k_0 = 1 + \alpha$$
$$= 1 + (1 - D_S) = 2 - D_S,$$

giving the updated product

$$D_S \cdot k_0 = (1 - \alpha)(1 + \alpha) = 1 - \alpha^2,$$

which is closer to 1 than D_S. Now set

$$k_1 = 1 + \alpha^2 = 1 + (1 - D_S k_0),$$

giving the updated product

$$D_S \cdot k_0 \cdot k_1 = (1 - \alpha^2)(1 + \alpha^2) = 1 - \alpha^4.$$

Continuing, set

$$k_2 = 1 + \alpha^4 = 1 + (1 - D_S k_0 k_1),$$

so that the updated product becomes

$$D_S \cdot k_0 \cdot k_1 \cdot k_2 = (1 - \alpha^4)(1 + \alpha^4) = 1 - \alpha^8, \text{ etc.}$$

This process continues until the desired number of iterations has been reached or until the updated product $D_S \cdot k_0 \cdot k_1 \cdot k_2 \cdots = 1$.

Notice that each k_j is 1 plus the radix complement of the product of D_S and all the k factors to that point. This can be generalized mathematically as

$$k_j = 1 + \left[D_S \cdot \prod_{i=0}^{j-1} k_i \right]_{rC}. \quad (2.27)$$

2.9 BINARY ARITHMETIC

Consider the following simple example for the division operation, which will be carried out in both decimal and binary:

$$\frac{D_D}{D_S} = \begin{bmatrix} 0.375 \\ 0.500 \end{bmatrix}_{10} = \begin{bmatrix} 0.011 \\ 0.100 \end{bmatrix}_2.$$

EXAMPLE 2.31 In decimal:

$$\alpha = 1 - D_S = 0.5$$
$$k_0 = 1 + \alpha = 1.5$$
$$D_S k_0 = 0.75$$
$$k_1 = 1 + (0.75)_{10C} = 1.25$$
$$D_S k_0 k_1 = 0.9375.$$
$$k_2 = 1 + (0.9375)_{10C} = 1.0625$$
$$D_S k_0 k_1 k_2 = 0.99609375$$
$$k_3 = 1 + (0.99609375)_{10C} = 1.00390625$$
$$D_S k_0 k_1 k_2 k_3 = 0.999984741.$$

Therefore, after four iterations the quotient becomes

$$Q = D_D k_0 k_1 k_2 k_3 = 0.749988556$$
$$= 0.749989 \text{ rounded to six places } (10^{-6}).$$

Note that a fifth iteration with $k_4 = 1.000015259$ produces $Q = 0.750000$ rounded to six places.
In binary:

$$\alpha = 1 - D_S = 0.1$$
$$k_0 = 1 + \alpha = 1.1$$
$$D_S k_0 = 0.11$$
$$k_1 = 1 + (0.11)_{2C} = 1.01$$
$$D_S k_0 k_1 = 0.1111$$
$$k_2 = 1 + (0.1111)_{2C} = 1.0001$$
$$D_S k_0 k_1 k_2 = 0.11111111$$
$$k_3 = 1 + (0.11111111)_{2C} = 1.00000001$$
$$D_S k_0 k_1 k_2 k_3 = 0.1111111111111111.$$

Therefore, at the end of four iterations the quotient is

$$Q = D_D k_0 k_1 k_2 k_3 = 0.1011111111111111010_2$$
$$= 0.75_{10} \text{ after rounding.}$$

In 32-bit FPN notation the quotient Q would be given as

$$\underbrace{0}_{S} \; \underbrace{01111111}_{E} \; \underbrace{01111111111111010\cdots 00}_{M},$$

where the most significant 1 bit in the mantissa M has been omitted in agreement the IEEE normalized FPN notation discussed in Section 2.8.

In the simple example just given the value of α was determined by the value of D_S. Because this is a quadratic convergence algorithm, the process starts off slowly, particularly if the divisor is close to $\frac{1}{2}$. The division calculations in computers can be speeded up by using ROM look-up tables to determine α based on the divisor value. Furthermore, it is common practice to fix the number of iterations and then deal with rounding problem.

With some simplifying assumptions the following algorithm generalizes the quadratic convergence process for iterative division:

Algorithm 2.13: $Q = D_D \div D_S$

(1) Set D_S to normalized FPN form, retain the MSB 1 in the mantissa, and adjust the exponent as required by the FPN notation.
(2) Calculate $\alpha = 1 - D_S$ by using Algorithm 2.9.
(3) Set $k_0 = 1 + \alpha$ and calculate $(D_S k_0)_{2C}$ by using Algorithms 2.10 and 2.6.
(4) Set $k_1 = 1 + (D_S k_0)_{2C}$ and calculate $(D_S k_0 k_1)_{2C}$ as in step (3).
(5) Repeat steps (1) through (4) for $k_j = 1 + \left[D_S \cdot \prod_{i=0}^{j-1} k_i \right]$ for all $j = 2, 3, \ldots$.
(6) Calculate $Q = D_D \prod_{i=0}^{j-1} k_i$ when $D_S \prod_{i=0}^{j-1} k_i = 1$.

2.9.6 BCD Addition and Subtraction

Compared to binary arithmetic, BCD arithmetic is more complex, particularly with regard to hardware implementation. This is true since not all possible four-bit binary number states correspond to the BCD number system. The six number patterns 1010, 1011, 1100, 1101, 1110, and 1111 are not valid BCD states, as is indicated in Table 2.2.

BCD Addition BCD addition is similar to unsigned binary addition, except that a correction must be made any time a sum exceeds $9_{10} = 1001_2$. Summation begins with the least significant digit (LSD) and ends with the most significant digit (MSD). If the sum exceeds 1001 for any given digit, that sum is corrected by adding $6_{10} = 0110_2$ with a carry of 0001 to the next MSD. The following example illustrates the addition process for two-decade BCD integers $A_{BCD} = A_{10} A_1$ and $B_{BCD} = B_{10} B_1$ represented in three-decade form:

2.9 BINARY ARITHMETIC   63

EXAMPLE 2.32

$$
\begin{array}{rl}
056_{10} \rightarrow & 0000\ 0101\ 0110_{BCD} \quad \text{Augend } A_{BCD} \\
+069_{10} \rightarrow & +0000\ 0110\ 1001_{BCD} \quad \text{Addend } B_{BCD} \\
\hline
125_{10} & 0000\ 1011\ 1111 \quad\quad \text{Sum} \\
& \quad\ 0110\ 0110 \quad\quad\quad \text{Correction} \\
& 1\ 1111\ 110 \quad\quad\quad\quad \text{Carries} \\
\hline
& 0001\ 0010\ 0101_{BCD} \quad \text{Result} = 125_{10}
\end{array}
$$

The following algorithm generalizes the process just given for operands having both integers and fractions.

Algorithm 2.14: $A_{BCD} + B_{BCD}$

(1) Set BCD operands in descending decade order, $A_{BCD} = \cdots A_{100} A_{10} A_1 A_{.1} A_{.01} \cdots$ and $B_{BCD} = \cdots B_{100} B_{10} B_1 B_{.1} B_{.01} \cdots$ such that the MSDs for A and B are 0000 (null).
(2) Set $i = LSD$ for matching operand decades.
(3) If $A_i + B_i > 1001$ by Algorithm 2.8, add 0110 to that result and carry 0001 over to the next MSD. If $A_i + B_i < 1001$, add 0000 to the result and carry 0000 to the next MSD.
(4) Repeat steps (2) and (3) for matching decades in the order of $10^1 i, 10^2 i, 10^3 i, \ldots,$ MSD.

BCD Subtraction Negative BCD numbers are most conveniently represented in 10's complement (10C). This permits negative BCD numbers to be added according to Algorithm 2.14. The result, if negative, will be represented in 10C form requiring negation to obtain the true value.

EXAMPLE 2.33

$$
\begin{array}{l}
08.25_{10} \rightarrow\ \ 08.25_{10}\ \rightarrow\ \ 0000\ 1000\ .\ 0010\ 0101_{BCD} \\
-13.52_{10} \rightarrow +86.48_{10C} \rightarrow +1000\ 0110\ .\ 0100\ 1000_{BCD)_{10C}} \\
\hline
-05.27_{10} \quad\quad 94.73_{10C} \quad\quad\quad 1000\ 1110\ .\ 0110\ 1101 \quad \text{Sum} \\
\quad\quad\quad\quad\quad\quad\quad\quad +\ \ \ \ \ \ 0110\ \ \ \ \ \ \ \ \ \ 0110 \quad\quad \text{Correction} \\
\quad\quad\quad\quad\quad\quad\quad\quad\quad\ \ \ 1\ 11\quad\quad\quad 1\ 1 \quad\quad\quad \text{Carries} \\
\hline
\quad\quad\quad\quad\quad\quad\quad\quad\quad 1001\ 0100\ .\ 0111\ 0011_{BCD)_{10C}} \quad \text{Result}
\end{array}
$$

The true (absolute) value of the result is found by negation to be

$$94.73_{10C})_{10C} = 05.27_{10} \text{ or } 0000\ 0101\ .\ 0010\ 0111_{BCD}.$$

Note that to convert directly from the *BCD* form, $BCD)_{10C} = N_{2C} + 1010 = \bar{N}_2 + 1_{LSB} + 1010$ for the LSD but thereafter is $BCD)_{10C} = \bar{N}_2 + 1010 = N_{9C}$, discarding any carry overflow in either case.

The following algorithm generalizes the process of BCD subtraction.

Algorithm 2.15: $A_{BCD} + B_{BCD)_{10C}}$ or $A_{BCD)_{10C}} + B_{BCD)_{10C}}$

(1) Convert any negative decimal number to its 10's complement (10C) by Algorithm 2.5 with $r = 10$.

(2) Represent each operand in BCD form.
(3) Carry out steps (1) through (4) of Algorithm 2.14. If the result is negative, the true value is found by negation: $[(result)_{10C}]_{10C} = (result)_{BCD}$. If the result is positive, that result is the true value.

2.9.7 Floating-Point Arithmetic

Up to this point the arithmetic operations have involved fixed-point representation in which all bits of a binary number were represented. In many practical applications the numbers may require many bits for their representation. In Section 2.8 the floating-point number (FPN) system was discussed for just that reason. Now it is necessary to deal with the arithmetic associated with the FPN system, namely addition, subtraction, multiplication, and division.

FPN Addition and Subtraction Before two numbers can be added or subtracted one from the other, it is necessary that they have the same exponent. This is equivalent to aligning their radix points. From Eq. (2.17) for radix 2, consider the following two FPNs:

$$X = M_X \cdot 2^{E_X}$$

and

$$Y = M_Y \cdot 2^{E_Y}.$$

Now, if for example $E_X \geq E_Y$, then Y is represented as $M'_Y \cdot 2^{E'_Y}$, where

$$M'_Y = \cdot \underbrace{00 \cdots 0}_{\substack{E_X - E_Y \\ \text{zeros}}} f_{-1} f_{-2} \cdots f_{-m} \quad \text{and} \quad E'_Y = E_Y + (E_X - E_Y) = E_X,$$

so that $X + Y = (M_X + M'_Y) \cdot 2^{E_X}$ or $X - Y = (M_X - M'_Y) \cdot 2^{E_X}$, etc. Here, $M_Y = \cdot f_{-1} f_{-2} \cdots f_{-m}$ originally, but is now adjusted so that the exponents for both operands are the same. The addition or subtraction of the fractions M_X and M'_Y is carried out according to Algorithm 2.8 or Algorithm 2.9, respectively.

Consider the following examples of FPN addition:

EXAMPLE 2.34 — Addition

$$\begin{array}{r} 145.500_{10} \\ +27.625_{10} \\ \hline 173.125_{10} \end{array} \quad \rightarrow \quad \begin{array}{l} 10010001.100_2 = X \\ \underline{00011000.101_2 = Y} \end{array}$$

Comparing and equalizing the exponents E_X and E_Y gives

$$145.500_{10} = .10010001100 \times 2^8$$
$$27.625_{10} = .00011011101 \times 2^8.$$

2.9 BINARY ARITHMETIC

In FPN notation the addition operation becomes

$$
\begin{array}{r|ccc}
 & S & E & M \\
\hline
145.500 \rightarrow & 0 & 10000111 & 10010001 1000\cdots 00 \\
+27.625 \rightarrow & +0 & 10000111 & 00011011 1010\cdots 00 \\
\hline
173.125 & 0 & 10000111 & 10101101 0010\cdots 00 \\
\end{array}
$$

where the exponents given in excess 127 form are $127 + 8 = 135$ as discussed in Section 2.7. To represent the result in normalized FPN notation, the most significant 1 bit of the mantissa is omitted, yielding the result 0 10000111 01011010 0100$\cdots$00 for the sign, exponent, and mantissa fields, respectively.

EXAMPLE 2.35 — Subtraction in 2's complement

$$
\begin{array}{r|ccc}
 & S & E & M \\
\hline
-145.500 \rightarrow & 0 & 10000111 & 01101110 1000\cdots 00 \\
+27.625 \rightarrow & +0 & 10000111 & 00011011 1010\cdots 00 \\
\hline
-117.875 & 0 & 10000111 & 10001010 0010\cdots 00 \\
\end{array}
$$

The true value is obtained by negation of the mantissa (taking its 2's complement), giving the sign magnitude result 1 10000111 01110101 1110$\cdots$00, which is -117.875_{10}. In normalized FPN notation the MSB 1 bit of the mantissa would be omitted, giving the 2's complement result 1 10000111 00010100 0010$\cdots$00.

FPN Multiplication and Division In some respects multiplication and division of FPNs is simpler than addition and subtraction from the point of view of the decision-making problem. Multiplication and division operations can be carried out by using the standard methods for such operations without the need to compare and equalize the exponents of the operands. The following generalizations of these processes illustrates the procedure involved.

The product of two operands X and Y in radix r is represented as follows:

$$
\begin{aligned}
P &= X \times Y \\
&= (M_X \cdot r^{E_X}) \times (M_Y \cdot r^{E_Y}) \\
&= (M_X \times M_Y) \cdot r^{(E_X + E_Y)} \\
&= M_P \cdot r^{E_P},
\end{aligned}
$$

where the exponents are added following Algorithm 2.8 while the mantissas are multiplied by using Algorithm 2.10. The addition and multiplication of signed numbers is covered by Algorithms 2.9 and 2.11, respectively.

Similarly, for division in radix r the quotient is given by

$$
\begin{aligned}
Q &= X \div Y \\
&= (M_X \cdot r^{E_X}) \div (M_Y \cdot r^{E_Y}) \\
&= (M_X \div M_Y) \cdot r^{(E_X - E_Y)}.
\end{aligned}
$$

Here, the exponents are subtracted (added) in radix 2 (binary) by using the same algorithms as for addition, namely Algorithms 2.8 and 2.9. The division of the mantissas, on the other hand, is best accomplished by using the quadratic convergence iterative divide method discussed in Subsection 2.9.5, since the mantissas are usually represented in normalized FPN form. The following examples illustrate the multiplication and division of FPNs.

EXAMPLE 2.36 — FPN signed-magnitude multiplication

$$
\begin{array}{r}
-7.25_{10} \rightarrow \\
\times 4.50_{10} \rightarrow \\
\hline
-32.625_{10}
\end{array}
\quad
\begin{array}{c}
S \\
1 \\
+0 \\
\hline
1
\end{array}
\quad
\begin{array}{c}
E \\
10000010 \\
+10000010 \\
\hline
10000101
\end{array}
\quad
\begin{array}{c}
M \\
11101000\cdots00 \\
\times 10010000\cdots00 \\
\hline
00000000\cdots00 \\
000000000\cdots00 \\
0000000000\cdots00 \\
00000000000\cdots00 \\
111010000000\cdots00 \\
0000000000000\cdots00 \\
00000000000000\cdots00 \\
111010000000000\cdots00 \\
\hline
1000001010000000\cdots00
\end{array}
$$

23-*bit representation*

The result, given in normalized signed-magnitude 32-bit FPN form, is

$$1\ 10000101\ 000001010000000\cdots 0_{FPN} = -32.625_{10},$$

where the MSB 1 bit of the mantissa is omitted. Note that the mantissa has a magnitude $.10000010100\cdots 00 \times 2^6 = 100000.101$.

EXAMPLE 2.37 — FPN signed-magnitude division

$$\frac{4.5}{-0.625} \Rightarrow -\frac{0.1001 \times 2^3}{0.1010} = \frac{X}{Y}$$

In FPN notation the operands are

$$X = 0\ 10000010\ 100100\cdots 00$$

$$Y = 1\ 01111111\ 101000\cdots 00.$$

Division of the mantissas $M_X/M_Y = D_D/D_S$ by Algorithm 2.13:

$$\alpha = 1 - D_S = 1 - .101000\cdots 00 = .011000\cdots 00$$

$$k_0 = 1 + \alpha = 1.011000\cdots 00$$

$$D_S k_0 = .11011100\cdots 00$$

$$k_1 = 1 + (D_S k_0)_{2C} = 1.00100100\cdots 00$$
$$D_S k_0 k_1 = .11111010111100\cdots 00$$
$$k_2 = 1 + (D_S k_0 k_1)_{2C} = 1.00000101000100\cdots 00$$
$$D_S k_0 k_1 k_2 = .11111111111001100101111$$
$$k_3 = 1 + (D_S k_0 k_1 k_2)_{2C} = 1.00000000000110011010001.$$

After four iterations the quotient is given by

$$Q = (D_D k_0 k_1 k_2 k_3) = .11100110011001100011000 \times 2^3,$$

which is truncated to 23 bits. The quotient has a decimal value of

$$(0.899996755 \times 2^3)_{10} = 7.1999740450_{10}.$$

In normalized FPN signed-magnitude 32 bit form the quotient is given by

$$Q = 1\ 10000010\ 11001100110011000110000,$$

where the MSB 1 bit of the mantissa has been omitted as discussed in Section 2.8. Note that the subtraction of exponents $E_X - E_Y$ is $130 - 127 = 003_{10}$ or 10000010_2 in excess 127 code. The sign bits are added in binary giving $S_X + S_Y = 0 + 1 = 1$, where any carry (in this case 0) is discarded.

Algorithm 2.16: Signed-Magnitude $(X \times Y)_{FPN}$ or $(X \div Y)_{FPN}$

(1) Set operands X and Y in IEEE normalized FPN form (see Section 2.7).
(2) Add the exponents E_X and E_Y according to Algorithms 2.8 or 2.9.
(3) If $X \times Y$, then multiply mantissa fractions according to Algorithm 2.10.
(4) If $X \div Y$, then divide mantissas according to Algorithm 2.13.
(5) Add the sign bits $S_X + S_Y$ and discard the carry.
(6) Set result in IEEE normalized FPN form.

2.9.8 Perspective on Arithmetic Codes

It should seem clear to the reader that certain arithmetic operations are more easily executed than others, depending on whether or not the operands are signed and depending on the number code used to carry out the operation. Table 2.9 is intended to show the general degree of difficulty of certain arithmetic operations relative to the arithmetic code (signed-magnitude, 2's complement, etc.) used.

Not indicated in Table 2.9 are the subdivisions within a given type of arithmetic operation. For example, no distinction is made between a direct division (restoring) algorithm and an iterative divide (nonrestoring) algorithm, which may differ significantly with regard to difficulty — the latter being easier for operands represented in FPN notation. As a general rule,

Table 2.9 Arithmetic codes vs the degree of difficulty of arithmetic operations

Arithmetic Operation	Unsigned Numbers	Signed-Magnitude	Two's Complement
Unsigned addition	Easy	Easy	Easy
Signed addition/subtraction	—	Difficult	Easy
Unsigned multiplication	Fairly difficult	Fairly difficult	Difficult
Signed multiplication	—	Fairly difficult	Difficult
Unsigned division	Difficult	Difficult	Very difficult

addition/subtraction of signed numbers involves relatively simple arithmetic manipulations of the operands compared to multiplication; and division requires more decision-making steps than multiplication. Also not shown in Table 2.9 are the 1's complement, BCD, and XS3 number codes, since they are not commonly used in computer numeric operations. Finally, a direct correspondence is implied between degree of difficulty and the hardware requirements to carry out a given arithmetic operation.

2.10 OTHER CODES

Most binary codes of value in digital design fall into one or more of the following ten categories:

Weighted binary codes	Unit distance codes
Unweighted binary codes	Reflective codes
Biased codes	Number codes
Decimal codes	Alphanumeric codes
Self-complementing codes	Error detecting codes

The previous sections have introduced examples of weighted binary codes, number codes, biased codes, and decimal codes. Number codes are those such as 2's and 1's complement that are used in addition/subtraction operations. Now, other codes (excluding alphanumeric codes) will be briefly discussed so as to provide a foundation for the developments in later chapters.

2.10.1 The Decimal Codes

Shown in Table 2.10 are seven decimal (10 state) codes that can be classified as either weighted or unweighted codes. All but one of these codes is weighted as indicated in the table. A weighted code can be converted to its decimal equivalent by using positional weights in a polynomial series as was done for the BCD code (1) discussed in Subsection 2.4.1. Code (2), the XS3 code, is a biased-weighted code considered in Subsection 2.4.1 and in Section 2.7. An unweighted code, such as code (7), cannot be converted to its decimal equivalent by any mathematical weighting procedure.

Not all weighted decimal codes are natural in the sense that their code weights cannot be derived from positive powers of 2 as is done for codes (1) and (2). Codes (3) through (6) in Table 2.10 are of this type. Code weights such as -1, -2, 5, and 6 cannot be generated

2.10 OTHER CODES

Table 2.10 Weighted and unweighted decimal codes

Dec. Value	Weighted codes						Unweighted
	(1) (BCD)	(2) (XS3)	(3) 2421	(4) 84-2-1	(5) 86421	(6) 51111	(7) Creeping Code
0	0000	0011	0000	0000	00000	00000	00000
1	0001	0100	0001	0111	00001	00001	10000
2	0010	0101	0010	0110	00010	00011	11000
3	0011	0110	0011	0101	00011	00111	11100
4	0100	0111	0100	0100	00100	01111	11110
5	0101	1000	1011	1011	00101	10000	11111
6	0110	1001	1100	1010	01000	11000	01111
7	0111	1010	1101	1001	01001	11100	00111
8	1000	1011	1110	1000	10000	11110	00011
9	1001	1100	1111	1111	10001	11111	00001

by any positive integer power of 2, but they can still serve as code weights. As an example, consider how decimal 5 is represented by code (4):

$$\text{Decimal equivalent} = 5$$

$$\text{84-2-1 code representation} = (1 \times 8) + (0 \times 4) + [1 \times (-2)] + [1 \times (-1)]$$

$$= 1011.$$

Note that there may be more than one combination of weighted bits that produce a given state. When this happens, the procedure is usually to use the fewest 1's. For example, decimal 7 can be represented by 00111 in code (5), 86421 code, but 01001 is preferred. An exception to this rule is the 2421 code discussed next.

Codes (2), (3), and (4) are examples of codes that have the unusual property of being *self-complementing*. This means that the 1's complement of the code number is the code for the 9's complement of the corresponding decimal number. In other words, the 1's complement of any state N (in decimal) is the same as the $(9 - N)$ state in the same self-complementing code. As an example, the 1's complement of state 3 in XS3 (0110) is state 6 (1001) in that same code. The 1's and 9's complement number codes were discussed in Subsection 2.6.3 and are presented in Tables 2.7 and 2.8, respectively.

2.10.2 Error Detection Codes

There is another class of weighted or semiweighted codes with the special property that their states contain either an even number or an odd number of logic 1's (or 0's). Shown in Table 2.11 are four examples of such codes. This unique feature make these codes attractive as error-detecting (parity-checking) codes. Notice that both the 2-out-of-5 code (semiweighted) and the biquinary code (weighted 50 43210) must have two 1's in each of their 10 states and are, therefore, even-parity codes. In contrast, the one-hot code (weighted 9876543210) is an odd-parity code, since by definition it is allowed to have only a single 1 for any given state. Code (d) is no more than the BCD code with an attached odd parity-generating bit, P.

Table 2.11 Error detection codes

Decimal Value	(a) Even Parity 2-out-of-5 (74210)	(b) Even Parity Biquinary 50 43210	(c) Odd Parity One-Hot 9876543210	(d) Odd Parity BCD P8421
0	11000	01 00001	0000000001	10000
1	00011	01 00010	0000000010	00001
2	00101	10 00100	0000000100	00010
3	00110	10 01000	0000001000	10011
4	01001	01 10000	0000010000	00100
5	01010	10 00001	0000100000	10101
6	01100	10 00010	0001000000	10110
7	10001	10 00100	0010000000	00111
8	10010	10 01000	0100000000	01000
9	10100	10 10000	1000000000	11001

The advantage of using an error-detecting code is that single-bit errors (those most likely to occur) are easily detected by a parity detector placed at the receiving end of a data bus. If a single error occurs, the parity is changed (odd-to-even or vice versa) and further processing can be delayed until the error is corrected. On the other hand, if two errors occur, the error cannot be detected by any simple means.

2.10.3 Unit Distance Codes

The last class of codes that will be discussed here are called *unit distance* codes, so called because only *one* bit is permitted to change between any two of their states — recall that in natural binary, adjacent states may differ by one or more bits. Three examples of unit distance codes are given in Table 2.12: (1) a decimal code, (2) a *reflective* unit distance code called *Gray code,* and (3) an XS3 decimal Gray code formed from the inner 10 states of code (2). The reflective character of the Gray and XS3 Gray codes are easily revealed by the fact that all bits except the MSB are mirrored across an imaginary plane located midway in the 16 states and 10 states, respectively, as indicated by the dashed lines. The unit distance property of the Gray code will be used in logic function graphics discussed at length in Chapter 4. Also, the unit distance and reflective character of the Gray code make it uniquely suitable as a position indicator code for rotating disks and shafts. Encoding errors produced by rotational irregularities can be detected and corrected by the use of such a code.

Although only a 4-bit Gray code is represented in Table 2.12, it should be noted that a Gray code of any number of bits is possible. Also, there are other unit distance codes that can be generated with any number of bits but they will most likely not be reflective.

2.10.4 Character Codes

The most common character code is called *ASCII* (pronounced "as-key"), the acronym for *American Standard Code for Information and Interchange.* ASCII code represents each character as a 7-bit binary string, hence a total of $2^7 = 128$ characters, and is given in Table 2.13. This code encodes numerals, punctuation characters, all upper- and lowercase

Table 2.12 Unit distance codes: (1) a decimal code (nonreflective); (2) four-bit Gray code (reflective); (3) XS3 Gray decimal code (reflective)

Decimal Value	(1) Decimal Code	(2) 4-Bit Gray Code	(3) XS3 Gray Decimal Code
0	0000	0000	0010
1	0001	0001	0110
2	0011	0011	0111
3	0010	0010	0101
4	0110	0110	0100
5	1110	0111	1100
6	1111	0101	1101
7	1101	0100	1111
8	1100	1100	1110
9	0100	1101	1010
10	—	1111	—
11	—	1110	—
12	—	1010	—
13	—	1011	—
14	—	1001	—
15	—	1000	—

Table 2.13 ASCII character code

$a_3a_2a_1a_0$	Row (Hex)	$a_6a_5a_4$ (column)							
		000 0	001 1	010 2	011 3	100 4	101 5	110 6	111 7
0000	0	NUL	DLE	SP	0	@	P	`	p
0001	1	SOH	DC1	!	1	A	Q	a	q
0010	2	STX	DC2	"	2	B	R	b	r
0011	3	ETX	DC3	#	3	C	S	c	s
0100	4	EOT	DC4	$	4	D	T	d	t
0101	5	ENQ	NAK	%	5	E	U	e	u
0110	6	ACK	SYN	&	6	F	V	f	v
0111	7	BEL	ETB	'	7	G	W	g	w
1000	8	BS	CAN	(	8	H	X	h	x
1001	9	HT	EM	)	9	I	Y	i	y
1010	A	LF	SUB	*	:	J	Z	j	z
1011	B	VT	ESC	+	;	K	[	k	{
1100	C	FF	FS	,	<	L	\	l	|
1101	D	CR	GS	-	=	M	]	m	}
1110	E	SO	RS	.	>	N	^	n	~
1111	F	SI	US	/	?	O	_	o	DEL

alphabet letters, and a variety of printer and typewriter control abbreviations. An eighth bit (not shown) is often used with the ASCII code for error detection purposes.

Another common character code is known as *EBCDIC* (pronounced "ebb-see-dick"), the acronym for *extended BCD interchange code*. It uses 8-bit BCD strings so as to encode a 256-character set.

FURTHER READING

Literature on number systems and arithmetic is extensive. Many journal articles and most texts on digital logic design cover these subjects to one extent or another. Portions of this chapter regarding number systems are taken from contributions by Tinder to *The Electrical Engineering Handbook,* cited here. Recognized classic treatments of number systems and arithmetic include those of Garner, Hwang, and Knuth. The IEEE publication on the standard for floating-point arithmetic is also frequently cited. These references, together with recent texts covering the subject areas, are cited here.

[1] H. L. Garner, "Number Systems and Arithmetic," in *Advances in Computers*, Vol. 6. Academic Press, New York, 1965, pp. 131–194.
[2] K. Hwang, *Computer Arithmetic*. John Wiley & Sons, New York, 1978.
[3] IEEE, *IEEE Standard for Binary Floating-Point Arithmetic* (ANSI/IEEE Std 754-1985) The Institute of Electrical and Electronic Engineers, New York, 1985.
[4] D. E. Knuth, *The Art of Computer Programming: Seminumerical Algorithms,* Vol. 2. Addison-Wesley, Reading, MA, 1969.
[5] V. P. Nelson, H. T. Nagle, B. D. Carroll, and J. D. Irwin, *Digital Logic Circuit Analysis and Design*. Prentice Hall, Englewood Cliffs, NJ, 1995.
[6] L. H. Pollard, *Computer Design and Architecture*. Prentice Hall, Englewood Cliffs, NJ, 1990.
[7] A. W. Shaw, *Logic Circuit Design*. Saunders College Publishing, Fort Worth, TX, 1993.
[8] R. F. Tinder, *Digital Engineering Design: A Modern Approach*, Prentice Hall, Englewood Cliffs, NJ, 1991.
[9] R. F. Tinder, "Number Systems," in *The Electrical Engineering Handbook*, 2nd ed. (R. C. Dorf, Ed.). CRC Press, 1997, pp. 1991–2006.
[10] C. Tung, "Arithmetic," *Computer Science* (A. F. Cardenas *et al.,* Eds.), Chapter 3. Wiley-Interscience, New York, 1972.
[11] J. F. Wakerly, *Digital Design Principles and Practice,* 2nd ed. Prentice Hall, Englewood Cliffs, NJ, 1994.

PROBLEMS

Note: Use Tables P2.1, P2.2, and P2.3 as needed in working the following problems.

2.1 Convert the following decimal numbers to binary:
  (a) 5
  (b) 14
  (c) 39
  (d) 107.25
  (e) 0.6875

PROBLEMS

2.2 Convert the following binary numbers to decimal by using the method of positional weights:
(a) 0110
(b) 1011
(c) 11001
(d) 11011001.11
(e) 0.01011

2.3 Convert the decimal numbers of Problem 2.1 to BCD.

2.4 Convert the binary numbers of Problem 2.2 to BCD. To do this, add trailing and leading 0's as required.

2.5 Convert the following BCD numbers to binary:
(a) 00010011
(b) 01010111
(c) 0101000110
(d) 1001000.00100101
(e) 0.100001110101

2.6 Convert the decimal numbers in Problem 2.1 to XS3.

2.7 Convert the BCD numbers in Problem 2.5 to XS3.

2.8 Convert the binary numbers in Problem 2.2 to BCH.

2.9 Convert the BCD numbers in Problem 2.5 to BCO.

2.10 Convert the following numbers to binary:
(a) 613520_8
(b) $2FD6A25B_{16}$
(c) 11110011100.011_{XS3}
(d) 6!

2.11 Convert the following fractions as indicated:
(a) $0.534_{10} \to N_2$ rounded to 8 bits.
(b) $0.3DF2_{16} \to N_2$ rounded to 8 bits.
(c) $0.534_{10} \to N_{16}$ terminated at 4 digits.
(d) $0.5427_8 \to N_2$ rounded to 8 bits.

2.12 Convert the following numbers to *signed-magnitude* binary form:
(a) $+56.25_{10}$
(b) -94.625_{10}
(c) $-7BD.5_{16}$
(d) $+125_8$
(e) -0110101.10011_{BCD}

2.13 Give the radix complement representation for the following numbers:
(a) The 10's complement of 47.63_{10}

(b) The 2's complement of 011011101.1101_2

(c) The 8's complement of 501.74_8

(d) The 16's complement of $AF3.C8_{16}$

2.14 Represent the following numbers in IEEE normalized FPN_2 form:
   (a) 1101011.1011_2
   (b) $+27.6875_{10}$
   (c) -145.500_{10}

2.15 Add the following binary additions and verify in decimal:
   (a) $10 + 11$
   (b) $101 + 011$
   (c) $10111 + 01110$
   (d) $101101.11 + 011010.10$
   (e) $0.1100 + 1.1101$

2.16 Carry out the following binary subtraction operations in 2's complement and verify in decimal:
   (a) $01100 - 00101$
   (b) $0111011 - 0011001$
   (c) $01001000 - 01110101$
   (d) $010001.0101 - 011011.1010$
   (e) $00.011010 - 01.110001$

2.17 Repeat Problem 2.16 in 1's complement.

2.18 Carry out the following binary multiplication operations and verify in decimal:
   (a) 11×0101
   (b) 11101×1111011
   (c) 1001.10×11101.11
   (d) 110.011×1101.0101
   (e) 0.1101×0.01111

2.19 Carry out the following complement multiplications and verify in decimal:
   (a) $00000111 \times -00001101$
   (b) 110×-11101 ($k = 5$)
   (c) -11.01×101.11 ($k = 5$)
   (d) 111.111×-1.101 ($k = 6$)
   (Hint: Consider switching minuend and subtrahend operands if it yields less work.)

2.20 Find the quotient for each of the following division operations by using the binary equivalent of the familiar "pencil-and-paper" method used in long division of decimal numbers. Show work details.
   (a) $1100 \div 100$
   (b) $111111 \div 1001$
   (c) $11001.1 \div 011.11$ (Carry out quotient to the 2^{-2} bit and give the remainder)
   (d) $100 \div 1010$ (Carry out quotient to the 2^{-6} bit and give remainder)

2.21 Use the direct quadratic convergence method to obtain the quotient for the following fractions. To do this, use Eqs. (2.26) and (2.27).
(a) $(0.25 \div 0.75)_{10}$ in decimal. Find Q after three iterations and rounded to 10^{-5}.
(b) $(0.01 \div 0.11)_2$ in binary. Compare Q after two iterations rounded to 2^{-8} with Q after three iterations rounded to 2^{-16}. For comparison, use decimal values derived from the binary results.

2.22 Carry out the following hexadecimal operations and verify in decimal:
(a) $1A8 + 67B$
(b) $ACEF1 + 16B7D$
(c) $1273_{16} - 3A8$
(d) $89_{16} \times 1A3$
(e) $A2 \times 15BE3$
(f) $1EC87 \div A5$ (Hint: Use decimal $\leftrightarrow$ hex methods with Table P2.3.)

2.23 Convert the following decimal numbers to BCD with the MSDs null (0000), then carry out the indicated arithmetic in BCD by using Algorithms 2.14 and 2.15 in Subsection 2.9.6:
(a) $049_{10} + 078_{10}$
(b) $168.6_{10} + 057.5_{10}$
(c) $093_{10} - 067_{10}$
(d) $034.79_{10} - 156.23_{10}$

2.24 Perform the FPN arithmetic indicated below. To do this follow the examples in Subsection 2.9.7.
(a) $135.25_{10} + 54.625_{10}$
(b) $54.625_{10} - 135.25_{10}$
(c) $3.75_{10} \times 5.0625_{10}$
(d) $4.50_{10} \times (-2.3125_{10})$
(e) $6.25 \div (-0.375_{10})$

Note: Use the sign-magnitude FPN system for parts (d) and (e) following Examples 2.36 and 2.37.

2.25 To add XS3 numbers, a correction by either adding or subtracting 0011 is necessary depending on whether or not a 1 carry is generated. Study, then write an algorithm for the addition in XS3 numbers.

2.26 Prove that a self-complementing unit-distance code is not possible.

2.27 An inspection of the binary and Gray codes in Tables 2.1 and 2.12 indicates a unique relationship between these codes. Examine these codes and devise a simple algorithm that will permit direct "pencil-and-paper" conversion between them, binary-to-Gray or vice versa.

2.28 Decipher the following ASCII code. It is given in hexadecimal, MSD first.

57 68 61 74 69 73 79 6F 75 72 6E 61 6D 65 3F

Table P2.1 Powers of 2

2^n	n	2^{-n}
1	0	1.0
2	1	0.5
4	2	0.25
8	3	0.125
16	4	0.062 5
32	5	0.031 25
64	6	0.015 625
128	7	0.007 812 5
256	8	0.003 906 25
512	9	0.001 953 125
1 024	10	0.000 976 562 5
2 048	11	0.000 488 281 25
4 096	12	0.000 244 140 625
8 192	13	0.000 122 070 312 5
16 384	14	0.000 061 035 156 25
32 768	15	0.000 030 517 578 125
65 536	16	0.000 015 258 789 062 5
131 072	17	0.000 007 629 394 531 25
262 144	18	0.000 003 814 697 265 625
524 288	19	0.000 001 907 348 632 812 5
1 048 576	20	0.000 000 953 674 316 406 25
2 097 152	21	0.000 000 476 837 158 203 125
4 194 304	22	0.000 000 238 418 579 101 562 5
8 388 608	23	0.000 000 119 209 289 550 781 25
16 777 216	24	0.000 000 059 604 644 775 390 625
33 554 432	25	0.000 000 029 802 322 387 695 312 5
67 108 864	26	0.000 000 014 901 161 193 847 656 25
134 217 728	27	0.000 000 007 450 580 596 923 828 125
268 435 456	28	0.000 000 003 725 290 298 461 914 062 5
536 870 912	29	0.000 000 001 862 645 149 230 957 031 25
1 073 741 824	30	0.000 000 000 931 322 574 615 478 515 625
2 147 483 648	31	0.000 000 000 465 661 287 307 739 257 812 5
4 294 967 296	32	0.000 000 000 232 830 643 653 869 628 906 25

Table P2.2 Hexadecimal addition table

	0	1	2	3	4	5	6	7	8	9	A	B	C	D	E	F
0	0	1	2	3	4	5	6	7	8	9	A	B	C	D	E	F
1	1	2	3	4	5	6	7	8	9	A	B	C	D	E	F	10
2	2	3	4	5	6	7	8	9	A	B	C	D	E	F	10	11
3	3	4	5	6	7	8	9	A	B	C	D	E	F	10	11	12
4	4	5	6	7	8	9	A	B	C	D	E	F	10	11	12	13
5	5	6	7	8	9	A	B	C	D	E	F	10	11	12	13	14
6	6	7	8	9	A	B	C	D	E	F	10	11	12	13	14	15
7	7	8	9	A	B	C	D	E	F	10	11	12	13	14	15	16
8	8	9	A	B	C	D	E	F	10	11	12	13	14	15	16	17
9	9	A	B	C	D	E	F	10	11	12	13	14	15	16	17	18
A	A	B	C	D	E	F	10	11	12	13	14	15	16	17	18	19
B	B	C	D	E	F	10	11	12	13	14	15	16	17	18	19	1A
C	C	D	E	F	10	11	12	13	14	15	16	17	18	19	1A	1B
D	D	E	F	10	11	12	13	14	15	16	17	18	19	1A	1B	1C
E	E	F	10	11	12	13	14	15	16	17	18	19	1A	1B	1C	1D
F	F	10	11	12	13	14	15	16	17	18	19	1A	1B	1C	1D	1E

Table P2.3 Hexadecimal multiplication table

	0	1	2	3	4	5	6	7	8	9	A	B	C	D	E	F
0	0	0	0	0	0	0	0	0	0	0	0	0	0	0	0	0
1	0	1	2	3	4	5	6	7	8	9	A	B	C	D	E	F
2	0	2	4	6	8	A	C	E	10	12	14	16	18	1A	1C	1E
3	0	3	6	9	C	F	12	15	18	1B	1E	21	24	27	2A	2D
4	0	4	8	C	10	14	18	1C	20	24	28	2C	30	34	38	3C
5	0	5	A	F	14	19	1E	23	28	2D	32	37	3C	41	46	4B
6	0	6	C	12	18	1E	24	2A	30	36	3C	42	48	4E	54	5A
7	0	7	E	15	1C	23	2A	31	38	3E	46	4D	54	5B	62	69
8	0	8	10	18	20	28	30	38	40	48	50	58	60	68	70	78
9	0	9	12	1B	24	2D	36	3E	48	51	5A	63	6C	75	7E	87
A	0	A	14	1E	28	32	3C	46	50	5A	64	6E	78	82	8C	96
B	0	B	16	21	2C	37	42	4D	58	63	6E	79	84	8F	9A	A5
C	0	C	18	24	30	3C	48	54	60	6C	78	84	90	9C	A8	B4
D	0	D	1A	27	34	41	4E	5B	68	75	82	8F	9C	A9	B6	C3
E	0	E	1C	2A	38	46	54	62	70	7E	8C	9A	A8	B6	C4	D2
F	0	F	1E	2D	3C	4B	5A	69	78	87	96	A5	B4	C3	D2	E1

CHAPTER 3

Background for Digital Design

3.1 INTRODUCTION

The contents of this chapter are considered all important to the reader's understanding of the remainder of this text and, hence, to an understanding of modern digital design methods. In this chapter the reader will learn mixed logic notation and symbology, Boolean algebra, and the reading and construction of logic circuits. Besides becoming the industrial standard, mixed logic notation and symbology, once learned, offers a remarkably simple, direct means of reading and constructing logic circuits and timing diagrams. Use will be made of the CMOS logic family to develop this symbology. Other logic families, such as NMOS and TTL, are discussed in Appendix A.

This chapter goes beyond the usual treatment of Boolean algebra to present what is called XOR algebra, an extension of Boolean algebra that deals with functions that have become very important in circuit design, particularly in arithmetic circuit design. CMOS realizations of XOR functions have, in a special sense, revolutionized thinking along these lines, making the use of such functions much more appealing to the logic designer.

3.2 BINARY STATE TERMINOLOGY AND MIXED LOGIC NOTATION

Digital systems are switching devices that operate in only one of two possible states at any given time, but that can be switched back and forth from one state to another at very high speed (millions of times per second). The two states are high voltage (HV) and low voltage (LV). The LV and HV levels are usually taken as 0 V and 2 to 5 V, respectively, for common CMOS logic circuits.

To design a useful digital device, meaningful logic names must be assigned to the inputs and outputs of a logic circuit so that their physical interpretation in terms of voltage levels can be made unambiguously. This requires the use of a notation that can easily *bridge the gap* between the logic domain in which the device is designed, and the physical domain in which the device is to be operated. The following subsection defines this notation.

3.2.1 Binary State Terminology

A state is said to be *active* if it is the condition for causing something to happen. And for every active state there must exist one that is *inactive*. In the binary (base 2) system of 1's and 0's, these descriptors take the following meaning:

> Logic 1 is the ACTIVE state
> Logic 0 is the INACTIVE state

Thus, in the *logic domain*, logic 1 is assigned to the active condition while logic 0 is assigned to the inactive condition. This will always be so.

A symbol that is attached to the name of a signal and that establishes which physical state, HV or LV, is to be the active state for that signal, is called the *activation level indicator*. The activation level indicators used in this text are

> (H) meaning ACTIVE HIGH
> (L) meaning ACTIVE LOW

Thus, a line signal LOAD(H) is one for which the active state occurs at high voltage (HV), and LOAD(L) is one for which the active state occurs at low voltage (LV). This is illustrated in Fig. 3.1. Here, the name LOAD is the physical waveform output of a digital device, and LOAD(H) and LOAD(L) are equivalent logical interpretations of that physical waveform. Notice that logic waveforms are rectangular (i.e., with zero rise and fall times), whereas physical waveforms must have finite rise and fall times. Finite rise and fall times are a consequence of the fact that changes in the physical state of anything cannot occur instantaneously. Logic level transitions, on the other hand, are nonphysical and occur abruptly at the active and inactive transition points of the physical waveform, as indicated by the vertical dotted lines in Fig. 3.1. Also, the physical waveforms in Fig. 3.1 have amplitudes measured in terms of voltage whereas logic waveforms have amplitudes indicated by the logic levels 0 and 1. Labels such as LOAD(H) or LD(H) and LOAD(L) or LD(L) are commonly referred to as *polarized mnemonics*. The word "polarized" refers to the use of activation level indicator symbols, (H) and (L). Thus, LD(L) means LOAD active (or asserted) low, and LD(H) refers to LOAD active (or asserted) high.

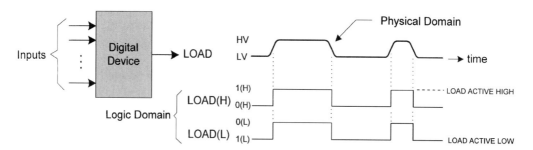

FIGURE 3.1
Mixed logic interpretation of a physical waveform showing a digital device with its voltage waveform and the positive and negative logic interpretations of the waveform.

3.2 BINARY STATE TERMINOLOGY AND MIXED LOGIC NOTATION

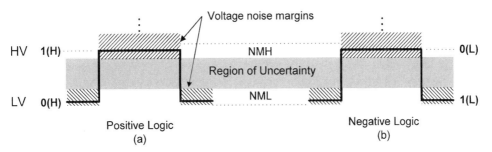

FIGURE 3.2
Logic interpretation of voltage waveforms. (a) Positive logic. (b) Negative logic.

What results from an inspection of the physical and logic waveforms in Fig. 3.1 is the important conclusion

$$\begin{aligned} \text{HV corresponds to } 1(H) &= 0(L) \\ \text{LV corresponds to } 0(H) &= 1(L). \end{aligned} \quad (3.1)$$

Relations (3.1) represent the essence of *mixed logic notation* and are used throughout this text for the purpose of converting from the logic domain to the physical domain or vice versa. Thus, the physical state of HV is represented by either 1(H) or 0(L) in the logic domain while the physical state of LV is represented by either 0(H) or 1(L) in the logic domain. The expression "mixed logic" applies to the use of both the *positive logic* and *negative logic* systems within a given application.

The positive and negative logic systems, which follow from Eqs. (3.1), are presented in Fig. 3.2. Here, the two systems are shown on logic waveform pulses similar to those shown in Fig. 3.1. The high noise margin (NMH) and low noise margin (NML) are included as a reminder that their inner boundaries are also the inner limits of the logic states (1 and 0) as well as the outer limits of the uncertainty region. A signal whose value lies in the uncertainty region cannot be taken as either logic 1 or logic 0.

The digital device shown in Fig. 3.3 illustrates the use of *polarized mnemonics* in the mixed logic digital system. Shown here are two inputs, LD(H) and CNT(L), and one output, DONE(H). The input LD(H) is said to arrive from a positive logic source (active high) while CNT(L) arrives from a negative logic source (hence, active low). The output DONE(H) is delivered to the next stage as a positive logic source (active high). LD and CNT,

FIGURE 3.3
Polarized mnemonics applied to a digital device.

82 CHAPTER 3 / BACKGROUND FOR DIGITAL DESIGN

which represent LOAD and COUNT, respectively, are meaningful abbreviations called mnemonics.

3.3 INTRODUCTION TO CMOS TERMINOLOGY AND SYMBOLOGY

Complementary MOSFET (CMOS) switching circuits are composed of n-type MOSFETs (NMOS for short) and p-type MOSFETs (PMOS). As a help in reading and constructing CMOS switching circuits, the simplified symbols and ideal equivalent circuits for both types are given in Fig. 3.4. Thus, for either, the OFF condition is always an *open circuit* while the ON condition is always a *short circuit*. But the voltage levels causing the ON and OFF conditions for NMOS and PMOS are opposite; hence, they are called *complementary*. Notice that the voltage to produce the ON or OFF condition is always applied to the gate, G, and that the drain-to-source is either nonconducting ($I_{Drain} = 0$) for the OFF condition or conducting ($V_{DS} = 0$) for the ON condition. Use of Fig. 3.4 makes reading and construction of CMOS circuits very easy. However, knowledge of which terminal is the drain and which is the source is important only when configuring at the transistor circuit layout level.

Proper CMOS circuit construction requires that the NMOS and PMOS sections be positioned as shown in Fig. 3.5. The reason for this particular configuration is that NMOS passes

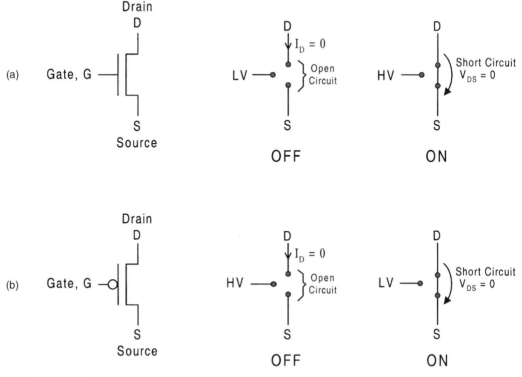

FIGURE 3.4
Symbols and ideal equivalent circuits for n and p MOSFETs: (a) NMOS. (b) PMOS.

3.4 LOGIC LEVEL CONVERSION: THE INVERTER

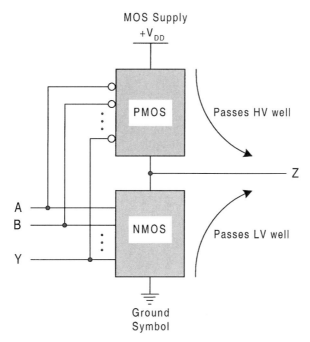

FIGURE 3.5
Proper PMOS and NMOS placement for generalized CMOS gate configurations.

LV well but not HV. Conversely, PMOS passes HV well but not LV. The proper placement of the NMOS and PMOS sections results in a sharp, relatively undistorted waveform. Inverting this configuration would require that the NMOS and PMOS sections pass voltage levels that they do not pass well, resulting in a distortion of the voltage waveform. Therefore, the PMOS section is always placed on the HV end with the NMOS on the LV side, as in Fig. 3.5.

3.4 LOGIC LEVEL CONVERSION: THE INVERTER

When a positive logic source is converted to a negative logic source, or vice versa, *logic level conversion* is said to occur. The physical device that performs logic level conversion is called the *inverter*. Shown in Fig. 3.6a is the CMOS version of the inverter. It is a CMOS inverter because it is composed of both NMOS and PMOS cast in the complementary configuration of Fig. 3.5. The physical truth table, shown in Fig. 3.6b, is easily understood by referring to Fig. 3.4. The logic interpretations and *conjugate logic symbols* that derive from the physical truth table are shown in Figs. 3.6c and 3.6d. The conjugate logic circuit symbols are used to indicate the logic level conversion $X(H) \rightarrow X(L)$ or $X(L) \rightarrow X(H)$ depending on where the *active low indicator bubble* is located. The designation "conjugate" indicates that the symbols are interchangeable, as they must be since they are derived from the same physical device (the inverter).

The CMOS inverter is used here for the purpose of developing the concept of logic level conversion. However, there are versions of the inverter that belong to logic families other than the CMOS family. These include the NMOS and TTL families, all of which yield

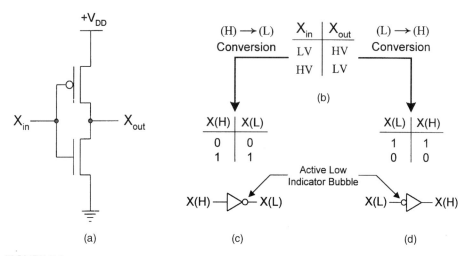

FIGURE 3.6
The inverter, its I/O behavior, and its two logic interpretations. (a) The CMOS transistor circuit. (b) Physical truth table. (c) Active-high-to-active-low conversion and logic circuit symbol. (d) Active-low-to-active-high conversion and logic circuit symbol.

the physical truth table and logic interpretations given in Figs. 3.6b, 3.6c, and 3.6d. More detailed information on these logic families is presented in Appendix A.

3.5 TRANSMISSION GATES AND TRI-STATE DRIVERS

A MOS transistor switch that functions as a *passive* (non-amplifying) switching device and that does not invert a voltage signal is called a *transmission gate* or *pass transistor*. Logic circuits composed of transmission gates are called *steering logic* and are discussed in Section 6.9. Shown in Fig. 3.7 are the circuit symbols and equivalent circuits for the NMOS, PMOS, and CMOS transmission gates. Here, it can be seen that the ON condition in Fig. 3.7b permits an input signal X_i to be transferred to the output; hence, $X_o = X_i$. Conversely, the OFF condition disconnects the output from the input, allowing no signal to be transferred. Notice that the CMOS transmission gate requires complementary "enable" inputs, EN and $\overline{EN}$, to the NMOS and PMOS gates, respectively. This simply means that when one enable input is at high voltage (HV) the other must be at low voltage (LV) and vice versa.

As indicated earlier, an NMOS switch passes LV well but not HV, the reverse being true for a PMOS switch. Consequently, some distortion of the transmitted waveform is to be expected in NMOS and PMOS transmission gates operated in the transfer mode (ON condition). The CMOS switch, on the other hand, combines the best of both worlds, thereby minimizing waveform distortion.

An *active* (restoring) switching device that operates in either a transfer or disconnect mode is called a *tri-state driver* or *tri-state buffer*. If in the transfer mode it is designed to invert, it is called an inverting tri-state driver. These devices are called "tri-state" or "three-state" because they operate in one of three states — logic 0, logic 1, or high-impedance (Hi-Z) state. In the Hi-Z or disconnect state the tri-state driver is functionally "floating," as if it were not there. Tri-state drivers are used to interface various IC devices to a common

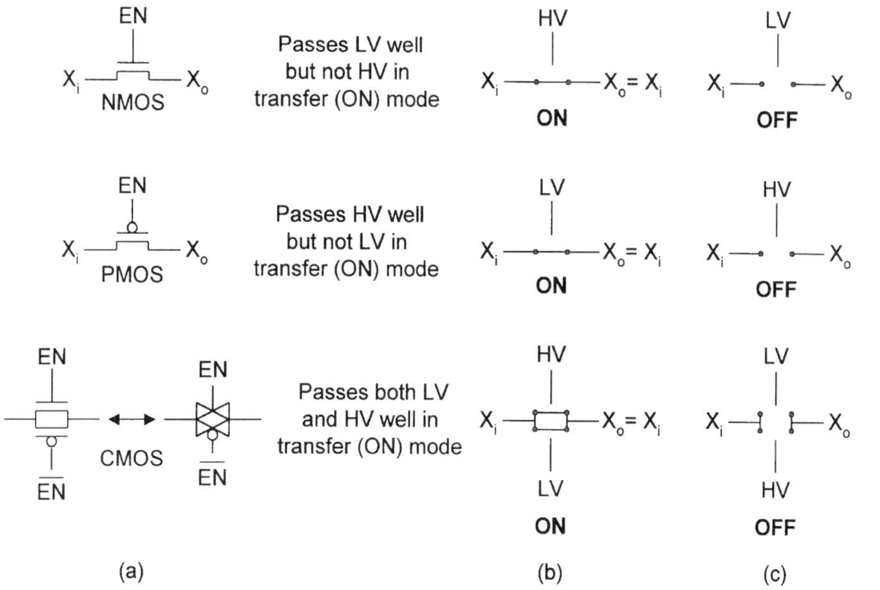

3.5 TRANSMISSION GATES AND TRI-STATE DRIVERS

FIGURE 3.7
Transmission gate circuit symbols and their idealized equivalent circuits. (a) Simplified circuit symbols for NMOS, PMOS, and CMOS transmission gates. (b) ON (transfer) mode equivalent circuit. (c) OFF (disconnect) mode equivalent circuit.

data bus so that the devices will not interfere with each other. By this means, tri-state drivers permit multiple signal sources to share a single line if only one of the signals is active at any given time. These drivers also serve as a controlled enable on the output of some devices. Note that the term "tri-state" is a trademark of National Semiconductor Corporation. Thus, the use of the term "tri-state" in this text acknowledges NSC's right of trademark. The terms tri-state and three-state are often used interchangeably.

Shown in Fig. 3.8 are four types of CMOS tri-state drivers constructed from the equivalent of two or three inverters. They differ in the activation levels of the control input, C, and the output, X_o, indicated by the logic circuit symbols. The choices are inverting or noninverting tri-state drivers with active high or active low control inputs, as provided in Figs. 3.8a–d. The buffering (driving) strength is the same for all tri-state drivers, however. This is so because during the transfer stage the outputs X_o are connected to either the supply $+V_{DD}$ or to ground depending on the noninverting or inverting character of the driver. For example, in the case of the inverting tri-state driver of Fig. 3.8c, a control $C = HV$ connects the output X_o to $+V_{DD}$ if the input is $X_i = LV$ or connects X_o to ground if $X_i = HV$. Thus, in the transfer mode, the transistors of a tri-state driver serve as transmission gates, thereby permitting an input signal to be enhanced (or refreshed); hence the meaning of the term driver. Of course, in the disconnect mode the tri-state driver produces a very large impedance (Hi-Z) between its input and output, virtually disconnecting the input from the output.

Note that the conjugate logic circuit symbols are provided for each tri-state driver shown in Fig. 3.8 and that these symbols are interchangeable — as they must be, since they are derived from the same physical device (the tri-state driver). The idea here parallels that of the inverter and its conjugate logic circuit symbols shown in Fig. 3.6. Symbol $\bar{X}$ appearing

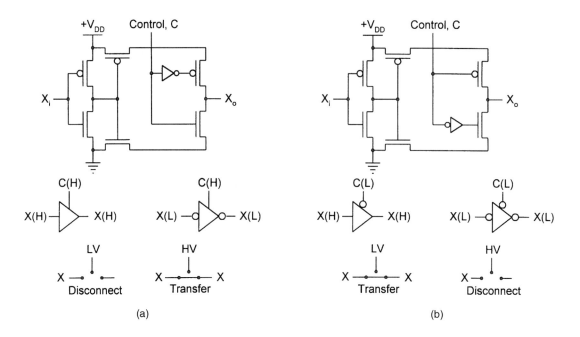
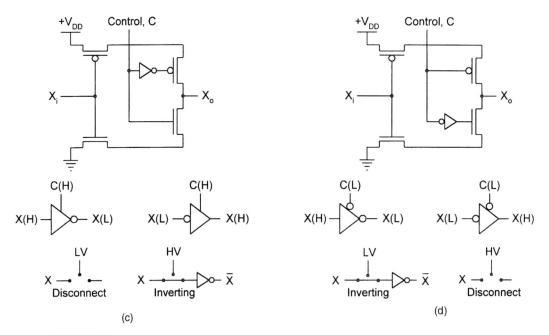

FIGURE 3.8
CMOS tri-state drivers, conjugate circuit symbols, and ideal equivalent circuits. (a) Noninverting tri-state driver with active high control, C. (b) Noninverting tri-state driver with active low control. (c) Inverting tri-state driver with active high control. (d) Inverting tri-state driver with active low control.

on the output of an inverting tri-state driver in the transfer mode indicates an inverted voltage signal. Thus, if X is at LV, then $\bar{X}$ is at HV and vice versa.

Buffers, or *line drivers* as they are sometimes called, may be composed of a series of inverters or gates used as inverters, or they may be simply a tri-state driver operated in the transfer mode. Remember, it is the function of a line driver to boost and sharpen signals that might otherwise degrade below switching levels or be distorted. The mixed logic circuit symbols for buffers are given in Fig. 3.20a.

3.6 AND AND OR OPERATORS AND THEIR MIXED-LOGIC CIRCUIT SYMBOLOGY

There are just two binary logic operations that underlie all of logic design and Boolean algebra (after George Boole, 1815–1864, English mathematician). These are the AND and OR operations. The following are the operator symbols (or connectives) that are used for AND and OR:

$$(\cdot) \to AND \qquad (+) \to OR$$

So, if one writes $X \cdot Y$, XY, or $(X)(Y)$, it is read as X AND Y. Note that the AND operator $(\cdot)$ is also called the *Boolean product* (or *intersection*) and may be represented by the alternative symbol $(\wedge)$. Thus, $X \cdot Y = X \wedge Y$ is the intersection or Boolean product of X and Y. In contrast, $X + Y$ is read as X OR Y. The operator $(+)$ is often called the *Boolean sum* (or *union*) and may be represented by the alternative symbol $(\vee)$. Hence, $X + Y = X \vee Y$ is the union or Boolean sum of X and Y.

By using the two Boolean operators, an endless variety of Boolean expressions can be represented. Simple examples are expressions such as

$$F = X + Y \cdot Z \qquad \text{and} \qquad G = X \cdot (Y + Z).$$

The first is read as F equals X OR (Y AND Z). In this expression the Boolean quantity $Y \cdot Z$ must first be evaluated before it is "ORed" with X. The second expression is read as G equals X AND (Y OR Z). In this case the quantity $(Y + Z)$ must first be evaluated before it can be "ANDed" with X. Thus, the hierarchy of Boolean operation is similar to that of Cartesian algebra for multiplication and addition.

3.6.1 Logic Circuit Symbology for AND and OR

The meanings of the AND and OR operators (functions) are best understood in terms of their logic circuit symbols. Shown in Fig. 3.9 are the distinctively shaped logic circuit symbols commonly used to represent the AND and OR operators, which may have multiple inputs and a single output. The functional descriptions of these symbols are stated as follows:

> The output of a logic AND circuit symbol is active if, and only if, all inputs are active.

> The output of a logic OR circuit symbol is active if one or more of the inputs are active.

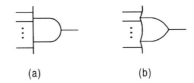

FIGURE 3.9
Distinctive logic circuit symbols for the binary operators. (a) AND symbol. (b) OR symbol.

The functional descriptions may be taken as the *definitions for AND and OR*. Remember that the term active implies logic 1.

The distinctively shaped symbols in Fig. 3.9 represent the functional descriptions for AND and OR and provide the logical interpretation for a variety of physical devices called *gates*. That is, each gate must satisfy the logical AND or logical OR functional description. By definition: *A gate is an interconnection of electronic switches and other circuit elements arranged in such a way as to produce the electrical equivalent of a logic operation.* The inputs and outputs of a gate are measured in terms of voltages (LV or HV), whereas the inputs and outputs of a logic symbol, as in Fig. 3.9, are expressed in terms of logic 1 or logic 0 together with the appropriate activation level indicators, (H) and (L).

3.6.2 NAND Gate Realization of Logic AND and OR

The physical device shown in Fig. 3.10a is a two-input NAND gate. NAND is short for NOT-AND. Because this version of NAND gate complies with the generalized CMOS gate configuration in Fig. 3.5, it is called a CMOS NAND gate. The physical truth table for this

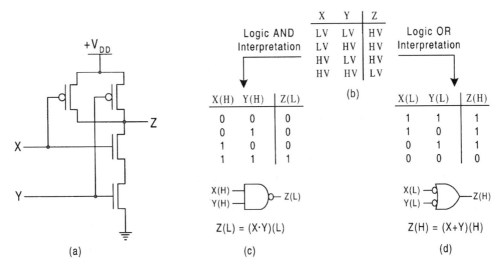

FIGURE 3.10
The two-input NAND gate, its I/O behavior, and its two logic interpretations. (a) CMOS transistor circuit. (b) Physical truth table. (c) Logic AND interpretation and circuit symbol. (d) Logic OR interpretation and circuit symbol.

3.6 AND AND OR OPERATORS AND THEIR MIXED-LOGIC CIRCUIT SYMBOLOGY   89

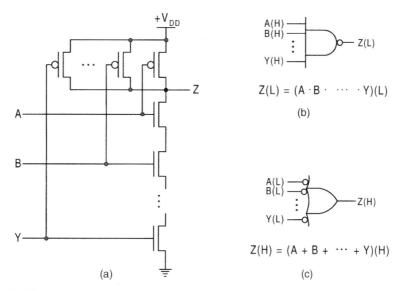

FIGURE 3.11
Multiple input NAND gates and logic circuit symbols. (a) CMOS logic circuit. (b) AND interpretation. (c) OR interpretation.

or any NAND gate is given in Fig. 3.10b. It is easily generated by applying the ON and OFF equivalent circuits for NMOS and PMOS given in Fig. 3.4.

The two logical AND and OR interpretations of the NAND gate and their mixed logic circuit symbols are given in Figs. 3.10c and 3.10d. They, too, apply to NAND gates belonging to logic families other than CMOS, as explained further in Appendix A. Notice that by applying Eqs. (3.1) the truth tables for the AND and OR interpretations satisfy the definitions for AND and OR given earlier in connection with Fig. 3.9—no other combination of activation level symbols applied to inputs X and Y satisfies these definitions. But both logic circuit symbols represent (in the logic domain) the physical NAND gate, since both are derived from it. Thus, one symbol (c) performs the AND operation with active low output, while the other symbol (d) performs the OR operation with active low inputs. The symbols are interchangeable in a logic circuit and, for that reason, are called *conjugate NAND gate symbols* even though, strictly speaking, they are only logic circuit symbols.

Multiple input CMOS NAND gates result by adding more PMOS in parallel and an equal number of NMOS in series, as shown in Fig. 3.11. The logic circuit symbols and output expressions shown in Figs. 3.11b and 3.11c result. The number of inputs is usually limited to eight or fewer, mainly because of an increase in resistance of the series N-MOSFETs, each of which has a small ON channel resistance associated with it. Therefore, too many inputs causes an increase in gate propagation delay and a degradation of the signal. The number of inputs that a gate can have is called the *fan-in*. For example, a four-input NAND gate would have a fan-in of 4.

3.6.3 NOR Gate Realization of Logic AND and OR

The transistor circuit for the two-input CMOS NOR gate is shown in Fig. 3.12a. NOR is short for NOT-OR. The physical truth table and the AND and OR logical interpretations

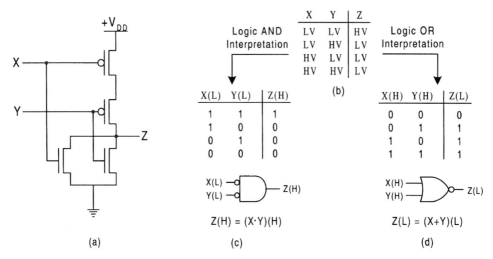

FIGURE 3.12
The two-input NOR gate, its I/O behavior and its two logic interpretations. (a) CMOS transistor circuit. (b) Physical truth table. (c) Logic AND interpretation and circuit symbol. (d) Logic OR interpretation and circuit symbol.

that derive from it are given Figs. 3.12b, 3.12c, and 3.12d, respectively, and these also apply to NOR gates belonging to other logic families, as discussed in Appendix A. As before, the physical truth table is obtained by applying the equivalent circuits given in Fig. 3.4 to the transistors in Fig. 3.12a. The AND and OR logic interpretations in parts (c) and (d) derive from the application of Eqs. (3.1) to the physical truth table and are observed to agree with the definitions of AND and OR given in connection with Fig. 3.9 — again, no other combination of activation level symbols applied to inputs X and Y satisfies these definitions. Thus, there results two logic circuit symbols, one performing the AND operation with active low inputs (c) and the other performing the OR operation with active low output. Since the logic symbols are interchangeable, they are called *conjugate NOR gate symbols*.

Multiple input NOR gates are produced by adding more PMOS in series and an equal number of NMOS in parallel, as indicated in Fig. 3.13a. The logic symbols for multiple input NOR gates are shown in Figs. 3.13b and 3.13c. As in the case of multiple NAND gates, there exists a practical limit to the number of NOR gate inputs (fan-in) because of the channel resistance effect. Thus, too many inputs to a NOR gate will increase the gate propagation delay and degrade the signal.

When fan-in restrictions become a problem, a gate *tree* structure (e.g., a NOR gate tree) can be used. A gate tree is a combination of like gates that form a multilevel array (see Fig. 4.49). Thus, a gate tree composed of several NOR gates can replace a multiple-input NOR gate when the number of inputs exceeds the fan-in limit for that gate.

3.6.4 NAND and NOR Gate Realization of Logic Level Conversion

Inherent in any NAND or NOR gate is the ability to function as an inverter. Thus, under the proper conditions, the NAND or NOR gate can perform the equivalent of logic level conversion as in Fig. 3.6. Shown in Figs. 3.14 and 3.15 are the various input connections

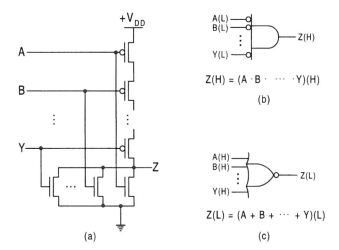

FIGURE 3.13
Multiple input NOR gates and logic circuit symbols. (a) CMOS logic circuit. (b) AND interpretation. (c) OR interpretation.

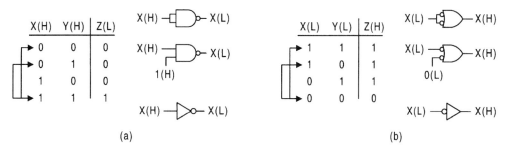

FIGURE 3.14
Nand gate realization of logic level conversion and equivalent symbology. (a) (H) → (L) conversion. (b) (L) → (H) conversion.

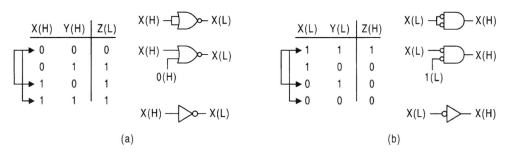

FIGURE 3.15
NOR gate realization of logic level conversion and equivalent symbology. (a) (H) → (L) conversion. (b) (L) → (H) conversion.

that permit this to happen. These input connections result directly from the logic truth tables in Figs. 3.10 and 3.12. The arrows on the left side of each truth table are given to draw attention to those portions of the table that dictate how the connections to the NAND and NOR gates must be made. For example, the extreme upper and lower arrows in Fig. 3.14a indicate that the two inputs to the NAND gate can be connected for the $X(H) \to Z(L) = X(L)$ conversion. The lower two arrows indicate that the same conversion can be achieved by setting $Y(H) = 1(H)$. It is not likely that a NAND or NOR gate would be used as a replacement for an inverter if the latter were available, but the substitution is permissible if the need is there. Obviously, the NAND or NOR gate inverter is more costly (spacewise) and is slower than the inverter in Fig. 3.6.

3.6.5 The AND and OR Gates and Their Realization of Logic AND and OR

NAND and NOR CMOS gates are natural electrical realizations of the AND and OR logic operators, but the AND and OR CMOS gates are not. This can be understood if one recalls that a transistor switch is, by its nature, an inverter. Thus, it might be expected that proper CMOS realizations of NOT-AND and NOT-OR would be simpler (by transistor count) than the equivalent CMOS realizations of AND and OR, and this is the case.

Shown in Fig. 3.16a is the transistor circuit for the CMOS version of the two-input AND gate. It is seen to be composed of the NAND gate followed by an inverter, hence NAND-NOT or NOT-AND-NOT. By application of Eqs. (3.1), the physical truth table for the AND gate, given in Fig. 3.16b, yields the AND and OR interpretations shown in Figs. 3.16c, and 3.16d. From these interpretations there results the two *conjugate AND gate symbols*, one performing the AND operation with active high inputs and output (c) and the other performing the OR operation with active low inputs and output as indicated by the active

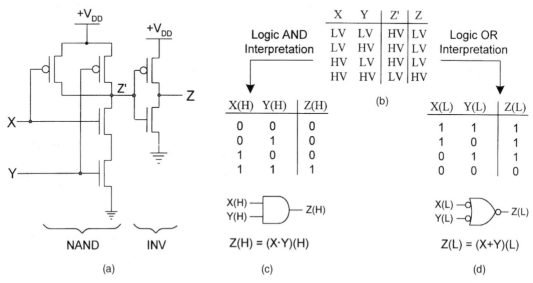

FIGURE 3.16

The two-input AND gate, its I/O behavior, and its two logic interpretations. (a) CMOS transistor circuit. (b) Physical truth table. (c) Logic AND interpretation and circuit symbol. (d) Logic OR interpretation and circuit symbol.

3.6 AND AND OR OPERATORS AND THEIR MIXED-LOGIC CIRCUIT SYMBOLOGY 93

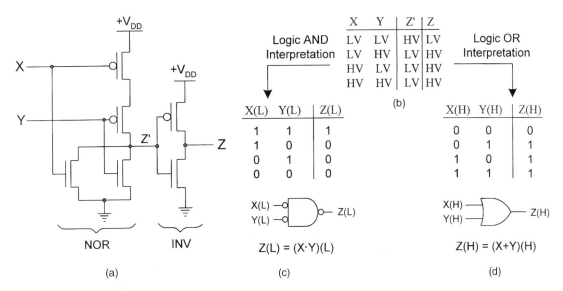

FIGURE 3.17
The two-input OR gate, its I/O behavior, and its two logic interpretations. (a) CMOS transistor circuit. (b) Physical truth table. (c) Logic AND interpretation and circuit symbol. (d) Logic OR interpretation and circuit symbol.

low indicator bubbles. The logic interpretations and mixed logic circuit symbols also apply to the AND gate of any logic family.

The CMOS version of the two-input OR gate, its physical truth table, and its two logic interpretations are shown in Fig. 3.17. Again, Eqs. (3.1) and the functional descriptions associated with Fig. 3.9 have been applied to the physical truth table, Fig. 3.17b, to yield the AND and OR interpretations and the mixed logic circuit symbols presented in Figs. 3.17c and 3.17d. The two logic circuit symbols are interchangeable and hence are *conjugate OR gate symbols*. One symbol performs the AND operation and has active low inputs and output, while the other performs the OR operation and has active high inputs and output. As before, the truth table and logic AND and OR interpretations apply also to an OR gate of any logic family.

Multiple input CMOS AND and OR gates are possible by combining the transistor circuit in either Fig. 3.11 or Fig. 3.13 with an inverter. The conjugate gate symbols for AND and OR that result are shown in Figs. 3.18 and 3.19. The same limitations on numbers of inputs that apply to CMOS NAND and NOR gates also apply to CMOS AND and OR gates.

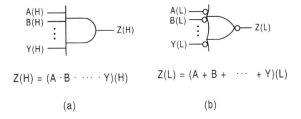

FIGURE 3.18
Logic symbols for multiple input AND gates. (a) AND interpretation. (b) OR interpretation.

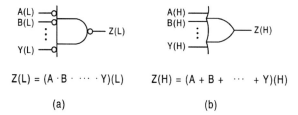

$$Z(L) = (A \cdot B \cdot \cdots \cdot Y)(L) \qquad Z(H) = (A + B + \cdots + Y)(H)$$

(a) \hspace{4cm} (b)

FIGURE 3.19
Logic circuit symbols for multiple input OR gates. (a) AND interpretation. (b) OR interpretation.

AND and OR gates can be configured by "flipping" end-for-end the NAND and NOR gates shown in Figs. 3.10a and 3.12a, respectively, such that the NMOS occupy the HV end while the PMOS reside at the LV end. However, to do this requires that the NMOS pass HV, which they do not do well, and that the PMOS pass LV, which they do not do well. Thus, although such flipped configurations logically satisfy the AND and OR interpretations for the respective gates, their output signals would be somewhat distorted. For minimum output signal distortion the PMOS and NMOS portions for any gate should be configured as in Fig. 3.5.

3.6.6 Summary of Logic Circuit Symbols for the AND and OR Functions and Logic Level Conversion

For reference purposes, a summary is now provided for the mixed logic symbology that has been covered so far. Shown in Fig. 3.20 are the conjugate mixed logic circuit symbols together with the physical gate names they represent. The conjugate mixed logic circuit symbols for the inverter and buffer are given in Fig. 3.20a. Notice that the conjugate pairs of logic circuit symbols in Fig. 3.20b are split into two groups, one group performing the AND function and the other performing the OR function. The buffer, not previously discussed, is included here for completeness. It functions as an amplifier to boost the signal to meet

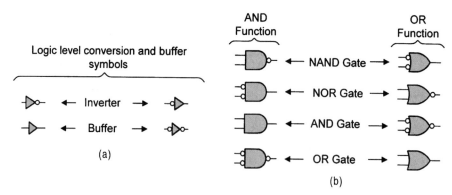

FIGURE 3.20
Summary of conjugate mixed logic circuit symbols and the gates they represent. (a) Logic level conversion and buffer symbols. (b) AND and OR function symbols.

fan-in requirements. For reference purposes, the ANSI/IEEE Standard logic symbols for gates are provided in Appendix C.1.

3.7 LOGIC LEVEL INCOMPATIBILITY: COMPLEMENTATION

The eight conjugate gate symbols in Fig. 3.20b perform one of two logic operations, AND or OR, regardless of the presence or absence of the active low indicator bubbles that serve to associate the symbols to the physical gates from which they are derived. However, the presence or absence of an active low indicator bubble on the input to a given circuit symbol indicates the activation level of the input, (L) or (H), that is "required" by that gate. Thus, the presence of active low indicator bubbles on the inputs to a logic symbol requires that all inputs arrive from negative logic sources while the absence of bubbles requires that the inputs arrive from positive logic sources. When these requirements are met the inputs are said to have *logic compatibility* with the logic symbol.

But suppose an input signal arrives at the input to a logic symbol with an activation level that is of opposite polarity to that required by the logic circuit symbol. When this happens a condition of *logic incompatibility* exists, and this requires that the signal name in the output be complemented.

The operation of *complementation* is defined by the following important relations applied to a logic function α:

$$\alpha(L) = \bar{\alpha}(H) \quad \text{and} \quad \alpha(H) = \bar{\alpha}(L) \tag{3.2}$$

such that

$$\begin{aligned}(\alpha \cdot \bar{\alpha})(H) &= 0(H) \\ (\alpha \cdot \bar{\alpha})(L) &= 0(L) \\ (\alpha + \bar{\alpha})(H) &= 1(H) \\ (\alpha + \bar{\alpha})(L) &= 1(L).\end{aligned} \tag{3.3}$$

The overbar is read as "the complement of." Thus, in the logic domain a logic function α ANDed with its complement $\bar{\alpha}$ is logic 0, or the function ORed with its complement is logic 1.

In Fig. 3.21 are four typical examples of input logic level incompatibility each requiring the complementation of the incompatible input name in the output expression. Note that this is indicated in two ways. In Fig. 3.21a, Eqs. (3.2) are applied directly to satisfy the logic level compatibility requirements of the logic symbol. In Fig. 3.21b, an incompatibility slash "/" is placed on the input line containing the logic incompatibility as a visual reminder that a logic level incompatibility exists and that the input name must be complemented in the output expression.

The pairs of logic circuit symbols in Figs. 3.21a and 3.21b are conjugate symbol forms as in Fig. 3.20. Because these conjugate circuit symbols are interchangeable, their output expressions are equal and are representative of a set of such equations called the *DeMorgan relations*. This subject will be considered further in Section 3.10.

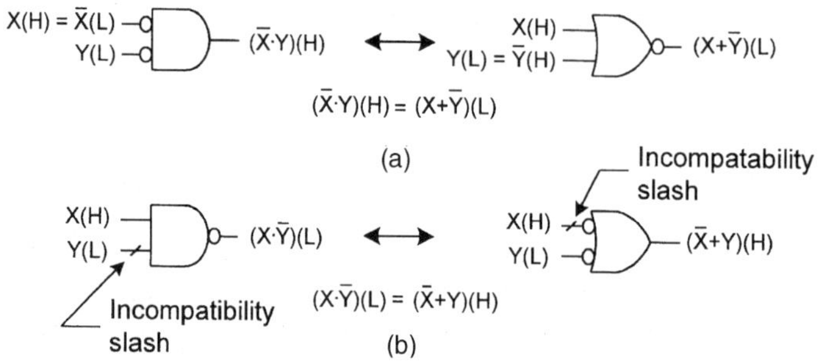

FIGURE 3.21
Examples of logic level incompatibility. (a) Applications of Eqs. (3.2). (b) Use of the incompatibility slash (/) as an alternative.

Earlier it was stated that the only logic function of the inverter is to perform logic level conversion, $(H) \to (L)$ or $(L) \to (H)$, and this is true. But to what purpose? The answer is simply stated:

The logical function of the inverter is to create or remove an input logic level incompatibility depending on the output function requirements of the logic symbol to which the input is attached.

Consider two examples of logic level conversion in Figs. 3.22a and 3.22b. Here, NAND and OR logic realizations of logic functions require the use of inverters to create and remove logic level incompatibilities, respectively.

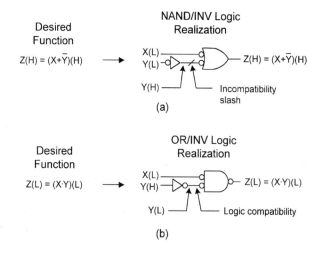

FIGURE 3.22
Examples of logic level conversion. (a) Creation of a logic incompatibility with active low inputs. (b) Removal of a logic incompatibility with inputs $X(L)$ and $Y(H)$.

3.8 READING AND CONSTRUCTION OF MIXED-LOGIC CIRCUITS

The very simple procedures that are necessary to construct and read mixed-logic circuits are now demonstrated by more complex examples. Consider the function $F(H)$:

$$F(H) = [\overline{A} \cdot B + \overline{B} \cdot C](H)$$

with OR output stage and AND input stages indicated.

Notice that this function is formed by ORing together two ANDed terms and is read as F equals A "bar" AND B ORed with B "bar" AND C, all active high. The logic circuit for this function is shown in Fig. 3.23, where it is assumed that the inputs arrive active high (H), that is, from positive logic sources. Two logic realizations are shown, one NAND/INV logic and the other AND/OR/INV logic, both yielding the function $F(H)$. Thus, by complementing between the AND and OR stages (area enclosed by dotted lines), the physical realization is altered but without changing the original function — logic level compatibility has been preserved. Observe that an incompatibility slash ("/") is placed on a given symbol input as a reminder that an input logic incompatibility requires complementation of the input name in the output expression. In Figs. 3.23c and 3.23d are two additional means of representing the function F — namely, the truth table and logic waveforms. Here, a binary input sequence is assumed and no account is taken of the path delays through the gates and inverters.

A second more complex example is shown in Fig. 3.24, where a function $Z(L)$ has been implemented by using NAND/NOR/INV logic in (a) and by using AND/OR/INV

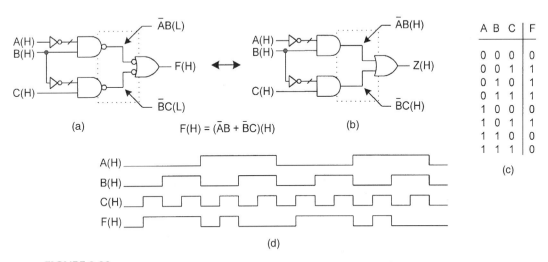

FIGURE 3.23
Examples of the reading, construction, and waveform analysis of a logic circuit. (a) NAND/INV and (b) AND/OR/INV logic realizations of the function $F(H)$ with active high inputs. (c) Truth table for the function F. (d) Logic waveforms for the function F assuming a binary input pattern.

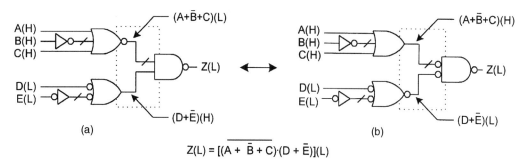

FIGURE 3.24
Logic realizations of the function Z(L) with inputs A(H), B(H), C(H), D(L), and E(L). (a) NAND/NOR/INV logic. (b) AND/OR/INV.

logic in (b). In this example the ORing input stages receive inputs that are assumed to arrive as A(H), B(H), C(H), D(L), and E(L). Here again, by complementing between the AND and OR stages (dotted boxes), the physical realization is changed without altering the original function. Notice that incompatibilities exist on inputs and between ORing and ANDing stages requiring (in each case) complementation of the signal name in the output expression as indicated by the "/" symbol.

Reading vs Construction of Logic Circuits Implied by Figs. 3.23 and 3.24 is the procedure to be followed when reading or constructing a logic circuit:

> *The reading of a logic circuit always begins with the inputs and ends at the output, hence "input-to-output."*
> *Construction of a logic circuit begins at the output stage and continues to the inputs, hence "top down."*

One must not begin construction of a mixed-logic circuit until the activation levels of the inputs are known and the output and input stage operators have been identified. If a circuit has been presented in positive logic form (no mixed logic symbology), it is advisable for the reader to convert the circuit to mixed logic form before reading it. This will speed up the reading process and minimize the probability for error.

3.9 XOR AND EQV OPERATORS AND THEIR MIXED-LOGIC CIRCUIT SYMBOLOGY

Certain functions consisting of AND and OR operations occur so often in digital logic design that special names and operator symbols have been assigned to them. By far the most common of these are the *exclusive or* (XOR) and *equivalence* (EQV) functions represented by the following operator symbols:

$\oplus \rightarrow XOR$, meaning *"one or the other but not both equivalent."*
$\odot \rightarrow EQV$, meaning *"both the same (equivalent)."*

3.9 XOR AND EQV OPERATORS

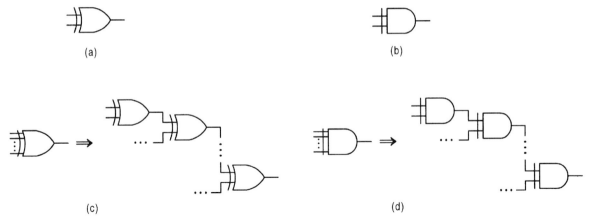

FIGURE 3.25
Distinctive logic circuit symbols for XOR and EQV. (a) The XOR function circuit symbol. (b) The EQV function circuit symbol. (c) and (d) illustrate the meaning of multiple input symbols.

Thus, if one writes $X \oplus Y$, it is read as X XOR Y; $X \odot Y$ is read as X EQV Y. The EQV operator is also known as XNOR (for EXCLUSIVE NOR), a name that will *not* be used in this text.

Like the AND and OR functions, the XOR and EQV functions are best understood in terms of the logic circuit symbols representing them. Figures 3.25a and 3.25b give the commonly used XOR and EQV circuit symbols for which the following functional descriptions apply:

> *The output of a logic XOR circuit symbol is active if one or the other of two inputs is active but not both active or inactive — that is, if the inputs are not logically equivalent.*

> *The output of a logic EQV circuit symbol is active if, and only if, both inputs are active or both inputs are inactive — that is, if both inputs are logically equivalent.*

A circuit symbol for either XOR or EQV consists of *two and only two* inputs. Multiple input XOR or EQV circuit symbols are understood to have the meaning indicated in Figs. 3.25c and 3.25d and are known as *tree forms*.

The *defining relations* for XOR and EQV are obtained from the functional descriptions just given. In Boolean sum-of-products and Boolean product-of-sums form these defining relations are

$$A \oplus B \equiv \bar{A} \cdot B + A \cdot \bar{B} = (\bar{A} + \bar{B}) \cdot (A + B) \tag{3.4}$$

and

$$A \odot B \equiv \bar{A} \cdot \bar{B} + A \cdot B = (\bar{A} + B) \cdot (A + \bar{B}). \tag{3.5}$$

In words, the XOR function in Eq. (3.4) is active if only one of the two variables in its defining relation is active but not both active or both inactive. Thus, $A \oplus B = 1$ if only one

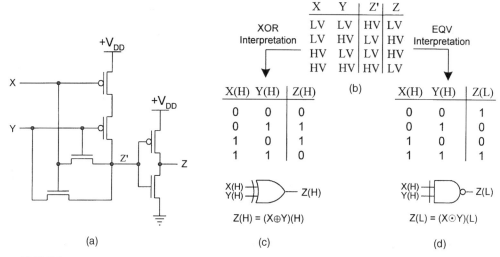

FIGURE 3.26
The XOR gate, its I/O behavior, and its two logic interpretations. (a) A CMOS transistor circuit. (b) Physical truth table. (c) Logic XOR interpretation. (d) Logic EQV interpretation.

of the two variables takes logic 1 at any given time. Conversely, the EQV function in Eq. (3.5) is active only if both variables in its defining relation are active or both are inactive. In this case $A \odot B = 1$ only if both variables are logically equivalent (both logic 1 or both logic 0).

3.9.1 The XOR and EQV Functions of the XOR Gate

Shown in Fig. 3.26a is one of several CMOS versions of the XOR gate. This version makes rather clever use of NMOS and PMOS totaling six transistors. Notice that the output stage is an inverter that acts not only to invert the signal, but also as a *gain element* to boost the signal. This is important since the remainder of the circuit is composed of NMOS and PMOS *transmission gates* (Fig. 3.7) that lack the ability to amplify. In Fig. 3.26b is the physical truth table for the XOR gate. Observe that all but the $X, Y = LV, LV$ input conditions produce a Z' voltage that is the voltage level from one or both inputs. This is characteristic of logic gates composed of pass transistors.

Presented in Figs. 3.26c and 3.26d are the XOR and EQV logic interpretations of the XOR gate together with their distinctively shaped circuit symbols. The logic truth tables for these interpretations derive from the defining relations given by Eqs. (3.4) and (3.5), respectively, and from Eqs. (3.1). Observe that the XOR symbol with active high inputs and output is interchangeable with an EQV symbol with active low output. Thus, it follows that $(X \oplus Y)(H) = (X \odot Y)(L)$.

3.9.2 The XOR and EQV Functions of the EQV Gate

A version of the CMOS EQV gate is shown in Fig. 3.27a. It is obtained from the XOR version in Fig. 3.26a by "complementing" the MOS transistors in the Z' circuit to obtain the Z'' circuit in Fig. 3.27a. Notice that all input conditions except $X, Y = HV, HV$ produce a Z'' output directly from one or both of the inputs. Also note that Z'' is an XOR output,

3.9 XOR AND EQV OPERATORS   101

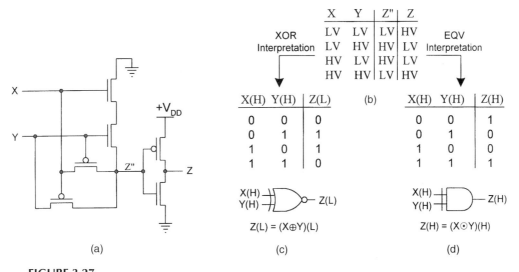

FIGURE 3.27
The EQV gate, its I/O behavior, and its two logic interpretations. (a) A CMOS transistor circuit. (b) Physical truth table. (c) Logic XOR interpretation. (d) Logic EQV interpretation.

whereas Z' is an EQV output, and that they are the inverse of each other. In each case an inverter is added to invert the signal as well as to buffer the output Z.

The physical truth table for the EQV gate and its two logic interpretations are given in parts (b), (c), and (d) of Fig. 3.27. The same procedure used for the XOR gate is used here to obtain the logic truth tables and circuit symbols for the EQV gate. In this case, the XOR symbol with active low output is interchangeable with the EQV symbol with active high inputs and output resulting in the relation $(X \oplus Y)(L) = (X \odot Y)(H)$.

3.9.3 Multiple Gate Realizations of the XOR and EQV Functions

The CMOS transistor circuits for the XOR and EQV functions given in Figs. 3.26 and 3.27 represent the most efficient use of MOSFETs for such purposes. However, there are occasions when such MOS implementations of these functions are not possible. One example is the use of programmable logic arrays (PLAs), as discussed in Section 7.3, to implement arithmetic-type circuits discussed in Chapter 8. PLAs are devices that must use two-level gate forms to implement XOR or EQV functions — XOR or EQV gates are not commonly part of the PLA architecture. Shown in Fig. 3.28 are four multiple-gate realizations of the XOR and EQV functions. The circuits in Figs. 3.28a and 3.28b have been derived from the defining relations for XOR and EQV given by Eqs. (3.4) and (3.5), respectively and are suitable for two-level circuit design. The three-level circuits in Figs. 3.28c and 3.28d are not suitable for two-level circuit design. These three-level circuits result from derivatives of the defining relations:

$$A \cdot (\overline{AB}) + B(\overline{AB}) = A\bar{B} + \bar{A}B \qquad \text{XOR form} \qquad (3.4a)$$

$$(A + \bar{A}\bar{B})(B + \bar{A}\bar{B}) = (A + \bar{B})(\bar{A} + B) \qquad \text{EQV form} \qquad (3.4b)$$

The applications of CMOS AND-OR-invert and OR-AND-invert gates to the implementation of XOR and EQV functions are given later in Subsection 7.7.1. Such CMOS

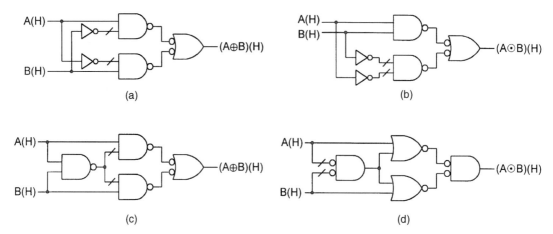

FIGURE 3.28
Multiple gate realizations of the XOR and EQV functions. (a, b) Two-level NAND implementations of the XOR and EQV functions. (c, d) Three-level NAND and NOR implementations of the XOR and EQV functions.

realizations of these functions are shown to be closer to a single level of path delay rather than two levels.

3.9.4 The Effect of Active Low Inputs to the XOR and EQV Circuit Symbols

An interesting property of the XOR and EQV logic symbols is the fact that when the two inputs are of opposite polarity the output function is complemented, but when the inputs are of the same polarity (both active high or both active low) the function remains unchanged. This is illustrated by the four examples in Fig. 3.29. Thus, a single incompatibility complements the function (changing an XOR output function to an EQV output function or vice versa), whereas a double incompatibility or no incompatibility retains the output function. These results may be proven by altering the appropriate logic truth table in Figs. 3.26 and 3.27 to agree with the input activation levels indicated for each logic circuit symbol.

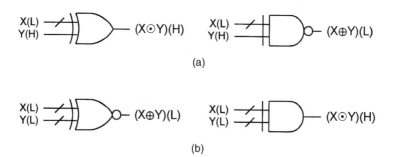

FIGURE 3.29
Effect of active low inputs to XOR and EQV logic circuit symbols. (a) Single logic incompatibilities. (b) Dual logic incompatibilities.

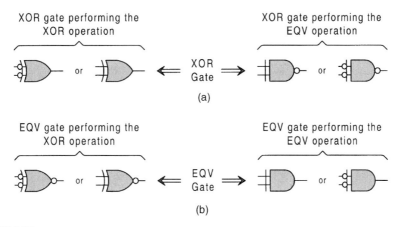

FIGURE 3.30
Summary of conjugate mixed logic circuit symbols for the XOR and EQV gates. (a) XOR gate symbols. (b) EQV gate symbols.

3.9.5 Summary of Conjugate Logic Circuit Symbols for XOR and EQV Gates

For reference purposes the logic circuit symbols representing the XOR and EQV gates are summarized here. Shown in Fig. 3.30a are the four conjugate forms for the XOR gate and in (b) the four conjugate forms for the EQV gate. The conjugate logic circuit symbol forms are interchangeable, as indicated by the two-way arrows. This information can be very useful when synthesizing logic circuits containing XOR or EQV functions. Notice that dual logic low indicator bubbles, representing double incompatibilities, have no effect on the logic function. The reader is referred to Fig. 3.29b for reinforcement of this fact.

3.9.6 Controlled Logic Level Conversion

In Section 3.4 the concept of logic level conversion was introduced in connection with the inverter. Here, the subject of logic level conversion is revisited as it relates to the XOR or EQV gate. Another interesting and useful property of the XOR and EQV gates is that they can be operated in either one of two modes: the *inverter mode* or the *transfer mode*. These modes are illustrated in Fig. 3.31, where exclusive use is made of the XOR symbol to represent the XOR and EQV gates. In Fig. 3.31a the XOR interpretation of the EQV gate is used for $(H) \to (L)$ logic level conversion or for logic transfer depending on the logic level of the controlling input. Notice that the buffer symbol is used to represent the transfer mode. These two modes are easily deduced from the truth table given at left in Fig. 3.31a. Similarly, in Fig. 3.31b, the XOR interpretation of the XOR gate is used for the $(L) \to (H)$ conversion mode or for the logic transfer mode depending on the logic level of the controlling input. Here again, these results are easily deduced from the truth table to the left in Fig. 3.31b, which has been altered from that in Fig. 3.26c to account for the active low inputs.

The positive logic interpretation of an XOR gate used as a *controlled inverter* is given in Fig. 3.31c. This is included to add greater understanding of the XOR gate and its operation. Although all three cases in Fig. 3.31 physically represent *controlled inversion*, it is common to find controlled inverters represented as in Fig. 3.31c. A typical example is in the design

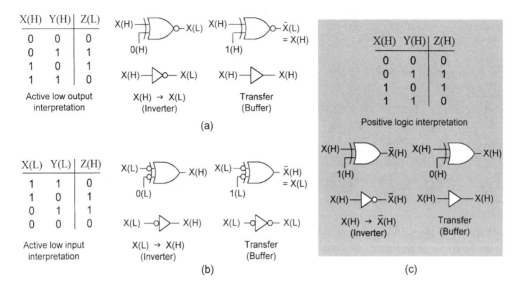

FIGURE 3.31
Controlled logic level conversion. (a) The EQV gate used for (H) → (L) conversion and logic transfer. (b) The XOR gate in mixed logic notation used for (L) → (H) conversion. (c) Positive logic interpretation of the XOR gate used as a controlled inverter.

of the adder/subtractor featured in Fig. 8.9. In making the transition from Fig. 3.31b to Fig. 3.31c, it should be recalled that complementation of both inputs to an XOR or EQV circuit symbol leaves the output function unaltered. Notice that the inverter and buffer symbols in Fig. 3.31 are the same as those given in Fig. 3.20a.

3.9.7 Construction and Waveform Analysis of Logic Circuits Containing XOR-Type Functions

As an extension of Section 3.8, the reading and construction of a *multilevel* logic circuit containing an XOR function is demonstrated by the NAND/XOR/NOR/INV circuit in Fig. 3.32a representing the function $Y = \bar{A} \oplus BC + \bar{B}\bar{C}$. A multilevel logic function is one that has more than two gate path delays from input to output. In this case there are three levels of path delay. Here, an XOR gate performs the XOR operation to yield the $(\bar{A} \oplus BC)(H)$ input to the NOR output stage performing the OR operation. The waveform for $\bar{B}\bar{C}(H)$ is obtained by ANDing the complement of the $B(H)$ waveform with the complement of the $C(H)$ waveform by using a NOR gate to perform the AND operation. Thus, there are three logic incompatibilities, one for the $A(H)$ input and the other two for the $B(H)$ and $C(H)$ inputs, but inverters are not needed to create these logic level incompatibilities.

Presented in Figs. 3.32b and 3.32c are the truth table and logic waveforms for the circuit in Fig. 3.32a. The inputs are arbitrarily given in binary sequence, and the output waveforms from the intermediate stages are given to reveal the advantage of the mixed logic method. No account is taken of the propagation delays of the gates and inverters. Notice that the $(\bar{A} \oplus BC)(H)$ and $\bar{B}\bar{C}(H)$ logic signals are logically compatible with the requirements of the ORing operation of the NOR gate output stage. If complementation is carried out within the dashed box, the waveforms for the resulting $(\bar{A} \oplus BC)(L)$ and $\bar{B}\bar{C}(L)$ signals would

3.10 LAWS OF BOOLEAN ALGEBRA

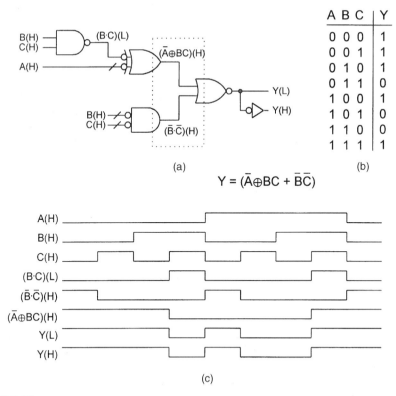

FIGURE 3.32
(a) NAND/NOR/XOR/INV logic circuit, (b) truth table, and (c) logic waveforms for the function Y with active high inputs and mixed logic outputs.

remain the same as those shown for the active high signals, but with opposite activation levels. However, the physical nature of the devices required for implementation would now become NAND/EQV/OR/AND/INV as taken from input to output.

Another interesting facet of the mixed logic method is the fact that an inverter on the output stage permits the generation of *mixed rail output* signals. The $Y(L)$ and $Y(H)$ signals constitute the mixed rail outputs and have waveforms that are identical if account is not taken of the propagation delay through the inverter. In the physical domain, however, the voltage waveforms represented by $Y(L)$ and $Y(H)$ would be the inverse of one another.

3.10 LAWS OF BOOLEAN ALGEBRA

To design a digital circuit that will perform a given function, it may be necessary to manipulate and combine the various switching variables in certain ways that are in agreement with mathematical logic. Use of the laws of Boolean algebra makes these manipulations and combinations relatively simple. This special mathematical logic, named in recognition of the English mathematician George Boole, can be rigorously and eloquently presented by using axioms, theorems, and corollaries. However, for our purposes there is no need for such a formal approach. The laws of Boolean algebra are relatively few and can be deduced from the truth tables for NOT, AND, and OR. In this section these laws are developed exclusively within the logic domain with only passing reference to activation levels.

3.10.1 NOT, AND, and OR Laws

NOT Laws The *unary operator* NOT is the logic equivalent of complementation and connotes inversion in the sense of supplying the lack of something. Although NOT is purely a logic concept and complementation arises more from a physical standpoint, the two terms, NOT and complementation, will be used interchangeably following established practice.

The truth table for NOT is the positive logic interpretation of the physical truth table given in Fig. 3.6b. It is from this truth table that the NOT laws are derived.

NOT Truth Table

X	$\bar{X}$
0	1
1	0

NOT Laws

$\bar{0} = 1$
$\bar{1} = 0$
$\bar{\bar{X}} = X$

(3.6)

The NOT operation, like complementation, is designated by the overscore (or "bar"). A double bar (or double complementation) of a function, sometimes called *involution*, is the function itself, as indicated in Eqs. (3.6).

As examples of the applications of the NOT laws, suppose that $X = A\bar{B}$. Then the function $\bar{X} = \overline{A\bar{B}}$ is read as A AND B bar the quantity complemented, and $\bar{\bar{X}} = \overline{\overline{A\bar{B}}} = A\bar{B}$. Or, if $Y = 0$, then $\bar{Y} = \bar{0} = 1$, etc. Finally, notice that Eqs. (3.2) can be generated one from the other by involution — even in mixed logic notation. Thus, $\bar{\alpha}(L) = \bar{\bar{\alpha}}(H) = \alpha(H)$, and so on.

AND Laws The AND laws are easily deduced by taking the rows two at a time from the truth table representing the logic AND interpretation of the AND gate given in Fig. 3.16c. Thus, by taking Y equal to logic values 0, 1, X, and $\bar{X}$, the four AND laws result and are given by Eqs. (3.7).

AND Truth Table

X	Y	$X \cdot Y$
0	0	0
0	1	0
1	0	0
1	1	1

AND Laws

$X \cdot 0 = 0$
$X \cdot 1 = X$
$X \cdot X = X$
$X \cdot \bar{X} = 0$

(3.7)

To illustrate the application of the AND laws, let X be the function $X = A + \bar{B}$ so that $(A + \bar{B}) \cdot 0 = 0$, $(A + \bar{B}) \cdot 1 = A + \bar{B}$, $(A + \bar{B}) \cdot (A + \bar{B}) = A + \bar{B}$, and $(A + \bar{B}) \cdot \overline{(A + \bar{B})} = 0$. These laws are valid regardless of the complexity of the function X, which can represent any multivariable function.

3.10 LAWS OF BOOLEAN ALGEBRA

OR Laws The four OR laws are deduced from the logic OR interpretation of the OR gate given in Fig. 3.17d by assigning to Y the values 0, 1, X, and $\bar{X}$ and are given by Eqs. (3.8). The OR laws are illustrated by letting X represent the function $X = \bar{B}C$. Then, according to the OR laws, $\bar{B}C + 0 = \bar{B}C$, $\bar{B}C + 1 = 1$, $\bar{B}C + \bar{B}C = \bar{B}C$, and $\bar{B}C + \overline{\bar{B}C} = 1$. Here again, use has been made of a multivariable function X to demonstrate the applicability of a fundamental Boolean law, the OR law.

$$
\begin{array}{c|c}
\textbf{OR} & \\
\textbf{Truth Table} & \textbf{OR Laws} \\
\begin{array}{cc|c} X & Y & X+Y \\ \hline 0 & 0 & 0 \\ 0 & 1 & 1 \\ 1 & 0 & 1 \\ 1 & 1 & 1 \end{array} & \longrightarrow \quad \begin{array}{l} X + 0 = X \\ X + 1 = 1 \\ X + X = X \\ X + \bar{X} = 1 \end{array}
\end{array} \tag{3.8}
$$

Notice that the AND and OR laws are easily verified by substituting 0 and 1 for the multivariable function X in the examples just given, and then comparing the results with the AND and OR truth tables.

3.10.2 The Concept of Duality

An inspection of the AND and OR laws reveals an interesting relationship that may not be obvious at first glance. If the 1's and 0's are interchanged while the AND and OR operators, $(\cdot)$ and $(+)$, are interchanged, the AND laws generate the OR laws and vice versa. For reference purposes, the interchange of 1's and 0's simultaneously with the interchange of operators is represented by the double arrows $(\leftrightarrow)$ as follows:

$$
\left\{ \begin{array}{c} 0 \leftrightarrow 1 \\ (\cdot) \leftrightarrow (+) \\ \odot \leftrightarrow \oplus \end{array} \right\}
$$

This simultaneous interchange of logic values and operators is called logic *duality*. The duality between the AND and OR laws is given by Eqs. (3.9).

$$
\begin{array}{ccc}
\textbf{AND Laws} & & \textbf{OR Laws} \\
\begin{array}{l} X \cdot 0 = 0 \\ X \cdot 1 = X \\ X \cdot X = X \\ X \cdot \bar{X} = 0 \end{array} & \begin{array}{c} \text{By} \\ \longleftrightarrow \\ \text{Duality} \end{array} & \begin{array}{l} X + 0 = X \\ X + 1 = 1 \\ X + X = X \\ X + \bar{X} = 1 \end{array}
\end{array} \tag{3.9}
$$

Perhaps the best way to demonstrate duality is by the two dual sets

$$(A \odot B)[A\bar{B} + \bar{A}B] = 0 \quad \longleftrightarrow \quad (A \oplus B) + [(A + \bar{B}) \cdot (\bar{A} + B)] = 1$$

and

$$X \odot (\bar{X} + Y) = X \cdot Y \quad \longleftrightarrow \quad X \oplus (\bar{X} \cdot Y) = X + Y,$$

where the double arrow ($\longleftrightarrow$) again represents the duality relationship of the set. For each dual set it can be seen that an operator in the left side equation has been replaced by its dual in the right side while the logic 0 and 1 (in the first dual set) are interchanged. Note that the two equations in a given set are *not* algebraically equal — they are duals of each other. However, a dual set of equations are complementary if an equation is equal to logic 1 or logic 0 as in the first example. Such is not the case for the second set. The concept of duality pervades the entire field of mathematical logic, as will become apparent with the development of Boolean algebra.

3.10.3 Associative, Commutative, Distributive, Absorptive, and Consensus Laws

The associative, commutative, distributive, absorptive, and consensus laws are presented straightforwardly in terms of the multivariable functions X, Y, and Z to emphasize their generality, but the more formal axiomatic approach is avoided for the sake of simplicity. These laws are given in a *dual* form that the reader may find useful as a mnemonic tool:

Associative Laws
$$\left. \begin{array}{l} (X \cdot Y) \cdot Z = X \cdot (Y \cdot Z) = X \cdot Y \cdot Z \\ (X + Y) + Z = X + (Y + Z) = X + Y + Z \end{array} \right\} \quad (3.10)$$

Commutative Laws
$$\left. \begin{array}{l} X \cdot Y \cdot Z = X \cdot Z \cdot Y = Z \cdot X \cdot Y = \cdots \\ X + Y + Z = X + Z + Y = Z + X + Y = \cdots \end{array} \right\} \quad (3.11)$$

Distributive Laws
$$\left. \begin{array}{ll} (X \cdot Y) + (X \cdot Z) = X \cdot (Y + Z) & \text{Factoring Law} \\ (X + Y) \cdot (X + Z) = X + (Y \cdot Z) & \text{Distributive Law} \end{array} \right\} \quad (3.12)$$

Absorptive Laws
$$\left. \begin{array}{l} X \cdot (\bar{X} + Y) = X \cdot Y \\ X + (\bar{X} \cdot Y) = X + Y \end{array} \right\} \quad (3.13)$$

Consensus Laws
$$\left. \begin{array}{l} (X \cdot Y) + (\bar{X} \cdot Z) + (Y \cdot Z) = (X \cdot Y) + (\bar{X} \cdot Z) \\ (X + Y) \cdot (\bar{X} + Z) \cdot (Y + Z) = (X + Y) \cdot (\bar{X} + Z) \end{array} \right\} \quad (3.14)$$

Notice that for each of the five sets of laws, duality exists whereby the AND and OR operators are simultaneously interchanged. The dual set of distributive laws in Eqs. (3.12) occur so often that they are sometimes given the names *factoring law* and *distributive law* for the first and second, respectively. The factoring law draws its name from its similarity to the factoring law of Cartesian algebra.

Although rigorous proof of these laws will not be attempted, they are easily verified by using truth tables. Shown in Figs. 3.33 and 3.34 are the truth table verifications for the AND form of the associative law and the factoring law. Their dual forms can be verified in a similar manner.

Proof of the commutative laws is obtained simply by assigning logic 0 and logic 1 to the X's and Y's in the two variable forms of these laws and then comparing the results with the AND and OR truth tables given by Eqs. (3.7) and (3.8), respectively.

The distributive law can also be verified by using truth tables. However, having verified the factoring law, it is simpler to prove this law with Boolean algebra by using the factoring

3.10 LAWS OF BOOLEAN ALGEBRA

Decimal	X	Y	Z	X·Y	Y·Z	(X·Y)·Z	X·(Y·Z)	X·Y·Z
0	0	0	0	0	0	0	0	0
1	0	0	1	0	0	0	0	0
2	0	1	0	0	0	0	0	0
3	0	1	1	0	1	0	0	0
4	1	0	0	0	0	0	0	0
5	1	0	1	0	0	0	0	0
6	1	1	0	1	0	0	0	0
7	1	1	1	1	1	1	1	1

FIGURE 3.33
Truth table for the AND form of the associative laws in Eqs. (3.10).

law together with the AND and OR laws. This is done in the following sequence of steps by using square brackets to draw attention to those portions where the laws indicated on the right are applied:

$$[(X+Y)(X+Z)] = [X \cdot (X+Z)] + [Y \cdot (X+Z)] \quad \text{Factoring law (applied twice)}$$
$$= [X \cdot X] + (X \cdot Z) + (Y \cdot X) + (Y \cdot Z) \quad \text{AND law } (X \cdot X = X)$$
$$= [X + (X \cdot Z) + (Y \cdot X)] + (Y \cdot Z) \quad \text{Factoring law}$$
$$= X \cdot [1 + Z + Y] + (Y \cdot Z) \quad \text{OR law } (1 + Z + Y = 1)$$
$$= X + (Y \cdot Z).$$

In similar fashion the second of the absorptive laws is proven as follows:

$$X + \bar{X}Y = [(X + \bar{X})(X + Y)] \quad \text{Distributive and OR laws}$$
$$= 1 \cdot (X + Y) \quad \text{AND law } (1 \cdot (X+Y) = X+Y)$$
$$= X + Y.$$

The remaining absorptive law is easily proved by first applying the factoring law followed by the AND law $X \cdot \bar{X} = 0$. Duality can also be used as a validation of one form once its dual is proven.

Decimal	X	Y	Z	X·Y	X·Z	Y+Z	(X·Y)+(X·Z)	X·(Y+Z)
0	0	0	0	0	0	0	0	0
1	0	0	1	0	0	1	0	0
2	0	1	0	0	0	1	0	0
3	0	1	1	0	0	1	0	0
4	1	0	0	0	0	0	0	0
5	1	0	1	0	1	1	1	1
6	1	1	0	1	0	1	1	1
7	1	1	1	1	1	1	1	1

FIGURE 3.34
Truth table for the factoring law given in Eqs. (3.12).

The first of the consensus laws in Eqs. (3.14) is proven by applying the OR and factoring laws:

$$XY + \bar{X}Z + YZ = XY + \bar{X}Z + [(X + \bar{X})YZ] \quad \text{OR law and factoring law}$$
$$= XY + \bar{X}Z + [XYZ + \bar{X}YZ] \quad \text{Factoring law}$$
$$= [XY(1+Z)] + [\bar{X}Z(1+Y)] \quad \text{Factoring law (applied twice); OR law}$$
$$= XY + \bar{X}Z.$$

Proof of the second of the consensus laws follows by duality.

3.10.4 DeMorgan's Laws

In the latter half of the nineteenth century, the English logician and mathematician Augustus DeMorgan proposed two theorems of mathematical logic that have since become known as DeMorgan's theorems. The Boolean algebraic representations of these theorems are commonly known as DeMorgan's laws. In terms of the two multivariable functions X and Y, these laws are given in dual form by

$$\textbf{DeMorgan's Laws} \quad \left\{ \begin{array}{c} \overline{X \cdot Y} = \bar{X} + \bar{Y} \\ \overline{X + Y} = \bar{X} \cdot \bar{Y} \end{array} \right\} \quad (3.15)$$

More generally, for any number of functions, the DeMorgan laws take the following form:

$$\overline{X \cdot Y \cdot Z \cdot \cdots \cdot N} = \bar{X} + \bar{Y} + \bar{Z} + \cdots + \bar{N}$$

and $\qquad (3.15a)$

$$\overline{X + Y + Z + \cdots + N} = \bar{X} \cdot \bar{Y} \cdot \bar{Z} \cdot \cdots \cdot \bar{N}.$$

DeMorgan's laws are easily verified by using truth tables. Shown in Fig. 3.35 is the truth table for the first of Eqs. (3.15).

Application of DeMorgan's laws can be demonstrated by proving the absorptive law $X + \bar{X}Y = X + Y$:

$$X + \bar{X}Y = \overline{\overline{X \cdot (\overline{\bar{X}Y})}} = \overline{\bar{X} \cdot (X + \bar{Y})} = \overline{\bar{X} \cdot X + \bar{X} \cdot \bar{Y}} = \overline{\bar{X} \cdot \bar{Y}} = X + Y.$$

Notice that the double bar over the term $X + \bar{X}Y$ is a NOT law and does not alter the term. Here, DeMorgan's laws are first applied by action of the "inner" bar followed by simplification under the "outer" bar. Final application of DeMorgan's law by application

X	Y	$X \cdot Y$	$\overline{X \cdot Y}$	$\bar{X}$	$\bar{Y}$	$\bar{X} + \bar{Y}$
0	0	0	1	1	1	1
0	1	0	1	1	0	1
1	0	0	1	0	1	1
1	1	1	0	0	0	0

FIGURE 3.35
Truth table for DeMorgan's Law $\overline{X \cdot Y} = \bar{X} + \bar{Y}$.

3.11 LAWS OF XOR ALGEBRA

of the outer bar takes place only after simplification. As a general rule, DeMorgan's laws should be applied to a function only after it has been sufficiently reduced so as to avoid unnecessary Boolean manipulation.

3.11 LAWS OF XOR ALGEBRA

The laws of XOR algebra share many similarities with those of conventional Boolean algebra discussed in the previous section and can be viewed as a natural extension of the conventional laws. Just as the AND and OR laws are deduced from their respective truth tables, the XOR and EQV laws are deduced from their respective truth tables in Figs. 3.26c and 3.27d and are given by Eqs. (3.16) together with their truth tables:

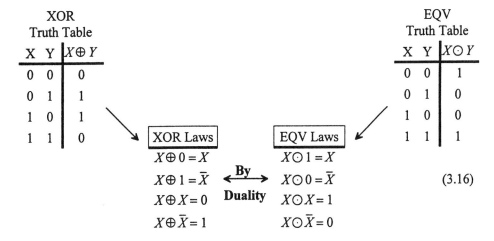

Here, the dual relationship between the XOR and EQV laws is established by interchanging the 1's and 0's while simultaneously interchanging the XOR and EQV operators, as indicated by the double arrow.

The associative and commutative laws for EQV and XOR follow from the associative and commutative laws for AND and OR given by Eqs. (3.10) and (3.11) by exchanging operator symbols: $\odot$ for $(\cdot)$ and $\oplus$ for $(+)$. The distributive, absorptive, and consensus laws of XOR algebra follow from their AND/OR counterparts in Eqs. (3.12), (3.13), and (3.14) by replacing the appropriate $(+)$ operator symbols with the $\oplus$ operator symbols, and by replacing the appropriate $(\cdot)$ symbols with the $\odot$ symbol, but not both in any given expression. In similar fashion, DeMorgan's laws in XOR algebra are produced by substituting $\odot$ for $(\cdot)$ and $\oplus$ for $(+)$ in Eqs. (3.15). These laws are presented as follows in dual form and in terms of variables X, Y, and Z, which may represent single or multivariable functions:

Associative Law
$$\begin{cases} (X \odot Y) \odot Z = X \odot (Y \odot Z) = X \odot Y \odot Z \\ (X \oplus Y) \oplus Z = X \oplus (Y \oplus Z) = X \oplus Y \oplus Z \end{cases} \qquad (3.17)$$

Commutative Laws
$$\begin{cases} X \odot Y \odot Z = X \odot Z \odot Y = Z \odot X \odot Y = \cdots \\ X \oplus Y \oplus Z = X \oplus Z \oplus Y = Z \oplus X \oplus Y = \cdots \end{cases} \qquad (3.18)$$

Distributive Laws
$$\begin{cases} (X \cdot Y) \oplus (X \cdot Z) = X \cdot (Y \oplus Z) & \text{Factoring Law} \\ (X + Y) \odot (X + Z) = X + (Y \odot Z) & \text{Distributive Law} \end{cases} \quad (3.19)$$

Absorptive Laws
$$\begin{cases} X \cdot (\bar{X} \oplus Y) = X \cdot Y \\ X + (\bar{X} \odot Y) = X + Y \end{cases} \quad (3.20)$$

Consensus Laws
$$\begin{cases} (X \cdot Y) \oplus (\bar{X} \cdot Z) + (Y \cdot Z) = (X \cdot Y) \oplus (\bar{X} \cdot Z) \\ (X + Y) \odot (\bar{X} + Z) \cdot (Y + Z) = (X + Y) \odot (\bar{X} + Z) \end{cases} \quad (3.21)$$

DeMorgan's Laws
$$\begin{cases} \overline{X \odot Y} = \bar{X} \oplus \bar{Y} = X \oplus Y \\ \overline{X \oplus Y} = \bar{X} \odot \bar{Y} = X \odot Y \end{cases}. \quad (3.22)$$

Notice that each of the six sets of equations are presented in dual form. Thus, by interchanging AND and OR operators while simultaneously interchanging EQV and XOR operators, duality of the set is established. The first of the distributive laws given in Eqs. (3.19) can be termed the *factoring law* of XOR algebra owing to its similarity with the factoring law of Cartesian algebra and that of Eqs. (3.12).

Generalizations of DeMorgan's XOR laws follow from Eqs. (3.15a) and (3.22) and are given by

$$\overline{X \odot Y \odot Z \odot \cdots \odot N} = \bar{X} \oplus \bar{Y} \oplus \bar{Z} \oplus \cdots \oplus \bar{N}$$

and (3.22a)

$$\overline{X \oplus Y \oplus Z \oplus \cdots \oplus N} = \bar{X} \odot \bar{Y} \odot \bar{Z} \odot \cdots \odot \bar{N}.$$

Verification of the associative, commutative, and distributive laws is easily accomplished by using truth tables. For example, the second of the distributive laws in Eqs. (3.19) is verified by the truth table in Fig. 3.36. Here, Eq. (3.5) is used together with the OR laws [Eqs. (3.8)] to show the identity of the terms $(X + Y) \odot (X + Z)$ and $X + (Y \odot Z)$.

The distributive laws may also be proven by using Boolean algebra. For example, the factoring law of Eqs. (3.19) is proven by applying the defining relation of the XOR function

X	Y	Z	$X+Y$	$X+Z$	$Y \odot Z$	$(X+Y) \odot (X+Z)$	$X + (Y \odot Z)$
0	0	0	0	0	1	1	1
0	0	1	0	1	0	0	0
0	1	0	1	0	0	0	0
0	1	1	1	1	1	1	1
1	0	0	1	1	1	1	1
1	0	1	1	1	0	1	1
1	1	0	1	1	0	1	1
1	1	1	1	1	1	1	1

FIGURE 3.36
Truth table for the XOR distributive law given in Eqs. (3.19).

3.11 LAWS OF XOR ALGEBRA

given by Eq. (3.4) and by using the AND and OR laws of Eqs. (3.9):

$$
\begin{aligned}
[(XY) \oplus (XZ)] &= [(\overline{XY})](XZ) + (XY)[(\overline{XZ})] && \text{Eq. (3.4) and Eq. (3.15)} \\
&= [(\bar{X} + \bar{Y})(XZ)] + [(XY)(\bar{X} + \bar{Z})] && \text{Factoring law [Eqs. (3.12)]} \\
&= [\bar{X}XZ + X\bar{Y}Z] + [X\bar{X}Y + XY\bar{Z}] && \text{AND and OR laws [Eqs. (3.9)]} \\
&= [X\bar{Y}Z + XY\bar{Z}] && \text{Factoring law [Eqs. (3.12)]} \\
&= X[\bar{Y}Z + Y\bar{Z}] && \text{Eq. (3.4)} \\
&= X(Y \oplus Z).
\end{aligned}
$$

In these equations, the square brackets [ ] are used to draw attention to those portions where the laws or equations indicated on the right are to be applied in going to the next step. Equation (3.4) refers to the defining relation for XOR given by $X \oplus Y = \bar{X}Y + X\bar{Y}$.

The absorptive laws are also easily proven by using Boolean algebra. Beginning with the first of Eqs. (3.20), there follows

$$
\begin{aligned}
X \cdot [(\bar{X} \oplus Y)] &= X \cdot (\bar{X}\bar{Y} + XY) && \text{Eq. (3.4)} \\
&= [X \cdot (\bar{X}\bar{Y} + XY)] && \text{Factoring law [Eqs. (3.12)]} \\
&= [X \cdot \bar{X}\bar{Y} + X \cdot XY] && \text{AND and OR laws [Eqs. (3.9)]} \\
&= XY,
\end{aligned}
$$

where the square brackets [ ] are again used to draw attention to those portions where the laws or equations indicated on the right are to be applied. The second of Eqs. (3.20) is proven by the following sequence of steps:

$$
\begin{aligned}
X + [(\bar{X} \odot Y)] &= X + (X\bar{Y}) + (\bar{X}Y) && \text{Eq. (3.5)} \\
&= [X + (X\bar{Y})] + \bar{X}Y && \text{Factoring law [Eqs. (3.12)]} \\
&= [X(1 + \bar{Y}) + \bar{X}Y] && \text{OR and AND laws} \\
&= [X + \bar{X}Y] && \text{Absorptive law [Eqs. (3.13)]} \\
&= X + Y.
\end{aligned}
$$

Notice that in the foregoing proofs, use is tacitly made of the important dual relations

$$
\begin{aligned}
\bar{X} \oplus Y &= X \oplus \bar{Y} = \overline{X \oplus Y} = X \odot Y \\
\bar{X} \odot Y &= X \odot \bar{Y} = \overline{X \odot Y} = X \oplus Y.
\end{aligned}
\tag{3.23}
$$

These relations are easily verified by replacing the variable (X or Y) by its complement ($\bar{X}$ or $\bar{Y}$) in the appropriate defining relation, (3.4) or (3.5).

An inspection of Eqs. (3.23) reveals what should already be understood — that complementing one of the connecting variables complements the function, and that the complement of an XOR function is the EQV function and vice versa. A generalization of this can be stated as follows:

In any string of terms interconnected only with XOR and/or EQV operators, an odd number of complementations (variable or operator complementations) complements the function, whereas an even number of complementations preserves the function.

To illustrate, consider a function F consisting of a string of four multivariable terms W, X, Y, and Z interconnected initially by XOR operators:

$$F = W \oplus X \oplus Y \oplus Z = W \oplus \bar{X} \oplus Y \odot Z$$
$$= W \odot \bar{X} \oplus \bar{Y} \oplus \bar{Z} = W \odot X \odot Y \oplus Z = \cdots$$
$$\bar{F} = W \oplus \bar{X} \oplus Y \oplus Z = W \oplus \bar{X} \oplus Y \odot \bar{Z} \qquad (3.24)$$
$$= W \odot X \odot Y \odot Z = \bar{W} \odot X \oplus \bar{Y} \oplus Z = \cdots$$

An examination of Eqs. (3.24) reveals that there are 64 possible expressions representing F and 64 for $\bar{F}$, all generated by repeated applications of Eqs. (3.23). The number 64 is derived from combinations of seven different objects taken an even or odd number at a time for F and $\bar{F}$, respectively.

Application of Eqs. (3.24) is illustrated by

$$A \odot (A \odot D + \bar{C}) \odot \bar{B} = A \odot [(A \oplus D)C] \odot B$$
$$= A \oplus [(A \oplus D)C] \oplus B,$$

where the original function has been converted from one having only EQV operators and two complemented variables to one having only XOR operators with no complemented variables. The two alternative forms (right side) differ from each other by only two complementations. Notice also that the first alternative form involved applications of DeMorgan's laws given by Eqs. (3.15) and (3.22).

3.11.1 Two Useful Corollaries

Interesting and useful relationships result between XOR algebra and conventional Boolean algebra by recognition of the following two dual corollaries, which follow directly from the definitions of the XOR and EQV operations:

COROLLARY I   *If two functions, α and β, never take the logic 1 value at the same time, then*

$$\alpha \cdot \beta = 0 \quad \text{and} \quad \alpha + \beta = \alpha \oplus \beta \qquad (3.25)$$

and the logic operators $(+)$ and $(\oplus)$ are interchangeable.

COROLLARY II   *If two functions, α and β, never take the logic 0 value at the same time, then*

$$\alpha + \beta = 1 \quad \text{and} \quad \alpha \cdot \beta = \alpha \odot \beta \qquad (3.26)$$

and the logic operators $(\cdot)$ and $(\odot)$ are interchangeable.

Corollary I requires that α and β each be terms consisting of ANDed variables called *product terms* (p-terms) and that they be *disjoint*, meaning that the two terms never take logic 1 simultaneously. By duality, Corollary II requires that α and β each be terms consisting of ORed variables called *sum terms* (s-terms) and that they be *conjoint*, meaning that the two terms never take logic 0 simultaneously. The subject of these corollaries will be revisited in Section 5.5 where their generalizations will be discussed.

The most obvious application of Corollaries I and II is in operator interchange as demonstrated by the following four examples:

$$[1] \quad AB + \bar{B}C = (AB) \oplus (\bar{B}C),$$

where $\alpha = AB$, $\beta = \bar{B}C$, and $\alpha \cdot \beta = 0$ by Corollary I.

$$[2] \quad (\bar{A} + B + X) \cdot (A + B + C + Y) = (\bar{A} + B + X) \odot (A + B + C + Y),$$

where $\alpha = (\bar{A} + B + X)$, $\beta = (A + B + C + Y)$ and $\alpha + \beta = 1$ according to Corollary II.

$$[3] \quad a + \bar{b} \oplus bc = a + \bar{b} + bc = a + \bar{b} + c,$$

where Corollary I has been applied followed by the absorptive law in Eqs. (3.13).

$$[4] \quad (X\bar{Y}) \odot (\bar{X} + Y + \bar{Z}) = (X\bar{Y})(\bar{X} + Y + \bar{Z}) = X\bar{Y}\bar{Z}$$

Here, Corollary II is applicable since $\overline{X\bar{Y}} = \bar{X} + Y$, and the result follows by using the AND and OR laws given by Eqs. (3.9).

3.11.2 Summary of Useful Identities

The laws of XOR algebra have been presented in the foregoing subsections. There are several identities that follow directly or indirectly from these laws. These identities are useful for function simplification and are presented here in *dual* form for reference purposes.

$$\begin{cases} \bar{X} \oplus Y = X \oplus \bar{Y} = \overline{X \oplus Y} = \bar{X} \odot \bar{Y} = X \odot Y \\ \bar{X} \odot Y = X \odot \bar{Y} = \overline{X \odot Y} = \bar{X} \oplus \bar{Y} = X \oplus Y \end{cases} \quad (3.27)$$

$$\begin{cases} X \oplus \bar{X} = X \odot X = 1 \\ X \odot \bar{X} = X \oplus X = 0 \end{cases} \quad (3.28)$$

$$\begin{cases} 1 \oplus X = \bar{X} \\ 0 \odot X = \bar{X} \end{cases} \quad \begin{cases} 0 \oplus X = X \\ 1 \odot X = X \end{cases} \quad (3.29)$$

$$\begin{cases} X\bar{Y} \oplus X = \bar{X}Y \oplus Y = XY \\ (X + \bar{Y}) \odot X = (\bar{X} + Y) \odot Y = X + Y \end{cases} \quad (3.30)$$

$$\begin{cases} \bar{X}Y \oplus X = X\bar{Y} \oplus Y = 1 \oplus \bar{X}\bar{Y} \\ (\bar{X} + Y) \odot X = (X + \bar{Y}) \odot Y = 0 \odot (\bar{X} + \bar{Y}) \end{cases} \quad (3.31)$$

$$\begin{cases} (XY) \oplus (X + Y) = X \oplus Y \\ (X + Y) \odot (XY) = X \odot Y \end{cases} \quad (3.32)$$

$$\begin{cases} XY + YZ + XZ = XY \oplus YZ \oplus XZ \\ (X + Y)(Y + Z)(X + Z) = (X + Y) \odot (Y + Z) \odot (X + Z) \end{cases} \quad (3.33)$$

Note that in these identities, either X or Y or both may represent multivariable functions or single variables of any polarity (i.e., either complemented or uncomplemented).

By applying the laws and corollaries previously given, the first of the two identities of Eqs. (3.33) is proven as follows (Boolean laws used are given to the right):

$$XY + YZ + XZ = XY(\bar{Z} + Z) + (\bar{X} + X)YZ + X(\bar{Y} + Y)X \quad \text{OR Laws}$$
$$\text{[Eqs. (3.8)]}$$
$$= XYZ + XY\bar{Z} + XYZ + \bar{X}YZ + XYZ + X\bar{Y}Z \quad \text{Eqs. (3.19) and}$$
$$\text{OR Laws}$$
$$= XYZ \oplus XY\bar{Z} \oplus (\bar{X}YZ + X\bar{Y}Z) \quad \text{Corollary I}$$
$$= XY \oplus (Y \oplus X)Z \quad \text{Eqs. (3.19), OR}$$
$$\text{Law, Eq. (3.4)}$$
$$= XY \oplus YZ \oplus XZ. \quad \text{Eq. (3.8)}$$

Proof of the second identity of Eqs. (3.33) follows by duality, that is, simply by interchanging all (+) with (·) operators while simultaneously interchanging all $\oplus$ with $\odot$ operators. The generalization of this identity is given by

$$[WXY \cdots + WXZ \cdots + WYZ \cdots + XYZ \cdots + \cdots]$$
$$= [WXY \cdots \oplus WXZ \cdots \oplus WYZ \cdots \oplus XYZ \cdots \oplus \cdots],$$

which also has its dual formed by the simultaneous interchange of the operators.

This concludes the treatment of Boolean algebra. While not intended to be an exhaustive coverage of the subject, it is adequate for the needs of digital design as presented in this text. Additional references on Boolean algebra are available in the list of further reading that follows.

3.12 WORKED EXAMPLES

EXAMPLE 3.1 Given the waveforms (heavy lines) at the top of Figs. 3.37a and 3.37b, draw the two waveforms for the two terminals below each.

EXAMPLE 3.2 Complete the physical truth table in Fig. 3.38b for the CMOS logic circuit given in Fig. 3.38a. Name the gate and give the two conjugate logic circuit symbols for this gate in part (c).

EXAMPLE 3.3 The logic circuit in Fig. 3.39 is a redundant circuit, meaning that excessive logic is used to implement the function Z(H). (a) Name the physical gates that are used in the logic circuit in Fig. 3.39. (b) Read the circuit in mixed-logic notation and express the results in reduced, polarized Boolean form at nodes W, X, Y, and Z.

(a) (1) NAND, (2) NOR, (3) NOR, (4) OR, (5) AND, (6) NOR

(b) $W(H) = A\bar{B}(H)$

$X(L) = \bar{B}\bar{C}(L)$

$Y(L) = (\bar{C} + D)(L)$

$Z(H) = \bar{W}XY(H) = \overline{(A\bar{B})}(\bar{B}\bar{C})(\bar{C} + D)(H) = (\bar{A} + B)(\bar{B}\bar{C})(\bar{C} + D)(H)$
$= (\bar{A} + B)(\bar{B}\bar{C} + \bar{B}\bar{C}D)(H)$
$= \bar{A}\bar{B}\bar{C}(H)$

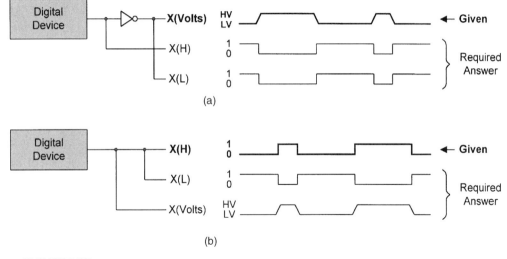

FIGURE 3.37
Physical (voltage) waveforms and mixed-logic notation. (a) Effect of logic level conversion. (b) Absence of logic level conversion.

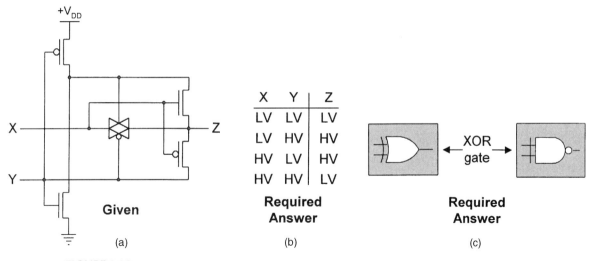

FIGURE 3.38
Physical truth table and logic circuit symbols for a CMOS logic gate. (a) Given logic circuit. (b) Physical truth table for an XOR gate. (c) Conjugate logic circuit symbols for the XOR gate.

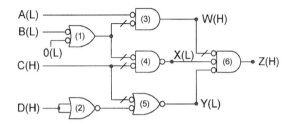

FIGURE 3.39
A redundant logic circuit.

117

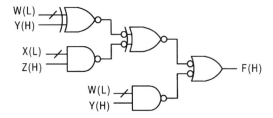

FIGURE 3.40
Logic circuit for the function given in Example 3.4.

EXAMPLE 3.4 Use three NAND gates and two EQV gates (nothing else) to implement the following function *exactly* as written:

$$F(H) = [(\bar{W} \oplus Y) \oplus (\bar{X}Z) + \bar{W}Y](H)$$

The solution is shown in Fig. 3.40.

EXAMPLE 3.5 A simple function of three inputs is given by the following expression:

$$F(H) = (\bar{A}B + C)(H).$$

(a) Construct the logic circuit by using AND/NOR/INV logic. Assume that the inputs arrive active high.
(b) Construct the CMOS circuit for the function given in Part (a).
(c) Obtain the physical truth table for the circuit of Part (b).
(d) Obtain the positive logic truth table for the circuit of Part (b).

The solutions to Example 3.5 are given in Fig. 3.41. Notice that PMOS and NMOS are organized according to Fig. 3.5, and that the PMOS section generates the complement of that of the NMOS section, hence the complementary MOS. Also note that the output of the A inverter is connected to both the PMOS and NMOS inputs of the complementary sections for F.

EXAMPLE 3.6 Use the laws of Boolean algebra, including XOR algebra and corollaries, to reduce each of the following expressions to their *simplest* form. Name the law(s) in each step.

[1] $\bar{A} + A\bar{B}\bar{C} + \overline{(B + C)} = \bar{A} + A\bar{B}\bar{C} + \bar{B}\bar{C}$ DeMorgan's law [Eqs. (3.15)]
$\qquad\qquad\qquad\qquad\quad = \bar{A} + \bar{B}\bar{C} + \bar{B}C$ Absorptive law [Eqs. (3.13)]
$\qquad\qquad\qquad\qquad\quad = \bar{A} + \bar{B}(\bar{C} + C)$ Factoring law [Eqs. (3.12)]
$\qquad\qquad\qquad\qquad\quad = \bar{A} + \bar{B}$ AND and Or laws [Eqs. (3.7) and (3.8)]

[2] $(a + b)(\bar{a} + c)(\bar{a} + \bar{c}) = (a + b)(\bar{a} + c \cdot \bar{c})$ Distributive law [Eqs. (3.12)]
$\qquad\qquad\qquad\qquad\quad = (a + b)\bar{a}$ AND and OR laws [Eqs. (3.7) and (3.8)]
$\qquad\qquad\qquad\qquad\quad = \bar{a}a + \bar{a}b$ Factoring law [Eqs. (3.12)]
$\qquad\qquad\qquad\qquad\quad = \bar{a}b$ AND and OR laws

3.12 WORKED EXAMPLES 119

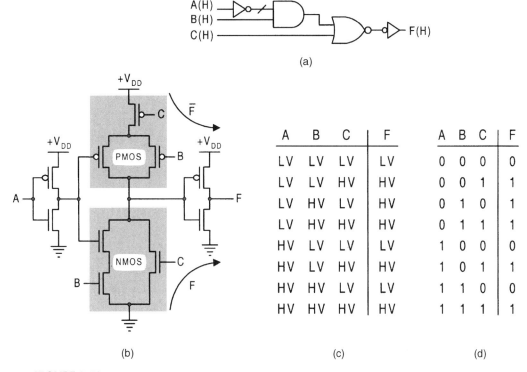

FIGURE 3.41
Circuit and truth table representations for the function F of Example 3.5. (a) Logic circuit. (b) CMOS circuit. (c) Physical truth table. (d) Logic truth table.

$$
\begin{aligned}
{[3]}\ (X+Y)(X+\bar{Z})[Y(X+\bar{Z})+\bar{Y}] &= (X+Y\bar{Z})[Y(X+\bar{Z})+\bar{Y}] & &\text{Distributive law} \\
& & &\text{[Eqs. (3.12)]} \\
&= (X+Y\bar{Z})[X+\bar{Y}+\bar{Z}] & &\text{Absorptive law} \\
& & &\text{[Eqs. (3.13)]} \\
&= X+Y\bar{Z}\bar{Y}+Y\bar{Z}\bar{Z} & &\text{Distributive law} \\
&= X+Y\bar{Z} & &\text{AND and OR laws}
\end{aligned}
$$

$$
\begin{aligned}
{[4]}\ \overline{(b\oplus\bar{c})+\overline{(ab)}(\overline{\bar{a}+c})} &= \overline{(b\odot c)+(\bar{a}+\bar{b})(a\bar{c})} & &\text{Eqs. (3.23); DeMorgan's laws} \\
& & &\text{[Eqs. (3.15)]} \\
&= \overline{(b\odot c)+\bar{a}a\bar{c}+a\bar{b}\bar{c}} & &\text{Factoring law [Eqs. (3.12)]} \\
&= \overline{(bc+\bar{b}\bar{c}+a\bar{b}\bar{c})} & &\text{Eq. (3.5); AND and OR laws} \\
&= \overline{(bc+\bar{b}\bar{c})} & &\text{Factoring law; AND and OR laws} \\
&= b\oplus c & &\text{Eqs. (3.5) and (3.23)}
\end{aligned}
$$

$$
\begin{aligned}
{[5]}\ (\bar{X}+Y)\odot(X\oplus Y) &= (\bar{X}+Y)\odot(\bar{X}Y+X\bar{Y}] & &\text{Eq. (3.4)} \\
&= (\bar{X}+Y)(\bar{X}Y+X\bar{Y}) & &\text{Corollary II [Eq. (3.26)]} \\
&= \bar{X}Y+\bar{X}Y & &\text{Factoring law; AND and OR laws} \\
&= \bar{X}Y & &\text{OR laws}
\end{aligned}
$$

[6] $(A + \bar{B} + \bar{A}C) \odot (\bar{A}B + C) = (A + \bar{B} + \bar{A}C)(\bar{A}B + C)$ Corollary II
$= (A + \bar{B} + C)(\bar{A}B + C)$ Absorption [Eqs. (3.13)]
$= C + (A + \bar{B})(\bar{A}B)$ Distributive law [Eqs. (3.12)]
$= C$ Factoring law; AND and OR laws

[7] $\bar{a}c + (\bar{a} + b) \odot (\bar{a} + bc) = \bar{a}c + (a\bar{b}) \oplus (\bar{a} + bc)$ DeMorgan's law [Eqs. (3.15); Eqs. (3.23)]
$= \bar{a}c + (a\bar{b}) + (\bar{a} + bc)$ Corollary I [Eq. (3.25)]
$= a\bar{b} + \bar{a} + bc$ Factoring law; AND and OR laws
$= \bar{a} + \bar{b} + bc$ Absorption [Eqs. (3.13)]
$= \bar{a} + \bar{b} + c$ Absorption

[8] $w\bar{x}\bar{y} + wx\bar{z} + \bar{w}x\bar{z} + w\bar{y}\bar{z} + x\bar{z} = w\bar{x}\bar{y} + wx\bar{z} + \bar{w}x\bar{z}$ Consensus law [Eqs. (3.14)]
$= w\bar{x}\bar{y} + x\bar{z}(w + \bar{w})$ Factoring law [Eqs. (3.12)]
$= w\bar{x}\bar{y} + x\bar{z}$ Or laws

[9] $A \oplus B \oplus (A + B) = A \oplus [\bar{B} \oplus (\bar{A}\bar{B})]$ Eqs. (3.27)
$= A \oplus [\bar{B}(1 \oplus \bar{A})]$ XOR Factoring law [Eqs. (3.19)]
$= A \oplus (A\bar{B})$ Eqs. (3.29)
$= A(1 \oplus \bar{B})$ XOR Factoring law
$= AB$ Eqs. (3.29)

[10] $f = d \oplus \bar{b}c\bar{d} \oplus ab\bar{d} \oplus cd \oplus a\bar{d} \oplus \bar{a}\bar{b}c\bar{d}$
$= [d \oplus cd] \oplus [ab\bar{d} \oplus a\bar{d}] \oplus [\bar{b}c\bar{d} \oplus \bar{a}\bar{b}c\bar{d}]$ Rearranging terms
$= [d(1 \oplus c)] \oplus [a\bar{d}(b \oplus 1)] \oplus [\bar{b}c\bar{d}(1 \oplus \bar{a})]$ XOR Factoring law [Eqs. (3.19)]
$= \bar{c}d \oplus a\bar{b}\bar{d} \oplus a\bar{b}c\bar{d}$ Repeated applications of Eqs. (3.29)
$= \bar{c}d \oplus [a\bar{b}\bar{d}(1 \oplus c)]$ XOR Factoring law [Eqs. (3.19)]
$= \bar{c}d \oplus a\bar{b}\bar{c}\bar{d}$ Application of Eqs. (3.29)

Notice that the gate/input tally of f has been reduced from 10/24 to 3/8 in the final expression. Application of Corollary I further reduces f to $(ab\bar{c} + \bar{c}d)$.

FURTHER READING

Additional reading on the subject of mixed logic notation and symbology can be found in the texts of Comer, Fletcher, Shaw and Tinder.

[1] D. J. Comer, *Digital Logic and State Machine Design*, 3rd ed. Saunders College Publishing, Fort Worth, TX, 1995.

[2] W. I. Fletcher, *An Engineering Approach to Digital Design*. Prentice Hall, Englewood Cliffs, NJ, 1980.
[3] A. W. Shaw, *Logic Circuit Design.* Sanders College Publishing, Fort Worth, TX, 1993.
[4] R. F. Tinder, *Digital Engineering Design: A Modern Approach*. Prentice Hall, Englewood Cliffs, NJ, 1991.

Virtually every text on digital or logic design provides some coverage of Boolean algebra. The texts of McCluskey and Dietmeyer are noteworthy for their coverage of both conventional Boolean algebra and XOR algebra including a very limited treatment of the Reed–Muller expansion theorem.

[5] D. L. Dietmeyer, *Logic Design of Digital Systems*, 2nd ed. Allyn and Bacon, Boston, MA, 1978.
[6] E. J. McCluskey, *Logic Design Principles*. Prentice-Hall, Englewood Cliffs, NJ, 1986.

A more formal treatment of XOR algebra can be found in the work of Fisher.

[7] L. T. Fisher, "Unateness Properties of AND-EXCLUSIVE OR," *IEEE Trans. on Computers* **C-23**, 166–172 (1974).

A brief history of Boolean algebra is provided in Chapter 2 of Hill and Peterson.

[8] F. J. Hill and G. R. Peterson, Digital Logic and Microprocessors, John Wiley, NY, 1984.

CMOS logic, which is emphasized in this text, is adequately covered by Weste and Eshraghian in Chapter 1 and portions of Chapter 5. But an excellent coverage of experimental work on various XOR and EQV gates on the MOS transistor level is given by Wang et al.

[9] N. H. E. Weste and K. Eshraghian, *Principles of CMOS VLSI Design*, Addison-Wesley, Reading, MA, 1985.
[10] J. Wang, S. Fang, and W. Feng, "New Efficient Designs for EXOR and XNOR Functions on the Transistor Level," *IEEE Journal of Solid-State Circuits* **29**(7), 780–786 (1994).

PROBLEMS

3.1 Define the following:
  (a) Mixed logic
  (b) Polarized mnemonic
  (c) Logic level conversion
  (d) Active and inactive states
  (e) Inverter
  (f) Gate

3.2 Identify the gate appropriate to each of the physical truth tables in Fig. P3.1. Note: It may be necessary to search this chapter for the answers.

A B	Y		A B	Y		A B	Y		A B	Y		A B	Y
LV LV	HV		LV LV	LV		LV LV	LV		LV LV	HV		LV LV	LV
LV HV	HV		LV HV	HV		LV HV	LV		LV HV	LV		LV HV	HV
HV LV	HV		HV LV	HV		HV LV	LV		HV LV	LV		HV LV	HV
HV HV	LV		HV HV	HV		HV HV	HV		HV HV	LV		HV HV	LV
(a)			(b)			(c)			(d)			(e)	

FIGURE P3.1

3.3 By using a sketch, indicate how the CMOS inverter of Fig. 3.6a can be converted to a two-transistor noninverting switch. What would be the disadvantage (if any) of such a device?

3.4 Given the waveforms from the two logic devices in Fig. P3.2, sketch the waveforms for X(voltage), X(L), Y(H), and Y(voltage). Keep the logic and voltage levels as shown.

3.5 With reference to Problem 3.4, explain the differences between the logic levels for the X(H) and X(L) waveforms and those for the Y(H) and Y(L) waveforms. Do these differences represent a contradiction in the definition of positive and negative logic? Explain.

3.6 Use the inverter, its I/O behavior, and the logic circuit symbols in Fig. 3.6 to explain how the PMOS indicator bubble in the inverter circuit is related to the active low indicator bubble appearing on the inverter symbols.

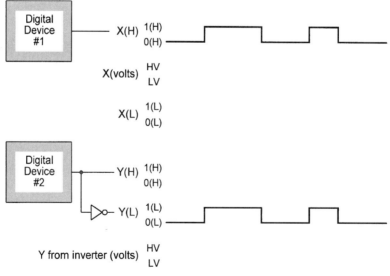

FIGURE P3.2

3.7 Use the definitions of positive and negative logic and Fig. 3.6 as a guide to construct the physical truth table and its two mixed logic truth tables for each of the tri-state drivers in Fig. 3.8 as listed below. Remember to include both inputs X_i and the control C. Use the letter D to represent the disconnect mode.
  (a) Noninverting tri-state driver with $C(H)$.
  (b) Noninverting tri-state driver with $C(L)$.
  (c) Inverting tri-state driver with $C(H)$.
  (d) Inverting tri-state driver with $C(L)$.

3.8 By adding an inverter to each, reconfigure the tri-state drivers in Figs. 3.8c and 3.8d so that they become noninverting tri-state enable switches with driver capability. Give the circuit symbols and ideal equivalent circuits for these two reconfigured tri-state switches.

3.9 Reconfigure the NAND gate in Fig. 3.10a by flipping it end-for-end such that the two series NMOS are on the $+V_{DD}$ (HV) end and the two parallel PMOS on the ground (LV) end.
  (a) Construct the physical and mixed logic truth tables for this reconfigured CMOS circuit. Is this a valid gate form and, if so, what logic function does it perform? (Hint: Compare with Fig. 3.16.)
  (b) What, if any, are the disadvantages of this new configuration? Explain.

3.10 Repeat Problem 3.9 for the NOR gate in Fig. 3.12a, but with the two parallel NMOS on the HV end and the series PMOS on the LV end. (Hint: Compare with Fig. 3.17.)

3.11 Explain why the AND and OR gates of Figs. 3.16 and 3.17 cannot be used for logic level conversion as is done for the NAND and NOR gates of Figs. 3.14 and 3.15.

3.12 Write the logic expressions for the action indicated by the situations given below. Use mnemonics or abbreviations where appropriate.
  (a) Bob will go fishing in a boat only if the boat does not leak and if it is not windy. Otherwise, he will fish from the bank, but only if the fish are biting.
  (b) A laboratory class consists of five students (A, B, C, D, and E) each from a different discipline. An experiment has been assigned that must be carried out with any one of the following combinations of students:

  A and C but not D
  A or B but not both (see Section 3.9)
  D but only if E is present

  (c) A robot is activated only if a majority of its three switches (X, Y, and Z) are turned ON and is deactivated if a majority of its three switches are turned OFF.

3.13 Archie (A), Betty (B), Cathy (C), and David (D) may attend a school dance, but will dance only with the opposite sex and then only under the following conditions: Archie will dance with either Betty or Cathy. However, Cathy will dance with Archie only if Betty is not present at the dance. David will dance only with Betty. Obtain the logic expression representing the active state of dancing for A, B, C, and D.

3.14 Use a minimum number of gates and inverters to implement the functions below with NAND/INV logic. Give the gate/input tally for each logic circuit, excluding inverters. Implement the function exactly as presented — make no alterations. Assume that all inputs arrive from positive logic sources. Use the inverters for logic level conversion.

(a) $Z(H) = (X\bar{Y} + W)(H)$
(b) $F(H) = [\overline{\bar{A}D} + (B + \bar{E})](H)$
(c) $g(L) = (\bar{w}\bar{y} + x + z)(L)$
(d) $G(L) = [(AB + \bar{C})(\overline{D + E})](L)$
(e) $Y(H) = [\overline{(a + \bar{b}c)} \cdot (\bar{d} + e)](H)$

3.15 Repeat Problem 3.14 by using NOR/INV logic. Assume that all inputs arrive from negetive logic sources.

3.16 Repeat Problem 3.14 by using AND/OR/INV logic. Assume that all inputs arrive from positive logic sources.

3.17 Use three NOR gates (nothing else) to implement the function $Y(H)$ below exactly as written. Assume the inputs arrive as follow: $A(H)$, $B(H)$, $C(H)$, $D(L)$, and $E(L)$.

$$Y(H) = [(\overline{\bar{A}D}) \cdot (B + C + \bar{E})](H)$$

3.18 Use three NAND gates (nothing else) to implement the function $Z(H)$ below exactly as written. Assume the inputs arrive as follow: $A(H)$, $B(H)$, $C(H)$, $D(L)$, and $E(L)$.

$$Z(H) = [(\overline{\bar{A} + D}) + (BC\bar{E})](H)$$

3.19 Name the gates used in each of the logic circuits shown in Fig. P3.3 and give the mixed logic expression at each node in mixed logic notation. Use Figs. 3.20, 3.23, and 3.24 as a guide if needed.

3.20 The CMOS circuits in Fig. P3.4 perform specific logic functions. Construct the physical and mixed logic truth tables for each circuit, indicate what logic function it performs and give its two conjugate logic circuit symbols. Note that $\bar{B}$ is the inverse voltage of B.

3.21 Use two NOR gates and one XOR gate (nothing else) to implement the function $Y(H)$ below exactly as written. Assume the inputs arrive as $A(H)$, $B(H)$, $C(H)$, $D(L)$, and $E(L)$.

$$Y(H) = [(A \oplus D) \cdot (B + C + \bar{E})](H)$$

3.22 Use three NAND gates, one EQV gate, and one inverter (nothing else) to implement the function $G(H)$ below exactly as written. Assume the inputs all arrive from positive

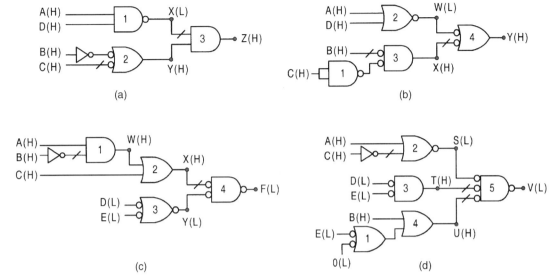

FIGURE P3.3

logic sources.

$$G(H) = [(X\bar{Y}) \oplus \bar{Z} + XYZ](H)$$

3.23 Use three NAND gates and two EQV gates (nothing else) to implement the function $F(H)$ below exactly as written. Assume the inputs arrive as $W(L)$, $X(L)$, $Y(H)$, and $Z(H)$.

$$F(H) = [(\bar{W} \oplus Y) \oplus (\bar{X}Z) + \bar{W}Y](H)$$

3.24 Unused inputs must not be left dangling. Instead, they must be tied to other inputs, or be connected to HV or LV depending on the logic operations involved. Implement the following functions with the logic indicated.
  (a) A four-input NOR gate performing the $(A\bar{B})(H)$ operation with inputs $A(L)$ and $B(H)$.
  (b) A three-input NAND gate performing the $X(H) \rightarrow X(L)$ logic level conversion operation.
  (c) An XOR gate performing the controlled $X(L) \rightarrow X(H)$ logic level conversion operation.
  (d) A four-input AND gate performing the $(\bar{A} + B)(L)$ operation with inputs $A(H)$ and $B(L)$.

3.25 Construct the truth table and the mixed logic waveforms for the functions below by using a binary input sequence in alphabetical order, all inputs active high. Use Table 2.1 in Section 2.3 if needed, and follow the format of Fig. 3.32 in constructing the waveforms. Do not take into account the propagation delays through gates and inverters.

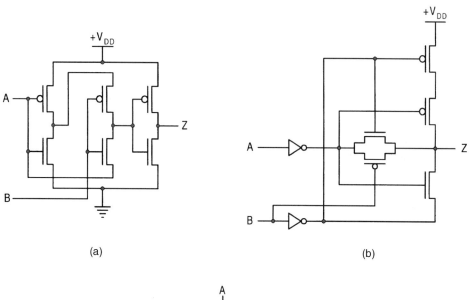

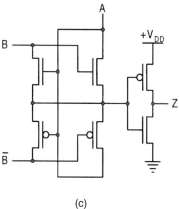

FIGURE P3.4

(a) Function $Z(H)$ in Fig. P3.3a.
(b) Function $G(H)$ in Problem 3.22.
(c) Function $F(H)$ of Problem 3.23.

3.26 Reduce the following expressions to their simplest form and name the Boolean laws used in each step. (Note: Follow the procedure of Examples 3.6 in Section 3.12.)

(a) $ab(c + \bar{b}) + a\bar{b}$
(b) $(X + \bar{Y})(XZ + \bar{Y})$ (Hint: First use the distributive law.)
(c) $\overline{A + \overline{AC} + B}$
(d) $(x + y)(x + \bar{z})[y(x + \bar{z}) + \bar{y}]$
(e) $AB + \bar{A}CD + BC + \bar{A}C$

PROBLEMS

3.27 Reduce the following expressions to their simplest form, but *do not* name the Boolean laws used in each step.

(a) $\bar{A} + AB\bar{C} + \overline{\bar{A} + C}$
(b) $(a\bar{c}d + a\bar{d})(a\bar{d} + \bar{c}d)$
(c) $(WX + \bar{Y} + \bar{W})(\bar{W}X + \bar{Y} + WX)$
(d) $(x + y)(\bar{x} + z)(y + z)$
(f) $\overline{(A + \overline{BC})(A\bar{B} + \overline{ABC})}$ (Hint: Simplify under short complementation bars first.)
(e) $\bar{a} + b + a(\overline{b + bc}) + (b + \bar{c}) \cdot ab\bar{c}d$
(f) $(\bar{A} + B + \bar{C} + D)(\bar{A} + \bar{C} + D)(\bar{A} + \bar{B} + D)$ (Hint: First use consensus.)

3.28 Reduce the following expressions to their simplest form and name the Boolean laws used in each step.

(a) $\overline{(a \oplus b + \bar{b})}(a + b)$
(b) $(XY) \oplus (X + \bar{Y})$
(c) $x \odot y \odot (xy)$
(d) $[(X + Y) \odot (\bar{X} + Z)] + X$
(e) $[(\bar{A} + B) \cdot \bar{C}] \oplus [\bar{A} + B + AC]$ (Hint: Find a way to use Corollary II.)

3.29 Reduce the following expressions to their simplest form, but *do not* name the Boolean laws used in each step.

(a) $\overline{\bar{A} + A \oplus B} + \bar{A}B$
(b) $\{\bar{S} + [S \oplus (S\bar{T})]\}(H) = [?](L)$
(c) $(X + Y) \odot (X \oplus Y)$
(d) $(a \odot b) \oplus (a\bar{b})$
(e) $(\bar{x} + \bar{y})\overline{(x \oplus y + \bar{y})}$
(f) $[1 \oplus (\bar{1} + \overline{0.1}) + 1 \odot 0](H) = [?](L)$

3.30 Use the laws of Boolean algebra, including the XOR laws, identities, and corollaries given by Eqs. (3.17) through (3.33), to prove whether the following equations are true (T) or false (F). *Do not* name the laws used.

(a) $X \odot (\bar{X} + Y) = X\bar{Y}$
(b) $ab(\overline{b + bc}) + b\bar{c} + ab\bar{c}d = bc$
(c) $\bar{A} \oplus \bar{B} \oplus (A\bar{B}) = \bar{A}B$
(d) $X \oplus (\bar{X}Y) = X + (\bar{X} \odot Y)$
(e) $[(A\bar{B})(A \odot B)](L) = \bar{A}B(H)$
(f) $AX\bar{Y} + A\bar{X}Y + \bar{A}Y = (AX) \oplus Y$ (Hint: First apply Corollary I.)

3.31 Use whatever conjugate gate forms are necessary to obtain a gate-minimum implementation of the following functions exactly as written (do not alter the functions):

(a) $F(H) = \{[A \oplus B] \cdot [(B\bar{C}) \odot D]\}(H)$ with inputs as $A(H)$, $B(H)$, $C(L)$, and $D(L)$.
(b) $K(L) = [A \oplus C \oplus (\overline{BD}) \oplus (\bar{A}\bar{B}C\bar{D})](L)$ with inputs from negative logic sources.

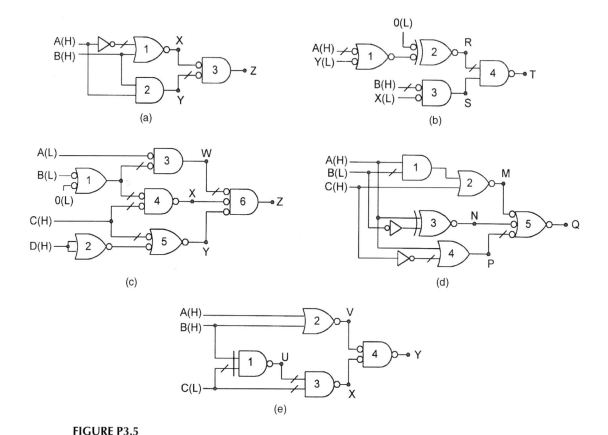

FIGURE P3.5

3.32 Use NOR/XOR/INV logic to implement the function below exactly as written by using the fewest number of gates and inverters possible, assuming the inputs A and B arrive active low and inputs X and Y arrive active high.

$$Z(H) = \{[\bar{X} \odot (\bar{A} + Y)] \cdot \bar{B}\}(H)$$

3.33 A room has two doors and a light that is controlled by three switches, A, B, and C. There is a switch beside each door and a third switch in another room. The light is turned on (LTON) any time an odd number of switches are closed (active). Find the function LTON(H) and implement it with a gate-minimum circuit. Assume that the inputs are all active high.

3.34 The logic circuits shown in Fig. P3.5 are redundant circuits, meaning that they contain more logic than is necessary to implement the output function. Identify each numbered gate and give the *simplest* mixed logic result at each node indicated. To do this, it will be necessary to use the various laws of Boolean algebra together with mixed logic notation.

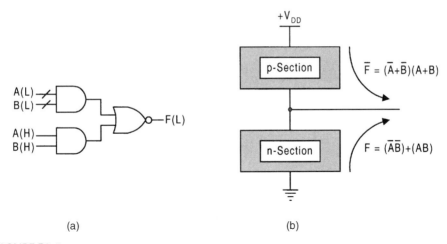

FIGURE P3.6

3.35 By following Subsection 3.10.2, write the dual forms for the functions $Y(H)$, $G(H)$, and $Z(H)$ in Problems 3.21, 3.22, and 3.32.

3.36 Use the laws of XOR algebra and identities given in Eqs. (3.17) through (3.33) to reduce the following function to its simplest (gate-minimum) form:

$$F = D \oplus B \oplus BD \oplus BCD \oplus A \oplus AD \oplus AC \oplus ACD \oplus AB.$$

3.37 The mixed logic circuit for the multiple gate realization of the XOR function $F(L) = (\bar{A}\bar{B} + AB)(L) = (A \odot B)(L) = (A \oplus B)(H)$ is shown in Fig. P3.6a, together with its CMOS organization in Fig. P3.6b. It derives from the defining relations given by Eqs. (3.4) and (3.5). Construct the CMOS circuit (excluding inverters) for this function by using the proper placement of the PMOS and NMOS as indicated in Figs. 3.5 and P3.6b. Also, construct the physical and logic truth tables for this function.

3.38 The logic circuit for the function $Y(H) = [A(B + CD)](H)$ is given in Fig. P3.7.
(a) Assuming that inputs A, B, C, and D arrive from positive logic sources, construct the CMOS circuit for the function $Y(H)$.
(b) Obtain the physical and positive logic truth table for this function.

3.39 Shown in Fig. P3.8 is a CMOS circuit having three inputs and one output.

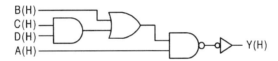

FIGURE P3.7

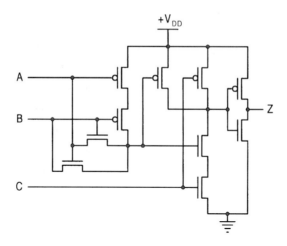

FIGURE P3.8

(a) Construct the physical truth table for this circuit taking into account all possible combinations of LV and HV inputs.
(b) If the inputs and outputs are all assumed to be active high, find the logic function for $Z(H)$ and its logic circuit.

3.40 The CMOS circuit in Fig. P3.9 is an example of a *gate-matrix layout*. The circuit has four inputs, A, B, C, and D, and one output Z. Note that X indicates an internal connection.

(a) Construct the physical truth table for this circuit taking into account all possible combinations of LV and HV inputs.
(b) If the inputs and outputs are all assumed to be active high, find the logic function for $Z(H)$ and construct the logic circuit for $Z(H)$.

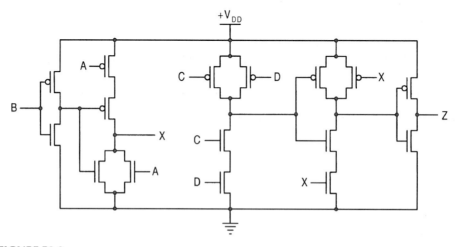

FIGURE P3.9

CHAPTER 4

Logic Function Representation and Minimization

4.1 INTRODUCTION

A given logic function can be represented in a variety of different forms, and often one of these forms proves to be the best for the application under consideration. It is the purpose of this chapter to consider the different forms of logic representation. It is also the purpose of this chapter to consider the reduction and minimization of these different forms. Knowing how to reduce or minimize a logic function is important so as to reduce design area, power consumption, and cost by eliminating unnecessary hardware. Also, the minimized function often reveals information that is not readily apparent from a nonminimized form. In short, the information in this chapter is essential to good design practices and specifically to an understanding of the remainder of this text.

4.2 SOP AND POS FORMS

Without specific mention of it, SOP and POS forms have already been used in the discussions of Chapter 3. Typical examples are the defining relations for XOR and EQV given by Eqs. (3.4) and (3.5) in Section 3.9, where each is given in both SOP and POS form. To understand what is meant by SOP and POS, the AND operation is taken as the *Boolean product* and the OR operation represents the *Boolean sum*. Thus, SOP means *sum-of-products* while POS denotes *product-of-sums*. These definitions will be used throughout the text.

4.2.1 The SOP Representation

Consider the function of three variables given by the Boolean expression

$$f(A, B, C) = \bar{A}B + BC + \underbrace{A\bar{B}\bar{C}}_{minterm} . \tag{4.1}$$

The function in Eq. (4.1) is written in *sum-of-products* (SOP) form, meaning ORing of ANDed terms also called *p-terms* (product-terms). Although there are three p-terms in this expression, only the term $A\bar{B}\bar{C}$ is called a *minterm*. A minterm is defined as follows:

> **Minterm:** *Any ANDed term containing all the variables of a function in complemented or uncomplemented form.*

Use will be made of the symbol

$$m_i = m_i(A, B, C, \ldots) \quad (4.2)$$

to represent the *i*th minterm of a function. Notice that two of the three p-terms in Eq. (4.1) cannot be minterms by this definition.

To simplify minterm representation, a shorthand notation is used and is based on the following *minterm code*:

> **MINTERM CODE**
> *Complmented variables:* logic 0
> *Uncomplented variables:* logic 1

Once the logic 0's and 1's have been assigned to all variables in a given minterm, a minterm code is established where the subscript in m_i becomes the decimal equivalent of the binary code formed by the logic state assignments. For example, the minterm in Eq. (4.1) is represented by

$$A\bar{B}\bar{C} = m_4,$$
$$100$$

since the binary of 100 has a decimal value of 4. A complete minterm code table for four variables is given in Fig. 4.1. A similar minterm code table can be constructed for any number of variables.

A function composed completely of a logical sum of minterms is said to be in *canonical SOP form*. A typical example is given by the following expressions, where use has been made of the minterm code shorthand notation and the operator symbol $\sum$ to represent the logical sum of minterms:

$$Y(A, B, C) = \underbrace{\bar{A}\bar{B}\bar{C}}_{000} + \underbrace{\bar{A}BC}_{011} + \underbrace{ABC}_{111} + \underbrace{A\bar{B}\bar{C}}_{100} + \underbrace{AB\bar{C}}_{110}$$

$$= m_0 + m_3 + m_7 + m_4 + m_6$$

$$= \sum m(0, 3, 4, 6, 7).$$

A reduced SOP function such as that in Eq. (4.1) can be expanded to canonical form by applying the factoring law and the AND and OR laws given in Section 3.10. This is

4.2 SOP AND POS FORMS

SOP Term	Binary	Decimal	m_i	SOP Term	Binary	Decimal	m_i
$\bar{A}\bar{B}\bar{C}\bar{D}$	0000	0	m_0	$A\bar{B}\bar{C}\bar{D}$	1000	8	m_8
$\bar{A}\bar{B}\bar{C}D$	0001	1	m_1	$A\bar{B}\bar{C}D$	1001	9	m_9
$\bar{A}\bar{B}C\bar{D}$	0010	2	m_2	$A\bar{B}C\bar{D}$	1010	10	m_{10}
$\bar{A}\bar{B}CD$	0011	3	m_3	$A\bar{B}CD$	1011	11	m_{11}
$\bar{A}B\bar{C}\bar{D}$	0100	4	m_4	$AB\bar{C}\bar{D}$	1100	12	m_{12}
$\bar{A}B\bar{C}D$	0101	5	m_5	$AB\bar{C}D$	1101	13	m_{13}
$\bar{A}BC\bar{D}$	0110	6	m_6	$ABC\bar{D}$	1110	14	m_{14}
$\bar{A}BCD$	0111	7	m_7	$ABCD$	1111	15	m_{15}

FIGURE 4.1
Minterm code table for four variables.

demonstrated by expanding Eq. (4.1) as follows:

$$f(A, B, C) = \bar{A}B + BC + A\bar{B}\bar{C}$$
$$= \bar{A}B(\bar{C} + C) + (\bar{A} + A)BC + A\bar{B}\bar{C}$$
$$= \bar{A}B\bar{C} + \bar{A}BC + \bar{A}BC + ABC + A\bar{B}\bar{C}$$
$$= m_2 + m_3 + m_3 + m_7 + m_4$$
$$= \sum m(2, 3, 4, 7). \qquad (4.3)$$

Note that the OR law $X + \bar{X} = 1$ has been applied twice and that the two identical minterms $\bar{A}BC$ are combined according to the OR law $X + X = X$.

The canonical truth table for Eqs. (4.3), shown in Fig. 4.2, is easily constructed from the minterm code form. However, the truth table can also be constructed directly from the original reduced form given by Eqs. (4.1). Notice that a logic 1 is placed in the f column each time an $\bar{A}B = 01$ occurs, each time a BC occurs, and for $A\bar{B}\bar{C}$. Thus, construction

A B C	f	
0 0 0	0	
0 0 1	0	
0 1 0	1	m_2
0 1 1	1	m_3
1 0 0	1	m_4
1 0 1	0	
1 1 0	0	
1 1 1	1	m_7

FIGURE 4.2
Truth table for Eq. (4.3).

of the truth table from a reduced form permits a simple means of obtaining the canonical representation without having to use the Boolean manipulation given by Eqs. (4.3).

4.2.2 The POS Representation

An alternative means of representing a logic expression is to cast it in *product-of-sums* (POS) form, meaning the ANDing of ORed terms, also called s-terms (sum-terms). An example of POS representation is given by the function

$$f(A, B, C, D) = (A + \bar{B})\underbrace{(\bar{A} + \bar{B} + C + \bar{D})}_{Maxterm}(B + \bar{C} + D) \tag{4.4}$$

where, of the three s-terms, only the term $(\bar{A} + \bar{B} + C + \bar{D})$ is called a *maxterm*.

A maxterm is defined as follows:

> **Maxterm:** Any ORed term containing all the variables of a function in complemented or uncomplemented form.

The symbol

$$M_i = M_i(A, B, C, \ldots) \tag{4.5}$$

will be used to represent the ith maxterm of a function.

Maxterm representation can be simplified considerably by using the *maxterm code*:

> **MAXTERM CODE**
> Complemented variable: logic 1
> Uncomplemented variable: logic 0

The assignment of the logic 1's and 0's in this manner to all variables in each maxterm establishes the maxterm code, where the subscript in M_i is the decimal equivalent of the binary number formed by the logic state assignments. The maxterm code table for four variables is given in Fig. 4.3. Use of this table is illustrated by maxterm in Eq. (4.4),

$$\begin{array}{cccc} \bar{A} + \bar{B} + C + \bar{D} = M_{13} \\ 1 & 1 & 0 & 1 \end{array}$$

where $1101_2 = 13_{10}$.

A comparison of the minterm and maxterm code tables in Figs. 4.1 and 4.3 indicates that

$$M_i = \bar{m}_i$$

and \hfill (4.6)

$$m_i = \bar{M}_i,$$

4.2 SOP AND POS FORMS

POS Term	Binary	Decimal	M_i	POS Term	Binary	Decimal	M_i
A+B+C+D	0000	0	M_0	$\bar{A}$+B+C+D	1000	8	M_8
A+B+C+$\bar{D}$	0001	1	M_1	$\bar{A}$+B+C+$\bar{D}$	1001	9	M_9
A+B+$\bar{C}$+D	0010	2	M_2	$\bar{A}$+B+$\bar{C}$+D	1010	10	M_{10}
A+B+$\bar{C}$+$\bar{D}$	0011	3	M_3	$\bar{A}$+B+$\bar{C}$+$\bar{D}$	1011	11	M_{11}
A+$\bar{B}$+C+D	0100	4	M_4	$\bar{A}$+$\bar{B}$+C+D	1100	12	M_{12}
A+$\bar{B}$+C+$\bar{D}$	0101	5	M_5	$\bar{A}$+$\bar{B}$+C+$\bar{D}$	1101	13	M_{13}
A+$\bar{B}$+$\bar{C}$+D	0110	6	M_6	$\bar{A}$+$\bar{B}$+$\bar{C}$+D	1110	14	M_{14}
A+$\bar{B}$+$\bar{C}$+$\bar{D}$	0111	7	M_7	$\bar{A}$+$\bar{B}$+$\bar{C}$+$\bar{D}$	1111	15	M_{15}

FIGURE 4.3
Maxterm code table for four variables.

revealing a complementary relationship between minterms and maxterms. The validity of Eqs. (4.6) is easily demonstrated by the following examples:

$$\bar{m}_5 = \overline{A\bar{B}C} = \bar{A} + B + \bar{C} = M_5$$

and

$$\bar{M}_{12} = \overline{\bar{A} + \bar{B} + C + D} = AB\bar{C}\bar{D} = m_{12},$$

where use has been made of DeMorgan's laws given by Eqs. (3.15a).

A function whose terms are *all* maxterms is said to be given in *canonical POS form* as indicated next by using maxterm code.

$$f(A, B, C) = \underbrace{(A + B + \bar{C})}_{001} \cdot \underbrace{(\bar{A} + B + \bar{C})}_{101} \cdot \underbrace{(\bar{A} + B + C)}_{100} \cdot \underbrace{(A + B + C)}_{000}$$

$$= M_1 \cdot M_5 \cdot M_4 \cdot M_0$$

$$= \prod M(0, 1, 4, 5)$$

Note that the operator symbol $\prod$ is used to denote the ANDing (Boolean product) of maxterms M_0, M_1, M_4, and M_5.

Expansion of a reduced POS function to canonical POS form can be accomplished as indicated by the following example:

$$f(A, B, C) = (A + \bar{C})(\bar{B} + \bar{C})(\bar{A} + B + C)$$

$$= (A + B\bar{B} + \bar{C})(\bar{A}A + \bar{B} + \bar{C})(\bar{A} + B + C)$$

$$= \underbrace{(A + \bar{B} + \bar{C})}_{M_3}\underbrace{(A + B + \bar{C})}_{M_1}\underbrace{(\bar{A} + \bar{B} + \bar{C})}_{M_7}\underbrace{(A + \bar{B} + \bar{C})}_{M_3}\underbrace{(\bar{A} + B + C)}_{M_4}$$

$$= \prod M(1, 3, 4, 7). \tag{4.7}$$

Here, use is made of multiple applications of the distributive, AND, and OR laws in the

A B C	f	
0 0 0	1	m_0
0 0 1	0 →	M_1
0 1 0	1	m_2
0 1 1	0 →	M_3
1 0 0	0 →	M_4
1 0 1	1	m_5
1 1 0	1	m_6
1 1 1	0 →	M_7

FIGURE 4.4
Truth table for Eqs. (4.8).

form of $(X + Y)(X + \bar{Y}) = X$. Notice that the AND law $M_3 \cdot M_3 = M_3$ is applied since this maxterm occurs twice in the canonical expression.

The results expressed by Eq. (4.7) are represented by the truth table in Fig. 4.4, where use is made of both minterm and maxterm codes. Function f values equal to logic 1 are read as minterms, while function values equal to logic 0 are read as maxterms. From this there emerges the result

$$f(A, B, C) = \sum m(0, 2, 5, 6) = \prod M(1, 3, 4, 7), \quad (4.8)$$

which shows that a given function can be represented in either canonical SOP or canonical POS form. Moreover, this shows that *if one form is known, the other is found simply by using the missing code numbers from the former.*

By applying DeMorgan's laws given by Eqs. (3.15a), it is easily shown that the complement of Eqs. (4.8) is

$$\bar{f}(A, B, C) = \prod M(0, 2, 5, 6) = \sum m(1, 3, 4, 7). \quad (4.9)$$

This follows from the result

$$\bar{f} = \overline{\sum m(0, 2, 5, 6)} = \overline{m_0 + m_2 + m_5 + m_6}$$
$$= \bar{m}_0 \cdot \bar{m}_2 \cdot \bar{m}_5 \cdot \bar{m}_6$$
$$= M_0 \cdot M_2 \cdot M_5 \cdot M_6$$
$$= \prod M(0, 2, 5, 6) = \sum m(1, 3, 4, 7)$$

A similar set of equations exist for $\bar{f} = \overline{\prod M(1, 3, 4, 7)}$. Equations (4.8) and (4.9), viewed as a set, illustrate the type of interrelationship that always exists between canonical forms.

There is more information that can be gathered from the interrelationship between canonical forms. By applying the OR law, $X + \bar{X} = 1$, and the OR form of the commutative laws to Eqs. (4.8) and (4.9), there results

$$f + \bar{f} = \sum m(0, 2, 5, 6) + \sum m(1, 3, 4, 7)$$
$$= \sum m(0, 1, 2, 3, 4, 5, 6, 7)$$
$$= 1.$$

Generally, the Boolean sum of all 2^n minterms of a function is logic 1 according to

$$\sum_{i=0}^{2^n-1} m_i = 1. \qquad (4.10)$$

Similarly, by using the AND law, $X \cdot \bar{X} = 0$, and the AND form of the commutative laws, there results

$$f \cdot \bar{f} = \prod M(1, 3, 4, 7) \cdot \prod M(0, 2, 5, 6)$$
$$= \prod M(0, 1, 2, 3, 4, 5, 6, 7)$$
$$= 0.$$

Or generally, the Boolean product of all 2^n maxterms of a function is logic 0 according to

$$\prod_{i=0}^{2^n-1} M_i = 0. \qquad (4.11)$$

Equations (4.10) and (4.11) are dual relations by the definition of duality given in Subsection 3.10.2.

To summarize, the following may be stated:

> *Any function ORed with its complement is logic 1 definite, and any function ANDed with its complement is logic 0 definite — the form of the function is irrelevant.*

4.3 INTRODUCTION TO LOGIC FUNCTION GRAPHICS

Graphical representation of logic truth tables are called *Karnaugh maps (K-maps)* after M. Karnaugh, who, in 1953, established the map method for combinational logic circuit synthesis. K-maps are important for the following reasons: (1) K-maps offer a straightforward method of identifying the minterms and maxterms inherent in relatively simple minimized or reduced functions. (2) K-maps provide the designer with a relatively effortless means of function minimization through pattern recognition for relatively simple functions. These two advantages make K-maps extremely useful in logic circuit design. However, it must be pointed out that the K-map method of minimization becomes intractable for very large complex functions. Computer assisted minimization is available for logic systems too complex for K-map use. The following is a systematic development of the K-map methods.

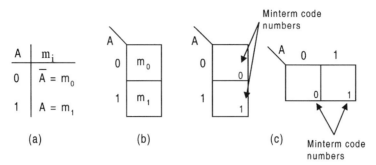

FIGURE 4.5
(a) Minterm code table for one variable and (b) its graphical equivalent. (c) Alternative formats for first order K-maps showing minterm positions.

4.3.1 First-Order K-maps

A first-order K-map is the graphical representation of a truth table of one variable and is developed from the minterm code table shown in Fig. 4.5a. The minterm positions in a first-order K-map are shown in Fig. 4.5b, leading to the alternative formats for a first-order K-map given in Fig. 4.5c. The number in the lower right-hand corner of a K-map in Fig. 4.5c indicates the position into which a minterm with that code number must be placed.

Consider the three functions given by the truth tables in Fig. 4.6a. Notice that all information contained within a given truth table is present in the corresponding K-map in Fig. 4.6b and that the functions are read as $f_1 = X$, $f_2 = \bar{X}$, and $f_3 = 1$ from either the truth tables or K-maps. Thus, a logic 1 indicates presence of a minterm and a logic 0 (the absence of a minterm) is a maxterm.

4.3.2 Second-Order K-maps

A second-order K-map is the graphical representation of a truth table for a function of two variables and is developed from the minterm code table for two variables given in Fig. 4.7a. The graphical equivalent of the minterm code table in Fig. 4.7a is given in Fig. 4.7b, where the minterm code decimal for each of four m_i is the binary equivalent of the *cell coordinates*

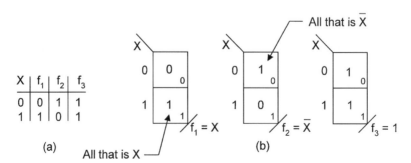

FIGURE 4.6
(a) Truth table and (b) first order K-maps for functions f_1, f_2, and f_3 of one variable X.

4.3 INTRODUCTION TO LOGIC FUNCTION GRAPHICS

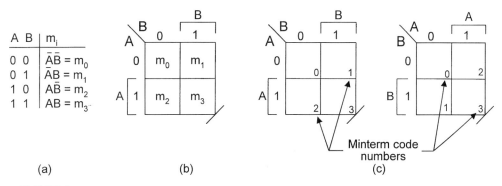

FIGURE 4.7
(a) Minterm code table for two variables and (b) its graphical equivalent. (c) Alternative formats for second-order K-maps showing minterm positions.

(read in alphabetical order AB) of the cell into which that minterm is placed. From these figures there result the two alternative K-map formats shown in Fig. 4.7c, where the number in the lower right-hand corner of each cell is the decimal equivalent of the coordinates for that cell given in binary.

As examples, functions f_1 and f_2 of two variables (X and Y) are represented by truth tables in Fig. 4.8a and by K-maps in Fig. 4.8b. Function f_1 is shown to have two minterms and two maxterms while function f_2 has three minterms and one maxterm. From the truth tables the functions can be read in SOP form as

$$f_1(X, Y) = \sum m(1, 3) = \bar{X}Y + XY = Y$$

and (4.12)

$$f_2(X, Y) = \sum m(0, 2, 3) = \bar{X}\bar{Y} + X\bar{Y} + XY = X + \bar{Y}.$$

However, by combining ("looping out") adjacent minterms these results are immediately obvious as indicated in Fig. 4.8b.

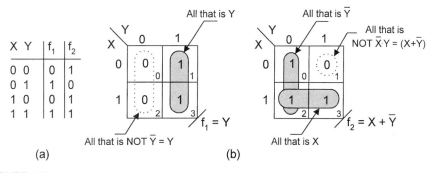

FIGURE 4.8
(a) Truth tables for functions f_1 and f_2. (b) K-maps for functions f_1 and f_2, showing minimum SOP cover (shaded) and POS cover (dashed loops).

The same results could have been obtained by reading the maxterms in the K-maps of Fig. 4.8b. Thus, in maxterm code the canonical and minimum POS forms become

$$f_1 = \prod M(0, 2) = (X + Y)(\bar{X} + Y) = Y$$

and   (4.13)

$$f_2 = \prod M(1) = (X + \bar{Y}),$$

where for f_1 the distributive law in Eqs. (3.12) has been applied. Another way of looking at the results given by Eqs. (4.13) is to read groups of adjacent maxterms. For f_1 the two adjacent maxterms (0's) are read as "all that is NOT $\bar{Y}$" or $\bar{\bar{Y}} = Y$. Similarly, for function f_2 the 0 in cell 1 is read as "all that is NOT $\bar{X}Y$" or simply $\overline{\bar{X}Y} = X + \bar{Y}$.

4.3.3 Third-Order K-maps

In dealing with functions of three variables, a suitable graphical representation and K-map format must be decided. One choice would be to use a three-dimensional graphic having one axis for each variable. However, such a graphical representation would be difficult to construct and equally difficult to read. A much better choice would be to maintain the domain concept in two dimensions. To do this requires the use of two variables for one axis. Shown in Fig. 4.9a is the graphical representation for the minterm code table of three variables as deduced from Fig. 4.1. Again, the minterm positions are those given by the coordinates of the cells read in alphabetical order XYZ. From this there results the two alternative formats for a third-order K-map given in Fig. 4.9b, where the minterm code numbers in decimal are shown in the lower right-hand corners of the cells.

Notice that the two-variable axes in the third-order K-maps of Fig. 4.9b are laid out in 2-bit *Gray code*, a unit distant code featured in Subsection 2.10.3. This is important so that each cell along the two-variable axis is surrounded by logically adjacent cells. The result is that the Y and Z domains in Fig. 4.9 are maintained intact. Notice that in Fig. 4.9a the logic adjacency along the two-variable axis is continuous as though the K-map were formed into a cylinder about the X axis (orthogonal to the YZ axis). Had the YZ axis been laid out in binary, the Z domain would be split into two separate sections, making map plotting and reading difficult. For this reason all axes of two or more variables are laid out in Gray code so as to maximize axis coherency.

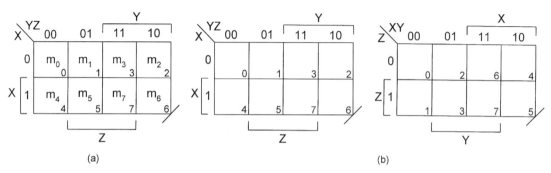

FIGURE 4.9
(a) Minterm positions and (b) alternative formats for third-order K-maps.

4.3 INTRODUCTION TO LOGIC FUNCTION GRAPHICS   141

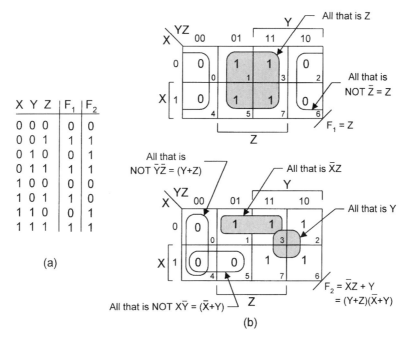

FIGURE 4.10
Truth tables for functions F_1 and F_2. (b) K-map representations for functions F_1 and F_2 showing minimum SOP cover (shaded) and minimum POS cover (unshaded).

To illustrate the application of third-order K-maps, two simple functions are presented in Fig. 4.10. Here, the truth tables for functions F_1 and F_2 are presented in Fig. 4.10a and their K-map representations together with minimum cover are given in Fig. 4.10b. Notice that the 1's and 0's are placed at the proper coordinates within the K-maps, in agreement with the truth table. From the K-maps, the canonical and minimum SOP forms for functions F_1 and F_2 are read as

$$F_1(X, Y, Z) = \sum m(1, 3, 5, 7) = \bar{X}\bar{Y}Z + \bar{X}YZ + X\bar{Y}Z + XYZ$$
$$= Z$$
$$F_2(X, Y, Z) = \sum m(1, 2, 3, 6, 7) = \bar{X}\bar{Y}Z + \bar{X}Y\bar{Z} + \bar{X}YZ + XY\bar{Z} + XYZ$$
$$= \bar{X}Z + Y.$$
(4.14)

By grouping minterms in Fig. 4.10b, the minimum SOP expressions, $F_1 = Z$ and $F_2 = \bar{X}Z + Y$, become immediately apparent.

The 0's in the K-maps of Fig. 4.10b can be given in canonical and minimum POS forms:

$$F_1(X, Y, Z) = \prod M(0, 2, 4, 6)$$
$$= (X + Y + Z)(X + \bar{Y} + Z)(\bar{X} + Y + Z)(\bar{X} + \bar{Y} + Z)$$
$$= Z$$
(4.15)
$$F_2(X, Y, Z) = \prod M(0, 4, 5) = (X + Y + Z)(\bar{X} + Y + Z)(\bar{X} + Y + \bar{Z})$$
$$= (Y + Z)(\bar{X} + Y),$$

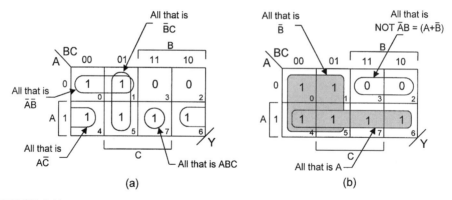

FIGURE 4.11
(a) K-map for the reduced function Y of Eq. (4.16). (b) K-map showing minimum SOP and POS cover for function Y.

as read in maxterm code. The minimum POS results are easily read from the K-maps of Fig. 4.10b by combining adjacent 0's as indicated. Thus, F_1 is read as "all that is NOT $\bar{\bar{Z}}$" or $\bar{\bar{Z}} = Z$. Similarly, F_2 can be read as "all that is NOT $\bar{Y}\bar{Z} + X\bar{Y}$" or $\overline{\bar{Y}\bar{Z} + X\bar{Y}} = (Y + Z)(\bar{X}+Y)$. Notice that the distributive law in Eqs. (3.12) is applied as $(Y+Z)(\bar{X}+Y) = Y+\bar{X}Z$, demonstrating that the SOP and POS forms for F_2 are algebraically equal, as they must be. The minimum POS results given by Eqs. (4.15) can also be obtained by applying the Boolean laws given in Section 3.10, but with somewhat more effort. For example, F_1 is minimized to give the result Z after three applications of the distributive law in Eqs. (3.12) together with the AND and OR laws.

The use of third-order K-maps is further illustrated by placing the reduced SOP function

$$Y = ABC + A\bar{C} + \bar{B}C + \bar{A}\bar{B} \tag{4.16}$$

into the third-order K-map in Fig. 4.11a. Then by grouping adjacent minterms (shaded loops) as in Fig. 4.11b, there results the minimum expression for Eq. (4.16),

$$Y = A + \bar{B}. \tag{4.17}$$

As expected, the same results could have been obtained by grouping the adjacent maxterms (0's) in Fig. 4.11b, which is equivalent to saying "all that is NOT $\bar{A}B$" or $\overline{\bar{A}B} = A + \bar{B}$.

Other information may be gleaned from Fig. 4.11. Extracting canonical information is as easy as reading the minterm code numbers in the lower right-hand corner of each cell. Thus, the canonical SOP and canonical POS forms for function Y are given by

$$Y = \sum m(0, 1, 4, 5, 6, 7)$$
$$= \bar{A}\bar{B}\bar{C} + \bar{A}\bar{B}C + A\bar{B}\bar{C} + A\bar{B}C + AB\bar{C} + ABC$$

or (4.18)

$$Y = \prod M(2, 3)$$
$$= (A + \bar{B} + C)(A + \bar{B} + \bar{C})$$

as read in minterm code and maxterm code, respectively.

4.3 INTRODUCTION TO LOGIC FUNCTION GRAPHICS   143

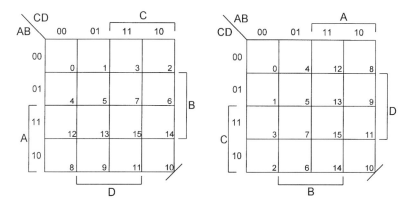

FIGURE 4.12
Alternative formats for fourth-order K-maps.

4.3.4 Fourth-Order K-maps

At this point it is expected that the reader is familiar with the formats for first-, second-, and third-order K-maps. Following the same development, two alternative formats for fourth-order K-maps are presented in Fig. 4.12, where use of the minterm code table in Fig. 4.1 is implied and where A is the MSB and D is the LSB. Here, both two-variable axes have logic coordinates that are unfolded in Gray code order so that all juxtaposed minterms (those separated by any single domain boundary) are logically adjacent. Notice that each cell in the K-maps of Fig. 4.12 has a number assigned to it that is the decimal equivalent of the binary coordinates for that cell (read in the order ABCD), and that each cell has four other cells that are logically adjacent to it. For example, cell 5 has cells 1, 4, 7, and 13 logically adjacent to it.

Just as a third-order K-map forms an imaginary cylinder about its single variable axis, a fourth-order K-map whose axes are laid out in Gray code will form an imaginary toroid (doughnut-shaped figure), the result of trying to form two cylinders about perpendicular axes. Thus, cells (0, 8) and (8, 10) and (1, 9) are examples of logically adjacent pairs, while cells (0, 2, 8, 10) and (0, 1, 4, 5) and (3, 7, 11, 15) are examples of logically adjacent groups of four.

To illustrate the application of fourth-order K-maps, consider the reduced SOP function

$$F(A, B, C, D) = \bar{A}C\bar{D} + \bar{C}D + A\bar{B}C\bar{D} + \bar{B}\bar{C}\bar{D} + \bar{A}BCD \qquad (4.19)$$

and its K-map representation in Fig. 4.13a. By grouping logically adjacent minterms as in Fig. 4.13b, a minimum SOP result is found to be

$$F_{SOP} = \bar{A}BC + \bar{C}D + \bar{B}\bar{D}. \qquad (4.20)$$

Notice that the original function in Eq. (4.19) requires six gates, whereas the minimum result in Eq. (4.20) requires only four gates. In both cases the gate count includes the final ORing operation of the p-terms. The minimum POS cover for function F is obtained by grouping the logically adjacent 0's as in Fig. 4.13c, giving

$$F_{POS} = (\bar{B} + C + D)(\bar{A} + \bar{B} + \bar{C})(B + \bar{C} + \bar{D}), \qquad (4.21)$$

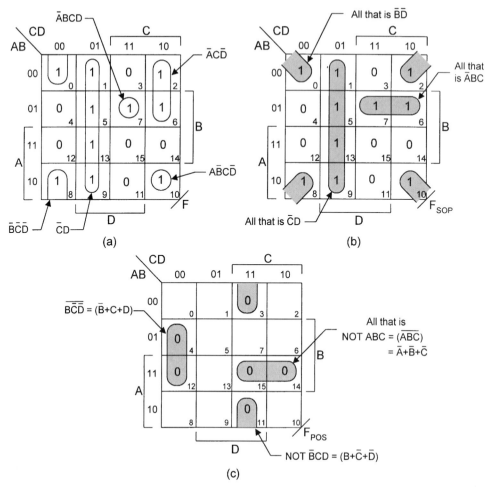

FIGURE 4.13
(a) K-map for the reduced SOP function in Eq. (4.19). (b) K-map showing minimum SOP cover for the function F. (c) K-map showing minimum POS cover for the function F.

which is saying "all that is NOT ($B\bar{C}\bar{D} + ABC + \bar{B}CD$" as indicated in Fig. 4.13c. The gate tally for Eq. (4.21) is four, including the final ANDing of s-terms, which is less than the original function in Eq. (4.19). Canonical minterm and maxterm representations are easily determined by reading the 1's and 0's in the K-maps of Fig. 4.13 to give

$$F = \sum m(0, 1, 2, 5, 6, 7, 8, 9, 10, 13)$$
$$= \prod M(3, 4, 11, 12, 14, 15). \qquad (4.22)$$

4.4 KARNAUGH MAP FUNCTION MINIMIZATION

Use of the K-map offers a simple and reliable method of minimizing (or at least greatly reducing) logic expressions. In fact, this is the most important application of K-maps. In

4.4 KARNAUGH MAP FUNCTION MINIMIZATION

Section 4.3, simple examples serve to demonstrate how K-maps can be used to extract both canonical and minimum SOP and POS forms depending on whether 1's or 0's are read. Now it is necessary to present certain important information that was not explicitly stated earlier, but was implied.

It should be clear from Section 4.3 that each line or edge of a K-map forms the boundary between two complementary domains. As a result, minterms or maxterms that are separated by a line or edge are logically adjacent and can be combined to form a reduced function. The following rule generalizes this point:

Reduction Rule

Each variable domain boundary crossed in an adjacent group (looping) requires the absence of that variable in the reduced term.

Thus, a pair of logically adjacent minterms or maxterms crosses one domain boundary and eliminates the domain variable in the reduced term; a grouping of four logical adjacencies crosses two domain boundaries and eliminates the two domain variables in the reduced function. In this way 2^n logic adjacencies ($n = 1, 2, 3, \ldots$) can be extracted (looped out) to produce a reduced $(N - n)$-variable term of an N-variable function.

To help ensure a minimized result from K-map extraction, thereby avoiding possible costly redundancies, the following loop-out protocol is recommended but not required:

Loop-out Protocol

Loop out the largest 2^n group of logically adjacent minterms or maxterms in the order of increasing $n = 0, 1, 2, 3, \ldots$.

When following this protocol, single isolated minterms or maxterms (*monads*), if present, should be looped out first. This should be followed by looping out groups of any two logically adjacent minterms or maxterms (*dyads or duads*) that cannot be looped out in any other way. The process continues with groups of four logic adjacencies (*quads*), then groups of eight (*octads*), etc. — always in groups of 2^n logic adjacencies.

As an example of the pitfalls that can result from failure to follow the loop-out protocol, consider the function represented in the K-map of Fig. 4.14. Instinctively, one may be tempted to loop out the quad (dashed loop) because it is so conspicuous. However, to do so creates a redundancy, since all minterms of that grouping are covered by the four dyads shown by the shaded loops.

Since K-maps are minterm-code based, minimum POS cover can be extracted directly, avoiding the "NOT" step indicated in Figs. 4.8, 4.10, 4.11, and 4.13, by using the following procedure:

Simplified POS Extraction Procedure

Take the *union* (ORing) of the complemented domains in which the 2^n groups of logically adjacent maxterms exist.

Groups of minterms or maxterms other than 2^n groups (e.g., groups of three, five, six, and seven) are forbidden since such groups are *not* continuously adjacent. Examples of such

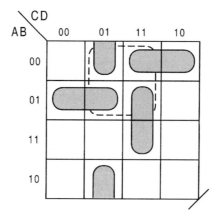

FIGURE 4.14
Minimum cover (shaded) by using the loop-out protocol avoids the redundant quad (dashed loop).

forbidden groups are presented in Fig. 4.15, which has been "crossed out" to indicate that such groupings are not allowed.

4.4.1 Examples of Function Minimization

Just as canonical forms can be read from a K-map in two ways (SOP and POS), so also can a function be read from a K-map in either minimum SOP form or minimum POS form. To illustrate, consider the function

$$G(A, X, Y) = \sum m(0, 3, 5, 7), \qquad (4.23)$$

which is mapped and minimized in Fig. 4.16. Noting that the 1's are looped out as two dyads and a monad as are the 0's, there results

$$G_{SOP} = \bar{A}\bar{X}\bar{Y} + XY + AY \qquad \text{(minimum SOP cover)}$$

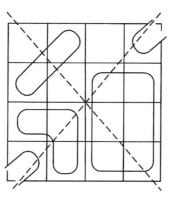

FIGURE 4.15
Examples of forbidden (non-2^n) groupings of minterms and maxterms.

4.4 KARNAUGH MAP FUNCTION MINIMIZATION

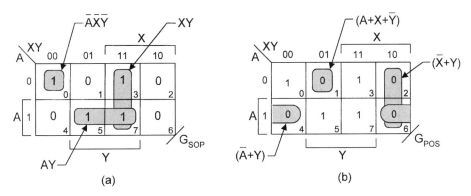

FIGURE 4.16
K-maps for Eq. (4.23). (a) Minimum SOP cover. (b) Minimum POS cover.

and

$$G_{POS} = (A + X + \bar{Y})(\bar{X} + Y)(\bar{A} + Y) \quad \text{(minimum POS cover)}.$$

Application of the laws of Boolean algebra shows that the SOP and POS minima are algebraically equal:

$$(A + X + \bar{Y})(\bar{X} + Y)(\bar{A} + Y) = (A + X + \bar{Y})(\bar{A}\bar{X} + \bar{X}Y + \bar{A}Y + Y)$$
$$= (A + X + \bar{Y})(\bar{A}\bar{X} + Y)$$
$$= \bar{A}\bar{X}\bar{Y} + XY + AY.$$

As a second example, consider the reduced function given in POS form:

$$Y(A, B, C, D) = (A + \bar{B} + \bar{D})(B + \bar{C})(A + \bar{C} + D)(\bar{A} + \bar{B} + C + \bar{D}). \quad (4.24)$$

To map this function, one simply maps the 0's in maxterm code, as indicated in Fig. 4.17a. The representation in Fig. 4.17a is not minimum. However, after the maxterms are regrouped, a minimum POS representation is shown in Fig. 4.17b. Notice that the dyad $M(5, 13)$ crosses the A boundary, permitting $(A\bar{A} + \bar{B} + C + \bar{D}) = (\bar{B} + C + \bar{D})$ as the reduced s-term. Similarly, the quad $M(2, 3, 6, 7)$ crosses the B and D boundaries to yield $(A + B\bar{B} + \bar{C} + D\bar{D}) = (A + \bar{C})$. Also, the quad $M(2, 3, 10, 11)$ crosses the A and D boundaries, eliminating these variables to give $(B + \bar{C})$ as the reduced s-term.

The minimum SOP cover for the function Y of Eq. (4.24) is shown in Fig. 4.17c and consists of one dyad and two quads. The dyad $m(14, 15)$ crosses the D boundary, permitting $ABC(D + \bar{D}) = ABC$, while the quad $m(0, 4, 8, 12)$ crosses the A and B boundaries, yielding $(A + \bar{A})(B + \bar{B})\bar{C}\bar{D} = \bar{C}\bar{D}$. Likewise, the quad $m(0, 1, 8, 9)$ crosses the A and D boundaries to give $\bar{B}\bar{C}$ as the reduced p-term. The minimized results that are extracted from Figs. 4.17b and 4.17c are now presented as

$$Y_{POS} = (\bar{B} + C + \bar{D})(A + \bar{C})(B + \bar{C}) \quad \text{Minimum POS cover}$$

and

$$Y_{SOP} = ABC + \bar{C}\bar{D} + \bar{B}\bar{C}, \quad \text{Minimum SOP cover}$$

148 CHAPTER 4 / LOGIC FUNCTION REPRESENTATION AND MINIMIZATION

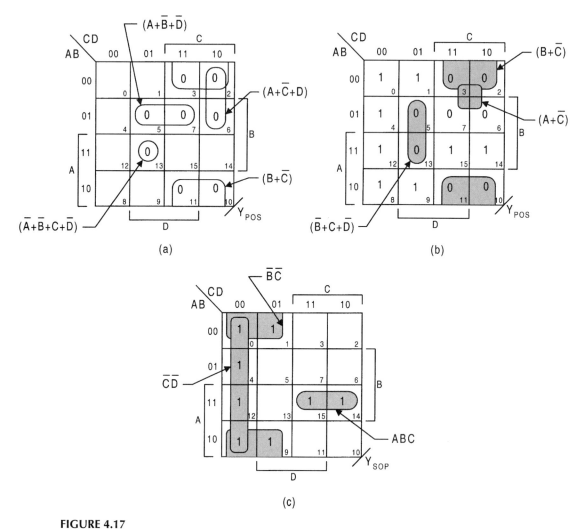

FIGURE 4.17
(a) K-map representing the cover as given by Eq. (4.24). (b) Minimum POS cover. (c) Minimum SOP cover.

which are shown to be algebraically equal if one carries out the required algebraic manipulations.

4.4.2 Prime Implicants

Groups of 2^n minterms or maxterms that cannot be combined with other 2^n groups in any way to produce terms of fewer variables are called *prime implicants* (PIs). The loop-out protocol described in the previous section offers a procedure for achieving minimum cover by systematically extracting PIs in the order of increasing $n(n = 0, 1, 2, 3\ldots)$. But the task of achieving minimum cover following the loop-out protocol (or any procedure for that matter) is not quite as straightforward as one might believe. Difficulties can arise when optional and redundant groupings of adjacent minterms or maxterms are present. To deal

4.4 KARNAUGH MAP FUNCTION MINIMIZATION

with these problems, it will be helpful to identify the following three subsets of PIs:

- *Essential Prime Implicants* (EPIs): Single-way PIs that must be used to achieve minimum cover
- *Optional Prime Implicants* (OPIs): Optional-way PIs that are used for alternative minimum cover
- *Redundant Prime Implicants* (RPIs): Superfluous PIs that cannot be used if minimum cover is to result.

Any grouping of 2^n adjacencies is an *implicant*, including a single minterm or maxterm, but it may not be a PI. For example, a solitary quad EPI contains eight RPIs, four monads, and four dyads, none of which are PIs.

To illustrate a simple mapping problem with optional coverage, consider the function

$$Z(A, B, C, D) = \sum m(2, 4, 6, 8, 9, 10, 11, 15), \quad (4.25)$$

which is mapped in Fig. 4.18a. Noting first the minterm adjacencies that form the three dyads (no monads exist) and the single quad, there results the SOP minimum expression

$$Z_{SOP} = ACD + \bar{A}B\bar{D} + \begin{Bmatrix} \bar{A}C\bar{D} \\ \bar{B}C\bar{D} \end{Bmatrix} + A\bar{B}, \quad (4.26)$$

which has three EPI p-terms (two dyads and one quad), and two OPI dyads indicated in braces. The minterm m_2 can be covered in two ways to form the OPI dyads $m(2, 6)$ and $m(2, 10)$ shown with dashed loops in Fig. 4.18a. Remember that when one OPI is selected

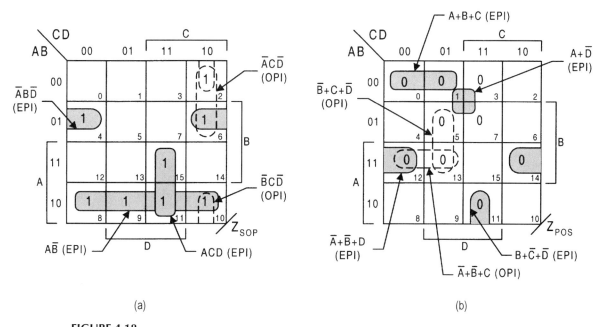

(a)   (b)

FIGURE 4.18
K-maps showing EPIs and OPIs for the function Z in Eq. (4.25). (a) SOP minimum cover. (b) POS minimum cover.

to be an EPI, the other OPI becomes redundant (an RPI). In similar fashion a minimum POS cover is extracted as shown in Fig. 4.18b, giving the result

$$Z_{POS} = (A + B + C)(B + \bar{C} + \bar{D})(\bar{A} + \bar{B} + D) \begin{Bmatrix} (\bar{A} + \bar{B} + C) \\ (\bar{B} + C + \bar{D}) \end{Bmatrix} (A + \bar{D}), \quad (4.27)$$

which is seen to have four dyads (including one of two OPIs) and one quad. The maxterm M_{13} can be looped out in two ways (dashed loops in Fig. 4.18b) to form the OPI dyads $M(12, 13)$ and $M(5, 13)$ represented by the bracketed s-terms in Eq. (4.27).

4.4.3 Incompletely Specified Functions: Don't Cares

In the design of logic circuits *nonessential* minterms or maxterms may be introduced so as to simplify the circuit. Such nonessential minterms or maxterms are called *don't cares* and are represented by the symbol

$$\phi = Min/Max = don't\ care.$$

Thus, the don't care can be taken as logic 0 or logic 1, take your choice. The symbol ϕ can be thought of as a logic 0 with a logic 1 superimposed on it.

Don't cares can arise under the following two conditions:

- When certain combinations of input logic variables can never occur, the output functions for such combinations are nonessential and are assigned don't cares.
- When all combinations of input logic variables occur but certain combinations of these variables are irrelevant, the output functions for such combinations are assigned don't cares.

As an example of the second condition, the BCD number system discussed in Subsection 2.4.1 has 10 4-bit binary patterns for decimal integers 0 through 9. Thus, there are six 4-bit patterns, representing decimal integers 10 through 15 that are never used — that is, we "don't care" about them. Accordingly, the don't care symbol ϕ can be assigned to any output generated by one of the six nonessential 4-bit patterns. This will be demonstrated in Subsection 6.5.2 for conversion from BCD to XS3 decimal codes.

Consider the three-variable function

$$f(A, B, C) = \underbrace{\sum m(1, 3, 4, 7)}_{\substack{\text{Essential} \\ \text{minterms}}} + \underbrace{\phi(2, 5)}_{\substack{\text{Nonessential} \\ \text{minterms} \\ (don't\ cares)}} \quad (4.28)$$

written in canonical SOP form showing essential minterms and nonessential minterms (don't cares). The K-maps representing minimum SOP and POS cover are shown in Figs. 4.19a and 4.19b, giving the results

$$F_{SOP} = A\bar{B} + C$$
$$F_{POS} = (A + C)(\bar{B} + C). \quad (4.29)$$

4.4 KARNAUGH MAP FUNCTION MINIMIZATION

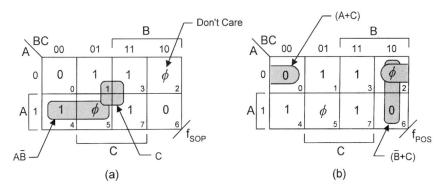

FIGURE 4.19
K-maps for Eq. (4.28) showing EPIs containing don't cares. (a) Minimum SOP cover. (b) Minimum POS cover.

Notice that the don't cares ϕ_2 and ϕ_5 are purposely used differently to obtain the minimum SOP and POS expressions of Eqs. (4.29). The result is that the F_{SOP} and F_{POS} expressions are algebraically equal since there is no *shared use* of don't cares between the two functions ($\phi_5 = 1$ and $\phi_2 = 0$ in both cases). Thus, F_{SOP} can be produced by algebraically manipulating F_{POS}. Had no use been made of the two don't cares in Fig. 4.19, the results would be quite different, namely $F_{SOP} = A\bar{B}\bar{C} + \bar{A}C + BC$ and $F_{POS} = (A + B + C)(\bar{A} + \bar{B} + C)$, which are logically equivalent but not algebraically equal.

As a second example consider the four-variable function given in canonical POS form showing essential and nonessential maxterms:

$$Y(A, B, C, D) = \underbrace{\prod M(0, 1, 2, 4, 6, 9, 11, 15)}_{\substack{\text{Essential} \\ \text{maxterms}}} \cdot \underbrace{\phi(3, 8, 10, 12)}_{\substack{\text{Nonessential} \\ \text{maxterms} \\ \text{(don't cares)}}}. \quad (4.30)$$

In Fig. 4.20 the 0's and ϕ's of Eq. (4.30) are mapped in maxterm code, and the minimum covers for Y_{POS} and Y_{SOP} are shown by the shaded loops in Figs. 4.20a and 4.20b, respectively. The resulting minimum POS and SOP expressions for Eq. (4.30) are

$$\begin{aligned} Y_{POS} &= (\bar{A} + \bar{C} + \bar{D})(A + D)B \\ Y_{SOP} &= \bar{A}BD + B\bar{C}D + A\bar{D} \end{aligned} \quad (4.31)$$

Again it is noted that these expressions are logically equivalent. However, they are algebraically unequal because of the shared use of don't cares (ϕ_8 and ϕ_{10}) in the loop-out process. Notice also that Y_{SOP} contains OPIs $B\bar{C}D$ and $AB\bar{C}$ with $\bar{A}BD$ as an EPI, since minterm m_{13} can be looped out in two ways (with m_5 or with ϕ_{12}). Similarly, OPIs $\bar{A}BD$ and $\bar{A}CD$ result if $B\bar{C}D$ is an EPI, since minterm m_7 can be looped out in two ways (with m_5 and with ϕ_3). No OPIs exist for Y_{POS}.

The Gate/Input Tally vs Cardinality of a Function Throughout this text use will be made of the ratio of the gate tally to the input tally (*gate/input tally*) as a measure of function complexity in terms of hardware cost. Input tallies include both external and internal inputs

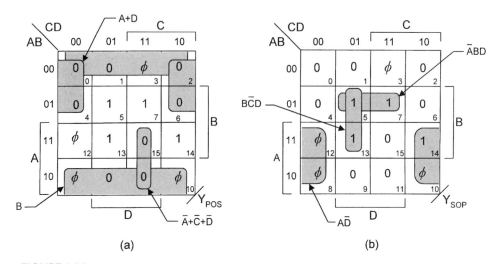

FIGURE 4.20
K-maps for Eq. (4.30) containing don't cares showing (a) minimum POS cover and (b) minimum SOP cover containing OPIs for minterms in cells 7 and 13 but not shown.

(connections) to gates. Gate tallies are weighted more heavily than input tallies. Inverters can be included in the gate/input tally of a given function only if the activation levels of the inputs are known. Unless otherwise stated, the gate/input tallies will be given exclusive of inverters. An inverter is considered to be a gate with one input.

The number of p-terms or s-terms representing a function is called the *cardinality* of the function. Thus, the cardinality of a cover represents the number of prime implicants of the function, and a minimum cardinality (or cover) consists only of EPIs. When significant numbers of don't cares are present in a function, there may exist several alternative covers of minimum cardinality that may differ in gate/input tally.

As an example of the use of the gate/input tally and cardinality, consider the minimized expressions in Eqs. (4.29). Here, F_{SOP} has a gate/input tally of 2/4, whereas the gate/input tally for F_{POS} is 3/6, both exclusive of inverters and both with a minimum cardinality of 2. Thus, the SOP expression is the simpler of the two. However, this may not always be true. Taking a gate and input count of Eqs. (4.31) reveals that the gate/input tally for Y_{POS} is 3/8 while that for Y_{SOP} is 4/11, again both exclusive of possible inverters. Thus, in this case, the POS expression is the simpler hardware-wise, but both expressions have a minimum cardinality of 3. Notice that a single variable EPI contributes to the cardinality count of the function but not to the gate tally.

4.5 MULTIPLE OUTPUT OPTIMIZATION

Frequently, logic system design problems require optimization of multiple output functions, all of which are functions of the same input variables. For complex systems this is generally regarded as a tedious task to accomplish without the aid of a computer, and for this reason computer programs have been written to obtain the optimum cover for multioutput functions of many variables. Examples of such computer programs are discussed in Appendix B.1. In this section a simple approach to this process will be presented but limited to two or

4.5 MULTIPLE OUTPUT OPTIMIZATION

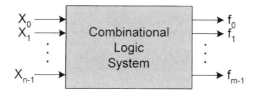

FIGURE 4.21
Block diagram for an *n*-input/*m*-output combinational logic system.

three outputs, each output being limited to four variables or less. Beyond these limitations, computer-aided optimization is recommended.

Consider the *n*-input/*m*-output system illustrated by the block diagram in Fig. 4.21. Suppose the object is to minimize each of the *m* output functions in such a way as to make use of as many of the shared terms between them as possible, thus optimizing the combinational logic of this system. The recommended procedure is given in four steps that follow.

Multiple-Output Minimization Procedure

Step 1. Obtain the canonical SOP or POS forms. If necessary, K-maps can be used for this purpose.

Step 2. AND the canonical SOP forms or OR the canonical POS forms in some systematic way (for example, $f_1 \cdot f_2, f_2 \cdot f_3, f_3 \cdot f_4, \ldots$, or $f_1 + f_2, f_2 + f_3, f_3 + f_4, \ldots$) and map each ANDed or ORed expression separately, looping out all shared PIs (common terms).

Minterm ANDing rules:

$$\begin{aligned} m_i \cdot m_i &= m_i \\ m_i \cdot m_j &= 0 \quad (i \neq j) \\ m_i \cdot \phi_i &= m_i \\ \phi_i \cdot \phi_i &= \phi_i \\ m_i \cdot \phi_j &= \phi_i \cdot \phi_j = 0 \quad (i \neq j) \end{aligned} \qquad (4.32)$$

Maxterm ORing rules:

$$\begin{aligned} M_i + M_i &= M_i \\ M_i + M_j &= 1 \quad (i \neq j) \\ M_i + \phi_i &= M_i \\ \phi_i + \phi_i &= \phi_i \\ M_i + \phi_j &= \phi_i + \phi_j = 1 \quad (i \neq j) \end{aligned} \qquad (4.33)$$

Step 3. Make a table of the results of step 2 giving all shared PIs in literal form.

Step 4. From K-maps of the original functions, loop out the shared PIs given in step 3, then loop out the remaining EPIs following the loop-out protocol with

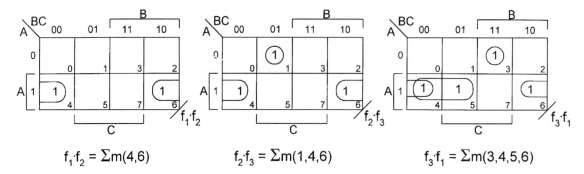

$f_1 \cdot f_2 = \Sigma m(4,6)$ $f_2 \cdot f_3 = \Sigma m(1,4,6)$ $f_3 \cdot f_1 = \Sigma m(3,4,5,6)$

FIGURE 4.22
ANDed functions, their K-maps, and minimum cover for Eqs. (4.34).

one exception. If the adjacencies of the shared PIs are part of a larger 2^n grouping of adjacencies, use the larger grouping, but only if it leads to simpler form.

For simple systems the four-step procedure just given can be shortened considerably by simply comparing the K-maps of the functions. Often the adjacency patterns lead to an immediate recognition of the shared PIs that should be included for optimum cover.

To illustrate the four-step procedure given previously, consider the system of three outputs, each a function of three variables:

$$\begin{cases} f_1(A, B, C) = \sum m(0, 3, 4, 5, 6) \\ f_2(A, B, C) = \sum m(1, 2, 4, 6, 7) \\ f_3(A, B, C) = \sum m(1, 3, 4, 5, 6) \end{cases}. \quad (4.34)$$

Equations (4.34) satisfy step 1 of the multiple-output minimization procedure. Then, moving on to step 2, the ANDed functions are given in Fig. 4.22, together with their respective K-maps and minimum cover. The minimum cover in each ANDed K-map indicates the common terms that must be included in the optimized expressions for the three-output system.

The table of shared PIs for each of the ANDed forms and the appropriate transfer of these shared PIs into the K-maps of the original functions are given in Fig. 4.23, in agreement with steps 3 and 4 of the multiple-output minimization procedure. Notice that the dyad $A\bar{C}$ is common to all three ANDed functions, as is evident from the ANDed function $f_1 \cdot f_2 \cdot f_3 = m(4, 6)$ indicated in the table of shared PIs of Fig. 4.23.

By looping out the shared PIs first in Fig. 4.23 followed by the remaining EPIs, there result the optimal expressions

$$\begin{cases} f_1 = \bar{A}BC + A\bar{C} + A\bar{B} + \bar{B}\bar{C} \\ f_2 = \bar{A}\bar{B}C + A\bar{C} + AB + B\bar{C} \\ f_3 = \bar{A}BC + \bar{A}\bar{B}C + A\bar{C} + A\bar{B} \end{cases}. \quad (4.35)$$

Notice that the dyad $m(1, 3)$ in the f_3 K-map is avoided, hence also an individual minimum for f_3, so that the expression for f_3 can be completely generated from the terms in f_1 and f_2, the optimal solution. The optimum gate/input tally is 10/28 for this system of three outputs, each output having a cardinality of 4.

4.5 MULTIPLE OUTPUT OPTIMIZATION

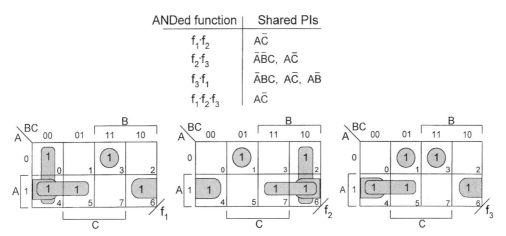

FIGURE 4.23
Table of shared PIs and the K-maps for the functions f_1, f_2, and f_3, showing optimal cover for the three-output system of Eqs. (4.34).

As a second example, consider the output functions for the four-input/two-output logic system represented by Eqs. (4.36) and by the block diagram in Fig. 4.24. It is required that this system be optimized with respect to both POS cover and SOP cover following the four-step multiple output minimization procedure and that the results be compared to determine which, if either, is the more optimum. The optimized system is to be implemented with either NOR/INV logic or NAND/INV logic.

$$f_1(A, B, C, D) = \prod M(1, 2, 3, 4, 5, 9, 10) \cdot \phi(6, 11, 13)$$
$$= \sum m(0, 7, 8, 12, 14, 15) + \phi(6, 11, 13)$$
$$f_2(A, B, C, D) = \prod M(2, 5, 9, 10, 11, 15) \cdot \phi(3, 4, 13, 14)$$
$$= \sum m(0, 1, 6, 7, 8, 12) + \phi(3, 4, 13, 14)$$

(4.36)

Optimized POS Cover. ORing of the canonical forms of Eqs. (4.36) yields

$$f_1 + f_2 = \prod M(2, 3, 4, 5, 9, 10, 11) \cdot \phi(13),$$

where use has been made of the ORing rules given by Eqs. (4.33) at the beginning of this

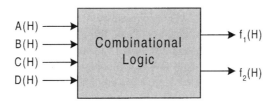

FIGURE 4.24
Block diagram for a four-input/two-output combinational logic system represented by Eqs. (4.36).

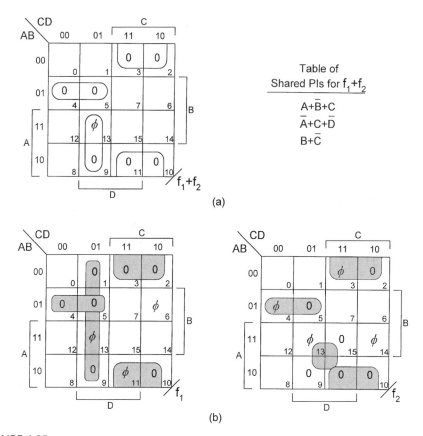

FIGURE 4.25

Multioutput POS optimization for the system represented by Eqs. (4.36) and Fig. (4.24). (a) K-map and shared PIs for $(f_1 + f_2)$. (b) K-maps showing optimal POS cover for functions f_1 and f_2.

section. Figure 4.25a gives the K-map and table of shared s-term PIs for $f_1 + f_2$. The final step involves looping out the individual functions by following the loop-out protocol in such a manner as to incorporate as many shared PIs as necessary to achieve optimum cover for the two outputs. Reading the K-maps in Fig. 4.25b for f_1 and f_2 produces the results

$$\begin{cases} f_1 = (A + \bar{B} + C)(B + \bar{C})(C + \bar{D}) \\ f_2 = (A + \bar{B} + C)(B + \bar{C})(\bar{A} + \bar{D}) \end{cases}, \quad (4.37)$$

which yields a combined gate/input tally of 6/15 exclusive of possible inverters. Notice that the shared PI dyad $(\bar{A} + C + \bar{D})$ is covered by the quads $(C + \bar{D})$ and $(\bar{A} + \bar{D})$ in the expressions for f_1 and f_2, respectively. Thus, the optimum coverage for both f_1 and f_2 is, in this case, that of the individual minimum forms. This is not usually the case, as is demonstrated next for the optimum SOP results. Note that if strict use had been made of all the shared PIs in the table of Fig. 4.25a together with a required dyad for each output, the combined gate/input tally would become 7/22, significantly greater than that of Eqs. (4.37).

4.5 MULTIPLE OUTPUT OPTIMIZATION

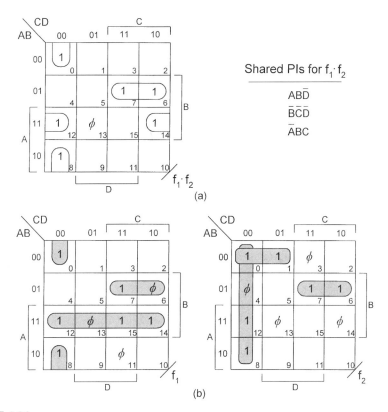

FIGURE 4.26
Multioutput SOP optimization for the system represented by Eqs. (4.36) and Fig. 4.24. (a) K-map and shared PIs for $f_1 \cdot f_2$. (b) K-maps showing optimal SOP cover for functions f_1 and f_2.

Optimized SOP Cover. ANDing the canonical SOP forms of Eqs. (4.36) by using the ANDing rules given by Eqs. (4.32) produces

$$f_1 \cdot f_2 = \sum m(0, 6, 7, 8, 12, 14) + \phi(13).$$

The K-map for $f_1 \cdot f_2$ and the table of shared p-term PIs is given in Fig. 4.26a. The K-maps in Fig. 4.26b show the optimized cover for the two-function system. The results are

$$\begin{cases} f_1 = \bar{A}BC + \bar{B}\bar{C}\bar{D} + AB \\ f_2 = \bar{A}BC + \bar{A}\bar{B}\bar{C} + \bar{C}\bar{D} \end{cases}, \qquad (4.38)$$

which represent a combined gate/input tally of 7/19 making use of only one of the three shared PIs. Here, shared PI dyads $AB\bar{D}$ and $\bar{B}\bar{C}\bar{D}$ are rejected in favor of quads AB and $\bar{C}\bar{D}$ in the f_1 and f_2 K-maps, respectively. Notice that function f_1 is not an individual minimum, but combined with the individual minimum for function f_2 results in an optimized system. An individual minimum for function f_1 is achieved by replacing the shared PI $m(6, 7)$ with the quad $m(6, 7, 14, 15)$ in Fig. 4.26b. When combined with the individual minimum for function f_2, there results a gate/input tally of 8/21, which is not optimal. Also, note that

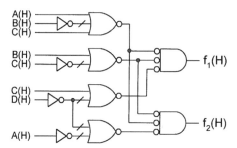

FIGURE 4.27
NOR/INV logic circuit for the optimized POS system of Fig. 4.25.

making use of all shared PIs in the table of Fig. 4.26a together with the required additional p-term cover yields a combined gate/input tally of 7/22.

Comparing the POS and SOP results with optimum system covers of cardinality 4 and 5, respectively, it is clear that the POS result is the more optimum (gate/input tally of 6/15 or 10/19 including inverters). Shown in Fig. 4.27 is the optimal NOR/INV logic implementation of the POS results given by Eqs. (4.37).

The simple search method used here to obtain optimum results becomes quite tedious when applied to multiple output systems more complicated than those just described. For example, a four-input/four-output SOP optimization problem would require at least 10 ANDed fourth-order K-maps, including one for each of six ANDed pairs. For systems this large and larger it is recommended that a computer optimization program (Appendix B) be used, particularly if a guaranteed optimum cover is sought. Optimum cover, as used here, means the least number of gates required for implementation of the multiple output system. Obviously, the number of inverters required and fan-in considerations must also be taken into account when appraising the total hardware cost.

4.6 ENTERED VARIABLE K-MAP MINIMIZATION

Conspicuously absent in the foregoing discussions on K-map function minimization is the treatment of function minimization in K-maps of lesser order than the number of variables of the function. An example of this would be the function reduction of five or more variables in a fourth-order K-map. In this section these problems are discussed by the subject of *entered variable* (EV) *mapping,* which is a "logical" and very useful extension of the conventional (1's and 0's) mapping methods developed previously.

Properly used, EV K-maps can significantly facilitate the function reduction process. But function reduction is not the only use to which EV K-maps can be put advantageously. Frequently, the specifications of a logic design problem lend themselves quite naturally to EV map representation from which useful information can be obtained directly. Many examples of this are provided in subsequent chapters. In fact, EV (entered variable) K-maps are the most common form of graphical representation used in this text.

If N is the number of variables in the function, then map entered variables originate when a conventional Nth-order K-map is compressed into a K-map of order $n < N$ with terms of $(N - n)$ variables entered into the appropriate cells of the nth-order K-map. Thus,

4.6 ENTERED VARIABLE K-MAP MINIMIZATION

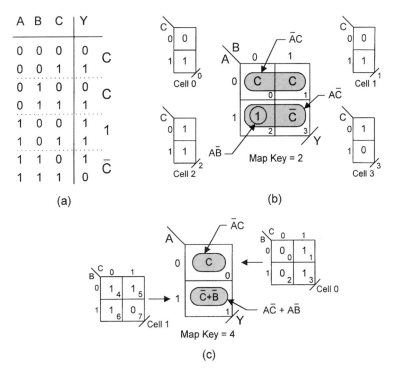

FIGURE 4.28
(a) Truth table for function Y in Eq. (4.39) showing subfunctions for a first-order map compression. (b), (c) Second and first-order EV K-maps showing submaps and minimum SOP cover extracted in minterm code.

each cell of the nth-order K-map becomes a submap of order $(N - n)$, hence K-maps within K-maps.

To illustrate, consider the three-variable function

$$Y(A, B, C) = \sum m(1, 3, 4, 5, 6), \qquad (4.39)$$

which has been placed in a truth table and mapped into a second-order EV K-map, as shown in Figs. 4.28a and 4.28b. The subfunctions indicated to the right of the truth table are also represented as first-order submaps corresponding to the cells 0, 1, 2, and 3 in the EV K-map of Fig. 4.28b. The minimum cover is then obtained by looping out the cell entries, as shown by the shaded loops, giving the minimum result

$$Y_{SOP} = \bar{A}C + A\bar{C} + A\bar{B}. \qquad (4.40)$$

Notice that the term $A\bar{C}$ covers only the $\bar{C}$ in the $1 = C + \bar{C}$ of cell 2. This requires that the C in the 1 be covered by one of the two OPIs, $A\bar{B}$ or $\bar{B}C$, and the former is chosen.

The same result can be obtained from a second-order compression if the expression of Eq. (4.39) is compressed into a first-order K-map. This is done in Fig. 4.28c, where B and C are now the EVs. The minimum cover is indicated by the shaded loops, yielding the expression in Eq. (4.40). The OPI $\bar{B}C$ is not easily seen in the first-order EV K-map, but can be found by observing the 1's representing $\bar{B}C$ in the two submaps shown in Fig. 4.28c.

Map Key It has already been pointed out that each cell of the compressed nth-order K-map represents a submap of order $(N-n)$ for an $N > n$ variable function. Thus, each submap covers 2^{N-n} possible minterms or maxterms. This leads to the conclusion that any compressed nth-order K-map, representing a function of $N > n$ variables, has a Map Key defined by

$$\text{Map Key} = 2^{N-n} \quad N > n \qquad (4.41)$$

The Map Key has the special property that when multiplied by a cell code number of the compressed nth-order K-map there results the code number of the first minterm or maxterm possible for that cell. Furthermore, the Map Key also gives the maximum number of minterms or maxterms that can be represented by a given cell of the compressed nth-order K-map. These facts may be summarized as follows:

Conventional K-map: Map Key = 1 (no EVs, 1's and 0's only)
First-order compression K-map: Map Key = 2 (one EV)
Second-order compression K-map: Map Key = 4 (two EVs)
Third-order compression K-map: Map Key = 8 (three EVs), etc.

As an example, the first-order compressed K-map in Fig. 4.28b has a Map Key of $2^{3-2} = 2$. So each of its cells represents two possible minterms (first-order submaps) beginning with minterm code number equal to (Map Key = 2) × (Cell Number). This is evident from an inspection of the truth table in Fig. 4.28a. Similarly, the second-order compression in Fig. 4.28c has a Map Key of $2^{3-1} = 4$. Therefore, each cell represents four possible minterms represented by the conventional second-order submaps shown to the sides of Fig. 4.28c.

The compressed K-maps in Fig. 4.28 can also be read in maxterm code as indicated by the shaded loops in Fig. 4.29. In this case the logic 1 in cell 2 must be excluded. The result for either the first-order or second-order compressed K-maps is

$$Y_{POS} = (\bar{A} + \bar{B} + \bar{C})(A + C). \qquad (4.42)$$

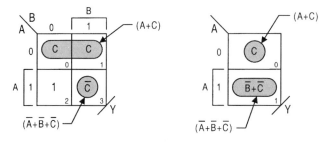

FIGURE 4.29
Second- and first-order EV K-maps showing minimum POS cover for function Y extracted in maxterm code.

4.6 ENTERED VARIABLE K-MAP MINIMIZATION   161

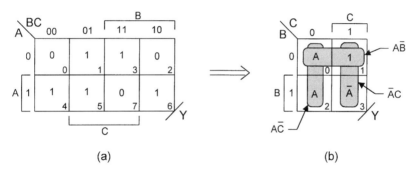

FIGURE 4.30
(a) Conventional K-map for function Y of Eq. (4.39). (b) Second-order EV K-map with entered variable A showing minimum cover for Y as extracted in minterm code.

That Y_{POS} in Eq. (4.40) and Y_{SOP} in Eq. (4.42) are algebraically equal is made evident by carrying out the following Boolean manipulation:

$$(\bar{A} + \bar{B} + \bar{C})(A + C) = \bar{A}C + A\bar{C} + [A\bar{B} + \bar{B}C],$$

where the two p-terms in brackets are OPIs, thereby rendering one to be redundant.

In the second-order K-maps of Figs. 4.28 and 4.29, C is taken to be the EV. However, any of the three variables could have been chosen as the EV in the first-order compression K-maps. As an example, variable A is the EV in Fig. 4.30, where the columns in the conventional K-map of (a) form the submaps of the cells in the compressed K-map of Fig. 4.30b. Minimum cover extracted in minterm code then yields the same result as Eq. (4.40). Or, if extracted in maxterm code, Eq. (4.42) would result. Thus, one concludes that the choice of EVs in a compressed K-map does not affect the extracted minimum result.

Reduced but nonminimum functions can be easily compressed into EV K-maps. This is demonstrated by mapping the four-variable function

$$X = B\bar{C}D + \bar{A}\bar{B} + A\bar{C}\bar{D} + \bar{A}BCD + AB\bar{C} \qquad (4.43)$$

into the third-order EV K-maps shown in Fig. 4.31, where the Map Key is 2. Here, D is the EV and $1 = (D + \bar{D})$. Figure 4.31a shows the p-terms (loops) exactly as presented in Eq. (4.43). However, regrouping of the logic adjacencies permits minimum SOP and POS cover to be extracted. This is done in Figs. 4.31b and 4.31c, yielding

$$\begin{aligned} X_{SOP} &= \bar{A}D + A\bar{C} + \bar{A}\bar{B} \\ X_{POS} &= (A + \bar{B} + D)(\bar{A} + \bar{C}), \end{aligned} \qquad (4.44)$$

where the expressions for X_{SOP} and X_{POS} represent gate/input tallies of 4/9 and 3/7, respectively, excluding possible inverters.

The four-variable function X in Eq. (4.43) can also be minimized in a second-order EV K-map. Shown in Fig. 4.32 is the second-order compression and minimum SOP and POS cover for this function, giving the same results as in Eqs. (4.44). Notice that after covering the D in cell 1 of Fig. 4.32a, it is necessary to cover all that remains in cell 0 by looping out the 1 as an island to give $\bar{A}\bar{B}$. In this case the 1 has the value $1 = C + \bar{C} = D + \bar{D}$. Clearly, the 1 in cell 0 cannot be used in extracting minimum cover in maxterm code.

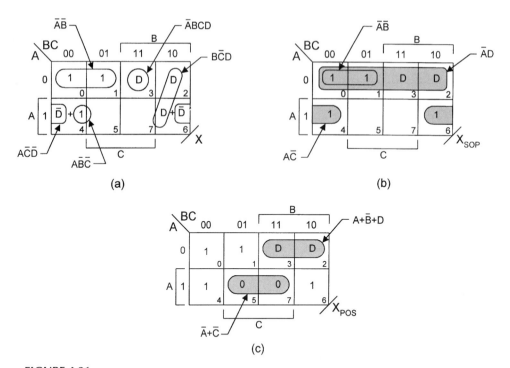

FIGURE 4.31
(a) First-order compression plot of the function X in Eq. (4.43) showing original p-terms. (b) Minimum SOP cover. (c) Minimum POS cover.

4.6.1 Incompletely Specified Functions

The EV mapping method is further illustrated by compressing the incompletely specified function

$$f(A, B, C, D) = \sum m(3, 6, 9, 10, 11) + \phi(0, 1, 4, 7, 8) \tag{4.45}$$

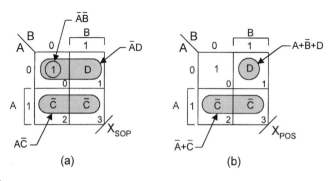

FIGURE 4.32
Second-order compressions of the function X showing (a) minimum SOP cover and (b) minimum POS cover.

4.6 ENTERED VARIABLE K-MAP MINIMIZATION

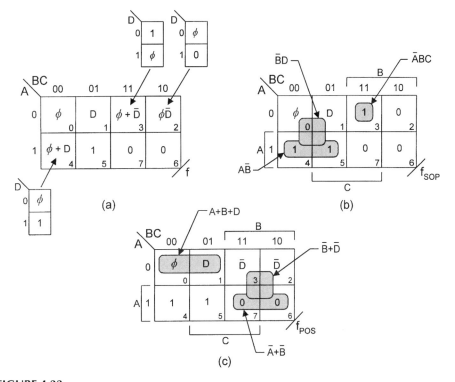

FIGURE 4.33
(a) First-order compression plot and submaps for the function f in Eq. (4.45). (b) Minimum SOP cover and (c) minimum POS cover.

into the third-order K-map in Fig. 4.33a, a first-order compression with a Map Key of 2. Here, the subfunctions are presented in their simplest form yet preserving all canonical information. In Figs. 4.33b and 4.33c are shown the minimum SOP and POS covers for this function, which produce the expressions

$$f_{SOP} = \bar{B}D + \bar{A}BC + A\bar{B}$$
$$f_{POS} = (A + B + D)(\bar{B} + \bar{D})(\bar{A} + \bar{B}),$$
(4.46)

both of which have a gate/input tally of 4/10. In extracting the minimum expressions of Eqs. (4.46), the loop-out protocol is first applied to the entered variable D and then applied to the 1's or 0's.

Some observations are necessary with regard to Fig. 4.33 and Eqs. (4.46). First, these expressions are logically equivalent but are *not* algebraically equal. The reason is that the don't cares ϕ_4 and ϕ_7 in cells 2 and 3 are used differently for the f_{SOP} and f_{POS}. For example, $(\phi_7 + \bar{D})_{SOP} = 1$ for $\phi_7 = 1$ but $(\phi_7 + \bar{D})_{POS} = \bar{D}$, since, in this case, $\phi_7 = 0$. Second, the extraction process involved some techniques in dealing with ϕ's that have not been discussed heretofore. These techniques are set off for reference purposes by the following:

Remember:

- Treat the don't care (ϕ) as an *entered variable* — which it is.
- In simplifying incompletely specified subfunctions, apply the absorptive laws:

$$X + \phi \bar{X} = X + \phi$$
$$X \cdot (\phi + \bar{X}) = \phi X.$$

- Subfunctions of the type $(\phi + X)$ have an essential SOP component but no essential POS component. (Proved by substituting the set $\{0, 1\}$ for ϕ.)
- Subfunctions of the type ϕX have an essential POS component but no essential SOP component. (Proved by substituting the set $\{0, 1\}$ for ϕ.)

Concluding this section is the function

$$Z(A, B, C, D) = \prod M(2, 4, 7, 11, 12, 14, 15)$$
$$= \sum m(0, 1, 3, 5, 6, 8, 9, 10, 13), \qquad (4.47)$$

which is represented by the second-order EV K-maps in Fig. 4.34, where C and D are the EVs and the Map Key is 4. This example is interesting because of the XOR function in cell 1, which must be represented by both the SOP and POS defining relations, given in Eqs. (3.4), so as to extract minimum SOP and POS cover. To assist the reader in identifying the subfunctions, second-order conventional submaps in C and D axes are shown for each cell. Thus, the subfunction for cell 0 is $\sum m(0, 1, 3) = \bar{C} + D$, while that for cell 1 is

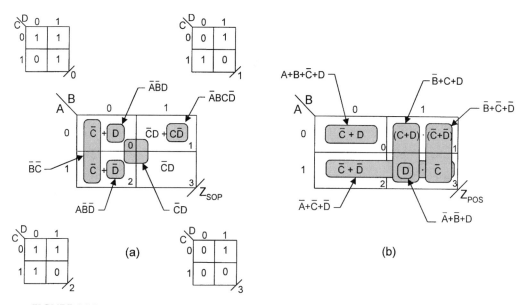

FIGURE 4.34
Second-order EV K-maps and submaps for Eqs. (4.47) showing (a) minimum SOP cover and (b) minimum POS cover.

$\sum m(5, 6) = C \oplus D = C\bar{D} + \bar{C}D = (C + D)(\bar{C} + \bar{D})$. The minimum SOP and POS results are given by

$$Z_{SOP} = \bar{A}BC\bar{D} + \bar{A}\bar{B}D + AB\bar{D} + \bar{C}D + \bar{B}\bar{C}$$
$$Z_{POS} = (A + B + \bar{C} + D)(\bar{B} + \bar{C} + \bar{D})(\bar{B} + C + D)(\bar{A} + \bar{B} + D)(\bar{A} + \bar{C} + \bar{D}).$$
(4.48)

From the results depicted in Fig. 4.34, certain conclusions are worth remembering and are set off by the following:

- In minterm code, subfunctions of the type XY are subsets of forms of the type $X + Y$.
- In maxterm code, subfunctions of the type $X + Y$ are subsets of forms of the type XY.

What this means is that subfunctions of the type XY can be looped out from terms of the type $X + Y$ to produce reduced SOP cover. For reduced POS cover, subfunctions of the type $X + Y$ can be looped out from terms of the type XY (there are more 0's in XY than in $X + Y$). For example, in Fig. 4.34 $\bar{C}D$ is looped out of both $\bar{C} + D$ and $\bar{C} + \bar{D}$ to contribute to minimum SOP cover. However, in Fig. 4.34b both $C + D$ and $\bar{C} + \bar{D}$ are looped out of $\bar{C}D$, leaving $\bar{C} + D$ to be covered by $\bar{A} + \bar{B} + D$.

4.7 FUNCTION REDUCTION OF FIVE OR MORE VARIABLES

Perhaps the most powerful application of the EV mapping method is the minimization or reduction of functions having five or more variables. However, beyond eight variables the EV method could become too tedious to be of value, given the computer methods available. The subject of computer-aided minimization tools is covered in Appendix B.

Consider the function

$$F(A, B, C, D, E) = \sum m(3, 11, 12, 19, 24, 25, 26, 27, 28, 30), \quad (4.49)$$

which is to be compressed into a fourth-order K-map. Shown in Fig. 4.35 is the first-order compression (Map Key = 2) and minimum SOP and POS cover for the five variable function in Eqs. (4.49). The minimized results are

$$F_{SOP} = BC\bar{D}\bar{E} + \bar{C}DE + AB\bar{E} + AB\bar{C}$$
$$F_{POS} = (A + \bar{D} + E)(\bar{C} + \bar{E})(B + E)(A + C + D)(B + D),$$
(4.50)

which have gate input tallies of 5/17 and 6/17, respectively. Thus, the SOP result is the simpler of the two. Also, since there are no don't cares involved, the two expressions are algebraically equal. Thus, one expression can be derived from the other by Boolean manipulation.

A more complex example is presented in Fig. 4.36, where the six-variable function

$$Z(A, B, C, D, E, F)$$
$$= \sum m(0, 2, 4, 6, 8, 10, 12, 14, 16, 20, 23, 32, 34, 36, 38, 40,$$
$$42, 44, 45, 46, 49, 51, 53, 54, 55, 57, 59, 60, 61, 62, 63) \quad (4.51)$$

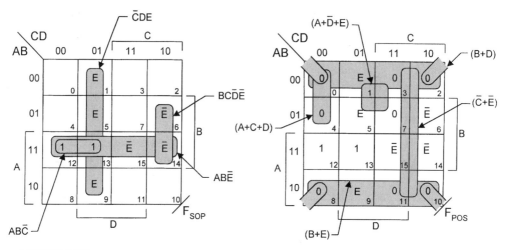

FIGURE 4.35
Minimum SOP and POS cover for the five-variable function given by Eq. (4.49).

is compressed into a fourth-order K-map, a second-order compression (Map Key = 4). The minimum SOP and POS cover is indicated by the shaded loops in Figs. 4.36a and 4.36b and yield the following minimum expressions for function Z:

$$Z_{SOP} = B\bar{C}DEF + \bar{A}\bar{C}E\bar{F} + ACD\bar{E} + ADE\bar{F} + ABF + \bar{B}\bar{F}$$

$$Z_{POS} = (A + \bar{B} + \bar{E} + F)(A + E + \bar{F})(A + D + \bar{F})(\bar{A} + \bar{B} + C + E + F) \quad (4.52)$$
$$\cdot (\bar{A} + \bar{B} + D + F)(B + D + \bar{F})(B + \bar{E} + \bar{F})(B + C + \bar{F})(A + \bar{B} + \bar{C}).$$

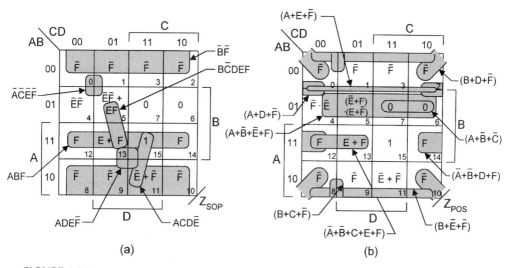

FIGURE 4.36
Fourth-order EV K-maps for the six-variable function Z in Eq. (4.51), showing (a) minimum SOP cover and (b) minimum POS cover.

4.7 FUNCTION REDUCTION OF FIVE OR MORE VARIABLES 167

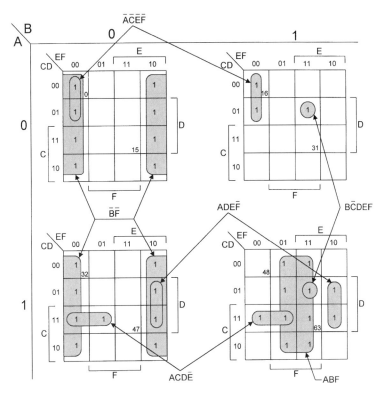

FIGURE 4.37
An $A/B \| CD/EF$ format for the six-variable function of Eq. (4.51) and Fig. 4.36a, showing minimum SOP cover.

Notice that the SOP form of the equivalence function, $E \odot F = \bar{E}\bar{F} + EF$ given by Eq. (3.5), is used in cell 5 to extract minimum cover in minterm code and that the POS form $(\bar{E} + F) \cdot (E + \bar{F})$ is required for extraction of minimum cover in maxterm code. Note also that the loop-out protocol is applied first to the EVs and then to the 1's (in minterm code) and 0's (in maxterm code) as "clean-up" operations. This protocol procedure is recommended to avoid possible redundancy.

There are other K-map formats that can be used to extract reduced or minimum cover for a given function. Consider again the six-variable function given in Eq. (4.51). Presented in Fig. 4.37 is the $A/B \| CD/EF$ format for the conventional (1's and 0's) mapping of this function where only minterm code extraction is considered. Observe that extraction of the EPIs takes on a three-dimensional (3-D) character in a 2-D layout, which can be somewhat perplexing.

As a final example, the format of Fig. 4.37 is used to deal with the following incompletely specified function of eight variables:

$$Z(a, b, c, d, e, f, S, T) = \sum m(16, 18, 20, 22, 24, 26, 28, 30, 48, 50, 52, 54, 56, 58, 60,$$
$$62, 98, 99, 102, 103, 106, 107, 110, 111, 160-191,$$
$$225-227, 229-231, 233-235, 237-239,$$
$$241, 243, 245-247, 248, 250, 252, 254, 255)$$
$$+ \phi(0-15, 32-47, 64-79, 112-159, 192-207). \quad (4.53)$$

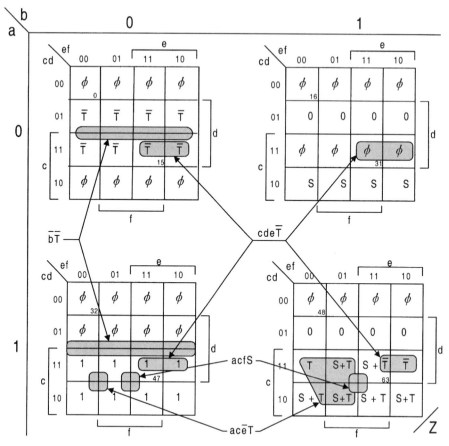

FIGURE 4.38
An $a/b\|cd/ef$ EV format for an eight-variable function Z given by Eq. (4.53).

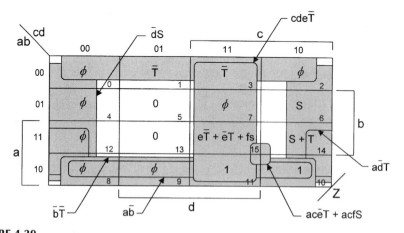

FIGURE 4.39
ab/cd EV K-map showing minimum SOP cover for the eight-variable function represented in Fig. 4.38 and by Eq. (4.53).

168

4.8 MINIMIZATION ALGORITHMS AND APPLICATION

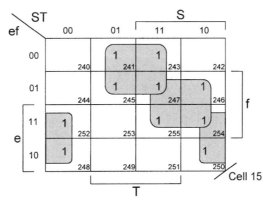

Cell 15 = $\Sigma m(241, 243, 245\text{-}7, 248, 250, 252, 254, 255)$

FIGURE 4.40
Submap for Cell 15 of Fig. 4.39 showing minimum subfunction cover.

Presented in Fig. 4.38 is the second-order compression of this function (Map Key = 4) by using the format $a/b\|cd/ef$, where S and T are the EVs. The minimized result, as extracted from Fig. 4.38, is given by

$$Z_{SOP} = cde\bar{T} + ac\bar{e}T + acfS + a\bar{d}T + \bar{d}S + \bar{b}\bar{T} + a\bar{b}, \qquad (4.54)$$

where, for clarity's sake, only the loopings for the first three and sixth p-terms are shown. Here again the p-terms are given in the order determined by the loop-out protocol first for the EVs then for the 1's as a "clean-up" operation. Note that the term $a\bar{b}$ covers all the 1's and don't cares in cells 32 through 47 of Fig. 4.38, but is not shown.

Next, the function of Eq. (4.53) is compressed into the fourth-order K-map of Fig. 4.39, a fourth-order compression (Map Key = 16). The same minimum result given by Eq. (4.54) is easily obtained from Fig. 4.39 as indicated by the shaded loopings. To help understand the entry in Cell 15 of Fig. 4.39, a submap for this cell is provided in Fig. 4.40. The last line of essential minterms in Eq. (4.53) pertains to Cell 15 and to Fig. 4.40.

4.8 MINIMIZATION ALGORITHMS AND APPLICATION

Tabular methods for function minimization have been devised that can be implemented by a computer and can therefore be used to minimize functions having a large number of input variables. One such method has become known as the *Quine–McCluskey* (Q-M) algorithm. Typical of these methods, the Q-M algorithm first finds the prime implicants (PIs) and then generates the minimum cover. Another important minimization algorithm is a heuristic-type algorithm called *Espresso*. This section will provide a description of these two algorithms together with simple illustrative applications.

4.8.1 The Quine–McCluskey Algorithm

To understand the Q-M algorithm, it is helpful to review the tabular format and notation that is unique to it. In Fig. 4.41 is shown the Q-M notation that will be used in the two examples

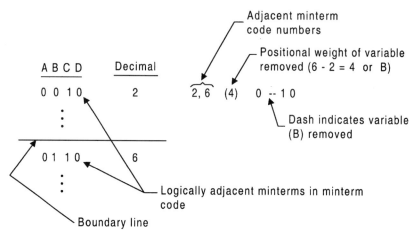

FIGURE 4.41
Quine–McCluskey (Q-M) notation for PI determination.

that follow. Notice that the Q-M notation uses minterm code, minterm code numbers, and positional weights for PI determination.

EXAMPLE 1 Consider the minimization of the incompletely specified function

$$Y(A, B, C, D) = \sum m(0, 1, 4, 6, 8, 14, 15) + \phi(2, 3, 9). \tag{4.55}$$

In the Q-M algorithm the ϕ's are treated as essential minterms, and minterm sets k are compared with sets $(k + 1)$ in a linear and exhaustive manner. The first step in the application of the Q-M algorithm is presented in Fig. 4.42. Here, a check mark ($\sqrt{}$) indicates that an implicant is covered by a PI in the column to the right and, therefore, cannot itself be a PI. Thus, unchecked terms in columns 4 and 6 are the indicated PIs and those that are lined out are redundant.

The second step in the application of the Q-M method is the identification of the essential prime implicants (EPIs). Presented in Fig. 4.43 is a table of the PIs (taken from Fig. 4.42) vs the essential minterms in Eq. (4.55). The check mark ($\sqrt{}$) within the table indicates that a given minterm is covered by a PI. The EPIs are selected from a minimum set of PIs that cover all of the essential minterms of the function Y in Eq. (4.55) and are presented in Eq. (4.56):

$$Y = ABC + \bar{B}\bar{C} + \bar{A}\bar{D}. \tag{4.56}$$

This result can be easily verified by the conventional K-map extraction method described in Section 4.4.

EXAMPLE 2 In this example a minimum POS result is required for the incompletely specified function

$$F(W, X, Y, Z) = \sum m(0, 1, 4, 5, 11, 12, 13, 14, 15) + \phi(2, 7, 9) \tag{4.57}$$

$$= \prod M(3, 6, 8, 10) \cdot \phi(2, 7, 9). \tag{4.58}$$

4.8 MINIMIZATION ALGORITHMS AND APPLICATION

No. of 1's							
0	0000 0	0,1 (1)	$\bar{A}\bar{B}\bar{C}$ -- ✓	0,1,2,3	$\bar{A}\bar{B}$ -- --	0 0 -- --
1	0001 1	0,2 (2)	$\bar{A}\bar{B}$ -- $\bar{D}$ ✓	0,1,8,9	-- $\bar{B}\bar{C}$ --	-- 0 0 --
	0010 2	0,4 (4)	$\bar{A}$ -- $\bar{C}\bar{D}$ ✓	~~0,2,1,3~~	~~$\bar{A}\bar{B}$ -- --~~	
	0100 4	0,8 (8)	-- $\bar{B}\bar{C}\bar{D}$ ✓	0,2,4,6	$\bar{A}$ -- -- $\bar{D}$	0 -- -- 0
	1000 8	1,3 (2)	$\bar{A}\bar{B}$ -- D ✓	~~0,4,2,6~~	~~$\bar{A}$ -- -- $\bar{D}$~~	
2	0011 3	1,9 (8)	-- $\bar{B}\bar{C}$ D ✓	~~0,8,1,9~~	~~-- $\bar{B}\bar{C}$ --~~	
	0110 6	2,3 (1)	$\bar{A}\bar{B}$ C -- ✓			
	1001 9	2,6 (4)	$\bar{A}$ -- C $\bar{D}$ ✓			
3	1110 14	4,6 (2)	$\bar{A}$ B -- $\bar{D}$ ✓			
4	1111 15	8,9 (1)	A $\bar{B}\bar{C}$ -- ✓			
		6,14 (8)	-- B C $\bar{D}$	6,14	-- B C $\bar{D}$	
		14,15 (1)	A B C --	14,15	A B C --	

✓ Indicates that an implicant is covered by a Prime Implicant in the columns to the right.

FIGURE 4.42
Determination of PIs for the 1's in the function Y of Eq. (4.55).

To do this, the 0's of Eq. (4.58) will be treated as 1's, as required by the Q-M algorithm, to yield $\bar{F}_{POS}$ in minimum SOP form. Then, application of DeMorgan's law, given by Eqs. (3.15), yields the results $\bar{\bar{F}}_{POS} = F_{POS}$ by involution. Here, the ϕ's in Eq. (4.58) are treated as essential minterms, not as nonessential maxterms. Shown in Fig. 4.44 is the tabular determination of PIs for the 0's, treated as 1's, in the maxterm form of function F given by Eq. (4.58).

The final step is to tabulate the PIs of Fig. 4.44 with the maxterms (now treated as minterms) in Eq. (4.58) to obtain the EPIs for the function $\bar{F}_{POS}$. This is done in Fig. 4.45,

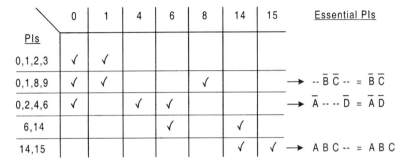

FIGURE 4.43
Table of PIs (from Fig. 4.42) vs minterms for the function Y of Eq. (4.55) showing the resulting EPIs.

```
        No.
        of 1's
              ⎧  0010  2      2,3 (1)    W̄ X Y --  ✓     2,3,6,7      W̄ -- Y --
           1  ⎨
              ⎩  1000  8      2,6 (4)    W̄ -- Y Z̄  ✓    ~~2,6,3,7~~  ~~W̄ -- Y --~~
              ⎧  0011  3      2,10 (8)   -- X̄ Y Z̄  ✓     2,10         -- X̄ Y Z̄
              ⎪  0110  6      8,9 (1)    W X̄ Ȳ --  ✓     8,9          W X̄ Ȳ --
           2  ⎨  1001  9      8,10 (2)   W X̄ -- Z̄  ✓     8,10         W X̄ -- Z̄
              ⎪                ─────────
              ⎩  1010  10     3,7 (4)    W̄ -- Y Z  ✓
           3 { 0111  7       6,7 (1)    W̄ X Y --  ✓
```

✓ Indicates that an implicant is covered by a Prime Implicant in the columns to the right.

FIGURE 4.44
Tabular determination of PIs for the 0's (treated as 1's) in function F of Eq. (4.58).

giving the final results

$$\bar{F}_{POS} = \bar{W}Y + W\bar{X}\bar{Z}$$

$$\bar{\bar{F}}_{POS} = F_{POS} = (W + \bar{Y})(\bar{W} + X + Z). \qquad (4.59)$$

Notice that the PI (2, 3, 6, 7) is the EPI $\bar{W}Y$ and that the remaining maxterms (treated as minterms) are covered by the PI (8,10), the minimum set of PIs covering all minterms.

Had the Q-M algorithm been applied to Eq. (4.57), the minimum SOP result would be

$$F_{SOP} = \bar{W}\bar{Y} + WX + WZ, \qquad (4.60)$$

which is algebraically equal to the POS result of Eq. (4.59). The reason for this is that the application of the Q-M algorithm uses the three ϕ's in the same way for the two cases, a feature of the Q-M method. As a general rule, this is rarely the case for SOP and POS minimized forms of incompletely specified functions obtained by other methods.

PIs	3	6	8	10	Essential PIs
2,3,6,7	✓	✓			→ W̄ -- Y -- = W̄Y
2,10				✓	
8,9			✓		
8,10			✓	✓	→ W X̄ -- Z̄ = W X̄ Z̄

FIGURE 4.45
Table of PIs (from Fig. 4.44) vs maxterms treated as minterms for function F of Eq. (4.58) showing the essential PIs.

4.8.2 Cube Representation and Function Reduction

The *cube notation* is commonly used in CAD programs and, in fact, is the notation that is used in the computer implementation of the Q-M algorithm described in Subsection 4.8.1. In this notation an n-dimensional cube has 2^n vertices formed by the intersection of n dimensional lines. Most commonly one thinks of a cube as three-dimensional (a 3-cube) having $2^3 = 8$ vertices. But the concept is much more general, extending to n dimensions that generally cannot be easily visualized by a geometrical figure.

Cube representation is usually based on minterm code. Thus, the minterms of a switching function can be mapped onto the 2^n vertices of an n-dimensional cube such that each pair of adjacent vertices differ by exactly one bit position. As an example, consider implicants (2, 3) and (6, 7) listed in the Q-M example of Fig. 4.44. In minterm code cube notation, these implicants would be represented as (0010, 0011) and (0110, 0111), respectively. Reduction of these implicants to PI (r-cube) form occurs between adjacencies (adjacent vertices) as follows:

$$0010 + 0011 = 001- = \bar{W}\bar{X}Y \quad \text{and} \quad 0110 + 0111 = 011- = \bar{W}XY$$

or, finally,

$$(001-) + (011-) = 0-1- = \bar{W}Y,$$

where 0 represents the complemented variable, 1 is the uncomplemented variable, and the "$-$" symbol represents an irrelevant input variable (representing both 1 and 0). Thus, in general, an r-cube of an n-variable function is produced by combining 2^r adjacent minterms, thereby eliminating r variables in a function reduction process.

4.8.3 Qualitative Description of the Espresso Algorithm

The two-level minimization algorithm called *Espresso* belongs to a class of minimization algorithms that use heuristic logic methods as opposed to the linear exhaustive PI search of the Q-M method. In effect, all heuristic methods group, expand and regroup adjacent minterms over a number of iterations until an optimal or near-optimal grouping, called the *irredundant set*, is found. The exact strategies used and the order in which they are used depends on the particular algorithm.

Though a detailed description of the Espresso algorithm is beyond the scope of this text, the principal steps involved can be qualitatively understood by the K-maps in Fig. 4.46. Here, the four basic steps of the Espresso algorithm are represented by four fourth-order K-maps labeled ORIGINAL, REDUCE, EXPAND and IRREDUNDANT. The ORIGINAL function, plotted in Figure 4.46a, is the graphical equivalent to the PI table of the Q-M method since it represents the largest number of prime implicants, that is, six PIs. The original function is then regrouped to form a smaller (REDUCED) number of prime implicants (four PIs) in Fig. 4.46b and then EXPANDED (RESHAPED) to form four PIs by eliminating two PIs. Notice that the cardinality is preserved in the REDUCE-to-EXPAND step. Finally, an IRREDUNDANT set is found by regrouping and eliminating yet another PI, resulting in only three EPIs. This irredundant set is said to have minimum *cardinality*, that is, minimum cover.

The Espresso algorithm just described qualitatively is usually called Espresso-II. Since its inception, various improvements have been made, adding to the speed and multiple-output

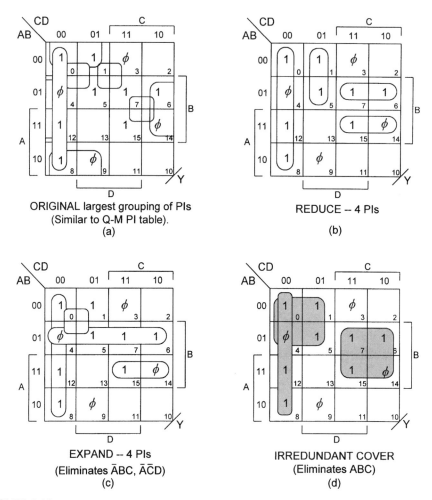

FIGURE 4.46
Four-variable K-maps for function Y illustrating four steps of the Espresso algorithm: (a) ORIGINAL, (b) REDUCE, (c) EXPAND, and (d) IRREDUNDANT COVER.

capability. These improved Espresso algorithms include the two referred to as Espresso-EXACT and Espresso-MV. A detailed description of these and other closely related algorithms can be found in references cited at the end of this chapter.

4.9 FACTORIZATION, RESUBSTITUTION, AND DECOMPOSITION METHODS

Beyond the SOP or POS minimum result, it is possible to further reduce the gate/input tally, reduce the number of inverters, and reduce the gate fan-in requirements for some functions by using a technique called *factoring* or *factorization*. The results of factorization lead to *multilevel* forms that are *hybrids*, since they cannot be classified as either purely SOP or POS. A multilevel logic realization of a function is one involving more than two levels of

4.9 FACTORIZATION, RESUBSTITUTION, AND DECOMPOSITION METHODS

gate path delay excluding possible inverters. The logic circuits considered to this point are classified as two-level.

4.9.1 Factorization

The method of factorization is illustrated by using a simple example. Consider the minimum SOP function

$$F = AB\bar{C} + AD + \bar{B}D + \bar{C}D, \tag{4.61}$$

which requires a gate/input tally of 5/13 excluding inverters. This requires four ANDing operations with a maximum of three inputs per gate, and one ORing operation requiring four inputs. Also, if it is assumed that the inputs arrive active high, two inverters are required, bringing the total gate/input tally to 7/15. Now suppose that it is desirable to limit the fan-in to a maximum of three inputs per gate and to eliminate the need to use inverters in generating the complements of B and C variables. This can be accomplished by factorization of Eq. (4.61) in the following manner:

$$\begin{aligned} F &= AB\bar{C} + AD + \bar{B}D + \bar{C}D \\ &= AB(\bar{B} + \bar{C}) + AD + (\bar{B} + \bar{C})D \\ &= AB(\overline{BC}) + AD + (\overline{BC})D. \end{aligned} \tag{4.62}$$

The term $AB\bar{C}$ is factored as $AB\bar{C} = AB(\bar{B} + \bar{C})$. Notice that if the function of Eq. (4.62) is implemented as a NAND circuit, a gate/input tally of 5/12 would result with a maximum fan-in of 3 with no inverters.

An interesting feature of the factorization method is that there is usually more than one way in which to factor a given function. So it is with Eq. (4.61), which can be factored as a three-level hybrid form in the following alternative way:

$$\begin{aligned} F &= AB\bar{C} + AD + \bar{B}D + \bar{C}D \\ &= AB\bar{C} + D(A + \bar{B} + \bar{C}). \end{aligned}$$

If implemented with NAND gates, the gate/input becomes 4/10 plus two inverters, assuming that the inputs arrive active high.

The factorization method can be extended to multioutput systems of the type considered in Section 4.5. The process is illustrated by the following simple example where three optimized functions are factored as indicated:

$$\begin{aligned} f_1 &= \bar{A}\bar{B} + \bar{A}\bar{C} + AB = \bar{A}(\bar{B} + \bar{C}) + AB = \bar{A}(\overline{BC}) + AB \\ f_2 &= \bar{A}\bar{B}C + \bar{B}\bar{C} + AB = \bar{B}(\bar{A} + \bar{C}) + AB = \bar{B}(\overline{AC}) + AB \\ f_3 &= \bar{B}C + \bar{A}BC + B\bar{C} = C(\bar{B} + \bar{C}) + B(\bar{A} + \bar{C}) = C(\overline{BC}) + B(\overline{AC}). \end{aligned} \tag{4.63}$$

Here, terms in expressions for f_2 and f_3 are factored as $\bar{A}\bar{B}C + \bar{B}\bar{C} = \bar{B}(\bar{A}C + \bar{C}) = \bar{B}(\bar{A} + \bar{C}) = \bar{B}(\overline{AC})$ and $\bar{B}C = C(\bar{B} + \bar{C}) = C(\overline{BC})$. With NAND logic and assuming the inputs arrive active high, the total gate/input tally for the factored expressions is 12/20, including one inverter, with fan-in requirements of two inputs per gate. In comparison, the original optimized SOP expressions require a gate/input tally of 14/30, including three inverters, and have a maximum fan-in requirement of 3.

The factorized expressions of Eqs. (4.63) are three-level functions, whereas the original SOP expressions are two-level. This brings up other aspects of the optimization problem, namely the *design area* (real estate usage) vs *delay* (performance), as discussed in Section 4.10.

4.9.2 Resubstitution Method

The Boolean resubstitution method possesses a close resemblance to polynomial division and works to generate multilevel functions that have improved fan-in (hence improved area) requirements. The process of resubstitution begins by finding a good, if not optimal, divisor P in the expression

$$F = PQ + R, \tag{4.64}$$

where F is the dividend, Q is the quotient, and R is the remainder. Heuristic algorithms exist that can accomplish this, but they are complex and fall outside the scope of this text. However, an attempt will be made to illustrate the resubstitution method with a simple example. Consider the minimized SOP five-variable function

$$F = AB\bar{E} + AB\bar{C}\bar{D} + CD\bar{E} + \bar{A}C\bar{E} + \bar{A}\bar{B}CD + \bar{A}\bar{B}E + \bar{C}\bar{D}E. \tag{4.65}$$

Noting the appearance of AB, $\bar{A}\bar{B}$, CD, $\bar{C}\bar{D}$, E, and $\bar{E}$ in six of the seven p-terms, the divisor is chosen to be $P = AB + CD + E$. The process continues by repeating three steps for each of the seven p-terms:

Step 1. Select term $AB\bar{E}$.
Step 2. AND (Boolean multiply) $AB\bar{E} \cdot P = AB\bar{E} + ABCD\bar{E} + AB\bar{E}E = AB\bar{E}$.
Step 3. Delete AB in $AB\bar{E} \cdot P$ to yield term $\bar{E} \cdot P$.

Step 1. Select term $AB\bar{C}\bar{D}$.
Step 2. AND $AB\bar{C}\bar{D} \cdot P = AB\bar{C}\bar{D} + AB\bar{C}\bar{D}E = AB\bar{C}\bar{D}$.
Step 3. Delete AB in $AB\bar{C}\bar{D} \cdot P$ to yield term $\bar{C}\bar{D} \cdot P$.

Repeat Steps 1, 2, and 3 for the remaining five terms in the order given by Eq. (4.65):

$CD\bar{E} \cdot P = ABCD\bar{E} + CD\bar{E} = CD\bar{E}$. Delete CD in $CD\bar{E} \cdot P$ to yield $\bar{E} \cdot P$.
$\bar{A}C\bar{E} \cdot P = 0$. Thus, no literals can be deleted in $\bar{A}C\bar{E} \cdot P$.
$\bar{A}\bar{B}CD \cdot P = \bar{A}\bar{B}CD + \bar{A}\bar{B}CDE = \bar{A}\bar{B}CD$. Delete CD in $\bar{A}\bar{B}CD \cdot P$ yield $\bar{A}\bar{B} \cdot P$.
$\bar{A}\bar{B}E \cdot P = \bar{A}\bar{B}CDE + \bar{A}\bar{B}E = \bar{A}\bar{B}E$. Delete E in $\bar{A}\bar{B}E \cdot P$ to yield $\bar{A}\bar{B} \cdot P$.
$\bar{C}\bar{D}E \cdot P = AB\bar{C}\bar{D}E + \bar{C}\bar{D}E = \bar{C}\bar{D}E$. Delete E in $\bar{C}\bar{D}E \cdot P$ to yield $\bar{C}\bar{D} \cdot P$.

In the preceding set of steps it should be observed that the only literals that can be deleted are those that appear as p-terms in the divisor P. Also, it should be noted that the choice of divisor P is somewhat arbitrary, since there are other combination of terms that can be used in the resubstitution process.

The final results of resubstitution are expressed by the partition

$$F = \bar{A}\bar{B}P + \bar{C}\bar{D}P + \bar{E}P + \bar{A}C\bar{E}$$
$$= PQ + R, \tag{4.66}$$

4.9 FACTORIZATION, RESUBSTITUTION, AND DECOMPOSITION METHODS   177

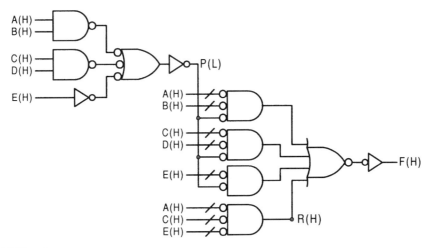

FIGURE 4.47
NAND/NOR/INV realization of the partitioned function given by Eq. (4.66).

where $P = AB + CD + E$, $Q = \bar{A}\bar{B} + \bar{C}\bar{D} + \bar{E}$ and $R = \bar{A}\bar{C}\bar{E}$. Function F, expressed by Eqs. (4.66), represents four levels of path delay, as shown implemented by NAND/NOR/INV logic in Fig. 4.47 where it is assumed that all inputs arrive active high. Notice that the gate/input tally is now 11/25, including three inverters, and that only one gate has a fan-in of 4. If a fan-in limitation of 4 is also applied to the original two-level SOP expression in Eq. (4.65), a three-level circuit results having a gate/input tally of 14/35, including five inverters, and four gates with a fan-in of 4. Thus, the partitioned function of Fig. 4.47 has an improved design area factor but not necessarily an improved performance. A discussion of the design area vs performance factors is given in Section 4.10.

The resubstitution method just described bears similarity to portions of some heuristic two-level minimization algorithms such as Espresso II, qualitatively described in Subsection 4.8.3. In particular, the introduction of a new literal, divisor term P in step 2 and the subsequent deletion of literals in step 3 of resubstitution is a generalization of the REDUCE and EXPAND processes in Espresso II. In these processes, Espresso seeks to add literals existing in one product term of the original expression to other candidate terms so that implicants covered by a given expanded implicant can be deleted. Thus, by repeated introduction of divisor P followed by deletions of redundant terms, the resubstitution process seeks a more optimum result, not unlike the heuristic processes in Espresso.

4.9.3 Decomposition by Using Shannon's Expansion Theorem

Shannon's expansion theorem states that any Boolean function of n variables $f(x_{n-1}, \ldots x_2, x_1, x_0)$ can be decomposed into the SOP form

$$\begin{aligned}
f(x_{n-1}, \ldots, x_2, x_1, x_0) &= \bar{x}_i f(x_{n-1}, \ldots, x_{i+1}, 0, x_{i-1}, \ldots, x_2, x_1, x_0) \\
&\quad + x_i f(x_{n-1}, \ldots, x_{i+1}, 1, x_{i-1}, \ldots, x_2, x_1, x_0) \\
&= \bar{x}_i f^{SOP}_{\bar{x}_i} + x_i f^{SOP}_{x_i}
\end{aligned} \qquad (4.67)$$

or into its dual POS form

$$f(x_{n-1}, \ldots, x_2, x_1, x_0) = [\bar{x}_i + f(x_{n-1}, \ldots, x_{i+1}, 1, x_{i-1}, \ldots, x_2, x_1, x_0)]$$
$$\cdot [x_i + f(x_{n-1}, \ldots, x_{i+1}, 0, x_{i-1}, \ldots, x_2, x_1, x_0)]$$
$$= [\bar{x}_i + f_{\bar{x}_i}^{POS}] \cdot [x_i + f_{x_i}^{POS}], \qquad (4.68)$$

where $f_{\bar{x}_i}^{SOP}$ and $f_{x_i}^{SOP}$ are the cofactors for $\bar{x}_i$ and x_i in Eq. (4.67), and $f_{\bar{x}_i}^{POS}$ and $f_{x_i}^{POS}$ are the cofactors for $\bar{x}_i$ and x_i in Eq. (4.68).

Proof of Eqs. (4.67) and (4.68) is easily obtained by setting $x_i = 1$ and then $x_i = 0$ and observing that in each case the surviving cofactors are identical to the left side of the respective equation. For example, setting $x_i = 1$ in Eq. (4.67) leads to

$$f(x_{n-1}, \ldots, x_{i+1}, 1, x_{i-1}, \ldots, x_2, x_1, x_0) = f(x_{n-1}, \ldots, x_{i+1}, 1, x_{i-1}, \ldots, x_2, x_1, x_0),$$

since $\bar{x}_i = 0$ when $x_i = 1$.

Multiple applications of Eqs. (4.67) and (4.68) are possible. For example, if decomposition is carried out with respect to two variables, x_1 and x_0, Eq. (4.67) becomes

$$f(x_{n-1}, \ldots, x_2, x_1, x_0) = \bar{x}_1 \bar{x}_0 f(x_{n-1}, \ldots, x_2, 0, 0) + \bar{x}_1 x_0 f(x_{n-1}, \ldots, x_2, 0, 1)$$
$$+ x_1 \bar{x}_0 f(x_{n-1}, \ldots, x_2, 1, 0) + x_1 x_0 f(x_{n-1}, \ldots, x_2, 1, 1)$$
$$= m_0 f(x_{n-1}, \ldots, x_2, \underline{m_0}) + m_1 f(x_{n-1}, \ldots, x_2, \underline{m_1})$$
$$+ m_2 f(x_{n-1}, \ldots, x_2, \underline{m_2}) + m_3 f(x_{n-1}, \ldots, x_2, \underline{m_3}),$$

or generally for decomposition with respect to $(x_{k-1}, \ldots, x_2, x_1, x_0)$,

$$f(x_{n-1}, \ldots, x_2, x_1, x_0) = \sum_{i=0}^{2^k-1} m_i(x_{n-1}, \ldots, x_2, x_1, x_0) \cdot f(x_{n-1}, \ldots, x_k, \underline{m_i}). \qquad (4.69)$$

Here, m_i are the canonical ANDed forms of variables x_j taken in ascending minterm code order from $i = 0$ to $(2^k - 1)$, and $\underline{m_i}$ represents their corresponding minterm code. As an example, decomposition with respect to variables (x_2, x_1, x_0) gives

$$f(x_{n-1}, \ldots, x_2, x_1, x_0) = \bar{x}_2 \bar{x}_1 \bar{x}_0 f(x_{n-1}, \ldots, x_3, 0, 0, 0)$$
$$+ \bar{x}_2 \bar{x}_1 x_0 f(x_{n-1}, \ldots, x_3, 0, 0, 1) + \cdots$$

for $k = 3$.

In similar fashion, the dual of Eq. (4.69) is the generalization of Eq. (4.68) given by

$$f(x_{n-1}, \ldots, x_2, x_1, x_0) = \prod_{i=0}^{2^k-1} [M_i(x_{n-1}, \ldots, x_2, x_1, x_0) + f(x_{n-1}, \ldots, x_k, \underline{M_i}), \qquad (4.70)$$

where now M_i represents the canonical ORed forms of variables x_j and $\underline{M_i}$ represents their corresponding maxterm code.

4.9 FACTORIZATION, RESUBSTITUTION, AND DECOMPOSITION METHODS

As a practical example of the application of Shannon's expansion theorem, consider the function

$$F(A, B, C, D) = \sum m(1, 3, 4, 5, 9, 10, 13, 14, 15), \qquad (4.71)$$

which is represented in the truth table of Fig. 4.48a and in the K-map of Fig. 4.48b. Applying Eq. (4.69) for decomposition with respect to variables C and D gives the cofactors

$$\begin{cases} F_0 = (A, B, 0, 0) = \bar{A}B \\ F_1 = (A, B, 0, 1) = 1 \\ F_2 = (A, B, 1, 0) = A \\ F_3 = (A, B, 1, 1) = A \odot B \end{cases},$$

from which the function F can be written as

$$F_{CD}(A, B, C, D) = \bar{C}\bar{D}(\bar{A}B) + \bar{C}D(1) + C\bar{D}(A) + CD(A \odot B)$$
$$= \bar{A}B\bar{C}\bar{D} + \bar{C}D + AC\bar{D} + (A \odot B)CD,$$

which could have been deduced directly from an inspection of the truth table or K-map in Fig. 4.48.

But the variables about which the function is to be decomposed are a matter of choice. If it is required that the function F be decomposed with respect to variables A and B, the result would be

$$F_{AB}(A, B, C, D) = \bar{A}\bar{B}(D) + \bar{A}B(\bar{C}) + A\bar{B}(C \oplus D) + AB(C + D),$$

which, like the previous result, can be read directly from either the truth table or the K-map. Note that decompositions of the type just described can be very useful in implementing

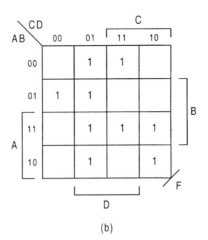

FIGURE 4.48
Truth table (a) and K-map (b) for the function F given by Eq. (4.71).

functions of a large number of variables by using devices with lesser input capability. The use of multiplexers discussed in Section 6.2 offers a good example of this fact.

The process of decomposition can be reversed to yield a purely SOP or purely POS expression from the decomposed expression. This is illustrated by factoring out $\bar{A}, A, \bar{B}$, and B in turn from F_{AB} to yield the SOP expression

$$\begin{aligned} F_{AB} &= \bar{A}\bar{B}(D) + \bar{A}B(\bar{C}) + A\bar{B}(C \oplus D) + AB(C + D) \\ &= \bar{A}[\bar{B}D + B\bar{C}] + A[\bar{B}(C \oplus D) + B(C + D)] \\ &= \bar{A}\bar{B}[D] + \bar{A}B[\bar{C}] + A\bar{B}[\bar{C}D + C\bar{D}] + AB[C + D] \\ &= \bar{A}\bar{B}D + \bar{A}B\bar{C} + A\bar{B}\bar{C}D + A\bar{B}C\bar{D} + ABC + ABD, \end{aligned}$$

where $C \oplus D = \bar{C}D + C\bar{D}$ follows from Eq. (3.4). A cursory inspection of the SOP form of F_{AB} verifies its agreement with Fig. 4.48.

4.10 DESIGN AREA VS PERFORMANCE

It is common to observe an inverse relationship between design area and performance (delay). That is, circuit realizations with improved design area commonly suffer from poorer performance and vice versa. It is known that CMOS gate performance decreases (i.e., delay increases) with increasing numbers of inputs (fan-in). The larger the fan-in, the greater is the path delay through the gate. As an example, consider the function of Eq. (4.65). It has a cardinality of 7 that must be ORed. Shown in Fig. 4.49a are four alternative ORing configurations for seven inputs. It is expected that there exists a *trade-off* between design area and delay for these four configurations, as illustrated in Fig. 4.49b.

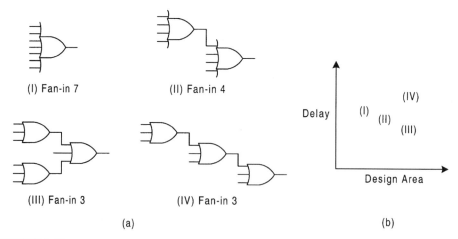

FIGURE 4.49
Area/delay trade-off for the seven-input function of Eq. (4.65). (a) Four alternative ORing configurations. (b) Possible area/delay trade-off points for configurations (I), (II), (III), and (IV), showing effect of treeing and cascading configurations.

4.12 WORKED EV K-MAP EXAMPLES 181

Here, *tree* configuration (III) is expected to show the least delay, but at the cost of greater design area. Tree configuration (II) would seem to have the most favorable area/delay trade-off, while gate (I) and *cascade* configuration (IV) are expected to have the least favorable trade-off. A dual set of ANDing operations would show the same area/delay trade-offs.

4.11 PERSPECTIVE ON LOGIC MINIMIZATION AND OPTIMIZATION

The EV mapping methods described in Sections 4.6 and 4.7 are useful up to three or four orders of map compression. However, with increasing compression order beyond third order, the gap usually widens between the reduced forms obtained and the absolute minimum result. This is especially true if reduced or minimized subfunctions are used to extract cover from such EV K-maps. For this reason a practical limit of four orders of K-map compression (eight variables) is set, and use of reduced or minimum subfunctions is highly recommended. The use of submaps can narrow the gap between a reduced result and one that is an absolute or exact minimum. This fact is implied by the simple examples given in the sections on EV mapping methods.

Beyond four orders of compression in fourth-order K-maps, the use of computer algorithmic methods for logic minimization becomes necessary. But even these computer programs have their limitations, particularly with regard to multiple output systems having a large number of input variables. It is an established fact that the generalized optimal solution for an n-variable function is impossible. The reason for this is that 2^n minterms must be dealt with in some manner or another. Minimization problems in the class referred to as $\mathcal{P}$ are called *tractable* problems — those for which an optimum or near-optimum solution is possible. Those that are *intractable* belong to the class of problems referred to as $\mathcal{NP}$-complete. The search for faster, more robust algorithms to optimize very large multiple output systems continues. These algorithms are most likely to be of the heuristic type. Though the Q-M linear tabular method is useful in introducing readers to the subject of logic minimization, it is of little practical importance given the much-improved heuristic methods now in use.

Finally, it must be said that the SOP (or POS) minimization of a function may not be an end in itself. Section 4.9 demonstrates that optimization may continue beyond minimization by techniques such as factorization and resubstitution that generate multilevel functions. To do this, however, brings into play other factors such as area/delay trade-offs. Thus, there emerge two approaches to function optimization from which the designer must choose: *Optimize design area under delay constraints* or *optimize delay under design area constraints*. It is unlikely that a system can be optimized with respect to both design area and delay, although it may be possible to come close to this for some systems.

4.12 WORKED EV K-MAP EXAMPLES

EXAMPLE 4.1 Compress the following four-variable function into a third-order K-map and extract minimum SOP and minimum POS cover from it.

$$f(A, B, C, S) = \sum m(2, 3, 5, 6, 7, 10, 12, 13, 15)$$
$$= \prod M(0, 1, 4, 8, 9, 11, 14). \quad (4.72)$$

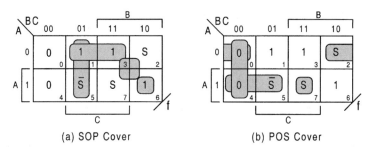

FIGURE 4.50
(a) Minimum SOP cover and (b) minimum POS cover for the function f.

The compressed third-order K-maps representing SOP and POS cover are shown in Fig. 4.50. Applying the loop-out protocol first to the EV and then to the 1's and 0's gives

$$f_{SOP} = \bar{B}C\bar{S} + BS + AB\bar{C} + \bar{A}C$$
$$f_{POS} = (\bar{A} + \bar{B} + \bar{C} + S)(\bar{A} + B + \bar{S})(A + C + S)(B + C)$$

EXAMPLE 4.2 A four-variable function Z containing don't cares is shown in the compressed third-order K-map of Fig. 4.51. Two first-order submaps for cells 4 and 6 are also shown to demonstrate that the don't care (ϕ) is treated as an EV, which it is.

(a) Represent the function Z in canonical SOP and POS form by using coded notation.

Noting that the Map Key is $2^{4-3} = 2$, the results can be written directly in canonical SOP and POS form by counting by 2's or by making use of first-order submaps in D, and by applying the minterm and maxterm codes, respectively. For example, cell 3 represents m_6 or M_7, cell 4 represents $(\phi m_8 + m_9)$ or ϕM_8, and so on. Proceeding in this manner, the results are given by

$$Z(A, B, C, D) = \sum m(0, 1, 5, 6, 9, 10, 11) + \phi(2, 3, 8, 13)$$
$$= \prod M(4, 7, 12, 14, 15) \cdot \phi(2, 3, 8, 13), \qquad (4.73)$$

where knowing one canonical form yields the other through observation of the missing numbers in the former.

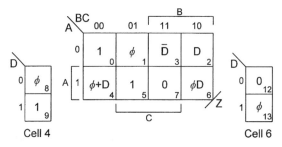

FIGURE 4.51
Compressed K-map for Example 4.2 showing sample first-order submaps.

4.12 WORKED EV K-MAP EXAMPLES 183

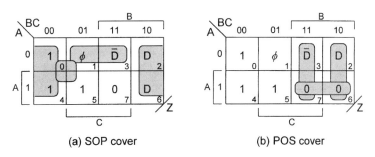

FIGURE 4.52
(a) Minimum SOP cover and (b) minimum POS cover for function Z of Example 4.2.

(b) Extract minimum SOP and minimum POS cover for the function Z.

In Fig. 4.52 are the third-order K-maps showing the minimum SOP and minimum POS cover for the function Z. Notice that the subfunction in cell 6 is interpreted differently in the SOP and POS K-maps.

From reading this cover, the results are

$$Z_{SOP} = \bar{A}C\bar{D} + \bar{C}D + \bar{B}$$
$$Z_{POS} = (\bar{B} + C + D)(\bar{B} + \bar{C} + \bar{D})(\bar{A} + \bar{B}),$$

which are seen to be logically equivalent but not algebraically equal. Notice that the 1's in the SOP K-map are looped out as the octad $\bar{B}$ by using $\phi_8 = 1$ in cell 4 of Fig. 4.51 to give $\phi_8 + D = 1$. Also, note that the 0 in cell 6 of the POS K-map in Fig. 4.51 is looped out as the quad $\bar{A} + \bar{B}$ by using $\phi_{13} = 0$ to give $\phi_{13}D = 0$. Thus, ϕ_{13} is used as a 1 for minimum SOP extraction but as a 0 for minimum POS extraction, meaning that the SOP and POS expressions cannot be algebraically equal.

EXAMPLE 4.3 A four-variable function $F(A, B, C, D)$ containing don't cares is compressed into the truth table given in Fig. 4.53.

(a) Represent the function F in a second-order K-map, and express F in canonical SOP and POS form by using coded notation.
(b) By proper interpretations of the don't care subfunctions, loop out the minimum SOP and POS cover from the second-order K-map and give the gate/input tallies for each.

A	B	F
0	0	C·D
0	1	C·(ϕ + $\bar{D}$)
1	0	(ϕ + C + D)
1	1	0

FIGURE 4.53
Compressed truth table for a function F of four variables.

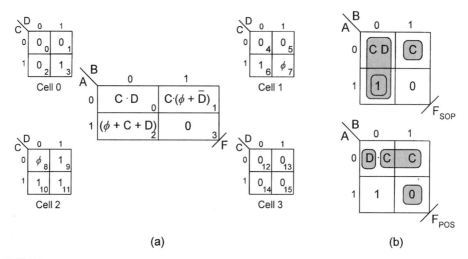

FIGURE 4.54
(a) Second-order compressed K-map and its submaps for the four-variable function given in the EV truth table of Fig. 4.53. (b) EV K-maps showing minimum SOP cover and minimum POS cover.

(a) The simplest means of obtaining the canonical forms from Fig. 4.53 is to use a second-order K-map. Shown in Fig. 4.54a is the second-order compressed K-map together with its submaps for a Map Key of $2^{4-2} = 4$. By reading the submaps directly, the canonical forms become

$$F = \sum m(3, 6, 9, 10, 11) + \phi(7, 8)$$
$$= \prod M(0, 1, 2, 4, 5, 12, 13, 14, 15) \cdot \phi(7, 8). \quad (4.74)$$

(b) The compressed second-order K-maps for the function F are given in Fig. 4.54b. From these K-maps the minimum SOP and minimum POS expressions are found to be

$$F_{SOP} = \bar{B}CD + \bar{A}BC + A\bar{B}$$
$$F_{POS} = (A + B + D)(A + C)(\bar{A} + \bar{B}),$$

with gate/input tallies of 4/11 and 4/10, respectively, excluding possible inverters. Notice that the minimum SOP and POS cover results from these K-maps by taking $\phi_7 = 1$ to give $C(\phi_7 + \bar{D}) = C$ in cell 1, and by taking $\phi_8 = 1$ to give $(\phi_8 + C + D) = 1$ in cell 2. Because the don't cares, ϕ_7 and ϕ_8, are used in the same way (no shared use) in both K-maps of Fig. 4.54b, the minimum SOP and POS expressions are algebraically equal.

EXAMPLE 4.4 A five-variable function f is given in the canonical form:

$$f(A, B, C, D, E) = \sum m(3, 9, 10, 12, 13, 16, 17, 24, 25, 26, 27, 29, 31). \quad (4.75)$$

(a) Use a fourth-order EV K-map to minimize this function in both SOP and POS form.
A compression of one order requires that the Map Key be 2. Therefore, each cell of the fourth-order EV map represents a first-order submap covering two possible minterm or

4.12 WORKED EV K-MAP EXAMPLES

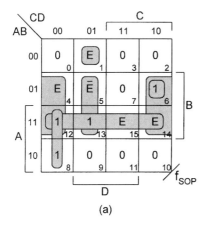
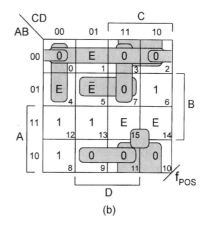

FIGURE 4.55
Fourth-order EV K-maps for the five-variable function f in Eq. (4.75). (a) Minimum SOP cover and (b) minimum POS cover.

maxterm positions. In Fig. 4.55 are the fourth-order EV K-maps showing minimum SOP and minimum POS cover for which the expression are

$$f_{SOP} = \bar{A}\bar{B}\bar{C}DE + B\bar{C}D\bar{E} + B\bar{D}E + ABE + \bar{A}BC\bar{D} + AC\bar{D}$$

$$f_{POS} = (A + C + D + E)(A + \bar{B} + \bar{D} + \bar{E})(\bar{A} + \bar{C} + E)(A + B + E)(\bar{A} + B + \bar{D})$$
$$\cdot (A + \bar{C} + \bar{D})(A + B + D)(B + \bar{C}).$$

Notice that the loop-out protocol is applied first to the EVs and then to the 1's or 0's as a "cleanup" operation, a practice that should always be followed. Also, notice that for the POS result, the term $(B + \bar{D} + E)$ is an OPI for the term $(A + B + E)$.

(b) Find the minimum SOP cover for the five-variable function in Eq. (4.75) by using conventional (1's and 0's) $A\|BC/DE$ format K-map similar to that used for a six-variable function in Fig. 4.37.

Shown in Fig. 4.56 is the conventional (1's and 0's) K-map indicating minimum SOP cover.

EXAMPLE 4.5 Map the reduced function in Eq. (4.76) into a fourth-order K-map and extract minimum SOP and POS cover. Give the gate/input tally for each result, exclusive of possible inverters.

$$Y = A\bar{B}\bar{C}\bar{D}E + ABCD + ABDE + BC\bar{D}\bar{E} + \bar{A}B\bar{C}\bar{D}\bar{E} + \bar{A}BDE + \bar{A}BC\bar{E} \quad (4.76)$$
$$+ \bar{B}C\bar{D}E + \bar{A}CDE + A\bar{B}CE$$

The function of Eq. (4.76) is mapped into the fourth-order K-map shown in Fig. 4.57, and the minimum SOP and minimum POS covers are indicated with shaded loops. The resulting minimum expressions are given by

$$Y_{SOP} = A\bar{B}\bar{D}E + \bar{A}B\bar{D}\bar{E} + BC\bar{E} + BDE + \bar{B}CE$$

$$Y_{POS} = (\bar{B} + D + \bar{E})(C + \bar{D} + E)(B + E)(\bar{A} + C + E)(B + C + \bar{D})(A + B + C),$$

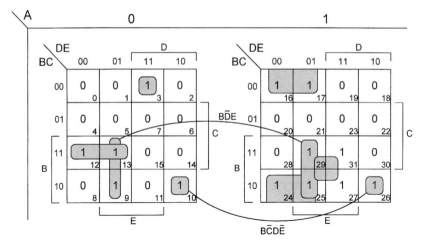

FIGURE 4.56
Conventional (1's and 0's) format of the type $A\|BC/DE$ for the function f in Eq. (4.75).

which represent gate/input tallies of 6/22 and 7/23, respectively, for the Y_{SOP} and Y_{POS} results, exclusive of inverters. Notice that the 1's in the SOP K-map are covered by the quads $BC\bar{E}$ and BDE, and that the 0 in cell 12 of the POS K-map is covered by the quads $(\bar{B}+D+\bar{E})$ and $(\bar{A}+C+E)$.

EXAMPLE 4.6 Compress the following function into a second-order K-map and extract minimum SOP and POS cover:

$$Z(A, B, C, D) = \prod M(2, 4, 7, 11, 12, 14, 15)$$
$$= \sum m(0, 1, 3, 5, 6, 8, 9, 10, 13). \tag{4.77}$$

In Fig. 4.58 are the second-order EV K-maps and submaps showing minimum SOP and

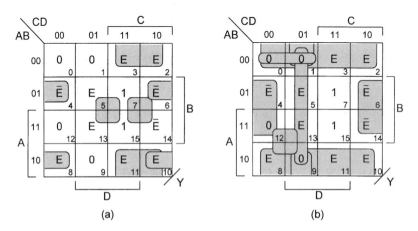

FIGURE 4.57
(a) Minimum SOP cover and (b) minimum POS cover for the function Y in Eq. (4.76).

4.12 WORKED EV K-MAP EXAMPLES 187

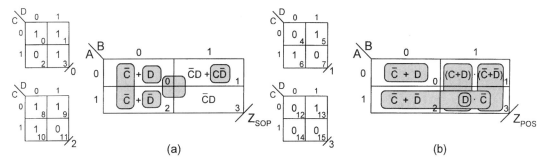

FIGURE 4.58
Second-order EV K-maps and submaps for the function Z in Eq. (4.77) showing (a) minimum SOP cover and (b) minimum POS cover.

minimum POS cover for the function Z with the results given by

$$Z_{SOP} = \bar{A}BC\bar{D} + \bar{A}\bar{B}D + A\bar{B}\bar{D} + \bar{B}\bar{C} + \bar{C}D$$

$$Z_{POS} = (A + B + \bar{C} + D)(\bar{B} + \bar{C} + \bar{D})(\bar{B} + C + D)(\bar{A} + \bar{B} + D)(\bar{A} + \bar{C} + \bar{D})$$

Notice how easy it is to read a subfunction when accompanied by a submap. Thus, the SOP term $\bar{C}D$ is easily observed to be present in each of the four submaps of Fig. 4.58a. Similarly, $\bar{C}D$, read as a POS term in Fig. 4.58b, is seen to contain both the $(C + D)$ and $(\bar{C} + \bar{D})$ terms by a cursory inspection of the submaps.

EXAMPLE 4.7 Compress the following six-variable function into a fourth-order EV K-map and extract minimum SOP and minimum POS cover for it:

$$W(A, B, C, D, E, F) = \sum m(4, 6, 8, 9, 10, 11, 12, 13, 14, 15, 20, 22, 26, 27, 30, 31,$$
$$36, 38, 39, 52, 54, 56, 57, 60, 61). \quad (4.78)$$

Compressing a six-variable function into a fourth-order K-map requires a Map Key of $2^{6-4} = 4$, hence four possible minterms per K-map cell. This is a second-order compression meaning that each cell of the K-map contains subfunctions from a second-order K-map. Shown in Fig. 4.59 are the fourth-order K-maps for the function W in Eq. (4.78) where the EVs are E and F. The minimum covers for the SOP and POS functions are indicated by shaded loops and yield

$$W_{SOP} = A\bar{B}\bar{C}DE + ABC\bar{E} + \bar{C}D\bar{F} + \bar{A}CE + \bar{A}\bar{B}$$

$$W_{POS} = (A + \bar{B} + \bar{C} + E)(C + E + \bar{F})(\bar{B} + C + \bar{F})(A + C + \bar{F})(\bar{A} + \bar{C} + \bar{E})$$
$$\cdot (\bar{A} + B + \bar{C})(C + D),$$

which represent gate/input tallies of 6/23 and 8/28, respectively. Note that the loop-out protocol is applied first to the EVs and then to the 1's and 0's as cleanup operations, a procedure that should always be followed. Observe also that these expressions are algebraically equal since no don't cares are involved.

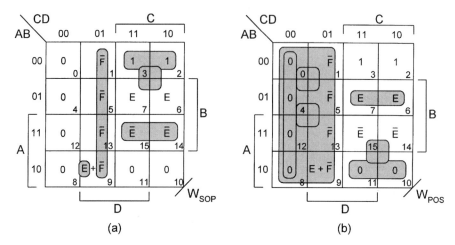

FIGURE 4.59
Fourth-order EV K-maps for the six-variable function W in Eq. (4.78) showing (a) minimum SOP cover and (b) minimum POS cover.

FURTHER READING

Nearly every text on digital or logic design provides some coverage of conventional (1's and 0's) K-map methods. Typical among these are texts of Hill and Peterson; Nelson, Nagle, Carroll and Irwin; and Roth.

[1] F. J. Hill and G. R. Peterson, *Digital Logic and Microprocessors*. John Wiley & Sons, New York, 1984.
[2] V. P. Nelson, H. T. Nagle, B. D. Carroll, and J. D. Irwin, *Digital Logic Circuit Analysis and Design*. Prentice Hall, Englewood Cliffs, NJ, 1995.
[3] C. H. Roth, *Fundamentals of Logic Design*, 4th ed., West, St. Paul, MN, 1992.

References on entered variable (EV) K-map methods are limited to only a few texts. The best sources appear to be the texts of Comer, Shaw, and Tinder.

[4] D. J. Comer, *Digital Logic and State Machine Design*, 3rd ed., Sanders College Publishing, Fort Worth, TX, 1995.
[5] A. W. Shaw, *Logic Circuit Design*. Sanders College Publishing, Fort Worth, TX, 1993.
[6] R. F. Tinder, *Digital Engineering Design: A Modern Approach*. Prentice Hall, Englewood Cliffs, NJ, 1991.

A classic contribution to logic minimization generally, is found in the text of Kohavi. There, can be found early treatment of the algebraic, tabular, and mapping methods. Entered variable K-map methods are not treated in this text.

[7] Z. Kohavi, *Switching and Finite Automata Theory*. McGraw-Hill, New York, 1978.

The two logic minimization algorithms briefly discussed in this chapter, the Quinn–McCluskey method and the Espresso approach (which is a heuristic algorithm), have been cited in countless publications. A few representative sources of these methods are presented here. Included are some of the original references as well as some of the more current ones, which often provide useful summaries of the methods.

[8] E. J. McCluskey, *Logic Design Principles*. Prentice-Hall, Englewood Cliffs, NJ, 1986.
[9] E. J. McCluskey, "Minimization of Boolean Functions," *Bell Syst. Tech. J.* 35(5), 1417–1444 (1956).
[10] W. V. Quine, "The Problem of Simplifying Truth Functions," *Am. Math Monthly* 59(8), 521–531 (1952).
[11] R. K. Brayton, G. Hachtel, C. McMullen, and A. Sangiovanni-Vincentelli, *Logic Minimization Algorithms for VLSI Synthesis*. Kluwer Academic Publishers, Boston, 1984.
[12] R. Rudell and A. Sangiovanni-Vincentelli, "Multiple-valued Minimization for PLA Optimization," *IEEE Transactions on CAD/CAS* **CAD-6**(5), 727–750 (1987).
[13] R. K. Brayton, P. C. McGeer, J. V. Sanghavi, and A. L. Sangiovanni-Vincentelli, "A New Exact Minimizer for Two-Level Logic Synthesis," in *Logic Synthesis and Optimization* (T. Sasao, Ed.). Kluwer Academic Publishers, Boston, 1993.

References on the factorization, resubstitution, and decomposition methods of optimization of multilevel circuits are fairly numerous but are set in fairly advanced notation. Perhaps the most useful are those found in texts by De Micheli, Kohavi, and Dietmeyer, and in the reference book edited by Sasao. Advanced preparation by the reader is recommended for use of these references. The text of De Micheli also has useful discussions of the area/delay trade-off factors.

[14] G. De Micheli, *Synthesis and Optimization of Digital Circuits*. McGraw-Hill, New York, 1994.
[15] D. L. Dietmeyer, *Logic Design of Digital Systems,* 2nd ed., Allyn and Bacon, Boston, 1971.
[16] M. Fujita, Y. Matsunaga, Y. Tamiya, and K.-C. Chen, "Multi-Level Logic Minimization of Large Combinational Circuits by Partitioning," in *Logic Synthesis and Optimization* (T. Sasao, Ed.). Kluwer Academic Publishers, Boston, 1993.
[17] Z. Kohavi, *Switching and Finite Automata Theory*. McGraw-Hill, New York, 1978.

PROBLEMS

4.1 Expand each of the following expressions into canonical (literal) form by using the appropriate Boolean laws:
(a) $e(a, b) = a + \bar{b}$
(b) $f(x, y) = x + \bar{x}y$
(c) $g(A, B, C) = A\bar{B}C + \bar{A}BC + AB + BC + \bar{A}B\bar{C}$
(d) $h(X, Y, Z) = (X + Y)(\bar{X} + \bar{Y} + Z)(Y + \bar{Z})(\bar{X} + Y + \bar{Z})$
(e) $E(A, B, C, D) = (\bar{A} + \bar{B}C)(B + D)(\bar{A} + C + D)(A + \bar{B} + C + \bar{D})(\bar{B} + D)$
(f) $F(w, x, y, z) = wxy\bar{z} + \bar{w}\bar{x}z + x\bar{y}z + w\bar{x}yz + xz + w\bar{x}\bar{y}\bar{z} + \bar{w}x\bar{y}z$
(g) $G(a, b, c, d,) = (a + \bar{b} + c + \bar{d})(b + \bar{c} + \bar{d})(\bar{a} + b)(\bar{b} + d)(\bar{a} + c + d)$
(h) $H(V, W, X, Y) = VW\bar{X}\bar{Y} + \bar{X}Y + WX\bar{Y} + V\bar{W}XY + VXY + \bar{V}\bar{W}XY + \bar{W}\bar{X}\bar{Y}$

4.2 Place each of the three-variable functions below in a canonical truth table and in a conventional (1's and 0's) K-map. Place the variables on the K-map axes in alphabetical

order beginning with the ordinate (vertical) axis, as has been done throughout this text.
(a) $P(A, B, C) = (A + B + \bar{C})(A + \bar{B} + \bar{C})(A + B + C)(\bar{A} + B + \bar{C})(\bar{A} + \bar{B} + \bar{C})$
(b) $Q(a, b, c) = \sum m(1, 2, 4, 5, 6)$
(c) $W(a, b, c) = a\bar{b}c + \bar{a}\bar{b}\bar{c} + ab\bar{c} + abc + a\bar{b}\bar{c}$
(d) $X(A, B, C) = \prod M(0, 1, 2, 6, 7)$
(e) $Y(w, x, y) = wx + \bar{x}\bar{y} + w(x \oplus y) + \bar{w}\bar{y}$ (Hint: Expand first.)
(f) $Z(A, B, C) = (A + B) \odot (\bar{A}C) + A\bar{B}$ (Hint: First construct a truth table with input A.)
(g) $F(X, Y, Z) = XY \oplus YZ \oplus XZ + X\bar{Y}$ [Hint: See Eq. (3.33).]

4.3 Place each of the four-variable functions below in a canonical truth table and in a conventional (1's and 0's) K-map. Place the variables on the K-map axes in alphabetical order beginning with the ordinate (vertical) axis, as has been done throughout this text.
(a) $R(u, v, w, x) = \sum m(0, 2, 3, 7, 8, 9, 10, 11, 13)$
(b) $S(a, b, c, d) = (a + \bar{b})(\bar{a} + b\bar{c})(\bar{b} + \bar{c})(\bar{a} + \bar{b} + c)$
(c) $T(W, X, Y, Z) = YZ + \bar{W}\bar{X}Y + WX\bar{Y}\bar{Z} + \bar{X}\bar{Y}Z + \bar{W}\bar{Y}Z + WXY\bar{Z} + X\bar{Y}Z$
(d) $U(A, B, C, D) = \prod M(0, 5, 8, 9, 11, 12, 15)$
(e) $V(a, b, c, d) = \sum m(0, 4, 5, 7, 8, 9, 13, 15)$
(f) $W(u, v, w, x) = [(v + w) \odot x](u + w)(u + \bar{v})(u + x)$
(g) $X(A, B, C, D) = (A \oplus B)\bar{C}\bar{D} + \bar{B}\bar{C}D + BCD + (\bar{A} + \bar{B})C\bar{D} + \bar{A}B(C \odot D)$
(Hint: First construct a truth table for CD, then map the result into a 1's and 0's K-map.)
(h) $F(W, X, Y, Z) = (X \oplus Z) \oplus [W(Y \oplus Z)] + X\bar{Y}\bar{Z}$
(Hint: First construct a truth table for WX, then map the result into a 1's and 0's K-map.)

4.4 Place each function of Problem 4.1 into a conventional (1's and 0's) K-map and extract canonical (coded) SOP and POS expressions from that K-map.

4.5 Minimize each function of Problem 4.2 in both SOP and POS form with a third-order K-map. By using the gate/input tally (exclusive of possible inverters) determine which is simpler, the SOP or POS expression. Do not implement with logic gates.

4.6 Minimize each function of Problem 4.3 in both SOP and POS form with a fourth-order K-map. By using the gate/input tally (exclusive of possible inverters), determine which is simpler, the SOP or POS expression. Do not implement with logic gates.

4.7 The following three-variable functions are incompletely specified functions, that is, they contain don't cares. By using a third-order K-map, minimize each function in both SOP and POS form *with and without* the use of the don't cares in each case. Identify any OPIs that may be present.
(a) $e(A, B, C) = \sum m(0, 1, 2, 7) + \phi(3, 5)$
(b) $f(X, Y, Z) = \prod M(3, 4, 6) \cdot \phi(0, 2)$
(c) $g(a, b, c) = \sum m(0, 1, 5, 7) + \phi(2, 4)$
(d) $h(x, y, z) = \prod M(3, 4, 5) \cdot \phi(0, 1, 2)$
(e) $i(X, Y, Z) = \sum m(0, 5) + \phi(1, 2, 3, 7)$

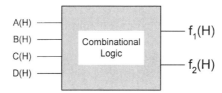

FIGURE P.4.1

4.8 The following four-variable functions are incompletely specified functions — they contain don't cares. Use a conventional (1's and 0's) K-map to minimize each function in both SOP and POS form and, with the help of the gate/input tally (exclusive of possible inverters), indicate which is simpler. Also, identify any OPIs that may be present.
(a) $E(a, b, c, d) = \sum m(6, 11, 12, 13, 14) + \phi(0, 1, 2, 3, 4, 5)$
(b) $F(A, B, C, D) = \prod M(0, 3, 6, 11, 13, 15) \cdot \phi(5, 8, 10, 14)$
(c) $G(W, X, Y, Z) = \sum m(0, 4, 6, 8, 9, 10, 11, 14, 15) + \phi(1, 5)$
(d) $H(w, x, y, z) = \prod M(1, 2, 3, 9, 10, 14) \cdot \phi(11, 13)$
(e) $I(A, B, C, D) = \sum m(4, 5, 7, 12, 14, 15) + \phi(3, 8, 10)$
(f) $J(a, b, c, d) = \prod M(0, 1, 2, 5, 7, 9) \cdot \phi(4, 6, 10, 13)$

4.9 Find the optimum cover (either SOP or POS) for the following four-input/two-output system (see Fig. P4.1). Base your choice on the total gate/input tally (including inverters) for the system. Assume the inputs and outputs are all active high. Do not construct the logic circuit.

$$f_1 = \sum m(0, 2, 4, 5, 9, 10, 11, 13, 15)$$
$$f_2 = \sum m(2, 5, 10, 11, 12, 13, 14, 15)$$

4.10 Three functions, each of three inputs, are given in canonical SOP form. Follow the discussion in Section 4.5 and find the optimized SOP minimum for the three functions taken as a system. Give the total gate/input tally for the system, exclusive of inverters.

$$f_1(A, B, C) = \sum m(1, 3, 5, 6, 7)$$
$$f_2(A, B, C) = \sum m(0, 1, 3, 6)$$
$$f_3(A, B, C) = \sum m(0, 5, 7)$$

4.11 Two functions, each of four variables, are given in canonical SOP form. Follow the discussion in Section 4.5 and find the optimized SOP and POS minima for the two functions taken as a system. By using the gate/input tally, exclusive of inverters, indicate which is simpler, the SOP result or the POS result.

$$F_1(A, B, C, D) = \sum m(7, 8, 10, 14, 15) + \phi(1, 2, 5, 6)$$
$$F_2(A, B, C, D) = \sum m(1, 5, 7, 8, 11, 14, 15) + \phi(2, 3, 10)$$

4.12 The two four-variable functions shown are presented in canonical POS form. Follow the discussion in Section 4.5 and find the optimized SOP and POS minima for the two functions taken as a system. Use the gate/input tally, including inverters, to determine which is simpler, the SOP result or the POS result. Implement the simpler of the two forms in either NAND/INV or NOR/INV logic. Assume that the inputs and outputs are all active high.

$$g_1(A, B, C, D) = \prod M(0, 3, 4, 11, 12, 13, 15) \cdot \phi(2, 5, 6)$$
$$g_2(A, B, C, D) = \prod M(0, 1, 9, 12, 13) \cdot \phi(2, 3, 4, 10)$$

4.13 Given below is a set of three functions, each of four variables. Follow the discussion in Section 4.5 and find the optimized SOP and POS minima for the three functions taken as a system. Use the gate/input tally, excluding inverters, to determine which is simpler, the SOP result or the POS result. [Hint: In determining the shared PIs, don't forget to include the ANDed and ORed functions $(y_1 \cdot y_2 \cdot y_3)$ and $(y_1 + y_2 + y_3)$.]

$$y_1(a, b, c, d) = \sum m(0, 1, 2, 5, 7, 8, 10, 14, 15)$$
$$y_2(a, b, c, d) = \sum m(0, 2, 4, 5, 6, 7, 10, 12)$$
$$y_3(a, b, c, d) = \sum m(0, 1, 2, 3, 4, 6, 8, 9, 10, 11)$$

4.14 Extract minimum SOP and POS expressions (cover) from the K-maps shown in Fig. P4.2. Where appropriate, application of the loop-out protocol discussed in Section 4.4 will help to avoid redundancy.

4.15 Following the discussion in Section 4.6, compress each function in Problem 4.2 into a second-order K-map (Map Key = 2) and extract minimum SOP and POS cover. Use the LSB variable as the entered variable (EV).

4.16 Following the discussion in Section 4.6, compress each function in Problem 4.3 into a third-order K-map (Map Key = 2) and extract minimum SOP and POS cover. Use the LSB variable as the entered variable (EV).

4.17 Following the discussion in Section 4.6, compress each function in Problem 4.7 into a second-order K-map (Map Key = 2) and extract minimum SOP and POS cover. Use the LSB variable as the entered variable (EV).

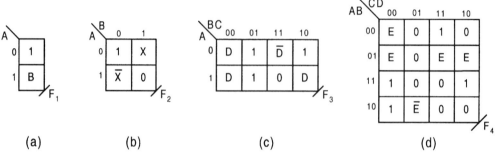

FIGURE P.4.2

4.18 Following the discussion in Section 4.6, compress each function in Problem 4.8 into a third-order K-map (Map Key = 2) and extract minimum SOP and POS cover. Use the LSB variable as the entered variable (EV).

4.19 Following the discussion in Section 4.7, compress each of the following five-variable functions into a fourth-order K-map (Map Key = 2) and extract minimum SOP and POS cover. Use the LSB variable as the entered variable (EV).
  (a) $q(A, B, C, D, E) = \prod M(0, 1, 2, 5, 14, 16, 17, 18, 19, 21, 26, 27, 30)$
  (b) $r(A, B, C, D, E) = A\bar{B}CE + \bar{A}CDE + \bar{B}C\bar{D}E + \bar{A}BC\bar{E} + \bar{A}BDE$
    $+ \bar{A}B\bar{C}\bar{D}\bar{E} + BC\bar{D}\bar{E} + ABD + B\bar{C}D + ABCDE$
  (c) $s(A, B, C, D, E) = \sum m(0, 2, 4, 5, 7, 10, 13, 15, 21, 23, 24, 25, 28, 29, 30)$
  (d) $t(A, B, C, D) = (A + B + D + \bar{E})(\bar{B} + \bar{C} + \bar{D} + E)(\bar{A} + \bar{B} + E)(\bar{A} + C + D + \bar{E})$
    $\cdot (\bar{B} + C + D)(B + C + \bar{D} + E)(A + B + \bar{C} + \bar{D} + \bar{E})(\bar{B} + \bar{C} + D + E)$
    $\cdot (A + \bar{B} + C)(\bar{B} + C + \bar{D})(B + \bar{C} + D + \bar{E})$

4.20 Minimize each function of Problem 4.19 in both SOP and POS by using a conventional (1's and 0's) K-map. To do this follow the example in Fig. 4.56.

4.21 Following the discussion in Section 4.6, compress each function in Problem 4.2 into a first-order K-map (Map Key = 4) and extract a minimum SOP and POS expression for each. Use the last two significant bit variables as EVs.

4.22 Following the discussion in Section 4.6, compress each function in Problem 4.3 into a second-order K-map (Map Key = 4) and extract a minimum SOP and POS expression for each. Use the last two significant bit variables as EVs.

4.23 Following the discussion in Section 4.6, compress each function in Problem 4.7 into a first-order K-map (Map Key = 4) and extract a minimum SOP and POS expression for each. Use the last two significant bit variables as EVs.

4.24 Following the discussion in Section 4.6, compress each function in Problem 4.8 into a second-order K-map (Map Key = 4) and extract a minimum SOP and POS expression for each. Use the last two significant bit variables as EVs.

4.25 Compress each function in Problem 4.19 into a third-order K-map (Map Key = 4) and extract a minimum SOP and POS expression for each. Use the last two significant bit variables as EVs.

4.26 Shown in Fig. P4.3 are two functions, F and Z, each of four variables, that have been compressed into third-order K-maps. (Hint: It will help to first simplify the

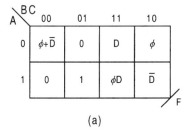
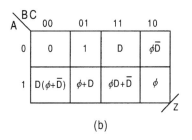

FIGURE P.4.3

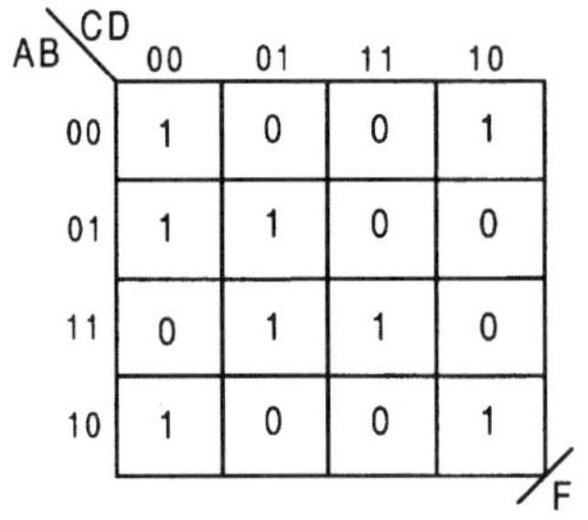

FIGURE P.4.4

subfunctions in cells 4 and 7 of function Z by applying the laws of Boolean algebra while treating the ϕ as an entered variable.)

(1) By following the example in Fig. 4.33a, construct the first-order submap for each of the eight cells in each K-map.
(2) Give the canonical SOP and POS expressions in *code* form for each function.
(3) Extract the minimum SOP and POS forms from each third-order K-map, keeping in mind the discussion on the use of don't cares in Subsection 4.6.1.

4.27 Compress the fourth-order K-map in Fig. P4.4 into a second-order K-map (Map Key = 4) and loop out minimum SOP and POS expressions according to the following format:
(a) Set A, B as the axis variables.
(b) Set B, C as the axis variables.
(c) Set A, C as the axis variables.

4.28 Compress the following function into a fourth-order K-map (Map Key = 4) and extract minimum SOP and POS forms. By using the gate/input tally (exclusive of inverters), indicate which form is simpler.

$$Y(A, B, C, D, E, F) = \prod M(0, 1, 5, 7, 9, 15, 16, 18, 21, 24, 29, 31, 35, 37, 39,$$
$$40, 45, 49, 50, 56, 58, 60, 61, 63)$$

4.29 Shown in Fig. P4.5 is a function of six variables that has been compressed into a third-order K-map, hence a third-order compression (Map Key = 8).

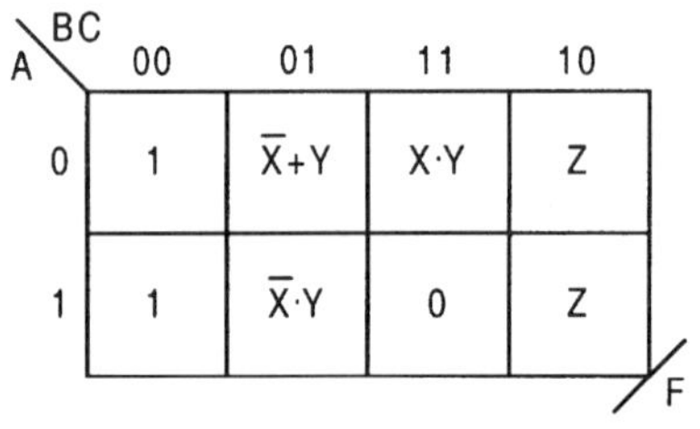

FIGURE P.4.5

(a) From the third-order K-map, write the canonical coded SOP and POS for this function.

(b) Use this K-map to extract the minimum SOP and POS expressions for this function.

4.30 Find the minimum SOP and POS expressions (cover) for each of the following sub-functions and give the cell location of each sub-function in the fourth-order K-map.
(a) $P(A, B, C, D, E, F, G) = \sum m(33, 34, 36, 38) + \phi(32, 29)$
(b) $Q(a, b, c, d, e, f, g, h) = \sum m(114, 116, 118, 122, 124, 126)$
(c) $R(A, B, C, D, E, F, G) = \prod M(105, 107, 108, 109, 110)$
(d) $S(a, b, c, d, e, f, g, h) = \prod M(176, 181, 182, 183, 184, 189, 191)$
    $\cdot \phi(177, 185, 190)$

4.31 Minimize each of the following functions in both SOP and POS form by using the Quine–McCluskey (Q-M) algorithm discussed in Section 4.8.
(a) $f(w, x, y) = \sum m(0, 1, 3, 5, 7)$
(b) $g(a, b, c) = \prod M(2, 3, 4, 6)$
(c) $F(W, X, Y, Z) = \sum m(0, 2, 4, 5, 6, 8, 10, 11, 13, 14)$
(d) $G(A, B, C, D) = \prod M(1, 2, 3, 5, 7, 9, 11, 12, 14)$

4.32 Minimize each of the functions of Problem 4.7 in both SOP and POS form by using the Quine–McCluskey (Q-M) algorithm discussed in Section 4.8. Keep in mind the manner in which the Q-M algorithm treats don't cares.

4.33 Minimize each of the functions of Problem 4.8 in both SOP and POS form by using the Quine–McCluskey (Q-M) algorithm discussed in Section 4.8. Keep in mind the manner in which the Q-M algorithm treats don't cares.

4.34 Use the method of *factorization* to obtain a gate-minimum SOP and POS result for the following two-level functions. Find the gate/input tally (including inverters) for each and compare the results with the two-level minimum forms. Assume the inputs all arrive from positive logic sources. (Hint: First minimize the functions in two-level form and then apply the factorization method.)
(a) $Y = \bar{A}B + B\bar{D} + \bar{A}C + ABC + AC\bar{D}$
(b) $F = A\bar{B}\bar{D}\bar{E} + AB\bar{C}\bar{E} + C\bar{D}E + BCDE + ABCD + (A \odot C)(B + \bar{D})$

4.35 Use the *resubstitution method* discussed in Subsection 4.9.2 to obtain a gate minimum for each of the following functions. Compare the gate/input tally (excluding invertors) of the result with that for the two-level minimum. Also, comment on fan-in and inverter requirements for each, and on the gate propagation delay level for each. Assume that all inputs are active high. (Hint: First obtain the two-level SOP minimum expression, then plan to use the suggested divisor given for each.)
(a) $F(W, X, Y, Z) = \sum m(0, 4, 5, 7, 10, 13, 14, 15)$ (Use divisor $X + \bar{Z}$)
(b) $G(A, B, C, D) = \sum m(0, 1, 2, 3, 4, 9, 10, 11, 13, 14, 15)$
                                                          (Use divisor $\bar{A} + C + D$)
(c) $H(W, X, Y, Z) = \prod M(0, 2, 4, 6, 9)$ (Your choice of divisor)

4.36 Decompose each function in Problem 4.31 by applying *Shannon's expansion theorem* discussed in Subsection 4.9.3. Try at least two sets of two-variable axes about which each expansion is to be performed. Compare the best expansion result for each with its two-level K-map minimum result.

4.37 Use BOOZER, the logic minimizer bundled with this text, to verify or compare (if applicable) the results with any of the previously stated problems. For example, Problem 4.37/4.6c would require use of BOOZER to minimize Problem 4.3c, since Problem 4.6 refers to Problem 4.3. [Hint: To obtain a minimum POS result by using BOOZER, map the function, complement each cell of the K-map, enter the results into BOOZER and minimize as an SOP function, and then complement the BOOZER result. Noter that either entered variables (EVs) or 1's and 0's can be entered into the BOOZER algorithm — your choice.]

CHAPTER 5

Function Minimization by Using K-map XOR Patterns and Reed–Muller Transformation Forms

5.1 INTRODUCTION

In this chapter it will be shown how simple "pencil-and-paper" methods can be used to extract gate-minimum multilevel logic designs not yet possible by any conventional method, including the use of CAD techniques. The methods described here make possible multilevel IC designs that occupy much less real estate than would be possible for an equivalent two-level design, and often with little or no sacrifice in speed — an advantage for VLSI design.

There are a variety of approaches to logic function minimization, which can be divided into two main categories: two-level and multilevel approaches. Chapter 4 was devoted primarily to the two-level approach to minimization. Combining entered variable (EV) subfunctions and the XOR patterns (described in the following section) in a K-map extraction process is a special and powerful form of multilevel function minimization. Used with two-level logic forms (AND and OR functions) this multilevel minimization approach leads to XOR/SOP, EQV/POS, and hybrid forms that can represent a substantial reduction in the hardware not possible otherwise. XOR/SOP and EQV/POS forms are those connecting p-terms (product terms) with XOR operators or s-terms (sum-terms) with EQV operators, respectively. Hybrid forms are those containing a mixture of these.

Another approach to multilevel logic minimization involves the use of Reed–Muller transformation forms (discussed in Sections 5.5 through 5.12) that are partitioned (broken up) into tractable parts with the assistance of entered variable Karnaugh maps (EV K-maps). The process is called the contracted Reed–Muller transformation (CRMT) minimization method and is expressly amenable to classroom (or pencil-and-paper) application. General information covering the subjects associated with Reed–Muller minimized logic synthesis are cited in Further Reading at the end of this chapter.

The word *level* (meaning level of a function) refers to the number of gate path delays from input to output. In the past the XOR gate (or EQV gate) has been viewed as a two-level device, meaning two units of path delay as implied by the defining relations for XOR and EQV given by Eqs. (3.4) and (3.5). But the emergence of CMOS IC technology has moved the XOR and EQV gates close to single-level gates with respect to compactness and speed, as is evident from Figs. 3.26 and 3.27. The term *multilevel*, as used in this text, means the use of XOR and/or EQV gates together with two-level logic to form multiple levels of path delay as measured from input to output.

The concept of minimization, as used in this text, is presented in terms of three degrees. A *minimum* result is one that yields the lowest gate/input tally for a particular method used, for example, a two-level minimum result, but may not be the lowest possible. An *exact* minimization designates a result that has the fewest p-terms possible in an expression or the fewest s-terms possible in an expression. An *absolute* minimum expression is one that has the lowest possible gate/input tally considering all possible methods of minimization. Thus, an absolute minimum is a *gate/input-tally minimum* (or simply *gate-minimum*) and is usually the result of a specific or unique method of minimization. As a reminder, the gate/input tally (defined in Subsection 4.4.3) will usually be given exclusive of possible inverters. Only when the input activation levels are known can the gate/input tally include the inverter count.

Where appropriate to do so, reference will be made to the defining relations for XOR and EQV given by Eqs. (3.4) and (3.5) and to the XOR and EQV laws, corollaries, and identities presented in Section 3.10. Reference will also be made to minterm code (logic 0 for a complemented variable and logic 1 for an uncomplemented variable), and to maxterm code which is the dual of minterm code as discussed in Section 4.2. The EV K-map methods used in this chapter may be considered as an extension of the conventional methods discussed in Sections 4.6 and 4.7.

5.2 XOR-TYPE PATTERNS AND EXTRACTION OF GATE-MINIMUM COVER FROM EV K-MAPS

There are four types of XOR patterns that can be easily identified in EV K-maps:

1. *Diagonal patterns*
2. *Adjacent patterns*
3. *Offset patterns*
4. *Associative patterns*

References will frequently be made to the so-called *XOR-type patterns* in EV K-maps. These are references to the diagonal, adjacent, offset, and associative patterns listed above and are found only in compressed K-maps. A kth-order K-map compression results when an N-variable function is represented in an nth-order K-map — that is, $k = N - n$. Of the XOR-type patterns, only the offset pattern requires third and higher order K-maps for its appearance. K-maps used in the following discussions are all minterm code based, but are used to extract gate-minimum functions in both minterm code and maxterm code.

Simple examples of the first three patterns are shown in Fig. 5.1a, where a six-variable function has been compressed into a third-order EV K-map. Empty cells 0 and 2 in Fig. 5.1a are to be disregarded so as to focus attention on the patterns: The diagonal pattern formed by cells 1 and 4 is read in minterm code as $\bar{B}X(A \oplus C)$ or in maxterm code as $B + X + A \odot C$.

5.2 XOR-TYPE PATTERNS

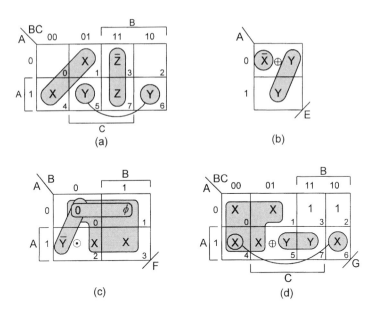

FIGURE 5.1
Examples of XOR patterns in compressed K-maps. (a) Diagonal, adjacent, and offset patterns. (b), (c), (d) Associate patterns.

Notice that the diagonal pattern lies in the $\bar{B}$ domain (B domain in maxterm code) "for all that is X," and that the defining relations for XOR and EQV, Eqs. (3.4) and (3.5), are used for cells 1 and 4 to give $A \oplus C$ and $A \odot C$, respectively, for minterm code and maxterm code. The adjacent pattern is formed by cells 3 and 7 and is read $BC(A \odot Z)$ in minterm code or as $\bar{B} + \bar{C} + A \odot Z$ in maxterm code. Here, the adjacent pattern lies at the intersection of domains B and C in minterm code ($\bar{B} + \bar{C}$ in maxterm code), and again the defining relations for XOR and EQV are used to obtain the minterm and maxterm extraction, respectively. The offset pattern is formed by cells 5 and 6 and is read in minterm code as $AY(B \oplus C)$ and in maxterm code as $\bar{A} + Y + B \odot C$. In this case, the offset pattern lies in the A domain ($\bar{A}$ in maxterm code) "for all that is Y," and the defining relations, Eqs. (3.4) and (3.5), are used for cells 5 and 6 to obtain $B \oplus C$ and $B \odot C$, respectively. Throughout this discussion it is assumed that any entered variable, for example X, Y, or Z, may represent a single variable or a multivariable function of any complexity.

Each of the three XOR-type patterns extracted from Fig. 5.1a has a gate/input tally of 2/5 (excluding inverters). The *gate/input tally* is a measure of logic circuit cost (in hardware and real estate) and is defined in Subsection 4.4.3. The gate count is, of course, the more significant of the two tallies and the input tally includes the inputs to all gates in the logic circuit. Unless stated otherwise, the gate/input tally will exclude inverters and their inputs. By comparison, the two-level logic gate/input tally for each of the patterns in Fig. 5.1a is 3/8.

The associative patterns shown in Figs. 5.1b, 5.1c, and 5.1d may combine with any or all of the other three patterns to form *compound patterns*. For this reason the associative patterns require special consideration and will be dealt with separately in the following subsection.

5.2.1 Extraction Procedure and Examples

Before illustrating the extraction process by example, it will be instructive to outline the extraction procedure. In this procedure, reference will be made to minterm and maxterm codes for clarification purposes. Since all K-maps are minterm code based, extraction of EQV/POS cover from them requires that the K-map domains be complemented, but not the entered variables. Extraction of XOR/SOP cover follows conventional procedure. The following six-step extraction procedure applies generally to all four types of XOR-type patterns.

Extraction Procedure

Step I. Identify the type of EV XOR pattern that exists in the K-map. A diagonal pattern requires identical K-map cell entries in diagonally located cells. An adjacent pattern requires complementary K-map cell entries in logically adjacent cells. An offset pattern requires identical cell entries in cells whose coordinates differ by two bits (a Hamming distance of 2). Associative patterns require terms associated by an XOR or EQV connective in at least one cell.

Step II. Write down the K-map domains in which the XOR pattern exists and any subfunctions that are the same in the pattern. Remember that in maxterm code the domains are complemented, whereas in minterm code they are not.

Step III. Extract the XOR pattern of type 1, 2, or 3 that exists by using the defining SOP or POS relations for XOR and EQV given by Eqs. (3.4) and (3.5). Associative patterns, of type 4, are extracted in a manner similar to the extraction of EV s- and p-terms as discussed in Section 4.6. Thus, associative patterns with XOR connectives are extracted in minterm code while those with EQV connectives are extracted in maxterm code. Compound associative patterns involve some combination of associative pattern with one or more of the other three patterns. They may also include the intersection (ANDing) of patterns or the union (ORing) of patterns. In all cases involving an associative pattern, the associating connective must be preserved in the resulting expression.

Step IV. Extract any remaining two-level SOP or POS cover that may exist.

Step V. Combine into SOP or POS form the results of steps I through IV. The resulting expression may be altered as follows: Single complementation of an XOR/EQV-associated term complements the XOR or EQV connective while double complementation of the associated terms retains the original connective.

Step VI. If necessary, test the validity of the extraction process. This can be done by introducing the K-map cell coordinates into the resulting expression. Generation of each cell subfunction of the K-map validates the extraction procedure.

Examples The simplest associative patterns are formed between XOR-associated or EQV-associated variables and like variables in adjacent cells. Three examples are presented in Figs. 5.1b, 5.1c, and 5.1d, all representing second-order K-map compressions (two EVs). For the first-order EV K-map, shown in Fig. 5.1b, the function E is read in minterm code as

$$E_{XOR/SOP} = (\bar{A} \cdot \bar{X}) \oplus Y = (A + X) \odot Y \tag{5.1}$$

5.2 XOR-TYPE PATTERNS

and is seen to be a two-level function. Here, according to step III of the extraction procedure, the associative XOR pattern is extracted in minterm code in SOP form with $\bar{X}$ located in the $\bar{A}$ domain, hence $\bar{A} \cdot \bar{X}$. The $E_{XOR/SOP}$ form can be converted to the $E_{EQV/POS}$ form by double complementation as required by Eqs. (3.24), or can be read in maxterm code directly from the K-map.

The function F in the second-order K-map of Fig. 5.1c is read in maxterm code, according to step III and is given by

$$F_{EQV/POS} = [(B + \bar{Y}) \odot X] \cdot A, \tag{5.2}$$

which is a three-level function. In this case the EQV connective associates the $\bar{Y}$ in cells 0 and 2 (hence $B + \bar{Y}$ in maxterm code) with the X in all four cells. The remaining POS cover in cell 0 is extracted with the don't care (ϕ) in cell 1 by ANDing the previous result with A as required by step IV in the extraction procedure.

The function G in the third-order EV K-map, shown in Fig. 5.1d, is also read in maxterm code. Here, the EQV connective associates the X's in cells 0, 1, 4, and 5 (thus, $B + X$ in maxterm code) with the Y's in cells 5 and 7 (hence, $\bar{A} + \bar{C} + Y$), giving the result

$$G_{EQV/POS} = [(B + X) \odot (\bar{A} + \bar{C} + Y)](\bar{A} + C + X), \tag{5.3}$$

which is also a three-level function. The term $(\bar{A} + C + X)$ removes the remaining POS cover in cells 4 and 6, as required by step IV.

For comparison purposes the two-level minimum results for E_{SOP}, F_{POS}, and G_{POS} are

$$E_{SOP} = XY + \bar{A}\bar{X}\bar{Y} + AY \tag{5.4}$$

$$F_{POS} = (X + Y)(B + \bar{X} + \bar{Y})(\bar{B} + X)A \tag{5.5}$$

$$G_{POS} = (B + X + Y)(\bar{A} + \bar{C} + \bar{X} + \bar{Y})(\bar{A} + C + X)$$
$$\cdot (A + B + X)(\bar{A} + \bar{B} + X + \bar{Y}). \tag{5.6}$$

The use of associative patterns often leads to significant reduction in hardware compared to the two-level SOP and POS forms. For example, function $E_{XOR/SOP}$ has a minimum gate/input tally of 2/4 compared to 4/10 for E_{SOP}, the two-level SOP minimum form. The gate/input tally for $F_{EQV/POS}$ is 3/6 compared to 4/11 for the F_{POS} expression, and function $G_{EQV/POS}$ has a minimum gate/input tally of 4/12 compared to 6/22 for G_{POS}, the two-level POS minimum result, all excluding inverters.

XOR patterns may be combined very effectively to yield gate-minimum results. Shown in Fig. 5.2a is a second-order compression where diagonal, adjacent, and offset patterns are associated in minterm code by the XOR operator in cell 1. Here, the defining relation for XOR, given in Eqs. (3.4), is applied to the diagonal pattern (cells 1 and 4) in the $\bar{B}$ domain for all that is X to yield $\bar{B}X(A \oplus C)$. This pattern is then associated with the intersection (ANDing) of the adjacent pattern $(A \odot Y)$ and the offset pattern $(B \oplus C)$ in cells 1, 2, 5, and 6 to give the gate-minimum, three-level result

$$H_{XOR/SOP} = [\bar{B}X(A \oplus C)] \oplus [(A \odot Y)(B \oplus C)] \tag{5.7}$$

with a gate/input tally of 6/13. The defining relation for EQV, given in Eqs. (3.5), is used

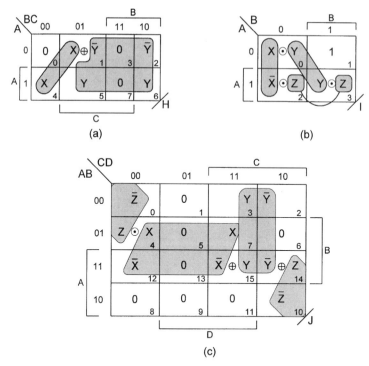

FIGURE 5.2
Examples of complex XOR patterns. (a) Combined XOR-type patterns. (b), (c) Compound associative patterns.

for the adjacent pattern ($\bar{Y}$ in the $\bar{A}$ domain and Y in the A domain), while Eqs. (3.4) are applied to the offset pattern ($\bar{Y}$ and Y in the $\bar{B}C$ domain, and $\bar{Y}$ and Y in the $B\bar{C}$ domain). Notice that the 0's in cells 3 and 7 play no role in this offset/adjacent pattern even though they are included in the shaded loop covering this pattern.

For comparison purposes, the two-level minimum result is

$$H_{SOP} = \bar{A}\bar{B}C\bar{X}\bar{Y} + A\bar{B}\bar{C}X + A\bar{B}CY + \bar{B}CXY + \bar{A}B\bar{C}\bar{Y} + AB\bar{C}Y, \qquad (5.8)$$

which has a gate/input tally of 7/31. Comparison with Eq. (5.7) makes it clear that the three-level result provides a better gate-minimum result but not necessarily a better performance. To evaluate the relative performance of the two approaches, fan-in restrictions and gate propagation delays would have to be established.

Compound (interconnected) associative patterns are also possible and may lead to gate-minimum functions, although often of a higher level (hence slower) than those where there is no interconnection between associative patterns. Two examples are given in Figs. 5.2b and 5.2c, both third-order compressions (hence three EVs). Function I is extracted in maxterm code, yielding the four-level, gate-minimum result

$$I_{EQV/POS} = [B + (A \oplus X)] \odot [Y + (A \oplus B)] \odot [\bar{A} + Z], \qquad (5.9)$$

5.2 XOR-TYPE PATTERNS

which has a gate/input tally of 7/14. Extraction involves the association of an adjacent pattern and a diagonal pattern with the three EQV connectives. The adjacent pattern in domain B (cells 0 and 2) requires the use of Eqs. (3.5) to give $[B + (A \oplus X)]$. This is associated with the diagonal pattern in cells 0 and 3, by using Eqs. (3.5), "for all that is Y" to give $[Y + (A \oplus B)]$, but is also associated with the cell 3 connective in domain $\bar{A}$ for all that is Z. Notice that the terms in square brackets are commutative. For comparison purposes the two-level POS result for function I is given by

$$I_{POS} = (A + B + X + \bar{Y})(A + B + \bar{X} + Y)(\bar{A} + B + X + Z)(\bar{A} + B + \bar{X} + \bar{Z})$$
$$\times (\bar{A} + \bar{B} + Y + \bar{Z})(\bar{A} + \bar{B} + \bar{Y} + Z) \qquad (5.10)$$

and has a gate/input tally of 7/30.

The function J in Fig. 5.2c is extracted in minterm code, giving the four-level, gate-minimum result

$$J_{XOR/SOP} = [\bar{D}(B \odot Z)(A \odot C)] \oplus [B(A \oplus X)(C \odot D)] \oplus [C(D \odot Y)(A \odot B)] \qquad (5.11)$$

with a gate/input tally of 11/25. This function is extracted as three sets of two intersecting patterns, all associated by the three XOR connectives. The "Z" set consists of adjacent and diagonal patterns where application of Eqs. (3.5) yields $(B \odot Z)$ and $(A \odot C)$, respectively, which intersect (AND) in the $\bar{D}$ domain. The "X" set consists of adjacent and offset patterns that are read as $(A \oplus X)$ and $(C \odot D)$, by application of Eqs. (3.4) and (3.5), and that intersect in the B domain. Here, as in Fig 5.2a, the 0's (now in cells 5 and 13) are disregarded in the development of the offset/adjacent pattern. Finally, the "Y" set also consists of adjacent and offset patterns such that the application of Eqs. (3.5) yields $(D \odot Y)$ and $(A \odot B)$, respectively, which intersect in the C domain. As in the previous example, the terms in square brackets of Eq. (5.11) are commutative. In comparison, the two-level SOP minimum for function J is given by

$$J_{SOP} = ABCD\bar{X}\bar{Y} + ABC\bar{D}YZ + \bar{A}\bar{B}C\bar{D}\bar{Z} + \bar{A}\bar{B}CDY + \bar{A}\bar{B}CD\bar{Y} + \bar{A}BCDX + AB\bar{C}\bar{D}\bar{X}$$
$$+ A\bar{B}C\bar{D}\bar{Z} + \bar{A}\bar{C}\bar{D}X\bar{Z} + B\bar{C}\bar{D}\bar{X}Z + BCDXY + AC\bar{D}\bar{Y}\bar{Z} \qquad (5.12)$$

and has a gate/input tally of 13/74. Again, the gate/minimum advantage of the multilevel function over its two-level counterpart in Eq. (5.12) is evident.

Both four-level functions, $I_{EQV/POS}$ and $J_{XOR/SOP}$, are easily verified by introducing in turn the coordinates for each cell into the particular expression. For example, if one sets $ABCD = 1111$ in Eq. (5.11), the subfuction $\bar{X} \oplus Y$ is generated for cell 15 as required by Fig. 5.2c. Generation of the subfunctions in each cell validates the extraction process.

The gate/input tallies for all six functions represented previously are given exclusive of inverters. When account is taken of the inverters required for inputs assumed to arrive active high, the gate/input tally differentials between the multilevel results and the two-level results increases significantly. These gate/input tallies from previous examples are compared in Table 5.1, where all inputs are assumed to arrive active high.

There are other factors that may significantly increase the gate/input tally and throughput time differentials between multilevel and standard two-level SOP and POS minimum forms. These include gate fan-in restrictions and static hazard cover considerations. Static hazards

Table 5.1 Gate/input tallies including *inverters* for functions E, F, G, H, I, and J represented as multilevel logic forms and as two-level logic forms

Function	E	F	G	H	I	J
Multilevel	2/4	4/7	5/13	7/14	8/15	12/36
Two-level	7/13	7/14	11/21	12/36	12/35	20/81

are a type of timing defect that will be discussed at length in Chapter 9. The term *fan-in* refers to the number of inputs required by a given gate. For logic families such as CMOS, propagation delay is increased significantly with increasing numbers of gate inputs, and it is here where the multilevel XOR forms often have a distinct advantage over their two-level counterparts. For example, the largest number of inputs to any gate in the implementation of function $J_{XOR/SOP}$ is 3, whereas for the two-level function J_{SOP} it is 12. Thus, depending on how such a function is implemented, the gate/input tally and throughput time differentials between the multilevel and two-level results could increase significantly. An example of how multiple output optimization considerations may further increase the gate/input tally differential between the multilevel and two-level approaches to design is given in Section 8.8.

5.3 ALGEBRAIC VERIFICATION OF OPTIMAL XOR FUNCTION EXTRACTION FROM K-MAPS

Verification of the multilevel XOR forms begins by direct K-map extraction of the function in SOP or POS form by using minterm code for XOR connectives and maxterm code for EQV connectives. It then proceeds by applying Corollary I [Eq. (3.25)] or Corollary II [Eq. (3.26)] together with commutivity, distributivity, and the defining relations for XOR and EQV given by Eqs. (3.18), (3.19), (3.4), and (3.5).

As an example, consider the function H in Fig. 5.2a, which is extracted in minterm code. Verification of this function is accomplished in six steps:

$$H = \bar{A}\bar{B}C(X \oplus \bar{Y}) + A\bar{B}\bar{C}X + \bar{A}B\bar{C}\bar{Y} + \bar{A}BCY + AB\bar{C}Y \quad (1) \text{ From K-map}$$

$$= [\bar{A}\bar{B}C(X \oplus \bar{Y})] \oplus (A\bar{B}\bar{C}X) \oplus (\bar{A}B\bar{C}\bar{Y})$$
$$\oplus (\bar{A}BCY) \oplus (AB\bar{C}Y) \quad (2) \text{ By Eq. (3.25)}$$

$$= (\bar{A}\bar{B}CX) \oplus (\bar{A}\bar{B}C\bar{Y}) \oplus (A\bar{B}\bar{C}X) \oplus (\bar{A}B\bar{C}\bar{Y})$$
$$\oplus (\bar{A}BCY) \oplus (AB\bar{C}Y) \quad (3) \text{ By Eqs. (3.19)}$$

$$= [\bar{B}X\{(\bar{A}C) \oplus (A\bar{C})\}] \oplus [\bar{B}C\{(\bar{A}\bar{Y}) \oplus (AY)\}]$$
$$\oplus [B\bar{C}\{(\bar{A}\bar{Y}) \oplus (AY)\}] \quad (4) \text{ By Eqs. (3.19)}$$

$$= [\bar{B}X(A \oplus C)] \oplus [\bar{B}C(A \odot Y)] \oplus [B\bar{C}(A \odot Y)] \quad (5) \text{ By Eqs. (3.25),}$$
$$\text{(3.19), (3.4), and (3.5)}$$

$$= [\bar{B}X(A \oplus C)] \oplus [(A \odot Y)(B \oplus C)] \quad (6) \text{ By Eqs. (3.19)}$$
$$\text{and (3.4)}$$

Notice that in going from step 3 to step 4 the commutative law of XOR algebra is used.

5.4 K-MAP PLOTTING AND ENTERED VARIABLE XOR PATTERNS

As a second example, consider function I in Fig. 5.2b, which has been extracted in maxterm code. Verification of this function is also accomplished in six steps:

$$
\begin{aligned}
I &= (A + B + X \odot Y)(\bar{A} + B + \bar{X} \odot Z)(\bar{A} + \bar{B} + Y \odot Z) & &\text{(1) From K-map} \\
&= (A + B + X \odot Y) \odot (\bar{A} + B + \bar{X} \odot Z) \odot (\bar{A} + \bar{B} + Y \odot Z) & &\text{(2) By Eq. (3.26)} \\
&= (A + B + X) \odot (A + B + Y) \odot (\bar{A} + B + \bar{X}) \odot (\bar{A} + B + Z) \\
&\quad \odot (\bar{A} + \bar{B} + Y) \odot (\bar{A} + \bar{B} + Z) & &\text{(3) By Eqs. (3.19)} \\
&= [B + (A + X) \odot (\bar{A} + \bar{X})] \odot [Y + (A + B) \odot (\bar{A} + \bar{B})] \\
&\quad \odot [\bar{A} + Z + (B \odot \bar{B})] & &\text{(4) By Eqs. (3.19)} \\
&= [B + (A + X)(\bar{A} + \bar{X})] \odot [Y + (A + B)(\bar{A} + \bar{B})] \odot [\bar{A} + Z] & &\text{(5) By Eq. (3.26)} \\
&= [B + (A \oplus X)] \odot [Y + (A \oplus B)] \odot [\bar{A} + Z] & &\text{(6) By Eqs. (3.4)}
\end{aligned}
$$

In going from step 3 to step 4, commutivity was applied before application of Eqs. (3.19). Also, in step 4, $B \odot \bar{B} = 0$.

5.4 K-MAP PLOTTING AND ENTERED VARIABLE XOR PATTERNS

At the onset let it be understood that one does not usually hunt for applications of the XOR pattern minimization methods described here. It is possible to do this, as the example in this section illustrates, but it is more likely that such methods would be applied to EV XOR patterns that occur naturally in the design of a variety of combinational logic devices. Examples of these include a 2×2 bit "fast" multiplier, comparator design, Gray-to-binary code conversion, XS3-to-BCD code conversion, dedicated ALU design, binary-to-2's complement conversion, and BCD to 84-2-1 code conversion, to name but a few, most covered in later chapters. EV XOR patterns may also occur quite naturally in the design of some state machines as, for example, the linear feedback shift register counters discussed in Subsection 12.4.3.

EV K-map plotting for the purpose of extracting a gate-minimum cover by using XOR patterns is not an exact science, and it is often difficult to find the optimum K-map compression involving specific EVs, hence specific K-map axis variables. However, for some functions it is possible to plot the map directly from the canonical form, as illustrated by the example that follows. For some relatively simple functions, the K-map plotting process can be deduced directly from the canonical expression. Consider the simple function given in canonical code form:

$$f(W, X, Y, Z) = \Sigma m(1, 2, 3, 6, 7, 8, 11, 12, 13). \tag{5.13}$$

Shown in Fig. 5.3 are the conventional (1's and 0's) K-map and the second-order compression (two EVs) K-map derived directly from the conventional K-map. The two-level SOP minimum and the multilevel XOR/SOP gate-minimum forms are

$$f_{SOP} = \bar{W}\bar{X}Z + WX\bar{Y} + W\bar{Y}\bar{Z} + \bar{X}YZ + \bar{W}Y \tag{5.14}$$

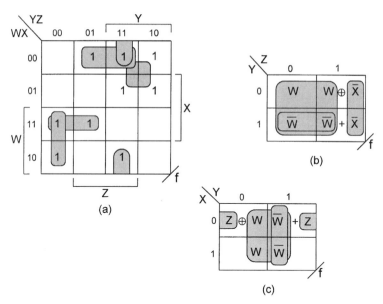

FIGURE 5.3
Compressed K-maps for extraction of gate-minimum XOR forms. (a) Conventional K-map for function f in Eq. (5.13). Second-order compression K-map deduced from K-map in (a) showing XOR patterns. (c) Alternative second-order K-map.

and

$$f_{XOR/SOP} = [(Y \oplus W)] \oplus (\bar{X}Z) + \bar{W}Y, \quad (5.15)$$

which have gate/input tallies of 6/19 and 5/10, respectively. The second-order K-map in Fig. 5.3b is deduced from the K-map in Fig. 5.3a by observing that $W \odot X$ exists in the $YZ = 01$ column, with W and $\bar{W}$ located in adjacent YZ columns. Thus, by taking Y and Z as the axis variables and W and X as the EVs for the compressed K-map, the XOR patterns appear, allowing one to easily extract gate-minimum results.

Notice that the $\bar{W}$ in the EV K-map of Fig. 5.3b must be looped out a second time to give the $\bar{W}Y$ term in Eq. (5.15). This is necessary because cover remains in $\bar{W} + \bar{X}$ after the associative pattern involving W and $\bar{X}$ in cell 1 has been extracted. That is, only $\bar{W} \oplus \bar{X}$ has been looped out of $\bar{W} + \bar{X}$, making it necessary to cover either $\bar{W}$ or $\bar{X}$ a second time. This is easily verified by introducing the coordinates of the cell 3 ($Y = 1$, $Z = 1$) into Eq. (5.15). Without the term $\bar{W}Y$ the subfunction $\bar{W} + \bar{X}$ cannot be generated. The residual cover in $\bar{W} + \bar{X}$ can also be looped out of cell 3 by extracting $\bar{X}YZ$ and using it in place of $\bar{W}Y$ in Eq. (5.15).

Only in one other compressed K-map are the gate-minimum XOR patterns and Eq. (5.15) results obvious, and that is shown in Fig. 5.3c. In all four other compressed K-map possibilities, those having axes W/X, X/Z, W/Z, and W/Y, the XOR patterns shown in Figs. 5.3b and 5.3c disappear, making a gate-minimum extraction without extensive Boolean manipulation very difficult if not impossible. Notice that the compressed K-map in Fig. 5.3c is easily obtained from that of Fig. 5.3a by introducing the coordinate values for X and Y into Fig. 5.3b to obtain the subfunctions in terms of W and Z.

For complex functions involving five or more variables, the process of generating a gate-minimum result by using XOR EV patterns becomes increasingly more a matter of trial and error as the number of variables increases. Again, the application of the EV XOR pattern approach to design is left more to the natural occurrence of such patterns than it is to the hunt-and-choose method. However, if it is known that XOR patterns occur naturally in some functions and if one is familiar with conventional (1's and 0's) K-map methods for five or more variables, it is possible to deduce a compressed K-map that will yield XOR/SOP or EQV/POS forms, but that may not necessarily represent gate-minimum results.

To overcome the obvious problem of dealing with complex K-map XOR patterns in functions having more the five variables, an algebraic approach can be used, a subject that is discussed at length in the remaining sections of this chapter.

5.5 THE SOP-TO-EXSOP REED–MULLER TRANSFORMATION

A generalization of Corollary I (Subsection 3.11.1) can be expressed in canonical form as

$$F_n(x_0, x_1, \ldots, x_{n-1}) = \sum_{i=0}^{2^n-1}(m_i \cdot f_i)$$

$$= \bigoplus_{i=0}^{2^n-1}(m_i \cdot f_i)$$

$$= (m_0 f_0) \oplus (m_1 f_1) \oplus (m_2 f_2) \oplus \cdots \oplus (m_{2^n-1} f_{2^n-1}), \quad (5.16)$$

where the 2^n m_i represent minterms read in minterm code, and the f_i represent their respective coefficients whose values derive from the binary set $\{0, 1\}$. The m_i symbols represent minterms that are, by definition, *mutually disjoint*, since only one minterm can be active (logic 1) for the same values of inputs. For this reason, it is permissible to interchange the OR and XOR operators as in Eq. (5.16). Thus, Eq. (5.16) expresses a transformation of an SOP expression to an EXSOP (EXclusive OR-sum-of-products) expression. Notice that if all f_i are logic 1, then $F_n = 1$.

By setting $\bar{x}_i = x_i \oplus 1$, all $\bar{x}_i$ are eliminated from the EXSOP form of Eq. (5.16) and a function of positive polarity results. Then after considerable Boolean manipulation involving multiple applications of the XOR form of the distributive law given by Eqs. (3.19), the EXSOP expression of Eq. (5.16) is recast as the *Reed–Muller expansion* in the EXSOP form

$$F_n(x_0, x_1, \ldots, x_{n-1}) = g_0 \oplus g_1 x_{n-1} \oplus g_2 x_{n-2} \oplus g_3 x_{n-2} x_{n-1}$$
$$\oplus g_4 x_{n-3} \cdots \oplus g_{2^n-1} x_0 x_1 \cdots x_{n-1}, \quad (5.17)$$

where the g_i are called the Reed–Muller (R-M) coefficients for a positive polarity (uncomplemented f_i) R-M expansion (PPRME). Each R-M coefficient is the set

$$g_i = \bigoplus_{j \subseteq i} f_j. \tag{5.18}$$

obtained from the subnumbers of i by replacing m 1's with 0's in 2^m possible ways in the binary number corresponding to decimal i:

$$g_0 = f_0 \qquad \qquad \cdots 000$$
$$g_1 = \oplus f(1,0) = f_1 \oplus f_0 \qquad \cdots 001 \cdots 000$$
$$g_2 = \oplus f(2,0) = f_2 \oplus f_0 \qquad \cdots 010 \cdots 000$$
$$g_3 = \oplus f(3,2,1,0) = f_3 \oplus f_2 \oplus f_1 \oplus f_0 \qquad \cdots 011 \cdots 010 \cdots 001 \cdots 000$$
$$g_4 = \oplus f(4,0) = f_4 \oplus f_0 \qquad \cdots 100 \cdots 000$$
$$g_5 = \oplus f(5,4,1,0) = f_5 \oplus f_4 \oplus f_1 \oplus f_0 \qquad \cdots 101 \cdots 100 \cdots 001 \cdots 000$$
$$\vdots$$
$$g_{2^n-1} = \bigoplus_{i=0}^{2^n-1} f_i.$$

Note that any g_i in Eq. (5.18) is 1 if an *odd* number of f coefficients are logic 1, but is 0 if an even number of f coefficients are logic 1. If a Karnaugh map (K-map) of F_n is available, the values for the g_i are easily determined by counting the 1's in the map domains defined by the 0's in the binary number representing i in g_i. For example, the value of g_5 is found by counting the 1's present in the $\bar{x}_0\bar{x}_2$ domain for a function $F_4 = (x_0x_1x_2x_3)$. Thus, $g_5 = 1$ if an odd number of 1's exists or $g_5 = 0$ otherwise. Similarly, to determine the logic value for g_8 one would count the number of 1's present in the $\bar{x}_1\bar{x}_2\bar{x}_3$ domain for the same function, etc. All terms in the PPRME expansion whose g coefficients are logic 0 are disregarded.

5.6 THE POS-TO-EQPOS REED–MULLER TRANSFORMATION

The dual of Eqs. (5.16) is the generalization of Corollary II (Subsection 3.11.1) and is expressed as

$$F_n(x_0, x_1, x_2, \ldots, x_{n-1}) = \prod_{i=0}^{2^n-1}(M_i + f_i)$$
$$= \bigodot_{i=0}^{2^n-1}(M_i + f_i)$$
$$= (M_0 + f_0) \odot (M_1 + f_1) \odot (M_2 + f_2)$$
$$\odot \cdots \odot (M_{2^n-1} + f_{2^n-1}), \tag{5.19}$$

where the $2^n M_i$ in Eq. (5.19) represent maxterms read in maxterm code, and the f_i represent their respective coefficients whose values derive from the binary set $\{0, 1\}$. The $2^n M_i$ maxterms are *mutually conjoint* since only one maxterm can be inactive (logic 0) for the same values of inputs. For this reason it is permissible to interchange the AND and EQV operators in Eq. (5.19). Thus, Eq. (5.19) expresses the transformation of a POS expression to an EQV-product-of-sums (EQPOS) expression. Note that if all f_i are logic 0, then $F_n = 0$.

Setting $x_i = \bar{x}_i \odot 0$ eliminates all x_i from Eq. (5.19), resulting in a negative polarity expression for the function F_n, which is simplified by multiple applications of the EQV form of the distributive law given by Eqs. (3.19). The result is the Reed–Muller expansion in the EQPOS form

$$F_n(x_0, x_1, x_2, \ldots, x_{n-1}) = g_0 \odot (g_1 + \bar{x}_{n-1}) \odot (g_2 + \bar{x}_{n-2}) \odot (g_3 + \bar{x}_{n-2} + \bar{x}_{n-1})$$
$$\odot (g_4 + \bar{x}_{n-3}) \odot \cdots \odot (g_{2^n-1} + \bar{x}_0 + \bar{x}_1 + \cdots + \bar{x}_{n-1})$$
(5.20)

where the g_i are now the R-M coefficients for an EQPOS (negative polarity) R-M expansion.

Each EQPOS R-M coefficient is the set

$$g_i = \bigodot_{i \subseteq j} f_j \quad (5.21)$$

obtained from the subnumbers of i by replacing m 1's with 0's in 2^m possible ways in the binary number corresponding to decimal i, as in Eq. (5.18). Thus, the array of g_i for an EQPOS expansion is the same as that for an EXSOP expansion except that the $\oplus$ operator is replaced by the $\odot$ operator. In Eq. (5.21) any g_i is 0 if an odd number of f coefficients are logic 0, but is 1 otherwise. Again, the use of a K-map can be helpful in obtaining the values by counting the 0's within a given domain similar to the procedure explained earlier for the case of the EXSOP expansion. Thus, any g_i is 0 if an odd number of 0's exist within a given domain defined by the 0's in the binary number. All terms in the R-M EQPOS expansion whose g coefficients are logic 1 are ignored.

5.7 EXAMPLES OF MINIMUM FUNCTION EXTRACTION

In this section two examples of minimum function extraction are presented that bear the same relationship to each other as do the conventional and EV K-map methods—that is, one is based on canonical forms (conventional method) while the other is equivalent to the use of entered variables in K-maps (called the CRMT method).

A SIMPLE EXSOP EXAMPLE Consider the function

$$F_3 = \bar{A}BC + A\bar{B} + A\bar{C} = \sum m(3, 4, 5, 6)$$
$$= \bigoplus m(3, 4, 5, 6), \quad (5.22)$$

where, for this example, $f_3 = f_4 = f_5 = f_6 = 1$ and $f_0 = f_1 = f_2 = f_7 = 0$. Therefore, the g_i are found as follows:

$$g_0 = f_0 = 0 \qquad\qquad g_4 = \oplus f_4(4, 0) = 1$$
$$g_1 = \oplus f_1(1, 0) = 0 \qquad g_5 = \oplus f_5(5, 4, 1, 0) = 0$$
$$g_2 = \oplus f_2(2, 0) = 0 \qquad g_6 = \oplus f_6(6, 4, 2, 0) = 0$$
$$g_3 = \oplus f_3(3, 2, 1, 0) = 1 \qquad g_7 = \oplus f_7(7\text{–}0) = 0.$$

Here, the notation (7–0) means (7, 6, 5, 4, 3, 2, 1, 0). From Eq. (5.17) the result is an exact minimum given directly as

$$F_3 = BCg_3 \oplus Ag_4 = BC \oplus A, \tag{5.23}$$

which is a much simplified result compared to the original function and has a gate/input tally of 2/4. This function is said to be a positive polarity R-M expression or PPRME.

The same result is achieved, but with less effort, if the variables of function F_3 are partitioned into two distinct sets: a disjoint set of *bond variables* (called the *bond set*) and a free set of variables (called the *free set*), both chosen from the set {A, B, C} and recast into a contracted (reduced) form for application of Eqs. (5.16) and (5.17). Here, {A, B} is chosen as the *bond set* to be coupled with the remaining (orthogonal) *free set*, {C}. In this particular case, any combination of bond and free sets would achieve the same desired result with equal ease. When a function is recast into bond and free sets it is said to be in a contracted Reed–Muller transformation (CRMT) form. For this example, the CRMT form of function F_3 is

$$F_{AB} = (\bar{A}B)C + (A\bar{B}) + (AB)\bar{C} = (\bar{A}B)C \oplus (A\bar{B}) \oplus (AB)\bar{C}$$
$$= (\bar{A}\bar{B})f_0 \oplus (\bar{A}B)f_1 \oplus (A\bar{B})f_2 \oplus (AB)f_3$$
$$= g_0 \oplus Bg_1 \oplus Ag_2 \oplus ABg_3, \tag{5.24}$$

where the subscript in F_{AB} identifies the bond set {A, B}. Notice that use was made of $A\bar{B}\bar{C} + A\bar{B}C = A\bar{B}$ and that all terms of the bond set are mutually disjoint. Now, the f coefficients are $f_0 = 0$, $f_1 = C$, $f_2 = 1$, and $f_3 = \bar{C}$. Therefore, the resulting CRMT coefficients become

$$g_0 = f_0 = 0 \qquad\qquad g_2 = f_2 \oplus f_0 = 1$$
$$g_1 = f_1 \oplus f_0 = C \qquad g_3 = f_3 \oplus f_2 \oplus f_1 \oplus f_0 = \bar{C} \oplus 1 \oplus C \oplus 0 = 0.$$

Introducing these coefficients into the CRMT expression gives the result

$$F_2 = g_0 \oplus Bg_1 \oplus Ag_2 \oplus ABg_3$$
$$= 0 \oplus BC \oplus A \oplus 0$$
$$= BC \oplus A \tag{5.25}$$

as before.

This simple example has been given to illustrate the application of the CRMT method of function minimization. The examples that follow are designed to establish the foundation

5.7 EXAMPLES OF MINIMUM FUNCTION EXTRACTION

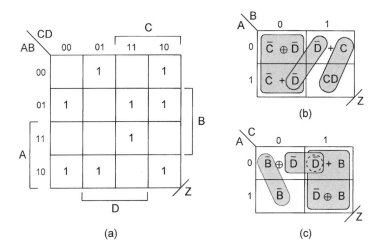

FIGURE 5.4
(a) Conventional K-map for function Z. (b), (c) Compressed EV K-maps for function Z representing bond sets {A, B} and {A, C} showing minimum cover by using XOR-type patterns.

for a tractable CRMT minimization procedure, one that is suitable for classroom (or pencil-and-paper) application.

A MORE COMPLEX EXSOP EXAMPLE   Consider the function and its canonical R-M transformation

$$Z_4(A, B, C, D) = \sum m(1, 2, 4, 6, 7, 8, 9, 10, 15)$$
$$= \bigoplus m(1, 2, 4, 6, 7, 8, 9, 10, 15). \quad (5.26)$$

In Fig. 5.4a is shown the conventional K-map for function Z. In Figs. 5.4b and 5.4c are shown the second-order compressed K-maps of function Z for bond sets {A, B} and {A, C}, respectively, which are representative of the six possible bond sets for two variables.

Considering first the bond set {A, B}, as depicted in Fig. 5.4b, and noting that the cell entries are the f coefficients, the function Z_4 is recast into the following CRMT form:

$$Z_{AB} = (\bar{A}\bar{B})f_0 \oplus (\bar{A}B)f_1 \oplus (A\bar{B})f_2 \oplus (AB)f_3$$
$$= (\bar{A}\bar{B})(C \oplus D) \oplus (\bar{A}B)(C + \bar{D}) \oplus (A\bar{B})(\bar{C} + \bar{D}) \oplus (AB)CD$$
$$= g_0 \oplus Bg_1 \oplus Ag_2 \oplus ABg_3 \quad (5.27)$$

for which the CRMT coefficients are

$g_0 = f_0 = C \oplus D$
$g_1 = \oplus f(1, 0) = (C + \bar{D}) \oplus C \oplus D = \bar{C}D \oplus \bar{C} \oplus D = \bar{C}\bar{D} \oplus D = \bar{C} \oplus CD = 1 \oplus C\bar{D}$
$g_2 = \oplus f(2, 0) = (\bar{C} + \bar{D}) \oplus C \oplus D = CD \oplus C \oplus \bar{D} = C\bar{D} \oplus \bar{D} = \bar{C}\bar{D}$
$g_3 = \oplus f(3, 2, 1, 0) = CD \oplus (\bar{C} + \bar{D}) \oplus 1 \oplus C\bar{D} = CD \oplus CD \oplus 0 \oplus C\bar{D} = C\bar{D}$,

where use has been made of the XOR form of Eqs. (3.31), and the XOR DeMorgan identities given in Eqs. (3.27) from which, for example, there results $(C + \bar{D}) \oplus C = \bar{C}D \oplus \bar{C}$ in g_1. Introducing these coefficients into Eq. (5.27) and simplifying by using the XOR form of Eqs. (3.30) gives the minimum result

$$Z_{AB} = C \oplus D \oplus B \oplus BC\bar{D} \oplus A\bar{C}\bar{D} \oplus ABC\bar{D}$$
$$= B \oplus C \oplus D \oplus A\bar{C}\bar{D} \oplus \bar{A}BC\bar{D}, \qquad (5.28)$$

which is a three-level function with a gate/input tally of 6/15.

Repeating the same procedure for Fig. 5.4c and bond set $\{A, C\}$, the function Z is recast into the CRMT form

$$Z_{AC} = (\bar{A}\bar{C})f_0 \oplus (\bar{A}C)f_1 \oplus (A\bar{C})f_2 \oplus (AC)f_3$$
$$= (\bar{A}\bar{C})(B \oplus D) \oplus (\bar{A}C)(B + \bar{D}) \oplus (A\bar{C})\bar{B} \oplus (AC)(B \oplus \bar{D})$$
$$= g_0 \oplus g_1 C \oplus g_2 A \oplus g_3 AC, \qquad (5.29)$$

where the g coefficients become

$$g_0 = B \oplus D$$
$$g_1 = (B + \bar{D}) \oplus B \oplus D = \bar{B}D \oplus \bar{B} \oplus D = \bar{B}\bar{D} \oplus D = \bar{B} \oplus BD = 1 \oplus B\bar{D}$$
$$g_2 = \bar{B} \oplus B \oplus D = \bar{D}$$
$$g_3 = B \oplus \bar{D} \oplus \bar{B} \oplus 1 \oplus B\bar{D} = \bar{D} \oplus B\bar{D} = \bar{B}\bar{D}.$$

Then by introducing these g coefficients into Eq. (5.29) and simplifying with the XOR form of Eqs. (3.30), there results the exact minimum

$$Z_{AC} = B \oplus D \oplus \bar{B}C\bar{D} \oplus CD \oplus A\bar{D} \oplus A\bar{B}C\bar{D}$$
$$= B \oplus A\bar{D} \oplus \bar{C}D \oplus \bar{A}\bar{B}C\bar{D}, \qquad (5.30)$$

which is a three-level function with a gate/input tally of 6/14 excluding possible inverters.

The results for Z_{AB} and Z_{AC} are typical of the results for the remaining four two-variable bond sets $\{C, D\}$, $\{B, D\}$, $\{A, D\}$, and $\{B, C\}$. All yield gate/input tallies of 6/14 or 6/15 and are three-level functions. Thus, in this case, the choice of bond set does not significantly affect the outcome, but the effort required in achieving an exact minimum may vary with the choice of the bond set. No attempt is made to use single- or three-variable bond sets for function Z.

A comparison is now made between the CRMT minimization method and other approaches to the minimization of function Z. Beginning with the canonical R-M approach and from Fig. 5.4a, the f coefficients easily seen to be

$$f_1 = f_2 = f_4 = f_6 = f_7 = f_8 = f_9 = f_{10} = f_{11} = f_{14} = f_{15} = 1 \quad \text{and}$$
$$f_0 = f_3 = f_5 = f_{12} = f_{13} = 0.$$

5.7 EXAMPLES OF MINIMUM FUNCTION EXTRACTION

Then from Eq. (5.16) the R-M g_i coefficients are evaluated as follows:

$$g_0 = f_0 = 0 \qquad\qquad g_8 = \oplus f(8, 0) = 1$$
$$g_1 = \oplus f(1, 0) = 1 \qquad\qquad g_9 = \oplus f(9, 8, 1, 0) = 1$$
$$g_2 = \oplus f(2, 0) = 1 \qquad\qquad g_{10} = \oplus f(10, 8, 2, 0) = 1$$
$$g_3 = \oplus f(3, 2, 1, 0) = 0 \qquad\qquad g_{11} = \oplus f(11 - 8, 3 - 0) = 1$$
$$g_4 = \oplus f(4, 0) = 1 \qquad\qquad g_{12} = \oplus f(12, 8, 4, 0) = 0$$
$$g_5 = \oplus f(5, 4, 1, 0) = 0 \qquad\qquad g_{13} = \oplus f(13, 12, 9, 8, 5, 4, 1, 0) = 0$$
$$g_6 = \oplus f(6, 4, 2, 0) = 1 \qquad\qquad g_{14} = \oplus f(14, 12, 10, 8, 6, 4, 2, 0) = 1$$
$$g_7 = \oplus f(7 - 0) = 1 \qquad\qquad g_{15} = \oplus f(15 - 0) = 1.$$

Note that the g_i coefficients are immediately realized by counting 1's within the domains of the conventional K-map shown in Fig. 5.4a. Thus, $g_{13} = 0$ since an even number of 1's exist in the $\bar{C}$ domain (determined from 1101), or $g_9 = 1$ because an odd number of 1's exist in the $\bar{B}\bar{C}$ domain (from 1001). Disregarding the $g = 0$ coefficients, there results the positive polarity R-M expression and its simplified mixed polarity form

$$Z_4 = Dg_1 \oplus Cg_2 \oplus Bg_4 \oplus BCg_6 \oplus BCDg_7 \oplus Ag_8 \oplus ADg_9 \oplus ACg_{10} \oplus ACDg_{11}$$
$$\oplus ABCg_{14} \oplus ABCDg_{15}$$
$$= D \oplus C \oplus B \oplus BC \oplus BCD \oplus A \oplus AD \oplus AC \oplus ACD \oplus ABC \oplus ABCD$$
$$= D \oplus C \oplus B \oplus BC\bar{D} \oplus A\bar{D} \oplus AC\bar{D} \oplus ABC\bar{D}$$
$$= B \oplus C \oplus D \oplus A\bar{C}\bar{D} \oplus \bar{A}BC\bar{D}, \qquad (5.31)$$

which is a three-level function having a gate/input tally of 6/15 excluding possible inverters. The function in Eq. (5.31) is seen to be the same as that in Eq. (5.28), but it is not an exact minimum. Here, multiple applications of the XOR identities in Eqs. (3.30) have been applied to excise terms.

Other comparisons are now made between the CRMT method and the EV K-map and conventional K-map methods presented in Sections 4.6 and 4.4. From Figs. 5.4b and 5.4c, the minimum cover extraction by using XOR type patterns (shown by loops) gives

$$Z_{K\text{-map } AB} = \bar{B}(C \oplus D) + \bar{D}(A \oplus B) + BCD \qquad (5.32)$$

and

$$Z_{K\text{-map } AC} = (\bar{B}\bar{C}) \oplus (\bar{A}\bar{D}) + C(B \oplus \bar{D}) \qquad (5.33)$$

representing three-level functions with gate/input tallies of 6/14 and 6/12, respectively, excluding possible inverters. The function $Z_{K\text{-map } AC}$ is a gate/input-tally minimum for function Z. The results in Eqs. (5.32) and (5.33) are hybrid forms classified as mixed AND/OR/EXSOP expressions. The two-level SOP minimum, obtained from Fig. 5.4a, offers another comparison and is

$$Z = \bar{B}\bar{C}D + \bar{A}B\bar{D} + BCD + \bar{B}C\bar{D} + AB\bar{C} \tag{5.34}$$

with a gate/input tally of 6/20, again excluding inverters.

Comparing the results for the four methods used to minimize function Z given by Eqs. (5.28) through (5.34), it is clear that the CRMT results in Eqs. (5.28) and (5.30) are competitive with the other methods. However, only the CRMT result in Eq. (5.30) is an exact EXSOP minimum result. As will be demonstrated further by other examples, the CRMT and EV K-map methods of minimization tend to yield results that are typically competitive with or more optimum than those obtained by the other methods, including computerized two-level results. These observations are valid for relatively simple expressions amenable to classroom methods. No means are yet available for making a fair comparison of the CRMT approach with related computer algorithmic methods.

AN EQPOS EXAMPLE Consider the four variable function G and its canonical R-M transformation

$$G(W, X, Y, Z) = \prod M(0, 1, 6, 7, 8, 10, 13, 15) = \bigodot M(0, 1, 6, 7, 8, 10, 13, 15), \tag{5.35}$$

which follows from Eqs. (5.19). The conventional (1's and 0's) K-map for this function is shown in Fig. 5.5a. Begin with the CRMT minimization method applied to bond set $\{W, X\}$ as depicted in Fig. 5.5b, which is a second-order compression of the function G. From Eqs. (5.20) and (5.21) and for bond set $\{W, X\}$, this function is represented in the

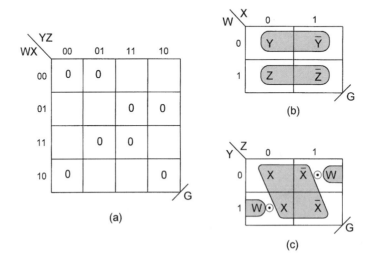

FIGURE 5.5
(a) Conventional K-map for function G. (b), (c) Compressed EV K-maps of function for bond sets $\{W, X\}$ and $\{Y, Z\}$ showing minimum cover by using XOR-type patterns.

5.7 EXAMPLES OF MINIMUM FUNCTION EXTRACTION

negative-polarity CRMT form

$$G_{WX} = (W + X + f_0) \odot (W + \bar{X} + f_1) \odot (\bar{W} + X + f_2) \odot (\bar{W} + \bar{X} + f_3)$$
$$= (W + X + Y) \odot (W + \bar{X} + \bar{Y}) \odot (\bar{W} + X + Z) \odot (\bar{W} + \bar{X} + \bar{Z})$$
$$= g_0 \odot (\bar{X} + g_1) \odot (\bar{W} + g_2) \odot (\bar{W} + \bar{X} + g_3), \tag{5.36}$$

which are read in maxterm code. From Eq. (5.21) the g coefficients become

$$g_0 = Y \qquad\qquad g_2 = Z \odot Y$$
$$g_1 = \bar{Y} \odot Y = 0 \qquad g_3 = Z \odot \bar{Z} \odot 0 = 1.$$

Introducing these coefficients into Eq. (5.36) yields the absolute minimum EQPOS expression

$$G_{WX} = Y \odot \bar{X} \odot (\bar{W} + Z \odot Y)$$
$$= Y \odot \bar{X} \odot (\bar{W} + Y) \odot (\bar{W} + Z)$$
$$= \bar{X} \odot (W + Y) \odot (\bar{W} + Z) \tag{5.37}$$

that is seen to be a three-level function with a gate/input tally of 4/8.

The CRMT minimization process is repeated for the bond set $\{Y, Z\}$ as depicted in Fig. 5.5c. The CRMT expression now becomes

$$G_{YZ} = (Y + Z + g_0) \odot (Y + \bar{Z} + g_1) \odot (\bar{Y} + Z + g_2) \odot (\bar{Y} + \bar{Z} + g_3)$$
$$= (Y + Z + X) \odot (Y + \bar{Z} + W \odot \bar{X}) \odot (\bar{Y} + Z + W \odot X) \odot (\bar{Y} + \bar{Z} + \bar{X})$$
$$= g_0 \odot (\bar{Z} + g_1)(\bar{Y} + g_2)(\bar{Y} + \bar{Z} + g_3) \tag{5.38}$$

for which the g coefficients are found to be

$$g_0 = f_0 = X$$
$$g_1 = \odot f(1, 0) = W \odot \bar{X} \odot X = \bar{W}$$
$$g_2 = \odot f(2, 0) = W \odot X \odot X = W$$
$$g_3 = \odot f(3, 2, 1, 0) = \bar{X} \odot W \odot X \odot \bar{W} = 1,$$

where use is made of $g_1 = \odot f(1, 0) = \bar{W}$ in the last term for g_3. Then, introducing these coefficients into Eq. (5.38) gives the absolute minimum result

$$G_{YZ} = X \odot (\bar{W} + \bar{Z}) \odot (W + \bar{Y}), \tag{5.39}$$

which is again a three-level function with a gate/input tally of 4/8, inverters excluded.

The same result is, in this case, obtained by minimizing a canonical R-M expansion of Eq. (5.35), which becomes

$$G_4(W, X, Y, Z) = \bigodot M(0, 1, 6, 7, 8, 10, 13, 15)$$
$$= g_0 \odot (M_1 + g_1) \odot (M_2 + g_2) \odot (M_3 + g_3) \odot \cdots \odot (M_{15} + g_{15}).$$

From the conventional K-map in Fig. 5.5a and counting 0's within a given domain, the g coefficients are found to be $g_0 = g_2 = g_4 = g_9 = g_{10} = 0$ with all the rest being logic 1. Introducing these values in the R-M expansion gives the minimum result

$$\begin{aligned}G_4 &= 0 \odot \bar{Y} \odot \bar{X} \odot (\bar{W} + \bar{Z}) \odot (\bar{W} + \bar{Y}) \\ &= 0 \odot \bar{X} \odot (\bar{W} + \bar{Z}) \odot (W + \bar{Y}) \\ &= X \odot (\bar{W} + \bar{Z}) \odot (W + \bar{Y}),\end{aligned} \quad (5.40)$$

as before in Eq. (5.39).

The result previously obtained for G_{YZ} can also be obtained by using the CRMT approach in a somewhat different way. The plan is to obtain the result for the SOP CRMT expansion ("for the 0's") and then complement that result to produce the EQPOS CRMT expansion form. Complementing each of the four EV cell entries in Fig. 5.5b gives

$$\begin{aligned}\bar{G}_{YZ}(\text{EPOS}) &= (\bar{Y}\bar{Z})\bar{X} + (\bar{Y}Z)(W \oplus \bar{X}) + (Y\bar{Z})(W \oplus X) + (YZ)X \\ &= g_0 \oplus Zg_1 \oplus Yg_2 \oplus YZg_3,\end{aligned} \quad (5.41)$$

with g values

$$g_0 = f_0 = \bar{X} \qquad\qquad g_2 = \oplus f(2,0) = W \oplus X \oplus \bar{X} = \bar{W}$$
$$g_1 = \oplus f(1,0) = W \oplus \bar{X} \oplus \bar{X} = W \qquad g_3 = \oplus f(3-0) = X \oplus W \oplus X \oplus W = 0,$$

where use is made of $g_1 = \oplus f(1,0) = W$ in the last term for g_3. Introducing these values into Eq. (5.41) gives

$$\bar{G}_{YZ}(\text{EPOS}) = \bar{X} \oplus ZW \oplus Y\bar{W},$$

resulting in the EQPOS expression

$$\begin{aligned}G_{YZ} &= \overline{\bar{X} \oplus ZW \oplus Y\bar{W}} \\ &= X \odot (\bar{W} + \bar{Z}) \odot (W + \bar{Y}),\end{aligned} \quad (5.42)$$

where an odd number of complementations (operators and operands) have been performed to complement the function. Notice that the f coefficients are also the complements of those required for the EQPOS expansion, as they must be, since the cells of the EV K-map in Fig. 5.5b were complemented.

It is interesting to compare the results just obtained for G with those read from the EV K-maps in Figs. 5.5b and c, and with two-level POS minimization. Following the procedure given by [3, 4], the results for G_{WX} and G_{YZ} are read directly in maxterm code from the K-maps (see K-map loopings) as

$$G_{K\text{-map }WX} = [W + (X \oplus Y)][\bar{W} + (X \oplus Z)] \quad (5.43)$$

and

$$G_{K\text{-map }YZ} = [W + (Y \odot Z)] \odot (X \oplus Z)] \tag{5.44}$$

with gate/input tallies of 5/10 and 4/8, respectively. Note that reading a K-map in maxterm code requires that the domains (not the entered variables) be complemented, since the K-maps are minterm-code based [3]. In comparison, the two-level minimum result from Fig. 5.5a is

$$G = (W + X + Y)(W + \bar{X} + \bar{Y})(\bar{W} + \bar{X} + \bar{Z})(\bar{W} + X + Z), \tag{5.45}$$

which has a gate/input tally of 5/16 excluding possible inverters.

Notice that all CRMT minimization results, the canonical R-M minimization result, and one EV K-map result for G all represent three-level functions with minimum gate/input tallies of 4/8 (excluding possible inverters). In comparison, the best two-level result that can be obtained for G yields a gate/input tally of 5/16, again excluding inverters. Notice also that the two-level result requires that four s-terms be ANDed in the output stage, whereas all other results mentioned earlier have a fan-in limit of two. Increased fan-in can slow the throughput of a circuit, particularly in CMOS, as was discussed in Subsections 3.6.2 and 3.6.3, and in Section 4.10.

5.8 HEURISTICS FOR CRMT MINIMIZATION

A given minimization method can yield a guaranteed exact minimum for a function if, and only if, an exhaustive search is carried out. Applied to the CRMT method this involves finding the optimum (bond set)/(free set) combination for the optimal reduction process of minimization required by the CRMT method. As the number of inputs to a function increases, the task of performing an exhaustive search becomes more difficult, eventually requiring computer algorithmic means. Even then, an intractable problem eventually ensues when the number of inputs becomes excessively large for the minimization algorithm used. When this happens, the minimization problem is known to be $\mathcal{NP}$-complete (see Section 4.11).

Fortunately, variation in the choice of bond set for CRMT minimization often results in little or no difference in minimum gate/input tally for a minimized function. However, the effort required to achieve a minimum result may vary considerably with bond set choice. Thus, if a guaranteed absolute minimum is not required, alternative choices of bond set should yield an acceptable minimum, but with some limits placed on the number of bond set variables. By the pencil-and-paper method this means that for practical reasons the number of bond set variables should not exceed four for most applications of the CRMT method. The limit on the total number of variables is placed between eight and twelve depending on one's ability to use entered variables. In any case, experience in the application of the laws and identities of XOR algebra is an invaluable asset in achieving a minimized result.

Given these preliminary comments, the following procedure should be helpful in applying the CRMT "hand" minimization method to functions of 12 variables or less:

Step 1. Choose a bond set and construct an entered variable (EV) K-map with the K-map axes as the bond set. The starting point in this process can be a canonical SOP or

POS expression, a conventional (1's and 0's) K-map, or a truth table. The cell subfunctions of the EV K-map become the f_i coefficients in the CRMT form of Eq. (5.16) or (5.19). Thus, the entered variables make up the free set. As a caveat, try to avoid bond sets that generate f coefficients like $\cdots Z(X+Y)\cdots$, since such coefficients do not produce simple g coefficients. Note that an EV truth table, such as that in Fig. 8.26, will also suffice for the purpose of the CRMT minimization method if the table-heading variables are taken as the bond set variables.

Step 2. For the choice of bond set used, obtain a set of minimum CRMT g_i coefficients from Eq. (5.18) or (5.21) by using the EV K-map cell entries as the f_i coefficients and by applying Eqs. (3.19). If alternative minimum expressions exist for a given g coefficient, choose among these for the "best" one in consideration of steps 3 and 4 that follow. Thus, if an exact minimum result is required for a given bond set, an exhaustive search for an optimum g set must be carried out.

Step 3. Recast the function in positive or negative CRMT form by using the g set from step 2 in Eq. (5.17) or (5.20).

Step 4. Reduce the results of STEP (3) by applying the laws and identities given by Eqs. (3.19) and (3.27)–(3.33). Keep in mind that identical EXSOP terms in the form $\cdots \oplus X \oplus X \oplus \cdots$ or EQPOS terms in the form $\cdots \odot X \odot X \odot \cdots$ can be excised immediately in their respective CRMT expressions.

Step 5. If an exact minimum result is required, an exhaustive search must be carried out by repeating Steps (1) through (4) for all possible bond sets. As a practical matter for pencil-and-paper application of the CRMT method, the exhaustive search process should not be conducted on functions exceeding five variables. For example, a five-variable function would have to be partitioned into 10 two- or 10 three-variable bond sets in addition to the partitioning for the remaining bond sets. Of course, if an exact minimum is not required, most any choice of CRMT bond set will yield an acceptable minimum for many applications—one that may even be a near-exact minimum.

Step 6. Don't cares, if present, must be considered during the bond set selection process. This is usually done with the intent of reducing the complexity of the CRMT g coefficients, if not optimizing them. It is often the case that simple f coefficients (EV K-map cell entries such as 0, 1, X, or $X \oplus Y$) yield simple g coefficients that lead to near-exact minimizations. In any case, the presence of don't cares will complicate considerably an exhaustive search process.

Step 7. If more than one function is to be optimized, the procedure is to set up the CRMT forms separately as in steps 1–5 and then follow a systematic reduction process for each, taking care to use shared terms in an optimal fashion.

5.9 INCOMPLETELY SPECIFIED FUNCTIONS

Consider the five-variable function

$$f(a, b, c, d, e) = \sum m(1, 3, 4, 6, 9, 10, 12, 13, 18, 21, 23, 25)$$
$$+ \phi(0, 8, 11, 14, 15, 16, 24, 26, 27, 29, 30, 31), \quad (5.46)$$

where ϕ is the symbol representing don't cares (nonessential minterms). Shown in Fig. 5.6a is the conventional K-map for this function and in Fig. 5.6b its second-order EV K-map

5.9 INCOMPLETELY SPECIFIED FUNCTIONS

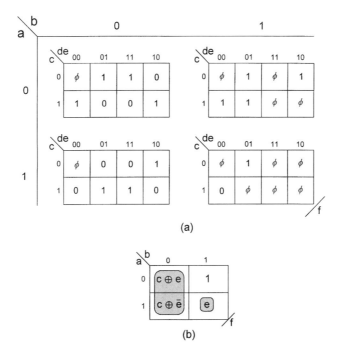

FIGURE 5.6
K-maps for the five-variable function f given in Eq. (5.46). (a) Conventional K-map array for bond set $\{a, b\}$. (b) Third-order compression of function f for bond set $\{a, b\}$ showing minimum cover by using the adjacent XOR pattern in maxterm code.

(third-order compression) for bond set $\{a, b\}$. Recasting this function into the CRMT form of Eqs. (5.16) and (5.17) gives

$$\begin{aligned}
f_{ab} &= \bar{a}\bar{b}f_0 \oplus \bar{a}bf_1 \oplus a\bar{b}f_2 \oplus abf_3 \\
&= \bar{a}\bar{b}(c \oplus e) \oplus \bar{a}b \oplus a\bar{b}(c \oplus \bar{e}) \oplus abe \\
&= g_0 \oplus bg_1 \oplus ag_2 \oplus abg_3
\end{aligned} \qquad (5.47)$$

with g values

$$g_0 = c \oplus e \qquad\qquad g_2 = c \oplus \bar{e} \oplus c \oplus e = 1$$
$$g_1 = 1 \oplus c \oplus e = \bar{c} \oplus e = c \oplus \bar{e} \qquad g_3 = e \oplus c \oplus \bar{e} \oplus c \oplus \bar{e} = e.$$

Here, don't cares are set as $\phi_0 = \phi_{24} = \phi_{26} = \phi_{30} = 0$ with the remainder equal to logic 1. Introducing the values into Eq. (5.47) and simplifying yields the mixed polarity result

$$\begin{aligned}
f_{ab} &= c \oplus e \oplus b\bar{c} \oplus be \oplus a \oplus abe \\
&= a \oplus e \oplus c \oplus b\bar{c} \oplus \bar{a}be \\
&= a \oplus e \oplus (b+c) \oplus \bar{a}be \\
&= a \oplus \bar{e} \oplus \bar{b}\bar{c} \oplus \bar{a}be,
\end{aligned} \qquad (5.48)$$

which is a three-level minimum with a gate/input tally of 5/11 excluding possible inverters. No attempt is made to examine bond sets other than $\{a, b\}$. Consequently, Eq. (5.48) cannot necessarily be regarded as an exact minimum.

The EQPOS CRMT form for bond set $\{a, b\}$ is obtained from Eqs. (5.20) and (5.21) and is

$$f_{ab} = g_0 \odot (\bar{b} + g_1) \odot (\bar{a} + g_2) \odot (\bar{a} + \bar{b} + g_3), \tag{5.49}$$

for which the g coefficients are

$$g_0 = c \oplus e = c \odot \bar{e} \qquad\qquad g_2 = \bar{c} \oplus e \odot c \oplus e = 0$$
$$g_1 = 1 \odot (c \odot \bar{e}) = c \odot \bar{e} = \bar{c} \odot e \qquad g_3 = e \odot e \odot c \odot c \odot \bar{e} = \bar{e}.$$

After introducing these coefficients into Eq. (5.49) there results the mixed polarity CRMT result

$$\begin{aligned}f_{ab} &= c \odot \bar{e} \odot (\bar{b} + c \odot \bar{e}) \odot \bar{a} \odot (\bar{a} + \bar{b} + \bar{e}) \\ &= \bar{a} \odot \bar{e} \odot c \odot (\bar{b} + c) \odot (\bar{b} + \bar{e}) \odot (\bar{a} + \bar{b} + \bar{e}) \\ &= \bar{a} \odot \bar{e} \odot (b + c) \odot (a + \bar{b} + \bar{e}).\end{aligned} \tag{5.50}$$

This is a three-level EQPOS minimum with a gate/input tally of 5/11.

Now it is desirable to compare the CRMT minimum forms of Eqs. (5.48) and (5.50) with the EV K-map and two-level results. Reading the loops of Fig. 5.6b in maxterm code (or the submaps in Fig. 5.6a) gives

$$f_{K\text{-}map} = (b + a \oplus c \oplus e)(\bar{a} + \bar{b} + e), \tag{5.51}$$

which is a four-level function having a gate/input tally of 5/11 excluding possible inverters. By comparison, the computer-minimized two-level POS minimum result is

$$f_{POS} = (a + b + \bar{c} + \bar{e})(\bar{a} + b + c + \bar{e})(\bar{a} + \bar{c} + e)(a + b + c + e) \tag{5.52}$$

and has a gate/input tally of 5/19. The SOP minimum result (not shown) has a gate/input tally of 7/22. No attempt is made to minimize function f by the EXSOP minimization approach, which is best accomplished by computer algorithmic means.

Figure 5.6 illustrates how the CRMT method can be applied to a five-variable function having a two-variable bond set, $\{a, b\}$ in this case. Shown in Fig. 5.7a is the conventional K-map array suitable for an eight variable function F having the bond set $\{w, x, y, z\}$, and in Fig. 5.7b its fourth-order compression, also for bond set $\{w, x, y, z\}$. These K-map formats also suggest a means by which functions with more than eight variables can be minimized by the CRMT method, providing one can deal with EV K-maps within EV K-maps. Large numbers of don't cares would greatly reduce the complexity of the K-map cell entries (and hence f coefficients) in Fig. 5.7b.

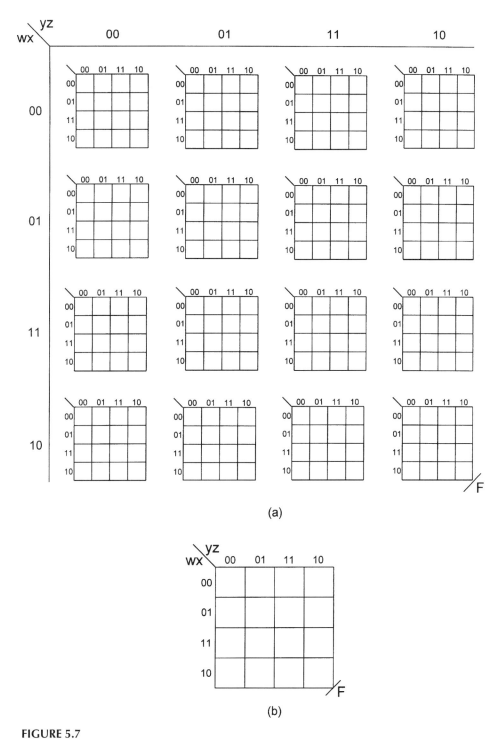

FIGURE 5.7
(a) K-map array for an eight variable function F having a bond set $\{w, x, y, z\}$. (b) K-map required for a fourth-order compression of function F having a bond set $\{w, x, y, z\}$.

5.10 MULTIPLE OUTPUT FUNCTIONS WITH DON'T CARES

The problem of selecting an optimum bond set is further complicated by the presence of don't cares in multiple output systems. Application of the CRMT minimization procedure given earlier to such systems is illustrated by minimizing the following two four-variable functions containing don't cares:

$$F(A, B, C, D) = \sum m(3, 6, 8, 9, 12, 15) + \phi(1, 4, 5, 11)$$

and (5.53)

$$H(A, B, C, D) = \sum m(1, 4, 7, 10, 12, 13) + \phi(2, 5, 6, 8, 11, 15).$$

The conventional fourth-order K-maps for functions F and H are shown in Fig. 5.8a. The don't cares are so chosen as to best meet the requirements of the CRMT minimization procedure for both functions, but with no guarantee of a two-function optimal result. The bond sets are arbitrarily chosen to be $\{C, D\}$ and $\{A, B\}$, respectively, for functions F and

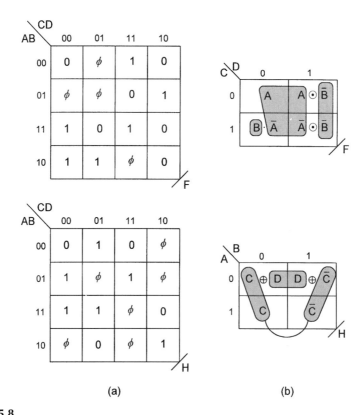

FIGURE 5.8
(a) Conventional K-maps for functions F and H of Eqs. (5.53) and (b) their second-order compressions for bond sets $\{C, D\}$ and $\{A, B\}$ showing minimum cover for each by using XOR-type patterns.

5.10 MULTIPLE OUTPUT FUNCTIONS WITH DON'T CARES

H, and their canonical SOP/EXSOP transformations now become

$$F = \sum m(3, 5, 6, 8, 9, 12, 15) = \bigoplus(3, 5, 6, 8, 9, 12, 15)$$

and (5.54)

$$H = \sum m(1, 2, 4, 7, 10, 11, 12, 13) = \bigoplus m(1, 2, 4, 7, 10, 11, 12, 13).$$

Here, for function F, the don't cares are choosen to be $\phi_1 = \phi_4 = \phi_{11} = 0$ and $\phi_5 = 1$. For function H the don't cares are assigned the values $\phi_5 = \phi_6 = \phi_8 = \phi_{15} = 0$ and $\phi_2 = \phi_{11} = 1$. The don't-care values are chosen in agreement with STEP (6) of the heuristics for CRMT minimization given previously. Thus, the CD columns and the AB rows have simple subfunctions of the type X and $X \oplus Y$ to improve chances for an optimum result.

Function F in Eqs. (5.53) is now recast as the contracted CRMT form

$$\begin{aligned} F_{CD} &= (\bar{C}\bar{D}f_0) \oplus (\bar{C}Df_1) \oplus (C\bar{D}f_2) \oplus (CDf_3) \\ &= (\bar{C}\bar{D})A \oplus (\bar{C}D)(A \oplus B) \oplus (C\bar{D})\bar{A}B \oplus (CD)(\bar{A} \oplus B) \\ &= g_0 \oplus Dg_1 \oplus Cg_2 \oplus CDg_3 \end{aligned} \quad (5.55)$$

for bond set $\{C, D\}$ and with CRMT coefficients

$$g_0 = A \qquad\qquad g_2 = \bar{A}B \oplus A = 1 \oplus \bar{A}\bar{B}$$
$$g_1 = A \oplus B \oplus A = B \qquad g_3 = \bar{A} \oplus B \oplus \bar{A}B \oplus B = \bar{A}\bar{B}.$$

Introducing these coefficients into Eq. (5.55) gives the minimized result for F_{CD}

$$\begin{aligned} F_{CD} &= A \oplus BD \oplus C \oplus \bar{A}\bar{B}C \oplus \bar{A}\bar{B}CD \\ &= A \oplus C \oplus BD \oplus \bar{A}\bar{B}C\bar{D}. \end{aligned} \quad (5.56)$$

Following the same procedure for function H_{AB}, there results

$$\begin{aligned} H_{AB} &= (\bar{A}\bar{B}f_0) \oplus (\bar{A}Bf_1) \oplus (A\bar{B}f_2) \oplus (ABf_3) \\ &= (\bar{A}\bar{B})(C \oplus D) \oplus (\bar{A}B)(\bar{C} \oplus D) \oplus (A\bar{B})C \oplus (AB)\bar{C} \\ &= g_0 \oplus Bg_1 \oplus Ag_2 \oplus ABg_3 \end{aligned} \quad (5.57)$$

for bond set $\{A, B\}$. From Fig. 5.8b and Eq. (5.18), the CRMT g coefficients become

$$g_0 = C \oplus D \qquad\qquad g_2 = C \oplus C \oplus D = D$$
$$g_1 = \bar{C} \oplus D \oplus C \oplus D = 1 \qquad g_3 = \bar{C} \oplus C \oplus 1 = 0,$$

which, when introduced into Eq. (5.57), give the absolute minimum result

$$\begin{aligned} H_{AB} &= C \oplus D \oplus B \oplus AD \\ &= C \oplus B \oplus \bar{A}D. \end{aligned} \quad (5.58)$$

The two CRMT optimized functions are now expressed together as

$$\begin{Bmatrix} F_{CD} = A \oplus C \oplus BD \oplus \bar{A}\bar{B}C\bar{D} \\ H_{AB} = C \oplus B \oplus \bar{A}D \end{Bmatrix}, \quad (5.59)$$

representing a three-level system with a combined gate/input tally of 8/18, but with no shared terms.

A comparison is now made with other approaches to the minimization of these functions. The EV K-map minimum results read directly from the cover (shown by the loopings) in Figs. 5.8b are

$$\begin{Bmatrix} F_{K\text{-}map} = [A \odot \bar{C} \odot (\bar{B} + \bar{D})](B + \bar{C} + D) \\ H_{K\text{-}map} = B \oplus C \oplus \bar{A}D \end{Bmatrix}, \quad (5.60)$$

representing a three-level system having a gate/input tally of 8/17 with no shared terms. Notice that function F is extracted (looped out) in maxterm code, whereas function H is extracted in minterm code. The computer-optimized two-level SOP result is

$$\begin{Bmatrix} F = A\bar{C}\bar{D} + \bar{A}B\bar{D} + ACD + \bar{B}D \\ H = \bar{B}C\bar{D} + \bar{A}\bar{C}D + \bar{A}B + B\bar{C} \end{Bmatrix}, \quad (5.61)$$

with a total gate/input tally of 10/29 excluding possible inverters.

For further comparison, these two functions are minimized together as a system by using canonical R-M forms. As a practical matter, an exhaustive search is not carried out on the choice of don't cares and, consequently, an exact EXSOP result cannot be guaranteed. However, a few trial-and-error attempts at minimization indicate that an exact or near-exact result is obtained for function F if all ϕ's are taken as logic 1, but that for function H the don't-care values are taken to be the same as those used by the CRMT method. Therefore, from the conventional K-maps in Fig. 5.8a there result the following canonical R-M forms: For F the R-M coefficients are

$$g_1 = g_4 = g_5 = g_7 = g_8 = g_9 = g_{10} = g_{11} = g_{12} = 1,$$

and for H they are

$$g_1 = g_2 = g_4 = g_9 = 1.$$

Introducing the g values for functions F and H into Equation (5.17) separately gives

$$F = D \oplus B \oplus BD \oplus BCD \oplus A \oplus AD \oplus AC \oplus ACD \oplus AB$$
$$= \bar{A}D \oplus AC\bar{D} \oplus B\bar{C}D \oplus A \oplus AB \oplus B$$
$$= \bar{A}D \oplus AC\bar{D} \oplus B\bar{C}D \oplus A\bar{B} \oplus B \quad (5.62)$$

for function F and

$$H = D \oplus C \oplus B \oplus AD = \bar{A}D \oplus C \oplus B \qquad (5.63)$$

for function H. Then, combining an optimum set of shared EXSOP terms results in the expressions

$$\begin{cases} F = [B \oplus \bar{A}D] \oplus A\bar{B} \oplus AC\bar{D} \oplus B\bar{C}D \\ H = [B \oplus \bar{A}D] \oplus C \end{cases}. \qquad (5.64)$$

This is a four-level system having a total gate/input tally of 9/20, including shared term $B \oplus \bar{A}D$.

Comparing results in Eqs. (5.60), (5.61), and (5.64) with those for the minimized CRMT forms in Eqs. (5.59) clearly shows that the CRMT method is competitive with the K-map and two-level minimization methods and illustrates the advantage of simplicity that the CRMT minimization approach has over that of the EXSOP minimization as a pencil-and-paper method.

5.11 K-MAP SUBFUNCTION PARTITIONING FOR COMBINED CRMT AND TWO-LEVEL MINIMIZATION

Any function can be partitioned in a manner that permits it to be minimized by a combination of the CRMT and two-level methods. Function partitioning for this purpose is best carried out within an EV K-map, hence subfunction partitioning. This partitioning process is significant because with K-map assistance it makes possible the selection of the most tractable (if not optimal) parts of a function for the combined two methods of minimization. This can be of great advantage for a multioutput function where shared term usage is important. There still remains the problem of knowing what is the "best" choice of function partitioning for optimal results. An absolute minimum result in the CRMT approach not only would require an exhaustive search of the best CRMT bond set minimum, but must be accompanied by an exhaustive two-level search. This is no easy task except for, perhaps, relatively simple functions. However, if an absolute minimum result is not sought, there may exist a variety of ways in which a given function can be partitioned without significant change in the cost (complexity) of the resulting minimized function. In any case, the combined minimum forms are classified as partitioned EXSOP/SOP forms or their dual EQPOS/POS.

As a simple example of K-map function partitioning, consider function Z_{AC} in Fig. 5.4c. Here, the literal $\bar{D}$ in cell 1 (see dashed loop in domain $\bar{A}C$) is separated out to be later ORed to the CRMT solution as the EPI $\bar{A}C\bar{D}$. After removal of the literal $\bar{D}$, the CRMT g coefficients become

$$g_0 = B \oplus D \qquad g_2 = \bar{B} \oplus B \oplus D = \bar{D}$$
$$g_1 = B \oplus B \oplus D = D \qquad g_3 = B \oplus \bar{D} \oplus \bar{B} \oplus D = 0.$$

Introducing these coefficients into Eq. (5.29) and adding the two level result $\bar{A}C\bar{D}$ yields

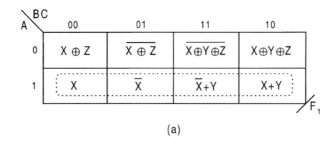

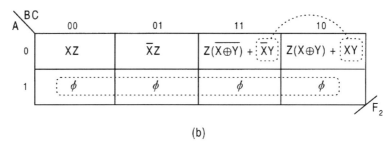

FIGURE 5.9
Combined CRMT and two-level minimization of a two-output system of six variables. (a) Third-order compression of function F_1, (b) third-order compression of function F_2.

the minimum result

$$Z_{AC} = [B \oplus D \oplus \bar{C}D \oplus A\bar{D}] + \bar{A}C\bar{D},$$
$$= [B \oplus CD \oplus A\bar{D}] + \bar{A}C\bar{D}, \qquad (5.65)$$

where Z_{AC} is now a four-level function with a gate/input tally of 6/13. Recall that the CRMT result in Eq. (5.30) is only a three-level function. The extra level in Eq. (5.65) compared to Eq. (5.30) is due to the OR operator of the mixed form.

A more interesting example is the EXSOP/SOP partitioning of the two function system shown in Figs. 5.9a and 5.9b. In this case, all entries in cells 100, 101, 111, and 110 for function F_1 are partitioned out (set to logic 0) for CRMT minimization, but are extracted optimally as shown by the two-level minimum cover. Similarly, for function F_2, terms $\bar{X}Y$ and XY in cells 011 and 010 are partitioned out of the CRMT minimization but are extracted optimally in SOP logic. Also, the don't cares in the F_2 K-map are set to logic 0 for both the CRMT and two-level minimizations.

The minimization process is now carried out on both the function F_1 and F_2 in such a manner as to make effective use of any shared terms that may occur. By using Figs. 5.9a and 5.9b, the function F_1 and F_2 are cast in the form of Eq. (5.17) to give

$$F_1, F_2 = g_0 \oplus Cg_1 \oplus Bg_2 \oplus BCg_3 \oplus Ag_4 \oplus ACg_5 \oplus ABg_6 \oplus ABCg_7. \qquad (5.66)$$

5.11 K-MAP SUBFUNCTION PARTITIONING FOR COMBINED CRMT

After partitioning, CRMT coefficients for function F_1 become

$$g_0 = X \oplus Z \qquad g_4 = X \oplus Z$$
$$g_1 = 1 \qquad g_5 = 1$$
$$g_2 = Y \qquad g_6 = Y$$
$$g_3 = 0 \qquad g_7 = 0.$$

The two-level minimization result for cells 100, 101, 111, and 110 is simply

$$A\bar{C}X + AC\bar{X} + ABY = A(X \oplus C + BY). \tag{5.67}$$

Introducing the g coefficients into Eq. (5.66) and adding the two-level result gives the mixed minimum result

$$F_1 = X \oplus Z \oplus C \oplus BY \oplus AX \oplus AZ \oplus AC \oplus ABY + A(X \oplus C + BY)$$
$$= (X \oplus C) \oplus \bar{A}Z \oplus \bar{A}BY \oplus A(X \oplus C) + A(X \oplus C + BY)$$
$$= \bar{A}[Z \oplus (X \oplus C) \oplus BY] + A(X \oplus C + BY). \tag{5.68}$$

Applying the same procedure to the partitioned F_2 function gives the CRMT g coefficients

$$g_0 = g_4 = XZ \qquad g_2 = g_6 = YZ$$
$$g_1 = g_5 = Z \qquad g_3 = g_7 = 0.$$

From the K-map for F_2, the two-level minimum result is easily seen to be

$$\bar{A}BC\bar{X}Y + \bar{A}B\bar{C}XY = \bar{A}BY(X \oplus C). \tag{5.69}$$

Now introducing the g coefficients into Eq. (5.66) and adding the two-level result yields an EXSOP/SOP minimum,

$$F_2 = XZ \oplus CZ \oplus BYZ \oplus AXZ \oplus ACZ \oplus ABYZ + \bar{A}BY(X \oplus C)$$
$$= \bar{A}XZ \oplus \bar{A}CZ \oplus \bar{A}BYZ + \bar{A}BY(X \oplus C)$$
$$= \bar{A}Z[(X \oplus C) \oplus (BY)] + \bar{A}(BY)(X \oplus C). \tag{5.70}$$

The combined two-function minimum result is now given by

$$\begin{cases} F_1 = \bar{A}[Z \oplus (X \oplus C) \oplus (BY)] + A(X \oplus C + BY) \\ F_2 = \bar{A}Z[(X \oplus C) \oplus (BY)] + \bar{A}(BY)(X \oplus C) \end{cases}, \tag{5.71}$$

which represents a five-level system with a combined gate/input tally of 11/24 excluding possible inverters.

The OR operators in Eqs. (5.71) add an extra level of path delay compared to forms that are exclusively EXSOP/SOP. This can be demonstrated by avoiding the partitioning of function F_1. When this is done the CRMT g coefficients become

$$g_0 = X \oplus Z \qquad g_4 = Z$$
$$g_1 = 1 \qquad g_5 = 0$$
$$g_2 = Y \qquad g_6 = Y \oplus (X + Y) \oplus X = Y \oplus \bar{X}\bar{Y} \oplus \bar{X} = Y \oplus \bar{X}Y = XY$$
$$g_3 = 0 \qquad g_7 = 0 \oplus 1 \oplus X\bar{Y} \oplus \bar{X}\bar{Y} = 1 \oplus \bar{Y} = Y.$$

Introducing these g coefficients into Eq. (5.66) gives the EXSOP/SOP result

$$F_1 = X \oplus Z \oplus C \oplus BY \oplus AZ \oplus ABXY \oplus ABCY$$
$$= (X \oplus C) \oplus \bar{A}Z \oplus BY \oplus ABY(X \oplus C)$$
$$= [\overline{ABY}(X \oplus C)] \oplus (\bar{A}Z) \oplus (BY), \qquad (5.72)$$

which is a four-level function with a gate/input tally of 7/15, exclusive of inverters. This compares to the mixed five-level function F_1 in Eqs. (5.71), which has a gate/input tally of 8/16.

Subfunction partitioning in maxterm code is equally effective in facilitating the minimization process. As a simple example, consider the function F_{CD} in Fig. 5.8b and the EQPOS CRMT form

$$F_{CD} = g_0 \odot (\bar{D} + g_1) \odot (\bar{C} + g_2) \odot (\bar{C} + \bar{D} + g_3), \qquad (5.73)$$

which follows Eq. (5.20). Proceeding with the CRMT minimization, with B partitioned out of the term $\bar{A} \cdot B$ in cell 10, gives the CRMT g coefficients

$$g_0 = A \qquad\qquad g_2 = \bar{A} \odot A = 0$$
$$g_1 = A \odot \bar{B} \odot A = \bar{B} \qquad g_3 = \bar{A} \odot \bar{B} \odot \bar{A} \odot \bar{B} = 1.$$

Introducing these coefficients into Eq. (5.73) and adding the two-level result gives

$$F_{CD} = [A \odot (\bar{B} + \bar{D}) \odot \bar{C}] \cdot (B + \bar{C} + D), \qquad (5.74)$$

which is exactly the same as the K-map minimum result in Eqs. (5.60).

Notice that the mixed CRMT/two-level method requires that the partitioning be carried out in either minterm or maxterm code form. Thus, if subfunctions of the type $X + Y$ are partitioned, the entire minimization process must be carried out in minterm code. Or, if terms such as $X \cdot Y$ are partitioned, the minimization process must be carried out in maxterm code. Note that either X or Y or both may represent multivariable functions or single literals of any polarity.

5.12 PERSPECTIVE ON THE CRMT AND CRMT/TWO-LEVEL MINIMIZATION METHODS

The main advantage of the CRMT method of function minimization lies in the fact that it breaks up the minimization process into tractable parts that are amenable to pencil-and-paper or classroom application. The CRMT minimization process can be thought of as consisting of three stages: the selection of a suitable bond set, the optimization of the CRMT g coefficients (for the chosen bond set), and the final minimization stage once the g coefficients have been introduced into the CRMT form. If an exact minimum is not required, a suitable bond set can be easily found, permitting the CRMT method to be applied to functions of as many as eight variables or more. Knowledge of the use of EV K-maps and familiarity with XOR algebra are skills essential to this process. A properly conducted CRMT minimization can yield results competitive with or more optimum than those obtained by other means.

It has been shown that minimization by the CRMT method yields results that are often similar to those obtained by the EV K-map method described in Section 5.4. This is particularly true when the EV K-map subfunctions are partitioned so as to take advantage of both the CRMT and two-level (SOP or POS) minimization methods. In fact, when subfunction partitioning is carried out in agreement with the minimum K-map cover (as indicated by loopings), the CRMT/two-level result is often the same as that obtained from the K-map. It is also true that when a function is partitioned for CRMT and two-level minimizations, an extra level results because of the OR (or AND) operator(s) that must be present in the resulting expression. Thus, a CRMT/two-level (mixed) result can be more optimum than the CRMT method (alone) only if the reduction in the gate/input tally of the CRMT portion of the mixed result more than compensates for the addition of the two-level part. At this point, this can be known only by a trial-and-error-method that is tantamount to an exhaustive search.

If an exact or absolute minimum CRMT result is sought, an exhaustive search must be undertaken for an optimum bond set. Without computer assistance this can be a tedious task even for functions of four variables, particularly if the function contains don't cares. Multiple-output systems further complicate the exhaustive search process and make computer assistance all the more necessary. One advantage of the mixed CRMT/two-level approach to function minimization is that each method can be carried out independently on more tractable parts.

FURTHER READING

Additional information on XOR algebra, XOR function extraction from K-maps, and logic synthesis with XOR and EQV gates can be found in the texts of Roth, Sasao (Ed.), and Tinder.

[1] C. H. Roth, *Fundamentals of Logic Design*, 4th ed. West, St. Paul, MN 1992 (Chapter 3).
[2] T. Sasao, "Logic Synthesis with XOR Gates," in *Logic Synthesis and Optimization* (T. Sasao, Ed). Kluwer, 1993 (see, e.g., Chapter 12).
[3] R. F. Tinder, *Digital Engineering Design: A Modern Approach.* Prentice Hall, 1991 (see, e.g., Chapter 3).

[4] R. F. Tinder, "Multilevel Logic Minimization by Using K-map XOR Patterns," *IEEE Trans. on Ed.* **38**(4), 370–375 (1995).

Earlier work on Reed–Muller expansions and the use of conventional K-map methods to obtain Reed–Muller coefficient values can be found in the work of Dietmeyer and Wu et al.

[5] D. L. Dietmeyer, *Logic Design of Digital Systems*. Allyn and Bacon, 1978 (Chapter 2).
[6] X. Wu, X. Chen, and S. L. Hurst, "Mapping of Reed–Muller Coefficients and the Minimization of Exclusive-OR Switching Functions," *Proc. IEE, Part E*, **129**, 15–20 (1982).

An excellent experimental study of the various XOR and EQV (XNOR) CMOS gate configurations can be found in the work of Wang, Fang, and Feng.

[7] J. Wang, S. Fang, and W. Feng, "New Efficient Designs for XOR and XNOR Functions on the Transistor Level," *IEEE Journal of Solid-State Circuits* **29**(7), 780–786 (1994).

Many approaches to the decomposition and minimization of multilevel (Reed–Muller) forms can be found in literature. A few representative works are cited below:

[8] D. Bochman, F. Dresig, and B. Steinbach, "A New Decomposition Method for Multilevel Circuit Design," The European Conference on Design Automation, Amsterdam, The Netherlands, 25–28 Feb. 1991, pp. 374–377.
[9] H. M. Fleisher and J. Yeager, "A Computer Algorithm for Minimizing Reed–Muller Canonical Forms," *IEEE Trans. Comput.* **36**(2), 247–250 (1987).
[10] J. M. Saul, "An Algorithm for the Multi-level Minimization of Reed–Muller Representations," *IEEE Int. Conf.on Computer Design:VLSI in ComputersandProcessors* (Cat. No. 91CH3040-3), pp. 634–637. IEEE Computer Soc. Press, Los Alamitos, CA, 1991.
[11] T. Sasao, "Logic Synthesis with XOR Gates," in *Logic Synthesis and Optimization* (T. Sasao, Ed.), Kluwer, 1993, pp. 259–285.
[12] N. Song and M. A. Perkowski, "EXORCISM-MV-2: Minimization of Exclusive Sum of Products Expressions for Multiple-valued Input Incompletely Specified Functions," Proc. of the 23rd International Symposium on Multiple-Valued Logic, ISMVL '93, Sacramento, CA, May 24–27, 1993, pp. 132–137.
[13] W. Wan and M. A. Perkowski, "A New Approach to the Decomposition of Incompletely Specified Functions Based on Graph-Coloring and Local Transformations and its Application to FPGA Mapping," Proc. of the IEEE EURO-DAC '92 European Design Automation Conference, Hamburg, Sept. 7–10, Hamburg, 1992, pp. 230–235.

PROBLEMS

Note: Most K-map minimization results of problems that follow can be verified by introducing the binary coordinates of each K-map cell into the resulting expression. Generation of each cell subfunction by this means validates the extraction results. In some cases, it may be necessary to construct a suitable EV K-map for this purpose. Also, to obtain correct answers for these problems, the reader will be required to make frequent use of the laws, corollaries, and identities of XOR algebra given in Section 3.11.

PROBLEMS

5.1 Compress the following function into a first-order K-map of axis A, and loop out a gate minimum expression by using XOR-type patterns in minterm code. Next, obtain the SOP minimum from the same K-map and compare the gate/input tallies for both the XOR and SOP forms. Finally, construct the logic circuits for the XOR and SOP results assuming that the inputs and output are all active high. What do you conclude from these comparisons?

$$E = \bar{A}\bar{X}\bar{Y} + \bar{A}XY + AY$$

5.2 The output F of a logic circuit is a function of three inputs A, B, and C. The output goes active under any of the following conditions as read in the order ABC:

All inputs are logic 1
An odd number of inputs are logic 1
None of the inputs are logic 1

(a) Construct a truth table for output function F and inputs ABC.
(b) Map the result in a second-order K-map and extract a gate-minimum expression by using XOR-type patterns.
(c) Further compress this function into a first-order K-map of axis A and again extract a gate-minimum expression by using XOR-type patterns. Compare the result with that of (b).
(d) Finally, place this function in a conventional (1's and 0's) K-map and extract minimum two-level SOP and POS logic expressions. By using the gate/input tally (exclusive of inverters), compare the results with those of (b) and (c). What do you conclude from this comparison?

5.3 Compress the following function into a *second-order* K-map with axes as indicated and extract a gate-minimum expression for each set of axes by using XOR patterns. Use the gate/input tally, exclusive of possible inverters, to compare this result with the minimum expressions for the two-level SOP and POS results. What do you conclude from this comparison? What is the gate delay level for the XOR pattern results? (Hint: It will be helpful to first plot this function into a conventional 1's and 0's K-map.)

$$F(W, X, Y, Z) = \sum m(0, 2, 5, 7, 9, 11, 12)$$

(a) Axes W, X
(b) Axes Y, Z
(c) Axes X, Y

5.4 Shown in Fig. P5.1 are six EV K-maps that contain XOR-type patterns and that represent two and three levels of compression. Use *minterm code* to loop out a gate-minimum cover for each by using XOR patterns (where appropriate). For comparison, loop out a minimum two-level SOP cover for each and compare their relative complexity by using the gate/input tally exclusive of possible inverters. Also, as part of the comparison, comment on the fan-in requirements for each.

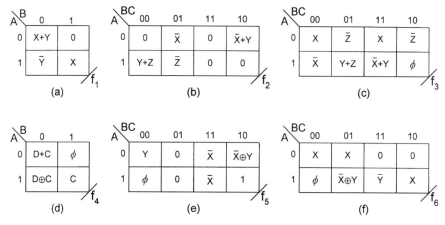

FIGURE P5.1

5.5 Use *maxterm code* and XOR-type patterns to loop out a gate-minimum expression for each of the five functions represented in Fig. P5.2. Give the gate/input tally for each and compare that value with the gate/input tallies for the two-level SOP and POS minimum expressions obtained from the same K-maps. [Hint: To obtain the two-level expressions from the K-maps in Figs. 5.2d and 5.2e, it will be necessary to expand the XOR and EQV subfunctions by using their defining relations given by Eqs. (3.4) and (3.5).]

5.6 Compress each of the following functions into a second-order K-map with axes A, B and loop out a gate-minimum expression for each by using XOR-type patterns where appropriate. Obtain the two-level SOP and POS minimum result and use the gate/input tally (exclusive of possible inverters) to compare the multi-level result. (Hint: Consider both minterm and maxterm codes when looping out XOR-type patterns for gate-minimum results.)

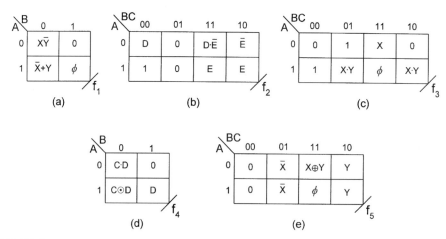

FIGURE P5.2

(a) $W(A, B, C, D) = \sum m(3, 6, 9, 12)$
(b) $X(A, B, C, D) = \prod M(2, 3, 4, 5, 7, 8, 9, 11, 14, 15)$
(c) $Y(A, B, C, D) = \sum m(1, 2, 4, 7, 11, 13, 14)$
(d) $Z(A, B, C, D) = \prod M(0, 3, 4, 6, 9, 10, 13)$

5.7 The following incompletely specified function contains XOR-type patterns:

$$G(A, B, C, D) = \prod M(0, 1, 2, 3, 8, 11, 12, 13) \cdot \phi(4, 5, 6, 7).$$

(a) Compress the following function into a second-order K-map of axes A, B and loop out a gate-minimum expression by using XOR-type patterns where appropriate. (Hint: Consider both minterm and maxterm codes and the best use of the don't cares when looping out XOR-type patterns for a gate-minimum result.)

(b) Use the same K-map to extract minimum SOP and POS expressions for this function. Compare the gate/input tallies (exclusive of possible inverters) for the XOR result with those for the SOP and POS results. What do you conclude from these comparisons?

(c) Construct the logic circuit for both the XOR result and the SOP result, assuming that the inputs and output are all active high.

5.8 Use XOR-type patterns to extract a gate-minimum expression for each of the three functions represented in Fig. P5.3. Use the gate/input tally (exclusive of inverters) to compare the multilevel result with that for the two-level SOP and POS minimum result. Note that compound XOR-type patterns may exist. [Hint: For f_2 and f_3, it will be necessary to make use of Eqs. (3.27).]

5.9 A computer program has been written that will yield a minimum solution to a combinational logic function, but only in SOP form. It accepts the data in either conventional (1's and 0's) form or in two-level EV *SOP form* — it does not recognize the XOR or EQV operators.

(1) Given the functions F_1 and F_2 represented by the EV K-maps in Fig. P5.4, extract a gate-minimum expression from each in maxterm code by using the pencil-and-paper method and XOR-type patterns.

(2) By following Example 2 in Section 4.8, outline the procedure required to "trick" the computer program into yielding a two-level minimum expression from the K-maps in Fig. P5.4, that can be easily converted to minimum POS form. (Hint: It will be necessary to complement the subfunction in each cell of the K-map and represent it in SOP form.)

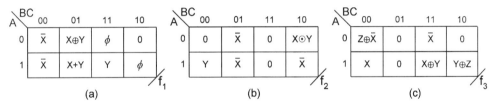

FIGURE P5.3

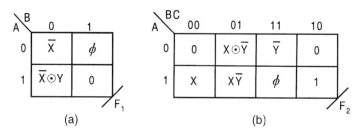

FIGURE P5.4

(3) Use the procedure in part 2 to obtain the two-level POS expression for function F_1 in Fig. P5.4a. Next, convert each cell of the original K-map to two-level POS sub-function form and extract a two-level POS minimum expression from it by using maxterm code. Should it agree with the results obtained by using the procedure of part 2? Explain.

(4) Repeat part 3 for function F_2 in Fig. P5.4b.

5.10 Repeat Problem 5.3 by using the CRMT method, taking each bond set as the axis indicated in the problem. Use the gate/input tally (exclusive of possible inverters) to compare the CRMT results with the two-level SOP minimum in each case.

5.11 Use the canonical Reed–Muller (R-M) approach to obtain an absolute minimum for the function F given in Problem 5.3. Compare the results with the two-level SOP minimum result by using the gate/input tally (exclusive of possible inverters).

5.12 Use the CRMT method to obtain an absolute minimum for the function G in Problem 5.7 by taking axes A, B as the bond set. Use the gate/input tally (exclusive of possible inverters) to compare the CRMT result with the two-level SOP minimum result.

5.13 Use the CRMT method to obtain an absolute minimum for each of the four functions given in Problem 5.6. Take axes A, B as the bond set for each. Construct the logic circuit for each CRMT minimum function assuming that all inputs and outputs are active high. Also, for comparison, construct the logic circuit for the minimum two-level SOP or POS minimum result, whichever is the simpler in each case.

5.14 Use the canonical R-M approach to obtain a gate-minimum for the four functions given in Problem 5.6. Then, by using the gate/input tally (exclusive of possible inverters), compare these results with the two-level SOP or POS minimum results, whichever is the simpler in each case.

5.15 (a) The following two functions are to be optimized together as a system by using the multiple-output CRMT method discussed in Section 5.10. To do this, collapse each function into a third-order K-map with axes A, B, C and then use the CRMT approach in minterm code to minimize each function while making the best use possible of shared terms. Plan to use $\{A, B, C\}$ as the bond set.

$$F_1(A, B, C, D, E) = \sum m(2, 3, 4\text{--}7, 9, 11, 12, 15, 21, 23, 25, 27)$$
$$F_2(A, B, C, D, E) = \sum m(4, 5, 10, 11, 13, 15\text{--}17, 20, 23\text{--}25, 30, 31)$$

(b) Obtain the optimized two-level SOP results for the two functions and compare them with the results of part (a) by using the gate/input tally (including inverters) assuming that the inputs and outputs are all active high.

(c) Construct the logic circuits for the circuits of parts (a) and (b).

5.16 (a) Use *subfunction partitioning* of the following function for CRMT/two-level minimization in minterm code. To do this, collapse this function into a third-order K-map of axes A, B, C and follow the discussion given in Section 5.11. Choose $\{A, B, C\}$ as the bond set for the CRMT portion.

$$F(A, B, C, D, E) = \sum m(4, 7, 10\text{--}12, 14, 16\text{--}19, 21, 23, 24\text{--}27, 28, 30)$$

(b) Without partitioning, use the CRMT method to obtain a gate-minimum for this function. Compare this result with that of (a) by using the gate/input tally exclusive of inverters.

5.17 A function F is to be activated by the use of three switches, A, B, and C. It is required that the function F be active iff a single switch is active. Thus, if any two or three of the switches are active the function must be inactive. Design a gate minimum circuit for the function F consisting of three XOR gates and an AND gate (nothing else). Assume that the inputs and output are all active high. (Hint: Apply the CRMT method.)

CHAPTER 6

Nonarithmetic Combinational Logic Devices

6.1 INTRODUCTION AND BACKGROUND

It is the purpose of combinational logic design to build larger, more sophisticated logic circuits by using the most adaptable and versatile building blocks available. The choice of discrete gates as the building blocks is not always a good one, owing to the complex nature of the circuits that must be designed and to the fact that there are integrated circuit (IC) packages available that are much more adaptable. It is the plan of this chapter to develop these building blocks and demonstrate their use in construction of larger combinational logic systems. Brief discussions of the various device performance characteristics and a design procedure are provided in advance of the logic device development.

6.1.1 The Building Blocks

It is well understood that the digital designer must be able to create combinational circuits that will perform a large variety of tasks. Typical examples of these tasks include the following:

Data manipulation (logically and arithmetically)
Code conversion
Combinational logic design
Data selection from various sources
Data busing and distribution to various destinations
Error detection

To implement circuits that will perform tasks of the type listed, the logic designer can draw upon an impressive and growing list of combinational logic devices that are commercially available in the form of IC packages called *chips*. Shown in Fig. 6.1 is a partial listing of the combinational logic chips, those that are of a nonarithmetic type (a) and those that are arithmetic in character (b). Only the devices in Fig. 6.1a will be considered in this chapter.

(a)	(b)
Non-Arithmetic Combinational Logic Devices	Arithmetic-Type Combinational Logic Circuits
Multiplexers (Data Selectors)	
Decoders/Demultiplexers	Adders
Priority Encoders	Subtractors
Code Converters	Arithmetic and Logic Units
Comparators	Multipliers
Parity Detectors	Dividers
Combinational Shifters	

FIGURE 6.1
Partial lists of available nonarithmetic IC devices and arithmetic IC devices.

6.1.2 Classification of Chips

IC chips for the devices of the type listed in Fig. 6.1 can be classified as small-scale integrated (SSI) circuits, medium-scale integrated (MSI) circuits, large-scale integrated (LSI) circuits, very-large-scale integrated (VLSI) circuits, and wafer-scale integrated (WSI) circuits. It has become customary to assign one of the preceding acronyms to a given IC circuit on the basis of the number of equivalent fundamental gates (meaning AND, OR, Inverter or NAND, NOR, Inverter) that are required to implement it. By one convention, these acronyms may be assigned the following gate count ranges:

SSI circuits: up to 20 gates
MSI circuits: 20 to about 200 gates
LSI circuits: 200 to thousands of gates
VLSI circuits: thousands to millions of gates

WSI chips might contain tens to hundreds of VLSI circuits. This classification scheme is obviously ineffective in revealing the true complexity of a given IC relative to the digital system in which it operates. For example, an LSI chip might be a 64-bit adder or it might be a moderately complex microprocessor. Thus, the reader should exercise caution when evaluating the complexity of a chip based on some count system. Finally, it is now common practice for logic designers to design chips for a limited, specific application. Such chips are called *application-specific* ICs, or ASICs, and may differ greatly from the usual commercial chips. ASICs can reduce total manufacturing costs and can often provide higher performance than is possible by combining commercially available devices.

6.1.3 Performance Characteristics and Other Practical Matters

The most desirable features a designer would want in a switching device, say, for integrated circuit applications are as follows:

- Fast switching speed
- Low power dissipation
- Wide noise margins

6.1 INTRODUCTION AND BACKGROUND

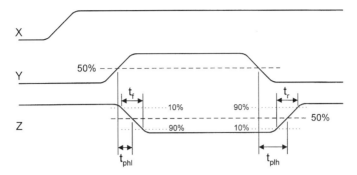

FIGURE 6.2
Voltage waveforms showing propagation delays and rise and fall times for a 2-input NAND gate with output Z as in Fig. 3.10.

- High fan-out capability
- High packing density
- Low cost

Although no single family or technology has all these desirable features, some may come close, at least for most of those listed above. A summary of these and other practical matters now follows.

Propagation Delay (Switching Speed) and Rise and Fall Times The propagation delay or switching speed of a device is the *measured output response to an input change*. Typically, a given logic circuit will have many outputs and many inputs with various input-to-output paths, each with a different path delay. Furthermore, propagation delays usually differ for output changes that are low-to-high (t_{plh}) compared to those that are high-to-low (t_{phl}), but both of which are measured from the 50% point of the input signal to the 50% point of the output response signal as illustrated in Fig. 6.2. The average propagation delay for a given input-to-output path is then given by

$$\tau_{p(avg)} = \frac{t_{plh} + t_{phl}}{2}, \quad (6.1)$$

where, typically, $t_{plh} > t_{phl}$. Since several input-to-output paths may be involved, the timing specifications given by manufacturers often include typical extremes in propagation delay data. A *minimum* value for τ_p is the smallest propagation delay that the logic device will ever exhibit; the *maximum* value is the delay that will "never" be exceeded. The maximum value is the one of most interest to designers since it is used to determine useful factors of safety. For modern CMOS, these values lie in the range of 0.1 to 10 ns. Also shown in Fig. 6.2 are the rise and fall times, t_r and t_f, as measured between the 10% and 90% marks of a given waveform.

Power Dissipation Logic devices consume power when they perform their tasks, and this power is dissipated in the form of heat, Joule heat. Of the various logic families,

CMOS devices consume the least power and then only during switching intervals—that is, dynamic power dissipation. Thus, CMOS power consumption is frequency dependent and may become quite large at very high frequencies. The bipolar families of devices (see Appendix A) consume the most power, mainly due to leakage current, but are much faster than CMOS. Other logic families tend to fall in between these two extremes.

A useful figure of merit for logic devices is called the *power–delay product* (PDP) given by

$$PDP = P_{Consumption} \times \tau_{p(avg)}, \tag{6.2}$$

which is the product of the power consumed by the device and its average propagation delay given by Eq. (6.1). The PDP of a device is sometimes called the *speed–power product* and is usually expressed in picojoules (1 mW × 1 ns = 1 pJ = 10^{-12} joules). Since it is desirable for a given logic device to have both a low power consumption and a small propagation delay, a low PDP is also desirable.

Noise Margins The noise margin of a logic device is the largest voltage that can be added to or subtracted from the logic voltage and still maintain the required logic level. The noise margins are defined as

$$\begin{cases} NML = V_{IL\max} - V_{OL\max} \\ NMH = V_{OH\min} - V_{IH\min} \end{cases} \tag{6.3}$$

and are shown in Fig. 3.2. The voltage parameters defined by manufacturers are expressed as follows:

$V_{IL\max}$ Maximum input voltage guaranteed to be recognized as LOW level.

$V_{OL\max}$ Maximum output voltage guaranteed to be recognized as LOW level.

$V_{OH\min}$ Minimum output voltage guaranteed to be recognized as HIGH level.

$V_{IH\min}$ Minimum input voltage guaranteed to be recognized as HIGH level.

As an example, typical values for high speed (HC) CMOS are $V_{IL\max} = 0.3V_{DD}$, $V_{IH\min} = 0.7V_{DD}$, with $V_{OL\max}$ being slightly above zero voltage and $V_{OH\min}$ being slightly below the supply level V_{DD}.

CMOS logic has always been considered as having good noise margins. However, in the low submicron ranges, CMOS noise margins have been reduced to relatively low values. The bipolar families are usually considered to have good noise margins. It is important that the noise margins of logic devices be wider than any noise transients that may occur so as to prevent unrecoverable errors in the output signals. Thus, noise margins may be regarded as the margins of safety within which digital systems must be operated if their behavior is to be predictable.

Fan-out and Fan-in Since the output from a switching device (for example, a gate) has a definite limit to the amount of current it can supply or absorb, there is a definite limit to the number of other switching devices that can be driven by a single output from that switch.

6.1 INTRODUCTION AND BACKGROUND

This limit is called the *fan-out* of a given device and is, in effect, the worst-case loading specification for that device. The fan-out limit is usually given in microamps (μA). If the fan-out limit of a device is exceeded, the signal can be degraded. MOS circuits are least affected by fan-out restrictions, whereas members of the bipolar families are dramatically affected by such restrictions. Propagation delay is essentially unaffected by fan-out limitations.

The maximum number of inputs permitted to control the operation of a digital device (usually a gate) is called the *fan-in*. Thus, a gate with four inputs has a fan-in of 4. In general for CMOS gates propagation delay increases with increasing fan-in. Fan-in and its consequences are discussed in Subsections 3.6.2 and 3.6.3 and in Section 4.10.

Cost The cardinality or cover of a function is a measure of the *cost* of that function. Design area is also a measure of the cost of a function and is called *area cost*. Thus, the cardinality or design area of a function can be given a monetary value, and this is what is of particular interest to manufacturers of digital devices. But there are more factors that contribute to the monetary cost of an IC. To one extent or another all of the factors previously mentioned directly or indirectly affect the cost of an IC. Appendix A gives qualitatively the performance characteristics as a measure of cost for commonly used IC logic families.

6.1.4 Part Numbering Systems

Typically, parts in data books are given specific part numbers indicative of the logic function they perform and the logic family to which they belong. Commercially available digital devices belonging to the CMOS and TTL (*transistor–transistor logic*) families are given the part prefix "74xnnn", where the "x" represents a string of literals indicating the logic family or subfamily and "nnn" is the part number. To understand this nomenclature the following literals are defined: H = High-speed, L = Low-power, A = Advanced, F = Fast, C = CMOS, and S = Schottky. For example, 74HC00 is a two-input high-speed CMOS NAND gate and a 74AS27 is a three-input advanced Schottky NOR gate. To avoid referring to any specific logic family or subfamily, the "x" descriptor is used along with the part number. For example, a 74x151 is an 8-to-1 multiplexer of a generic type, meaning that it belongs to any of the families for which the prefix "74···" is applicable. The TTL subfamilies designated 74nnn, 74Lnnn, and 74Hnnn have been made obsolete by the presence of modern Schottky subfamilies.

Another member of the bipolar family is called ECL for *emitter-coupled logic*. The ECL family is currently the fastest of the logic families but has an extremely high power consumption and high PDP. ECL parts are named either with a 5-digit number system (10nnn) or a 6-digit system (100nnn), depending on which subfamily is being referenced. In either case all part numbers "nnn" are always three digits in length, unlike those for CMOS and TTL families, which can be two or three digits in length. Appendix A qualitatively summarizes the performance characteristics of TTL, ECL, NMOS, and CMOS families.

6.1.5 Design Procedure

The design of any combinational logic device generally begins with the description of and specifications for the device and ends with a suitable logic implementation. To assist the reader in developing good design practices, the following six-step sequence is recommended:

Step 1: Understand the device. Describe the function of the device; then clearly indicate its input/output (I/O) specifications and timing constraints, and construct its block diagram(s).

Step 2: State any relevant algorithms. State all algorithms and/or binary manipulations necessary for the design. Include a general operations format if necessary.

Step 3: Construct the truth tables. From step 2, construct the truth tables that detail the I/O relationships. Truth tables are usually presented in positive logic form.

Step 4: Obtain the output functions. Map or use a minimization program to obtain any minimum or reduced expressions that may be required for the output functions.

Step 5: Construct the logic diagrams. Use either a gate or modular level approach (or both) to implement the logic expressions obtained in step 4. Implement from output to input, taking into account any mixed logic I/O conditions and timing constraints that may be required.

Step 6: Check the results. Check the final logic circuit by simulation before implementation as a physical device. Real-time tests of the physical device should be the final test stage.

This text follows the six-step sequence where appropriate and does so without specifically mentioning each step.

6.2 MULTIPLEXERS

There is a type of device that performs the function of selecting one of many data input lines for transmission to some destination. This device is called a *multiplexer* (*MUX* for short) or *data selector*. It requires n data select lines to control 2^n data input lines. Thus, a MUX is a 2^n-input/1-output device identified by the block diagram in Fig. 6.3a. Shown in Fig. 6.3b is the mechanical switch equivalent of the MUX. Notice that the function of the enable (EN = G) is to provide a disable capability to the device. Commercial MUX ICs usually come with active low enable, EN(L).

The general logic equation for the MUX of Fig. 6.3 can be expressed as

$$Y = \sum_{i=0}^{2^n-1}(m_i \cdot I_i) \cdot EN, \qquad (6.4)$$

where m_i represents the ith minterm of the data select inputs (e.g., $m_2 = \bar{S}_{n-1} \cdots \bar{S}_2 S_1 \bar{S}_0$). The validity of this equation will be verified in the following subsection on multiplexer design.

6.2.1 Multiplexer Design

The easiest and most "logical" way to design a MUX is to represent the MUX by a compressed, entered variable (EV) truth table. This is illustrated by the design of a 4-to-1

6.2 MULTIPLEXERS

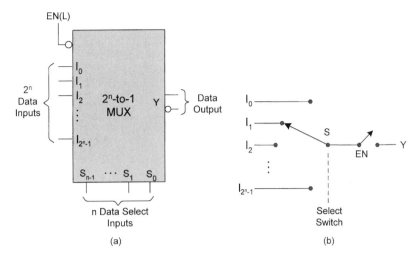

FIGURE 6.3
A 2^n-to-1 multiplexer (MUX) or data selector. (a) Block diagram symbol. (b) Mechanical switch equivalent.

(2^2-to-1) MUX with active low enable. Shown in Fig. 6.4 are the logic symbol, EV truth table, and K-map for the 4-to-1 MUX. From the K-map there result the output expressions given by

$$Y = \bar{S}_1 \bar{S}_0 I_0 EN + \bar{S}_1 S_0 I_1 EN + S_1 \bar{S}_0 I_2 EN + S_1 S_0 I_3 EN$$
$$= m_0 I_0 EN + m_1 I_1 EN + m_2 I_2 EN + m_3 I_3 EN$$
$$= \sum_{i=0}^{2^2-1} (m_i \cdot I_i) \cdot EN, \qquad (6.5)$$

representing four data input lines, two data select lines, and one output. The circuit for the 4-to-1 MUX is obtained directly from Eq. (6.5) and is presented in Fig. 6.4d, together with its shorthand circuit symbol given in Fig. 6.4e.

An $m \times 2^n$ input MUX can be produced by stacking m 2^n-to-1 MUXs with outputs to an m-to-1 MUX output stage. This is illustrated in Fig. 6.5, where four 8-to-1 (74x151) MUXs are stacked to produce a 32-to-1 MUX. Notice that this MUX can be disabled simply by using the EN(L) line to the output stage MUX. For an explanation of the part identification notation (e.g., 74x···), see Subsection 6.1.4.

Many variations of the stacked MUX configuration are possible, limited only by the availability of different MUX sizes. For example, two 16-to-1 MUXs combine to form a 32-to-1 MUX or four 4-to-1 MUXs combine to produce a 16-to-1 MUX. In the former case a 2-to-1 MUX must be used to complete the stack configuration, whereas in the latter case a 4-to-1 MUX is required. There are other ways in which to stack and package MUXs. One variation is illustrated in the discussion that follows.

More than one MUX can be packaged in an IC, and this can be done in a variety of ways. One configuration is illustrated by the design of the 74x153 4-input/2-bit MUX

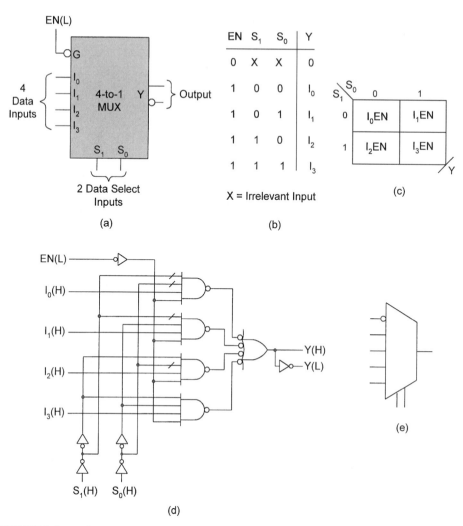

FIGURE 6.4
Design of a 4-to-1 MUX or data selector. (a) Block diagram symbol. (b) Compressed EV truth table. (c) MUX EV K-map. (d) Logic circuit for a 4-to-1 MUX with active low enable and mixed logic output. (e) Shorthand 4-to-1 MUX circuit symbol.

shown in Fig. 6.6. Reading directly from the EV truth table in Fig. 6.6a yields the output expressions

$$1Y = \bar{S}_1\bar{S}_0 1I_0 \cdot 1G + \bar{S}_1 S_0 1I_1 \cdot 1G + S_1\bar{S}_0 1I_2 \cdot 1G + S_1 S_0 1I_3 \cdot 1G$$
$$2Y = \bar{S}_1\bar{S}_0 2I_0 \cdot 2G + \bar{S}_1 S_0 2I_1 \cdot 2G + S_1\bar{S}_0 2I_2 \cdot 2G + S_1 S_0 2I_3 \cdot 2G,$$
(6.6)

which are implemented with NAND/INV logic in Fig. 6.6c.

The traditional logic symbol for the '153 MUX is given in Fig. 6.6b. Notice that there are two data select inputs that simultaneously control data selection to both outputs, and

6.2 MULTIPLEXERS   245

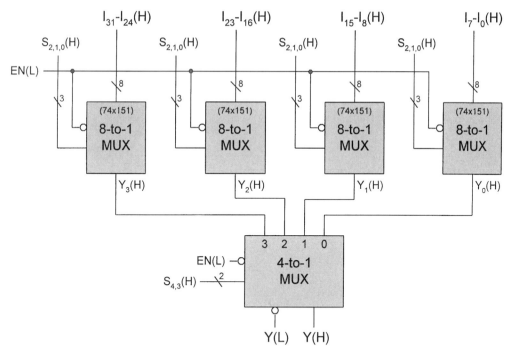

FIGURE 6.5
Four 8-to-1 MUXs and a 4-to-1 MUX combine to produce a 32-to-1 MUX having five data select inputs S_4, S_3, S_2, S_1, and S_0 and an active low enable.

that there are two independently operated enable inputs, $1G$ and $2G$, that enable or disable either or both of the MUXs.

6.2.2 Combinational Logic Design with MUXs

A MUX is a *function generator* and can be used to implement a function in canonical form or in compressed (reduced or minimum) form. To understand this, consider the function

$$Y(A, B, C, D) = \sum m(3, 4, 5, 6, 7, 9, 10, 12, 14, 15)$$
$$= \prod M(0, 1, 2, 8, 11, 13). \qquad (6.7)$$

If this function is implemented with a 16-to-1 MUX, then all inputs representing minterms in Eq. (6.7) are connected to logic 1 (HV) while all maxterms are connected to logic 0 (LV). In this case the data select variables are the four function variables. But if it is desirable to implement Eq. (6.7) with a 4-to-1 MUX, two levels of K-map compression are needed, as in Fig. 6.7b.

Notice that the data select variables, $S_1 = A$ and $S_0 = B$, form the axes of the MUX K-map in Fig. 6.7b and that the functions generated by the 4-to-1 MUX in Fig. 6.7 are

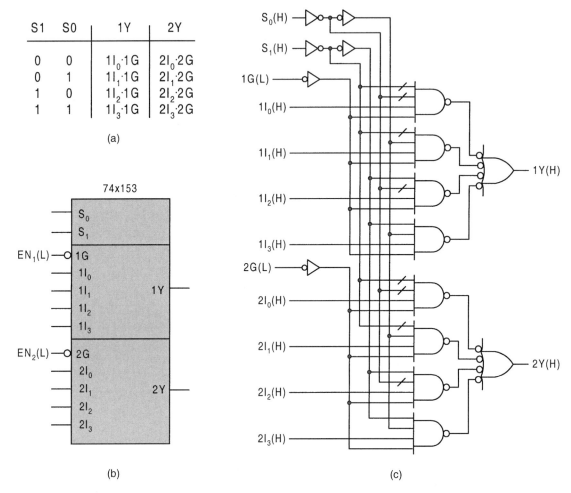

FIGURE 6.6
Design of the 74x153 4-input/2-bit MUX. (a) EV truth table. (b) Traditional logic symbol. (c) Logic diagram.

given by

$$\begin{cases} Y = [\bar{A}\bar{B}CD + \bar{A}B + A\bar{B}(C \oplus D) + AB(C + \bar{D})] \\ \text{or} \\ Y = [(A + B + CD)(\bar{A} + B + C \oplus D)(\bar{A} + \bar{B} + C + \bar{D})] \end{cases}, \quad (6.8)$$

both of which are three-level hybrid forms that are generated (H) or (L) from the $Y(H)$ and $Y(L)$ outputs of the MUX. The latter of the two may not seem obvious but is easily verified by extracting cover in maxterm code from the second-order K-map in Fig. 6.7b. Note also that if an 8-to-1 MUX is used to implement the function in Eq. (6.7), the K-map in Fig. 6.7a applies, where the data select variables are now A, B, and C. In this case inputs I_1 and I_4

6.2 MULTIPLEXERS

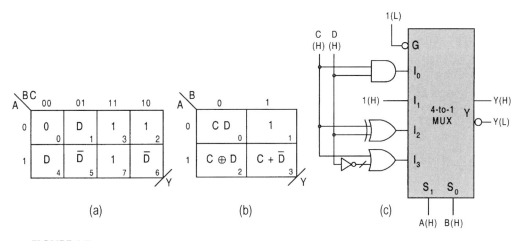

FIGURE 6.7
Implementation of the function in Eq. (6.7) by using a 4-to-1 MUX. (a), (b) First- and second-order K-map compressions for the function Y. (c) Logic circuit.

are connected to $D(H)$, inputs I_5 and I_6 are connected to $D(L)$, inputs I_2, I_3, and I_7 connect to $1(H)$, and I_0 goes to ground $0(H)$.

There still remains the problem of dealing with active low inputs to a MUX. The rules are simply stated:

For Active Low Data Select Inputs to MUXs

(1) Complement the MUX K-map axis of any active low data select input and renumber the K-map cells. The new cell numbers identify the MUX inputs to which they must be connected.

or

(2) Use an inverter on the input of an active low data select input and *do not* complement the MUX K-map axis.

For All Other Active Low Inputs

Active low nondata select inputs are dealt with as any combinational logic problem with mixed-logic inputs (see, e.g., Section 3.7). Therefore, *do not* complement any EV subfunction in a MUX K-map.

To illustrate the problem of mixed-logic inputs to a MUX, consider the function of Eq. (6.7) with inputs that arrive as $A(H)$, $B(L)$, $C(H)$, and $D(L)$. Implementation with a 4-to-1 MUX follows as in Fig. 6.8, where the B axis of the MUX K-map is complemented since no inverter is used on the $B(L)$ data select input line. Notice that no additional inverters are required when compared to the implementation of Fig. 6.7, and that the resulting outputs are identical to those of Eqs. (6.8). The use of an EQV gate in place of an XOR gate is a consequence of the $D(L)$ and the fact that only one inverter is used.

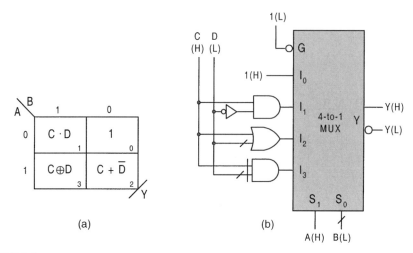

FIGURE 6.8
Implementation of the function in Eq. (6.7) with mixed-logic inputs by using a 4-to-1 MUX. (a) Second-order K-map compressions for the function Y showing the renumbered cells due to $B(L)$. (b) Logic circuit.

6.3 DECODERS/DEMULTIPLEXERS

A decoder is an n-input/2^n-output combinational logic device that has the function of activating one of the 2^n outputs for every unique binary input pattern of n bits. The circuit symbol for an n-to-2^n decoder is shown in Fig. 6.9a, where $I_{n-1} - I_0$ are the data inputs, $Y_{2^n-1} - Y_0$ are the outputs, and G is the enable. Each output is identified by the minterm code m_i of the binary input pattern it represents and can be represented, generally, as

$$Y_i = \sum_{i=0}^{2^n-1} m_i \cdot EN, \qquad (6.9)$$

where $m_0 = \bar{I}_{n-1} \cdots \bar{I}_2 \bar{I}_1 \bar{I}_0$, $m_1 = \bar{I}_{n-1} \cdots \bar{I}_2 \bar{I}_1 I_0$, $m_2 = \bar{I}_{n-1} \cdots \bar{I}_2 I_1 \bar{I}_0$, and so on. For this reason a decoder can be called a *minterm code generator*. Commercial decoders are available in a variety of sizes and packaged configurations, but most all feature active low outputs and an active low enable.

Shown in Fig. 6.9b is the same decoder used as a demultiplexer (DMUX). Now, the active low enable $EN(L)$ to the decoder becomes the single data input $I_{Data}(L)$ to the DMUX, and the data inputs $I_0, I_1, I_2, \ldots, I_{n-1}$ for the decoder become the data select inputs $S_0, S_1, S_2, \ldots, S_{n-1}$ for the DMUX. The outputs for the decoder and DMUX are the same if it is understood that I_i is replaced by S_i in Eq. (6.9). The active low outputs and active low enable are of particular importance when a decoder is used as a DMUX, since the DMUX is often paired with a MUX for data routing as explained later.

6.3.1 Decoder Design

Decoder design is illustrated by the design of a 3-to-8 decoder. Shown in Fig. 6.10 is the collapsed canonical I/O truth table for the enable input (EN), the three data inputs (I_2, I_1,

6.3 DECODERS/DEMULTIPLEXERS   249

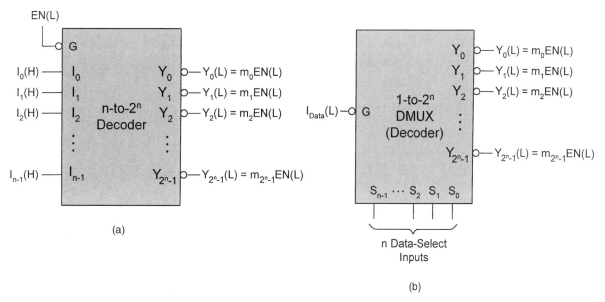

FIGURE 6.9
Generalization of a decoder/demultiplexer (DMUX). (a) An n-to-2^n decoder with an active low enable. (b) The decoder of (a) used as a 1-to-2^n DMUX with data input $I_{Data}(L)$.

and I_0) and the eight outputs $Y_7, \ldots, Y_2, Y_1,$ and Y_0. The truth table is termed a *collapsed truth table* because of the *irrelevant input* symbol X used to represent either logic 0 or logic 1. Thus, X X X in the first row of the table represents eight minterms in variables I_i. Notice that only one minterm code line is activated for each corresponding three-bit binary pattern that appears on the input with active EN.

Each output (Y_i) column in Fig. 6.10 represents a third-order K-map containing a single minterm ANDed with EN. However, it is not necessary to construct eight EV K-maps to obtain the eight output expression for Y_i, since this information can be read directly from

EN	I_2	I_1	I_0	Y_7	Y_6	Y_5	Y_4	Y_3	Y_2	Y_1	Y_0	
0	X	X	X	0	0	0	0	0	0	0	0	
1	0	0	0	0	0	0	0	0	0	0	1	$Y_0 = \bar{I}_1 \bar{I}_1 \bar{I}_0 \cdot EN$
1	0	0	1	0	0	0	0	0	0	1	0	$Y_1 = \bar{I}_1 \bar{I}_1 I_0 \cdot EN$
1	0	1	0	0	0	0	0	0	1	0	0	$Y_2 = \bar{I}_1 I_1 \bar{I}_0 \cdot EN$
1	0	1	1	0	0	0	0	1	0	0	0	$Y_3 = \bar{I}_1 I_1 I_0 \cdot EN$
1	1	0	0	0	0	0	1	0	0	0	0	$Y_4 = I_1 \bar{I}_1 \bar{I}_0 \cdot EN$
1	1	0	1	0	0	1	0	0	0	0	0	$Y_5 = I_1 \bar{I}_1 I_0 \cdot EN$
1	1	1	0	0	1	0	0	0	0	0	0	$Y_6 = I_1 I_1 \bar{I}_0 \cdot EN$
1	1	1	1	1	0	0	0	0	0	0	0	$Y_7 = I_1 I_1 I_0 \cdot EN$

X indicates an irrelevant input and represents either logic 0 or logic 1.

FIGURE 6.10
Collapsed truth table for a 3-to-8 decoder/demultiplexer with enable showing output expressions that derived directly from the truth table.

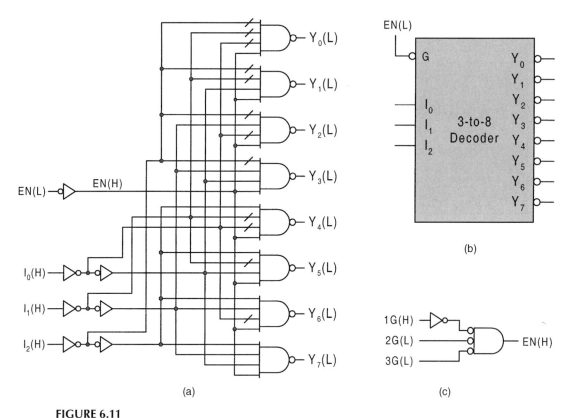

FIGURE 6.11
Implementation of the 3-to-8 decoder/demultiplexer in Fig. 6.10. (a) Logic circuit with active low outputs and active low enable. (b) Logic symbol. (c) An alternative three-enable configuration used by the commercial 74x138 decoder.

the truth table and is provided in the third column of Fig. 6.10. When the requirements of active low outputs and active low enable are introduced, the Y_i expressions for the 3-to-8 decoder/demultiplexer in Fig. 6.10 are implemented with NAND/INV logic as shown in Fig. 6.11a. Its logic symbol is given in Fig. 6.11(b).

A single enable input is used to enable or disable the decoder of Fig. 6.11. But other enable configurations are common. For example, a commercially available 3-to-8 decoder, the 74x138, has the same decoder logic as in Fig. 6.11a, except the commercial unit features three enable inputs as indicated in Fig. 6.11c. Multiple enable inputs permit greater versatility when controlling a given decoder from various sources.

Decoders can be stacked (cascaded) in hierarchical configurations to produce much larger decoders. This requires that the 2^m outputs of one decoder drive the EN controls of 2^n other decoders, assuming that all outputs of the leading decoders are used. As an example, four 3-to-8 decoders are enable/selected by a 2-to-4 decoder in Fig. 6.12 to produce a 5-to-32 decoder. Similarly, two 4-to-16 decoders can be stacked to produce a 5-to-32 decoder when enable/selected by a single inverter. Or cascading four 4-to-16 decoders produces a 6-to-64 decoder when enable/selected by a 2-to-4 decoder. Note that stacking any two decoders requires only an inverter acting as a 1-to-2 enable/select decoder.

6.3 DECODERS/DEMULTIPLEXERS

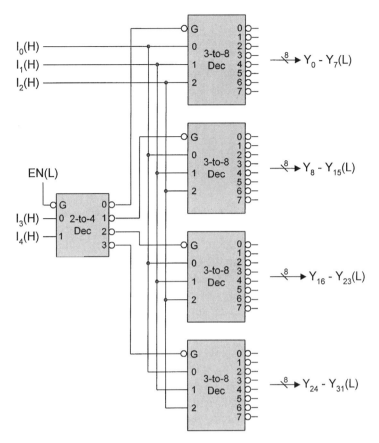

FIGURE 6.12
Stacking of four 3-to-8 decoders to produce a 5-to-32 decoder.

Decoders can be packaged in pairs such as the dual 2-to-4 74x139 decoder. Such a dual set will have two independent enable inputs, one for each or the two 2-to-4 decoders. The 2-to-4 decoder in Fig. 6.12 is actually one half of the 74x139, indicated as $\frac{1}{2}$74x139.

6.3.2 Combinational Logic Design with Decoders

Decoders can be used effectively to implement any function represented in canonical form. All that is needed is the external logic required to OR minterms for SOP representation or to AND maxterms for POS representation. As a simple example, consider the two functions in canonical form:

$$F(A, B, C) = \sum m(1, 3, 4, 7) \quad SOP$$
$$G(A, B, C) = \prod M(2, 3, 5, 6) \quad POS. \tag{6.10}$$

Assuming that the inputs arrive as $A(H), B(H), C(H)$, and that the outputs are delivered

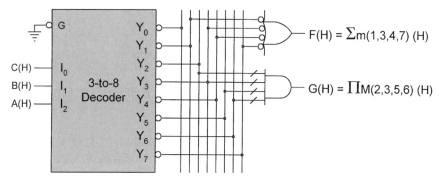

FIGURE 6.13
Decoder implementations of Eqs. (6.10) assuming inputs and outputs are all active high.

active high, these two functions are implemented as given in Fig. 6.13. To understand why function G is implemented with an AND gate, consider what is generated by the ANDing operation:

$$G = \bar{m}_2 \cdot \bar{m}_3 \cdot \bar{m}_5 \cdot \bar{m}_6 = M_2 \cdot M_3 \cdot M_5 \cdot M_6 = \prod M(2, 3, 5, 6).$$

If it is desirable to issue G active low, a NAND gate would be used in place of the AND gate. Or if F is to be issued active low, an AND would be needed in place of the NAND. Actually, to fully understand the versatile nature of function implementation with decoders, the reader should experiment by replacing the NAND and AND gates in Fig. 6.13 with a variety of gates, including treed XOR and EQV gates.

The problem of mixed-logic inputs can be dealt with in a manner similar to those issued to MUXs. The rules are similar to those for MUXs and are stated as follows:

For Mixed-Logic Data Inputs to Decoders

(1) Complement the bit of any active low input to a decoder and renumber the minterms accordingly.

or

(2) Use an inverter on the input line of an active low input and do not complement the bit.

Consider, as an example, the two functions of Eqs. (6.10) with inputs that arrive as $A(H)$, $B(L)$, and $C(L)$. Functionally, the mixed-logic forms of Eqs. (6.10) become

$$F_{SOP}[A(H), B(L), C(L)] = F_{SOP}[A, \bar{B}, \bar{C}](H)$$

and (6.11)

$$G_{POS}[A(H), B(L), C(L)] = G_{POS}[A, \bar{B}, \bar{C}](H).$$

6.3 DECODERS/DEMULTIPLEXERS

Then by Eqs. (6.11) and if inverters are not to be used, the B and C bits must be complemented:

$$m_0 = 000 \rightarrow 011 = m_3 \qquad m_4 = 100 \rightarrow 111 = m_7$$
$$m_1 = 001 \rightarrow 010 = m_2 \qquad m_5 = 101 \rightarrow 110 = m_6$$
$$m_2 = 010 \rightarrow 001 = m_1 \qquad m_6 = 110 \rightarrow 101 = m_5$$
$$m_3 = 011 \rightarrow 000 = m_0 \qquad m_7 = 111 \rightarrow 100 = m_4.$$

Thus, to accommodate the mixed-logic inputs, the two functions in Eqs. (6.10) must be connected to the decoder according to the renumbered functions

$$F[A(H), B(L), C(L)] = \sum m(0, 2, 4, 7) \quad \text{and}$$
$$G[A(H), B(L), C(L)] = \prod M(0, 1, 5, 6).$$

Of course, if inverters are used on the $B(L)$ and $C(L)$ inputs, no complementation is necessary and the functions are implemented according to Eqs. (6.10).

Decoders, used as demultiplexers (DMUXs), are simply reoriented so that the active low enable is the only data input. Now the I_i inputs become the data select inputs S_i as in Fig. 6.9b. Used in conjunction with MUXs, the MUX/DMUX system offers a means of time-shared bussing of multiple-line data X_i on a single line as illustrated in Fig. 6.14. Bussing data over large distances by using this system results in a significant savings on hardware, but is a relatively slow process.

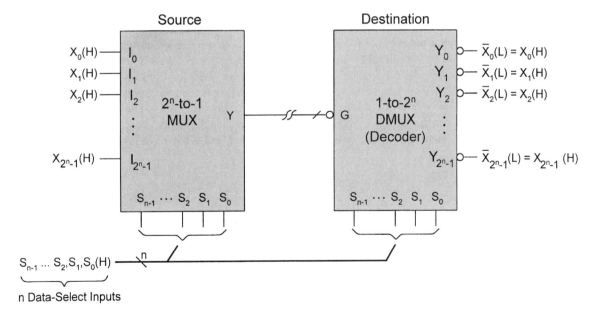

FIGURE 6.14
Generalization of a MUX/DMUX system for bussing data on $2^n - 1$ lines over a single time-shared line from the MUX source to the DMUX destination.

6.4 ENCODERS

By definition an *encoder* performs an operation that is the opposite to that of a decoder. That is, an encoder must generate a different output bit pattern (code) for each input line that becomes active. For a binary encoder, this requirement can be enforced only if one output binary word of n bits is associated with only one of 2^n "decimal" input lines $(0, 1, 2, 3, \ldots, 2^n - 1)$. Obviously, if more than one input line is active, the output becomes ambiguous in such an encoder. The ambiguity problem is overcome by prioritizing the input. When this is done the result is called a *priority encoder* (PE), which assigns a priority to each input according to a PE schedule for that encoder. Most encoders are PEs.

A priority encoder is generally an n-input/m-output ($n \leq 2^m$) device as indicated by the circuit symbol in Fig. 6.15. In addition to the n address inputs and m outputs, a commercial PE will usually have three other input and output lines that are used to cascade (stack) PEs: an *enable-in* input (*EI*), an *enable-out* output (*EO*), and a *group signal* output (*GS*). The purpose of the *GS* output is to indicate any legitimate encoding condition, meaning that *EI* is active concurrently with a single active address input. All inputs and outputs of a PE are active low as shown.

The design of a simple 3-input/2-output PE with cascading capability is illustrated in Fig. 6.16. Shown in Fig. 6.16a is the priority schedule and the collapsed I/O truth table for this encoder. The EV K-maps in Fig. 6.16b are plotted from the truth table, and the minimum cover yields the following output expressions:

$$Y_1 = I_2 EI + I_1 EI \qquad Y_0 = \bar{I}_1 I_0 EI + I_2 EI$$
$$EO = \bar{I}_2 \bar{I}_1 \bar{I}_0 EI \qquad GS = (I_2 + I_1 + I_0) EI = \overline{EO} \cdot EI. \tag{6.12}$$

These expressions are implemented with minimum logic in Fig. 6.16c for active low inputs and outputs as required in Fig. 6.15. Notice that this circuit represents a simple multioutput optimization, which is deduced by inspection of expressions for *EO* and *GS*.

The outputs *EO* and *GS* require special attention since their logic values have been specifically chosen to make cascading of PEs possible. When the address inputs for the nth stage are *all* inactive (logic 0), it is the function of EO_n to activate the $(n - 1)$th stage. This assumes that prioritization is assigned from highest active input (decimal-wise) to lowest. Therefore, *EO* can be active only for inactive address inputs and active *EI*. It is the function

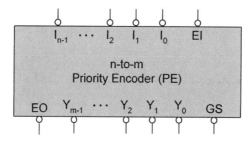

FIGURE 6.15
Logic symbol for an *n*-to-*m* priority encoder with cascading capability.

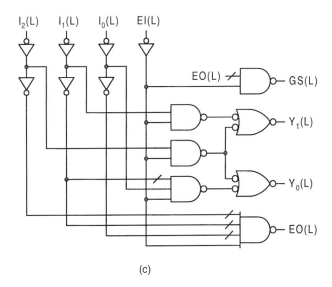

FIGURE 6.16
Design of a three-input priority encoder with cascading capability. (a) Priority schedule and collapsed truth table. (b) EV K-maps. (c) Minimized logic circuit.

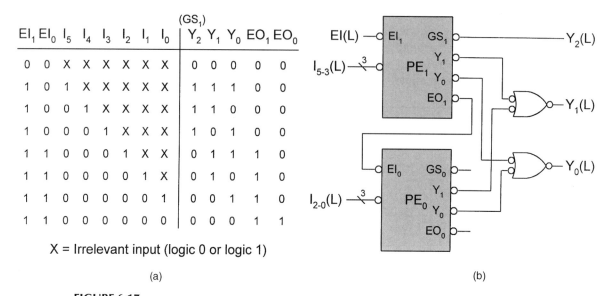

FIGURE 6.17
Two 3-to-2 PEs in cascade to produce a 6-to-3 PE. (a) Collapsed truth table. (b) Logic circuit.

of the *GS* output to go active for any valid encoding condition for that stage, meaning *EI* active concurrently with a single active address input, and to provide an additional output bit (MSB) for each stage added in cascade fashion.

Two 3-to-2 PEs of the type in Fig. 6.16 are shown stacked in Fig. 6.17 to form a 6-to-3 PE. The truth table for the two PE combination is given in Fig. 6.17a and includes the cascading bits *EI*, *EO*, and *GS* for PE_1 and PE_0. From Eqs. (6.12) and Fig. 6.17, the cascading bits for the system are expressed as

$$GS_1 = \overline{EO}_1 EI_1$$
$$EO_1 = \bar{I}_5 \bar{I}_4 \bar{I}_3 EI_1 = EI_0 \qquad (6.13)$$
$$EO_0 = \bar{I}_2 \bar{I}_1 \bar{I}_0 EI_0.$$

Notice that $GS_1 = Y_2$ indicates a valid encoding condition only for active inputs I_5, I_4, and I_3, while GS_0 (not shown) indicates a valid encoding condition only for active inputs I_2, I_1, and I_0. Output state 100 cannot occur according to Eqs. (6.12) and (6.13) and Fig. 6.17a.

Priority encoders of the type represented in Fig. 6.15 are commercially available as IC chips. Typically, they are produced in the 8-to-3 line size, such as the 74x148, which can be stacked to produce 16-to-4 line and 32-to-5 line PEs. Their applications include code conversion, code generation, and *n*-bit encoding in digital systems having a hierarchy of subsystems that must be prioritized.

Those PEs that do not have *EO* and *GS* outputs, and hence cannot be stacked, are also available commercially as ICs. Their design closely follows the design of PEs of the type in Fig. 6.16. They are commonly produced in 9-to-4 line size for use as BCD priority encoding, keyboard encoding, and range selection.

6.5 CODE CONVERTERS

Various codes are discussed at length in Section 2.10. On occasion it is necessary to convert one code to another. Devices that are designed to execute a code conversion are called *code converters*. Considering the many codes that are currently in use, it follows that there are a very large number of converters possible. Not taking into account any particular area of specialty, a few of the more common code conversions are as follows:

Binary-to-Gray conversion and vice versa
BCD-to-XS3 conversion and vice versa
Binary-to-BCD conversion and vice versa

6.5.1 Procedure for Code Converter Design

The following is a simple three-step procedure that will be followed in this text, often without reference to this subsection:

1. Generally, follow the design procedure in Subsection 6.1.5.
2. If conversion involves any of the decimal code input (e.g., BCD), only 10 states can be used. The six unused input states are called *false data* inputs. For these six states the outputs must be represented either by don't cares (ϕ's) or by some unused output state, for example all 1's. That is, if the requirement is for *false data rejection* (FDR), then the output states must correspond to at least one unused output state; if not, ϕ's are entered for the output states. Thus, FDR means that the outputs must never correspond to a used output state when any one of the six unused states arrive at the input terminals. If false data is not rejected, then the outputs corresponding to the six unused states can take on any logic values, including those of used output states.
3. If the input code is any other than binary or BCD and if EV K-maps are to be used in minimizing the logic, it is recommended that the input code be arranged in the order of ascending binary, taking care to match each output state with its corresponding input state.

6.5.2 Examples of Code Converter Design

To illustrate the code converter design process, four examples are presented. These examples are quite adequate since the conversion procedure varies only slightly from one conversion to another. The four examples are Gray-to-binary conversion, BCD-to-XS3 conversion, BCD-to-binary conversion, and BCD-to-seven-segment-display conversion. Of these, the last two are by far the most complex and perhaps the most important considering that binary, BCD, and the seven-segment display are commonly used in digital design and computer technology.

Gray-to-Binary Conversion The Gray-to-binary conversion table for 4-bit codes is given in Fig. 6.18a. Here, for convenience of plotting EV K-maps, the input Gray code and the corresponding output binary code have been rearranged such that the Gray code is given in ascending minterm code (compare Tables 2.1 and 2.12). The second-order EV

Gray A B C D	Binary A' B' C' D'	Gray A B C D	Binary A' B' C' D'
0 0 0 0	0 0 0 0	1 0 0 0	1 1 1 1
0 0 0 1	0 0 0 1	1 0 0 1	1 1 1 0
0 0 1 0	0 0 1 1	1 0 1 0	1 1 0 0
0 0 1 1	0 0 1 0	1 0 1 1	1 1 0 1
0 1 0 0	0 1 1 1	1 1 0 0	1 0 0 0
0 1 0 1	0 1 1 0	1 1 0 1	1 0 0 1
0 1 1 0	0 1 0 0	1 1 1 0	1 0 1 1
0 1 1 1	0 1 0 1	1 1 1 1	1 0 1 0

(a)

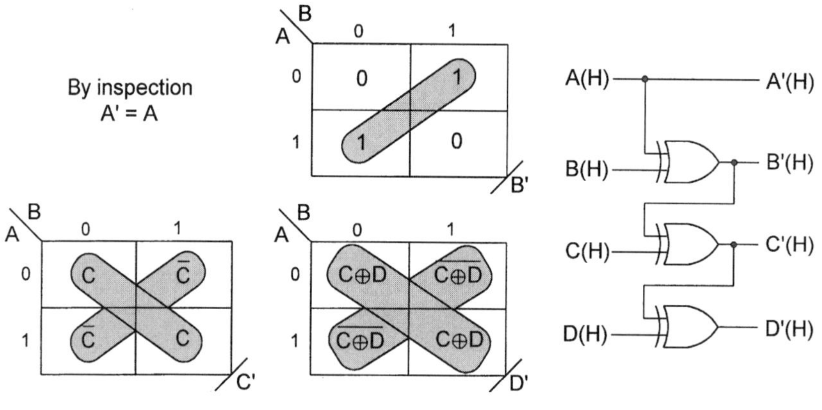

(b) (c)

FIGURE 6.18
Design of a 4-bit Gray-to-binary converter. (a) I/O truth table. (b) Output EV K-maps plotted from (a) showing minimum cover. (c) Resulting logic circuit according to Eqs. (6.14).

K-maps, shown in Fig. 6.18b, are plotted directly from the truth table and yield the minimum cover,

$$\begin{aligned} A' &= A \\ B' &= A \oplus B \\ C' &= A \oplus B \oplus C \\ D' &= A \oplus B \oplus C \oplus D, \end{aligned} \tag{6.14}$$

from which the logic circuit of Fig. 6.18c results. Noticing the trend in Eqs. (6.14), it is clear that an XOR gate is added in series fashion with each additional bit of the Gray-to-binary conversion. With this trend in mind any size Gray-to-binary converter can be implemented without the need to repeat the steps indicated in Fig. 6.18.

BCD-to-XS3 Conversion As a second example, consider the conversion between two decimal codes, BCD and XS3. Shown in Fig. 6.19a is the truth table for the BCD-to-XS3 conversion where, for this design, false data is *not* rejected. Thus, ϕ's become the output

6.5 CODE CONVERTERS

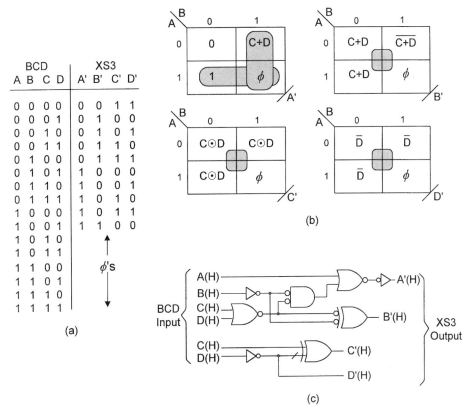

BCD A B C D	XS3 A' B' C' D'
0 0 0 0	0 0 1 1
0 0 0 1	0 1 0 0
0 0 1 0	0 1 0 1
0 0 1 1	0 1 1 0
0 1 0 0	0 1 1 1
0 1 0 1	1 0 0 0
0 1 1 0	1 0 0 1
0 1 1 1	1 0 1 0
1 0 0 0	1 0 1 1
1 0 0 1	1 1 0 0
1 0 1 0	
1 0 1 1	
1 1 0 0	φ's
1 1 0 1	
1 1 1 0	
1 1 1 1	

(a)

FIGURE 6.19
Design of the BCD-to-XS3 converter without FDR. (a) Truth table. (b) EV K-maps showing minimum cover. (c) Logic circuit according to Eqs. (6.15).

XS3 code patterns corresponding to the six unused input BCD states. The resulting EV K-maps for the output functions are given in Fig. 6.19b, from which the gate-minimum cover is extracted as

$$\begin{cases} A' = B(C+D) + A \\ B' = B \oplus (C+D) \\ C' = C \odot D \\ D' = \bar{D} \end{cases}, \quad (6.15)$$

representing a gate/input tally of 5/10, excluding inverters. The subfunctions for cell 2 in Fig. 6.19b result from an appropriate use of the φ's. Shown in Fig. 6.19c is the three-level logic circuit for this converter implemented with NOR/XOR/INV logic assuming that the inputs and outputs are all active high. The subject of mixed logic inputs to an XOR gate is discussed at length in Subsection 3.9.4.

Had FDR been a design objective for the BCD-to-XS3 converter, the φ's in Fig. 6.19a would have to be replaced by an unused (FDR) output state. If the FDR state is taken to be 1111, the K-maps of Fig. 6.19b are altered accordingly, and the resulting gate-minimum output expressions become

$$A' = B(C + D) + A$$
$$B' = B \oplus (C + D) + AB$$
$$C' = C \odot D + AC + AB \quad (6.16)$$
$$= C \oplus D + AC + AB$$
$$D' = \bar{D} + AC + AB,$$

now representing a gate/input tally of 10/22, excluding an inverter but including three shared PIs. Clearly, the FDR feature comes at a significant price in terms of hardware. FDR states other than the 1111 can be used, as for example the 0000 state, but all increase the cost in hardware even further.

Conversion between BCD and Binary The BCD and binary codes are two of the most widely used codes in digital design, so it is fitting that conversion between them be considered. The simplest approach is to establish workable algorithms to produce an expandable modular design. Even so, such tasks rank among the more difficult conversion problems. Consider, for example, that a two-digit BCD word converts to an 8-bit binary number, whereas an 8-bit binary word converts to a three-digit BCD representation.

Two algorithms will now be considered that make conversion between these two codes tractable and that lead to modular designs. In these algorithms shifting right by one bit is equivalent to dividing by 2 ($sr1 = \div 2$) and shifting left by one bit is equivalent to multiplying by 2 ($sl1 = \times 2$). Also, LSD and MSD refer to the least significant digit and most significant digit, respectively, for the BCD number. A detailed treatment of the binary and BCD number systems is provided in Sections 2.3 and 2.4, and the pencil-and-paper conversion method between the binary and BCD number systems is discussed in Subsection 2.4.1.

The first of the two algorithms, the BCD-to-binary conversion, requires that the BCD number first be placed in imaginary storage cells. For example, a two-decade BCD number will occupy eight imaginary storage cells. After this has been done, then the algorithm proceeds as follows:

Algorithm 6.1 BCD-to-Binary Conversion by the ($\div 2$)/(-3) Process

(1) Shift the BCD number to the right by one bit into the new LSD position, but keeping account of the bits shifted out of the LSD.
(2) Subtract 0011 from the LSD (or add 1101 to the LSD if in 2's complement) iff the new LSD number is greater than 7 (0111). After subtracting 3, shift right immediately even if the new LSD is greater than 7.
(3) Repeat steps (1) and (2) until the final LSD number can no longer be greater than decimal 7. The answer is now in binary.

Algorithm 6.1 is sometimes referred to as the shift-right/subtract 3 [or ($\div 2$)/(-3)] algorithm. The algorithm for binary-to-BCD conversion can be thought of as the mathematical dual of Algorithm 6.1. In this case the process could be called the shift-left/add 3 [or ($\times 2$)/($+3$)] algorithm. Begin by placing the binary number outside and to the right of the LSD positions, then proceed as follows:

6.5 CODE CONVERTERS

> **Algorithm 6.2 Binary-to-BCD Conversion by the (×2)/(+3) Process**
>
> (1) Shift the binary number to the left by one bit into the new LSD position.
> (2) Add 0011 to the new LSD iff the new LSD number is greater than 4 (0100). After adding 3, shift left immediately.
> (3) Repeat steps (1) and (2). When all binary bits have been shifted into digit positions, the process ceases and the answer is in BCD.

To design converters by either Algorithm 6.1 or 6.2, it is not necessary to resort to arithmetic means as implied by the algorithms. Rather, a modular approach is easily established by constructing a relatively simple truth table followed by appropriate minimization methods. The process is now illustrated by designing an 8-bit BCD-to-binary converter.

BCD-to-Binary Conversion A truth table matching BCD with the corresponding binary from 0 to 19 is given in Fig. 6.20a. Inherent in this truth table is the shift-right/subtract-3 Algorithm 6.1. The decimal range 0 to 19 is chosen to illustrate the process but is easily extended to 39, 79, or 99, etc. This truth table will be used to design a BCD-to-binary module that can be cascaded to produce any size converter. It is possible to use a decimal range of 0 to 99 for this purpose, but the size of the module is considered too large to be of value for this example.

Shown in Fig. 6.20b are the K-maps and minimum cover for four of the five output functions of the 2-digit BCD-to-binary module. The two-level minimum expressions for the BCD-to-binary module as read from the K-maps and truth table are

$$B_4 = D_4 D_2 D_1 + D_4 D_3$$
$$B_3 = D_4 \bar{D}_3 \bar{D}_1 + D_4 \bar{D}_2 D_1 + \bar{D}_4 D_3$$
$$B_2 = D_4 \bar{D}_2 D_1 + \bar{D}_4 D_2 + D_2 \bar{D}_1 \quad (6.17)$$
$$B_1 = D_4 \bar{D}_1 + \bar{D}_4 D_1 = D_4 \oplus D_1$$
$$B_0 = D_0 \quad \text{by inspection,}$$

which represent a gate/input tally of 11/27 excluding inverters but including one shared PI. The logic circuit for this module is given in Fig. 6.21a and is cascaded in Fig. 6.21b to produce the 8-bit BCD-to-binary converter. Notice that the four modules are cascaded in such a way as to satisfy the right-shift requirement of Algorithm 6.1. If expansion beyond 8 bits is needed, each additional bit requires that an additional module be added in cascade fashion.

BCD-to-Seven-Segment Display No discussion of code conversion is complete without including the BCD-to-seven-segment decoder (converter). Light-emitting diodes (LEDs) and liquid crystal displays (LCDs) are used extensively to produce the familiar Arabic numerals. The use of the BCD-to-seven-segment decoder is an important means of accomplishing this.

Shown in Fig. 6.22a is the seven-segment display format and the nine decimal digits that are produced by the decoder/display. The truth table for the converter is given in Fig. 6.22b and features a blanking input *BI*, but lacks FDR since ϕ's are assigned to the six unused input states.

The EV K-maps for the seven segment outputs are shown in Fig. 6.22c, where a gate-minimum POS cover for each has been looped out. This results in the minimum

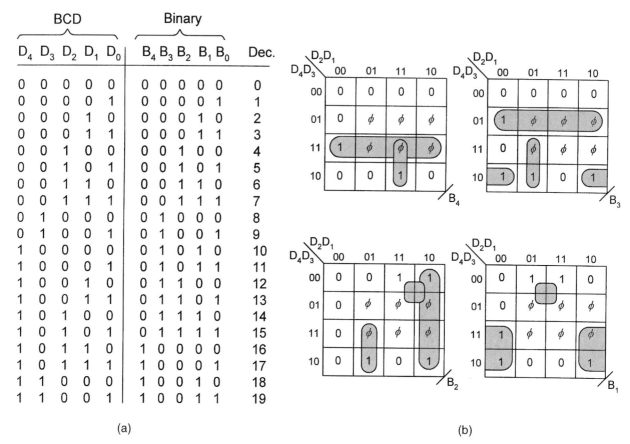

FIGURE 6.20
Design of an 8-bit BCD-to-binary converter. (a) Truth table for a 2-digit BCD-to-binary module. (b) K-maps plotted directly from the truth table showing minimum cover.

expressions

$$a = (A + B + C + \bar{D})(\bar{B} + D)$$
$$b = \bar{B} + C \odot D = \bar{B} + C \oplus \bar{D}$$
$$= (\bar{B} + C + \bar{D})(\bar{B} + \bar{C} + D)$$
$$c = (B + \bar{C} + D)$$
$$d = (B + C + \bar{D})(\bar{B} + C \oplus D) \quad\quad (6.18)$$
$$= (B + C + \bar{D})(\bar{B} + C + D)(\bar{B} + \bar{C} + \bar{D})$$
$$e = \bar{D}(\bar{B} + C)$$
$$f = (\bar{C} + \bar{D})(A + B + \bar{D})(B + \bar{C})$$
$$g = (A + B + C)(\bar{B} + \bar{C} + \bar{D}),$$

6.5 CODE CONVERTERS

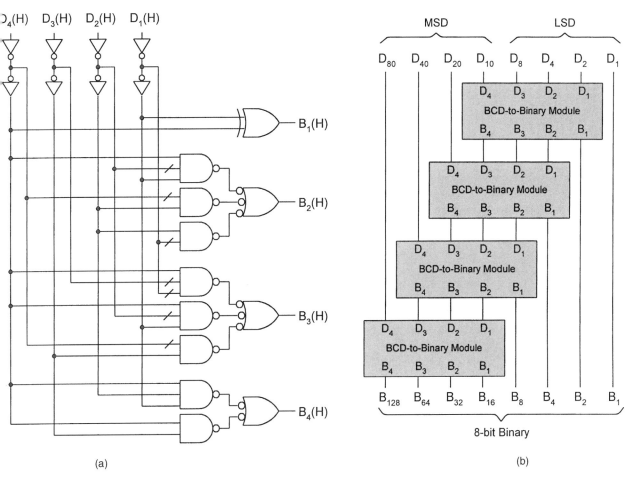

FIGURE 6.21
Implementation of the BCD-to-binary module and the 8-bit BCD-to-binary converter. (a) Logic circuit for the 2-digit module according to Eqs. (6.17). (b) Cascaded modules to produce the 8-bit converter.

which represent a POS gate/input tally of 20/53, excluding inverters. Notice that two of these seven expressions are also presented in hybrid (POS/XOR) form and that one, d, is a three-level expression if the XOR term is considered one level as in Fig. 3.26a.

A BCD-to-seven-segment converter uses either a *common cathode* or *common anode* LED display as shown in Fig. 6.23, but where the individual LEDs are arranged as in Fig. 6.22a. The common anode configuration requires that the outputs in Eqs. (6.18) be active low while the common cathode configuration requires that they be active high. For this example the common cathode configuration of LEDs is chosen, requiring that the expressions in Eqs. (6.18) be complemented and active low. When this is done, the resulting two-level SOP (active low) expressions become

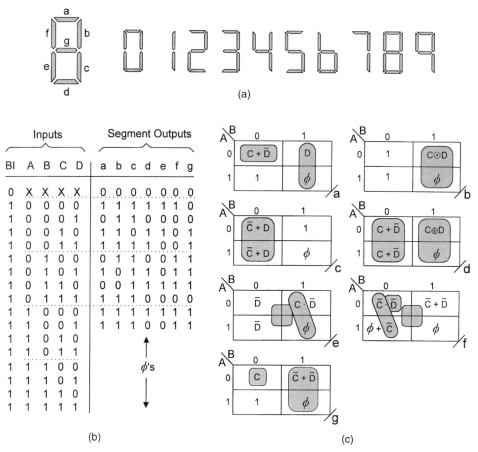

FIGURE 6.22
Gate-minimum design of a BCD-to-seven-segment converter. (a) Display format and the nine decimal digits. (b) Truth table with a blanking input and without FDR. (c) EV K-maps for the seven-segment outputs showing minimum cover, including optimal use of don't cares.

$$a(L) = [\bar{A}\bar{B}\bar{C}D + B\bar{D}](L)$$
$$b(L) = [B\bar{C}D + BC\bar{D}](L)$$
$$c(L) = [\bar{B}C\bar{D}](L)$$
$$d(L) = [\bar{B}\bar{C}D + BC\bar{D} + BCD](L) \qquad (6.19)$$
$$e(L) = [D + B\bar{C}](L)$$
$$f(L) = [CD + \bar{A}\bar{B}D + \bar{B}C](L)$$
$$g(L) = [\bar{A}\bar{B}\bar{C} + BCD](L)$$

and are implemented in Fig. 6.24, where the active low blanking input $BI(L)$ is realized by using inverting tri-state drivers with active low controls as in Fig. 3.8d. Thus, BI serves to

6.6 MAGNITUDE COMPARATORS

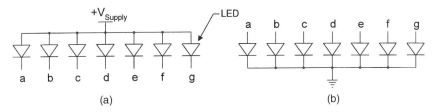

FIGURE 6.23
LED configurations for the BCD-to-seven-segment display converter. (a) Common anode. (b) Common cathode.

enable the decoder if $BI(L) = 1(L)$, or disable it if $BI(L) = 0(L)$. Notice that the common cathode configuration of LEDs in Fig. 6.23b requires the inputs to be $1(H) = HV$ to force the diode into forward bias (conducting mode). Thus, a $0(L)$ for any output in Eqs. (6.19) is a $1(H)$, which is the same as a 1 in the positive logic truth table of Fig. 6.22b. Coupling the decoder of Fig. 6.24 with the common anode configuration requires the use of noninverting tri-state drivers with active low controls as in Fig. 3.8b. In this case, each output in Eqs. (6.19) would be issued active high such that any $0(H)$ output (to the LED) forces a diode in Fig. 6.23a into forward bias. A commercial IC chip with logic suitable for coupling with the common anode LED configuration of Fig. 6.23a is the 74x49. Its logic differs somewhat from that of Fig. 6.24 because it generates the blanking condition in a different way — it uses a form of FDR — and it reverses the input lettering from MSB (D) to LSB (A).

The blanking feature shown in Fig. 6.25 is useful in removing leading zeros in integer displays and trailing zeros in fixed-point decimal displays. When the blanking feature is used in this way it is called *zero-blanking*. For example, 036.70 would appear if no zeros are blanked but would be 36.7 after zero-blanking. To accomplish the zero-blanking capability requires that additional logic be connected to the *BI* input. The idea here is that when the inputs to an MSD stage are zero, the zero-blanking logic must deactivate *BI* [$BI(L) = 0(L)$] but must not do so for intermediate zeros as, for example, in 40.7. ICs with this capability are designed with a *zero-blanking input* (*ZBI*) and a *zero-blanking output* (*ZBO*) so that when the decade stages are connected together, ZBO-to-ZBI, zero blanking can *ripple* in the direction of the radix point terminal. This is easily accomplished as illustrated in Fig. 6.25 for an integer display, where external logic is connected to the *BI* inputs of the BCD-to-seven-segment decoders of Fig. 6.24 such that *only* leading zeros are blanked in ripple fashion from MSD-to-LSD.

6.6 MAGNITUDE COMPARATORS

A device that determines which of two binary numbers is larger or if they are equal is called a *magnitude comparator* or simply *comparator*. A vending machine, for example, must engage a comparator each time a coin is inserted into the coin slot so that the desired item can be dispensed when the correct change has been inserted. The block diagram symbol for an *n*-bit comparator with cascading capability is given in Fig. 6.26. Here, it is to be understood that *gt* and $(A > B)$ represent *A* greater than *B*; *eq* and $(A = B)$ represent *A* equal to *B*; *lt* and

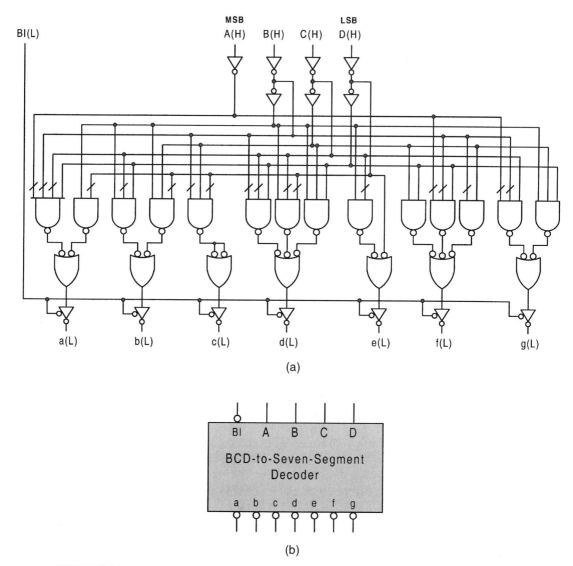

FIGURE 6.24

Logic circuit (a) and circuit symbol (b) for the BCD-to-seven-segment decoder according to Eqs. (6.19) and Fig. 6.22, featuring an active low blanking input *BI* implemented by using inverting three-state drivers with active low controls.

($A < B$) represent A less than B. For cascading purposes, the inputs *gt*, *eq*, and *lt* to the kth stage are the outputs ($A > B$, $A = B$, and $A < B$) from the next least significant $(k - 1)$th stage, while the corresponding outputs of the kth stage are the inputs (*gt*, *eq*, and *lt*) to the next most significant $(k + 1)$th stage. Thus, the magnitudes of the kth stage are more significant than those of the $(k - 1)$th stage, as the magnitudes of the $(k + 1)$th stage are more significant

6.6 MAGNITUDE COMPARATORS

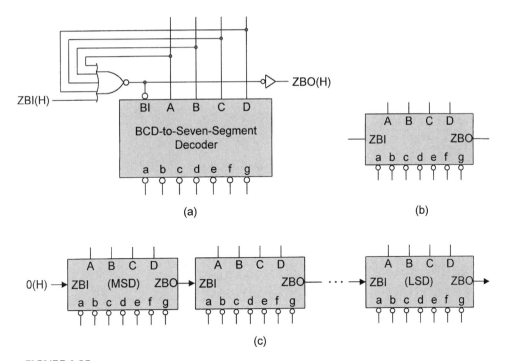

FIGURE 6.25
BCD-to-seven-segment decoding with zero-blanking capability. (a) External logic required for zero blanking. (b) Circuit symbol for zero-blanking decoder module. (c) Cascaded modules for integer representation showing rippled zero blanking of leading zeros from MSD-to-LSD stages.

than those of the kth stage, etc. Though seemingly trivial, these facts are important in establishing the proper truth table entries for comparator design, as the following examples will illustrate.

The design of a useful cascadable comparator begins with the 1-bit design. Shown in Fig. 6.27 are the EV truth table and EV K-maps for a cascadable 1-bit comparator. The

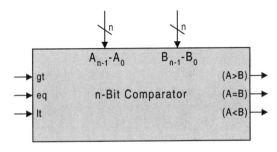

FIGURE 6.26
Circuit symbol for an n-bit comparator with cascading capability.

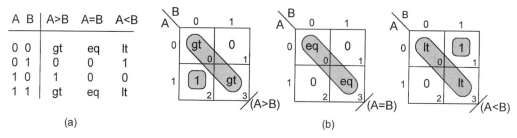

FIGURE 6.27
Design of a cascadable 1-bit comparator. (a) EV truth table. (b) Output EV K-maps showing minimum cover with XOR-type functions.

gate-minimum cover for each of the three outputs, as extracted from the K-maps, is

$$\begin{cases} (A > B) = gt(A \odot B) + A\bar{B} \\ (A = B) = eq(A \odot B) \\ (A < B) = lt(A \odot B) + \bar{A}B \end{cases}, \quad (6.20)$$

as given in three-level form. These represent a gate/input tally of 8/16, excluding inverters. Notice that in the truth table of Fig. 6.27a all three inputs, *gt*, *eq*, and *lt*, appear only when A and B are of equal magnitude and that logic 1 appears when one, A or B, dominates in magnitude.

In order to establish an important trend in the output expressions, one that can be used to establish the logic for any size comparator, it is helpful to construct the truth table and EV K-maps for a cascadable 2-bit comparator. These are provided in Fig. 6.28. As in the 1-bit comparator design, the EVs (*gt*, *eq*, and *lt*) in the truth table of Fig. 6.28a are the outputs from the next least significant stage, which explains why they appear only when A and B are of equal magnitude. The gate-minimum output cover given in Fig. 6.28b yields the following expressions in three-level form:

$$\begin{cases} (A > B) = gt\bar{A}_0\bar{B}_0(A_1 \odot B_1) + A_0(gt + \bar{B}_0)(A_1 \odot B_1) + A_1\bar{B}_1 \\ \quad\quad\quad\quad = gt(A_1 \odot B_1)(A_0 \odot B_0) + A_0\bar{B}_0(A_1 \odot B_1) + A_1\bar{B}_1 \\ (A = B) = eq\bar{A}_0\bar{B}_0(A_1 \odot B_1) + eqA_0B_0(A_1 \odot B_1) \\ \quad\quad\quad\quad = eq(A_1 \odot B_1)(A_0 \odot B_0) \\ (A < B) = \bar{A}_0(lt + B_0)(A_1 \odot B_1) + ltA_0B_0(A_1 \odot B_1) + \bar{A}_1B_1 \\ \quad\quad\quad\quad = lt(A_1 \odot B_1)(A_0 \odot B_0) + A_0B_0(A_1 \odot B_1) + \bar{A}_1B_1 \end{cases}. \quad (6.21)$$

These represent a gate/input tally of 11/29 (excluding inverters) with a maximum fan-in of 4. Notice that in arriving at the second equation for each of the outputs, use is made of the absorptive law, Eqs. (3.13), in the form $gt + \bar{B}_0 = gtB_0 + \bar{B}_0$ and $lt + B_0 = lt\bar{B}_0 + B_0$. In comparison, a two-level optimization of this comparator gives a gate/input tally of 21/60,

6.6 MAGNITUDE COMPARATORS

Dec. AB	A_1	A_0	B_1	B_0	A>B	A=B	A<B
00	0	0	0	0	gt	eq	lt
01	0	0	0	1	0	0	1
02	0	0	1	0	0	0	1
03	0	0	1	1	0	0	1
10	0	1	0	0	1	0	0
11	0	1	0	1	gt	eq	lt
12	0	1	1	0	0	0	1
13	0	1	1	1	0	0	1
20	1	0	0	0	1	0	0
21	1	0	0	1	1	0	0
22	1	0	1	0	gt	eq	lt
23	1	0	1	1	0	0	1
30	1	1	0	0	1	0	0
31	1	1	0	1	1	0	0
32	1	1	1	0	1	0	0
33	1	1	1	1	gt	eq	lt

(a)

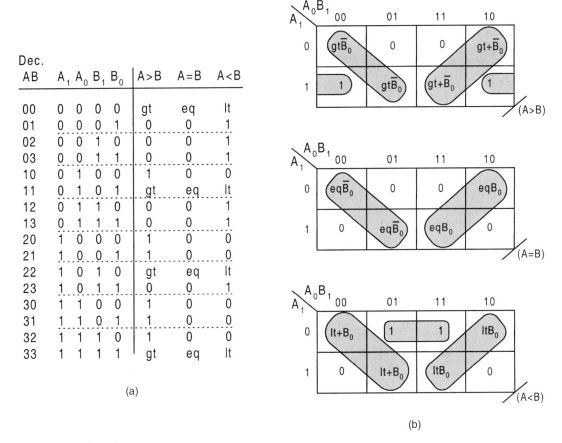

(b)

FIGURE 6.28
Design of a cascadable 2-bit comparator. (a) Compressed truth table. (b) EV K-maps plotted from (a) showing minimum cover involving XOR-type functions.

excluding inverters, and has a maximum fan-in of 7. Thus, the advantage of the multilevel gate-minimum form over that of the two-level implementation is evident. This is especially true if CMOS is used in the design of the comparator. Recall that the CMOS EQV gate in Fig. 3.27a has about the same gate path delay as a two-input NAND gate.

If $A = B$ in Eqs. (6.20) and (6.21), there results $(A > B) = gt$ and $(A < B) = lt$, each of which must be logic 0 for the $A = B$ condition from the next least significant stage. Therefore, it follows generally that

$$(A = B) = \overline{(A > B)} \cdot \overline{(A < B)}, \tag{6.22}$$

which yields $(A = B) = 0$ for any condition other than $A = B$.

The trend that emerges in the output expressions is easily deduced from an inspection of Eqs. (6.20) and (6.21), thereby permitting the three output expressions to be written for any number of bits. As an example, the output expressions for a cascadable 4-bit comparator become

$$\begin{cases} (A > B) = gt \cdot \prod_{i=0}^{3}(A_i \odot B_i) + A_0\bar{B}_0 \prod_{i=1}^{3}(A_i \odot B_i) + A_1\bar{B}_1 \prod_{i=2}^{3}(A_i \odot B_i) \\ \qquad\qquad + A_2\bar{B}_2(A_i \odot B_i) + A_3\bar{B}_3 \\ (A = B) = eq \cdot \prod_{i=0}^{3}(A_i \odot B_i) = \overline{(A > B)} \cdot \overline{(A < B)} \\ (A < B) = lt \cdot \prod_{i=0}^{3}(A_i \odot B_i) + \bar{A}_0 B_0 \prod_{i=1}^{3}(A_i \odot B_i) + \bar{A}_1 B_1 \prod_{i=2}^{3}(A_i \odot B_i) \\ \qquad\qquad + \bar{A}_2 B_2(A_i \odot B_i) + \bar{A}_3 B_3 \end{cases}, \quad (6.23)$$

where

$$\prod_{i=0}^{3}(A_i \odot B_i) = (A_0 \odot B_0)(A_1 \odot B_1)(A_2 \odot B_2)(A_3 \odot B_3)$$

and

$$\prod_{i=2}^{3}(A_i \odot B_i) = (A_2 \odot B_2)(A_3 \odot B_3), \text{ etc.}$$

Implementation of Eqs. (6.23) is given in Fig. 6.29, using three-level NAND/NOR/EQV logic and limiting fan-in to 5. Here, the gate/input tally is 23/67, excluding inverters.

The commercial chip that is equivalent to the cascadable 4-bit comparator of Fig. 6.29 is the 74xx85. Though the two differ somewhat in the logic circuit makeup, they function in exactly the same way. Either can be cascaded to form a comparator of any number of bits in multiples of four bits. Shown in Fig. 6.30 is an 8-bit comparator formed by cascading two 4-bit comparators in series. Notice that the inputs to the *least significant stage* are required to be at the fixed logic levels shown.

Combining three or more comparators in series suffers significantly in propagation delay. Much better is the series/parallel arrangement shown in Fig. 6.31. Here, six 4-bit comparators are combined to form a 24-bit comparator where no more than two comparators are placed in series and all inputs and outputs are active high. In this case, a dominant A or B word is picked up by one of the five stages in descending order of significance and is issued as either $(A > B)$ or $(A < B)$ by the output comparator. Words A and B of equal magnitude are picked up by the least significant stage and issued by the output comparator as $(A = B)$. The series/parallel comparator arrangement in Fig. 6.31 is justifiable on the basis of path delay arguments when the alternative is considered — namely, the series configuration of six 4-bit comparators. Notice that not all of the 24 bits of the comparator need be used. Any size A, B words up to 24 bits can be compared by using the comparator of Fig. 6.31. The only requirement is that words A and B be of equal length and that MSB inputs not in use be held at 0(H).

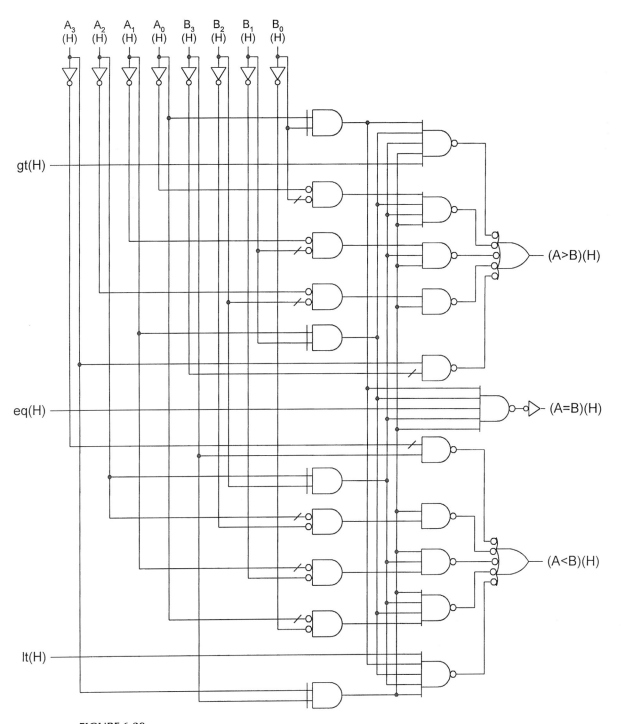

FIGURE 6.29
Three-level logic circuit for the cascadable 4-bit comparator by using NAND/NOR/EQV logic with a fan-in limit of 5.

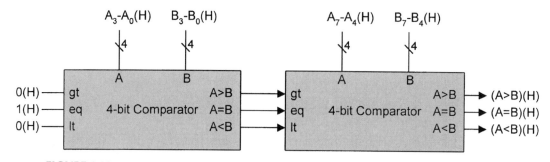

FIGURE 6.30
A cascadable 8-bit comparator created by combining two 4-bit comparators.

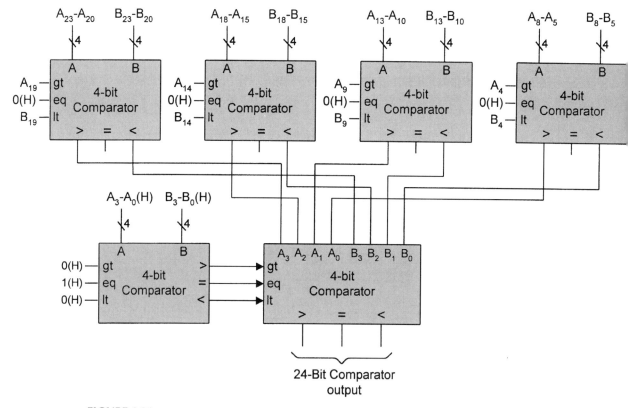

FIGURE 6.31
Speed-optimized series/parallel arrangement for a 24-bit comparator composed of six 4-bit comparators.

6.7 PARITY GENERATORS AND ERROR CHECKING SYSTEMS

A parity bit can be appended to any word of n bits to generate an even or an odd number of 1's (or 0's). A combinational logic device that generates even or odd parity is called a *parity generator*. A device that is used to detect even or odd parity is called a *parity detector*. To understand the concept of parity generation and detection, consider the following 8-bit words to which a ninth parity bit is appended as the LSB shown in brackets:

$\qquad$ 1 1 0 1 0 1 0 1 [1]: Even parity generation = Odd parity detection

$\qquad$ 1 1 0 1 0 1 0 1 [0]: Odd parity generation = Even parity detection

or

$\qquad$ 1 1 0 1 0 0 0 1 [1]: Odd parity generation = Even parity detection

$\qquad$ 1 1 0 1 0 0 0 1 [0]: Even parity generation = Odd parity detection.

Thus, parity generation includes the parity bit in the count of 1's, whereas parity detection excludes the parity bit in the count but uses the parity bit to identify the parity status (even or odd) of the word. The parity bit may be appended either at the LSB position, as in the examples just given, or at the MSB position.

XOR gates can be cascaded to produce n-bit parity circuits. As an example, consider the design of a 4-bit *even-parity generator module* shown in Fig. 6.32. The second-order EV K-maps in Fig. 6.32b follow directly from the truth table in Fig. 6.32a and yield the output expression

$$P_{Even\ Gen} = A \oplus B \oplus C \oplus D, \qquad (6.24)$$

which is implemented in Fig. 6.32c.

A B C D	$P_{Even\ Gen}$	A B C D	$P_{Even\ Gen}$
0 0 0 0	0	1 0 0 0	1
0 0 0 1	1	1 0 0 1	0
0 0 1 0	1	1 0 1 0	0
0 0 1 1	0	1 0 1 1	1
0 1 0 0	1	1 1 0 0	0
0 1 0 1	0	1 1 0 1	1
0 1 1 0	0	1 1 1 0	1
0 1 1 1	1	1 1 1 1	0

(a)

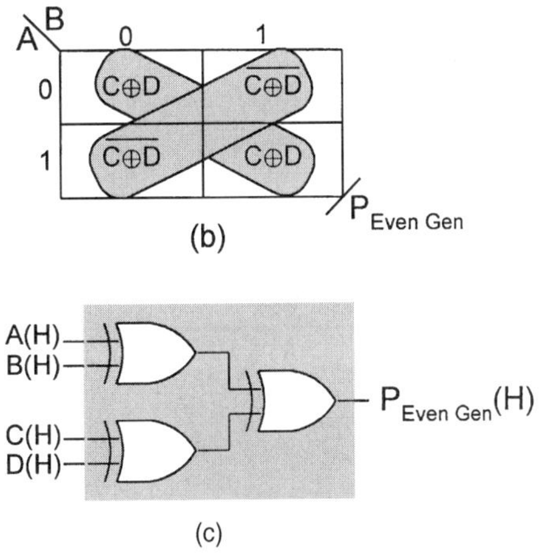

(b) EV K-map with entries $C \oplus D$, $\overline{C \oplus D}$, $\overline{C \oplus D}$, $C \oplus D$ yielding $P_{Even\ Gen}$

(c) Logic circuit: A(H), B(H) into XOR; C(H), D(H) into XOR; both outputs into XOR giving $P_{Even\ Gen}(H)$

FIGURE 6.32
Design of the 4-bit even parity generator module. (a) Truth table. (b) EV K-map. (c) Logic circuit according to Eq. (6.24).

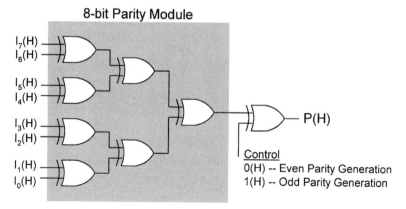

FIGURE 6.33

An 8-bit even/odd parity generator circuit with a control input.

This 4-bit parity generator module of Fig. 6.32 can be cascaded to form the 8-bit parity generator module shown in Fig. 6.33. Here, an additional XOR gate is added as a control to generate either an even or odd parity generate output. Notice that two 8-bit parity generator modules can be cascaded to produce a 16-bit module, and two 16-bit modules can be cascaded to produce a 32-bit module, etc., each with a control XOR gate to produce either an even or odd parity generation output. Note also that any 2^n-bit parity module is an n-level XOR circuit, excluding any control XOR gate. Thus, a 32-bit parity module represents a 5-level XOR circuit, a 64-bit module is a 6-level circuit, etc.

Error Checking Systems   Errors occur in digital systems for a variety of reasons. These reasons include logic noise, power supply surges, electromagnetic interference, and crosstalk due to physically close signals. Single-error detection in a digital system usually amounts to a parity checking system where the parities at the source and destination sites are compared. A typical single-error checking system for an 8-bit data bus is illustrated in Fig. 6.34. In this case an 8-bit parity generator of the type in Fig. 6.33 is used at the source to generate either an even or odd parity bit. The 9-bit parity detector at the destination is the 8-bit parity

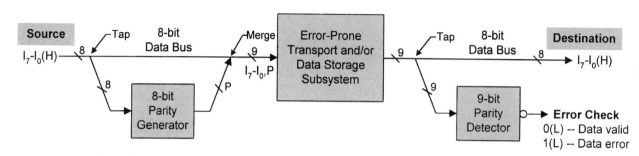

FIGURE 6.34

Error checking system for an 8-bit data bus by using an 8-bit parity generator at the source and a 9-bit parity detector at the destination site.

6.8 COMBINATIONAL SHIFTERS

generator module of Fig. 6.33 with an XOR gate and inverter as the output stage, hence an EQV gate, as implied in Fig. 6.34.

The parity checking scheme illustrated in Fig. 6.34 is valid for the detection of a single error in the 8-bit word. Actually, it is valid for any odd number of errors, but the probability that three or more errors will occur in a given word is near zero. What a single-bit parity checking system cannot do is detect an even number of errors (e.g., two errors). It is also true that the error checking system of Fig. 6.34 cannot correct any single error it detects. To do so would require detecting its location, which is no trivial task. To identify the location of an error bit requires multiple parity detection units on the submodular level down on the bit level, a significant cost in hardware. However, when this is done, an erroneous bit can be corrected. Memory systems in modern computers have such single-error correction capability.

6.8 COMBINATIONAL SHIFTERS

Shifting or rotating of word bits to the right or left can be accomplished by using combinational logic. Devices that can accomplish this are called *combinational shifters*, or *barrel shifters* if their function is only to rotate word bits. Shifters are used for bit extraction operations, transport, editing, data modification, and arithmetic manipulation, among other applications.

A general n-bit shifter is an $(n + m + 3)$-input/n-output device represented by the logic symbol in Fig. 6.35a, with the interpretations of the control inputs given in Fig. 6.35b. A shifter of this type accepts n data input bits $(I_{n-1} - I_0)$ and either passes these values straight through to the data outputs $(Y_{n-1} - Y_0)$, or shifts or rotates them by one or more bit positions to the right or left with 0 or 1 fill on command of the $m + 3$ control inputs. The control inputs consist of a rotate control (R), fill control (F), direction control (D), and m inputs $(A_{m-1} - A_0)$ to control the number of bit positions to be shifted or rotated—usually binary encoded to $0, 1, 2, 3, \ldots, p$ bit positions. For the shifter of Fig. 6.35, the control

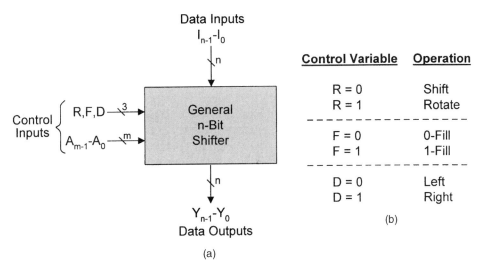

FIGURE 6.35
A general n-bit shifter. (a) Block diagram symbol. (b) Interpretation of the control inputs R, F, and D.

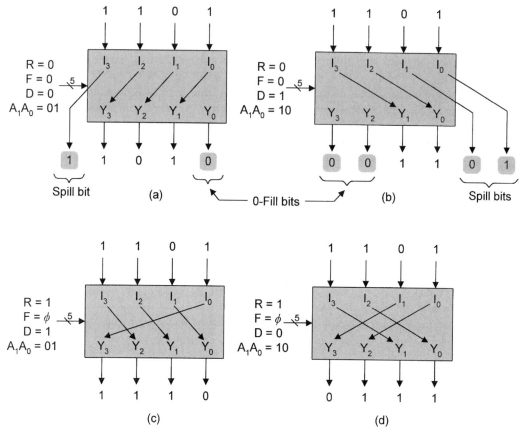

FIGURE 6.36
Examples of shift and rotation operations by using a general 4-bit shifter. (a) Shift left one bit (sl1) with 0-fill. (b) Shift right 2 bits (sr2) with 0-Fill. (c) Rotate right one bit (rr1). (d) Rotate left two bits (rl2).

input word is given the positional weights $RFDA_{m-1} \cdots A_0$ and is sometimes represented as $S_{k-1} \cdots S_1 S_0$, where $k = m + 3$.

The control variable ($R = 0$) causes the shifter in Fig. 6.35 to shift the input word (right or left) by p bits while, at the same time, causes Spill and 0- or 1-Fill by p bits. The control variable ($R = 1$) rotates the input word (right or left) by p bits with no Spill or Fill. These operations are illustrated in Fig. 6.36 for a 4-bit shifter with input word 1101. In Figs. 6.36a and 6.36b, the shifter is set for shifting operations, first shift left by one bit (sl1) and then shift right by two bits (sr2) with spill and 0-fill bits equal to the number of bits shifted. Note that the spill bits are always lost. In Figs. 6.36c and 6.36d, the shifter is set for rotation with settings for 1-bit right rotation (rr1) and 2-bits left rotation (rl2), respectively. Notice that rotation requires that end bits, normally spilled (discarded) in a shift operation, be rotated around to the opposite end of the shifter. The rotation mode could be called the barrel shifter mode, the word barrel implying "shift around."

The design of a 4-bit shifter capable of left shifting or rotation by up to three bits is shown in Fig. 6.37. To reduce the size of the table, the fill bit F is included as an entered variable in

6.8 COMBINATIONAL SHIFTERS

R	A_1	A_0	Y_3	Y_2	Y_1	Y_0	
0	0	0	I_3	I_2	I_1	I_0	Transfer
0	0	1	I_2	I_1	I_0	F	sl1
0	1	0	I_1	I_0	F	F	sl2
0	1	1	I_0	F	F	F	sl3
1	0	0	I_3	I_2	I_1	I_0	Transfer
1	0	1	I_2	I_1	I_0	I_3	rl1
1	1	0	I_1	I_0	I_3	I_2	rl2
1	1	1	I_0	I_3	I_2	I_1	rl3

(a)

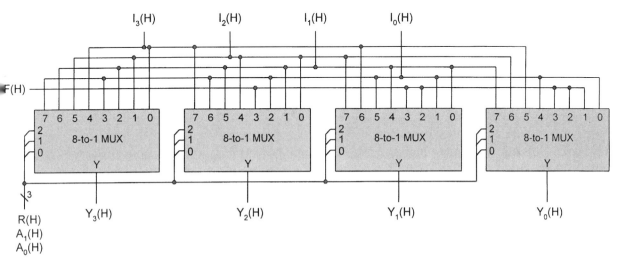

FIGURE 6.37
Design of a 4-bit general shifter that can shift or rotate left up to three bits with F fill. (a) EV Truth table. (b) Output K-maps suitable for 8-to-1 MUX implementation.

the truth table of Fig. 6.37a, and the absence of the direction variable D implies that $D = 0$ for left shift/rotate. Thus, there are four control variables R, F, A_1, and A_0, as indicated in Fig. 6.37a. The EV K-maps in Fig. 6.37b are plotted directly from the truth table.

The four outputs represented by the K-maps in Fig. 6.37b can be implemented in a variety of ways, including discrete logic. However, their form suggests an 8-to-1 MUX approach, which is the approach of choice for this example. Shown in Fig. 6.38 is the MUX implementation of the general 4-bit left shifter/rotator represented by Fig. 6.37.

FIGURE 6.38
MUX implementation of the general 4-bit left shifter/rotator with F (0 or 1) fill. The MUX select inputs are weighted $R A_1 A_0$ for $S_2 S_1 S_0$, respectively.

278   CHAPTER 6 / NONARITHMETIC COMBINATIONAL LOGIC DEVICES

Notice that in Fig. 6.38 the control variable D is missing, meaning that $D = 0$ for left shifting. Had the variable D been included for left or right shifting/rotating, fourth-order EV K-maps and 16-to-1 MUXs would have to be used if the architecture of Fig. 6.38 were to be retained. Also note that this shifter can be cascaded by connecting $Y_3(H)$ of one stage to $F(H)$ of the next most significant stage, etc.

6.9 STEERING LOGIC AND TRI-STATE GATE APPLICATIONS

Any of the combinational logic devices discussed to this point can be designed by using transmission gates together with inverters. When this is done the devices are classified as *steering logic*. A transmission gate lacks logic function capability by itself, but can be configured with other transmission gates to steer the logic signals in a manner that carries out a logic function. Inverters are necessarily included in most steering logic designs because transmission gates are passive and noninverting, as explained in Section 3.5.

As a simple example, consider a 4-to-1 MUX defined by the truth table in Fig. 6.39a. This device can be implemented easily with CMOS transmission gates and inverters and at

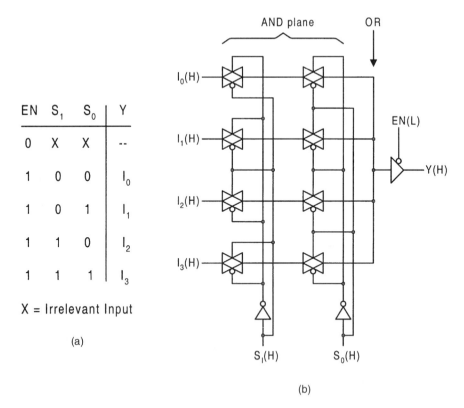

FIGURE 6.39
Transmission gate implementation of the 4-to-1 MUX. (a) Truth table. (b) Logic circuit with a tri-state buffered/enabled output.

a considerable savings in transistor count and design area, as indicated in Fig. 6.39b. Here, the MUX is enabled and buffered by using a tri-state driver with an active low control input. Thus, the disable condition $[EN(L) = 0(L)]$ is actually a disconnect state as indicated in Fig. 3.8b and is represented by the dash in the truth table of Fig. 6.39a. Notice that the AND plane, equivalent to the four four-input NAND gates in Fig. 6.4d, is constructed with only eight transmission gates, and that the OR operation is "wired" to the tri-state driver since only one line can be active at any given time. Consequently, the transmission gate design represents a significant savings in hardware cost and will be faster (shorter throughput) compared to the NAND gate design in Fig. 6.4d.

An important aspect of steering logic designs is that the transmission gates are non-restoring devices and must be buffered to prevent degradation of the signal. The following is offered as a "rule of thumb" in dealing with transmission gate designs:

> For signal-restoring purposes in logic designs that use transmission gates, plan to buffer each signal for every four transmission gates through which the signal must pass. CMOS transmission gates should be counted as two pass transistors.

The design of a 2-to-4 decoder is used here as another simple example of a CMOS transmission gate (TG) implementation of a combinational logic device. Shown in Fig. 6.40a is the truth table for this decoder and in Fig. 6.40b is the TG implementation of the decoder taken directly from the truth table. The outputs are shown enabled and buffered with inverting tri-state drivers having an active low control input as in Fig. 3.8d. If active high outputs are required, noninverting tri-state drivers of the type shown in Fig. 3.8b can be used. Notice that each "1" CMOS TG leading to a Y output must have a "0" TG associated with it and that all TGs have complementary EN inputs connected to them from an inverter. Thus, since there is a series of two "1" TGs per Y output, there are two "0" TGs for each output, making a total of 16 TGs or a total of 60 transistors, including inverters and tri-state buffer/enables. In comparison, the gatewise CMOS NAND implementation of the same decoder yields a transistor count of only 34. Though the transistor count for the TG design is greater than that for the gatewise implementation, the speed (throughput) should be comparable and perhaps a little faster for the TG design.

6.10 INTRODUCTION TO VHDL DESCRIPTION OF COMBINATIONAL PRIMITIVES

With ever-increasing complexity of digital systems, there comes a greater need for simulation, modeling, testing, automated design, and documentation of these systems. The challenge here is to make the English language readable by a computer for computer-aided design (CAD) purposes. Hardware description languages (HDLs) satisfy these requirements. VHSIC (for very high speed integrated circuit) is such an HDL. It was funded by the Department of Defense in the late 1970s and early 1980s to deal with the complex circuits of the time. However, the capabilities of the VHSIC language soon reached their limit, giving way to more advanced HDLs that could meet the challenges of the future. One important language that has emerged is known as VHSIC Hardware Description Language or simply VHDL. VHDL was first proposed in 1981 as a standard for dealing with very complex

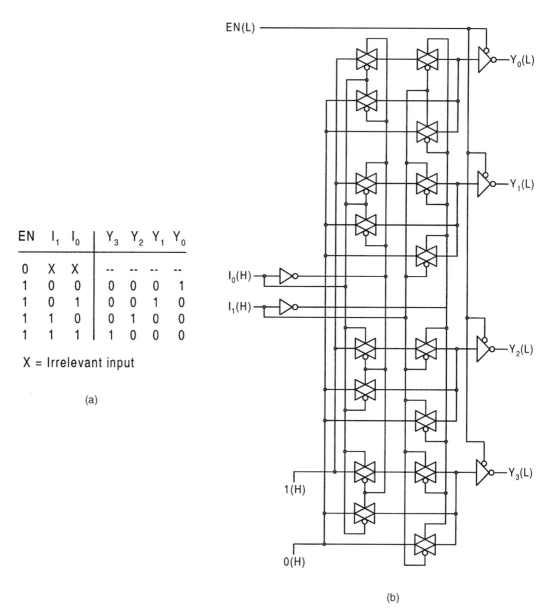

FIGURE 6.40
Transmission gate implementation of the 2-to-4 decoder. (a) Truth table. (b) Logic circuit with active low outputs produced by tri-state buffer/enables.

circuits. It has since gone through several revisions and in 1987 and 1993 was adopted as the IEEE 1076-1987 and 1076-1993 standards, respectively. The examples given in this section will use these standards.

VHDL is a powerful tool capable of top-down design descriptions covering various levels of abstraction ranging from the *behavior level*, to the *dataflow level*, to the *structural*

6.10 INTRODUCTION TO VHDL DESCRIPTION OF COMBINATIONAL PRIMITIVES

level. Examples of the behavior level of representation are truth tables and algorithmic descriptions. The structural level of representation typically includes various *primitives* together with the interconnections required to make a circuit. The primitives covered in this chapter include discrete gates, decoders, encoders, MUXs, comparators, parity generators, and combinational shifters. Other primitives, including those associated with arithmetic circuits, PLDs, and sequential circuits, will be considered in subsequent chapters.

Before illustrating the use of VHDL for some well known combinational primitives, it will be helpful to review some assignment statements relevant to behavioral modeling. For example,

a <= b;

is read as "*a* is assigned the value of *b*." As a second example,

Y <= I2 after 10 ns;

is interpreted as follows: "Y is assigned the value of I2 after 10 nanoseconds have elapsed." In these two examples, "<=" is an assignment operator that assigns a value to a signal. Another assignment operator is the ":=" operator. It is used to assign a value to a variable:

result := X;   or   delay := 4 ns;

Here, *result* and *delay* are variables and are assigned the values of X and 4 ns, respectively.

To illustrate, consider the VHDL behavioral description of the two-input NOR gate primitive given in Fig. 6.41a. In this case the **entity** is called *nor2* with inputs i1 and i2, and output, o1, as indicated in Fig. 6.41b. The architecture for *nor2* is arbitrarily called *nor2_behavior* and provides that "o1 is assigned the value i1 **nor** i2 [or **not**(i1 **or** i2)] after 5 nanoseconds." Thus, 5 ns is the propagation delay of the gate. Notice that all VHDL *keywords*, such as **entity**, **port**, and **end**, and the VHDL logical "**or**" are highlighted in bold for visual effect. Also, note that the entries are case insensitive. The keyword **port** is used to identify the input and output ports of the entity.

As a second example consider the VHDL behavioral description for a 4-to-1 MUX primitive given in Fig. 6.42. For this example an arbitrary delay is assigned (to the data selection process) the keywords **generic** (**del**:time), meaning that time delay (**del**) is determined by

```
entity nor2 is
    port (i1, i2: in bit; o1: out bit);
end nor2;

architecture nor2_behavior of nor2 is
begin
    o1 <= not (i1 or i2) after 5 ns;
end nor2_behavior;
```

(a)   (b)

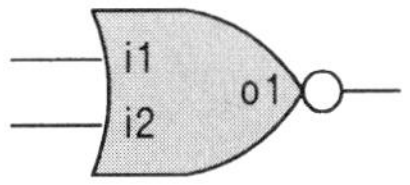

FIGURE 6.41
Behavioral model for a 2-input NOR gate. (a) VHDL entity declaration and behavioral description. (b) NOR gate circuit symbol with inputs, i1 and i2, and output, o1.

```
entity mux4 is
    generic (del: time);
    port (i0, i1, i2, i3: in bit_vector (0 to 3);
          o1: out bit_vector (0 to 3));
    group sel is (s1,s0);
end mux4;

architecture mux4_behavior of mux4 is
    begin
        o1 <= i0 after del when s1 = '0' and s0 = '0' else
              i1 after del when s1 = '0' and s0 = '1' else
              i2 after del when s1 = '1' and s0 = '0' else
              i3 after del when s1 = '1' and s0 = '1';
end mux4_behavior;
```

(a)   (b)

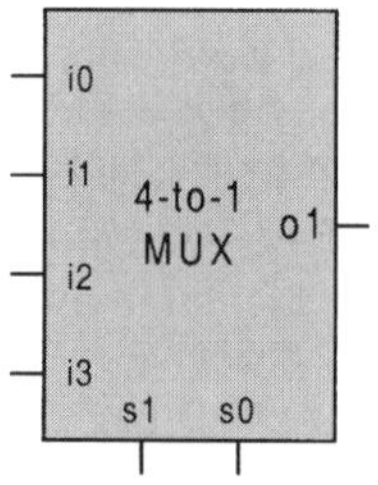

FIGURE 6.42
Behavioral model for a 4-to-1 MUX. (a) Entity declaration and behavioral description. (b) Logic symbol.

the logic and the environment and, therefore, is not given a specific value for any of the behavioral events. Again, as in Fig. 6.41, the VHDL keywords are highlighted as a visual effect for the reader. The logic symbol for the 4-to-1 MUX is given in Fig. 6.42b and indicates four data inputs (i3, i2, i1, i0), two data select inputs (s1 and s0), and a single output, o1, all active high.

Notice that the VHDL language grammar is to some extent intuitive. For example, **group** sel **is** (s1, s0) identifies a collection of named entities (s1, s0) as belonging to the group name "sel." Or, the third line under **architecture / begin** has the following meaning: Output o1 is assigned the value i2 after an arbitrary delay when the select inputs, s1 and s0, are 1 and 0, respectively, or else.... The behavioral model in Fig. 6.42 is but one of several VHDL description formats that could be used. The reader should experiment with others to gain experience in behavioral modeling.

The complete VHDL gate-level description of the 1-bit comparator is next offered as the final example in this chapter. In Figs. 6.43a and 6.43b are given the truth table and logic circuit symbol for the 1-bit comparator. The K-maps and gate-minimum cover for the bit comparator were given previously in Fig. 6.27b, resulting in the output expressions given by Eqs. (6.20). By using the factorization method presented in Subsection 4.9.1, Eqs. (6.20) are converted to two-level minimum form as follows:

$$a\_gt\_b = gt(a \odot b) + a\bar{b} = gta + gt\bar{b} + a\bar{b}$$
$$a\_eq\_b = eq(a \odot b) = eq\bar{a}\bar{b} + eqab \qquad (6.25)$$
$$a\_lt\_b = lt(a \odot b) + \bar{a}b = lt\bar{a} + ltb + \bar{a}b.$$

Here, gt, eq, and lt have the same meaning as is used in Section 6.6.

The logic circuit representing Eqs. (6.25) is shown in Fig. 6.43c, where the gate numbers and intermediate output functions, $im1$, $im2$, $im3$, ... are specified for each inverter and NAND gate. This is done for tracking purposes during the VHDL description that follows.

The VHDL gate-level description of the bit-comparator is divided into three parts: entity declaration, *behavioral* description, and the *structural* description. The average primitive

6.10 INTRODUCTION TO VHDL DESCRIPTION OF COMBINATIONAL PRIMITIVES 283

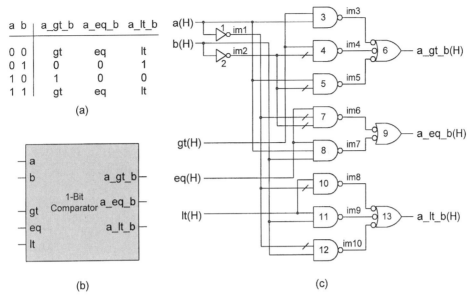

FIGURE 6.43
Design of the cascadable bit-comparator. (a) Truth table. (b) Logic circuit symbol. (c) Circuit diagram in NAND/INV logic according to Eqs. (6.25).

propagation delay, denoted as *avg_delay*, is computed in each *gate model* at the end of the VHDL description by using $(tplh+tphl)/2$, which is Eq. (6.1) discussed in Subsection 6.1.3. The propagation delays are defined in Fig. 6.2. As in the two previous examples, the VHDL keywords and logic operators are indicated in bold for visual effect. Two new keywords have been added: **generic map** associates constants within a portion of the VHDL description to constants defined outside that portion; **port map** associates port signals with a portion of the VHDL description to ports outside of that portion.

The following is the complete gate-level VHDL description of the 1-bit comparator given the name *bit_compare*:

entity bit_compare **is**
    generic (tplh1, tphl1, tplh2, tphl2, tplh3, tphl3: time);
    port (a, b, gt, eq, lt: **in** bit; a_gt_b, a_eq_b, a_lt_b: **out** bit);
end bit-compare;

architecture behave_comp **of** bit_compare **is**
    a_gt_b <= '1' **when** a > b **else** '0';
    a_lt_b <= '1' **when** a < b **else** '0';
    a_eq_b <= gt **if** gt <= '1' **else**
        eq **if** eq <= '1' **else**
        lt **if** lt <= '1';
end behave_comp;

```vhdl
architecture structure_comp of bit_compare is
    component inv
        generic (tplh, tphl: time); port (i1: in bit; o1: out bit);
    end component;
    component nand2
        generic (tplh, tphl: time); port (i1, i2: in bit; o1: out bit);
    end component;
    component nand3
        generic (tplh, tphl: time); port (i1, i2, i3: in bit; o1: out bit);
    end component;
    for all: inv use entity avg_delay_inv;
    for all: nand2 use entity avg_delay_nand2;
    for all: nand3 use entity avg_delay_nand3;
-- Intermediate signals must be declared:
    signal im1, im2, im3, im4, im5, im6, im7, im8, im9, im10: bit;
begin
    a_gt_b output
        gate1: inv generic map (tplh, tphl) port map (a, im1);
        gate2: inv generic map (tplh, tphl) port map (b, im2);
        gate3: nand2 generic map (tplh, tphl) port map (gt, a, im3);
        gate4: nand2 generic map (tplh, tphl) port map (gt, im2, im4);
        gate5: nand2 generic map (tplh, tphl) port map (a, im2, im5);
        gate6: nand3 generic map (tplh, tphl) port map (im3, im4, im5, a_gt_b);
    a_eq_b output
        gate7: nand3 generic map (tplh, tphl) port map (eq, im1, im2, im6);
        gate8: nand3 generic map (tplh, tphl) port map (eq, a, b, im7);
        gate9: nand2 generic map (tplh, tphl) port map (im6, im7, a_eq_b);
    a_lt_b output
        gate10: nand2 generic map (tplh, tphl) port map (lt, im1, im8);
        gate11: nand2 generic map (tplh, tphl) port map (lt, b, im9);
        gate12: nand2 generic map (tplh, tphl) port map (im1, b, im10);
        gate13: nand3 generic map (tplh, tphl) port map (im8, im9, im10, a_lt_b);
end structure_comp;
```

The following are the gate model descriptions for inv, nand2 and nand3:

```vhdl
entity inv is
    generic (tplh: time := 4ns; tphl: time := 2 ns);
    port (i1: in bit; o1: out bit);
end inv;
```

6.10 INTRODUCTION TO VHDL DESCRIPTION OF COMBINATIONAL PRIMITIVES   285

architecture avg_delay_inv **of** inv **is**
begin
    o1 <= **not** i1 **after** (tplh + tphl)/2;
end avg_delay1;

entity nand2 **is**
    generic (tplh: time := 6ns; tphl: time := 4 ns);
    port (i1, i2: **in** bit; o1: **out** bit);
end nand2;

architecture avg_delay_nand2 **of** nand2 **is**
begin
    o1 <= i1 **nand** i2 **after** (tplh + tphl)/2;
end avg_delay2;

entity nand3 **is**
    generic (tplh: time := 7ns; tphl: time := 5 ns);
    port (i1, i2, i3: **in** bit; o1: **out** bit);
end nand3;

architecture avg_delay_nand3 **of** nand3 **is**
    o1 <= **not** (i1 **and** i2 **and** i3) **after** (tplh + tphl)/2;
end avg_delay3;

In the preceding examples VHDL syntax has been applied sometimes without comment. There are relatively few syntax rules that need be followed to create proper VHDL descriptions of devices. The following are some of the more important examples of these syntax rules:

- VHDL is *not* case sensitive. Upper- or lowercase characters can be used as desired.
- Identifiers must begin with a letter and subsequent characters must be alphanumeric but may contain the underscore "_". For example, in Fig. 6.42 the identifiers are mux4 and mux4_behavior.
- The semicolon ";" is used to indicate the termination of a statement. For example: "**end** nand3;".
- Two dashes "--" are used to indicate the beginning of a *comment*. A comment in VHDL is not read by the compiler but serves as a message or reminder to the reader.

An interesting and useful feature of VHDL is that it supports what is called *operator overloading*. This feature permits custom operations to be defined. The following example illustrates how operator overloading can be used to define a new data type:

function "and" (l,r: std_logic_1164) **return** UX01 **is**

begin
   return((and_table)(l,r));
end "and";

.
.
.

architecture example **of** and_operation **is**
   signal Y,A,B: std_logic_1164;
begin
   Y <= A **and** B;
end example;

Here, std_logic_1164 refers to an IEEE standard logic package within which UX01 is a subtype for 4-valued logic systems. Thus, the operation "and" takes on the new meaning "and_table" contained in the standard package. Also, l and r (in line 1) are two of a class of *value kind attributes* that return the leftmost element index (l) or rightmost element index (r) of a given type or subtype.

Several operators, or *constructs* as they are sometimes called, have been used in the examples given previously. Although it is true that the VHDL language contains a large number of these constructs, only a few are necessary for most logic synthesis purposes. The most important of these are given in Fig. 6.44. Included are assignment, relational, logical,

---

Assignment Operators:
   <= Signal assignment   := Variable assignment

Relational Operators:
   = Equality   /= Inequality   > Greater than
   < Less than   <= Less than or equal >= Greater than or equal

Logical Operators:
   and AND   **or** OR   **nand** NAND   **nor** NOR
   xor XOR   **xnor** EQV   **not** Logical negation

Arithmetic Operators:
   + Addition   - Subtraction   * Multiplication   / Division

Shift Operators:
   sll Shift left logical   **srl** Shift right logical
   rol Rotate left logical   **ror** Rotate right logical

Miscellaneous Operators:
   ** Exponentiation   **abs** Absolute value   & Concatenation

---

FIGURE 6.44
Partial listing of important VHDL operators supported by the IEEE 1076-1993 standard.

Value	Description	Value	Description
'U'	Uninitialized	'X'	Unknown
'0'	Logic 0 (driven)	'1'	Logic 1 (driven)
'L'	Logic 0 (read)	'H'	Logic 1 (read)
'-'	Don't care	'Z'	High impedance

FIGURE 6.45
Eight logic data types supported by the IEEE 1076-1164 standard.

arithmetic, shift, and miscellaneous operators. Be aware that some of the operators have different meanings depending on the synthesis tool used.

The IEEE standard 1164 supports standard data types that allow multiple I/O values to be represented. As an example, the standard data type having eight values permit the accurate modeling of a digital circuit during simulation and is presented in Fig. 6.45. The word "driven" used in the description of data type characters '0' and '1' indicates that these logic values are assigned (or forced) to a signal (e.g., an output). The word "read" would apply to input logic values that must be read by a device. Note that each data type character must be enclosed in single quotes as, for example, 'X'.

VHDL is a large and complex language that is easy to learn at the beginning but difficult to master. It is particularly well suited to the design of very large systems, perhaps more so than any other HDL. Libraries of circuit elements can be easily built, used, and reused in a very effective and efficient manner, and this can be done at different levels of abstraction ranging from the block diagram level to the transistor level. In fact, one of VHDL's strengths is that it offers nearly unlimited use of reusable components and access to standard libraries such as the built-in IEEE 1076-1993 Standard and the IEEE 1076-1164 Standard. Used in the hands of a skilled designer, VHDL can greatly increase productivity as well as facilitate the move into more advanced tools (for example, simulators) and advanced target systems. Further Reading contains essential references for continued development in VHDL.

6.11 FURTHER READING

Most textbooks on digital design cover one or more of the performance characteristics related to digital design. Typical among these are the texts by Comer, McCluskey, Tinder, and Wakerly. The text by Wakerly covers these subjects in considerable detail and considers various logic families. The performance characteristics covering several of the more common logic families can be found in the *Electrical Engineering Handbook* (R. C. Dorf, Editor-in-Chief).

[1] D. J. Comer, *Digital Logic and State Machine Design*, 3rd ed. Sanders College Publishing, Fort Worth, TX, 1995.
[2] R. C. Dorf, Editor-inChief, *Electrical Engineering Handbook*, 2nd ed., CRC Press, Boca Raton, FL, 1997, pp. 1769–1790.
[3] E. J. McCluskey, *Logic Design Principles*. Prentice-Hall, Englewood Cliffs, NJ, 1986.
[4] R. F. Tinder, *Digital Engineering Design: A Modern Approach*. Prentice-Hall, Englewood Cliffs, NJ, 1991.

[5] J. F. Wakerly, *Digital Design Principles and Practices*, 2nd ed. Prentice-Hall, Englewood Cliffs, NJ, 1994.

The usual combinational logic devices such as MUXs, decoders, code converters, and comparators are covered adequately by most texts, including those just cited, but the texts by Tinder and Wakerly provide what is perhaps the best coverage of the group.

Steering logic seems to be covered adequately in only a few texts, among which are those of Hayes and Katz.

[6] J. P Hayes, *Introduction to Digital Logic Design*. Addison Wesley, Reading, MA, 1993.
[7] R. H. Katz, *Contemporary Logic Design*. Benjamin/Commings Publishing, Redwood City, CA, 1994.

There are numerous texts and reference books on VHDL. For instructional purposes the texts of Dewey, Navabi, Pellerin and Taylor, Perry, Roth, and Skahill are good choices. The texts by Pellerin and Taylor and by Skahill include CD-ROMs containing fully functional VHDL compilers. The text by Dewey is somewhat unusual in that it nicely combines digital design and analysis with VHDL.

[8] A. M. Dewey, *Analysis and Design of Digital Systems with VHDL*. PWS Publishint Co., Boston, 1997.
[9] Z. Navabi, *VHDL Analysis and Modeling of Digital Systems*. McGraw-Hill, New York, 1993.
[10] D. Pellerin and D. Taylor, *VHDL Made Easy*. Prentice Hall PTR, Upper Shaddle River, NJ, 1997.
[11] D. L. Perry, *VHDL*. McGraw-Hill, New York, 1991.
[12] C. H. Roth, Jr., *Digital Systems Design Using VHDL*. PWS Publishing Co., Boston, 1998.
[13] K. Skahill, *VHDL for Programmable Logic*. Addison-Wesley, Reading, MA, 1996.

The latest VHDL IEEE standard is the 1076-1993 standard. Standard 1076 has been augmented by standards 1164, 1076.3 and 1076.4. These latest standards are identified as follows:

Standard 1076-1993, *IEEE Standard VHDL Language Reference Manual*, IEEE, 1994.
Standard 1164-1993, *IEEE Standard Multivalue Logic System for VHDL Model Interoperability*, IEEE, 1993.
Standard 1076.3, *VHDL Synthesis Packages,* IEEE, 1995.
Standard 1076.4, *VITAL ASIC Modeling Specification*, IEEE, 1995.

These IEEE documents can be obtained from IEEE at the following address: IEEE Service Center, 445 Hoes Lane, PO Box 1331, Piscataway, NJ 08855-1331 (Phone: 1-800-678-IEEE).

PROBLEMS

6.1 The propagation delays for a state-of-the-art CMOS logic gate are calculated to be $\tau_{plh} = 0.25$ ns and $\tau_{phl} = 0.35$ ns with a power dissipation of 0.47 mW. Calculate the power–delay product (PDP) in picojoules predicted for this gate.

PROBLEMS

6.2 The voltage parameters for a high-speed CMOS (HC) gate are measured to be $V_{IL\,max} = 0.23V_{DD}$ and $V_{IH\,min} = 0.59V_{DD}$ with $V_{OL\,max} = 0.08V_{DD}$ and $V_{OH\,min} = 0.90V_{DD}$ for a supply of 2.8 V.

(a) Calculate the noise margins for this gate.

(b) Use the values calculated in part (a) to explain what they mean in relationship to the interpretation of logic 1 and logic 0 for this gate.

6.3 Construct a logic circuit that combines two 16-to-1 MUXs to form a 32-to-1 MUX. (Hint: Use an inverter to select the appropriate MUX.)

6.4 Use an 8-to-1 MUX to implement each of the following functions, assuming that all inputs and outputs are active high.

(a) $W(A, B, C) = \sum m(1, 2, 4, 5, 6)$
(b) $X(A, B, C) = A\bar{B}C + \bar{A}\bar{B}\bar{C} + AB\bar{C} + ABC + A\bar{B}\bar{C}$
(c) $Y(A, B, C) = \prod M(0, 1, 2, 6, 7)$
(d) $Z(A, B, C) = (A + B) \odot (\bar{A}C) + A\bar{B}$

6.5 Repeat Problem 6.4 but instead use a 4-to-1 MUX to implement each function. To do this use *minimum* external logic and the two most significant inputs as the data select variables.

6.6 Repeat Problem 6.5 assuming that only input B arrives from a negative logic source and that one inverter is permitted to be used on a data select input.

6.7 Use a 4-to-1 MUX to implement each of the following functions, assuming that all inputs and outputs are active high. It is required that minimum external logic to the MUX be used in each case, and that the data select inputs be A and B.

(a) $U(A, B, C, D) = \sum m(0, 4, 5, 7, 8, 9, 13, 15)$
(b) $V(A, B, C, D) = \prod M(0, 5, 8, 9, 11, 12, 15)$
(c) $W(A, B, C, D) = \sum m(4, 5, 7, 12, 14, 15) + \phi(3, 8, 10)$
(d) $X(A, B, C, D) = \prod M(0, 1, 2, 5, 7, 9) \cdot \phi(4, 6, 10, 13)$

6.8 Implement the following function by using the hardware indicated (nothing else). Assume that the inputs arrive as $A(H)$, $B(L)$, $C(H)$, and $D(L)$, and that the output is issued active high.

$$F(A, B, C, D) = \sum m(0, 1, 3, 4, 6, 8, 9, 10, 11, 12, 15)$$

Permitted hardware: One 4-to-1 MUX; one NAND gate; one XOR gate.

6.9 Implement the following function by using the hardware indicated (nothing else). Assume that the inputs arrive as $A(L)$, $B(H)$, $C(H)$, and $D(H)$, that the output is issued active high. (Hint: First find the absolute minimum expression for Z.)

$$Z = \bar{A}\bar{C}\bar{D} + \bar{A}\bar{B}\bar{C}D + (\bar{A} + B)C\bar{D} + (A \odot B)CD$$

Permitted hardware: One 2-to-1 MUX; one NAND gate; one AND gate.

6.10 (a) Configure a 6-to-64 decoder by using only 3-to-8 decoders.

(b) Configure a 6-to-64 decoder by using only 4-to-16 and 2-to-4 decoders.

6.11 Implement the function in Problem 4.28 by using a 16-to-1 MUX assuming that all inputs and the output are active high.

6.12 Repeat Problem 6.11 if only the *B* input is active low and *no* additional hardware (e.g., an inverter) is permitted.

6.13 Design a bitwise logic function generator that will generate any of the 16 possible logic functions. End with a single expression *F* that represents the 16 bitwise logic functions. To do this use a 4-to-1 MUX and nothing else. (Hint: Interchange the names for the data and data-select inputs to the MUX.)

6.14 Implement each function in Problem 6.4 by using a 3-to-8 decoder and the necessary external hardware, assuming that all inputs and outputs are active high.

6.15 Implement function *F* in Problem 6.8 by using a 4-to-16 decoder, one OR gate, and two NAND gates (maximum fan-in of 6), taking the input activation levels as given in Problem 6.8.

6.16 Repeat Problem 6.15 by replacing the 4-to-16 decoder with two 3-to-8 decoders and one inverter.

6.17 The function below is to have inputs that arrive as $A(H)$, $B(L)$, and $C(H)$, with an output $F(L)$.

$$F(A, B, C) = \sum m(0, 1, 6, 7)$$

(a) Implement this function by using a 3-to-8 decoder and one NAND gate (nothing else). Assume that the decoder has active low outputs. (Hint: Use the AND form of the two conjugate NAND gate circuit symbols to meet the requirement of an active low output.)

(b) Repeat part (a) by using two 2-to-4 decoders, a NAND gate, and one inverter (nothing else).

6.18 The circuit shown in Fig. P6.1 connects a decoder to a MUX. Analyze this circuit by finding $Y(H)$ in terms of inputs A, B, C, and D.

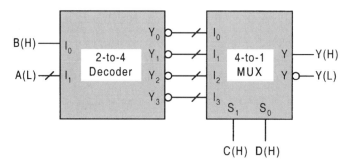

FIGURE P6.1

PROBLEMS

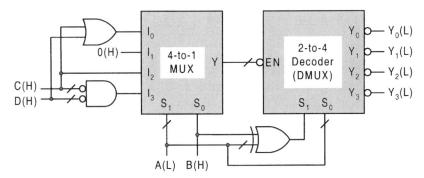

FIGURE P6.2

6.19 Analyze the network in Fig. P6.2 by obtaining the output logic expressions for $Y_0(L)$, $Y_1(L)$, $Y_2(L)$, and $Y_3(L)$ in terms of inputs A, B, C, and D. Assume that the DMUX is enabled if $EN(L) = 1(L)$.

6.20 Design an 8-input (I_7–I_0), noncascadable priority encoder (PE) that will issue all 2^n binary states. Thus, include an EN input but do not include the cascading bits EI, EO, and GS.

(a) Construct the collapsed priority schedule table for this PE by using X as the irrelevant input symbol. Assume that the logic value $EN = 1$ enables the PE. (Hint: There are three outputs.)

(b) Map the output functions into fourth-order EV K-maps with axes I_7, I_6, I_5, I_4 and extract minimum SOP expressions for these functions.

(c) Implement the results of part (b) by using NAND/INV logic assuming that all inputs and outputs are active low.

6.21 Repeat Problem 6.20 but with cascading bits EI, EO, and GS, all of which must be active low.

6.22 Use NOR/INV logic to design a noncascadable three-input priority encoder that will operate according to the following priority schedule:

Input C — Highest priority encoded as 10
Input A — Middle priority encoded as 01
Input B — Lowest priority encoded as 00
Assign 11 to the inactive state.

Assume that all inputs arrive active high and that the outputs are issued active low.

6.23 Design a Gray-to-BCD code converter by using four 4-to-1 MUXs and a gate-minimum external logic. The inputs arrive as $A(H)$, $B(L)$, $C(H)$, and $D(H)$, and the outputs are all active high.

6.24 Use three XOR gates and two OR gates (nothing else) to design a gate-minimum circuit for a 4-bit binary-to-2's complement (or vice versa) converter. Let B_3, B_2, B_1, B_0 and T_3, T_2, T_1, T_0 represent the 4-bit binary and 2's complement words, respectively, and assume that all inputs and outputs are active high. (Hint: Use second-order EV K-maps, constructed from the truth table, and XOR patterns. Factor T_3 from the SOP minimum expression and then apply the defining relation for XOR.)

6.25 Design a gate-minimum circuit for a BCD-to-creeping code converter assuming that all inputs and outputs are active high and that the inputs are restricted to the ten BCD states. To do this refer to Table 2.10 and plan to use second-order EV K-maps for convenience.

6.26 (a) Design an 8-bit binary-to-BCD converter based on Algorithm 6.2 in Subsection 6.5.2. To do this, first design the minimum NAND/INV logic for the converter module required to convert to BCD the binary numbers equivalent of decimal 0 through 19. Note that the shift-left/add-3 algorithm is inherent in the truth table for the converter module and that the LSB in binary is the same as the LSB in BCD. Next, cascade the modules as in Fig. P6.3 to carry out Algorithm 6.2 for the 8-bit converter. All inputs and outputs are assumed to be active high.

(b) Use the converter module of part (a) together with Fig. P6.3 to find the BCD for a binary number equivalent to 159_{10}.

6.27 Analyze the logic circuit for the BCD-to-seven-segment decoder in Fig. 6.24 by constructing a mixed-logic truth table for active high binary inputs equivalent to decimals 2_{10} and 9_{10}. Thus, the seven outputs must all be active low, suitable to drive the common cathode LED display in Fig. 6.23b.

6.28 (a) Following Fig. 6.28a, construct the truth table for a cascadable 3-bit comparator, but do not map it.

(b) By using only Eqs. (6.23), write the three output expressions for the 3-bit comparator.

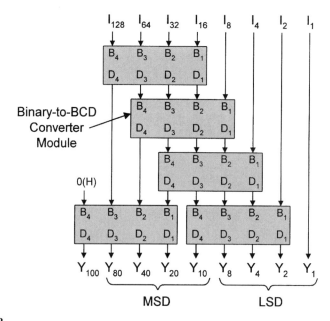

FIGURE P6.3

(c) Implement the results of part (b) by making use of Eq. (6.22). Is an *eq* input necessary?

6.29 Given the block symbol for a 4-bit cascadable comparator in Fig. 6.30 and with the appropriate gate-minimum NAND/EQV/INV external logic, design a 5-bit cascadable comparator.

6.30 Design a 4-bit even-parity detector with respect to logic 1 by using only XOR gates (nothing else). Show how this result can be used to produce a 4-bit odd-parity detector without adding additional hardware.

6.31 (a) Design a logic circuit that will detect a majority of seven inputs A, B, C, D, E, F, and G that are active at any one time, and this under the following conditions: Circuits 1 and 2 represent *majority functions* that must each detect a majority of its three inputs that are active. Thus, if any four or more of all seven inputs are active, the output *Majority Detect* will be active; otherwise the output will be inactive. To do this, use the "divide-and-conquer" approach. Referring to Fig. P6.4, construct truth tables for *identical* circuits 1 and 2 such that their outputs are active any time two or three of their inputs are active. Inputs A and D are the MSB inputs for each of the two circuits. Next, map each majority function (actually, one will do) from the truth tables and extract a gate-minimum cover. Finally, introduce the input $G = Z$ into the logic for circuit 3 such that the output, *Majority Detect*, is active iff the input conditions are met. End with a gate-minimum logic circuit that will contain XOR functions.

(Hints: Properly done, the output *Majority Detect* can be obtained directly without the use of a truth table. Note that if G is inactive, the output of circuit 3 can detect a majority of 4, 5, or 6 active inputs. However, with input G active, the output of circuit 3 can detect a majority of only 5 or 7 active inputs. To obtain a gate-minimum circuit for *Majority Detect* (seven gates and one inverter for Circuit 3, 15 total), it will be necessary to plot a fourth-order K-map with input G as the entered variable.)

(b) Repeat part (a) but without circuits 1 and 2. Thus, find the optimum two-level logic expression for the output *Majority Detect* with the seven inputs presented directly to circuit 3. To do this, plot a fourth-order EV K-map with EVs E, F, and G (a third-order compression with a Map Key of 8), then use the logic minimizer (e.g., BOOZER software bundled with this text) to obtain the result. Do not

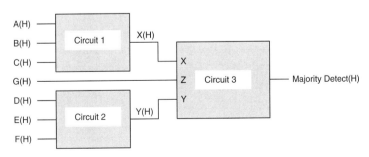

FIGURE P6.4

implement the function. [Hint: The easiest approach is to count the number of inputs (A, B, C, and D) that are active for each cell of the fourth-order K-map. Then enter the three EVs (in minimum subfunction form with the aid of a third-order K-map) in those cells as needed to bring the count of active inputs to four or more. If this is correctly done, 35 EPIs and no inverters will be required.]

(c) From the K-map in part (b), extract a gate-minimum expression by using XOR patterns. If this is correctly done, the gate count will be reduced to 18 (a four-level function) with no inverters required.

6.32 Design a 4-bit general shifter that can shift or rotate, right or left, up to three bits with F fill. To do this, do the following:

(a) Construct the truth table for this shifter.

(b) Use the truth table of part (a) to plot the fourth-order K-map for each of the four outputs.

(c) From the results of part (b), write the expression for each output in a form suitable for a 4-to-1 MUX implementation by taking the data select inputs as A_1 and A_0. Do not implement.

6.33 Find the gate-minimum external logic for a 4-bit shifter that will cause it to operate according to the following table. To do this, make use of XOR patterns. Show the resulting logic circuit required to be connected to the shifter assuming that all inputs and outputs are active high.

Shifter input	*Action*
Even 1 parity	Shift right 1 position with 0 fill
Odd 1 parity	Shift left 2 positions with 1 fill
1111 (Exception)	Transfer

(Hint: It is only necessary to know the logic for F, D, A_1, and A_0, which are the outputs for the truth table. Properly done, the minimum external circuitry will consist of three XOR gates, two NAND gates, and two inverters. Note that maxterm code must be used to extract A_0 from the K-map).

6.34 Combine two 4-to-1 steering logic MUXs, shown in Fig. 6.39b, to produce an 8-to-1 line MUX. To do this you are permitted to use a single additional CMOS inverter and a CMOS OR gate.

6.35 Write the VHDL behavioral description for the majority function $F = AB + BC + AC$. To do this, use the keywords that are defined in Section 6.10.

6.36 Write the VHDL behavioral and architecture descriptions for a 2-to-4 decoder with active high outputs. To do this follow the examples in Section 6.10.

CHAPTER 7

Programmable Logic Devices

7.1 INTRODUCTION

A class of devices called *programmable logic devices* (or PLDs) can be thought of as universal logic implementers in the sense that they can be configured (actually programmed) by the user to perform a variety of specific logic functions. So useful and versatile are these PLDs that one might question why any other means of design would ever be considered. Well, the answer is, of course, that there is a time and place for a variety of approaches to design — that is, no one single approach to design satisfies all possible problem situations. However, the option to use PLDs offers the logic designer a wide range of versatile devices that are commercially available for design purposes.

Some PLDs are made to perform only combinational logic functions; others can perform both combinational and sequential logic functions. This chapter will consider those PLDs capable of performing both combinational and sequential logic functions, but the exemplar applications will be limited to combinational logic design. Four commonly used PLDs considered here are the *read-only memory* (ROM) devices and their subgroups, the *field programmable logic array* (FPLA) devices, the *programmable array logic* (PAL) devices and their subgroups, and *field programmable gate arrays* (FPGAs) and subgroups. Other PLDs include *erasable programmable logic devices* (EPLDs), including *erasable programmable ROMs*, *generic array logic* (GAL) devices, and *field programmable logic sequencers* (FPLSs). Except for FPGAs, most of the PLDs mentioned have some commonality, namely a two-level AND/OR configuration. What is connected to the AND/OR network distinguishes one PLD from another. The development that follows attempts to illustrate the differences between these various PLDs and to provide a few useful examples of their application to combinational logic design.

7.2 READ-ONLY MEMORIES

A ROM is an n-input/m-output device composed of a nonprogrammable decoder (ANDing) stage and a programmable (OR) stage as illustrated in Fig. 7.1. Bit combinations of the n

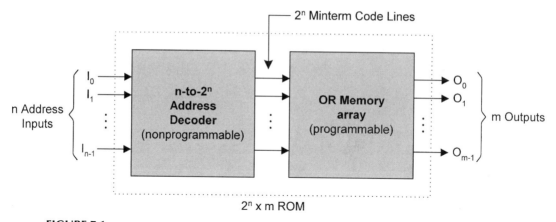

FIGURE 7.1
Block diagram showing the fixed address decoder (AND) stage and the programmable OR memory stage for a ROM of size 2^n words $\times$ m bits.

input variables are called *addresses*, and in ROMs there are 2^n minterm code lines, each representing a coded minterm on the output side of the decoder stage. Therefore, since there are n possible addresses to a ROM, there are 2^n possible *words* that can be stored in the ROM, each word being m bits in size. Any m-bit output word programmed into the ROM can be selected by the appropriate input address and is nonvolatile — it is stored permanently in the ROM.

The dimensions and size of an n-input/m-ouptut ROM are given by

$$\underbrace{2^n \times m}_{\text{Dimensions}} = \underbrace{(2^n)(m) \; bits}_{\text{Size}},$$

meaning that 2^n words, each of m bits, produce a total of $(2^n)(m)$ bits. The size of a ROM may be rounded off to the nearest integer power of 2 in K (10^3) bits of ROM. For example, an 8-input/4-output ROM is represented as a $2^8 \times 4 = 256 \times 4$ bit $= 1{,}024$ bit or 1Kbit ROM. The problem with the bit-roundoff form of representation is that knowledge of the dimensions is lost. Thus, a 1Kbit ROM could be any of the following: $2^7 \times 8 = 2^6 \times 16$, etc. This can be avoided by giving the ROM size in dimension ($2^n \times m$) form, which clearly specifies the number of inputs and outputs.

ROMs may differ in a variety of ways, but the main differences center about the manner in which they are programmed, and on whether or not they can be erased and reprogrammed and how this is accomplished. Members of the ROM family of PLDs may be divided into three main categories:

- Read-only memories (ROMs) — Mask programmable OR stage only
- Programmable ROMs (PROMs) — User programmable once
- Erasable PROMs (EPROMs) — User erasable many times

7.2 READ-ONLY MEMORIES

Mask-programmable ROMs are programmed during the fabrication process by selectively including or omitting the switching elements (transistors or diodes) that form the memory array stage of the ROM. Because the masking process is expensive, the use of mask-programmable ROMs is economically justifiable only if large numbers are produced to perform the same function.

When one or a few ROM-type devices are needed to perform certain functions, PROMs can be very useful. Most PROMs are fabricated with *fusible links* on all transistors (or diodes) in the OR memory stage, thereby permitting user programming of the ROM — a write-once capability. Shown in Fig. 7.2 is the circuit for an unprogrammed $2^n \times m$ PROM

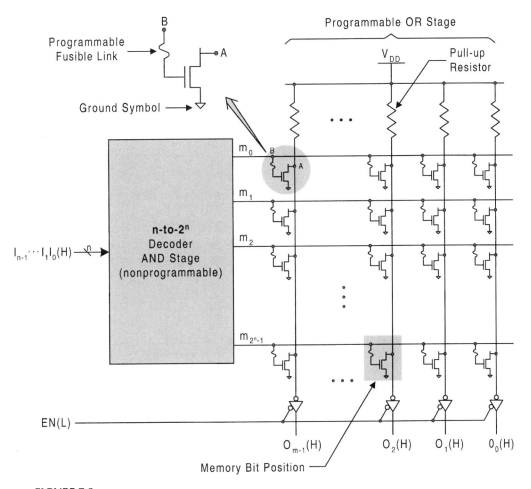

FIGURE 7.2
Logic circuit for an unprogrammed $2^n \times m$ PROM showing the nonprogrammable decoder (AND) section and programmable NMOS connections (fusible links) for each normally active bit location in the OR (memory) section. Tri-state drivers provide an enable capability.

consisting of a nonprogrammable decoder (AND) section and NMOS programmable connections (fusible links) on every memory bit location to form the programmable OR section. Also shown are inverting tri-state drivers (see Fig. 3.8d) on the PROM outputs to provide an active low enable control capability. The PROM chip is produced with all transistors "connected" such that each memory bit position is normally active high. Thus, when a decoder output line becomes active, all connected NMOS are turned ON, pulling those bit positions to 0(H), resulting in a 1(H) from each tri-state driver. A bit position is programmed when a fusible link is "blown," disconnecting that bit position. Disconnected bit positions cannot be pulled low and therefore, must remain at 1(H). If all bit positions on an OR column are disconnected by programming, the output is 0(H) from the tri-state driver. Or, if one or more OR column bit positions are left connected, the output will be a 1(H) if a decoder line to one of those bit positions becomes active — the OR function. The PROM of Fig. 7.2 is programmed (one time only permitted) with a PROM programmer by applying voltage pulses to target fusible links, causing disconnection of these bit positions.

The masking process of a mask-programmable ROM places NMOS connections at predetermined (programmed) memory bit positions. The positioned NMOS connections would look similar to those in Fig. 7.2, except their fusible links would be missing. Because the masking process is expensive, mask-programmable ROMs are used only for high-volume commercial applications.

Much more useful, generally, are the EPROMs, since they can be programmed, erased, and reprogrammed many times. These devices fall into two main categories: *ultraviolet erasable PROMs* (UVEPROMs) and *electrically erasable PROMs* (EEPROMs). In either case the technology is similar — use is made of *floating-gate NMOS transistors* at each memory bit location, as illustrated by the OR memory stage in Fig. 7.3.

Each transistor in Fig. 7.3 has two gates, a connected outer gate and an inner floating (unconnected) gate that is surrounded by a highly insulating material. Programming occurs when a high positive voltage is applied to the connected gate inducing a negative charge on the floating gate which remains after the high voltage is removed. Then, when a decoder line becomes active (HV), the negative charge prevents the NMOS from being turned ON, thereby maintaining a 1(H) at the memory bit position. This is equivalent to blowing a fusible link in Fig. 7.2. If all floating-gate NMOS in an OR column are so programmed, the output from the inverter is 0(H). But if a decoder line is active to any unprogrammed floating-gate NMOS, that bit position will be pulled to ground 0(H), causing the output to be 1(H) from the inverter — again, the OR function.

Erasure of a programmed floating-gate NMOS occurs by removing the negative charge on its floating gate. This charge can remain on the gate nearly indefinitely, but if the floating gate in a UVEPROM is exposed (through a "window") to ultraviolet light of a certain frequency, the negative charge on the floating gate is removed and erasure occurs. Similarly, if a voltage of negative potential is applied to the outer connected gate of an EEPROM, removal of the negative charge occurs. The technology for the UVEPROMs and EEPROMs differ somewhat as to the manner in which the floating gate is insulated; otherwise, they share much in common.

Technologies other than those just described are used to manufacture PROMs. For example, in *bipolar* PROMs, diodes with fusible links replace the NMOS in Fig. 7.2 with each diode conducting in the $A \rightarrow B$ direction (see blowup of the fusible link). Now, however,

7.2 READ-ONLY MEMORIES

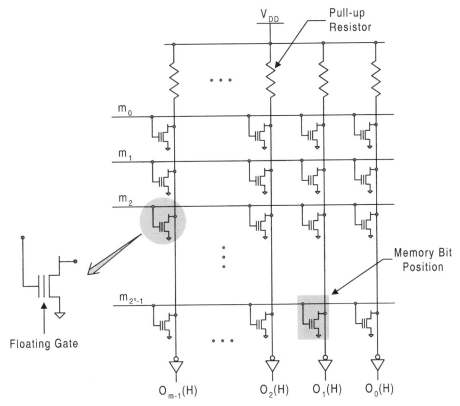

FIGURE 7.3
Logic circuit for an unprogrammed OR (memory) section of an EPROM illustrating the floating-gate NMOS transistor technology required for a program/erase/program cycle.

the decoder has active low outputs. Then, when a decoder output becomes active, 1(L) = 0(H), a bit position for a connected diode is pulled low, 0(H). A disconnected (programmed) diode maintains a 1(H) at the bit position when selected. If all diode bit positions in an OR column are disconnected, then the output for that column must be 0(H) from the inverting tri-state driver. However, if one (or more) of the bit position diodes in an OR column is left connected and is selected by an active low decoder line, the output will be 1(H) from the tri-state driver. The diode technology, used extensively in the early stages of ROM development, is now in less common use than MOS technology.

7.2.1 PROM Applications

The AND section of any ROM device is a nonprogrammable decoder, and since a decoder is a minterm code generator the following requirement must be met for ROM programming:

> To program a ROM, all input and output data must be represented in canonical form.

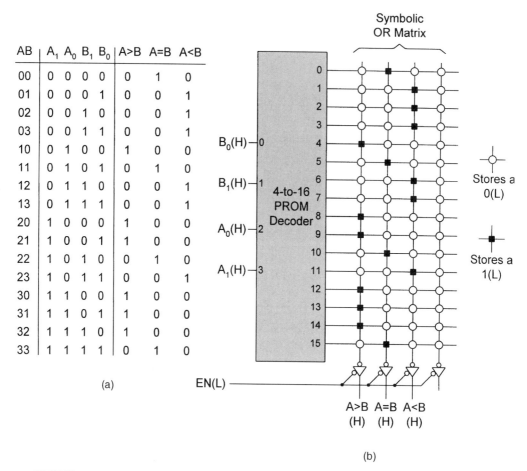

FIGURE 7.4
PROM implementation of a noncascadable 2-bit comparator. (a) Canonical truth table. (b) Decoder and symbolic representation of fusible bit position patterns in the OR memory section.

This fact is illustrated by considering the noncascadable 2-bit comparator represented by the canonical truth table in Fig. 7.4a. This truth table has been constructed from Fig. 6.28 by setting $gt = 1$ when $A > B$, $eq = 1$ when $A = B$, and $lt = 1$ when $A < B$, or by setting these inputs to zero otherwise. Shown in Fig. 7.4b is a $2^4 \times 4$ PROM that has been programmed to function as the 4-input/3-output noncascadable 2-bit comparator. Here, a symbolic notation is used so as to avoid the need to exhibit the details of the logic circuit, including the output inverters. For consistency with Fig. 7.2, tri-state drivers are added to the output with the understanding that a filled square (■) represents the storage of 1(L) and that an open circle (○) represents the storage of 0(L). Notice that one of the four output lines is left unused.

Programming a PROM that functions as the cascadable 2-bit comparator of Fig. 6.28 would require a PROM of dimensions $2^7 \times 3$, a significant increase in hardware. In fact, to do so might seem to be hardware overkill, considering that a 7-to-128 line decoder

7.3 PROGRAMMABLE LOGIC ARRAYS

section would be required. Recall that the three-level logic circuit for the cascadable 2-bit comparator in Fig. 6.28 has a gate/input tally of only 23/67.

7.3 PROGRAMMABLE LOGIC ARRAYS

Like the ROM, the PLA is an n-input/m-output device composed of an input ANDing stage and a memory (ORing) output stage. Unlike the ROM, however, both stages of the PLA are programmable, as indicated by the block diagram in Fig. 7.5. The AND matrix (array) generates the product terms (p-terms), while the OR matrix ORs the appropriate product terms together to produce the required SOP functions.

The dimensions of a PLA are specified by using three numbers:

$$\underset{\substack{\uparrow \\ \text{No. of} \\ \text{inputs}}}{n} \times \underset{\substack{\uparrow \\ \text{No. of} \\ \text{product} \\ \text{terms}}}{p} \times \underset{\substack{\uparrow \\ \text{No. of} \\ \text{outputs}}}{m}$$

The number p gives the maximum number of product terms (p-terms) permitted by the PLA. The magnitude of p is set by the PLA manufacturer based on expected user needs and is usually much less than 2^n, the decoder output of a ROM. For example, a PLA specified by dimensions $16 \times 48 \times 8$ would have 16 possible inputs and could generate 8 different outputs (representing 8 different SOP expressions) composed of up to 48 unique ORed p-terms. A p-term may or may not be a minterm. In contrast, a 16-input ROM could generate the specified number of outputs with up to $2^{16} = 65{,}536$ minterms.

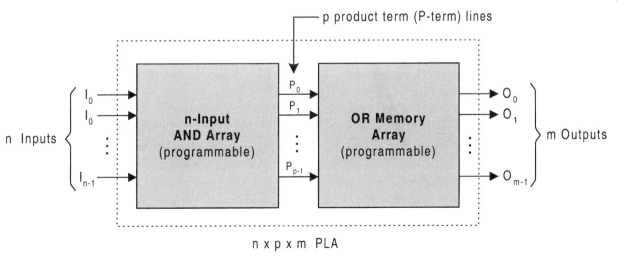

FIGURE 7.5
Block diagram for an n-input/m-output PLA showing the programmable AND and OR array stages and the p product-term lines.

The comparison here is 48 p-terms and 48 ORing connections for the unprogrammed PLA vs 65,536 minterms and 65,536 ORing connections for the unprogrammed ROM, a vast difference in hardware. Typically, commercial PLAs have inputs ranging in number from 8 to 16, with 20 to 60 addressable p-terms and up to 10 outputs that may or may not have controlled polarity by using XOR gates on the output lines (see Subsection 3.9.6). PLA IC chips of most any $n \times p \times m$ dimensions can be manufactured to user specifications.

PLAs, like ROMs, are constructed of interconnecting arrays of switching elements that perform the AND and OR operations. Members of the PLA family fall generally into two classes:

- Programmable logic arrays (PLAs) — Mask programmable AND and OR stages
- Field programmable logic arrays (FPLAs) — One-time programmable AND and OR stages

Thus, PLAs are programmed during fabrication in a manner similar to ROMs, while FPLAs are write-once programmed by the user.

Shown in Fig. 7.6 is the MOS version for an unprogrammed $n \times p \times m$ FPLA illustrating the programmable bit connections in both the AND and OR array sections. A given p-term row can become active, $P_i(H) = 1(H)$, iff all of the NMOS switches on the AND side are either disconnected or turned OFF. If any one of the NMOS in a p-term row is turned ON, then that row is pulled low, 0(H). An active p-term line causes a connected OR bit position in its path to be pulled low, resulting in a 1(H) from the output tri-state inverter. The buffer between the AND and OR stages is necessary to boost and sharpen the signal. Such a buffer could consist of two CMOS inverters. Notice that tri-state drivers provide an active low enable control on the OR stage.

The programming of a PLA is best understood by considering the 3-input/2-output FPLA segment in Fig. 7.7. Here, the single p-term line is programmed to generate $O_1(H) = 1(H)$ and $O_0(H) = 0(H)$ any time the p-term $\bar{I}_1 \cdot I_0$ becomes active. Notice that both NMOS bit positions for I_2 are disconnected together with those for the $I_1(L)$ and $I_0(H)$ lines. Thus, if I_1 is 0(H) and I_0 is 1(H), the p-term line $P_i(H)$ is forced active which, in turn, causes output $O_1(H)$ to become 1(H), but not output $O_0(H)$, whose bit position has been disconnected, causing it to become 0(H). In effect, disconnection (blowing) of a fusible link in the AND plane actually "makes the connection" of the p-term input (I_j or $\bar{I}_j$), and disconnection of a fusible link in the OR plane stores a $1(H) = 0(L)$. These facts may make it easier to understand the symbolic representations illustrated in Subsection 7.3.1.

7.3.1 PLA Applications

Unlike the programming of ROMs which require canonical data, PLAs require minimum or reduced SOP data. Thus, the most efficient application of an FPLA would be one for which the needed product terms fit within the FPLA's limited p-term capability. A good example is the FPLA implementation of the 4-bit combinational shifter of Fig. 6.37. Shown in Fig. 7.8 is the truth table (reproduced from Fig. 6.37a) and the EV K-maps and minimum cover for this shifter. The minimum output expressions for this shifter are obtained directly

7.3 PROGRAMMABLE LOGIC ARRAYS

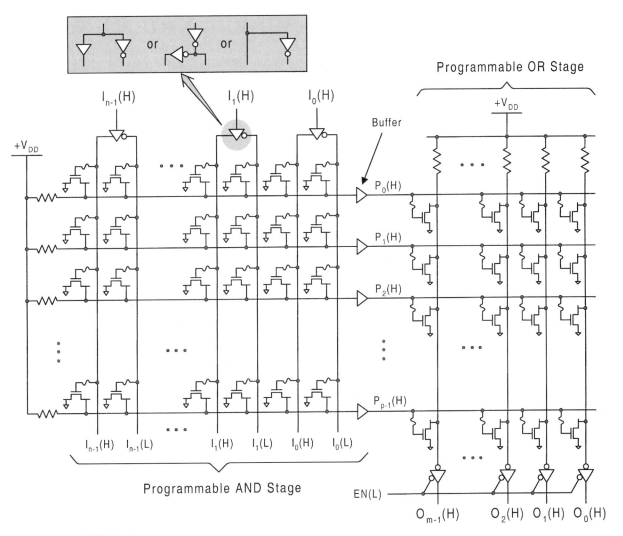

FIGURE 7.6
MOS version of an unprogrammed $n \times p \times m$ FPLA showing the programmable NMOS connections for normally active bit locations in the AND and OR (memory) sections, and showing tri-state outputs with active low enable controls.

from the K-maps and are given as follows:

$$
\begin{aligned}
Y_3 &= \bar{A}_1\bar{A}_0 I_3 + \bar{A}_1 A_0 I_2 + A_1\bar{A}_0 I_1 + A_1 A_0 I_0 \\
Y_2 &= \bar{A}_1\bar{A}_0 I_2 + \bar{A}_1 A_0 I_1 + A_1\bar{A}_0 I_0 + R A_1 A_0 I_3 + \bar{R} F A_1 A_0 \\
Y_1 &= \bar{A}_1\bar{A}_0 I_1 + \bar{A}_1 A_0 I_0 + R A_1\bar{A}_0 I_3 + R A_1 A_0 I_2 + \bar{R} F A_1 \\
Y_0 &= \bar{A}_1\bar{A}_0 I_0 + R\bar{A}_1 A_0 I_3 + R A_1\bar{A}_0 I_2 + R A_1 A_0 I_1 + \bar{R} F A_1 + \bar{R} F A_0 .
\end{aligned}
\tag{7.1}
$$

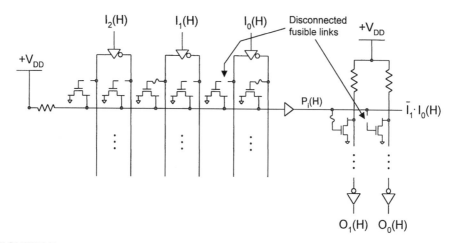

FIGURE 7.7
Portion of a 3-input/2-output FPLA programmed for a single p-term showing connected and disconnected NMOS bit positions in both AND and OR planes.

Having obtained the minimum expressions for the four outputs given by Eqs. (7.1), all that remains is to construct the *p-term table* from which the FPLA can be programmed. In this case an FPLA having minimum dimensions of $8 \times 19 \times 4$ will be required. Note that there are 20 p-terms but one ($\bar{R}FA_1$) is a shared PI. Presented in Fig. 7.9 is the p-term table for the 4-bit shifter. Notice that uncomplemented input variables are indicated by logic 1

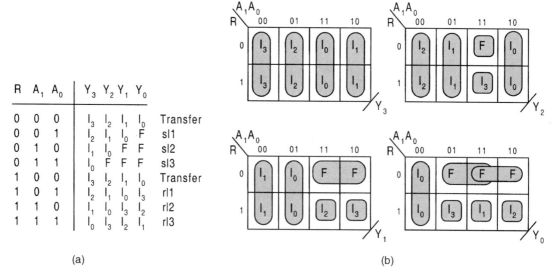

FIGURE 7.8
Design of a 4-bit general shifter that can shift or rotate left up to three bits with F fill. (a) EV truth table. (b) Output EV K-maps showing minimum two-level cover suitable for FPLA implementation.

7.3 PROGRAMMABLE LOGIC ARRAYS

p-Term		Inputs							Outputs			
	R	F	A_1	A_0	I_3	I_2	I_1	I_0	Y_3	Y_2	Y_1	Y_0
$\bar{A}_1\bar{A}_0 I_3$	–	–	0	0	1	–	–	–	1	0	0	0
$\bar{A}_1 A_0 I_2$	–	–	0	1	–	1	–	–	1	0	0	0
$A_1\bar{A}_0 I_1$	–	–	1	0	–	–	1	–	1	0	0	0
$A_1 A_0 I_0$	–	–	1	1	–	–	–	1	1	0	0	0
$\bar{A}_1\bar{A}_0 I_2$	–	–	0	0	–	1	–	–	0	1	0	0
$\bar{A}_1 A_0 I_1$	–	–	0	1	–	–	1	–	0	1	0	0
$A_1\bar{A}_0 I_0$	–	–	1	0	–	–	–	1	0	1	0	0
$RA_1 A_0 I_3$	1	–	1	1	1	–	–	–	0	1	0	0
$\bar{R}FA_1 A_0$	0	1	1	1	–	–	–	–	0	1	0	0
$\bar{A}_1\bar{A}_0 I_1$	–	–	0	0	–	–	1	–	0	0	1	0
$\bar{A}_1 A_0 I_0$	–	–	0	1	–	–	–	1	0	0	1	0
$RA_1\bar{A}_0 I_3$	1	–	1	0	1	–	–	–	0	0	1	0
$RA_1 A_0 I_2$	1	–	1	1	–	1	–	–	0	0	1	0
$\bar{R}FA_1$	0	1	1	–	–	–	–	–	0	0	1	1
$\bar{A}_1\bar{A}_0 I_0$	–	–	0	0	–	–	–	1	0	0	0	1
$R\bar{A}_1 A_0 I_3$	1	–	0	1	1	–	–	–	0	0	0	1
$RA_1\bar{A}_0 I_2$	1	–	1	0	–	1	–	–	0	0	0	1
$RA_1 A_0 I_1$	1	–	1	1	–	–	1	–	0	0	0	1
$\bar{R}FA_0$	0	1	–	1	–	–	–	–	0	0	0	1

FIGURE 7.9
P-term table for the 4-bit shifter represented in Fig. 7.8 and by Eq. (7.1).

and complemented input variables by logic 0, and the absence of an input to a p-term is indicated by a dash.

Presented in Fig. 7.10 is the symbolic representation of an $8 \times 20 \times 4$ FPLA programmed according to the p-term table in Fig. 7.9. The symbolism that is used avoids the need to provide details of the specific technology used in the FPLA. However, reference to Figs. 7.6 and 7.7 permits one to associate the symbolism of Fig. 7.10 with FPLA MOS technology. Notice the existence of the shared PI, $\bar{R}FA_1$, in the Y_1 and Y_0 output columns.

The symbolism of Fig. 7.10 is meant to be easily understood. The × in the AND plane signifies an input to a p-term and represents a disconnected fusible NMOS link in the sense of Fig. 7.7. The filled square (■) in the OR plane represents the programmed storage of 1(L) created by a connected fusible NMOS link in the sense of Fig. 7.7, and the open circle (○) in the OR plane indicates the programmed storage of 0(L) created by a disconnected fusible NMOS link. To assist the reader in deciphering this notation, representative p-terms are provided in Fig. 7.10 at the left and adjacent to their respective p-term lines.

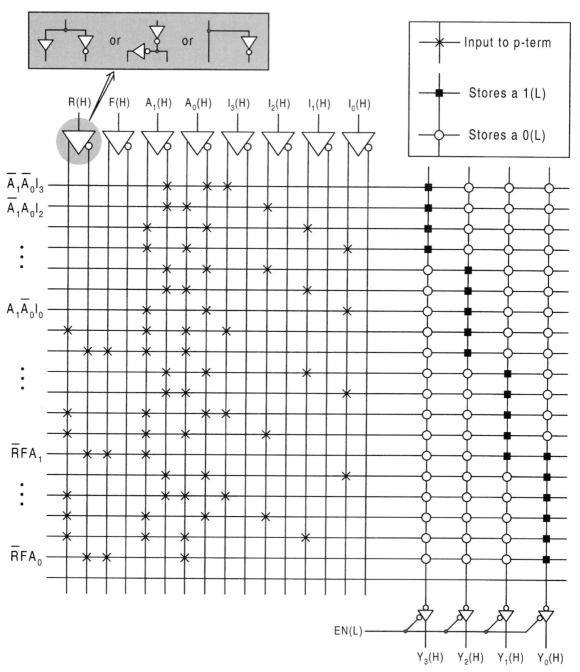

FIGURE 7.10
Symbolic representation of an 8 × 20 × 4 FPLA programmed according to the p-term table in Fig. 7.9 and showing the fusible bit position patterns in both the AND and OR planes. Representative p-terms are given to the left of their respective p-term lines.

7.4 PROGRAMMABLE ARRAY LOGIC DEVICES

Perhaps the most widely used PLD is the *programmable array logic* (PAL) device. The PAL device can be programmed only in the AND plane. For reference purposes the main differences between ROMs, PLAs and PAL devices may be stated as follows:

ROM devices — Programmable in the OR plane only
PLA devices — Programmable in both AND and OR planes
PAL devices — Programmable in the AND plane only

Thus, a PLA device may be thought of as a combination of ROM and PAL device programming characteristics. PAL devices, like PLAs, commonly include a variety of devices external to their OR planes, including XOR gates, AND–OR–invert logic, and registered outputs.

The acronym PAL is a registered trademark of Advanced Micro Devices, Inc. Therefore, hereafter it will be understood that use of the name PAL will acknowledge AMD's right of trademark for all devices that carry the name PAL.

PAL devices characteristically provide a fixed number of p-terms per output and cannot take advantage of shared PIs. This is illustrated by the unprogrammed $8 \times 20 \times 4$ PAL device shown in Fig. 7.11, which allows up to five p-terms per output. If the number of p-terms for a given output exceeds the number provided by the PAL device, the remaining p-terms can be given to another column output line, and the two ORed external to the OR plane. Thus, in the case of the 4-bit shifter, the $Y_0(H)$ output, requiring six p-terms, would have to have one or more of its p-terms given to a fifth output line (not shown) and the two lines ORed external to the OR plane. The MOS version of a basic PAL device would look like that for the PLA in Fig. 7.6, except that the NMOS bit positions in the OR stage would be permanently connected — no fusible links.

The basic PAL device simply employs an AND and OR section either in the form of AND/OR or AND/NOR as in Fig. 7.12a. However, the output logic of most PAL devices goes beyond the basic PAL. Shown in Fig. 7.12b is a segment of a PAL device that supports an *L-type* (logic-type) macrocell consisting of a controlled inverter (XOR gate), an AND-controlled output enable, and a feedback path. The feedback path is useful for cascading combinational logic functions, or for the design of asynchronous (self-timed) sequential machines covered in Chapter 14. In either case, one output function is fed back to become the input in the generation of another output function.

Macrocells containing on-chip flip-flops are also found in PAL devices, as illustrated in Fig. 7.12c. These are called *R-type* (registered-type) macrocells; they support feedback from a flip-flop output and a controlled tri-state driver/enable. Both the clock signal and the output enable signal can be supplied externally or can be generated from within the PAL device. PAL devices with R-type macrocells are useful in the design of synchronous (clock-driven) sequential machines, which are discussed at length in Chapters 10 through 13. The description of flip-flops and the details of their design and operation are presented in Chapter 10.

The versatility of PAL devices is improved significantly by the use of *V-type* (variable-type) macrocells such as that illustrated in Fig. 7.13. Now, the output signal generated from a 4-to-1 MUX can be combinational or registered depending on the data select inputs S_1 and S_0 whose logic values are set by programming fusible links. Thus, data select inputs

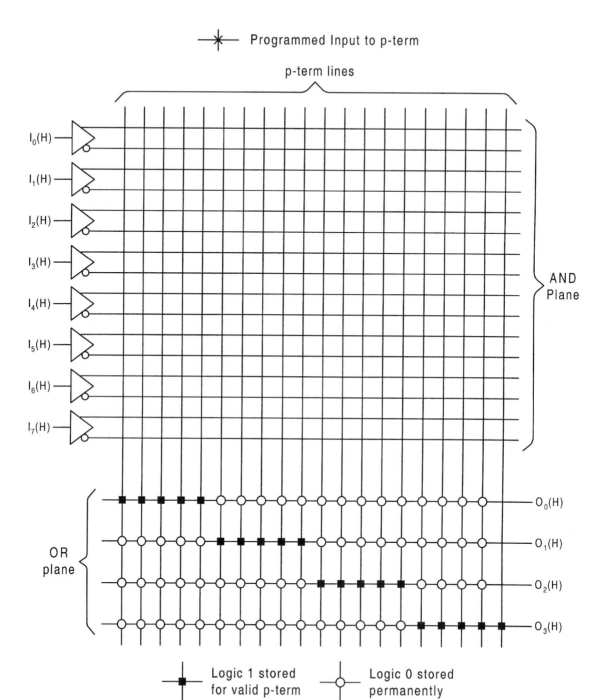

FIGURE 7.11
Symbolic representation of a basic unprogrammed 8 × 20 × 4 PAL with five p-term capability (unalterable) for each of the four outputs.

7.4 PROGRAMMABLE ARRAY LOGIC DEVICES

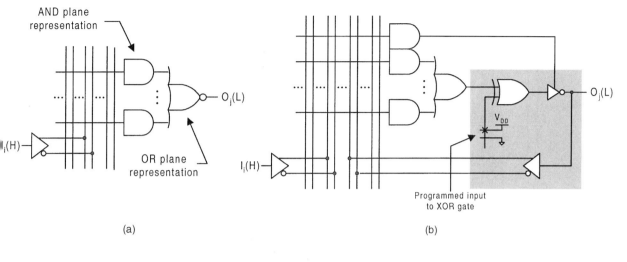

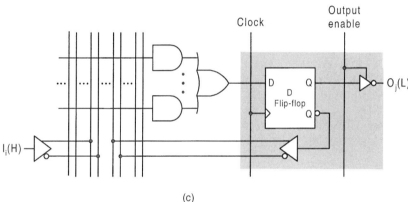

FIGURE 7.12
Logic equivalent segments of real PAL devices showing macrocell logic. (a) Basic I/O PAL. (b) L-type macrocell for a programmable I/O PAL device with controlled inverter output, feedback capability, and AND-controlled enable. (c) Macrocell for a registered (R-type) I/O PAL device with feedback from a flip-flop and with a controlled enable.

of $S_1, S_0 = 00$ or $S_1, S_0 = 01$ generate an active low or active high output (from the tri-state driver), respectively, directly from the AND/OR sections of the PAL device — hence, combinational. For data select inputs of $S_1, S_0 = 10$ or $S_1, S_0 = 11$, registered (flip-flop) outputs are generated active low or active high, respectively, from the inverting tri-state driver. Thus, the V-type macrocell combines the capabilities of the L-type and R-type macrocells of Fig. 7.12. But the V-type macrocell goes well beyond these capabilities and offers even more flexibility. A 2-to-1 MUX permits the active high and active low feedback signals to be generated either by the 4-to-1 MUX with its four options or directly by the active low output of the flip-flop. The data select input S to the 2-to-1 MUX is taken from the programmed data select input S_1 to the 4-to-1 MUX, as shown in Fig. 7.13. Because of the flexibility they offer, PAL devices with V-type macrocells are a popular choice for designers of both combinational logic or sequential machine design.

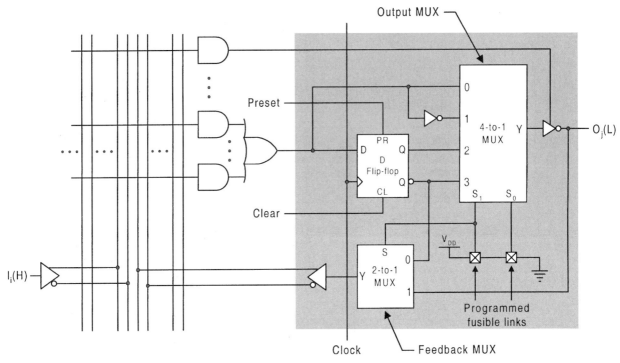

FIGURE 7.13
Output macrocell for V-type PAL devices showing I/O connections and cell logic. Source: *PAL Device Data Book*, Advanced Micro Devices, Inc., Sunnyvale, CA, 1992.

7.5 MIXED-LOGIC INPUTS TO AND OUTPUTS FROM ROMs, PLAs, AND PAL DEVICES

The subject of mixed-logic inputs to decoders was discussed in Subsection 6.3.2. There, two options were given for dealing with inputs that arrive both from positive and negative logic sources. These input rules are necessary since the decoder is normally an IC chip with no user access to its internal structure. Because ROMs, PLAs, and PALs also exist as IC chips, the same input rules also apply to them. For reference purposes, these input rules are stated as follows:

For Mixed-Logic Inputs to ROMs, PLAs, and PALs

(1) In the program table for a ROM, PLA or PAL device, complement each bit in the column of an active low input

or

(2) Use an inverter on the input line of an active low input to one of these PLDs and *do not* complement the bits in that column of the program table.

Clearly, these mixed-logic input rules are basically the same as those stated in Subsection 6.3.2 for decoders.

7.5 ROMs, PLAs, AND PAL DEVICES

The mixed-logic output rules for ROMs, PLAs, and PALs differ somewhat depending on the PLD in question. The mixed-logic output rules for ROMs are as follows:

For Mixed-Logic Outputs from ROMs

(1) In the ROM program table complement each bit in the column of any output from a ROM that is required to be active low,

or

(2) Use the inverter capability of the ROM, or add one externally, on that output line to generate the active low output and *do not* complement the bits in the output column.

Since ROMs accept only canonical (minterm) data, it is permissible, as an option, to complement the output column for an active low output. This is equivalent to complementing the conventional (1's and 0's) K-map for the active low output to yield

$$Y_{SOP}(L) = \bar{Y}_{SOP}(H), \tag{7.2}$$

which follows from the definition of complementation given by Eq. (3.2). But this is only possible for canonical data as in the case of a ROM. For example, suppose it is required that the output $(A = B)$ in Fig. 7.4 be generated active low. To do this, the $(A = B)$ output column can be complemented, interchanging the (○) and (■) symbols, or an inverter can be connected to the $(A = B)$ output line from the PROM.

The situation for PLA and PAL devices is much different from that for ROMs relative to active low outputs. Now, reduced or minimum forms constitute the output functions, and complementation of the output columns of active low outputs would result in radically different functions. Equation (7.2) does not apply to output column complementation for these PLDs. The rule for dealing with active low outputs from PLA and PAL devices is stated in the following way:

For Active Low Outputs from PLA and PAL Devices

If a given output line from a PLA or PAL device must be generated with an activation level different from that provided internally by the PLD, use *must* be made of an inverter added externally to that line. Complementation of an output column in the p-term table is *not* permitted.

As an example, consider the FPLA in Fig. 7.10, which has been programmed to function as a 4-bit shifter with F fill. Suppose that R arrives active low and that all outputs must be delivered to the next stage active low. To achieve this objective with minimum external logic, the R column is complemented and, if mixed logic outputs are not provided internal to the FPLA, inverters are placed on the output lines for Y_3, Y_2, Y_1, and Y_0. Complementation of the R column in the p-term table of Fig. 7.9 requires that the 1's and 0's be interchanged but leaving all dashes unaltered. Thus, the ×'s in the two R columns in Fig. 7.10 will be moved from the active high column to the active low column and vice versa. The meaning of the ×'s was explained previously in discussing the symbolism of Fig. 7.10.

7.6 MULTIPLE PLD SCHEMES FOR AUGMENTING INPUT AND OUTPUT CAPABILITY

Occasions arise when the I/O requirements of a design exceed the capabilities of the available PLD. When this happens the designer may have no alternative but to combine PLDs in some suitable fashion to meet the design requirements. To accomplish this requires the combined use of tri-state driver and *wired-OR* technologies, which permit nearly an unlimited number of outputs from different PLDs to be ORed together. The use of OR gates to accomplish this task would suffer the disadvantage of fan-in limitations and speed reduction.

Figures 7.2 and 7.6 illustrate the use of inverting tri-state drivers (see Fig. 3.8) with active low enable control EN(L) on the output lines from the PROM and FPLA, respectively. These tri-state drivers not only function to satisfy the logic level requirements of the device outputs, but also permit a type of PLD multiplexing based on the transfer and disconnect (high Z) modes of the tri-state drivers. The two principal modes of tri-state driver operation are illustrated in Figs. 7.14a and 7.14b, and the block diagrams for PLDs with active high tri-state driver outputs controlled by active high and active low enable controls are shown in Fig. 7.14c.

By using a multiplexed scheme involving a decoder and PLDs with active low tri-state enable controls, it is possible to increase the input capability beyond that of the stand-alone PLDs. Such a scheme is shown in Fig. 7.15, where a $(k-n)$-to-$2^{(k-n)}$ line decoder is used to select $2^{(k-n)}$ n-input PLDs, each with m outputs. The use of active low tri-state enables,

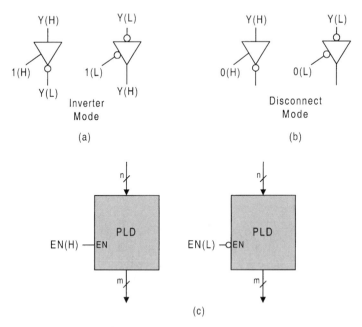

FIGURE 7.14
Tri-state driver/enables used in PLDs. (a) Transfer and (b) disconnect modes of operation for active high and active low tri-state drivers. (c) Block diagrams for a PLD with active high and active low tri-state driver/enables, and active high outputs.

7.6 MULTIPLE PLD SCHEMES FOR AUGMENTING INPUT AND OUTPUT

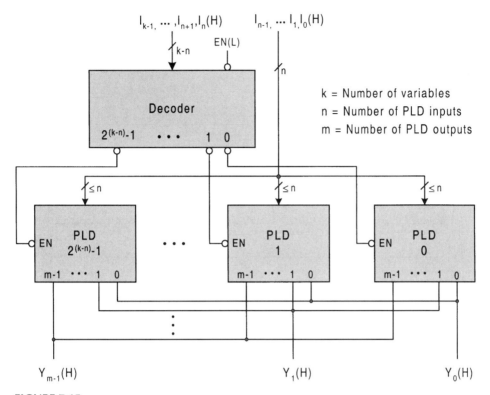

FIGURE 7.15
Multiplexed scheme by using a decoder and n-input PLDs to increase input variable capability from n to $k > n$.

as in Fig. 7.14, makes possible the wire-ORed connection of outputs shown in Fig. 7.15. This is so because the most significant $I_{k-1}, \ldots, I_n$ input bits activate only one of the $2^{(k-n)}$ n-input PLDs at any given time. As indicated in Fig. 7.15 by the notation $\leq n$, not all the available inputs to a given PLD need be used. Also, each PLD output need not be connected (wire-ORed) with the corresponding outputs from other PLDs to form *tri-state bus* lines.

Although Fig. 7.15 satisfies the need to augment the input capability of PLDs, it does not address the problem of limited output capability. When the number of output functions of a design exceeds the output capability of the PLDs in use, a parallel arrangement of the type shown in Fig. 7.16 can be used. This scheme is applicable to any stage of the multiplexed configuration of Fig. 7.15. It indicates that p PLDs, each of m outputs, yield a maximum of $(p \times m)$ possible outputs per stage, thereby increasing the output capability from m to $(p \times m)$. However, it is important to understand that *these outputs must not be wire-ORed together*, since they are from PLDs that are activated by the same decoder output — the PLDs in Fig. 7.16 are not multiplexed. Note that the PLDs need not have the same number of inputs or outputs, but the number of inputs is limited to n or less.

EXAMPLE 7.1 Suppose it is required to generate three output functions of 10 variables by using $2^8 \times 4$ PROMs. Since the number of input variables exceeds the number of PLD

314 CHAPTER 7 / PROGRAMMABLE LOGIC DEVICES

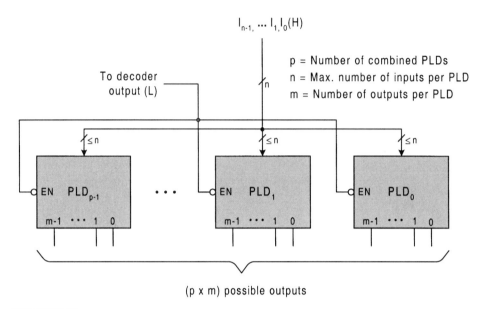

FIGURE 7.16
Scheme for increasing the output function capability from m to $(p \times m)$ for any stage of the PLD configuration shown in Fig. 7.15.

inputs by 2, a 2-to-4 decoder is required to selectively activate the four PROMs one at a time. Presented in Fig. 7.17 is the ROM program table that satisfies the requirements of this example. Since four PROMs are required ($2^{(k-n)} = 4$), the program table must be partitioned into four parts, each part being a program for a single PROM. Once programmed, the four PROMs are configured with the 2-to-4 decoder as illustrated in Fig. 7.18. The outputs are wire-ORed together as shown to generate the three outputs of 10 variables.

EXAMPLE 7.2 Augmentation of the input, output, and p-term capability of the PLDs in use is illustrated by the following example. Consider the implementation of 16 functions of 10 variables by using $8 \times 20 \times 4$ FPLAs subject to the following p-term constraints dictated by the function makeup:

Eight of the functions each require that $20 < p \leq 40$, while the remaining 8 require that $p \leq 20$. Here, p is the number of p-terms per function.

The p-term program table format for this system is given in Fig. 7.19. The p-terms are listed on the left side of the table, and the 10 FPLA inputs to the p-terms are indicated by using the notation given in Fig. 7.19. It is implied by the program table of Fig. 7.19 that two inputs, I_9 and I_8, to a 2-to-4 decoder are to be used to determine the FPLA assignment. With this in mind, one possible assignment scheme is as follows: The eight functions requiring 20 to 40 p-terms take decoder addresses of 00 or 01, while functions requiring 20 or fewer p-terms take decoder addresses of 10 or 11.

Implementation of the 16 functions of 10 variables, according to the decoder assignments just given, requires that six $8 \times 20 \times 4$ FPLAs be selected by the 2-to-4 decoder as shown in Fig. 7.20. Here, four of the functions, Y_3, Y_2, Y_1, and Y_0, are assigned to FPLA$_0$ and

PROM Inputs										PROM Outputs
I_9	I_8	I_7	I_6	I_5	I_4	I_3	I_2	I_1	I_0	$Y_2\ Y_1\ Y_0$
0	0	0	0	0	0	0	0	0	0	
					⋮					$PROM_0$
0	0	1	1	1	1	1	1	1	1	
0	1	0	0	0	0	0	0	0	0	
					⋮					$PROM_1$
0	1	1	1	1	1	1	1	1	1	
1	0	0	0	0	0	0	0	0	0	
					⋮					$PROM_2$
1	0	1	1	1	1	1	1	1	1	
1	1	0	0	0	0	0	0	0	0	
					⋮					$PROM_3$
1	1	1	1	1	1	1	1	1	1	

FIGURE 7.17
Partitioned PROM program table required to generate three output functions of 10 variables by using four $2^8 \times 4$ PROMs.

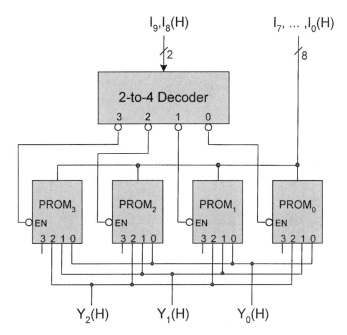

FIGURE 7.18
Multiplexed scheme required by the program table of Fig. 7.17 to generate three functions of 10 variables by using four $2^8 \times 4$ PROMs.

P-terms	FPLA Inputs I_9 I_8 : I_7 I_6 I_5 I_4 I_3 I_2 I_1 I_0	FPLA Outputs Y_{15} ... Y_3 Y_2 Y_1 Y_0	FPLA Assignment

FIGURE 7.19
P-term table format for a multiple FPLA scheme to generate 16 functions of 10 variables by using six $8 \times 20 \times 4$ FPLAs subject to the p-term conditions stated in Example 7.2.

FPLA$_2$; the remaining four, Y_7, Y_6, Y_5, and Y_4, are assigned to FPLA$_1$ and FPLA$_3$. Notice that the active low tri-state enables for FPLA$_0$ and FPLA$_1$ are connected together and to a single decoder output, as are those for FPLA$_2$ and FPLA$_3$. By wire-ORing the outputs in this manner, eight of the 16 output functions are each permitted to have up to 40 p-terms with the remaining eight functions limited to 20 p-terms.

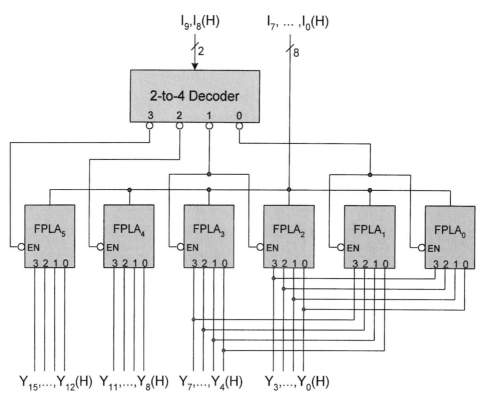

FIGURE 7.20
Scheme to generate 16 functions of 10 variables by using $8 \times 20 \times 4$ FPLAs subject to p-term requirements: 8 functions for which $20 < p \leq 40$, and 8 functions for which $p \leq 20$.

7.7 INTRODUCTION TO FPGAs AND OTHER GENERAL-PURPOSE DEVICES

The devices briefly discussed in this section vary in size from SSI to LSI circuit modules and range greatly in complexity from simple gate-level modules to those having a multiplicity of components, including flip-flops. Of the general purpose devices, the simplest are the AND–OR–invert or OR–AND–invert type devices with logic equivalent gate counts numbering typically in the three to five range. These SSI circuits are often used as building blocks in complex logic devices. Then, within the PLD spectrum of devices, the most complex class belongs to the field programmable gate array (FPGA) devices, which may contain a variety of primitive components, including discrete gates, MUXs, and flip-flops. Since this chapter deals only with combinational logic devices, the treatment here will deal mostly with those PLDs and general-purpose devices that are combinational in character.

The discussions to this point in the text have dealt only with combinational logic, consequently, the reader is not expected to grasp the significance of flip-flop operation in the output logic stage to some PLDs. The use of these "registered" PLDs will become evident in later chapters when sequential machines are discussed in detail. Treatment here will begin with simple general-purpose building block devices and will end with an introduction to the complex FPGAs.

7.7.1 AND–OR–Invert and OR–AND–Invert Building Blocks

Just as the XOR and EQV functions can be implemented by what amounts to one gate level of MOS transistors as in Figs. 3.26 and 3.27, so also can the AND–OR–invert and OR–AND–invert functions be implemented with just one gate level of transistors. Shown in Fig. 7.21a is the CMOS realization of the AND–OR–invert (AOI) gate. It is called a gate since it is a CMOS SSI circuit and has a propagation delay equivalent to that of a single NAND or NOR gate. There are many useful applications of this basic building block, including integration with much larger PLD logic blocks. The physical truth table and its mixed-logic interpretation are presented in Figs. 7.21b and 7.21c, respectively. The output logic expression can be read directly from the mixed-logic truth table and is

$$F(L) = [AB + CD](L), \tag{7.3}$$

which results in the logic equivalent circuit for the AOI gate shown in Fig. 7.21d.

As a simple example of the use of the AOI gate, consider the two-level active low EQV function given by

$$F(L) = (AB + \bar{A}\bar{B})(L).$$

This function can be implemented in Fig. 7.21d by connecting A to C and B to D via two inverters, a transistor count of 12. In comparison, the EQV gate of Fig. 3.27a requires only six transistors.

The CMOS realization of the OR–AND–invert (OAI) gate is the dual of that for the AND–OR–invert gate and is easily obtained by flipping the latter end-for-end while interchanging all NMOS with PMOS and vice versa. This is done in Fig. 7.22a. The same duality exists between the truth tables of Figs. 7.21 and 7.22, where H's and L's are interchange between physical truth tables and 1's and 0's are interchanged between mixed-logic truth tables. The output expression for the OAI gate is obtained directly from Fig. 7.22c by reading the 0's

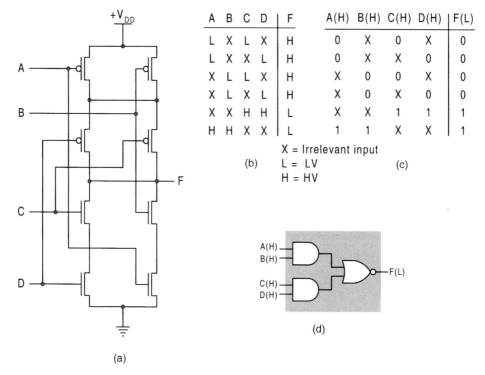

FIGURE 7.21
The AND–OR–invert (AOI) gate. (a) CMOS circuit. (b) Physical truth table. (c) Mixed-logic interpretation of the physical truth table. (d) Logic equivalent circuit for the AOI gate.

in the F column and complementing the results to give

$$F(L) = [(A + B)(C + D)](L), \tag{7.4}$$

which yields the logic equivalent circuit for the OAI gate shown in Fig. 7.22d.

As a simple example of the use of the OAI gate, consider the two-level active low XOR function given by

$$F(L) = (A + B)(\bar{A} + \bar{B})(L).$$

This function can be implemented in Fig. 7.22d by connecting A to C and B to D via two inverters for a total of 12 transistors, twice the transistor count of the XOR gate of Fig. 3.26a.

There are several variations on the AOI and OAI gate themes of Figs. 7.21 and 7.22. For example, adding an inverter to the output of each of these gates makes them AND–OR and OR–AND gates. More input stages can also be added to the AOI or OAI gates. Shown in Fig. 7.23 is the logic equivalent circuit for the 74x54, a 10-input AOI circuit. Clearly, the additional input capability adds to the versatility of the device.

7.7 INTRODUCTION TO FPGAs AND OTHER GENERAL-PURPOSE DEVICES

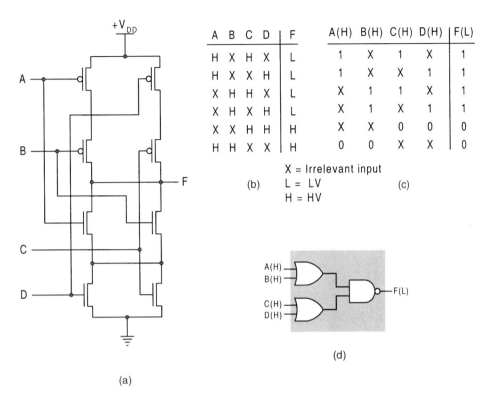

FIGURE 7.22
The OR–AND–invert (OAI) gate. (a) CMOS circuit. (b) Physical truth table. (c) Mixed-logic interpretation of the physical truth table. (d) Logic equivalent circuit for the OAI gate.

7.7.2 Actel Field Programmable Gate Arrays

As indicated previously, the architectures of FPGAs can be very complex and, in fact, are generally regarded as being at the high end in complexity of the programmable gate array (PGA) spectrum. The ACT-1 family of FPGAs (from Actel Corp.) discussed in this

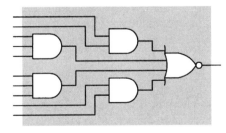

FIGURE 7.23
Logic equivalent circuit for the 10-input 74x54 AOI circuit.

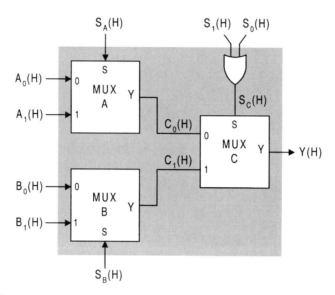

FIGURE 7.24
An ACT-1 family logic module. Source: *ACT Family Field Programmable Gate Array Databook*, Actel Corp., Sunnyvale, CA, 1991.

subsection are programmable by the end user. They represent some of the simpler PGAs in the sense that the logic components of these devices consist of relatively simple combinational structures. However, the programmable switching matrices required to interconnect the ACT-1 logic cells are relatively complex and cannot be reprogrammed.

Shown in Fig. 7.24 is an ACT-1 logic module consisting of three 2-to-1 MUXs and a 2-input OR gate for a total of eight inputs and one output. The output logic function is easily deduced from Fig. 7.24 to be

$$Y = (\overline{S_1 + S_0})C_0 + (S_1 + S_0)C_1$$
$$= (\overline{S_1 + S_0})(\bar{S}_A A_0 + S_A A_1) + (S_1 + S_0)(\bar{S}_B B_0 + S_B B_1). \qquad (7.5)$$

To achieve the status of an FPGA, Actel enmeshes hundreds of these modules in a matrix of programmable interconnections. A segment of the ACT-1 interconnect architecture is illustrated in Fig. 7.25, courtesy of the Actel Corporation. Here, dedicated vertical tracking lines connect with each input to and output from a logic module, while other vertical tracking lines function as feedthrough between channels. Connections between horizontal and vertical tracking lines are made by "blowing" the cross fuses at intersections in accordance with the required fuse pattern program. Since some wire tracks may be blocked by previously allocated tracks, a sequence of jogs from horizontal to vertical tracks and vice versa can be used to circumvent the blockage, thereby permitting connection to an appropriate logic module. Because of the versatility jog programming provides, the ACT-1 family of FPGAs achieves a gate-equivalence capability several times the number of logic modules. For example, the ACT-1 A1010, which has 295 logic modules, purports to have a gate equivalency of 1200.

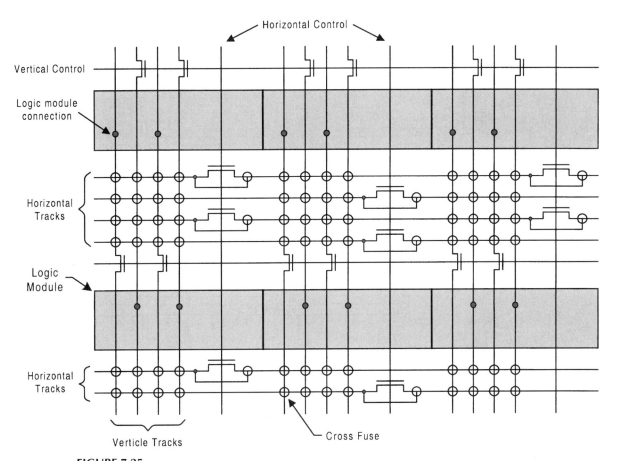

FIGURE 7.25
Interconnect architecture for ACT-1 family of FPGAs. Source: *ACT Family Field Programmable Gate Array Databook,* Actel Corp., Sunnyvale, CA, 1991.

7.7.3 Xilinx FPGAs

Field programmable gate arrays (FPGAs) offer the most design versatility of the PLDs considered so far, and their architecture differs markedly from those of PALs and Actel PLDs. The Xilinx FPGA family of IC devices consists of an array of configurable logic blocks (CLBs), I/O blocks (IOBs), and a switching interconnect matrix, as illustrated in Fig. 7.26. The low-end FPGAs support three kinds of interconnects: direct, general-purpose, and long-line interconnections. The *direct interconnects* (not shown in Fig. 7.26) connect CLBs to adjacent CLBs for localized applications with minimum propagation delay. The *general-purpose interconnects* connect CLBs to other CLBs via the horizontal and vertical interconnect lines and switching matrices. Finally, the *long-line interconnects* are reserved for signals that must be distributed to numerous CLBs and/or IOBs with minimum time–delay distribution (time skew) problems. The programming of sophisticated devices such as Xilinx FPGAs requires the use of dedicated software such as XACT by Xilinx, Inc. FPGAs

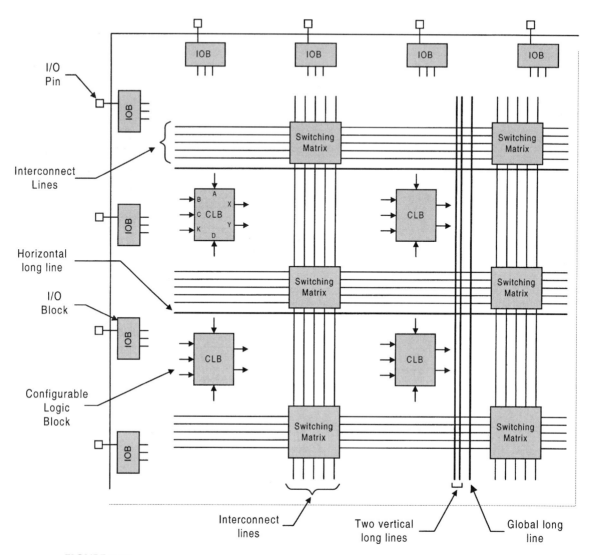

FIGURE 7.26
Segment of a Xilinx 2000 series logic cell architecture showing configurable logic blocks (CLBs), I/O blocks (IOBs), switching matrices, and vertical and horizontal interconnect lines. Source: *The Programmable Logic Data Book,* 1994. Xilinx Inc., San Jose, CA 95124.

by other manufacturers, which differ significantly from those of Xilinx, require their own dedicated software.

Though Xilinx no longer produces the series 2000 FPGAs, a general description of these devices is useful to introduce the subject of FPGAs. Shown in Fig. 7.27 is the Xilinx series 2000 FPGA logic cell, which consists of a combinational logic section, six MUXs, and a memory element (flip-flop). (Note: The reader is not expected to have a knowledge of flip-flops, which are covered at length in Chapter 10.) The combinational logic section can generate any single function of four variables, $F = G$, or any two functions, each of

7.7 INTRODUCTION TO FPGAs AND OTHER GENERAL-PURPOSE DEVICES

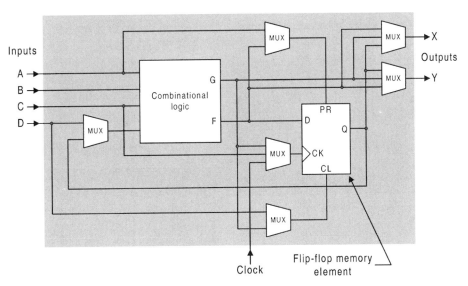

FIGURE 7.27
Logic structure for a Xilinx series 2000 logic cell block (LCB) with four data inputs, a clock input, and two outputs. Figure courtesy of Xilinx Inc., San Jose, CA.

three or fewer variables with separate F and G outputs. The functions are generated by a 16-bit programmable lookup table based on static RAM technology, thereby making reprogramming possible.

The simplest Xilinx IOB belongs to the 2000 series and is one consisting of two MUXs, a tri-state driver/enable, a noninverting buffer, and a flip-flop, as shown in Fig. 7.28. The I/O pin can function as either a dedicated input or a dedicated output, or it can be switched dynamically between the two. When the tri-state driver is disabled (disconnect mode), the I/O pin performs as a dedicated input. In this mode the 2-to-1 MUX selects either a buffered input from the flip-flop or one directly from the buffered input and delivers that input to the logic cell. When the tri-state driver is enabled (transfer mode), the I/O pin functions as a dedicated output from the logic cell. However, in this mode the 2-to-1 MUX can select the logic cell output via the buffer and return it to the logic cell as feedback, a bi-directional I/O condition.

The size and capability of the Xilinx FPGAs vary dramatically depending on the series and family to which the FPGA belongs. Shown in Table 7.1 are a few specifications for representative FPGAs of three families belonging to the XC4000 series. They range in CLB numbers from 100 for the XC4003E to 8464 for the XC40250XV. The XC40250XV features a gate range up to 500,000 (including RAM), has more than 100,000,000 transistors, and can operate at over 100 MHz. In comparison, the Pentium II microprocessor has 7,500,000 transistors but can operate at higher speeds.

Presented in Fig. 7.29 is the simplified architecture for the Xilinx XC4000 family of CLBs. Each CLB contains three function generators and two independent memory elements (flip-flops) that are triggered on either the rising edge or falling edge of the clock signal, depending on the logic level from the 2-to-1 MUXs. Multiplexers in the CLB map the four control inputs $C1, C2, C3,$ and $C4$ into the internal control signals $H1, Din, S/R,$ and CK_{EN} in any required manner. Combinational logic can be extracted directly from the three

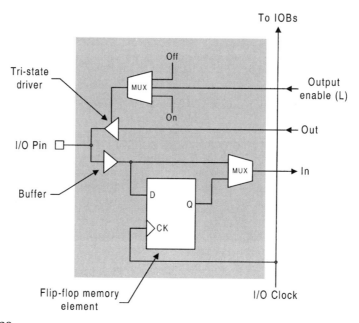

FIGURE 7.28
Logic structure for a relatively simple Xilinx I/O block (IOB). Figure courtesy of Xilinx Inc., San Jose, CA.

function generators at the X and Y outputs via the G', F' MUXs. By this means the CLB can implement any two functions of up to four variables, any function of five variables, or even some functions of up to nine variables. As an added advantage of the XC4000 series CLBs, function generators F' and G' contain dedicated arithmetic logic to increase the performance of the basic arithmetic and comparator operations.

The flip-flops generate outputs QX and QY from programmable data inputs via the 4-to-1 MUXs. The S/R control logic either presets ($PR = 1$, $CL = 0$) or clears ($CL = 1$, $PR = 0$) the flip-flops, depending on the logic level of the S/R input. A clock enable CK_{EN} input to the flip-flops permits the CLB to hold (store) data for an indefinite period of time. Again, it should be understood that the reader need not have a knowledge of flip-flops to obtain useful information from this subsection. The subject of flip-flops is covered at length in Chapter 10.

Table 7.1 Range of Xilinx FPGAs Belonging to the XC4000 CMOS Series

Technology (supply/scale factor — maximum frequency)	Product name	Maximum logic gates (no RAM)	Maximum RAM bits (no logic)	CLB matrix array	Number of flip-flops	Maximum user I/O
5 V/0.5 μ–66 MHz	XC4003E	3 K	3.2 K	$10 \times 10 = 100$	360	80
3.3 V/0.3 μ–80 MHz	XC4085XL	85 K	100 K	$56 \times 56 = 3{,}136$	7,168	448
2.5 V/0.25 μ–100 MHz	XC40250XV	250 K	271 K	$92 \times 92 = 8{,}464$	18,400	448

7.7 INTRODUCTION TO FPGAs AND OTHER GENERAL-PURPOSE DEVICES 325

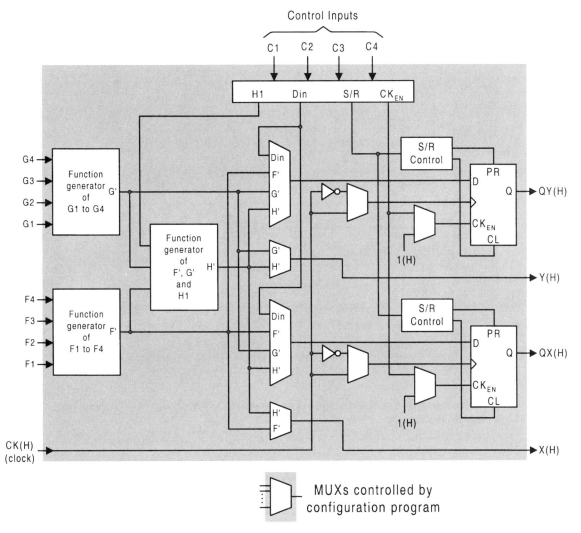

FIGURE 7.29
Simplified block diagram for the Xilinx XC4000 family of CLBs. Source: *The Programmable Logic Data Book*, Xilinx, Inc., 1994.

As expected, the I/O blocks (IOBs) for the XC4000 families of FPGAs are user-configurable, relatively complex, and versatile, just as are the CLBs. Shown in Fig. 7.30 is the simplified block diagram representing this family of IOBs. The buffered input signals from the I/O pin either are issued directly to the CLB (see inputs labeled I_1 and I_2), or are issued to the CLB from the flip-flop output Q after an optional delay, all determined by 2-to-1 MUXs. The delay can be used to eliminate the need for a data hold-time requirement at the external pin.

The output signals from the CLB can be inverted or not inverted and can either pass directly to the I/O pad or be stored in the memory element (flip-flop). The output enable OE signal acts on the tri-state driver/buffer to produce either a bidirectional I/O capability or a

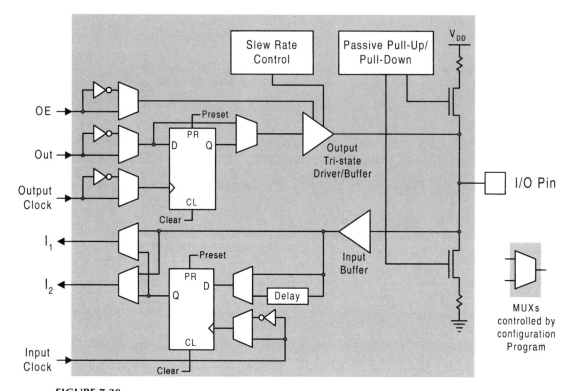

FIGURE 7.30
Simplified block diagram for the XC4000 series input/output block (IOB). Source: *The Programmable Logic Data Book,* Xilinx, Inc., 1994.

monodirectional input capability by imposing a high-impedance condition on the tri-state driver. The slew-rate control can also act on the tri-state driver, but does so to minimize the power consumption from bus transients when switching noncritical signals.

There are many other features and programmable options of the IOB. These include programmable pull-up and pull-down resistors that connect unused I/O pins (via NMOS) to either V_{DD} or ground to minimize power consumption. Separate input and output clock signals can be inverted or not inverted to produce rising- or falling-edge triggering of the flip-flops. Also, the flip-flops can be globally preset or cleared as is the case for the CLBs. Other characteristics of the IOBs, as well as those of the CLBs, are best understood by consulting Xilinx's *Programmable Logic Data Book* (see Further Reading at the end of this chapter).

The matrix of programmable interconnects for the Xilinx XC4000 families of FPGAs are significantly different from and more complex than those of the discontinued XC2000 series. For the XC4000 series there are three main types of interconnects: *single-length lines*, *double-length lines*, and *longlines*. A typical routing scheme for CLB connections to adjacent single-length lines is illustrated in Fig. 7.31. The switch matrix consists of six programmable transmission gates at the intersection of each single-length line as shown in the figure. The transmission gate configuration permits a given line signal to be routed

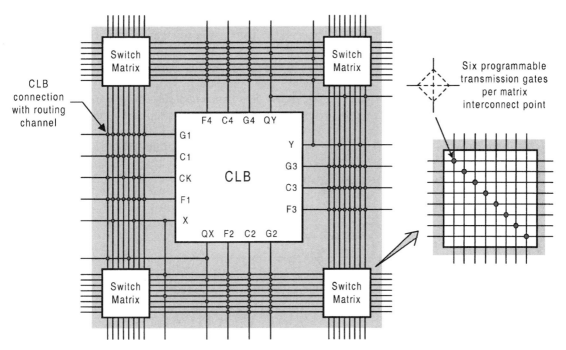

FIGURE 7.31
Interconnect scheme between CLB and routing channels consisting of single-length lines. Also shown are the six programmable transmission gate positions per interconnect point within each switch matrix. Source: *The Programmable Logic Data Book,* Xilinx, Inc., 1994.

in any one of three directions, in one of two transverse directions or on the same line, or along any combination of these. The single-length lines are used primarily for local network branching with fan-out greater than 1. The routing channels are designed to minimize path delay and power consumption, and their number between switch matrices varies with the size of the matrix.

Double-length lines (not shown) are those that interconnect every other switch matrix and are, therefore, twice as long as the single-length lines shown in Fig. 7.31. The double-length lines, which pass by two CLBs before entering a switch matrix, are grouped in pairs and offer the most efficient implementation of point-to-point interconnections of intermediate length. Any CLB input, except CK, can be driven from adjacent double-length lines, and the CLB outputs can be routed to nearby double-length lines in either horizontal and vertical directions.

The longlines (also not shown in Fig. 7.31) run the length of the array in both vertical and horizontal directions. They are designed to distribute signals of various types throughout the array with minimum time delay (skew) problems. Networks with high fan-out and time-critical signals are ideally suited for longline usage. A programmable switch located at the center of each intersecting longline permits a signal to be split into two independent signals, both traveling half the vertical or horizontal length of the array. Inputs to the CLB can be driven by adjacent longlines, but outputs from the CLB can be routed to adjacent longlines only via tri-state drivers. Programmable interconnect points between longlines and single-length lines are available. However, no interconnect points are provided between double-length lines and others.

Logic cell arrays (LCAs), such as the XC4000 series FPGAs, have the unique property that they can be reconfigured within a system and can even be dynamically altered to perform different functions at different times in a process. This family of devices contain 16×1 and 32×1 static RAM primitives that are user-configurable via look-up tables. Self-diagnosis, hardware for different environments, or dual-purpose applications — these exemplify the versatility that is inherent in reconfigurable LCAs. Properly used, such devices can minimize design effort and reduce costs. However, all of the above are possible only with the use of CAD help, the topic of Section 7.8. For much more complete information on Xilinx FPGAs, the reader is referred to Further Reading at the end of this chapter.

7.7.4 Other Classes of General-Purpose PLDs

To provide the reader with an overall perspective on the diversity of general-purpose PLDs and for reference purposes, the following additional classes of PLDs are offered as an extension of the devices discussed in the previous two subsections:

Generic Array Logic (GAL) Devices: Erasable MSI PLDs that may contain AOIs, XOR gates, and registers in the output stage for sequential machine design. GAL is a registered trademark of Lattice Semiconductor, Hillsboro, OR 97124.

Erasable programmable logic devices (EPLDs): Erasable CMOS-based devices whose macrocells typically contain discrete gates, MUXs, and registers (flip-flops) for sequential machine design. Some EPLDs may contain arithmetic logic units (ALUs). Both Altera and Xilinx Corporation offer EPLDs of various complexity. Detailed information on EPLDs are available from the *Altera Data Book*, Altera Corporation, and from *The Programmable Logic Data Book*, Xilinx Corporation, 1994.

Field programmable logic sequencers (FPLS): Similar to PAL and GAL devices that have output logic consisting of discrete gates and flip-flops. Detailed information on these devices is available from Phillips, *Programmable Logic Devices (PLD) Handbook*, Phillips Semiconductor, Sunnyvale, CA, 1994; and *Programmable Logic Data Handbook*, Signetics Co., Sunnyvale, CA, 1992.

7.8 CAD HELP IN PROGRAMMING PLD DEVICES

The programming of a PAL IC device, like that of a ROM or PLA, can be accomplished by a physical device (a programmer) that applies voltage pulses to target fusible links, causing disconnection of the bit positions as discussed in Section 7.2. The programmer may be a dedicated device or one of universal capability for ROMs, PLAs, and PAL devices, and it may even have erasure capability. In any case, the difficult part of the programming process is providing the instructions required to fit the Boolean expressions into the capability of the specific PLD device, for example, a PAL device, that may support an L-, R-, or V-type

7.8 CAD HELP IN PROGRAMMING PLD DEVICES

macrocell structure. This difficulty is overcome by proprietary software from manufacturers of PLD devices and PLD device programmers. A commonly used CAD package for this purpose is called ABEL (for Advanced Boolean Expression Language, a product of DATA I/O Corp., Redmond, WA). The ABEL compiler accepts I/O data in tabular form, minimizes it by using algorithms based on Espresso (discussed in Section 4.8), and generates a *fuse map* that can be read in one of several standard formats by the programming device: JEDEC (Joint Electron Device Engineering Council) format, ASCII hex format, Intel Hex format, etc. As the name implies, the fuse map (pattern) targets those fusible links that must be disconnected.

ABEL is not the only PLD programming compiler that is currently in use for PAL devices. A CAD software called PALASM (for PAL assembler) is commonly used to convert SOP Boolean expressions or truth tables into fuse maps for PAL devices. I/O pin specifications on the chip are required. Like ABEL, the PALASM compiler generates a fuse map in JEDEC format that can be read by the PAL programming device. PALASM is available without cost from AMD Inc. (Advanced Micro Devices, Inc.).

For Xilinx FPGA devices, dedicated and sophisticated CAE (computer-aided engineering) software called XACT (for Xilinx automated CAE tools) is available from the Xilinx Corp. XACT uses a three-step interrelated and iterative design process: *design entry, design implementation*, and *design verification*. For design entry at the gate level, the designer may begin with a schematic capture and labeling of the circuit to be implemented. To do this the designer can choose Viewlogic's Viewdraw, Mentor Graphics' Design Architect, or OrCAD's SDT, since these are all supported by the XACT development system. Design entry at the behavioral level (for example, Boolean equations or state machine descriptions) is best accomplished by the Xilinx-ABEL and X-BLOX products of Xilinx and other products by CAE vendors. Also, for complex systems, the use of high-level hardware description languages such as VHDL is recommended. Xilinx offers HDL interfaces for synthesis tools from synopsis and Viewlogic Systems. Mentor Graphics, Cadence Design Systems, and Exemplar Logic also offer HDL synthesis tools fashioned for Xilinx FPGAs.

One advantage of the Xilinx design environment is that the designers can combine schematic, text, gate-level, and behavioral-level descriptions at the design entry stage and then reuse such descriptions within the same system or in other systems at some future time. This is called *mix-mode design entry* and can be carried out by using different design entry tools, allowing the designer to choose the most effective and efficient design tool for each portion of the design.

Following the design entry comes the design implementation. Here, the FPGA design entry tools are mapped into the resources of the target device architecture, optimally selecting the routing channels that connect the CLBs and IOBs of the logic cell array. Although this can be accomplished automatically, the designer can and should exert control over the implementation process to minimize potential problems. For this purpose the Xilinx design environment provides an interactive, graphics-based editor that permits user manipulation of the logic and routing schemes for the FPGA device.

The design verification is the last part of the design process and necessarily follows the implementation stage. In-circuit testing, simulation at various levels, and static timing analysis are involved in the verification stage. This is accomplished by use of timing calculators, back-annotation capability, and static timing analyzers, which are available from Xilinx Corp. and various third-party vendors.

FURTHER READING

Any recent text will have some coverage of the basic PLDs: ROMs, PLAs, and PAL devices. However, there are a few texts that appear to cover these subjects better than most. The text by Nelson, Nagle, Carroll, and Irwin and that by Yarbrough appear to cover the basics rather well and extend their coverage into the industrial level. Other important books that deal strictly with digital design with PLDs are those of Pellerin and Holley, Lala, Bolton, and Carter. The text by Tinder appears to be the only one dealing with mixed logic inputs to and outputs from PLDs. For multiple PLD schemes for augmenting input and output capability, the text by Tinder and that by Ercegovac and Lang are recommended. Data handbooks are often a necessary source for detailed current information, and none is better than that for PAL devices by Advanced Micro Devices.

[1] M. Bolton, *Digital Systems Design with Programmable Logic*. Addison-Wesley, Reading, MA, 1990.
[2] J. W. Carter, *Digital Designing with Programmable Logic Devices*. Prentice Hall, Upper Saddle River, NJ, 1997.
[3] M. D. Ercegovac and T. Lang, *Digital Systems and Hardware/Firmware Algorithms*. John Wiley, New York, 1985.
[4] P. K. Lala, *Digital System Design Using Programmable Logic Devices*. Prentice Hall, Englewood Cliffs, NJ, 1990.
[5] V. P. Nelson, H. T. Nagle, B. D. Carroll, and J. D. Irwin, *Digital Logic Circuit Analysis and Design*. Prentice Hall, Englewood Cliffs, NJ, 1995.
[6] D. Pellerin and M. Holley, *Practical Design Using Programmable Logic*. Prentice Hall, Englewood Cliffs, NJ, 1991.
[7] *PAL Device Data Book*. Advanced Micro Devices, Inc., Sunnyvale, CA, 1992.
[8] R. F. Tinder, *Digital Engineering Design: A Modern Approach*. Prentice Hall, Englewood Cliffs, NJ, 1991.
[9] J. M. Yarbrough, *Digital Logic: Applications and Design*. West Puglishing Co., Minneapolis/St. Paul, 1997.

The subject of FPGAs is covered adequately by several recent texts and more extensively by the device manufacturers of these devices. For introductory material on FPGAs, the texts by Katz and Yarbrough (previously cited) are adequate. However, for current detailed information, no sources are better than the recent data books from Xilinx, Actel, and Altera. For the most recent information on the Xilinx CX4000XV family of FPGAs, the world wide web is perhaps the best source. GAL devices are covered by Lattice Semiconductor's data book. For EPLD component specifications and applications, the reader will find Intel's data book useful.

[10] *ACT Family Field Programmable Gate Array Databook*. Actel Corp., Sunnyvale, CA, 1991.
[11] *Altera Data Book*. Altera Corp., San Jose, CA, 1995.
[12] *GAL Data Book*. Lattice Semiconductor, Hillsboro, OR, 1992.
[13] http://www.xilinx.com/spot/virtexspot.htm
[14] R. H. Katz, *Contempory Logic Design*. The Benjamin/Commings Publishing Co., Inc., Redwood City, CA, 1994.
[15] *Programmable Gate Array Data Book*. Xilinx, Inc., San Jose, CA, 1995.
[16] *Programmable Logic Data Book*. Intel Corp., Santa Clara, CA, 1994.

[17] *The Programmable Logic Data Book*. Xilinx, Inc., San Jose, CA, 1996.
[18] *XACT, Logic Cell Array Macro Library*. Xilinx, Inc., San Jose, CA, 1992.

PROBLEMS

7.1 A $2^4 \times 4$ ROM is to be used to implement the following system of three functions, assuming that all inputs and outputs are active high.

$$y_1(a, b, c, d) = \sum m(0, 1, 2, 5, 7, 8, 10, 14, 15)$$
$$y_2(a, b, c, d) = \sum m(0, 2, 4, 5, 6, 7, 8, 10, 12)$$
$$y_3(a, b, c, d) = \sum m(0, 1, 2, 3, 4, 6, 8, 9, 10, 11)$$

(a) Construct the ROM program table for this system of three functions.

(b) From the program table of part (a), construct the symbolic representation of fusible links by following the example in Fig. 7.4.

7.2 A $2^4 \times 4$ ROM is to be used to design and implement a BCD-to-binary module that can be cascaded to produce any size of a BCD-to-binary converter (e.g., see Fig. 6.21).

(a) Construct the ROM program table for this module following Fig. 6.20.

(b) From the program table in part (a), construct a symbolic representation of the fusible links by referring to the example in Fig. 7.4.

7.3 A $2^4 \times 4$ PROM is to be used to implement the following system of three functions:

$$F_1(A, B, C) = \prod M(0, 2, 4)$$
$$F_2(A, B, C, D) = \prod M(3\text{–}12)$$
$$F_3(A, B, C) = \prod M(2, 3, 5, 7)$$

(a) Construct the program table for this ROM if the inputs arrive as $A(H)$, $B(H)$, $C(L)$, and $D(H)$, and the outputs are $F_1(H)$, $F_2(L)$, and $F_3(H)$. Use of inverters is *not* permitted.

(b) From the program table in part (a), construct a symbolic representation of the fusible links following the example in Fig. 7.4.

7.4 A multiplexed scheme of $2^6 \times 4$ EPROMs is to be used to design and implement a circuit that will convert 8-bit one-hot code (see Table 2.11) to 4-bit binary. Assume that all false data are rejected and indicated by binary 1000.

(a) In place of a ROM program table, give the coded canonical SOP forms for each binary output.

(b) Illustrate with a block diagram how the outputs of the multiplexed scheme must be wired-ORed to produce the four binary outputs of the converter. (Refer to Section 7.6 and Figs. 7.14 and 7.15 for a discussion of multiplexed schemes and wired-ORed connections.)

7.5 Design an XS3-to-Gray code converter by using a $4 \times 8 \times 4$ FPLA. Assume that all inputs and outputs are active high and that false data are *not* rejected.
   (a) Construct the minimized p-term table for this converter.
   (b) From the p-term table in part (a), construct the symbolic representation for the fusible links following the example in Fig. 7.10.

7.6 The following three functions are to be implemented by using a $4 \times 8 \times 4$ FPLA.

$$F_1 = A\bar{B} + \bar{A}B\bar{C} + \bar{B}C + AC$$
$$F_2 = A \oplus (\bar{B}C)$$
$$F_3(A, B, C) = \prod M(1, 3, 6)$$

   (a) Construct the minimized p-term table for the three functions. Assume that the inputs arrive as $A(H)$, $B(H)$, and $C(L)$ and that the outputs are issued as $F_1(H)$, $F_2(L)$, and $F_3(H)$.
   (b) From the p-term table in part (a), construct the symbolic representation for the fusible links following the example in Fig. 7.10. An inverter is permitted on the active low input.

7.7 A BCD-to-XS3 code converter is to be designed by using a $4 \times 12 \times 4$ PAL.
   (a) Construct the minimized p-term table for this converter. Assume that the inputs and outputs are all active low, and that all false data are encoded as 0000. Keep in mind that PALs cannot take advantage of shared PIs as can PLAs.
   (b) From the p-term table of part (a), construct the symbolic representation for the fusible links following the example in Fig. 7.11. Inverters may be used on the inputs.

7.8 A cascadable 2-bit comparator is to be designed by using a PAL.
   (a) Given the compressed truth table and EV K-maps in Fig. 6.28, find the minimum SOP logic expressions for the three outputs and construct the minimum p-term table from these expressions. Assume that all inputs and outputs are active high.
   (b) From the p-term table in part (a), determine the minimum size PAL that can be used and then construct the symbolic representation of the fusible links for this comparator. Include tri-state enables.

7.9 The three functions in Problem 7.1 are to be designed by using a $4 \times 16 \times 4$ PAL.
   (a) Construct the minimized p-term table for these three functions keeping in mind that a PAL cannot take advantage of shared PIs as can PLAs. Assume that the inputs arrive as $a(L)$, $b(L)$, $c(H)$, and $d(H)$, and that the outputs must be issued as $y_1(H)$, $y_2(L)$, and $y_3(H)$. Note that inverters are *not* permitted on the inputs.
   (b) From the program table of part (a), construct the symbolic representation of fusible links by following a form similar to that of Fig. 7.11, but with tri-state enables on the outputs.

7.10 The Actel (ACT-1) logic module, shown in Fig. 7.24, is embedded by the hundreds in Actel's FPGAs. This module is remarkably versatile in its ability to implement a large number of simple SOP functions active high, or POS functions active low. As examples, implement each of the following logic functions by using a single ACT-1 module assuming all inputs are active high: [Hint: Use Fig. 7.24, not Eq. (7.5), and plan to include 0's and 1's as inputs where needed.]

(a) $Y(H) = (\bar{A} + \bar{B})(H) = A \cdot B(L)$ A two-input NAND gate
(b) $Y(H) = \bar{A}\bar{B}C(H) = (A + B + \bar{C})(L)$ A p-term (or s-term)
(c) $Y(H) = (\bar{A}B + A\bar{B})(H) = [(\bar{A} + \bar{B})(A + B)](L)$ Defining relations for XOR
(d) $Y(A, B, C)(H) = \Sigma m(2, 3, 5, 7)(H)$ Canonical SOP function

7.11 The AOI and OAI gates in Figs. 7.21, 7.22, and 7.23 are versatile building blocks that can be used in the implementation of a variety of logic functions. Simple examples are given in Subsection 7.7.1. With a minimum of external logic, apply these AOI and OAI gates in creative ways to implement the following functions:

(a) Use one AOI gate (nothing else) to implement the expression for $(A = B)(H)$ given by Eq. (6.22).

(b) Use a minimum number of AOI gates to implement the three-function system in Problem 7.6 with input and output activation levels as stated. (Hint: For F_1, use Fig. 7.23)

(c) Use a minimum number of OAI gates to implement the three-function system in Problem 7.3 with input and output activation levels as stated. (Hint: For F_2, use the dual of Fig. 7.23.)

CHAPTER 8

Arithmetic Devices and Arithmetic Logic Units (ALUs)

8.1 INTRODUCTION

In this chapter digital circuits will be designed with electrical capabilities that can be interpreted as performing the basic arithmetic operations of binary numbers. The basic operations include

- Addition
- Subtraction
- Multiplication
- Division

Now, Boolean equations are uniquely defined so as to perform specific arithmetic operations, and the 1's and 0's, which have previously been used only as logic levels, must take on a numerical significance. The reader must keep in mind that an arithmetic circuit is only the electrical analog of the arithmetic operation it represents. In fact, it is the interpretation of the electrical circuit's behavior that bridges the gap between physical and logic domains.

The treatment of arithmetic circuits presented in this text is not intended to be a treatise on the subject. Rather, the intent is to introduce the subjects at both the beginning and intermediate-to-advanced levels utilizing, where necessary, appropriate algorithms for the basic operations in question. The subjects of arithmetic logic units (ALUs) and the application of dual-rail methods, which are covered in later sections, fall within the intermediate-to-advanced level of coverage and may not be expected to be part of a first-level course in digital design.

8.2 BINARY ADDERS

Because of the nature of binary and the requirements for arithmetic manipulation, approaches to basic adder design vary considerably depending on the form the manipulation takes. There are ripple-carry adders, carry-save, carry select, and carry-look-ahead adders, the last three being classified as "high-speed" adders. This list of four does not nearly cover the scope of adder design, nor does it establish the general character of these arithmetic devices. It does introduce the concept of computational speed as it relates to the addition process, however that is characterized.

8.2.1 The Half Adder

The *half adder* (HA) is the simplest of all the arithmetic circuits and may be regarded as the smallest building block for modular design of arithmetic circuits. The HA consists of two inputs, A and B, and two outputs, sum S and carry C, as indicated by the logic circuit symbol in Fig. 8.1a. The operation format and truth table for the HA are given in Fig. 8.1b and 8.1c. Here, A plus B yields the result CS, where carry is the MSB and sum is the LSB. When mapped as in Fig. 8.1d, the results for sum and carry are read directly as

$$\begin{Bmatrix} S = A \oplus B \\ C = A \cdot B \end{Bmatrix}. \tag{8.1}$$

The logic circuit for the HA is implemented by using Eqs. (8.1) and is given in Fig. 8.1e. Here, the choice is made to use the XOR gate in the implementation of the HA. However, use could have been made of the two- or three-level XOR realizations given in Fig. 3.28, or the transmission gate approach by using AOI or OAI gates as in Fig. 7.21 and 7.22. The XOR gate, given in Fig. 3.26, is the simplest CMOS design possible.

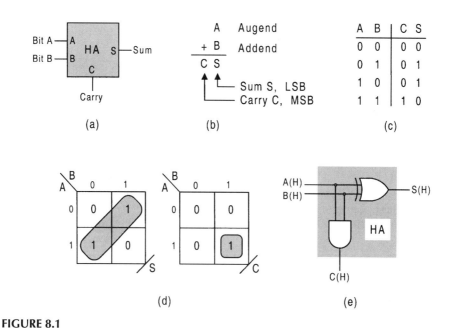

FIGURE 8.1
Design of the half adder (HA). (a) Block diagram for the HA. (b) Operation format. (c) Truth table for A plus B, showing carry C, and sum S. (d) K-maps for sum and carry. (e) Logic circuit for the HA.

8.2 BINARY ADDERS

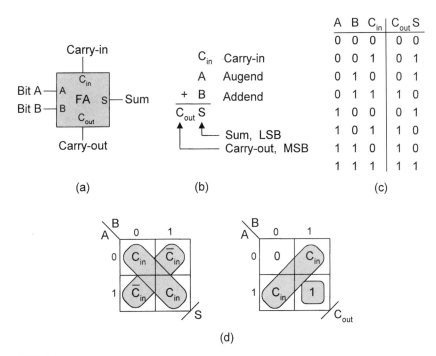

FIGURE 8.2
Design of the full adder (FA). (a) Block diagram for the FA. (b) Operation format for the FA. (c) Truth table for A plus B plus C_{in}, showing carry-out C_{out}, and sum S. (d) EV K-maps for sum and carry-out.

8.2.2 The Full Adder

The half adder (HA) just designed has severe limitations in modular design applications because it cannot accept a carry-in from the previous stage. Thus, the HA cannot be used for multiple bit addition operations. The limitations of the HA are easily overcome by using the *full adder* (FA) presented in Fig. 8.2. The FA features three inputs, A, B, and carry-in C_{in}, and two outputs, sum S and carry-out C_{out}, as indicated by the logic symbol and operation format in Figs. 8.2a and 8.2b. The truth table for A plus B plus C_{in} is given in Fig. 8.2c with outputs C_{out} and S indicated in the two columns on the right. As in the case of the HA, the inputs are given in ascending binary order. EV K-maps for sum and carry-out in Fig. 8.2d, showing diagonal XOR patterns, are plotted directly from the truth table and give the results

$$\left\{ \begin{aligned} S &= C_{in}(\overline{A \oplus B}) + \bar{C}_{in}(A \oplus B) \\ &= A \oplus B \oplus C_{in} \\ C_{out} &= C_{in}(A \oplus B) + AB \end{aligned} \right\}. \qquad (8.2)$$

Here, multioutput optimization has been used to the extent that the term $(A \oplus B)$ is used by both the S and C_{out} expressions. Recall that $A \oplus B \oplus C_{in} = (A \oplus B) \oplus C_{in}$.

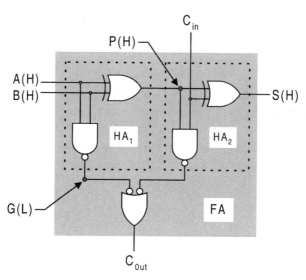

FIGURE 8.3
Logic circuit for the full adder (FA) showing it to be composed of two HAs and an OR gate.

The output expressions in Eqs. (8.2) are used to implement the FA shown in Fig. 8.3. As indicated, the FA is composed of two HAs and a NAND gate. The nodes $P(H)$ and $G(L)$ are indicated for later reference in Section 8.4. Notice that the NAND logic in Fig. 8.3 can be replaced by AND/OR logic but, for reasons to be discussed in Section 8.5, are kept as NAND.

8.2.3 Ripple-Carry Adders

The full adder (FA) can now be used to design a *ripple-carry adder*, sometimes called a *pseudoparallel adder* or simply *parallel adder*. An n-bit ripple-carry (R-C) adder is a $(2n+1)$-input/$(n+1)$-output combinational logic device that can add two n-bit binary numbers. The block diagram symbol and general operation format for this adder are presented in Figs. 8.4a and 8.4b, together with an illustration of the ripple-carry effect in Fig. 8.4c. The general operation format represents the familiar addition algorithm used in conventional arithmetic where carry from an addition operation is always to the next most significant stage. Notice that the subscripts are consistent with the powers of 2 to the left of the radix point in polynomial notion as is discussed in Section 2.3. Thus, the bits of each word representing a number are written in ascending order of positional weight from right to left. The addition process follows Algorithm 2.8 given in Section 2.9. Actually, the position of the radix point in the two numbers is arbitrary, since the adder has no means of sensing these positions. The only requirement is that the user of this adder make certain the radix points "line up" just as in conventional arithmetic. Thus, if significant bit positions exist to the right of the radix point for augend A and addend B, meaning that these numbers have a fraction component, then there must be an equal number of such positions for the two numbers, each of n bits total.

The modular design of the n-bit ripple-carry adder follows directly from Fig. 8.4. All that is required is that a series array of n FAs designated $FA_0, FA_1, \ldots, FA_{n-1}$, one for

8.2 BINARY ADDERS

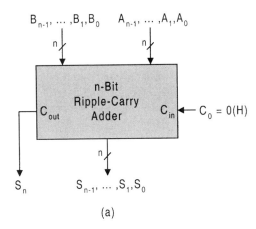

FIGURE 8.4
The n-bit ripple-carry adder. (a) Block diagram circuit symbol. (b) Operation format. (c) Ripple carry effect.

each bit, be connected such that the carry-out of one stage is the carry-in to the next most significant stage. This connection is shown in Fig. 8.5 where it is assumed that all inputs arrive active high. Notice that the condition of no carry to the initial stage (FA_0) is satisfied by $C_0 = 0(H)$ (ground) or by using a HA for this stage.

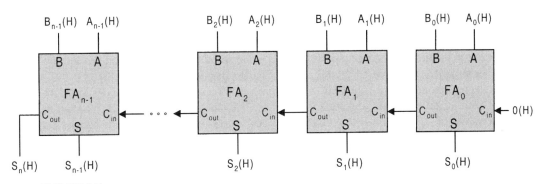

FIGURE 8.5
An n-bit ripple-carry adder implemented with n full adders.

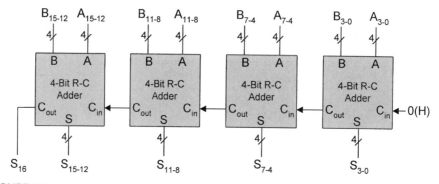

FIGURE 8.6
A 16-bit ripple-carry adder implemented with four 4-bit ripple-carry (R-C) adders.

An n-bit ripple-carry adder is more likely to be designed by using n m-bit adder modules rather than individual FAs. One example, represented in Fig. 8.6, features four 4-bit ripple-carry (R-C) adders in ripple-carry fashion to produce a 16-bit adder. Other examples include a 16-bit R-C adder composed of two 8-bit R-C adders, or a 32-bit implemented with four 8-bit R-C adders.

There are practical limitations associated with the operation of an R-C adder of the type presented in Figs. 8.5 and 8.6. To operate an R-C adder correctly, *all* input bits, for augend A and addend B, must be presented stably to the adders for a time equal to or greater than the time it takes for the carry signal to ripple from stage to stage across the entire adder. Thus, this ripple time determines the maximum frequency of input data presentation (to the adder stages) for a series of addition operations. Remember that an addition operation in an R-C adder is not complete until the carry signal passes through the last (MSB) stage of the adder. Discussions of ways to speed up the addition process will be considered later in Sections 8.4 and 8.5.

8.3 BINARY SUBTRACTORS

The electrical equivalent of the binary subtraction operation can be carried out by using full subtractors, just as the electrical equivalent of binary addition can be accomplished by using full adders. Although this is rarely done, the design of the *full subtractor* provides a good introduction to the more useful subject of adder/subtractors (devices that can serve in either capacity) and to the subject of parallel dividers considered in Section 8.7.

The design of the full subtractor (FS) follows in a manner similar to the design of the FA with, of course, some important differences. The inputs now become the minuend A, subtrahend B, and borrow-in B_{in}, and the outputs become the difference D and borrow-out B_{out}, which are shown by the block symbol in Fig. 8.7a. The operation format for the FS and truth table for $A - (B$ plus $B_{in})$ are given in Figs. 8.7b and 8.7c. Notice that $B_{out} = 1$ any time that $(B$ plus $B_{in}) > A$, and that the difference $D = 1$ any time the three inputs exhibit odd parity (odd number of 1's)—the same as for sum S in Fig. 8.2c.

8.3 BINARY SUBTRACTORS

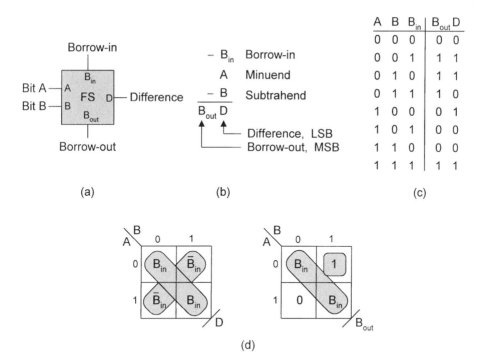

FIGURE 8.7
Design of the full subtractor (FS). (a) Block diagram for the FS. (b) Operation format. (c) Truth table for $A - (B \text{ plus } B_{in})$, showing outputs borrow-out B_{out} and difference D. (d) EV K-maps for difference and borrow-out.

The EV K-maps for difference and borrow-out are plotted directly from the truth table in Fig. 8.7c and are given in Fig. 8.7d, where again XOR diagonal patterns exist. The outputs, as read from the K-maps, are

$$\begin{cases} D = A \oplus B \oplus B_{in} \\ B_{out} = B_{in}\overline{(A \oplus B)} + \bar{A}B \end{cases}, \quad (8.3)$$

where it is recalled from Eqs. (3.23) that $A \odot B = \overline{A \oplus B}$. The FS can now be implemented from Eqs. (8.3) with the results shown in Fig. 8.8. Observe that the FS consists of two half subtractors (HSs) and that the only difference between a HS and a HA is the presence of two inverters in the NAND portion of the FS circuit. Use will be made of this fact in the next section dealing with the subject of adder/subtractors, devices that can perform either addition or sign-complement subtraction.

Full subtractors can be cascaded in series such that the borrow-out of one stage is the borrow-in to the next most significant stage. When this is done, a *ripple-borrow subtractor* results, similar to the ripple-carry adders of Figs. 8.5 and 8.6. However, the ripple-borrow subtractor suffers from the same practical limitation as does the ripple-carry adder—namely, that the subtraction process is not complete until the borrow signal completes

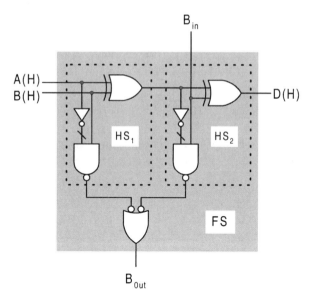

FIGURE 8.8
Logic circuit for the full subtractor (FS) showing it to be composed of two half subtractors (HSs) and a NAND gate.

its ripple path across the entire subtractor. As with the ripple-carry adder, the ripple-borrow subtractor is limited to applications not requiring high-speed calculations.

8.3.1 Adder/Subtractors

The adder/subtractor approach to arithmetic circuit design permits the use of one set of hardware to accomplish both addition- and subtraction-type operations. One means to accomplish this becomes evident from an inspection of the FA and FS circuits of Figs. 8.3 and 8.8. As was pointed out previously, the FA and FS differ only by two inverters in the AND/OR part of the circuit. If the inverters are replaced by a controlled inverters (XOR gates as in Fig. 3.31c), an adder/subtractor results. Thus, when the control input to the XOR gate is 1(H) the XOR gate functions as an inverter, creating an FS, or when the control input is 0(H) the device functions as an FA. When such FA/FS modules are connected in series as in Figs. 8.5 and 8.6, the result is a parallel adder/subtractor.

Another approach to adder/subtractor design makes use of 2's complement arithmetic as discussed in Subsection 2.9.2. The design of the adder/subtractor now follows by using the ripple-carry adder hardware of Fig. 8.5 together with XOR gates used as controlled inverters on the $B_i(H)$ inputs. The result is the n-bit adder/subtractor shown in Fig. 8.9. If the mode control input is set $\bar{A}/S(H) = 1(H)$, the operation is subtraction [A_i plus $(-B_i)$] in 2's complement and the final carry-out is discarded. However, if the add/subtract control is set $\bar{A}/S(H) = 0(H)$, the operation is addition and the final carry-out becomes the most significant sum bit S_n. For subtraction the final sum bit, S_{n-1}, is the sign bit, which can be positive (if 0) or negative (if 1) depending on the outcome. Notice that for subtraction a 1(H) is introduced into FA_0 as the initial carry-in C_0, a requirement of Eq. (2.14). Also, note

8.3 BINARY SUBTRACTORS   343

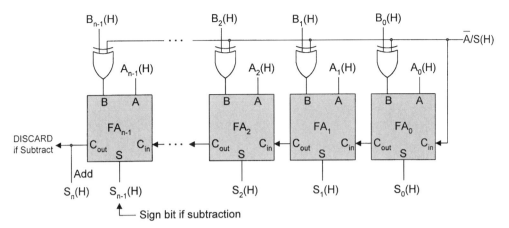

FIGURE 8.9
An *n*-bit adder/subtractor with ripple/carry implemented with full adders and XOR gates.

that if only subtraction is required, the XOR gates can be replaced by inverters. Remember, if the subtraction results in a negative number, that number is in 2's complement (see Algorithm 2.9).

8.3.2 Sign-Bit Error Detection

In Subsection 2.6.2 it is stated that the decimal range of representation for *n* integer bits in 2's complement is

$$-(r^{n-1}) \leq N_{10} \leq (r^{n-1} - 1) \quad \text{for } r = 2.$$

Table 2.7 illustrates this fact with a few examples in 8-bit binary, where the sign bit is the MSB. What this means with regard to 2's complement arithmetic is as follows: If any two positive numbers sum to a decimal value $N_{10} > (2^{n-1} - 1)$, the sign bit will be in error (thus negative). Similarly, if any two negative numbers sum to a decimal value $N_{10} < -(2^{n-1})$, the sign bit will again be in error (thus positive). If number *A* is positive and number *B* is negative, the sign bit will always be correct, assuming that both numbers are properly represented such that S_{n-1} is the sign bit for each. Now, with reference to Fig. 8.9, two negative numbers cannot be properly represented in 2's complement, so that leaves two positive numbers as the only possibility for sign-bit error in this adder/subtractor or in any ripple-carry adder. Two examples of 8-bit addition of positive numbers are illustrated in Fig. 8.10. In Fig. 8.10a the two numbers sum to decimal 128, resulting in a sign-bit overflow error because $128 > (2^{8-1} - 1) = 127$. The sign bit "1" indicates a negative number, which is an error. In Fig. 8.10b, the two numbers sum to a decimal value of 127, which is within the acceptable limit, and no sign-bit overflow error occurs.

It is useful to be able to detect a sign-bit overflow error in a ripple-carry adder or adder/subtractor so that corrective steps can take place. An inspection of Fig. 8.10 and the truth table for an FA in Fig. 8.2c indicates that a sign-bit overflow error can occur in the

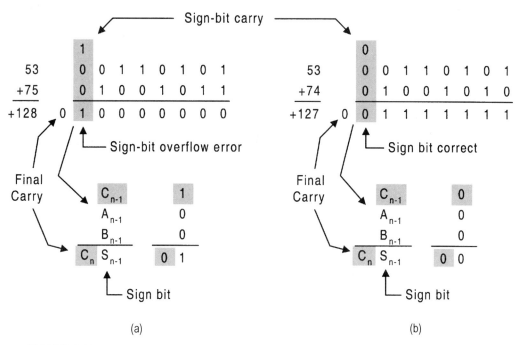

FIGURE 8.10
Eight-bit additions showing sign bits and final two carries. (a) Sign-bit overflow error created for sum exceeding decimal 127. (b) Correct sign bit for sum equal to 127.

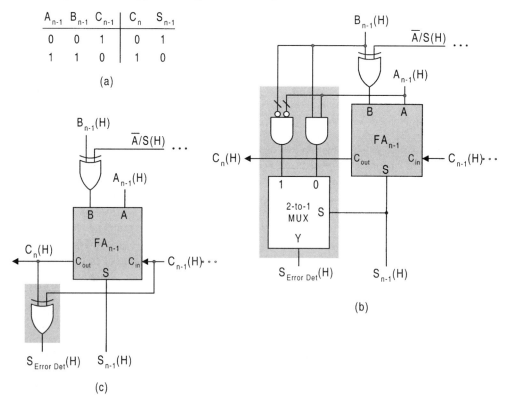

FIGURE 8.11
Overflow error detection circuits for sign-bit stage. (a) Truth table for overflow error in the sign-bit stage. (b) Overflow error detection circuit by using external data inputs and the sum output. (c) Alternative overflow error detection circuit by using input and output carries.

sum of two positive numbers only if $C_n \neq C_{n-1}$. Thus, a sign-bit overflow error detector can be implemented by

$$S_{Error\ Det} = C_n \oplus C_{n-1}, \qquad (8.4)$$

requiring that the sign-bit carry-in C_{n-1} be accessible, which it may not be for IC chips. Another approach permits a detector to be designed that depends only on the external inputs to and sum bit from the sign-bit stage. A further inspection of truth table for an FA in Fig. 8.2c indicates that a sign-bit overflow error can occur only if $A = B$ when $C_{out} \neq S$ for the $(n-1)$th stage. Shown in Fig. 8.11a is the truth table for the sign-bit overflow error conditions based on this fact. From this truth table there results the expression

$$S_{Error\ Det} = \bar{S}_{n-1}(A_{n-1}B_{n-1}) + S_{n-1}(\bar{A}_{n-1}\bar{B}_{n-1}), \qquad (8.5)$$

which permits the use of the 2-to-1 MUX implementation shown in Fig. 8.11b. For purposes of comparison, the implementation of Eq. (8.4) is given in Fig. 8.11c.

8.4 THE CARRY LOOK-AHEAD ADDER

The ripple-carry (R-C) adder discussed in Section 8.3 is satisfactory for most applications up to 16 bits at moderate speeds. Where larger numbers of bits must be added together at high speeds, fast adder configurations must be used. One clever, if not also intuitive, design makes use of the modular approach while reducing the propagation time of the R-C effect. The approach has become known as the *carry look-ahead (CLA) adder*. In effect, the CLA adder "anticipates" the need for a carry and then generates and propagates it more directly than does a standard R-C adder.

The design of the CLA adder begins with a generalization of Eqs. (8.2). For the ith stage of the ripple-carry adder, the sum and carry-out expressions are

$$\begin{cases} S_i = A_i \oplus B_i \oplus C_i \\ \quad = \text{Sum of the } i\text{th stage} \\ C_{i+1} = C_i(A_i \oplus B_i) + A_i B_i \\ \quad = \text{Carry-out of the } i\text{th stage} \end{cases} \qquad (8.6)$$

From the expression for C_{i+1} it is concluded that $C_{i+1} = 1$ is assured if $(A_i \oplus B_i) = 1$ and $C_i = 1$, or if $A_i B_i = 1$.

Next, it is desirable to expand Eqs. (8.6) for each of n stages, beginning with the 1st (first) stage. To accomplish this, it is convenient to define two quantities for the ith stage,

$$\begin{cases} P_i = A_i \oplus B_i = \text{Carry Propagate} \\ G_i = A_i \cdot B_i = \text{Carry Generate} \end{cases}, \qquad (8.7)$$

which are shown in Fig. 8.3 to be the intermediate functions $P(H)$ and $G(L)$ in the full adder. Introducing these equation into an expansion of Eqs. (8.6) gives

$$
\begin{aligned}
&\text{1st stage} && \left\{\begin{aligned} S_0 &= P_0 \oplus C_0 \\ C_1 &= P_0 C_0 + G_0 \end{aligned}\right\} \\
&\text{2nd stage} && \left\{\begin{aligned} S_1 &= P_1 \oplus C_1 \\ C_2 &= P_1 C_1 + G_1 \\ &= P_1 P_0 C_0 + P_1 G_0 + G_1 \end{aligned}\right\} \\
&\text{3rd stage} && \left\{\begin{aligned} S_2 &= P_2 \oplus C_2 \\ C_3 &= P_2 C_2 + G_2 \\ &= P_2 P_1 P_0 C_0 + P_2 P_1 G_0 + P_2 G_1 + G_2 \end{aligned}\right\} \\
& && \qquad\vdots \\
&n\text{th stage} && \left\{\begin{aligned} S_n &= P_n \oplus C_n \\ C_{n+1} &= P_n C_n + G_n \\ &= P_n P_{n-1} P_{n-2} \cdots P_0 C_0 + P_n P_{n-1} P_{n-2} \cdots \\ & \quad P_1 G_0 + \cdots + G_n \end{aligned}\right\}
\end{aligned} \qquad (8.8)
$$

To implement Eqs. (8.8), use is made of the carry look-ahead module shown in Fig. 8.12, which is deduced from Fig. 8.3 and Eqs. (8.7). Notice that the CLA module has but one half adder but has two additional outputs, carry generate G and carry propagate P. Whereas the S_i terms are produced within the each FA, the C_{i+1} terms must be formed externally by what is called the *carry generate/propagate (CGP) network*. It is the CGP network to which $G(L)$ and $P(H)$ must be connected in accordance with Eqs. (8.8).

Construction of an n-bit CLA adder requires the use of n CLA modules together with the CGP network according to Eqs. (8.8). This is done in Fig. 8.13 for the three least significant

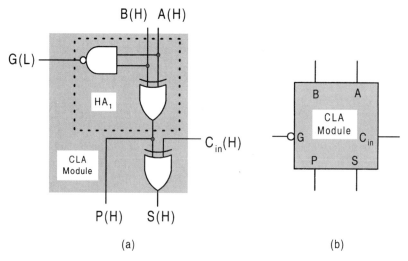

FIGURE 8.12
The carry-look-ahead (CLA) module. (a) Logic circuit deduced from Fig. 8.3 and Eqs. (8.7). (b) Logic circuit symbol.

8.4 THE CARRY LOOK-AHEAD ADDER

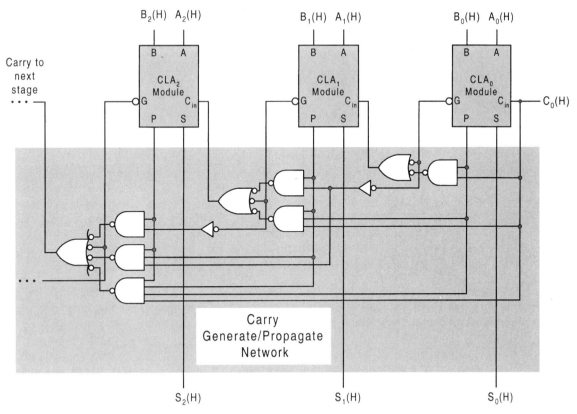

FIGURE 8.13
The three least significant stages of an n-bit carry look-ahead (CLA) adder showing the carry generate/propagate network.

bit stages of an n-bit CLA adder. Each of the CLA modules are of the type given in Fig. 8.12 and the shaded portion below the CLAs in Fig. 8.13 represents the CGP network as required by Eqs. (8.8). Notice that for $C_0(H) = 0(H)$, a requirement for proper operation of the adder, one NAND gate in each stage is disabled. This does not mean that such disabled NAND gates should be eliminated from the CGP network. All NAND gates shown in the CGP network become necessary if m-bit CLA adder modules are to be cascaded to form m n-bit CLA adders. It is also interesting to note that the CLA_0 module together with its CGP logic is exactly the same as an FA in Fig. 8.3. Furthermore, if the extra hardware is of no concern, all CLA modules can be replaced by FAs with G and P outputs.

An inspection of Fig. 8.13 makes it clear that the hardware requirement for the CGP network increases significantly as the number of stages increases. The gate count for the CLA adder in Fig. 8.13 is 5, 11, and 18 for the three stages 0, 1, and 2 as shown. A fourth stage would have a total gate count of 26 and a fifth stage would be 35, etc. In fact, the total number of gates required for an n-bit CLA is given by

$$\text{Total gate count} = \underbrace{(n^2 + 3n)/2}_{CGP\ network} + \underbrace{3n}_{CLA\ modules} = (n^2 + 9n)/2 \quad (8.9)$$

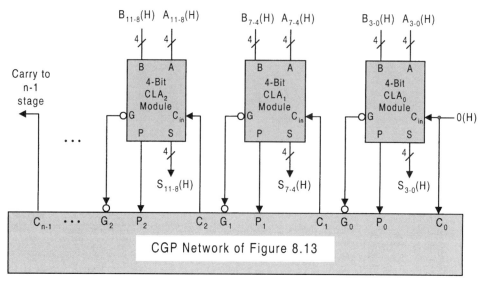

FIGURE 8.14
The three least significant stages of a carry look-ahead (CLA) adder showing the carry generate/propagate network between 4-bit modules.

This count is to be compared to $5n$ for a ripple-carry (R-C) adder of the type shown in Fig. 8.5. For example, an 8-bit CLA adder requires 68 gates, compared to 40 for the R-C adder, but the CLA adder is more than twice as fast. Furthermore, as the number of stages increases, so does the maximum fan-in requirement. For the CLA adder, the maximum fan-in is $(n + 1)$, compared to 2 for the R-C adder. The possibility of additional hardware to compensate for fan-in and fan-out limitations of the CLA adder must be weighed against the fact that the maximum propagation delay for the n-bit CLA adder of Fig. 8.13 is 4 gate delays, compared to $2n$ for the R-C adder.

The modular approach to CLA adder design can be taken one step further by using the CGP network for n-bit CLA adder modules to create a larger adder. To do this, one simply cascades m n-bit CLA modules in series to create an adder of $k = m \times n$ bits in size. This is demonstrated in Fig. 8.14, where 4-bit CLA adder modules are cascaded by using the CGP network shown in Fig. 8.13. The adder in Fig. 8.14 is called a *group CLA adder*. This group CLA adder configuration saves hardware, but at the expense of increased propagation delay. For example, a 16-bit CLA adder with 1-bit CLA modules requires 200 gates with 4 gate-levels of delay, 2 for generation of G and P and 2 for carry generation. In comparison, a 16-bit group CLA adder with 4-bit CLA modules requires 118 gates but has 6 gate-levels of delay (2 extra gate delays for carry). Note that the comparisons just made do not take into account any additional gates required to deal with fan-in and fan-out limitations.

The foregoing discussion suggests that there is a practical limit to the number of CLA adder stages that can be used without creating excessive amounts of hardware with the accompanying fan-in and fan-out restrictions and path delay problems. To overcome this limitation the concept of the group CLA adder configuration can again be used, but in a rather different way. For example, each 4-bit CLA stage in Fig. 8.14 can be replaced by 4-bit R-C FA stages or by 4-bit adder/subtractors stages as in Fig. 8.9 such that the MSB module of each stage is equipped with G and P outputs or is a CLA module of the type in

8.5 MULTIPLE-NUMBER ADDITION AND THE CARRY-SAVE ADDER   349

Fig. 8.12. In fact, the nature of the adder stage to be cascaded in CLA form is immaterial as long as the MSB module of each stage has G and P outputs to be connected to the CGP network. In any case combining R-C and CLA technologies reduces hardware requirements but, of course, does so at the expense of increasing path delay.

8.5 MULTIPLE-NUMBER ADDITION AND THE CARRY-SAVE ADDER

To add more than two binary numbers by the "conventional" methods discussed in Sections 8.3 and 8.4 requires that the addition process be carried out in stages, meaning that k operands are added in $k-1$ two-operand additions, each addition (following the first) being that of the accumulated sum with a operand. This staged addition operation is relatively slow since carries are propagated in each stage. However, addition of binary numbers is not limited to two number addition stages. Actually, many binary numbers can be added together in ways that reduces the total addition time. One way to accomplish this is to use a type of "fast" adder called an *iterative carry-save (ICS) adder*. The algorithm for iterative CS addition of more than two operands is stated as follows:

Algorithm 8.1: *Sum $S = A + B + C + D + E \cdots$ by Carry-Save Method*

(1) Set integers $A = A_{n-1}A_{n-2}\cdots A_1 A_0$, $B = B_{n-1}B_{n-2}\cdots B_1 B_0$, $C = C_{n-1}C_{n-2}\cdots C_1 C_0$, etc.
(2) Sum $S^0 = A + B + C = S^0_{n-1}S^0_{n-2}\cdots S^0_1 S^0_0$ exclusive of all carries-out C^1_o.
(3) Sum $S^1 = S^0 + D + C^1_o = S^1_{n-1}S^1_{n-2}\cdots S^1_1 S^1_0$ exclusive of all carries-out C^2_o, but with carries C^1_o shifted left by one bit.
(4) Sum $S^2 = S^1 + E + C^2_o = S^2_{n-1}S^2_{n-2}\cdots S^2_1 S^2_0$ exclusive of all carries C^3_o but with carries C^2_o shifted left by one bit.
(5) Continue process until last integer has been added and only two resultant operands remain: pseudosum S' and final CS carries C'.
(6) End with final sum $S = S' + C' = S_n S_{n-1}\cdots S_1 S_0$ by using either R-C or CLA addition.

The carry-save process described in Algorithm 8.1 applies to m integer operands and involves $m-2$ CS additions resulting in two operands: a pseudosum S' and the final CS carries C'. The final step (step 6 in Algorithm 8.1) adds the two operands S' and C' by using either a R-C adder or a CLA adder. The CS process (steps 1 through 5 in Algorithm 8.1) avoids the ripple-carry problem of R-C adders, resulting in a savings of addition time. For R-C adders of the type in Figs. 8.5 and 8.6, a time $(m-1)t_{R-C}$ would be required to add m integer operands. However, for addition of m integer operands by the CS method, a time $(m-2)t_{CS} + t_{R-C}$ is required, which represents a significant savings of time for $m > 3$, since $t_{CS} < t_{R-C}$ for a given number of addition operations. If CLA addition is involved, the time t_{R-C} can be replaced by t_{CLA} with less savings in time between the two methods.

The iterative CS process just described in Algorithm 8.1 is illustrated in Fig. 8.15a where four 4-bit integers, A, B, C, and D, are added. The ICS adder required for this process is shown in Fig. 8.15b, where use is made of FAs, a HA, and either a R-C adder or a CLA adder for the final sum operation.

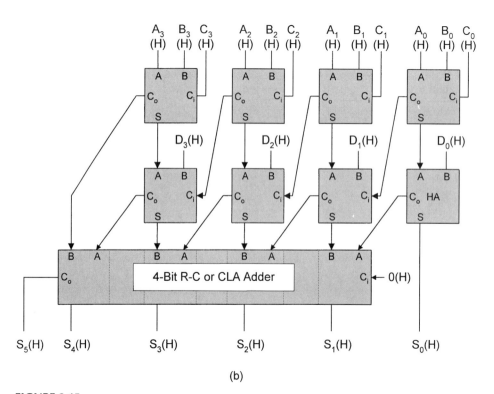

FIGURE 8.15
The carry-save (CS) method of addition. (a) CS addition of four 4-bit numbers. (b) The CS adder designed to add four 4-bit operands by using FAs a HA and a 4-bit R-C or CLA adder.

8.6 MULTIPLIERS

An $n \times m$ multiplier is an $(n + m)$–input/$(n + m)$–output device that performs the logic equivalent of an n-bit×m-bit binary multiplication. Shown in Fig. 8.16a is the block circuit symbol, and in Fig. 8.16b the operation format for a 4-bit × 4-bit multiplier that follows Algorithm 2.10, given in Subsection 2.9.4. As indicated in the operation format the

8.6 MULTIPLIERS

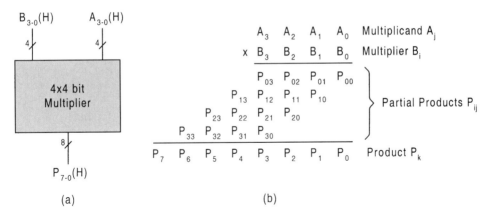

FIGURE 8.16
Characteristics of the 4×4 bit multiplier. (a) Block circuit symbol. (b) Operation format showing partial products and final product bits.

multiplication process consists of four multiplicand bits A_j, four multiplier bits B_i, partial product bits P_{ij}, and the final eight product bits P_k.

The 4 × 4 bit multiplication process, represented in Fig. 8.16b and by Example 2.27 in Subsection 2.9.4, can be implemented at the modular level by using HAs and FAs to perform the summation between the partial products and carries. Note that the partial products are generated by AND operations $P_{ij} = B_i \cdot A_j$, and that the product bits in each column of Fig. 8.16 are the sums of these partial products and the carries between stages as indicated by the XOR operations in Fig. 8.17.

Some explanation of Fig. 8.17 may be helpful. Recalling the sum expression for an FA in Eq. (8.2), it follows that k FAs can accommodate $2k$ XOR operators or $2k + 1$ variables in an XOR string such as that of Eq. (5.16). For $2k - 1$ XOR operators in the string, use can be made of one HA. Note that the C_{ij} symbols in Fig. 8.17 refer to caries from the ith to the jth sum stage.

The product bits P_i indicated in Fig. 8.17 can be computed by first generating the partial products followed by a summation of these partial products with the appropriate carries. This is done in Fig. 8.18, where six of the eight sum stages are shown separated by dotted

$$P_0 = P_{00}$$
$$P_1 = P_{01} \oplus P_{10}$$
$$P_2 = P_{02} \oplus P_{11} \oplus P_{20} \oplus C_{12}$$
$$P_3 = P_{03} \oplus P_{12} \oplus P_{21} \oplus P_{30} \oplus [C_{23} \oplus C_{23}']$$
$$P_4 = P_{13} \oplus P_{22} \oplus P_{31} \oplus [C_{34} \oplus C_{34}' \oplus C_{34}'']$$
$$P_5 = P_{23} \oplus P_{32} \oplus [C_{45} \oplus C_{45}' \oplus C_{45}'']$$
$$P_6 = P_{33} \oplus [C_{56} \oplus C_{56}']$$
$$P_7 = C_{67}$$

FIGURE 8.17
The summations of partial products and carries required to produce the product bits for the multiplication of two 4-bit operands shown in Fig. 8.16b.

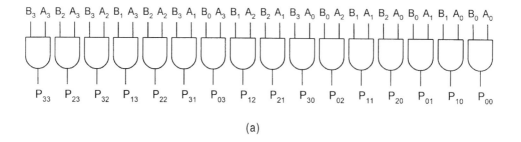

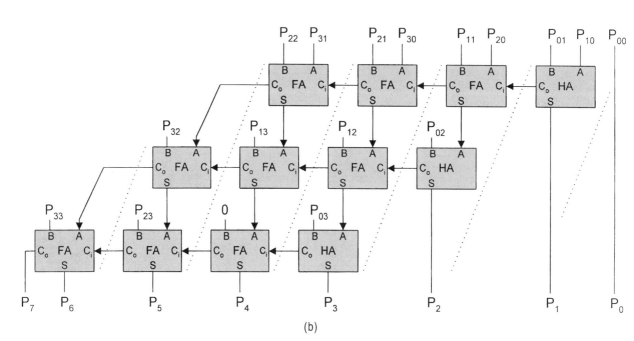

FIGURE 8.18
Implementation of the 4×4 bit multiplier of Fig. 8.17. (a) Generation of the partial products. (b) Use of FAs and HAs to sum the partial products and carries.

lines, the final stage being the carry $C_{67} = P_7$ and the initial stage being $P_{00} = P_0$. Although the string of XOR operations in each product expression of Fig. 8.17 can be computed in any arbitrary order, the carry configurations of Fig. 8.18 are so chosen as to make use of standard 4-bit R-C adders with HAs as initial stages. Now it is possible to replace the 4-bit R-C adders with 4-bit CLA adders at a significant savings in calculation time. Note that the number of carries C_{ij} for each stage in Fig. 8.17 agrees with those in Fig. 8.18.

The multiplication process can also be carried out by using the carry-save (CS) method of summing the partial products and carries in a manner similar to that of Fig. 8.15. However, in this case the operands must be summed by the iterative CS method expressed by Algorithm 8.1. Such a CS scheme, shown in Fig. 8.19, is a type of *array multiplier* called an *iterative CS multiplier*. Here, a 4-bit CLA adder is used for the final sum of the two operands, S^1 and C_0^2, as in Fig. 8.15.

8.7 PARALLEL DIVIDERS

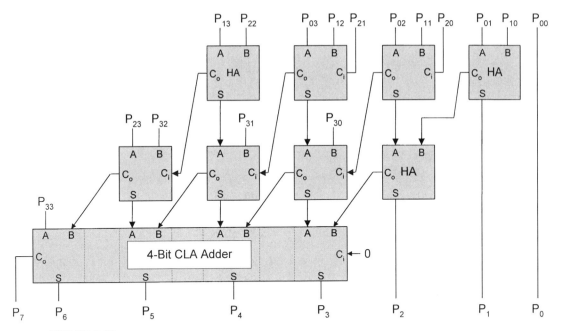

FIGURE 8.19
The iterative carry-save (ICS) method with FAs, HAs, and a 4-bit CLA adder used to multiply two 4-bit operands. Partial products P_{ij} are generated as in Fig. 8.18.

The iterative CS multiplier of Fig. 8.19 has the advantage of reduced computation time compared to the R-C approach illustrated in Fig. 8.18. For n rows and $2n - 1$ columns of partial products, the summation time for an iterative CS (ICS) multiplier is

$$t_{ICS} = (n-2)t_{FA} + t_{R-C}, \qquad (8.10)$$

where t_{FA} is the delay of a FA and t_{R-C} is the time required for the a R-C adder to complete the summation. In comparison, the summation time required for a multiplier that uses only R-C adders (as in Fig. 8.18), is

$$t_{RCA} = 2(n-1)t_{FA} + t_{R-C} \qquad (8.11)$$

Thus for large n, the iterative CS multiplier is about twice as fast as the one that uses only R-C adders. If CLA adders are used for both types of multipliers, the difference in speed between the two is reduced and may even shrink to near zero for certain values of n.

8.7 PARALLEL DIVIDERS

An $n \div m$ parallel divider is an $(n + m)$-bit/variable-bit output device that performs the electrical equivalent of the binary operation symbolized by $A \div B = Q$ with remainder R. As used here, A and B are the dividend and divisor operands, respectively, and Q is

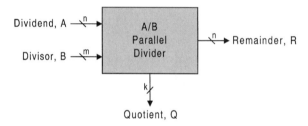

FIGURE 8.20
Block diagram symbol for an *n/m* parallel divider.

the quotient. The block diagram symbol for this divider is given in Fig. 8.20, where it is understood that $m \leq n$ for the binary integers of these operands.

The details of the logic circuitry for a divider depend on the algorithm used to execute the division operation. Recall that in Subsection 2.9.5, Algorithm 2.12 presented a division procedure that is close to the familiar *subtract-and-shift* procedure used in decimal long division. It is this same procedure that is used to design a parallel divider, but modified in the following way to accommodate hardware application:

1. Successively subtract the subtrahend *B* from the minuend *A* by starting from the MSB end of *A* and shifting 1 bit toward the LSB after each subtraction stage:
   (a) When the most significant (MS) borrow bit for the present stage is 0, the minuend for the next stage (remainder from the present stage) is the difference of the present stage.
   (b) When the MS borrow for the present stage is 1, the minuend for the next stage is the minuend of the present stage.
2. Complement the MS borrow bit for each stage and let it become the quotient bit for that stage.
3. Repeat steps 1 and 2 until the subtrahend *B* has been shifted to the LSB end of the minuend *A*. The final remainder *R* will be determined by the logic level of the MS borrow as in step 1a or 1b.

The procedure just described, a modification of Algorithm 2.12, is illustrated in Fig. 8.21a for a 5-bit dividend, $A = 10001$ and a 3-bit divisor $B = 011$. The result is $A \div B = Q$ with remainder *R*, where the 3-bit quotient is $Q = 101$ and the 5-bit remainder is $R = 00010$. Thus, in decimal $17 \div 3 = 5$ with remainder 2, which would be written as 17 and 2/3 or 17.66666.... Similarly, in Fig. 8.21b, $A = 11011_2 \, (27_{10})$, $B = 100_2 \, (4_{10})$ with the result $Q = 00110_2 \, (6_{10})$ and $R = 00011_2 \, (3_{10})$.

To design a parallel divider, the requirements of the subtract-and-shift process, illustrated in Fig. 8.21, must be met. First, the subtrahend must be successively subtracted from the minuend, and then shifted from the MSB end of the minuend toward its LSB end by one bit after each subtraction stage. This is easily accomplished by shifting the subtrahend presentation to an array of full subtractors (FSs). Second, the remainder *R* must be properly gated. Taking B_{out} to mean the MS borrow for a given stage, the division process requires that $R = D$ when $B_{out} = 0$ and $R = A$ for $B_{out} = 1$, where *D* and *A* are the difference and minuend, respectively, for the present stage. Shown in Fig. 8.22 are the truth table (a), EV K-map (b), and the subtractor module (c), together with its block symbol (d), that meet the

8.7 PARALLEL DIVIDERS

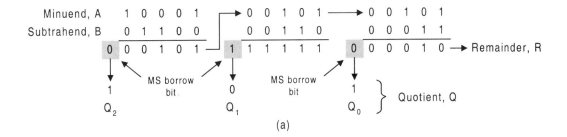

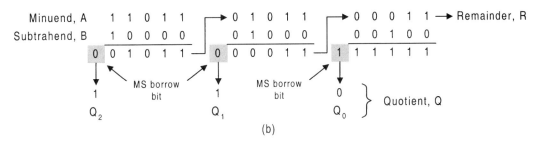

FIGURE 8.21
Two illustrations of the subtract-and-shift method of binary division for 5-bit dividends and 3-bit divisors. (a) The result is $Q = 00101\,(5_{10})$ with $R = 00010\,(2_{10})$ when $A = 10001\,(17_{10})$ and $B = 011\,(3_{10})$. (b) The result is $Q = 00110\,(6_{10})$ with $R = 00011\,(3_{10})$ when $A = 11011\,(27_{10})$ and $B = 100\,(4_{10})$.

remainder requirements just given. Notice that the expression $R = B_{out}A + \bar{B}_{out}D$ applies to a 2-to-1 MUX with inputs A and D and output R when B_{out} is the data select input.

All that remains is to construct an array of subtractor modules for the subtract-and-shift process required for the parallel divider. This is done in Fig. 8.23a for a dividend (minuend) of 5 bits and a divisor (subtrahend) of 3 bits. Here, the quotient outputs $Q_2(H)$, $Q_1(H)$, and $Q_0(H)$ are issued complemented from the active low outputs of inverter symbols defined in Fig. 7.6. The truth table for a full subtractor is given in Fig. 8.23b to assist the reader in analyzing this divider circuit.

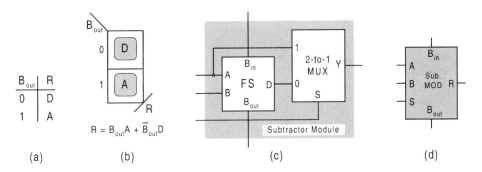

FIGURE 8.22
Design of the subtractor module for use in the design of a parallel divider. (a) Truth table for remainder R and final borrow-out. (b) EV K-map and gating logic for R. (c) The subtractor module by using a FS and a 2-to-1 MUX. (d) Circuit block symbol for the subtractor module in (c).

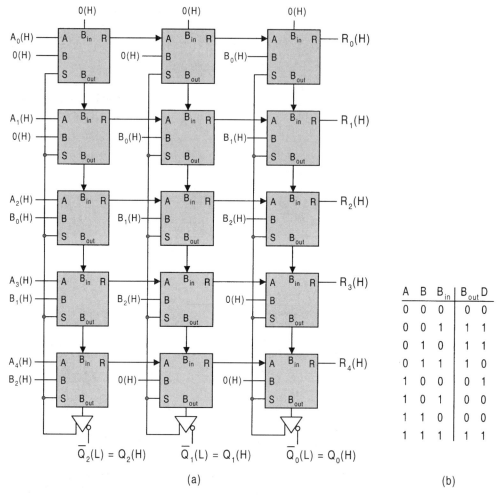

FIGURE 8.23
(a) Parallel divider for a 5-bit dividend, A, 3-bit divisor, B, a 3-bit quotient, Q, and a 5-bit remainder, R, designed with an array of 15 subtractor modules of the type shown in Fig. 8.22. (b) Truth table for a full subtractor.

The divider in Fig. 8.23 can be expanded to accommodate larger dividends and quotients by adding subtractor modules in both the Y- and X-directions, respectively. Referring to Fig. 8.20, the relationship between n, m, and k is given by

$$k = n - m + 1 \qquad (8.12)$$

for full usage of an $n \times k$ array of subtractor modules. For example, a 16-bit dividend ($n = 16$) and an 8-bit divisor ($m = 8$) can be used to generate a 9-bit quotient ($k = 9$) in a 16×9 array of subtractor modules. Or a 32-bit dividend and a 24-bit divisor can be used to generate a 9-bit quotient in a 32×9 array of subtractor modules. In all such cases the remainder is of n bits. It is also acceptable to use any lesser portion of a given array of subtractor modules to carry out a divide operation, but Eq. (8.12) must still apply to that

8.8 ARITHMETIC AND LOGIC UNITS 357

portion of the array that is used. However, to do this requires proper placement of 0's on modules that are not used. For example, if $m = n$, only one quotient bit is generated and 0's must be placed on all open inputs to the right of the MSB column of the array.

Dividers of the type shown in Fig. 8.23 can be classified as "fast" dividers. This is because they are asynchronous in nature, meaning that the results are generated as fast as the logic permits. However, it is characteristic of such circuits that with increasing operand size the hardware requirements increase rapidly making them impractical for many applications where space requirements are important. There are other methods of achieving the division operation with less hardware but at the expense of operation time, as expected. These methods require storage elements such as registers and fall outside the treatment of this text.

8.8 ARITHMETIC AND LOGIC UNITS

As the name implies, the *arithmetic and logic unit (ALU)* is a universal combinational logic device capable of performing both arithmetic and logic operations. It is this versatility that makes the ALU an attractive building block in the *central processing unit (CPU)* of a computer or microprocessor. It is the object of this section to develop the techniques required to design and cascade ALU devices following three very different approaches: the *dedicated ALU approach*, the *MUX approach*, and the *dual-rail approach* with completion signals.

The number and complexity of the operations that a given ALU can perform is a matter of the designer's choice and may vary widely from ALU to ALU, as will be demonstrated in this section. However, the choice of operations is usually drawn from the list in Fig. 8.24. Other possible operations include zero, unity, sign-complement, magnitude comparison, parity generation, multiplication, division, powers, and shifting. Multiplication, division, and related operations such as arithmetic shifting are complex and are found only in the most sophisticated ALU chips. Also, the AND, OR, and XOR operations are often applied to complemented and uncomplemented operands, making possible a wide assortment of such operations.

Presented in Fig. 8.25 is the block diagram symbol for a general n-bit slice ALU. This ALU accepts two n-bit input operands, $B_{n-1} \cdots B_1 B_0$ and $A_{n-1} \cdots A_1 A_0$, and a carry-in bit, C_{in}, and operates with them in some predetermined way to output an n-bit function, $F_{n-1} \cdots F_1 F_0$ and a carry-out bit, C_{out}. Here, the term *n-bit slice* indicates a partition of identical n-bit modules of stages that can be cascaded in parallel. Thus, an FA in an n-bit R-C adder could be called a 1-bit slice for that adder. Also, use of sign-complement arithmetic avoids the need for both carry and borrow parameters.

Arithmetic Operations	Logic Operations
Negation	Transfer
Increment	Complementation
Decrement	AND
Addition	OR
Subtraction	XOR (EQV)

FIGURE 8.24
Arithmetic and logic operations common to ALUs.

358 CHAPTER 8 / ARITHMETIC DEVICES AND ARITHMETIC LOGIC UNITS (ALUs)

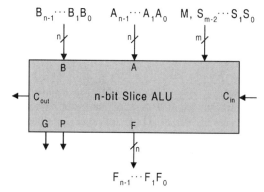

FIGURE 8.25
Block diagram symbol for a general 1-bit slice ALU with CLA capability.

The choice of operation between the two operands, A and B, is determined by the m mode/select inputs, $M, S_{m-2} \cdots S_1 S_0$, shown in Fig. 8.25. The mode input M sets the ALU for either an arithmetic or logic operation, and the function select inputs, $S_{m-2} \cdots S_1 S_0$, determine which particular operation, within the arithmetic or logic mode, is to be performed. Just as the carry-out bit C_{out} is required for cascading standard R-C arithmetic units, as in Fig. 8.5, the carry propagate and carry generate bits, P and G, are required for cascading carry look-ahead (CLA) units. In this section, ALUs with R-C and CLA capabilities are designed. Commercial ALU chips are available that have both of these features.

8.8.1 Dedicated ALU Design Featuring R-C and CLA Capability

The EV operation table in Fig. 8.26 represents a simple 1-bit slice ALU capable of performing four specific arithmetic functions and four specific logic functions, all on command of

	M	S_1	S_0	F	Operation*	C_{out}
Arithmetic Operations	0	0	0	$A \oplus C_{in}$	Transfer (LSB C_{in}= 0) or increment (LSB C_{in}= 1) of A	$A \cdot C_{in}$
	0	0	1	$\overline{A} \oplus C_{in}$	1's (LSB C_{in}= 0) or 2's (LSB C_{in}= 1) complement of A	$\overline{A} \cdot C_{in}$
	0	1	0	$A \oplus B \oplus C_{in}$	A plus B if LSB C_{in} = 0 or A plus B plus 1 if LSB C_{in} = 1	$C_{in}(A \oplus B) + A \cdot B$
	0	1	1	$\overline{A} \oplus B \oplus C_{in}$	B minus A if LSB C_{in} = 1 or $\overline{A}$ plus B if LSB C_{in} = 0	$C_{in}(A \oplus B) + \overline{A} \cdot B$
Logic Operations	1	0	0	A	Transfer A	0
	1	0	1	$\overline{A}$	Complement of A	0
	1	1	0	A + B	A OR B	0
	1	1	1	$\overline{A}$ + B	A complement OR A	0

* Subtraction operations assume 2's complement arithmetic.

FIGURE 8.26
Operation table for a simple 1-bit slice ALU showing output functions, F and C_{out}, for four arithmetic operations ($M = 0$) and four logic operations ($M = 1$).

8.8 ARITHMETIC AND LOGIC UNITS

the three mode/select inputs M, S_1, and S_0. On the right side of the table are the function expressions F, a brief description of each function operation, and carry-out expressions, C_{out}. The expressions for F and C_{out} are deduced from Eqs. (8.2) for the full adder (FA) together with Eqs. (3.23). Note that *false carry rejection* during the logic mode is realized by placing zeros in the C_{out} column for the four logic operations — the carry-out function has no relevance in a logic operation. Notice further that the two possible logic states for C_{in} lead to different interpretations for each of the four arithmetic operations. For example, $A \oplus C_{in}$ is the transfer of A if $C_{in} = 0$, but represents the increment of A if the LSB $C_{in} = 1$ ($B = 0$ is implied). Or, $\bar{A} \oplus C_{in}$ represents the 1's complement of A if LSB $C_{in} = 0$ but is the 2's complement of A if LSB $C_{in} = 1$ ($B = 1$ is implied). Subtraction operations by this ALU are carried out by 2's complement arithmetic as discussed in Subsection 2.9.2.

The dedicated ALU of Fig. 8.26 is now designed by using the EV K-map methods with XOR patterns as discussed in Section 5.2. Shown in Figs. 8.27a and 8.27b are the third-order EV K-maps for F and C_{out}, which are plotted directly from the operation table in Fig. 8.26 by using the mode/select inputs as the K-map axes. These third-order EV K-maps represent three orders of map compression because there are three EVs. By compressing the third-order K-maps by one additional order (hence now four orders of K-map compression), there results the second-order EV K-maps shown in Figs. 8.27c and 8.27d. Optimum cover is then extracted from these second-order K-maps by using associative XOR-type patterns, as indicated by the shaded loops. (See Section 5.2 for a discussion of associative patterns). From these K-maps there results

$$\begin{cases} F = (\bar{M}C_{in}) \oplus (A \oplus S_0) \oplus (S_1 B) + M S_1 B \\ C_{out} = (\bar{M}C_{in})[(A \oplus S_0) \oplus (S_1 B)] + \bar{M}(S_1 B)(A \oplus S_0) \end{cases}, \qquad (8.13)$$

which represent four-level logic with a collective gate/input tally of 10/22 excluding any inverters that may be required. Notice that several terms in Eqs. (8.13) are shared between

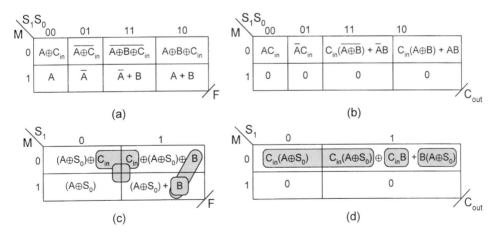

FIGURE 8.27
K-map representations for function F and carry-out C_{out} given in the operation table of Fig. 8.26 for a 1-bit slice ALU. (a), (b) Third-order EV K-maps plotted directly from Fig. 8.26 (c), (d) Second-order EV K-maps showing optimum cover for the two-output system.

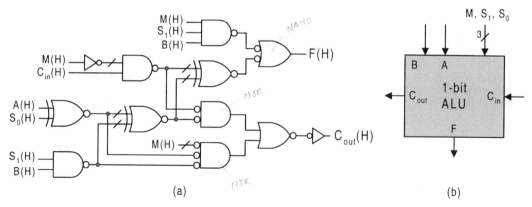

FIGURE 8.28

(a) Optimum gate-level implementation of the 1-bit slice ALU represented by the operation table of Fig. 8.26 and by Eqs. 8.13. (b) Block diagram symbol for the 1-bit ALU in (a).

the two functions, a result of the optimal organization of the operations in the table of Fig. 8.26.

Some explanation of Eqs. (8.13) is necessary. Referring to function F, the separate "island" loop for operand B in cell 3 of Fig. 8.27c is necessary to complete the cover for that cell. This is so because after $(A \oplus S_0)$ and B have been used to loop out the associative patterns, cover still remains in cell 3, as explained for a similar case in Section 5.4. The residual cover is extracted either by looping out operand B as an island to give the term MS_1B, or by looping out $(A \oplus S_0)$ in the M domain to give the term $M(A \oplus S_0)$. It is the former that is chosen for this design example.

Equations (8.13) can be implemented in a number of ways by using discrete logic. One means of accomplishing this is shown in Fig. 8.28, where use is made of NAND/NOR/EQV logic, assuming that all inputs and outputs are active high. Notice that this is a four-level circuit with a maximum fan-in of 3. Also, the reader is reminded that two active low inputs to an XOR gate or EQV gate retains the function, as discussed in Subsection 3.9.4.

An n-bit ripple-carry (R-C) ALU can be produced by cascading the 1-bit slice ALU of Fig. 8.28 in a manner similar to that used to produce an n-bit R-C adder from n FAs in Fig. 8.5. This is done in Fig. 8.29, but with the added requirement of connecting the three mode/select input to all stages as shown. It is also possible to construct an n-bit R-C ALU by cascading m-bit ALU modules in a manner similar to cascading configuration of R-C adders in Fig. 8.6.

The two functions in Eqs. (8.13) are not the only expressions that can be derived from the operation table for F and C_{out} in Fig. 8.26. Referring to Fig. 5.9, it can be seen that the functions F_1 and F_2 are exactly those for F and C_{out}, respectively, if the appropriate substitutions are made and if the don't cares in the F_2 K-map are each set to logic zero. Thus, from Eqs. (5.71) and (5.72) the two outputs for the 1-bit ALU now become either

$$\left\{ \begin{array}{l} F = \bar{M}[C_{in} \oplus (A \oplus S_0) \oplus (S_1 B)] + M(A \oplus S_0 + S_1 B) \\ C_{out} = \bar{M}C_i[(A \oplus S_0) \oplus (S_1 B)] + \bar{M}(S_1 B)(A \oplus S_0) \end{array} \right\} \quad (8.14)$$

8.8 ARITHMETIC AND LOGIC UNITS

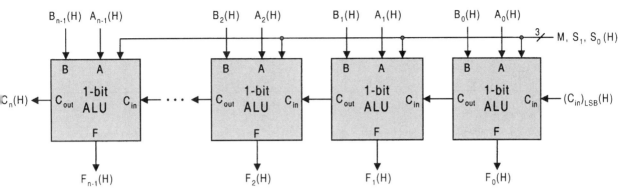

FIGURE 8.29
An n-bit R-C ALU produced by cascading n 1-bit slice ALUs of the type shown in Fig. 8.28.

or

$$\left\{ \begin{array}{l} F = [\overline{MS_1B}(A \oplus S_0)] \oplus (\bar{M}C_i) \oplus (S_1B) \\ C_{out} = \bar{M}C_i[(A \oplus S_0) \oplus (S_1B)] + \bar{M}(S_1B)(A \oplus S_0) \end{array} \right\}, \qquad (8.15)$$

depending on whether or not subfunction partitioning is used for a combined CRMT and two-level result for function F. The two outputs in Eqs. (8.14) represent a five-level system with a combined gate/input tally of 11/24, and those in Eqs. (8.15) represent a four-level system with a total gate/input tally of 11/25, both excluding possible inverters. Thus, the CRMT results in Section 5.11 are comparable but somewhat less optimal than those of Eqs. (8.13) extracted from K-maps.

The n-bit R-C ALU of Fig. 8.29, like the R-C adder, suffers a size limitation due to the ripple-carry effect, as discussed in Subsection 8.2.3. To overcome this limitation the carry look-ahead (CLA) feature can be coupled with the ALU design. (See Section 8.4 for a discussion of the CLA adder.) In Fig. 8.30 is the I/O table for the 1-bit slice ALU with CLA

	M	S_1	S_0	F	P	G
Arithmetic Operations	0	0	0	$A \oplus C_{in}$	A	0
	0	0	1	$\bar{A} \oplus C_{in}$	$\bar{A}$	A
	0	1	0	$A \oplus B \oplus C_{in}$	$(A \oplus B)$	$A \cdot B$
	0	1	1	$\overline{A \oplus B} \oplus C_{in}$	$\overline{(A \oplus B)}$	$\overline{A \cdot B}$
Logic Operations	1	0	0	A	0	0
	1	0	1	$\bar{A}$	0	0
	1	1	0	$A + B$	0	0
	1	1	1	$\overline{A + B}$	0	0

* Subtraction operations assume 2's complement arithmetic.

FIGURE 8.30
Operation table for the simple 1-bit slice ALU of Fig. 8.26 showing CLA output functions P and G based on Eqs. (8.7).

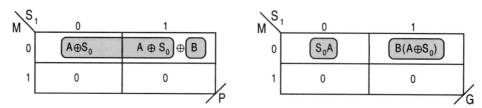

FIGURE 8.31
EV K-maps and minimum cover for carry propagate (P) and carry generate (G) parameters given in Fig. 8.30.

capability. Here, the C_{out} functions of Fig. 8.26 are replaced with those for carry propagate, P, and carry generate, G, which are derived from Eqs. (8.2) and (8.7). Logic 0's are placed in the P and G columns for the logic operations to indicate false carry rejection, as was done in the operation table of Fig. 8.26.

The design of the 1-bit slice ALU of Fig. 8.30 continues by plotting the P and G outputs in second-order K-maps as was the case for C_{out} in Fig. 8.27d. This is done in Fig. 8.31, where optimum cover is indicated with shaded loops for each of the outputs, yielding the results

$$\left\{\begin{array}{l} P = \bar{M}[(A \oplus S_0) \oplus (S_1 B)] \\ G = \bar{M}(S_1 B)(A \oplus S_0) + \bar{M}\bar{S}_1 S_0 A \end{array}\right\}. \tag{8.16}$$

Notice that a single associative pattern is used to extract optimum cover for output, P.

In completing the design of the 1-bit slice ALU with CLA capability, it must be remembered that it is the carry parameters in Eqs. (8.16) that take the place of the C_{out} expression in Eqs. (8.13). This is done by combining the carry expressions of Eqs. (8.16) with the expression for function F in Eqs. (8.13). The result is the logic circuit in Fig. 8.32a and its block diagram symbol in Fig. 8.32b for a 1-bit slice ALU with CLA capability.

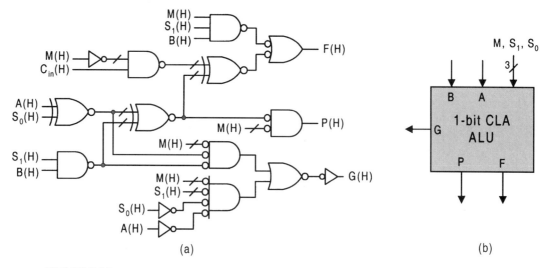

FIGURE 8.32
(a) Optimum gate-level implementation of the 1-bit slice ALU with CLA capability represented by the operation table of Fig. 8.30 and by Eqs. (8.16). (b) Block diagram symbol for the 1-bit ALU in (a).

8.8 ARITHMETIC AND LOGIC UNITS

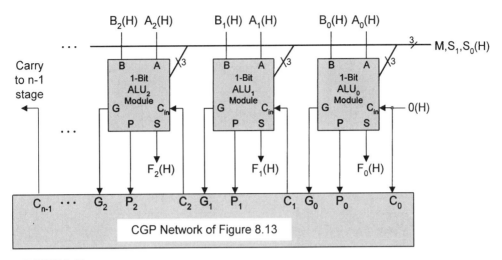

FIGURE 8.33
The three least significant stages of an *n*-bit carry look-ahead (CLA) ALU showing the block symbol for the carry generate/propagate network used between 1-bit modules.

The 1-bit slice ALU module in Fig. 8.32 can be cascaded by using the CLA carry generate/propagate (CGP) network given in Fig. 8.13. This is accomplished in Fig. 8.33, where the three least significant 1-bit stages are shown connected to the CGP network. Cascadable units of this type can be cascaded as groups to produce even larger units. This is done by connecting the carry-out of one *n*-bit stage to the carry-in of another, etc., with the mode/select inputs connected to all stages.

8.8.2 The MUX Approach to ALU Design

There are occasions when a designer requires an ALU that is fully programmable, that will perform a wide variety of tasks both arithmetic and logic, and that can be put to a circuit board with off-the-shelf components, or that can be designed for VLSI circuit applications. The MUX approach to ALU design can provide this versatility and computational power, which are difficult to achieve otherwise. In this subsection a programmable ALU (PALU) will be designed by the MUX approach that can perform the following operations on two *n*-bit operands:

1. Arithmetic operations
   (a) Add with carry, or increment
   (b) Subtract with borrow, or decrement
   (c) Partial multiply or divide steps
   (d) One's or 2's complement
2. Comparator operations
   (a) Equal
3. Bitwise logic operations
   (a) AND, OR, XOR, and EQV
   (b) Logic-level conversion
   (c) Transmit or complement data bit

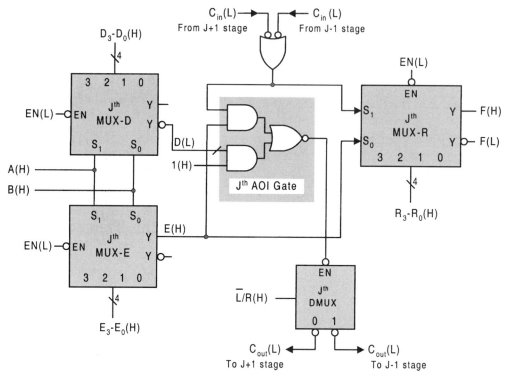

FIGURE 8.34
Implementation of the *J*th 1-bit slice PALU with shift left and shift right capability by using three 4-to-1 MUXs, a DMUX, a NAND gate, and an AOI gate.

4. Shift operations
   (a) Shift or rotate left or right by one bit

It is required that each 1-bit slice of the PALU have a resulting output function *F* and that the PALU be cascaded with respect to the carry-in and carry-out parameters, C_{in} and C_{out}. Also, it is required that the PALU operate with *false carry rejection*, that is, that C_{out} be disabled for all nonarithmetic operations.

The logic circuit diagram for the *J*th 1-bit slice PALU and its block diagram symbol are presented in Figs. 8.34 and 8.35. As can be seen, the PALU consists of three 4-to-1 MUXs, a 1-to-2 DMUX, a NAND gate, and an AOI gate (see Subsection 7.7.1 for a discussion of AOI gates). For reference purposes the MUXs are named

MUX-D ⇒ Disable Carry MUX — Output *D*
MUX-E ⇒ Extend Carry MUX — Output *E*
MUX-R ⇒ Result MUX — Output *F*

To help understand how the PALU is able to perform such a variety of tasks as listed earlier, it useful to write the Boolean expressions for the outputs from the three MUXs and from

8.8 ARITHMETIC AND LOGIC UNITS

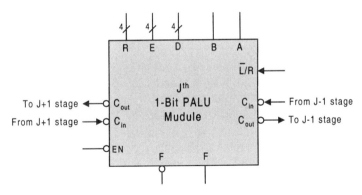

FIGURE 8.35
Block diagram symbol for the Jth 1-bit PALU module of Fig. 8.34.

the AOI gate. Referring to Fig. 8.34, these outputs are

$$\begin{cases} F = R = \sum_{i=3}^{0} = R_i m_i = R_3 C_{in} E + R_2 C_{in} \bar{E} + R_1 \bar{C}_{in} E + R_0 \bar{C}_{in} \bar{E} \\ E = \sum_{i=3}^{0} = E_i m_i = E_3 A B + E_2 A \bar{B} + E_1 \bar{A} B + E_0 \bar{A} \bar{B} \\ D = \sum_{i=3}^{0} = D_i m_i = D_3 A B + D_2 A \bar{B} + D_1 \bar{A} B + D_0 \bar{A} \bar{B} \end{cases} \quad (8.17)$$

and

$$C_{out} = C_{in} E + \bar{D}, \quad (8.18)$$

where the control address inputs R_i, E_i, and D_i are the coefficients of m_i that represent the canonical forms of input sets C_{in}, E or A, B. Thus, $R_2 m_2 = R_2 C_{in} \bar{E}$ or $E_0 m_0 = E_0 \bar{A} \bar{B}$, etc.

Notice that Eqs. (8.17) derive directly from Eq. (6.5), but where now the data select and data inputs are interchanged — that is, there are now four data select inputs (*control address inputs*) each for R_i, E_i and D_i, two data inputs, A and B. Referring to the K-map in Fig. 6.4c for a 4-to-1 MUX, it is easy to see that 16 possible functions in data variables S_1 and S_0 result by assigning 1's and 0's to the four variables I_3, I_2, I_1, and I_0 in the K-map, taking $EN = 1$. Thus, for programming purposes, the four components for each of the control address inputs (coefficients of m_i), R_i, E_i, and D_i, are encoded in the order of descending positional weight, but are given in decimal equivalent to represent any one of the 16 possible logic functions in variables, A and B. As examples, $D = A + \bar{B}$ is represented as $D = 1101_2 = 13_{10}$, $E = A\bar{B}$ if $E = 0100_2 = 4_{10}$, or $F = E \odot C_{in}$ is represented as $R = 1001_2 = 9_{10}$ and $F = E \oplus C_{in}$ when $R = 0110_2 = 6_{10}$, etc.

In Fig. 8.34 it can be seen that the MUX output signals, $D(L)$ and $E(H)$, together with carry-in C_{in}, drive the AOI gate, but that only the MUX output $E(H)$ combines with $C_{in}(H)$ in MUX-R to produce the resultant function F. The carry-out $C_{out}(L)$ from the AOI gate, on the other hand, is a function of E, D, and the carry-in input, C_{in}, as indicated in Eq. (8.18). This is necessary for the arithmetic operations that the PALU must perform. The $\bar{D}$ in Eq. (8.18) is used to disable C_{out} for nonarithmetic operations. If D is set equal to 0 for a given nonarithmetic operation, it follows that $\bar{D}=1$ and $C_{out}(L)=1(L)=0(H)$, which is interpreted as a carry-out disable for that operation. Thus, the carry-out disable feature of the PALU is equivalent to the false carry rejection used in the ALU operation tables of Figs. 8.26 and 8.30. In a sense, the output of MUX-D performs the same mode control operation as does M in the ALU of Figs. 8.28 and 8.29.

The operation of the PALU is illustrated by 12 examples in the operation table of Fig. 8.36 with the help of the n-bit PALU shown in Fig. 8.37. The first five operations in Fig. 8.36 are arithmetic while the last seven are logic. To understand the entries in this table, some explanation is necessary. The control address inputs, R, E, and D, are the binary coefficients in Eqs. (8.17) represented in decimal. Operations (1) through (4) are arithmetic operations that follow directly from Eqs. (8.2) for the FA and require specific values for the C_{in} to the LSB stage of the n-bit PALU in Fig. 8.37. For these arithmetic operations, the carry must propagate from LSB-to-MSB (left), which requires that the direction input be set to $\bar{L}/R = 0$. Note that this PALU cannot support CLA capability since not all carry operations are based on Eqs. (8.2).

Operation (1) requires that the operand magnitudes fall within the limits set in Subsection 8.3.1. Operation (2), A minus B, adds A to the 2's complement of B and requires

	Operation *	F	C_{out}	$\bar{L}/R$	R	E	D
(1)	A plus B (LSB C_{in} = 0)	$A \oplus B \oplus C_{in}$	$C_{in}(A \oplus B) + AB$	0	6	6	7
(2)	A minus B (LSB C_{in} = 1)	$A \oplus \bar{B} \oplus C_{in}$	$C_{in}(A \oplus \bar{B}) + A\bar{B}$	0	6	9	11
(3)	Increment B (LSB C_{in} = 1)	$B \oplus C_{in}$	$C_{in} \cdot B$	0	6	10	15
(4)	2' complement of A (LSB C_{in} = 1)	$\bar{A} \oplus C_{in}$	$C_{in} + \bar{A}$	0	6	3	3
(5)	A = B (LSB C_{in} = 0)	$(A \odot B)\bar{C}_{in}$	$C_{in} + A \oplus B$	0	2	9	9
(6)	$A \oplus B$	$A \oplus B$	1	ϕ	10	6	0
(7)	$A \cdot B$	$A \cdot B$	1	ϕ	10	8	0
(8)	$\bar{A} + B$	$\bar{A} + B$	1	ϕ	10	11	0
(9)	Complement A	$\bar{A}$	1	ϕ	10	3	0
(10)	Transfer B	B	1	ϕ	10	10	0
(11)	Shift A left 1 bit (Fill = LSB C_{in})	C_{in}	A	0	12	0	3
(12)	Shift B right 1 bit (Fill = MSB C_{in})	C_{in}	B	1	12	0	5

* Subtraction operations assume 2's complement arithmetic with LSB C_{in} = 1.

FIGURE 8.36
Twelve sample operations generated by the 1-bit slice PALU in Fig. 8.34 showing the shift direction input and the decimal values for the data select variables, R, E, and D.

8.8 ARITHMETIC AND LOGIC UNITS

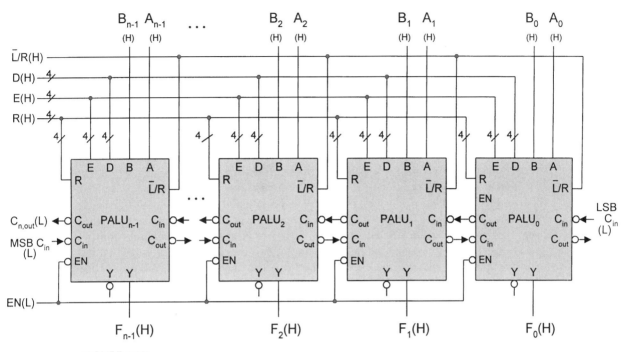

FIGURE 8.37
An n-bit PALU with operational characteristics given by Figs. 8.34, 8.35, and 8.36, and by Eqs. (8.17). and (8.18).

that LSB $C_{in} = 1$ in Fig. 8.37 and that the MSB be reserved for the sign bit as discussed in Section 8.3. For operation (2), the requirements of F and C_{out} are that $R = E \oplus C_{in} = 0110_2 = 6_{10}$, $E = A \oplus \bar{B} = A \odot B = 1001_2 = 9_{10}$, and that $D = \bar{A} + B = 1011 = 11_{10}$. Operation (3) simply requires that LSB $C_{in} = 1$ when $A = 0$, for which the requirements of function F are that $R = E \oplus C_{in} = 0110_2 = 6_{10}$, $E = B = 1010_2 = 10_{10}$, and $D = 1 = 1111 = 15_{10}$ so that $\bar{D} = 0$ in Eq. (8.18). The 2's complement of operation (4) follows Algorithm 2.5 represented by Eq. (2.14). The 2's complement operation sets LSB $C_{in} = 1$, then with $B = 1$ and A is complemented, $R = E \oplus C_{in} = 0110_2 = 6_{10}$, $E = \bar{A} = 0011_2 = 3_{10}$ and $D = \bar{A} = 0011_2 = 3_{10}$. Operations (6) through (10) are simple bitwise logic operations for which $F = R = E$ when $R = 1010_2 = 10_{10}$, and $C_{out} = 1$ and $\bar{L}/R = \phi$ for false carry rejection when $D = 0$.

Operation (5) is the comparator operation, $A = B$, considered either an arithmetic or logic operation. The requirement for this operation is as follows: If the two operands, A and B, are equal on a bitwise comparison basis, then $F_{n-1} = 1$ for an n-bit PALU with its LSB $C_{in}(L) = 0(L)$. Or if the operands are not equal, $F_{n-1} = 0$. Thus, $C_{out} = 0$ will ripple from the LSB stage to the MSB stage and all outputs will be $F_i = 1$. However, if any one of the bitwise comparisons yields an inequality, the carry $C_{out} = 1$ will ripple from that stage to the MSB stage and generate $F_{n-1} = 0$ at the MSB stage. Therefore, operation (5) requires that $R = \bar{C}_{in}E = 0010_2 = 2_{10}$, $E = A \odot B = 1001_2 = 9_{10}$, and $D = A \odot B = 1001$, which, when introduced into Eqs. (8.17) and (8.18), yields the results given in Fig. 8.36.

$$F = (A \odot B)\bar{C}_{in}$$

$$C_{out} = C_{in}E + D$$

$$= C_{in}(A \odot B) + A \oplus B$$

$$= C_{in} + A \oplus B$$

By comparing the result $F = (A \odot B)\bar{C}_{in}$ with Eqs. (6.20) for a 1-bit comparator, it becomes evident that $\bar{C}_{in} = eq$ is the $(A = B)$ output from the next least significant stage.

The remaining two operations, (11) and (12), are shift operations that can be considered both arithmetic and logic in nature. For example, operation (11) shifts word A by one bit to the left, which can be interpreted as a $\times 2$ operation, and hence a partial multiplication step. Referring to Section 8.6 and Algorithm 2.10, it is clear that a bit-by-bit left shift is a partial step used to execute the multiplication of two operands. This requires that the Jth function bit F_J receive the C_{in} from the $(J - 1)$th stage, and that the C_{out} of the Jth stage be the A input to that stage, for which $F = R = C_{in} = 1100_2 = 12_{10}$ and $D = \bar{A} = 0011_2 = 3_{10}$.

There are many other operations possible by the PALU that are not listed in the operation table of Fig. 8.36. As examples, operation (9) can be interpreted as the 1's complement of A according to Algorithm 2.7 as applied to binary, and operation (7) can be considered as a partial product required in the multiplication of two operands. Also, arithmetic operations other than the operations (1), (2), and (3) are possible.

There are a total of 16 bitwise logic operations that can be generated by the PALU, but only five are listed in Fig. 8.36. For reference purposes, the 16 logic functions in two operands, A and B, that can be generated by Eqs. (8.17) are summarized by

$$F = \sum_{i=3}^{0} F_i m_i = F_3 AB + F_2 A\bar{B} + F_1 \bar{A}B + F_0 \bar{A}\bar{B} \qquad (8.19)$$

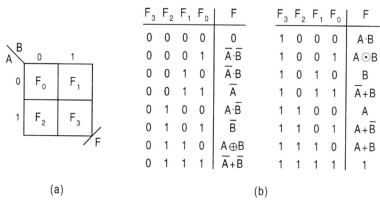

(a)   (b)

FIGURE 8.38
The 16 possible bitwise logic functions for operands A and B according to Eq. (8.19) (a) K-map representation. (b) Tabular representation.

8.9 DUAL-RAIL SYSTEMS AND ALUs WITH COMPLETION SIGNALS

and can be represented either by the K-map in Fig. 8.38a or by the table in Fig. 8.38b. The table is generated by assigning 1's and 0's to the four coefficients, F_3, F_2, F_1, and F_0. These functions will again be used by an even more versatile ALU, which is designed in the following section.

8.9 DUAL-RAIL SYSTEMS AND ALUs WITH COMPLETION SIGNALS

As implied in Section 8.8, ALUs are important because they make possible the use of the same device to perform many different operations, thereby increasing versatility while minimizing the need to combine different modules for those operations. Because of these advantages, ALUs are commonly found in processors where the specific operations are performed on command from the controller in the processor. Although these ALUs support a variety of arithmetic and logic operations and may include CLA capability, they typically have single rail carries (like those treated in Section 8.8) and cannot communicate to the processor when a given operation has been completed. To remedy this situation, completion signals are issued following worst-case delays that are associated with the various ALU operations.

This section features a programmable ALU (PALU) that will issue a final completion (DONE) signal immediately following the completion of any operation, no matter how complex or simple it is. This is a significant feature for an ALU, since arithmetic operations require more time to complete (because of the carry problem) than do bitwise logic operations. Used in a microprocessor, PALUs with DONE signals avoid the need to assign worst-case delays to the various ALU operations. Thus, whenever an operation (logic or arithmetic) is completed, a DONE signal is sent to the CPU (central processing unit), thereby permitting immediate execution of the next process without unnecessary delay.

Listed in Fig. 8.39 are the four modes of operation that the PALU can perform. As indicated, the PALU can perform bitwise logic operations ($M_1 M_0 = 01$), or it can perform left or right shift operations ($M_1 M_0 = 11$) on operand B. But it can also perform arithmetic operations on the result of either a logic or a shift operation, as indicated by mode controls $M_1 M_0 = 00$ and $M_1 M_0 = 10$, respectively. For example, any logic operation in Fig. 8.38 (e.g., $A \oplus B$) or shift in operand B can be added to or subtracted from operand A. With DONE signals issued following the completion of each process, it is clear that this PALU offers a higher degree of versatility than is available from the ALUs in Section 8.8.

An ALU will now be designed that is even more versatile than that of the MUX approach in Subsection 8.8.2. In addition, it is the goal of this section to develop the concepts

M_1	M_0	MODE
0	0	Arithmetic on Logic
0	1	Logic
1	0	Arithmetic on B-Shift
1	1	B-Shift (right or left)

FIGURE 8.39
Modes of PALU operation.

370 CHAPTER 8 / ARITHMETIC DEVICES AND ARITHMETIC LOGIC UNITS (ALUs)

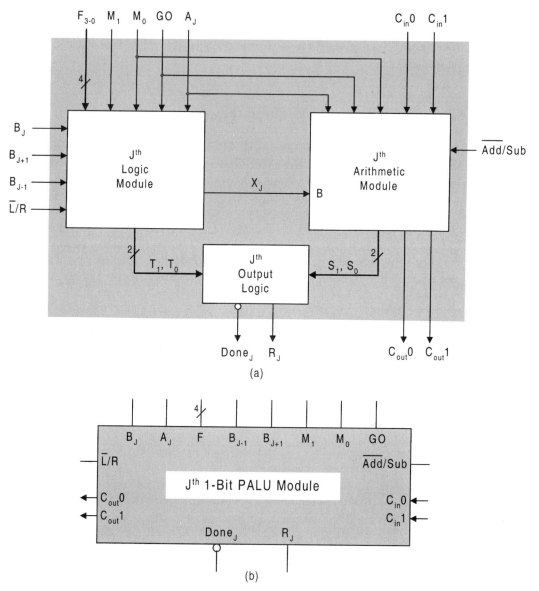

FIGURE 8.40
Block diagram structure (a) and block circuit symbol (b) for the Jth 1-bit PALU module capable of performing the modes of operation listed in Fig. 8.39 with completion (Done) signals.

of dual-rail systems and completion signals, and apply them to the design of a PALU with operational capability defined by Fig. 8.39. Also, both ripple-carry (R-C) and carry look-ahead (CLA) designs will be considered. The following develops the details of this unusual but versatile PALU.

Shown in Figs. 8.40a and 8.40b are the block diagram and logic circuit symbol for a 1-bit slice PALU that can perform the operations represented in Fig. 8.39 and issue a *Done* signal following the completion of each operation. An inspection of Fig. 8.40 indicates that there

8.9 DUAL-RAIL SYSTEMS AND ALUs WITH COMPLETION SIGNALS

are inputs specific to the Jth module and those that are global (applied to all modules in a cascaded system). Specific to the Jth PALU module are the operands inputs A_J, B_J, B_{J-1}, and B_{J-1}. The B_{J-1} input arrives from the B input to the next LSB stage in a cascaded configuration and is necessary for left shifting. Similarly, the B_{J+1} input arrives from the B input to the next MSB stage to permit right shifting. The input and output dual-rail carries shown in Fig. 8.40 are also specific to the Jth module and are defined as follows:

$$\left\{\begin{array}{l} C_{in}0 = \text{carry-in 0 to stage } J \text{ from stage } J-1 \\ C_{in}1 = \text{carry-in 1 to stage } J \text{ from stage } J-1 \\ C_{out}0 = \text{carry-out 0 to stage } J+1 \text{ from stage } J \\ C_{out}1 = \text{carry-out 1 to stage } J+1 \text{ from stage } J \end{array}\right\}. \tag{8.20}$$

The meaning here is that $C_{in}0 = 1$ when the carry-in to the Jth stage is logic 0, and $C_{in}1 = 1$ when the carry-in to the Jth is logic 1. Thus, both carry-in parameters cannot be active at the same time. Similarly, $C_{out}0 = 1$ when carry-out to the $(J+1)$th is logic 0, or $C_{out}1 = 1$ when the carry to the $(J+1)$th is logic 1, where again only one carry parameter can be active at any given time.

The global inputs to the PALU include the two mode control inputs, M_1 and M_0, the function generate signals, F_3, F_2, F_1, and F_0, a shift-direction input $\bar{L}/R$ (meaning right, "not left" when active), an add/subtract input $\overline{Add}/Sub$ (meaning subtract, not add when active), and a start signal called *GO*. The add/subtract control input operates the same as that use for the adder/subtractor design in Subsection 8.3.1, but only if the mode control $M_0 = 0$ according to Fig. 8.39. Also, the shift-direction control $\bar{L}/R$ is operational only if mode control $M_1 = 1$, as indicated in Fig. 8.39.

The two outputs, $Done_J$ and R_J, are specific to the Jth PALU mudule. When the Jth stage result of a bitwise logic operation or arithmetic operation is indicated by the output R_J, a completion signal $Done_J$ is issued. However, it is the requirement of an n-bit PALU design that a final (overall) completion signal, *DONE*, will not be issued until the *Done* signals from all n stages have become active. Thus, the results from those n stages must not be read until the final *DONE* has emerged.

Logic Module The logic module is responsible for carrying out both the 16 bitwise logic operations given by Eq. (8.19) and the shift left or right operation with 0 or 1 fill. (See Section 6.8 for details of a combinational shifter.) Presented in Fig. 8.41 are the output parameters, X_J, T_1, and T_0, for the Jth PALU logic module. The output function X_J, representing the mode control settings for logic and shift operations (according to Fig. 8.39), is given by the truth table and EV K-map in Figs. 8.41a and 8.41b. The dual-rail outputs from the logic module, T_1 and T_0, are defined in the truth table of Fig. 8.41c and represent only logic and shift modes — arithmetic operations are excluded.

The output function X_J is read directly from the EV K-map in Fig. 8.41b and is

$$X_J = \bar{M}_1 F_J + M_1(\overline{\bar{L}/R} \cdot B_{J-1} + \bar{L}/R \cdot B_{J+1}), \tag{8.21}$$

where the quantity $(\overline{\bar{L}/R} \cdot B_{J-1} + \bar{L}/R \cdot B_{J+1})$ represents the shift left or right operation but only when mode control $M_1 = 1$, as required by the mode control table in Fig. 8.39. Thus, right shift by one bit occurs when $\bar{L}/R = 1$ and left shift by one bit occurs when $\bar{L}/R = 0$. Function F_J represents the 16 possible bitwise logic operations, as in Eq. (8.19),

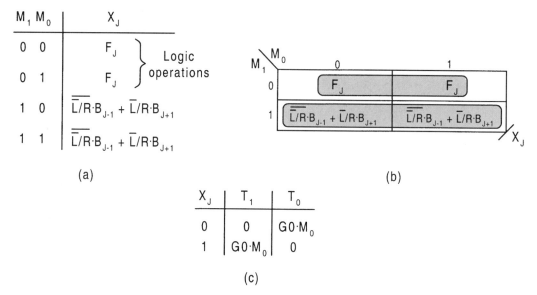

FIGURE 8.41
Requirements of output parameters for the Jth PALU logic module according to Fig. 8.39. (a) Truth table showing mode settings for logic and shift operations. (b) EV K-map for (a). (c) Dual-rail outputs representing only logic and shift operations from logic module.

and is given here for reference purposes:

$$F_J = F_3 AB + F_2 A\bar{B} + F_1 \bar{A}B + F_0 \bar{A}\bar{B}. \tag{8.22}$$

The dual-rail outputs from the Jth logic module, defined in Fig. 8.41c, are read directly from the truth table and are

$$\begin{Bmatrix} T_1 = GO \cdot M_0 \cdot X \\ T_0 = GO \cdot M_0 \cdot \bar{X} \end{Bmatrix}_J. \tag{8.23}$$

The meaning of these dual-rail parameters is as follows: For logic or shift operations, the mode control requirement is $M_0 = 1$ according to Fig. 8.39. Thus, for $GO = 1$ (start active), $T_1 = 1$ and $T_0 = 0$ if $X_J = 1$, or $T_1 = 0$ and $T_0 = 1$ if $X_J = 0$. The dual-rail outputs are necessary to generate completion signals following logic and shift operations.

Presented in Fig. 8.42 is the logic circuit for the Jth PALU logic module as required by Eqs. (8.21)–(8.23). The 4-to-1 MUX provides the 16 possible bitwise logic functions of the two operands, A_J and B_J, as represented by Eq. (8.22), but only if the mode control setting is $M_1 = 0$. The shift right/left portion of the circuit is activated when the mode setting is $M_1 = 1$. Then when $\bar{L}/R = 1$, B_{j+1} is received from the next MSB stage producing a 1-bit right shift, or if $\bar{L}/R = 0$, B_{j-1} is received from the next LSB stage, forcing a 1-bit left shift.

Arithmetic Module To design an arithmetic module with completion signal capability, it is necessary to first establish the concept of dual-rail carries. This is accomplished by rewriting the equations for a full adder as they pertain to the Jth 1-bit PALU arithmetic

8.9 DUAL-RAIL SYSTEMS AND ALUs WITH COMPLETION SIGNALS

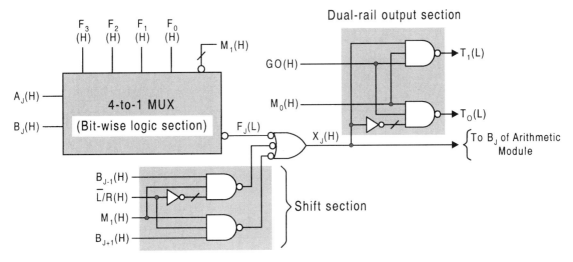

FIGURE 8.42
Logic circuit for the Jth PALU logic module based on Eqs. (8.21)–(8.23).

module. The sum expression from Eqs. (8.2) is restated here as

$$S = A \oplus B \oplus C_{in}$$

and is partitioned into the following two parts with respect to logic 1 and logic 0:

$$S_1 = (A \oplus B)\bar{C}_{in} + (A \odot B)C_{in}$$
$$= (A \oplus B)C_{in}0 + (A \odot B)C_{in}1$$

and (8.24)

$$S_0 = (A \oplus B)\bar{C}_{in} + (A \odot B)C_{in}$$
$$= (A \oplus B)C_{in}1 + (A \odot B)C_{in}0$$

Here, use has been made of Eq. (3.4) in the form $S = x\bar{y} + \bar{x}y$, where $x = A \oplus B$ and $y = C_{in}$. Thus, for S_1, carry-in is represented as $\bar{C}_{in} = C_{in}0$, and $C_{in} = C_{in}1$ whereas for S_0 the carry-in is represented as $\bar{C}_{in} = C_{in}1$ and $C_{in} = C_{in}0$. The split-rail sums, S_1 and S_0, in Eqs. (8.24), are summarized in Fig. 8.43a together with the dual-rail carry-outs, $C_{out}1$ and $C_{out}0$, as they are affected by the operands, A and B, and the dual-rail carry-ins, $C_{in}1$ and $C_{in}0$. Here, S_1 is active if the sum is logic 1 otherwise inactive, or S_0 is active if the sum is logic 0 otherwise inactive.

The carries have a similar meaning. An active $C_{in/out}1$ implies a carry (in or out) of logic 1, and an active $C_{in/out}0$ implies a carry (in or out) of logic 0. Thus, $C_{in/out}1$ and $C_{in/out}0$ cannot both be active (take logic 1) at the same time. Also, $C_{in/out}1 = C_{in/out}0 = 0$ is an indeterminate state of no carry. In effect, a three-state code representation is used for carry-in $\{C_{in}1, C_{in}0\}$, carry-out $\{C_{out}1, C_{out}0\}$, and sums $\{S_1, S_0\}$. This means that each of these three dual-rail pairs can assume one of the three state values taken from the set $\{00, 01, 10\}$. All three pairs are required to be set initially in logic state (00) and then transit to either state (01) or (10) following the activation of the start signal GO shown in Fig. 8.40. State (11) is not permitted.

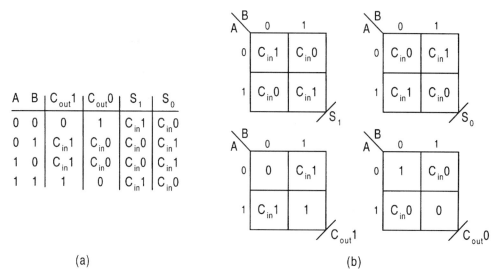

FIGURE 8.43
Requirements of a split-rail sum and carry system. (a) Truth table for sum and carry partitions. (b) EV K-maps plotted from the truth table in (a).

With these definitions in mind and by referring to Subsection 8.2.2 for the full adder, the carry partitions in the truth table of Fig. 8.43 can be understood. The dual-rail sum and carry-outs in Fig. 8.43a are plotted in the EV K-maps of Fig. 8.43b. From these K-maps there results the dual-rail carry-out expressions for the Jth stage,

$$\begin{cases} C_{out}1 = (A \oplus B)C_{in}1 + AB \\ C_{out}0 = (A \oplus B)C_{in}0 + \bar{A}\bar{B} \end{cases}. \tag{8.25}$$

Now, it is necessary to introduce the mode control M_0 (consistent with Fig. 8.39) and start input GO together with an add/subtract parameter $\alpha = B \oplus (\overline{Add/Sub})$ that replaces operand B in Eqs. (8.24) and (8.25). When this is done the dual-rail sum and carry-out parameters for Jth the module become

$$\begin{cases} S_1 = GO \cdot \bar{M}_0[(A \oplus \alpha)C_{in}0 + (A \odot \alpha)C_{in}1] \\ S_0 = GO \cdot \bar{M}_0[(A \oplus \alpha)C_{in}1 + (A \odot \alpha)C_{in}0] \\ C_{out}1 = GO \cdot \bar{M}_0[(A \oplus \alpha)C_{in}1 + A\alpha] \\ C_{out}0 = GO \cdot \bar{M}_0[(A \oplus \alpha)C_{in}0 + \bar{A}\bar{\alpha}] \end{cases}. \tag{8.26}$$

Use of the mode control M_0 avoids issuing a false sum from the arithmetic module and acts as a false data rejection feature during nonarithmetic operations. The XOR function $\alpha = B \oplus (\overline{Add/Sub})$ that replaces operand B permits the subtraction operation by using 2's complement arithmetic as is done for the adder/subtractor in Subsection 8.3.1. For addition, $\overline{Add/Sub} = 0$ passes operand B, but for subtraction $\overline{Add/Sub} = 1$ complements operand B, as required in 2's complement arithmetic.

8.9 DUAL-RAIL SYSTEMS AND ALUs WITH COMPLETION SIGNALS

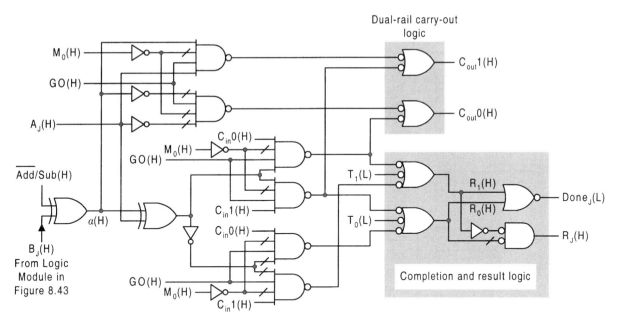

FIGURE 8.44
Logic circuit for the Jth arithmetic module represented by Eqs. (8.26) and (8.27), showing the completion ($Done_J$) and result (R_J) output logic that includes the dual-rail inputs T_1 and T_2 from the logic module and the dual-rail carry-out logic.

Shown in Fig. 8.44 is the logic circuit for the Jth arithmetic module as represented by Eqs. (8.26). In addition, there is included the completion ($Done_J$) and result (R_J) logic section that derives from equations that combine the results from the logic module with those from the arithmetic module. For the Jth PALU module, these equations are given by

$$\left\{ \begin{array}{l} R_1 = T_1 + S_1 \\ R_0 = T_0 + S_0 \\ \text{and} \\ R = R_1 \cdot \bar{R}_0 \\ Done = R_1 + R_0 \end{array} \right\}. \quad (8.27)$$

Notice that the result $R = R_1 \bar{R}_0$ is a resolved result required to issue a logic 1 when $R_1 = 1$ or a logic 0 when $R_0 = 1$ according to the definitions of the dual-rail components of R. However, a completion signal $Done = 1$ is issued in either case to indicate that a valid result is present. Thus, except for the initialization state $GO = 0$, for which $R_1 = R_0 = 0$, one or the other of R_1 or R_0 will be active, indicating a valid result, but they will never both be active simultaneously.

The cascading of 1-bit modules to form an n-bit PALU is easily accomplished as illustrated in Fig. 8.45. All that is required is to connect each set of dual-rail carry-outs from one stage to the next MSB stage and all global inputs to all stages. The shift inputs, B_{j-1} and

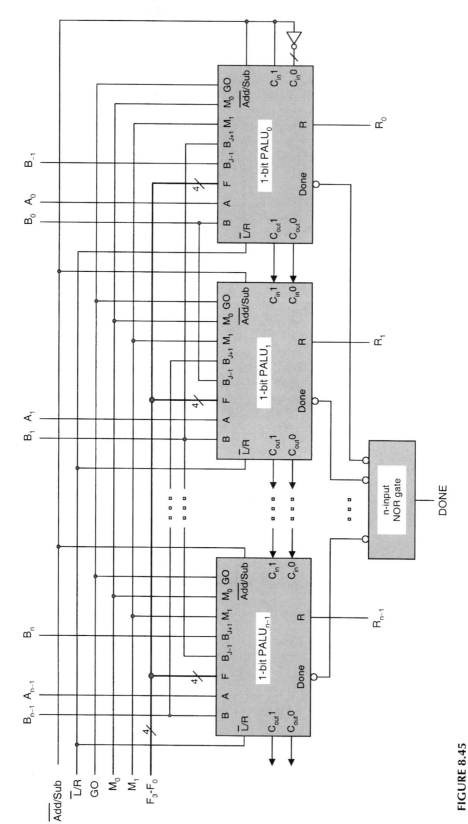

FIGURE 8.45

Logic circuit for an *n*-bit PALU with completion signal capability consisting of cascaded 1-bit modules of the type shown in Fig. 8.40. Note that all inputs and outputs are assumed to be active high and that the *n*-input NOR gate is of CMOS construction as illustrated in Fig. 8.46.

8.9 DUAL-RAIL SYSTEMS AND ALUs WITH COMPLETION SIGNALS

B_{j+1}, must be connected appropriately for left and right shifting as shown. Barrel shifting (rotation) right can be accomplished by connecting the R_0 output to the B_n input. Similarly, barrel shifting left results if the R_{n-1} output is connected to the B_{-1} input. Notice that the carry-in inputs, $C_{in}1$ and $C_{in}0$, to the LSB (PALU$_0$) stage are correctly initialized for add or 2's complement arithmetic (see Subsection 8.3.1 regarding adder/subtractors). Thus, if $\overline{Add}/Sub = 1$, required for subtraction by 2's complement, a logic 1 is carried in to the LSB stage ($C_{in}1 = 1$). Conversely, if $\overline{Add}/Sub = 0$ for addition, a logic 0 is carried in ($C_{in}0 = 1$).

The n-input NOR gate in Fig. 8.45 requires special consideration. This gate must AND the individual $Done$(L) signals to produce the final $DONE$(H) represented by the expression

$$DONE = \prod_{i=0}^{n-1} (Done)_i. \qquad (8.28)$$

Thus, the conjugate gate form shown in Fig. 3.13b must be used. With inputs to such a gate numbering more that four, there is the problem of fan-in as discussed in Section 4.10. The larger the fan-in, the greater is the path delay through the gate. In fact, there is a definite limit as to the number of inputs available in commercial NOR gate chips.

The fan-in problem is effectively eliminated by using the CMOS NOR gate construction shown in Fig. 8.46a. Here, the number of permissible inputs is practically unlimited with negligible effect on the path delay through the gate, which is essentially that of a two-input NOR gate. All $Done$ inputs must go to LV before the output $DONE$ can go to HV. So if one

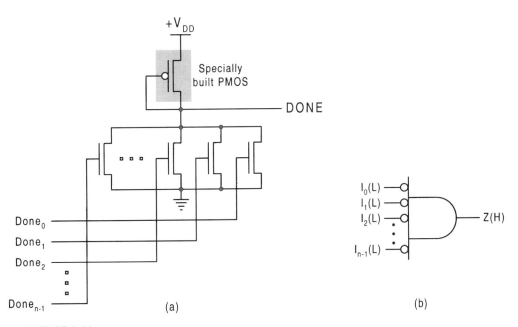

FIGURE 8.46
Multiple input NOR gate specifically designed to minimize fan-in-limitations. (a) CMOS circuit required for Fig. 8.45. (b) Generalized NOR gate symbol for this circuit.

or more of the inputs are at HV, the output is pulled low (to LV). Thus, the PMOS must be specially designed so that the Drain-to-Source resistance remains sufficiently high so as to minimize Drain current when one or more of the NMOS are turned ON. Here, the PMOS serves the same function as the depletion-mode NMOS in Fig. A.1 of Appendix A. Both function as pull-up transistors. Hence, the CMOS NOR gate of Fig. 8.46 could be replaced by the NMOS technology of Figure A.1. In any case, the NOR gate logic symbol for this circuit is given in Fig. 8.46b, which is the same as that in Fig. 3.13b.

8.9.1 Carry Look-Ahead Configuration

Like the R-C adder of Fig. 8.5, the PALU just described suffers a size limitation due to the ripple-carry effect. The carry propagation time increases with increase in the size of the PALU. To reduce significantly the propagation time, carry look-ahead (CLA) capabilities can be incorporated into a dual-rail PALU design. From Eqs. (8.7) the following definitions can be applied to the Jth 1-bit PALU module:

$$\left\{ \begin{array}{l} P = A \oplus \alpha = \text{Carry propagate} \\ P' = GO\bar{M}_0 \cdot P = \text{Modified carry propagate} \\ G_1 = GO\bar{M}_0 \cdot A\alpha = \text{Carry generate w/r to logic 1} \\ G_0 = GO\bar{M}_0 \cdot \bar{A}\bar{\alpha} = \text{Carry generate w/r to logic 0} \end{array} \right\}. \quad (8.29)$$

Here, G_1 and G_0 are the dual-rail carry generate parameters, and $\alpha = B \oplus \overline{(Add/Sub)}$ is the add/subtract parameter that replaces operand B in Eqs. (8.7). Introducing Eqs. (8.29) into the sum and carry-out expressions of Eqs. (8.26) yields

$$\left\{ \begin{array}{l} S_1 = (GO\bar{M}_0)PC_{in}0 + (GO\bar{M}_0)\bar{P}C_{in}1 \\ S_0 = (GO\bar{M}_0)PC_{in}1 + (GO\bar{M}_0)\bar{P}C_{in}0 \\ C_{out}1 = GO\bar{M}_0 PC_{in}1 + GO\bar{M}_0 A\alpha \\ \quad\quad = P'C_{in}1 + G_1 \\ C_{out}0 = GO\bar{M}_0 PC_{in}0 + GO\bar{M}_0 \bar{A}\bar{\alpha} \\ \quad\quad = P'C_{in}0 + G_0 \end{array} \right\}, \quad (8.30)$$

which are applied to the Jth 1-bit PALU module with CLA capability. As in Eqs. (8.26), the appearance of the mode control $\bar{M}_0$ in Eqs. (8.29) and (8.30) avoids issuing a false sum from the arithmetic module and acts as a false carry rejection feature during nonarithmetic operations. The carry-out expressions $C_{out}1 = P'C_{in}1 + G_1$ and $C_{out}0 = P'C_{in}0 + G_0$ can be expanded as in Eqs. (8.8) and, therefore, constitute the CGP network similar to that in Fig. 8.13 with P' replacing P in that network. Thus, all that remains in the design of the dual-rail PALU with CLA capability is to implement Eqs. (8.29) together with the sum expressions in Eqs. (8.30). Presented in Fig. 8.47a is the logic circuit for the arithmetic module of a 1-bit PALU with completion signal and CLA capability as required by Eqs. (8.27), (8.29), and (8.30). Remember that it is the modified carry propagate parameter P', not P, that combines with the dual-rail carry generate parameters, G_1 and G_0, to form the CGP network as in Fig. 8.13. The logic circuit symbol for this arithmetic module is given in Fig. 8.47b.

8.9 DUAL-RAIL SYSTEMS AND ALUs WITH COMPLETION SIGNALS

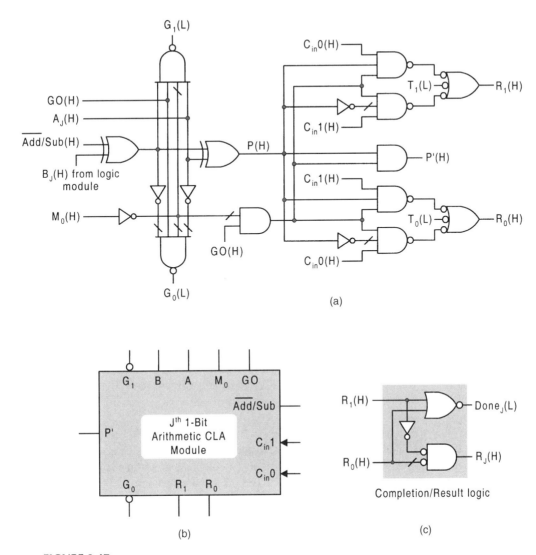

FIGURE 8.47
The Jth 1-bit arithmetic module for a PALU with completion signal and CLA capability according to Eqs. (8.29) and (8.30). (a) Logic circuit showing carry propagate, and dual-rail carry generate, carry inputs and result signals. (b) Block circuit symbol for the logic circuit in (a). (c) Completion/result logic circuit for combined logic and arithmetic modules.

The PALU can be completed by combining the logic module in Fig. 8.42 with the arithmetic module of Fig. 8.47. This requires that the completion signals from the logic module, T_1 and T_0, be combined with the completion/result signals from the arithmetic module, S_1 and S_0, to yield the *Done* and R signals as indicated in Fig. 8.47c. Further modifications and a significant increase in hardware are required to make the PALU symmetric with respect to both operands. In this case either an A or B shift would be possible with arithmetic operations performed on either.

The 1-bit PALU must now be cascaded in a manner similar to the CLA adder in Fig. 8.13, except that now one CGP network is needed for logic 1 (G_1, P') and another is needed for

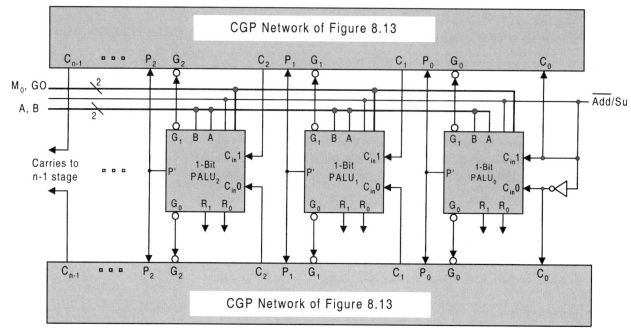

FIGURE 8.48
The three least significant bit stages of an n-bit arithmetic module for a PALU with CLA and completion signal capability showing the carry generate/propagate (CGP) network required for the dual-rail carries and carry generate parameters.

logic 0 (G_0, P'), as required for dual-rail carries. This is demonstrated in Fig. 8.48, where for simplicity only the arithmetic module is featured. Notice that the carry-in's for the LSB stage are properly initialized for addition or 2's complement arithmetic. Hence, for subtraction, $\overline{Add/Sub} = 1$ introduces a logic 1 into $C_{in}1$ and a logic 0 into $C_{in}0$, as required by Eq. (2.14). But for addition, $\overline{Add/Sub} = 0$ introduces a logic 0 into carry-in $C_{in}1$ and a logic 1 into $C_{in}0$ (meaning 0 carry-in). To complete the PALU design, it is necessary to include the logic module in Fig. 8.42 with the arithmetic module in Fig. 8.48 and combine the completion and result signals as indicated in Fig. 8.47c. It is also necessary to connect the X_J output from the logic module to the B_J input of the arithmetic module for each stage.

Clearly, the hardware commitment for the dual-rail CLA PALU increases considerably as the number of stages increases. For this reason it is recommended that the group CLA method be used on, say, 4-bit stages. For example, each of the 1-bit PALUs in Fig. 8.48 would be replaced by four 1-bit stages of a "conventional" type (i.e., without CLA capability) and then cascaded with the dual-rail CPG networks as in Fig. 8.48. This requires that only the MSB stage of each group of four need be altered, as in Fig. 8.47, to accommodate the CLA feature.

8.10 VHDL DESCRIPTION OF ARITHMETIC DEVICES

To recap what was stated or intimated in Section 6.10, the VHDL model of a circuit is called its *entity*, which consists of an *interface description* and an *architectural description*.

8.10 VHDL DESCRIPTION OF ARITHMETIC DEVICES

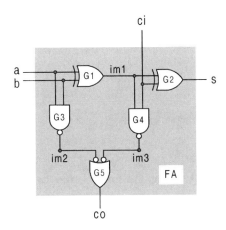

FIGURE 8.49
Logic circuit for the full adder showing inputs, outputs, and intermediate connection labels required for the VHDL description.

The interface of an entity specifies its circuit ports (inputs and outputs) and the architecture gives its contents (e.g., its interconnections). The architecture of an entity is divided into different levels of abstraction, which include its *structure, dataflow,* and *behavior*, the behavior description being the most abstract. The following VHDL description of the full adder, shown in Fig. 8.49, illustrates these three levels of abstraction (refer to Section 6.10 for a description of the key words used):

-- Declare entity:
entity full_adder_example **is**
    port (a, b, ci: **in** bit; s, co: **out** bit);
end full_adder_example;

-- Declare architecture beginning with structure:
architecture structure **of** full_adder_example **is**
    component XOR2
        port (x, y: **in** bit; z: **out** bit); -- declares XOR gate
    component NAND2
        port (x, y: **in** bit; z: **out** bit); -- declares NAND gate
end component;

signal im1, im2, im3: bit; -- declares intermediate signals
-- Declare interconnections between component gates:
begin
    G1: xor2 **port map** (a, b, im1);
    G2: xor2 **port map** (im1, ci, s);

```
        G3: nand2 port map (a, b, im2);
        G4: nand2 port map (im1, ci, im3);
        G5: nand2 port map (im2, im3, co);
end full_adder_example;
```

-- Declare dataflow:

```
architecture dataflow of full_adder_example is
begin
        s <= a xor b xor ci after 12 ns;
        co <= (a and b) after 10 ns or (im1 and ci) after 16 ns;
                                                -- 16 ns is worst case delay
end dataflow;
```

-- Declare behavior:

```
architecture behavior of full_adder_example is
begin
        process (a, b, ci)
            variable a, b, ci, s: integer;
            begin
                if a = '0' then a := 0; else a := 1;       -- converts a to integer
                end if
                if b = '0' then b := 0; else b := 1;       -- converts b to integer
                end if
                if ci = '0' then ci = 0; else ci := 1;     -- converts ci to integer
            s := a + b + ci                                -- computes sum of inputs
            case s is
                when 0 => s <= '0'; co <= '0';
                when 1 => s <= '1'; co <= '0';
                when 2 => s <= '0'; co <= '1';
                when 3 => s <= '1'; co <= '1';
            end case
        end process
end full_adder_example
```

In the full adder example just given, notice that all keywords are presented in bold type and that the symbol "=>" is read as "is the value of." Also note that the operators that are used are those listed in Fig. 6.44 of Section 6.10 and that the double dash "--" is used to indicate the beginning of a comment. The delay times given for *s* and *co* are based on average gate path delays of 6 ns for the XOR gate and 5 ns for the two-input NAND gate, as expressed in Eq. (6.1).

An important feature of VHDL is its modularity capability, which allows models to be reused in the description of other larger entities. A good example is the VHDL structural

8.10 VHDL DESCRIPTION OF ARITHMETIC DEVICES

description of a four-bit adder composed of four full adders described in the previous example. If Fig. 8.49 is used with reference to Figs. 8.4 and 8.5, the structure of the four-bit adder is given as follows:

entity four_bit_adder **is**

    port (a0, a1, a2, a3, b0, b1, b2, b3, ci: **in** bit; s0, s1, s2, s3, co: **out** bit;

end four_bit_adder

architecture connect_four **of** four_bit_adder **is**

    component full_adder

        port (a, b, ci: **in** bit; s, co: **out** bit);

    end component;

for all: full_adder **use entity** full_adder_example;

signal c1, c2, c3: bit

begin

    FA0: full_adder **port map** (a0, b0, ci, s0, c1);

    FA1: full_adder **port map** (a1, b1, c1, s1, c2);

    FA2: full_adder **port map** (a2, b2, c2, s2, c3);

    FA3: full_adder **port map** (a3, b3, c3, s3, co);

end connect_four

end four_bit_adder

Just given is an architectural description for the full-adder primitive, followed by that for a four-bit adder formed by cascading four full-adder primitives. However, within VHDL compilers, encapsulations of such primitives are provided so that they can be easily retrieved and used in the architectural descriptions of larger systems. Thus, for well-known primitives like those just considered, there is no need to construct the detailed architectural descriptions—this has already been accomplished for the convenience of the user. These primitives exist within standard logic packages. The IEEE 1164 standard library is an example, and its contents are made available by making the statements

library ieee
use std_logic_1164.all

Once a standard library is made available in the design description, use can be made of data types, functions, and operators provided by that standard. Standard data types include bit, bit_vector, integer, time, and others, and the operators are of the type given in Fig. 6.44. The standard package defined by the IEEE 1076 standard includes declarations for all the standard data types. For detailed information on these and other subjects related to standard libraries and packages the reader is referred to Further Reading.

FURTHER READING

Most recent texts give a fair account of the basic arithmetic devices, including the full adder, parallel adders, subtractors, adder/subtractors, and carry look-ahead adders. Typical examples are the texts of Comer, Ercegovac and Lang, Hayes, Katz, Pollard, Sandige, Tinder, Wakerly, and Yarbrough. The subject of multiple operand addition and the carry-save adder appears to be covered adequately only in texts by Ercegovac and Lang and by Tinder. Note that some of the listed devices may or may not be the strength of a given text.

[1] D. J. Comer, *Digital Logic and State Machine Design*, 3rd. ed., Sanders College Publishing, Fort Worth, TX, 1995.
[2] M. D. Ercegovac and T. Lang, *Digital Systems and Hardware/Firmware Algorithms*. John Wiley & Sons, New York, 1985.
[3] J. P Hayes, *Introduction to Digital Logic Design*. Addison Wesley, Reading, MA, 1993.
[4] R. H. Katz, *Contemporary Logic Design*. Benjamin/Commings Publishing, Redwood City, CA, 1994.
[5] L. H. Pollard, *Computer Design and Architecture*. Prentice Hall, Englewood Cliffs, NJ, 1990.
[6] R. S. Sandige, *Modern Digital Design*. McGraw-Hill, New York, 1990.
[7] R. F. Tinder, *Digital Engineering Design: A Modern Approach*. Prentice-Hall, Englewood Cliffs, NJ, 1991.
[8] J. F. Wakerly, *Digital Design Principles and Practices*, 2nd ed., Prentice-Hall, Englewood Cliffs, NJ, 1994.
[9] J. M. Yarbrough, *Digital Logic Applications and Design*. West Publishing Co., Minneapolis/St. Paul, 1997.

A few books adequately cover combinational multipliers. These include the texts by Ercegovac and Lang, Katz, Pollard, and Tinder, all previously cited. Of these, only the text by Tinder appears to cover combinational dividers.

A somewhat older text by Kostopoulos is unique in the sense that it covers a rather broad range of subjects relative to arithmetic methods and circuits. In addition to the usual coverage of the basic arithmetic methods and circuits, Kostopoulos provides a very good treatment of combinational multipliers and dividers, and a unique coverage of combinational square and square root binary circuits. The text also includes combinational BCD adders, subtractors, and multipliers.

[10] G. K. Kostopoulos, *Digital Engineering*. John Wiley & Sons, New York, 1975.

The subject of arithmetic and logic units (ALUs) is somewhat esoteric. Nevertheless, it is covered to one extent or another by a few well-known texts. These include those by Ercegovac and Lang, Hayes, Katz, Tinder, and Yarbrough, all previously cited. In addition, the text of Mead and Conway discusses an ALU suitable for processor application that is the starting point for the ALU treated in Subsection 8.8.2 of this text. Apparently, only the text by Tinder develops the dedicated multilevel ALU by using XOR/SOP logic. Coverage of dual-rail arithmetic systems, particularly ALUs, is difficult if not impossible to find in any text. The source on which this text is based is the thesis of Amar cited next. Here, dual-rail ALUs, multipliers, and dividers, all with completion signals, are discussed in detail.

[11] A. Amar, "ALUs, Multipliers and Dividers with Completion Signals," M.S. Thesis. School of Electrical Engineering and Computer Science, Washington State University, Pullman, WA, 1994.
[12] C. Mead and L. Conway, *Introduction to VLSI Systems*. Addison-Wesley, Reading, MA, 1980.

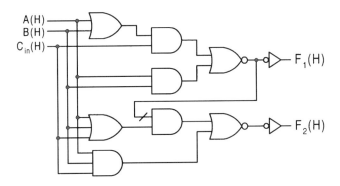

FIGURE P8.1

PROBLEMS

8.1 Use one half adder, one NOR gate, and one inverter (nothing else) to implement a circuit that will indicate the product of a positive 2-bit binary number and 3. Assume all inputs and outputs are active high. (Hint: Construct a truth table with AB as the inputs and Y_3, Y_2, Y_1, Y_0 as the outputs.)

8.2 Use a full adder (nothing else) to implement a circuit that will indicate the binary equivalent of $(x^2 + x + 1)$, where $x = AB$ is a 2-bit binary word (number). Assume that the inputs and outputs are all active high. (Hint: Construct a truth table with AB as the inputs and Y_3, Y_2, Y_1, Y_0 as the outputs.)

8.3 Design a three-input logic circuit that will cause an output F to go active under the following conditions:

All inputs are logic 1
An odd number of inputs are logic 1
None of the inputs are logic 1

To do this use *one* full adder and *two* NOR gates (nothing else). Assume that the inputs arrive active high and that the output is active low.

8.4 Prove that the logic circuit in Fig. P8.1 is that of a full adder. Also, prove that C_{out} is the majority function $AB + BC + AC$. (Hint: Make use of K-maps or truth tables to solve this problem.)

8.5 Use the logic circuit in Fig. P8.1 (exactly as it is given) to construct a staged CMOS implementation of the full adder. Refer to Fig. 3.41 for assistance if needed.

8.6 Use the symbol for a 3-to-8 decoder and two NAND gates (nothing else) to implement the full adder. Assume that the inputs arrive as $A(H)$, $B(H)$, and $C_{in}(L)$ and that the outputs are issued active high. (Hint: First construct the truth table for the full adder, taking into account the activation levels of the inputs and outputs.)

8.7 Without working too hard, design a 4-bit binary-to-2's complement converter by using half adders and inverters (nothing else). Use block symbols for the half adders and assume that the inputs and outputs are all active high.

8.8 Without working too hard, use *four* full adders (nothing else) to design a circuit that will convert XS3 to BCD. Assume that all inputs and outputs are active high. (Hint: Use 2's complement arithmetic.)

8.9 Analyze the adder/subtractor in Fig. 8.9 in 4 bits by adding or subtracting the binary equivalent of the numbers listed below. To do this, give the sum (or difference) and carry logic values at each stage.

(a) $A = 1101;\ B = 0111$ if $\bar{A}/S(H) = 0(H)$
(b) $A = 1101;\ B = 0111$ if $\bar{A}/S(H) = 1(H)$
(c) $A = 0110;\ B = 1101$ if $\bar{A}/S(H) = 1(H)$

8.10 Analyze the 3-bit carry look-ahead adder (CLA) in Fig. 8.13 by introducing the number given below. To do this, give the $G(L)$ and $P(H)$ logic values in addition to the sum and carry values.

(a) $A = 011;\ B = 110$
(b) $A = 111;\ B = 101$

8.11 (a) By using Algorithm 2.14 in Subsection 2.9.6 and two 4-bit ripple/carry adders, complete the design of the single-digit BCD adder in Fig. P8.2a, one that can be bit-sliced (cascaded) to produce an n-digit BCD adder. To do this, first find the minimum logic for the correction parameter X that will indicate when the sum is

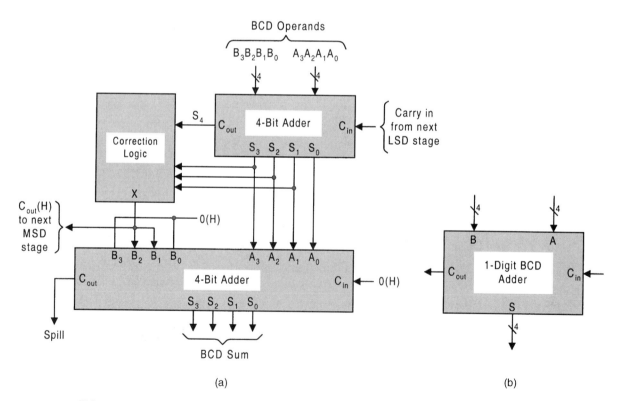

FIGURE P8.2

PROBLEMS

greater than 9 but less than 19. Then use one of two 4-bit adders (as shown) to add 0110 to the sum conditionally on X. Note the logic symbol for the 1-digit BCD adder in Fig. P8.2b. What restrictions, if any, are placed on the operands?

(b) Test the design of a two-digit BCD adder by adding the following two BCD numbers:

Test #1 $A = 9_{10}$; $B = 7_{10}$
Test #2 $A = 34_{10}$; $B = 19_{10}$

To do this, indicate the logic values for each sum, carry, and correction parameter X.

8.12 (a) Alter the design of Problem 8.11 with the appropriate external logic so as to create a one-digit BCD adder/subtractor that can be bit-sliced to form an n-digit BCD adder/subtractor, all with B as the subtrahend. (Hint: Add an enable $EN(H) = \bar{A}/S(H)$ mode control to the correction logic of Problem 8.11 and use 10's complement arithmetic to perform BCD subtraction operations. To do this, follow Fig. 8.9 with the appropriate external logic on the input for cascading purposes. Also, remember that a 1 carry cannot be generated in BCD subtraction. It is important to note that, in this case, the negative number is not converted to 10's complement as in Algorithm 2.15, which is an alternative means of BCD subtraction.)

(b) Test the design of a two-digit BCD adder/subtractor by introducing the following numbers and mode control:

Test #1 $A = 6_{10}$, $B = 29_{10}$, and $\bar{A}/S(H) = 0(H)$
Test #2 $A = 6_{10}$, $B = 29_{10}$, and $\bar{A}/S(H) = 1(H)$

To do this, indicate the logic values for each operand, sum, carry, mode control, and correction parameter. Note that if the result of a subtraction is a negative number, it will be represented in 10's complement, and that its true value can be found by negation of each digit separately.

8.13 Use two 4-bit binary adders and the necessary correction and external logic to design a one-digit XS3 adder (similar to BCD adder in Fig. P8.2) that can be cascaded to form an n-digit XS3 adder. To do this, use the following algorithm:

Add XS3 numbers by using binary addition: If there is no 1 carry from the 4-bit sum, correct that sum by subtracting 0011 (3_{10}). If a 1 carry is generated from the 4-bit sum, correct that sum by adding 0011. Remember that XS3 numbers less than 3 or greater than 12 are not used and that the sum of two XS3 numbers cannot exceed 24_{10}.

[Hint: First, find the minimum logic for the correction parameter X that will indicate when the sum is greater than 12 but less than 25, the range over which a 1 carry is generated. Also, controlled inverters (XOR gates) must be used for the addition or subtraction of 3_{10}.] What restrictions, if any, are placed on the operands?

8.14 (a) Alter the design of Problem 8.13 so as to create a one-digit XS3 adder/subtractor that can be cascaded to form an n-digit XS3 adder/subtractor, all with B as the subtrahend. Note that a 1 carry cannot be generated in XS3 subtraction. (Hint: An additional four controlled inverters are needed for the add/subtract operations.)

(b) Test the design of a two-digit XS3 adder/subtractor by introducing the following numbers and mode control:

Test #1 $A = 6_{10}$, $B = 29_{10}$, and $\bar{A}/S(H) = 0(H)$
Test #2 $A = 6_{10}$, $B = 29_{10}$, and $\bar{A}/S(H) = 1(H)$

To do this indicate the logic values for each operand, sum, carry, mode control, and correction parameter. Note that the decimal value of a negative XS3 number is found by subtracting ...0011 from the negated number and reading it as a BCD number.

8.15 In Fig. P8.3a is shown a network containing several combinational logic devices including a 4-bit ripple/carry adder.
(a) Complete the truth table in Fig. P8.3b.
(b) Use a decoder and a single OR gate to accomplish the result given in part (a). Assume that the decoder has active high outputs.

8.16 (a) Design a 4-bit noncascadable comparator by using two 4-bit subtractors and one NOR gate (nothing else). [Hint: It will be necessary to switch operands on one of the two subtractors. Also, in a subtractor, a final borrow-out of 1 indicates (minuend) < (subtrahend), but a final borrow-out of 0 indicates (minuend) ≥ (subtrahend). Thus, if both borrow-outs are logic 0, then the two numbers are equal. Note that a negative difference is given in 2's complement but taking into account $B_{in})_{LSB} = 0$.]

(b) Test the design in part (a) by using the following operands:

Test #1 $A = 1101$; $B = 0110$
Test #2 $A = 0110$; $B = 1101$
Test #3 $A = 1010$; $B = 1010$

(c) Show that the difference of the two operands can also be read from the circuit.

8.17 (a) By using Eqs. (8.8), complete the carry look-ahead adder (CLA) circuit in Fig. 8.13 for a cascadable 4-bit CLA adder unit. Thus, include the carry generate/propagate logic from the fourth stage.
(b) Test the results by adding the following numbers:

Test #1 $A = 0111$; $B = 0110$
Test #2 $A = 1101$; $B = 1010$

8.18 Analyze the carry-save circuit of Fig. 8.15b by introducing the three operands given in Fig. 8.15a into the circuit. To do this, give the logic values for each operand, sum, and carry.

8.19 Analyze the 4 × 4 binary multiplier in Fig. 8.18 by introducing the following operands into the circuit:
(a) $A = 1101$; $B = 0110$
(b) $A = 1001$; $B = 1011$

PROBLEMS 389

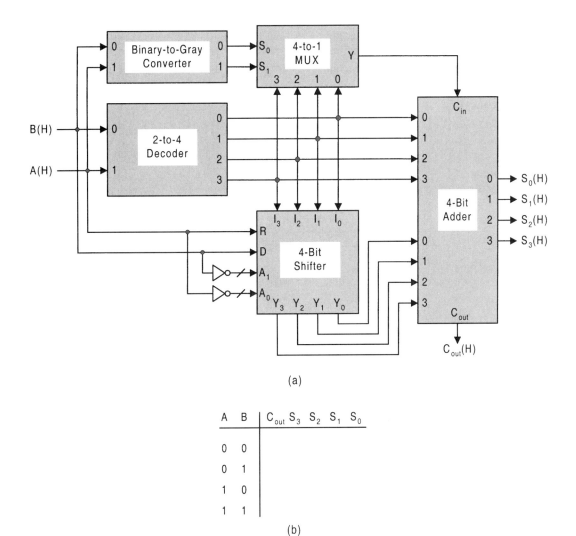

FIGURE P8.3

To do this, list all partial products, indicate the sum and carry values at each stage, and end with the final product values.

8.20 (a) Combine the 4 × 4 binary multiplier in Fig. 8.18 with the 8-bit binary-to-BCD converter in Fig. P6.3 to produce a one-digit BCD multiplier that can be bit-sliced to form an *n*-digit BCD multiplier. What restrictions must be placed on the operands?

(b) Test the results of part (a) by introducing the following operands into the resulting circuit of part (a):

Test #1 $A = 0111$; $B = 1000$
Test #2 $A = 1001$; $B = 0011$

	M	S_1	S_0	F	C_{out}	Operation*
Arithmetic Operations	0	0	0	$A \oplus B \oplus C_{in}$	$C_{in}(A \oplus B) + A \cdot B$	A plus B
	0	0	1	$A \oplus \bar{B} \oplus C_{in}$	$C_{in}(A \odot B) + A \cdot \bar{B}$	A minus B*
	0	1	0	$(A+B) \oplus C_{in}$	$C_{in} \cdot (A+B)$	A plus $\bar{A}B$
	0	1	1	$(\overline{AB}) \oplus C_{in}$	$C_{in} \cdot \bar{B} + A$	A plus $(A+\bar{B})$
Logic Operations	1	0	0	$A \oplus B$	ϕ	A XOR B
	1	0	1	$A \odot B$	ϕ	A EQV B
	1	1	0	$A + B$	ϕ	A OR B
	1	1	1	$\bar{A} \cdot \bar{B}$	ϕ	$\bar{A}$ AND $\bar{B}$

* Subtraction operations assume 2's complement arithmetic.

FIGURE P8.4

(c) Use a block symbol for the 1-digit BCD multiplier together with the block symbol for the binary-to-BCD converter of Fig. P6.3 to design a 2 × 2 BCD multiplier. To do this, form a array of 1-digit multipliers and connect them properly to a 4-digit BCD adder. Indicate the digit orders of magnitude (10^0, 10^1, 10^2, and 10^3) at all stages of the multiplier.

8.21 By using the results shown in Fig. 6.19, alter the design of the BCD multiplier of Problem 8.20 so as to produce a cascadable one-digit XS3 multiplier. (Hint: It will be necessary to design an XS3-to-BCD converter as an output device.)

8.22 With reference to Fig. 8.22, analyze the parallel divider shown in Fig. 8.23. To do this, introduce the operands $A = 11010$ and $B = 110$ and indicate on the logic circuit the logic value for each operand, borrow, remainder, and quotient.

8.23 Shown in Fig. P8.4 is the operation table for a cascadable one-bit arithmetic and logic unit (ALU) that has three mode/select inputs that control four arithmetic operations and four bitwise logic operations.
(a) Design this ALU by using a gate-minimum logic. Note that this design includes the use of compound XOR-type patterns similar to those used in Fig. 8.27. End with a logic circuit for both function F and C_{out}.
(b) Test the design of part (a) by introducing the following operands with $(C_{in})_{LSB} = \overline{Add/Sub}$. for arithmetic operations:

Tests #1 $A = 10$; $B = 11$ 2-Bit ALU; $MS_1S_0 = \begin{Bmatrix} 000 \\ 101 \end{Bmatrix}$

Tests #2 $A = 0100$; $B = 0111$ 4-Bit ALU; $MS_1S_0 = \begin{Bmatrix} 001 \\ 100 \end{Bmatrix}$

CHAPTER 9

Propagation Delay and Timing Defects in Combinational Logic

9.1 INTRODUCTION

To this point in the text, combinational logic circuits have been treated as though they were composed of "ideal" gates in the sense of having no propagation delay. Now it is necessary to take a step into the real world and consider that each gate has associated with it a propagation time delay and that, as a result of this delay, undesirable effects may occur.

Under certain conditions unwanted transients can occur in otherwise steady-state signals. These transients have become known as *glitches*, a term that derives from the German *glitsche*, meaning a "slip" (hence, the slang, error or mishap). A glitch is a type of *logic noise* that is undesirable because its presence in an output may initiate an unwanted process in a next-stage switching device to which that output is an input. In some circuits glitches can be avoided through good design practices; in other circuits they are unavoidable and must be dealt with accordingly.

There are three kinds of logic noise that occur in combinational logic circuits and that are classified as *hazards*.

Static hazards:
  Static 1-hazard (also called *SOP hazard*) — A glitch that occurs in an otherwise steady-state 1 output signal from SOP logic because of a change in an input for which there are two asymmetric paths (delay-wise) to the output.
  Static 0-hazard (also called *POS hazard*) — A glitch that occurs in an otherwise steady-state 0 output signal from POS logic because of a change in an input for which there are two asymmetric paths (delay-wise) to the output.

Static 1-Hazard
$1 \to 0 \to 1$

Static 0-Hazard
$0 \to 1 \to 0$

391

Dynamic hazards: Multiple glitches that occur in the outputs from multilevel circuits because of a change in an input for which there are three or more asymmetric paths (delay-wise) of that input to the output.

Dynamic 0→1→0→1 Hazard Dynamic 1→0→1→0 Hazard

Function hazards: A type of logic noise that is produced when two or more inputs to a gate are caused to change in close proximity to each other.

In this chapter the discussion will center on how these hazards occur and how they can be avoided or eliminated. Since the subject of hazards is also of considerable importance to sequential machine design, it with be revisited in subsequent chapters.

9.2 STATIC HAZARDS IN TWO-LEVEL COMBINATIONAL LOGIC CIRCUITS

A single glitch that is produced as a result of an asymmetric path delay through an inverter (or gate) is called a *static hazard*. The term "static" is used to indicate that the hazard appears in an otherwise steady-state output signal. Thus, a static hazard is not "stationary" or "motionless," as implied by the usual usage of the word static, but is quite unstationary and transient.

The best way to introduce static hazard detection and elimination in combinational logic is by means of simple examples. However, before proceeding further it will be helpful to define certain terms that are used in identifying static hazards in SOP or POS combinational logic circuits, and to provide a simple procedure for their elimination:

Coupled variable: An input variable that is complemented in one term of an output expression and uncomplemented in another term.
Coupled term: One of two terms containing *only one* coupled variable.
Residue: That part of a coupled term that remains after removing the coupled variable.
Hazard cover (or consensus term): The RPI required to eliminate the static hazard:
 AND the residues of coupled p-term to obtain the SOP hazard cover, or
 OR the residues of coupled s-terms to obtain the POS hazard cover.

Note that in either case the RPI (redundant prime implicant) is a result of the application of a consensus law given by Eqs. (3.14).

Static Hazard Detection and Elimination Static hazard detection involves identifying the coupled terms in an logic expression. Static hazard elimination occurs when the consensus p-term RPI is ORed to the SOP expression containing the static hazard, or when the consensus s-term RPI is ANDed to the POS expression containing the static hazard. Note that if the RPI is contained in a more minimum term, that term should be used.

9.2 STATIC HAZARDS IN TWO-LEVEL COMBINATIONAL LOGIC CIRCUITS

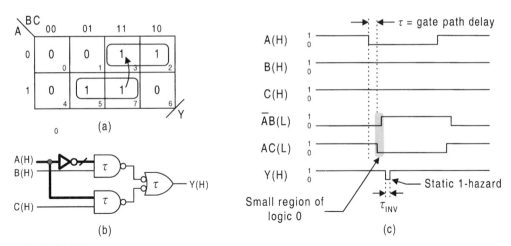

FIGURE 9.1
A static 1-hazard produced by a $1 \to 0$ change in input A. (a) K-map for Equation (9.1) showing loops for coupled terms and transition from state 111 to state 011. (b) Logic circuit for Equation (9.1) showing asymmetric paths for A. (c) Timing diagram for the circuit in (b) illustrating the production of the static 1-hazard after two gate path delays (2τ) following the change in A.

Consider the SOP function given by

$$Y = \bar{A}B + AC. \qquad (9.1)$$

$$\underset{011 \quad 111}{\phantom{Y = \bar{A}B + AC}}$$

Here, A is the coupled variable, $\bar{A}B$ and AC are the coupled terms, and the hazardous transition read in minterm code is $111 \to 011$ as indicated by the coupled terms in Eq. (9.1) and by the K-map in Fig. 9.1a. The logic circuit for Eq. (9.1) is given in Fig. 9.1b, where the two asymmetric paths of input A to the output are indicated by the heavier lines. With all inputs active high, the transition $111 \to 011$ produces a static 1-hazard after a $1 \to 0$ change in input A, as illustrated by the logic timing diagram in Fig. 9.1c. Thus, when the coupled terms are ORed, a small region of logic 0 (shaded region) creates the SOP hazard of magnitude equal to that through the inverter. The path delay through a NAND gate is designated by τ with input leads assumed to be ideal with zero path delay.

The ANDed residues of Eq. (9.1) is the RPI BC. When this is added to Eq. (9.1) there results

$$Y = \bar{A}B + AC + \underbrace{BC}_{\text{Hazard cover}}, \qquad (9.2)$$

which eliminates the hazard. This is demonstrated by the K-map, logic circuit, and timing diagram in Fig. 9.2. Notice that the hazard cover BC in the K-map of Fig. 9.2a is an RPI and that it covers the hazardous transition $111 \to 011$ indicated by the arrow. When this RPI is added to the original expression, as in Eq. (9.2), the result is the logic circuit in

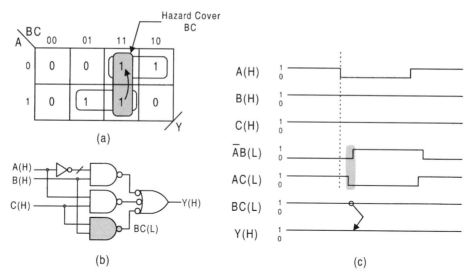

FIGURE 9.2
Elimination of the static 1-hazard by adding hazard cover. (a) K-map showing RPI that covers the hazardous transition 111 → 011. (b) Logic circuit that includes the shaded hazard cover gate $BC(L)$. (c) Timing diagram showing the elimination of the static hazard in Fig. 9.1 due to presence of the hazard cover term $BC(L)$.

Fig. 9.2b, which completely eliminates the static 1-hazard as illustrated in Fig. 9.2c. In fact, the hazard is eliminated regardless of the magnitude of the delay through the inverter — the hazard cannot form even if, for example, the delay is a ridiculous thousand times that of a NAND gate. It is also true that the results shown in Fig. 9.2 are valid if AND/OR/INV logic replaces the NAND/INV logic shown. In this case the coupled terms and hazard cover RPI would all be active high, but the waveforms would remain the same, the benefit of mixed-logic notation.

There is the question of how the activation level of the coupled variable A affects the results illustrated in Figs. 9.1 and 9.2. If A arrives active low, $A(L)$, then the inverter must be placed on the A line to the AC gate. The static 1-hazard is still formed, but as a result of a 011 → 111 transition following a 0 → 1 change in input A. This is illustrated in Fig. 9.3 for purposes of comparison with Figs. 9.1 and 9.2. Nevertheless, the static 1-hazard is eliminated by the hazard cover BC as shown in Fig. 9.3. Again, replacing the NAND/INV logic by AND/OR/INV logic would not alter the waveforms but would change the activation levels of the coupled terms and hazard cover to active high.

The forgoing discussion dealt with static hazards in SOP logic. The detection and elimination of static 0-hazards in POS combinational logic follows in similar but dual fashion. Consider the function

$$Y = (A + B)(\bar{A} + C), \tag{9.3}$$

with transition $000 \to 100$.

where A is again the coupled variable but now $(A + B)$ and $(\bar{A} + C)$ are the coupled terms.

9.2 STATIC HAZARDS IN TWO-LEVEL COMBINATIONAL LOGIC CIRCUITS 395

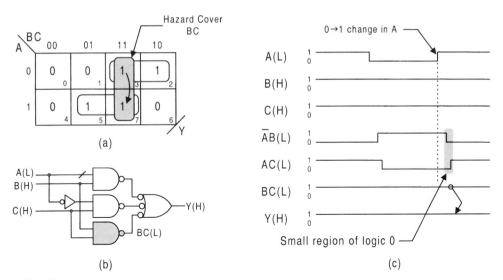

FIGURE 9.3
Elimination of the static 1-hazard for the case of an active low coupled variable. (a) K-map showing RPI that covers the hazardous transition $011 \to 111$. (b) Logic circuit that includes the shaded hazard cover gate $BC(L)$. (c) Timing diagram showing the elimination of the static 1-hazard due to presence of the hazard cover term $BC(L)$.

Read in maxterm code, assuming NOR/INV logic and active high inputs, the hazardous transition is $000 \to 100$ following a $0 \to 1$ change in A as shown in Fig. 9.4a. The logic circuit for Eq. (9.3) is given in Fig. 9.4b, where the two asymmetric paths for input A to the output are indicated by heavy lines. The static 0-hazard is formed as a result of the two

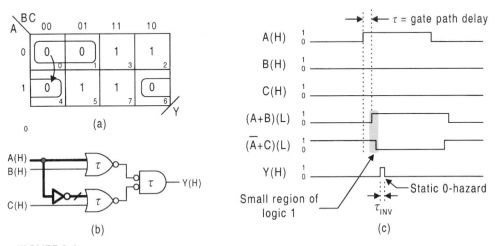

FIGURE 9.4
A static 0-hazard produced by a $0 \to 1$ change in input A. (a) K-map for Eq. (9.3) showing loops for coupled terms and transition from state 000 to state 100. (b) Logic circuit for Eq. (9.3) showing asymmetric paths for A. (c) Timing diagram for the circuit in (b) illustrating the production of the static 0-hazard after two gate path delays 2τ following the change in A.

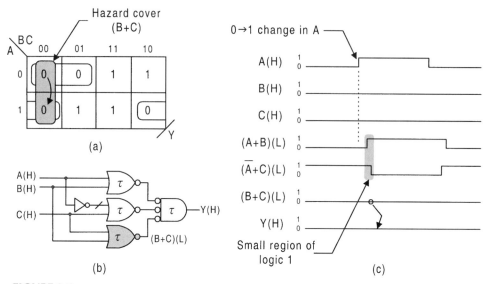

FIGURE 9.5
Elimination of the static-0 hazard by adding hazard cover. (a) K-map showing RPI that covers the hazardous transition 000 → 100. (b) Logic circuit that includes the shaded hazard cover gate $(B + C)(L)$. (c) Timing diagram showing the elimination of the static 0-hazard due to presence of the hazard cover term $(B + C)(L)$.

asymmetric paths (delay-wise) such that a small region of logic 1 exists when the waveforms for the coupled terms are ANDed as illustrated in Fig. 9.4c.

The static 0-hazard shown in Fig. 9.4c is eliminated by adding hazard cover to the function of Eq. (9.3). The ORed residues for function Y in Eq. (9.3) is $(B + C)$. When this is added (ANDed) to Eq. (9.3), there results

$$Y = (A + B)(\bar{A} + C) \cdot \underbrace{(B + C)}_{\text{Hazard cover}}, \tag{9.4}$$

which eliminates the static 0-hazard as illustrated in Fig. 9.5. Notice that $(B + C)$ covers the transition 000 → 100 as indicated by the arrow in the K-map of Fig. 9.5a, and that it is by definition an RPI. The logic circuit for Eq. (9.4) is given in Fig. 9.5b and is now hazard-free as illustrated in Fig. 9.5c. Here, it is seen that the small region of logic 1 that caused the static 0-hazard in Fig. 9.4 has been rendered ineffectual because of the ANDed RPI $(B + C)(L)$, which remains at 0(L) during the hazardous transition. Note that if only input A arrives active low $A(L)$, the static 0-hazard still forms, but as result of a 100 → 000 transition following a 1 → 0 change in A. Changing from NOR/INV logic to OR/AND/INV logic in Figs. 9.4 and 9.5 does not alter any of the conclusions drawn to this point. However, the activation levels of the coupled terms and hazard cover must be changed to active high in Fig. 9.5c, but leaving their waveforms unaltered.

Detection and elimination of static hazards in two-level combinational logic circuits is actually much simpler than would seem evident from the foregoing discussion. Actually, all that is necessary is to follow the simple procedure given next.

9.2 STATIC HAZARDS IN TWO-LEVEL COMBINATIONAL LOGIC CIRCUITS

Procedure for Detection and Elimination of Static Hazards in Combinational Logic Circuits

1. Identify the couple terms in an SOP or POS expression.
2. OR their consensus (RPI) P-terms to the SOP expression, or AND their consensus s-terms to the POS expression.
3. Reject any set of two terms containing more than one couple variable. Remember: *Only one variable is allowed to change in a static hazard transition*. All other variables *must* be constant.
4. Read the initial and final states from the coupled terms in a hazardous transition by using minterm code for SOP and maxterm code for POS.

As an example, consider the following minimum SOP function of four variables showing two hazard transitions together with the corresponding hazard cover for each:

$$F_{SOP} = B\bar{C}\bar{D} + \bar{A}\bar{C}D + \bar{B}C + \underbrace{\bar{A}\bar{B}D + \bar{A}B\bar{C}}_{Hazard\ cover}. \tag{9.5}$$

with transitions 0001 → 0011 and 0100 → 0101.

In this expression, NAND/INV logic is assumed with inputs that arrive active high. Here, the hazard cover $\bar{A}B\bar{C}$ is the ANDed residues (consensus p-term) of coupled terms $B\bar{C}\bar{D}$ and $\bar{A}\bar{C}D$, where D is the coupled variable. And hazard cover $\bar{A}\bar{B}D$ is the ANDed residues of couple terms $\bar{A}\bar{C}D$ and $\bar{B}C$, where C is the coupled variable. These static 1-hazard transitions are illustrated in the K-map of Fig. 9.6a by using arrows indicating a 1 → 0 change in the couple variable. The consensus terms can be seen to cover the hazard transitions (arrows).

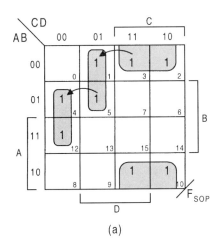
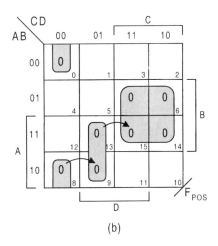

FIGURE 9.6
K-map representation of static hazards in the function F represented by Eq. (9.5) and by Eq. (9.6). (a) Static 1 hazard transitions. (b) Static 0 hazard transitions.

Terms $B\bar{C}\bar{D}$ and $\bar{B}C$ cannot be coupled terms and cannot produce a static hazard, since they contain more than one couple variable–they have no consensus term. These terms form what is called a *function hazard*, a subject that is discussed in a later section.

The procedure for detection and elimination of static 0-hazards in POS logic is the dual of that for the detection and elimination of static 1 hazards in SOP logic. As an example, consider the minimum POS representation of the function F together with the hazard transitions and hazard cover given in Eq. (9.6):

$$F_{POS} = (B + C + D)(\bar{A} + C + \bar{D})(\bar{B} + \bar{C}) \cdot \underbrace{(\bar{A} + B + C)(\bar{A} + \bar{B} + \bar{D})}_{Hazard\ cover}. \quad (9.6)$$

with transitions 1101 → 1111 and 1000 → 1001.

Here again, inputs C and D are the coupled variables where the inputs are assumed to arrive active high but with NOR/INV logic. Notice that the initial and final states are read from the coupled terms by using maxterm code and are exactly those shown by arrows in Fig. 9.6b, indicating a $0 \rightarrow 1$ change in the coupled variable. The hazard covers are the ORed residues (consensus s-terms) of the coupled terms and are ANDed to the original expression in Eq. (9.6). Note also that the s-terms $(B + C + D)$ and $(\bar{B} + \bar{C})$ are not coupled terms and cannot produce a static 0-hazard since they contain two coupled variables — remember, only one variable is allowed to change in the production of a static hazard. Also, if a coupled variable arrives active low with all other inputs active high, then the arrow indicating the hazard transition for that coupled variable must be reversed in Eq. (9.6).

As a final example, consider the function Z of five variables,

$$Z = B\bar{C}D + C D\bar{E} + \bar{A}D\bar{E} + ABE + \bar{A}\bar{B}C + \underbrace{\bar{A}BC\bar{E} + ABCD}_{Hazard\ cover}, \quad (9.7)$$

with transitions 11100 → 11101 and 01010 → 01000.

where the coupled variables are easily seen to be D and E. Assuming NAND/INV or AND/OR/INV logic and that all inputs arrive active high, the two hazard transitions are indicated by arrows in Eq. (9.7). Thus, read in minterm code, coupled terms $B\bar{C}D$ and $\bar{A}D\bar{E}$ produce a static 1-hazard during the transition $01010 \rightarrow 01000$ following a $1 \rightarrow 0$ change in D, while terms ABE and $CD\bar{E}$ generate a static 1-hazard during the transition $11101 \rightarrow 11100$ following a $1 \rightarrow 0$ change in E. Adding the ANDed residues of the coupled terms for each hazard transition gives the hazard-free expression in Eq. (9.7).

It should be clear to the reader that the detection and elimination of static hazards can be conducted without the aid of K-maps or logic circuits simply by following the four steps given previously. Exceptions to this rule are discussed in the next section, where diagrams of a rather different sort are used to simplify the process of identifying and eliminating static hazards in multilevel XOR-type functions.

9.3 DETECTION AND ELIMINATION HAZARDS IN MULTILEVEL XOR-TYPE FUNCTIONS

Conventional static hazard analysis used in two-level logic does not address the problem of hazards in multilevel XOR-type functions. This section presents a simple but general procedure for the detection and elimination of static hazards in these functions. It is shown that all static hazards can be eliminated with redundant cover derived by using a method based on lumped path delay diagrams (LPDDs), and that this method is of unrestricted applicability. The problems associated with dynamic hazards, as they relate to static hazard cover, are also considered.

Multilevel XOR-type functions of the type considered in this section find use in arithmetic circuits, such as dedicated arithmetic logic units, and in error detection circuits. These functions can be obtained from logic synthesis algorithms or from K-map extraction as was demonstrated in Chapter 5. If steady, clean outputs from these functions are required, it is necessary to remove any logic (hazard) noise that may be present.

Modern CMOS IC technology has produced XOR and EQV gates whose speed and compactness are close to those of other two-input gates (see, e.g., Figs. 3.26 and 3.27). This has made the use of XOR and EQV gates more practical and has led to the development of various methods of multilevel function implementation that take advantage of these gates. These implementations can produce gate-minimum results not possible with two-level logic. When fan-in restrictions on two-level implementations are considered, multilevel implementations become even more attractive.

The simpler multilevel functions include the XOR/SOP and EQV/POS forms. The XOR/SOP form connects p-terms with OR and XOR operators, while the EQV/POS form connects s-terms with AND and EQV operators. XOP and EOS forms are special cases of XOR/SOP and EQV/POS, respectively, and are considered to be special two-level forms of representation. The XOP form connects p-terms only with XOR operators, and the EOS form connects s-terms only with EQV operators. Multilevel functions more complex than XOR/SOP and EQV/POS are classified simply as compound multilevel forms for lack of a practical classification scheme.

As was stated in Section 9.2, a static hazard is a glitch in an otherwise steady-state output signal and is produced by two asymmetric paths from a single input. Figure 9.7 is a generalization of the condition that allows the static hazard to form in multilevel circuits. The coupled variable must be an input to the initial gate in each path (Gates 1 and 3 in

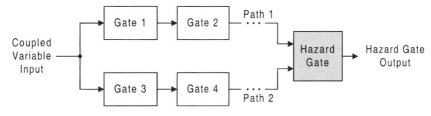

FIGURE 9.7
Alternative paths of the coupled variable to the hazard gate that is necessary for static hazard formation in a multilevel logic circuit.

Fig. 9.7). The signal must propagate through each path until both paths reach a single gate, called the *hazard gate*. If, for example, the signal requires more time to propagate through Path 1 than through Path 2, the signal from Path 2 will reach the hazard gate before the signal from Path 1. This asymmetry in path delay may cause a static hazard, depending on the nature of the hazard gate and the gates in the two paths. Here, an inverter is considered to be a gate and the hazard gate may or may not be the output gate. Also, lead delays are treated as ideal with negligible path delay, and each gate has an inherent delay associated with it that depends on its type, the number of inputs, and the technology used to implement it.

Static hazard analysis in multilevel functions is more complicated than in two-level logic. Nevertheless, the static hazard is formed in agreement with the requirements of Fig. 9.7. Each path may consist of any number of gates, and the gates in Path 1 may differ in several respects from those in Path 2. Thus, if

$$(\Sigma \text{ Path 1 delays}) \neq (\Sigma \text{ Path 2 delays}), \tag{9.8}$$

hazard formation is possible according to Fig. 9.7. Furthermore, in multilevel functions of the type considered in this section, the difference between the Path 1 and Path 2 delays may range from that of an inverter to one or more gate delays. Thus, the size (or strength) of the static hazard glitch in a multilevel logic circuit may be considerable. Whereas the static hazard glitch in two-level logic caused by an inverter may or may not cross the switching threshold, a static hazard glitch in a multilevel XOR-type circuit may be quite large and may easily cross the switching threshold.

9.3.1 XOP and EOS Functions

The simplest XOR/SOP or EQV/POS functions that produce static hazards are very similar to conventional two-level functions. If no more than one term in an SOP function can be active at any given time, the terms are mutually disjoint and the OR operators can be replaced with XOR operators as indicated by Corollary I in Subsection 3.11.1. The result is an XOP function. Hazards in an XOP function can be detected and eliminated by a method parallel to that described for SOP functions in Section 9.2.

As an example, consider the reduced function N in SOP and XOP form:

$$N_{SOP} = \bar{A}BC + ABD + A\bar{B}CD + A\bar{B}\bar{C}\bar{D} \tag{9.9}$$

$$N_{XOP} = \bar{A}BC \oplus ABD \oplus A\bar{B}CD \oplus A\bar{B}\bar{C}\bar{D}. \tag{9.10}$$

The p-terms are mutually disjoint, so the direct conversion from SOP to XOP is permitted. It follows from the conventional methods discussed in Section 9.2 that two static hazards will occur in N_{SOP} of Eq. (9.9): between coupled terms ABD and $\bar{A}BC$ on a $1111 \rightarrow 0111$ transition following a $1 \rightarrow 0$ change in A, and between coupled terms ABD and $A\bar{B}CD$ on a ($1111 \rightarrow 1011$) transition following a $1 \rightarrow 0$ change in B. Each hazard is caused by an inverter through which the coupled variable must pass. This inverter makes the two path delays (Fig. 9.7) unequal, allowing the hazards to form following a change in each coupled variable. Each hazard is eliminated by adding a consensus p-term consisting of the ANDed

9.3 DETECTION AND ELIMINATION HAZARDS

residues, giving the result

$$N_{SOP} = \bar{A}BC + ABD + A\bar{B}CD + A\bar{B}\bar{C}\bar{D} + \underbrace{(BCD + ACD)}_{Hazard\ cover} \qquad (9.11)$$

The method of hazard detection and elimination for N_{XOP} in Eq. (9.10) is quite similar to that of N_{SOP} in Eq. (9.11). However, a hazard can now occur in either direction between the coupled terms. Thus, for coupled terms ABD and $\bar{A}BC$ a static 1-hazard is produced on either a 1111 → 0111 or 0111 → 1111 transition following a 1 → 0 or 0 → 1 change in couple variable A. Similarly, for coupled terms ABD and $A\bar{B}CD$ a static 1-hazard can occur on either a 1111 → 1011 or 1011 → 1111 transition following a 1 → 0 or 0 → 1 change in coupled variable B. The complete details of these events are provided in Eq. (9.12), where the two hazards have been eliminated by the adding hazard cover. Note that each cover term is added to the function with an OR operator and not with an XOR operator. This is important because after hazard cover is added, the terms are no longer mutually disjoint. Therefore, adding the hazard cover by using an XOR operator would fundamentally alter the function.

$$N_{XOP} = \bar{A}BC \oplus ABD \oplus A\bar{B}CD \oplus A\bar{B}\bar{C}\bar{D} + \underbrace{(BCD + ACD)}_{Hazard\ cover} \qquad (9.12)$$

The bidirectionality of the XOP hazard production in Eq. (9.12) is due to the nature of the XOR gate. Unlike an OR gate or an AND gate, an XOR gate will produce an output change with *any* single input change. Furthermore, if both inputs to an XOR gate change simultaneously, the final output will be the same as the initial output. Therefore, if the two inputs change at different times but in close proximity to one another, a short glitch will occur, regardless of the state transition involved or the order in which the inputs change. However, on a 01 → 10 change in the two inputs to an OR gate, for example, the order in which the inputs change will determine whether or not a hazard occurs. This difference between XOR gates and OR gates is the reason a static hazard can be caused by a coupled variable change in either direction for an XOP function but in only one direction for an SOP function.

The timing diagram in Fig. 9.8 illustrates the behavior of N_{SOP} and N_{XOP} without and with hazard cover. At least one hazard cover signal is active during each hazard to prevent it from propagating to the final output signal. Note that hazards occur in N_{SOP} only on the falling edge of the coupled variable, but occur in N_{XOP} on both the rising and falling edge of the coupled variable.

A relationship similar to that between SOP and XOP functions exists between POS and EOS functions. If no more than one term in a POS function can be inactive at any given time, the terms are mutually conjoint and the AND operators can be replaced with EQV operators to form an EOS function. Hazards in the EOS function will be identical to those in POS, except that hazards in EOS will occur in both transitional directions. Hazard cover is formed by ORing the residues of the two coupled terms. An AND operator must be used to connect each hazard cover term to the original function.

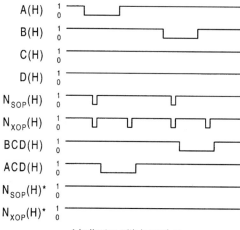

* Indicates with hazard cover

FIGURE 9.8

Timing diagram for functions N_{SOP} and N_{XOP} without and with (*) hazard cover in accordance with Eqs. (9.11) and (9.12).

As an example, consider the reduced POS and EOS forms of function L, which are the complements of function N in Eqs. (9.11) and (9.12), i.e., $L = \bar{N}$. These POS and EOS functions are represented by the following expressions, together with the hazard transitions and hazard cover for each:

$$L_{POS} = (A + \bar{B} + \bar{C})(\bar{A} + \bar{B} + \bar{D})(\bar{A} + B + \bar{C} + D)(\bar{A} + B + C + D)$$

$$\cdot \underbrace{(\bar{B} + \bar{C} + \bar{D})(\bar{A} + \bar{C} + \bar{D})}_{Hazard\ cover} \qquad (9.13)$$

$$L_{EOS} = (A + \bar{B} + \bar{C}) \odot (\bar{A} + \bar{B} + \bar{D}) \odot (\bar{A} + B + \bar{C} + D) \odot (\bar{A} + B + C + D)$$

$$\cdot \underbrace{(\bar{B} + \bar{C} + \bar{D})(\bar{A} + \bar{C} + \bar{D})}_{Hazard\ cover}. \qquad (9.14)$$

The coupled variables are A and B, and the coupled terms and hazard transitions are indicated by arrows. The hazard cover terms for both the POS and EOS forms are $(\bar{B} + \bar{C} + \bar{D})$ and $(\bar{A} + \bar{C} + \bar{D})$, each of which is the ORed residues of the respective coupled terms. Notice that in both cases the hazard cover terms are ANDed to the original expressions.

The timing diagram in Fig. 9.9 illustrates the behavior of L_{POS} and L_{EOS} without and with hazard cover. This behavior is similar to that of N_{SOP} and N_{XOP} in Fig. 9.8, but static

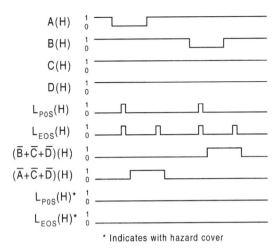

FIGURE 9.9
Timing diagram for functions L_{POS} and L_{EOS} without and with (*) hazard cover in accordance with Eqs. (9.13) and (9.14).

hazards are static 0 hazards in L_{POS} and L_{EOS} rather than the static 1 hazards as in N_{SOP} and N_{XOP}. Again notice that the static hazards in L_{EOS}, like those in N_{XOR}, are formed following both a $0 \rightarrow 1$ and $1 \rightarrow 0$ change in the coupled variable. It is this characteristic that distinguishes SOP and POS forms from XOP and EOS forms. The former types generate static hazards on a single change of the coupled variable, whereas the latter types generate static hazards on both $1 \rightarrow 0$ and $0 \rightarrow 1$ changes in the coupled variable.

9.3.2 Methods for the Detection and Elimination of Static Hazards in Complex Multilevel XOR-type Functions

Function minimization methods involving K-map XOR-type patterns and Reed–Muller transformation forms are considered in detail in Chapter 5. For certain functions these methods lead to gate-minimum forms that cannot be achieved by any other means. An inspection of these forms reveals that they are of the general form

$$F = (\alpha \oplus \beta) + \Gamma + \cdots, \tag{9.15}$$

where α, β, and Γ can be composed of SOP, POS, XOP, or EOS terms or some combination of these. The XOP and EOS functions discussed in Subsection 9.3.1 are a special case of Eq. (9.15), and the methods used there for the detection and elimination of static hazards parallel those used for two-level logic discussed in Section 9.2. However, these simple methods cannot be applied to the more complex functions considered in this subsection. Now, use must be made of special graphic methods to assist in the determination of path delay asymmetry according to Fig. 9.1.

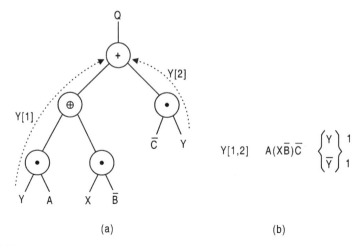

FIGURE 9.10
Hazard detection and hazard cover for function Q in Eq. (9.15). (a) LPDD showing two paths for input Y. (b) Path requirements of input Y to produce an active Q output and the hazard cover for the static 1 hazard.

Consider the five-variable function

$$Q = (AY) \oplus (X\bar{B}) + \bar{C}Y, \qquad (9.16)$$

which is a relatively simple form of Eq. (9.15). This function has a static 1-hazard for which the hazard cover cannot be readily identified by the standard methods used in Subsection 9.3.1. Other means must be used to identify the conditions for hazard formation and the cover that is necessary to eliminate it.

Shown in Fig. 9.10a is the *lumped path delay diagram* (LPDD) that is a graphical equivalent of the logic circuit for function Q in Eq. (9.16). Use of the LPDD makes possible a simple means of detecting and eliminating the static 1-hazard. However, some explanation of the use of this diagram is needed. The inputs are assumed to be active high, and inputs such as $\bar{B}$ and $\bar{C}$ imply the use of an inverter that is not shown. Two paths, $Y[1]$ and $Y[2]$, for variable Y are shown from input to output Q. The path requirements for input Y and $\bar{Y}$ that cause the output Q to be active are given in Fig. 9.10b. Thus, for Y inactive ($\bar{Y}$), path $Y[1]$ is enabled to cause $Q = 1$ if both A and $X\bar{B}$ are active. And for Y active, path $Y[2]$ is enabled to cause $Q = 1$ if $\bar{C}$ active ($C = 0$). The hazard cover is found by ANDing both enabling path requirements to give ($AX\bar{B}\bar{C}$) as indicated in Fig. 9.10b. But for Y inactive, input A is irrelevant to path $Y[1]$. Thus, the final result for function Q is given by

$$\begin{aligned} Q &= (AY) \oplus (X\bar{B}) + \bar{C}Y + AX\bar{B}\bar{C} \\ &= (AY) \oplus (X\bar{B}) + \bar{C}Y + X\bar{B}\bar{C}. \end{aligned} \qquad (9.17)$$

The timing diagram for function Q in Fig. 9.11 confirms the results presented in Fig. 9.10 and in Eq. (9.17). Notice that the hazard cover $\bar{B}\bar{C}X$ removes the static 1-hazard as indicated by $Q(H)^*$. The size (strength) of the static hazard, indicated by Δt, is the difference in path

9.3 DETECTION AND ELIMINATION HAZARDS

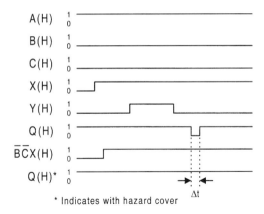

* Indicates with hazard cover

FIGURE 9.11
Timing diagram for functions Q without and with (*) hazard cover in agreement with Eq. (9.17).

delays $Y[1]$ and $Y[2]$ in Fig. 9.10a as it relates to Fig. 9.1. This path delay asymmetry may be expressed as

$$\Delta t = (t_{AND} + t_{XOR}) - (t_{AND}) = t_{XOR},$$

which is easily ascertained from an inspection of Fig. 9.10a.

That the hazard cover for Eq. (9.16) is $\bar{B}\bar{C}X$ and is independent of input A can be easily demonstrated by the use of a binary decision diagram (BDD) constructed from an inspection of Eq. (9.16). This BDD is given in Fig. 9.12, where the binary decisions begin with variable Y and end with an output (for Q) that is either logic 0 or logic 1. The requirement for static 1-hazard cover is that the decisions must lead to an active output Q. Thus, for $Y = 0$ the path to $Q = 1$ is enabled if $X = 1$ and $B = 0$ or $X\bar{B}$ and is enabled for $Y = 1$ if $C = 0$

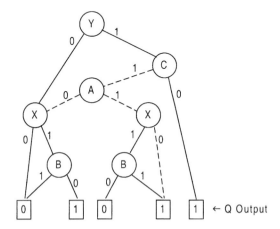

FIGURE 9.12
BDD for function Q in Eq. (9.16) showing the binary decisions required for an active output independent of input A.

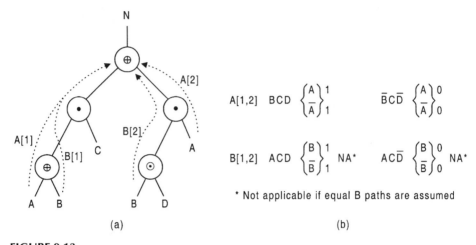

FIGURE 9.13
Hazard detection and hazard cover for function N in Eq. (9.18). (a) LPDD showing two paths for input A and two paths for input B. (b) Path requirements for inputs A and B required to produce an active N output, and the hazard covers necessary to eliminate the static 1 hazards.

or $\bar{C}$, all read in positive logic. ANDing the two enabling conditions gives the hazard cover $X\bar{B} \cdot \bar{C}$. Clearly, the binary decisions involving input A (dashed lines) are irrelevant to the enabling path conditions for an active output Q.

A BDD is a graphical representation of a set of binary-valued decisions, each of which ends with a result that is either logic 1 or logic 0. Thus, the BDD allows one to easily determine the output for any possible combination of input values. The BDD is used by starting with an input variable (top of the BDD) and proceeding in a downward path to an output logic value that corresponds to the value of the last input in that path. Thus, the final element, usually presented in a rectangle, is the logic value of the function (e.g., output Q in Fig. 9.12) for the input values used.

As a second and more complex example, consider the four-variable function

$$N = [(A \oplus B)C] \oplus [(B \odot D)A], \qquad (9.18)$$

which is a three-level form of Eqs. (9.9) and (9.10) obtained by the CRMT method discussed in Section 5.7 taking $\{A, C\}$ as the *bond set*. The LPDD, shown in Fig. 9.13, indicates that there is at least one static 1-hazard and one static 0-hazard associated with this function. Input A has two paths to the output Z, $A[1]$ and $A[2]$. Path $A[1]$ to output N is enabled if C is active, with B active or inactive depending on input A. Path $A[2]$ to output N is enabled if B and D are logically equivalent (via the EQV operator). Therefore, for input conditions BCD, output N is active for $A = 1$ via path $A[2]$ only and is active for $A = 0$ via path $A[1]$ only, indicating a static 1-hazard. However, for input conditions $\bar{B}C\bar{D}$ the output N is inactive for both A and $\bar{A}$: For $A = 1$ the output $N = 1 \oplus 1 = 0$, and for $A = 0$ the output $N = 0 \oplus 0 = 0$. When the output is inactive for both input conditions, a static 0-hazard is indicated. The result of this static hazard analysis is that BCD becomes the static 1-hazard cover that must be ORed to Eq. (9.18), and that $\overline{\bar{B}C\bar{D}} = (B + \bar{C} + D)$ is the static 0-hazard cover that must be ANDed to Eq. (9.18). When hazard cover is added, the final expression

9.3 DETECTION AND ELIMINATION HAZARDS

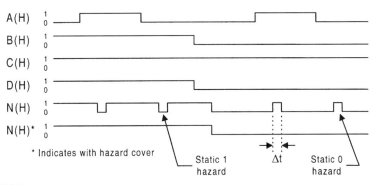

FIGURE 9.14
Timing diagram for function N in Eq. (9.18) showing static 1 and static 0 hazards and showing the result of adding hazard cover according to Eq. (9.19).

for N becomes

$$N = \{[(A \oplus B)C] \oplus [(B \odot D)A] + \underbrace{BCD\} \cdot (B + \bar{C} + D)}_{Hazard\ cover}, \quad (9.19)$$

which is now a five-level function. Note that the order in which the hazard cover is added is immaterial. Thus, Eq. (9.19) could have been completed by first ANDing $(B + \bar{C} + D)$ to the original expression followed by ORing BCD to the result.

The paths $B[1]$ and $B[2]$ are not expected to create static hazards, assuming that the XOR and EQV gates have nearly the same path delays. This is a good assumption if CMOS technology is used for their implementation as in Figs. 3.26 and 3.27. However, if the two B path delays are significantly asymmetric, then both static 1 and static 0 hazards would exist and would be eliminated by adding hazard covers in addition to those for the path A hazards (see Fig. 9.13b).

The timing diagram in Fig. 9.14 illustrates the results expressed by Eqs. (9.18) and (9.19) and by Fig. 9.13. The static 1-hazard occurs with changes in A under input conditions BCD, and static 0 hazards occur with changes in A under input conditions $\bar{B}C\bar{D}$. But when BCD is ORed to the expression in Eq. (9.18), the static 1-hazard disappears. Similarly, when $\bar{B}C\bar{D} = (B + \bar{C} + D)$ is ANDed to function N in Eq. (9.18), the static 0 hazards disappear. Notice that the strength of either type of static hazard in Fig. 9.14 is the difference in delay between the two paths expressed by

$$\Delta t = (t_{XOR} + t_{AND}) - t_{AND} = t_{XOR},$$

where each hazard is initiated after a delay of $(t_{XOR} + t_{AND})$ following a change in input A. This information is easily deduced from an inspection of Fig. 9.13.

The BDD for function N in Eq. (9.18) is given in Fig. 9.15. Once the coupled variables have been identified by the LPDD, the BDD can be used to obtain the hazard cover for both the static 1-hazard and static 0-hazard. An inspection of the binary decisions required to render $N = 1$ indicate a path BC for input condition $A = 0$ and a path BD for $A = 1$. When these two input paths are ANDed together the result is BCD, the enabling condition

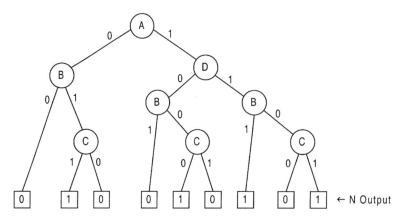

FIGURE 9.15
BDD for function N in Eq. (9.18) showing binary decisions required for static 1 and static 0 hazard formation.

for the static 1-hazard to form and, hence, also the hazard cover for that hazard. There are no other valid input paths for output $N = 1$, since their ANDing (intersection) is logic 0. In a similar fashion, the binary decisions required to produce $N = 0$ indicate a path $\bar{B}$ for $A = 0$ and $\bar{D}\bar{B}C$ for input condition $A = 1$. When these input path conditions are ANDed the result is $\bar{B}C\bar{D}$, which when complemented yields $(B + \bar{C} + D)$, the hazard cover for the static 0-hazard. All other ANDed input path combinations for $A = 1$ result in logic 0 and hence are invalid.

BDDs can be very useful in identifying the hazard cover(s) for a given coupled variable, which is best identified by first by using an LPDD. The difficulty is not in the reading of the BDD to obtain the hazard cover, but in its construction. The reader should appreciate the fact that constructing of a BDD from a Boolean expression of the type considered in this section is no trivial task. In contrast, the LPDD, which is essentially a logic circuit, is easily constructed from the Boolean expression. For this reason, LPDDs should be used for most hazard analyses, reserving the use of BDDs for the difficult cases where the hazard cover is not easily revealed by an inspection of the LPDD.

9.3.3 General Procedure for the Detection and Elimination of Static Hazards in Complex Multilevel XOR-Type Functions

The static 1 and static 0 hazards in N were detected and eliminated by following a procedure that is applicable to any function. The procedure consists of the following three steps:

Step I: *Use an LPDD to identify the two paths for each coupled variable whose path delays to the output differ according to Fig. 9.1.* A determination of the path delays is not always a straightforward task, since the technology used for the individual gates may not be known. Worse yet, integrated circuits may make such determination nearly impossible without empirical data.

Step II: *Find the hazard conditions and hazard cover for each coupled input variable in the LPDD by ANDing the variables that enable the two paths from the coupled variable to the output with those variables required to block (disable) all other paths.* The gates

9.3 DETECTION AND ELIMINATION HAZARDS   409

in the two enabling paths must not prevent the propagation of the coupled variable to the output stage. To accomplish this, other noncoupled variable inputs to AND operators must be active (logic 1), and other noncoupled variable inputs to OR operators must be inactive (logic 0). All other paths to the output stage must be blocked by proper selection of input activation levels. The use of a BDD is quite suitable for the purpose of finding hazard cover. Moreover, the BDD can also be used to simplify separately static 1-hazard cover and static 0-hazard cover. Note that static 1 and static 0 covers *must never* be simplified together.

For N_{XOP}, the LPDD (Fig. 9.13a) shows that C must be logic 1 so that the AND operator does not block path $A[1]$. The output from the EQV operator, which becomes an input to the AND operator in path $A[2]$, must also be logic 1. Inputs B and D must therefore be logically equivalent so as to enable path $A[2]$ to the output. There are no other paths to consider. Thus, the hazard cover is BCD for the static 1-hazard and $\overline{BCD} = (B + \bar{C} + D)$.

Step III: *Add the hazard cover to the original function by using an OR operator for static 1-hazard cover and by using an AND operator for a static 0-hazard cover.* The 1-hazard cover and the 0-hazard cover must be added separately but may be added in either order. In Eq. (9.19) the static 1-hazard cover is added to N before the static 0-hazard cover. If the 0-hazard cover were added to Eq. (9.19) before the 1-hazard cover, the result would be

$$N = \{[(A \oplus B)C] \oplus [(B \odot D)A] \cdot \underbrace{(B + \bar{C} + D)}_{Hazard\ cover}\} + BCD \qquad (9.20)$$

Eqs. (9.19) and (9.20) have gate/input tallies of 9/20, excluding inverters. There is often little advantage of one implementation over the other.

9.3.4 Detection of Dynamic Hazards in Complex Multilevel XOR-Type Functions

Up to this point the discussion has centered around the detection and elimination of static hazards in multilevel XOR-type functions. As it turns out these functions also have a propensity to form dynamic hazards, and there may or may not be a means of eliminating these defects. One example is the five-variable, four-level function given by

$$K = [B + (A \oplus X)] \odot \{[Y + (A \oplus B)] \odot [\bar{A} + Z]\}. \qquad (9.21)$$

The LPDD for this function, presented in Fig. 9.16a, reveals both static and dynamic hazards as indicated in Figs. 9.16b and 9.16c. The enabling path conditions for inputs A and B required to produce the static 1 and static 0 hazards are found in the same manner as for function N in Fig. 9.13. Obviously, static hazard analysis is considerably more involved than the analysis for function N. Furthermore, the additional logic required to eliminate all static hazards in function K is considerable. To eliminate both static 1 and 0 hazards requires that four p-terms be ORed to function K in Eq. (9.21) and by the ORing of three s-terms to that result, as required by Fig. 9.16b. The order in which the ORed and ANDed terms are added is immaterial.

The input conditions required to produce dynamic hazards in function K are given in Fig. 9.16c. Remember: *Dynamic hazards can exist in a given function only if there are three or more paths of an input to the output.* This condition is satisfied for input A as indicated. Notice that a dynamic hazard is identified in an LPDD when the enabling paths

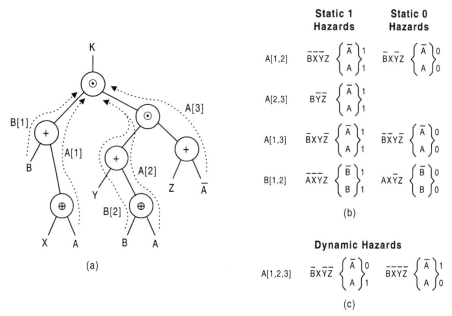

FIGURE 9.16
Hazard analysis of function K in Eq. (9.21). (a) LPDD showing three paths for input A and two paths for input B. (b) Path enabling requirements for A and B to produce static 1 and static 0 hazards. (c) Path A enabling requirements for the production of dynamic hazards in function K.

of the coupled variable to the output yield both a logic 1 and logic 0 as in Fig. 9.16c. This same information can be gleaned from a BDD, but, because of the difficulty in constructing the BDD, the LPDD approach is preferred.

The timing diagram for Eq. (9.21) is shown in Fig. 9.17, where dynamic hazards of the 1-0-1-0 and 0-1-0-1 types occur following $1 \to 0$ and $0 \to 1$ changes in coupled variable

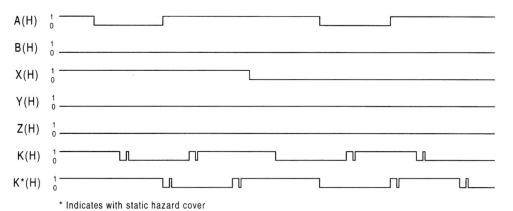

* Indicates with static hazard cover

FIGURE 9.17
Timing diagram for function K in Eq. (9.21), showing dynamic hazards produced without and with static hazard cover under input conditions given in Fig. 9.16c.

9.3 DETECTION AND ELIMINATION HAZARDS

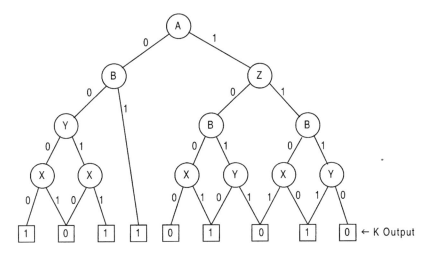

FIGURE 9.18
BDD for function K in Eq. (9.21) that can be used to show enabling paths of input A to output K as in Figs. 9.16b and 9.16c.

A, respectively. Notice that the dynamic hazards continue to occur even after input X is changed from logic 1 to logic 0. This is predictable from Fig. 9.16c, since the enabling paths for input A, $\bar{B}X\bar{Y}\bar{Z}$ and $\bar{B}\bar{X}\bar{Y}\bar{Z}$, are satisfied in both cases. Input conditions other than these enabling paths for A would not allow production of the dynamic hazards. As indicated by the K^* waveform in Fig. 9.17, static hazard cover cannot be used to eliminate a dynamic hazard.

The enabling paths for input A, shown in Figs. 9.16b and 9.16c, can also be deduced from the BDD in Fig. 9.18 for function K in Eq. (9.21). However, somewhat greater effort is needed to obtain this information from the BDD owing to its nature. For example, the enabling paths of A required to produce the dynamic hazards are seen to be $\bar{A}\bar{B}X\bar{Y}\bar{Z}$ and $A\bar{B}X\bar{Y}\bar{Z}$, yielding $K=0$ and $K=1$, respectively, with active X. Similarly, for $\bar{X}$ the enabling paths of A are observed to be $\bar{A}\bar{B}\bar{X}\bar{Y}\bar{Z}$ and $A\bar{B}\bar{X}\bar{Y}\bar{Z}$ for $K=1$ and $K=0$, respectively. The static 1 and static 0 hazards due to coupled variable A are deduced from the BDD in a similar manner.

A few points need to be remembered when using LPDD and/or BDD graphical means to obtain the enabling paths of coupled variables.

- The LPDD should be used to identify the coupled variable and any asymmetry that may exist in the alternative paths.
- The LPDD or a BDD can be used to deduce the enabling paths for that coupled variable.
- A BDD must be constructed for each coupled variable, whereas a single LPDD can be used for all coupled variables.
- Both logic values of the coupled variable must be considered when using either the LPDD or BDD, but only for the LPDD must account be taken of blocked paths.

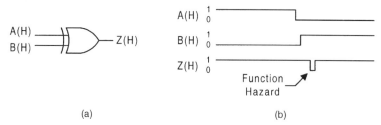

FIGURE 9.19
Demonstration of function hazard formation. (a) An XOR gate. (b) Timing diagram showing production of a function hazard when inputs A and B are changed in close proximity to each other.

9.4 FUNCTION HAZARDS

In the expression for Z_{SOP} given by Eq. (9.7), it is observed that pairs of terms such as $B\bar{C}D$ and $C\bar{D}\bar{E}$ or $\bar{A}\bar{D}\bar{E}$ and ABE each contain two couple variables. These pairs of terms are not coupled terms and cannot produce static hazards in the sense of Section 9.2. Also, their ANDed residues are always logic 0 — as are the ORed residues logic 1 for pairs of s-terms containing two (or more) coupled variables in a POS expression. But these pairs of terms can produce another type of hazard called a *function hazard*, which is also static in the sense that it occurs in an otherwise steady-state signal. Function hazards result when an attempt is made to change two or more coupled variables in close proximity to each other. Potential hazards of this type are very common. In fact, any two (or more) input gate can produce a function hazard if the two inputs are caused to change in close proximity to each other. As an example, consider a simple XOR gate in Fig. 9.19a. If the two inputs are changed close together as shown in Fig. 9.19b, a function hazard results. In effect, function hazards in most circuits can be avoided if care is taken not to permit the inputs to change too close together in time.

9.5 STUCK-AT FAULTS AND THE EFFECT OF HAZARD COVER ON FAULT TESTABILITY

If, by some means, an input to a logic gate becomes permanently stuck at logic 0 or logic 1, a single *stuck-at fault* is said to exist. Inadvertent shorted connections, open connections, or connections to the voltage supply can take place during the manufacture of a given device such as a gate. When this happens the device fails to operate correctly. Models have been created to test specifically for stuck-at faults in various logic devices. One such model has become known as the *single stuck-at fault model* and is regarded as the simplest and most reliable model to use. Here, exactly one line, say to a gate, is assumed to be fixed at a logic 1 or logic 0 and, therefore, cannot respond to an input signal. Testing for such faults in a complex combinational logic circuit is often complicated and may involve the application of elaborate testing procedures, the subject of which is beyond the scope of this text. For the reader wishing more information on fault models, test sets, design testability, and related subject matter, references are given in Further Reading at the end of this chapter.

Because a single input change to an XOR or EQV operator produces an output change, multilevel functions containing these operators can be more easily tested than their

9.5 STUCK-AT FAULTS AND THE EFFECT OF HAZARD COVER 413

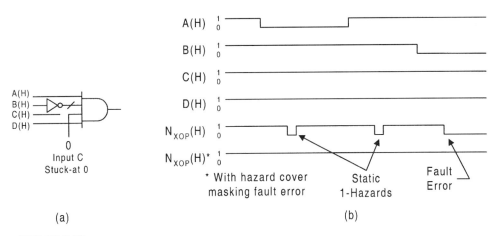

FIGURE 9.20
Effect of stuck-at fault on function N_{XOP} in Eq. (9.12). (a) AND representing the $A\bar{B}CD$ term and showing a stuck-at fault on input C. (b) Timing diagram showing effect of the stuck-at 0 fault and the masking effect of hazard cover.

two-level SOP or POS counterparts. This, of course, is one advantage in the use of XOP, EOS, and CRMT circuits discussed in Chapter 5. However, if static hazards must be eliminated in these circuits prior to fault testing, this advantage may be lessened or eliminated. Static hazard cover must always be redundant cover (i.e., not essential to function representation). Redundant cover can make stuck-at fault testing more difficult and may even mask an existing stuck-at fault. When considering the testability of a circuit, the designer must consider the effect of any static hazard cover needed.

As an example, consider function N_{XOP} in Eqs. (9.10) and (9.12) before and after the addition of static hazard cover. Suppose there is a stuck-at-0 fault at any input to the term $A\bar{B}CD$. This fault causes an output error on the input condition 1011. However, the addition of hazard cover ACD holds the output active and masks the presence of this fault. Thus, after hazard cover is added, one cannot test for this fault by the standard methods of observing the final output. The timing diagram in Fig. 9.20 illustrates the masking effect of hazard cover in the N_{XOP} function. This timing diagram can be easily understood if it is recalled that an odd number of 1's in an XOR string such as that for function N_{XOP} in Eqs. (9.10) and (9.12) yields a logic 1 for that function. Consequently, introducing the change $1111 \to 1011$ into these equations with and without the hazard cover ACD and with $C = 0$ results in the timing diagram shown in Fig. 9.20b. In conclusion it can be stated that *fault detection and location test sets* should be used prior to the addition of hazard cover; if not, some stuck-at faults may not be detected and located.

FURTHER READING

The subject of static hazards in two-level combinational logic circuits is covered adequately in texts by Breuer and Friedman, Katz, McCluskey, Tinder, Wakerly, and Yarbrough. Dynamic hazards and function hazards are also covered by McCluskey. However, there is no known published information on static hazards in circuits of the XOR type considered in this chapter.

[1] M. A. Breuer and A. D. Friedman, *Diagnosis and Reliable Design of Digital Systems.* Computer Science Press, 1976, pp. 10–13.
[2] R. H. Katz, *Contemporary Logic Design.* Benjamin/Cummings Publishing Co., Redwood City, CA, 1992.
[3] E. J. McCluskey, *Logic Design Principles.* Prentice-Hall, Englewood Cliffs, NJ, 1986.
[4] R. F. Tinder, *Digital Engineering Design: A Modern Approach.* Prentice-Hall, Englewood Cliffs, NJ, 1991.
[5] J. F. Wakerly, *Digital Design Principles and Practice.* Prentice-Hall, Englewood Cliffs, NJ, 1986.
[6] J. M. Yarbrough, *Digital Logic.* West Publishing Co., Minneapolis/St. Paul, 1997.

A discussion of the construction and application of binary decision diagrams (BDDs) is limited to a few texts, among which are those of De Micheli and Sasao (Ed.). However, more extensive information is available from published journal articles. Typical of these are articles by Akers and Bryant.

[7] S. Akers, "Binary Decision Diagrams," *IEEE Trans. on Computers,* **C-27,** 509–516 (1978).
[8] R. Bryant, "Graph-based Algorithms for Boolean Function Manipulation, " *IEEE Trans. on Computers* **C-35**(8), 677–691 (1986).
[9] G. De Micheli, *Synthesis and Optimization of Digital Circuits.* McGraw-Hill, New York, 1994.
[10] T. Sasao (Ed.), *Logic Synthesis and Optimization.* Kluwer Academic Publishers, Boston, 1993.

The subjects of fault detection and fault models are well covered by a number of texts and articles. For the beginning reader the text of Hayes does a commendable job. The text by McCluskey (previously cited) and that by Nelson, Nagle, Carroll, and Irwin are also recommended. For the advanced reader the texts by De Mecheli (previously cited) and Lala can be useful.

[11] J. P. Hayes, *Introduction to Digital Design.* Addison-Wesley, Reading, MA, 1993.
[12] P. K. Lala, *Fault Tolerant and Fault Testable Hardware Design.* Prentice-Hall, Englewood Cliffs, NJ, 1985.
[13] V. P Nelson, H. T. Nagle, B. D. Carroll, and J. D. Irwin, *Digital Logic Circuit Analysis and Design.* Prentice-Hall, Englewood Cliffs, NJ, 1995.

The following articles are noteworthy for their coverage of fault detection and testing, and of fault-tolerant systems:

[14] A. Chatterjee and M. A. d'Abreu, "The Design of Fault-Tolerant Linear Digital State Variable Systems: Theory and Techniques," *IEEE Trans. on Computers* **42**(7), 794–808 (1993).
[15] T. Lin and K. G. Shin, "An Optimal Retry Policy Based on Fault Classification," *IEEE Trans. on Computers* **43**(9), 1014–1025 (1994).
[16] B. Vinnakota and N. K. Jha, "Diagnosability and Diagnosis of Algorithm-Based Fault-Tolerant Systems," *IEEE Trans. on Computers* **42**(8), 924–937 (1993).

PROBLEMS

9.1 A function Y is represented in the K-map of Fig. P9.1. Refer to the examples in Section 9.2 and do the following:

(a) Loop out a minimum SOP and POS cover and then use arrows in separate K-maps to indicate the direction of the hazardous SOP and POS transitions that are present in this function.

(b) Find the hazard covers and combine them with the minimum SOP and POS expressions. Also, show these covers on the appropriate K-maps.

AB\CD	00	01	11	10
00	0	1	1	0
01	1	1	1	1
11	0	0	0	1
10	0	1	1	1

FIGURE P9.1

9.2 The following minimum SOP function contains both static 1-hazards and static 0-hazards:

$$F = \bar{A}B\bar{C} + BD + AC\bar{D} + \bar{B}\bar{C}\bar{D}.$$

(a) Map and loop out this function in a fourth-order K-map and indicate the direction of each SOP hazardous transitions on the map by using an arrow. Follow the example in Fig. 9.6a.

(b) Find the hazard covers and add them to the original expression above.

(c) By following the example in Eq. (9.5), confirm that the same information can be obtained directly from the minimum SOP expression just given.

(d) Repeat parts (a), (b), and (c) for the POS hazards (static 0-hazards) in the minimum POS expression for this function.

(e) Use the gate/input tally (exclusive of possible inverters) to compare the SOP and POS expressions with hazard cover included. Which result is simpler?

9.3 The following function contains a single static 1-hazard:

$$F_{SOP} = \bar{A}\bar{B}\bar{C} + AC + CD.$$

(a) From this expression (without using a K-map), determine the coupled terms, the initial and final states of the hazardous transition, and the hazard cover to be added to the expression. To do this, follow the example in Eq. (9.5).

(b) Use a timing diagram to show the development of this hazard, similar to the example in Fig. 9.1c. Then, by adding the hazard cover, show that the hazard is eliminated following the example in Fig. 9.2c. Assume that the inputs and output are all active high.

(c) Construct the logic circuit for F_{SOP} and include the hazard cover.

9.4 Map the expression in Problem 9.3 and extract a minimum expression for F_{SOP}. This function contains a single static 0-hazard.

(a) From this expression (without using a K-map), determine the coupled variable, coupled terms, the initial and final states of the hazardous transition, and the hazard cover to be added to the expression. To do this, follow the example in Eq. (9.6).

(b) Use a timing diagram to show the development of this hazard, similar to the example in Fig. 9.4c. Then by adding the hazard cover, show that the hazard is eliminated following the example in Fig. 9.5c. Assume that the inputs and output are all active high.

(c) Construct the logic circuit for F_{POS} and include the hazard cover.

9.5 Each of the following minimum or reduced functions contains one or more static hazards. For each function (without using a K-map), determine the coupled variable, coupled terms, the initial and final states of the hazardous transition, and the hazard cover to be added.

(a) $W = \bar{A}BC\bar{D} + \bar{B}\bar{C}\bar{D} + A\bar{C}D$

(b) $R = (U + W + \bar{X})(\bar{U} + \bar{V} + \bar{W})(\bar{V} + X)$

(c) $G = W\bar{X}\bar{Y} + XYZ + WY\bar{Z} + \bar{W}\bar{Y}$

(d) $T = (\bar{A} + \bar{B} + \bar{C})(\bar{A} + B + C)(A + D)(B + \bar{D})$

(e) $Y = \bar{w}y\bar{z} + xz + \bar{x}\bar{y}$

9.6 A five-variable function Z is represented in the second-order K-map of Fig. P9.2. It contains a single SOP hazard (static 1-hazard).

(a) Extract minimum SOP cover for this function and determine the coupled variable, coupled terms, the initial and final states of the hazardous transition, and the hazard cover to be added to the expression. To do this, follow the example in Eq. (9.5). (Hint: There are two possible coupled terms depending on how minimum cover is extracted.)

(b) Use a timing diagram to show the development of this hazard, similar to the example in Fig. 9.1c. Assume that all inputs are active high. Then by adding the hazard cover, show that the hazard is eliminated following the example in Fig. 9.2c.

A\B	0	1
0	$CD\bar{E}$	E
1	$\bar{E}$	$\bar{C}D+E$

Z

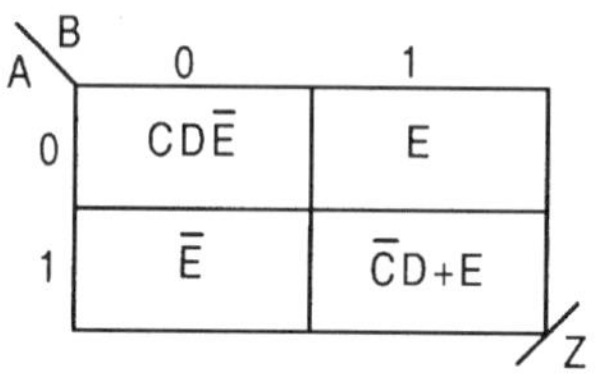

FIGURE P9.2

9.7 Find the minimum POS expression for the function in Fig. P9.2. How many static 0-hazards does this function have? What do you conclude as to the relative complexity of the SOP and POS expressions for this function when account is taken of hazard cover?

9.8 The following six-variable function has several static 1-hazards. Construct a table listing the coupled variable, coupled terms, initial and final states, and the hazard cover for each of the hazards.

$$F = \bar{A}\bar{B}C\bar{D}\bar{F} + \bar{A}BCE + ABCF + A\bar{B}C\bar{E} + D\bar{E}F + CD\bar{E}$$

9.9 The following multilevel functions have one or more static hazards. For each expression (without using a K-map), determine the coupled variable, coupled terms, the initial and final states of the hazardous transition (read in alphabetical order), and the hazard cover to be added to the expression. To do this, follow the examples in Eq. (9.12), (9.13), and (9.14), whichever is relevant to the particular function.

(a) $G = W\bar{X}\bar{Y} \oplus XYZ \oplus WY\bar{Z} \oplus \bar{W}\bar{Y}$

(b) $T = (\bar{A} + \bar{B} + \bar{C}) \odot (\bar{A} + B + C) \odot (A + D) \odot (\bar{B} + \bar{D})$

9.10 The following three-level XOR-type function has two static 1-hazards:

$$f = (Y \oplus W) \oplus (\bar{X}Z) + \bar{W}Y.$$

(a) Construct the lumped path delay diagram (LPDD) for this function. From the LPDD determine the hazard cover and initial and final states for each of the static hazards. Follow the example in Fig. 9.10. (Hint: Keep in mind the bidirectionality of the static hazards in XOR-type functions and read the states for function f in the order of $WXYZ$).

(b) By using two binary decision diagrams (BDDs), show the binary decisions required for each static 1-hazard formation.

9.11 The following three-level function has both a static 1-hazard and a static 0-hazard:

$$F = [X \oplus \bar{W}Y] \oplus (WZ).$$

(a) Construct the LPDD for this function (exactly as written). Then, determine the hazard cover and initial and final states for each of the static hazards. Read the states in alphabetical order. Follow the example in Fig. 9.13 and Eq. (9.19).

(b) Use a timing diagram to show the development of the two hazards, similar to the example in Fig. 9.14. Then by adding the hazard cover, show that the hazards are eliminated. Assume that all inputs and the output are active high.

(c) By using a binary decision diagram (BDD), show the binary decisions required for the static 1-hazard formation.

9.12 The following four-level function has three static 1-hazards, one static 0-hazard, and one dynamic hazard:

$$Y = [B \oplus (\bar{A}D)] \oplus [A\bar{B} \oplus AC\bar{D} \oplus B\bar{C}D]$$

(a) Construct an LPDD for this function (exactly as written) and find the hazard cover for each of the static hazards and the conditions required to produce the dynamic hazard. Indicate the initial and final states for each of the static hazards. Follow the example in Fig. 9.16. (Hint: No dynamic hazard exists due to a change in either input A or D, and the one that does exist is conditional on A. Also, look for the possibility that a static hazard may change from a static 1-hazard to a static 0-hazard depending on the order of change of one or two variables.)

(b) Use a BDD to show the enabling paths of the variable whose change is responsible for the dynamic hazard formation. Follow the example in Fig. 9.18.

(c) Use a timing diagram (simulation) to show the development of the dynamic hazard. Demonstrate that the dynamic hazard cannot be eliminated by adding static hazard cover.

(d) Demonstrate with both an LPDD and a timing diagram that the dynamic hazard can be eliminated by making the B paths to the output less asymmetric.

CHAPTER 10

Introduction to Synchronous State Machine Design and Analysis

10.1 INTRODUCTION

Up to this point only combinational logic machines have been considered, those whose outputs depend solely on the present state of the inputs. Adders, decoders, MUXs, PLAs, ALUs, and many other combinational logic machines are remarkable and very necessary machines in their own right to the field of logic design. However, they all suffer the same limitation. They cannot perform operations sequentially. A ROM, for example, cannot make use of its present input instructions to carry out a next-stage set of functions, and an adder cannot count sequentially without changing the inputs after each addition. In short, combinational logic devices lack *true memory*, and so lack the ability to perform sequential operations. Yet their presence in a sequential machine may be indispensable.

We deal with sequential devices all the time. In fact, our experience with such devices is so commonplace that we often take them for granted. For example, at one time or another we have all had the experience of being delayed by a modern four-way traffic control light system that is vehicle actuated with pedestrian overrides and the like. Once at the light we must wait for a certain sequence of events to take place before we are allowed to proceed. The controller for such a traffic light system is a fairly complex digital sequential machine.

Then there is the familiar elevator system for a multistory building. We may push the button to go down only to find that upward-bound stops have priority over our command. But once in the elevator and downward bound, we are likely to find the elevator stopping at floors preceding ours in sequence, again demonstrating a sequential priority. Added to these features are the usual safety and emergency overrides, and a motor control system that allows for the carrier to be accelerated or decelerated at some reasonable rate. Obviously, modern elevator systems are controlled by rather sophisticated sequential machines.

The list of sequential machines that touch our daily lives is vast and continuously growing. As examples, the cars we drive, the homes we live in, and our places of employment all use sequential machines of one type or another. Automobiles use digital sequential machines to control starting, braking, fuel injection, cruise control, and safety features. Most homes have automatic washing machines, microwave ovens, sophisticated audio and video devices of various types, and, of course, computers. Some homes have complex security, energy,

420 CHAPTER 10 / INTRODUCTION TO SYNCHRONOUS STATE MACHINE DESIGN

and climate control systems. All of these remarkable and now commonplace gifts of modern technology are made possible through the use of digital sequential machines.

The machines just mentioned are called *sequential machines*, or simply *state machines*, because they possess true memory and can issue time-dependent sequences of logic signals controlled by present and past input information. These sequential machines may also be *synchronous* because the data path is controlled by a *system clock*. In synchronous sequential machines, input data are introduced into the machine and are processed sequentially according to some algorithm, and outputs are generated — all regulated by a system clock. Sequential machines whose operation is clock independent (i.e., self-timed) are called *asynchronous sequential machines*, the subject of Chapters 14, 15, and 16.

Synchronous sequential machines and their design, analysis, and operation are the subjects covered in this chapter. Treatment begins with a discussion of the models used for these machines. This is followed by a discussion of an important type of graphic that is used to represent the sequential behavior of sequential machines and by a detailed development of the devices used for their memory. The chapter ends with the design and analysis of relatively simple state machines. The intricacies of design are numerous and require detailed consideration. For this reason they are discussed later, in Chapter 11.

10.1.1 A Sequence of Logic States

Consider that a synchronous sequential machine has been built by some means and that it is represented by the block symbol in Fig. 10.1a. Then suppose the voltage waveforms from its three outputs are detected (say with a waveform analyzer) and displayed as in

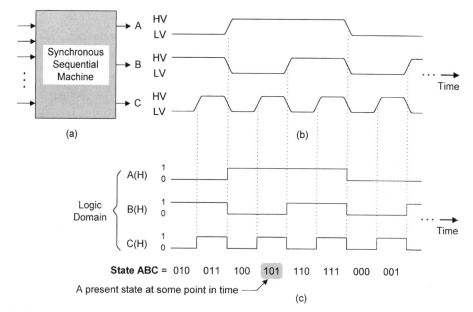

FIGURE 10.1

A sequence of logic events from a synchronous state machine. (a) Block diagram symbol and (b) output voltage waveforms. (c) Timing diagram representing the positive logic interpretation of the voltage waveforms and showing a sequence of logic states.

Fig. 10.1b. From these physical waveforms the positive logic waveforms are constructed and a sequence of logic states is read in the order *ABC* as shown in Fig. 10.1c. A group of logic waveforms such as these is commonly known as a *timing diagram*, and the sequence of logic states derived from these waveforms is seen to be a binary count sequence. Here, *A*, *B*, and *C* are called *state variables* because their values collectively define the *present state* of the machine at some point in time. Knowing the present state in a sequence of states also reveals the *next state*. Thus, in Fig. 10.1c, if state 101 is the present state, then 110 is the next state. This short discussion evokes the following definition:

> A *logic state* is a unique set of binary values that characterize the logic status of a sequential machine at some point in time.

A sequential machine always has a finite number of states and is therefore called a *finite state machine* (*FSM*) or simply *state machine*. Thus, if there are N state variables, there can be no more than 2^N states in the FSM and no fewer than 2. That is, for any FSM,

$$2 \leq (\text{number of states}) \leq 2^N$$

For example, a two-state FSM requires one state variable, a three- or four-state FSM requires two state variables, five- to eight-state FSMs require three state variables, etc. More state variables can be used for an FSM than are needed to satisfy the 2^N requirement, but this is done only rarely to overcome certain design problems or limitations. The abbreviation FSM will be used frequently throughout the remainder of this text.

To help understand the meaning of the various models used in the description and design of FSMs, four binary sequences of states are given in Fig. 10.2, each presenting a different feature of the sequence. The simple ascending binary sequence (a) is the same as that in Fig. 10.1. This sequence and the remaining three will be described as they relate to the various models that are presented in the following section.

10.2 MODELS FOR SEQUENTIAL MACHINES

Models are important in the design of sequential machines because they permit the design process to be organized and standardized. The use of models also provides a means of communicating design information from one person to another. References can be made to specific parts of a design by using standard model nomenclature. In this section the general model for sequential machines will be developed beginning with the most elemental forms.

Notice that each state in the sequence of Fig. 10.2a becomes the *present state* (*PS*) at some point in time and has associated with it a *next state* (*NS*) that is predictable given the PS. Now the question is: What logic elements of an FSM are required to do what is required in Fig. 10.2a? To answer this question, consider the thinking process we use to carry out a sequence of events each day. Whether the sequence is the daily routine of getting up in the morning, eating breakfast, and going to work, or simply giving our telephone number to someone, we must be able to remember our present position in the sequence to know what the next step must be. It is no different for a sequential machine. There must be a memory section, as in Fig. 10.3a, which generates the present state. And there must be a *next state logic* section, as in Fig. 10.3b, which has been programmed to know what the next state must be, given the present state from the memory. Thus, an FSM conforming to the model

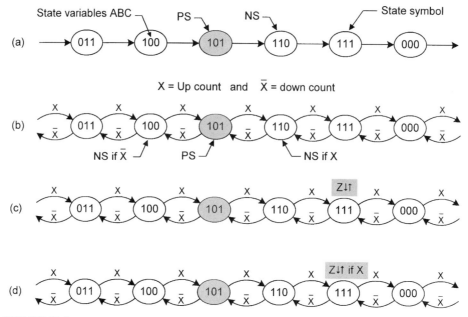

FIGURE 10.2
A sequence of states with present and next states based on Fig. 10.1. (a) A simple ascending binary sequence showing present state (PS) and next state (NS). (b) A bidirectional (up/down) binary sequence showing PS and NS depending on logic level of input X. (c) A bidirectional binary sequence with output Z in state 111. (d) A bidirectional sequence with output Z in state 111 conditional on input X (up-count).

of Fig. 10.3b is capable of performing the simple ascending binary sequence represented by Fig. 10.2a. However, to carryout the bidirectional binary sequence of Fig. 10.2b, a machine conforming to the *basic model* of Fig. 10.3c is required to have external input capability. As in the case of Fig. 10.2b, an input X would force the FSM to count up in binary, while $\bar{X}$ would cause it to count down in binary.

If it is necessary that the FSM issue an output on arrival in any given state, output-forming logic must be added as indicated in Fig. 10.4. This model has become known as *Moore's model* and any FSM that conforms to this model is often called a *Moore machine* in honor of E. F. Moore, a pioneer in sequential circuit design. For example, an FSM that can generate the bidirectional binary sequence in Fig. 10.2c is called a Moore FSM, since an output Z is unconditionally activated on arrival in state 111 (up arrow, ↑) and is deactivated on exiting this state (down arrow, ↓); hence the double arrow (↓↑) for the output symbol Z↓↑. Such an output could be termed a *Moore output*, that is, an output that is issued as a function of the PS only. The functional relationships for a Moore FSM are

$$\begin{cases} PS = f(NS) \\ NS = f'(IP, PS) \\ OP = f''(PS) \end{cases}, \quad (10.1)$$

where *IP* represents the external inputs and *OP* the outputs.

10.2 MODELS FOR SEQUENTIAL MACHINES

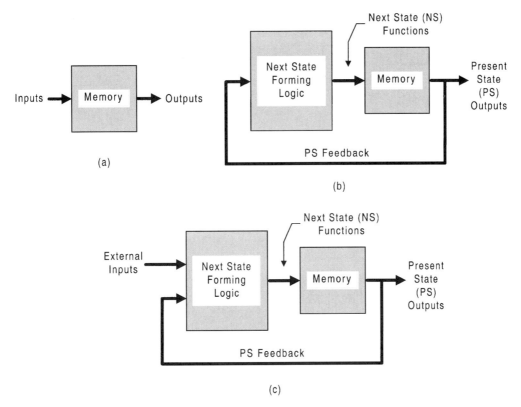

FIGURE 10.3
Development of the basic model for sequential machines. (a) The memory section only. (b) Model for an FSM capable of performing the sequence in Fig. 10.2a, showing the memory section and NS-forming logic. (c) The basic model for an FSM capable of performing the sequence in Fig. 10.2b when the external input is X.

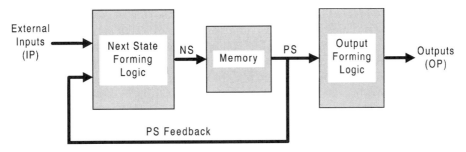

FIGURE 10.4
Moore's model for a sequential machine capable of performing the bidirectional binary sequence in Fig. 10.2c, showing the basic model in Fig. 10.3c with the added output-forming logic that depends only on the PS.

424 CHAPTER 10 / INTRODUCTION TO SYNCHRONOUS STATE MACHINE DESIGN

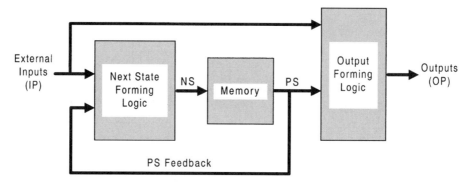

FIGURE 10.5
Mealy's (general) model for a sequential machine capable of performing the bidirectional binary sequence in Fig. 10.2d, showing the basic model in Fig. 10.3c with the added output-forming logic that depends on both IP and PS.

Now suppose it is necessary to issue an output conditional on an input as in the bidirectional binary sequence of Fig. 10.2d. This requires a model whose outputs not only depend on the PS, but also depend on the inputs, as illustrated in Fig. 10.5. Such a model is the *most general model* for state machines and is known as *Mealy's model* after G. H. Mealy, another pioneer in the field of sequential machines. Thus, an FSM that conforms to this model can be called a *Mealy machine* and would be capable of generating the bidirectional binary sequence of Fig. 10.2d, where the output is issued in state 111 but only if X is active (i.e., on an up count). Such an output could be termed a *Mealy output*, that is, an output that is issued conditional on an input. The functional relationships for a Mealy FSM are

$$\left\{ \begin{array}{l} PS = f(NS) \\ NS = f'(IP, PS) \\ OP = f''(IP, PS) \end{array} \right\}. \qquad (10.2)$$

As is evident from an inspection of Figs. 10.4 and 10.5, the only difference between a Mealy FSM and a Moore FSM is that the Mealy machine has one or more outputs that are conditional on one or more inputs. The Moore machine has no conditional outputs. Hereafter, reference made to a Mealy machine or a Moore machine will imply this difference. Similarly, outputs that are referred to as Mealy outputs will be those that are issued conditionally on one or more inputs, and outputs referred to as Moore outputs will be those that are issued unconditionally.

10.3 THE FULLY DOCUMENTED STATE DIAGRAM

In Fig. 10.2 a single input X is used to influence the sequential behavior of a binary sequence. A more complex example might involve several inputs that control the sequential behavior of the FSM. Such an FSM might be caused to enter one of several possible sequences (or routines), each with subroutines and outputs, all controlled by external inputs whose values

10.3 THE FULLY DOCUMENTED STATE DIAGRAM

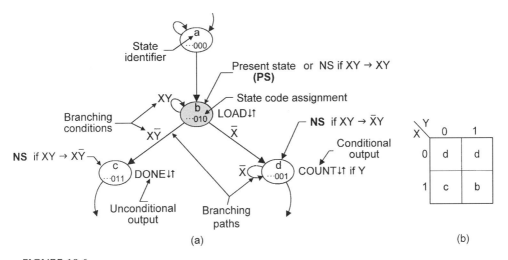

FIGURE 10.6
(a) Features of the fully documented state diagram section. (b) The input/state map for state b.

change at various times during the operation of the FSM. Obviously, some means must be found by which both simple and complex FSM behavior can be represented in a precise and meaningful way. The *fully documented state diagram* discussed in this section is one means of representing the sequential behavior of an FSM.

Presented in Fig. 10.6a is a portion of a state diagram showing the important features used in its construction. Attention is drawn to states identified as a, b, c, and d. Here, state b is the present state (PS) at some point in time and is given the state code assignment $\cdots 010$. Notice that state b branches to itself under the branching condition XY, the *holding condition* for that state, and that the next state (NS) depends on which input, X or Y, changes first. If X changes first, hence $XY \to \bar{X}Y$, the FSM will transit to the next state d, where it will hold on the input condition $\bar{X}$, where $\bar{X} = \bar{X}Y + \bar{X}\bar{Y}$. Or if Y changes first, $XY \to X\bar{Y}$, the FSM will transit to state c, where there is no holding condition.

The output notation is straightforward. There are two types of outputs that can be represented in a fully documented state diagram. Referring to state b in Fig. 10.6a, the output

$$\text{LOAD} \downarrow\uparrow$$

is an *unconditional (Moore) output* issued any time the FSM is in state b. The down/up arrows ($\downarrow\uparrow$) signify that LOAD becomes active (up arrow, $\uparrow$) when the FSM enters state b and becomes inactive (down arrow, $\downarrow$) when the FSM leaves that state. The order in which the arrows are placed is immaterial as, for example, up/down. The output DONE in state c is also an unconditional or Moore output. The second type of output, shown in state d of Fig. 10.6a and indicated by

$$\text{COUNT} \downarrow\uparrow \text{ if } Y,$$

is a *conditional output* that is generated in state d but only if Y is active — hence, COUNT

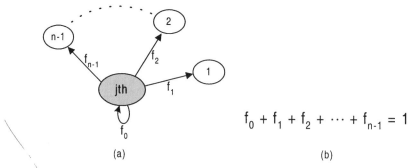

FIGURE 10.7
Application of the sum rule given by Eq. (10.3). (a) State diagram segment showing branching conditions relative to the jth state. (b) Application of the sum rule to the jth state in the state diagram segment.

is a *Mealy output* according to the Mealy model in Fig. 10.5. Thus, if input Y should toggle between active and inactive conditions while the FSM resides in state d, so also would the output COUNT toggle with Y.

The Sum Rule There are certain rules that must "normally" be followed for proper construction of state diagrams. One of these rules is called the *sum rule* and is stated as follows:

> The Boolean sum of all branching conditions from a given state must be logic 1.

With reference to Fig. 10.7, this rule is expressed mathematically as

$$\sum_{i=0}^{n-1} f_{i \leftarrow j} = 1, \qquad (10.3)$$

where $f_{i \leftarrow j}$ represents the branching condition from the jth to the ith state and is summed over n states as indicated in Fig. 10.7b. For example, if the sum rule is applied to state b in Fig. 10.6a, the result is

$$XY + \bar{X} + X\bar{Y} = 1,$$

since according to the absorptive law, Eq. (3.13), $\bar{X} + X\bar{Y} = \bar{X} + \bar{Y}$, which is the complement of XY. The graphical representation of the sum rule, as applied to state b, is shown in Fig. 10.6b and is called the *input/state map*. Had the sum rule not been satisfied, one or more branching conditions would not be accounted for and one or more of the cells in the input/state map of Fig. 10.6b would be vacant. If applied, unaccounted-for branching conditions can cause an FSM to malfunction.

The Mutually Exclusive Requirement While satisfying the sum ($\sum = 1$) rule is a necessary condition for state branching accountability in a fully documented state diagram, it is not sufficient to ensure that the branching conditions are nonoverlapping. The meaning of this can be stated as follows:

10.3 THE FULLY DOCUMENTED STATE DIAGRAM

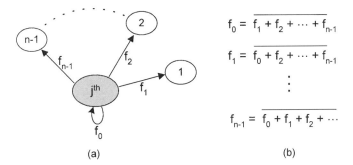

(a) (b)

FIGURE 10.8
Application of the mutually exclusive requirement given by Eq. (10.4). (a) State diagram segment showing branching conditions relative to the jth state. (b) Application of the mutually exclusive requirement to the state diagram segment in (a).

> *Each possible branching condition from a given state must be associated with no more than one branching path.*

With reference to Fig. 10.8a, this condition is expressed mathematically as

$$f_{i \leftarrow j} = \overline{\sum_{\substack{k=0 \\ k \neq i}}^{n-1} f_{k \leftarrow j}}, \qquad (10.4)$$

where each branching condition is seen to be the complement of the Boolean sum of those remaining as indicated in Fig. 10.8b. When applied to state b in Fig. 10.6, Eq. (10.4) gives the results $XY = \overline{\bar{X} + X\bar{Y}} = \overline{\bar{X} + \bar{Y}} = XY$ and $\bar{X} = \overline{XY + X\bar{Y}} = \bar{X}$, etc., clearly indicating that both the mutually exclusive requirement and the sum rule are satisfied. See Problem 10.24 for more on this subject.

Now consider the case shown in Fig. 10.9a, where the sum rule is obeyed but not the mutually exclusive requirement. In this case, the branching condition XY is associated

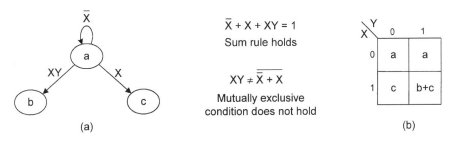

(a) (b)

FIGURE 10.9
(a) A portion of a state diagram for which the mutually exclusive condition does not hold. (b) Input/state map showing violation of the mutually exclusive requirement as applied to state a under branching condition XY.

with both the $a \to b$ and the $a \to c$ branching paths. Thus, if $\bar{X}Y \to XY$ while in state a, malfunction of the FSM is likely to occur. In Fig. 10.9b is the input/state map showing violation of Eq. (10.4) under input condition XY shared by branching paths $a \to b$ and $a \to c$. Thus, if the mutually exclusive requirement is to hold for a given state, the input/state map must not have cells containing more than one state identifier.

When Rules Can Be Broken There are conditions under which violation of the sum rule or of the mutual exclusion requirement is permissible. Simply stated, these conditions are as follows: If certain branching conditions can never occur or are never permitted to occur, they can be excluded from the sum rule and from the mutually exclusive requirement. This means that Eqs. (10.3) and (10.4) need not be satisfied for the FSM to operate properly. As an example, suppose that in Fig. 10.6 the branching condition is $\bar{Y}$ for branching path $b \to c$. Thus, the sum rule holds since $XY + \bar{X} + \bar{Y} = 1$. However, the branching condition $\bar{X}\bar{Y}$ is common to both the $b \to c$ and $b \to d$ branching paths with branching conditions $\bar{Y}$ and $\bar{X}$, respectively. Clearly, the mutually exclusive requirement of Eq. (10.4) is not satisfied, which is of no consequence if the input condition $\bar{X}\bar{Y}$ can never occur. But if the input condition $\bar{X}\bar{Y}$ is possible, then branching from state b under $\bar{X}\bar{Y}$ is ambiguous, leading to possible FSM malfunction. See Problem 10.24b for a more direct means of testing for the mutual exclusivity of branching conditions.

10.4 THE BASIC MEMORY CELLS

Developing the concept of memory begins with the basic building block for memory called the *basic memory cell* or simply *basic cell*. A basic cell plays a major role in designing a memory device (element) that will remember a logic 1 or a logic 0 indefinitely or until it is directed to change to the other value. In this section two flavors of the basic cell will be heuristically developed and used later in the design of important memory elements called *flip-flops*.

10.4.1 The Set-Dominant Basic Cell

Consider the wire loop in Fig. 10.10a consisting of a fictitious lumped path delay (LPD) memory element Δt and two inverters whose function it is to maintain an imaginary signal. The LPD memory element is the path delay for the entire wire loop including inverters concentrated (lumped) in Δt, hence the meaning of the word "fictitious." But since there is no external access to this circuit, introduction of such a signal into this circuit is not possible. This can be partially remedied by replacing one of the inverters with an NAND gate performing the OR operation as in Fig. 10.10b. Now, a *Set* ($0 \to 1$) can be introduced into the circuit if $S(L) = 1(L)$. This can be further understood by an inspection of the Boolean expression $Q_{t+1} = S + Q_t$ for the circuit in Fig. 10.10b, where the following definitions apply:

$$Q_{t+1} = \text{Next state}$$

$$Q_t = \text{Present state.}$$

10.4 THE BASIC MEMORY CELLS

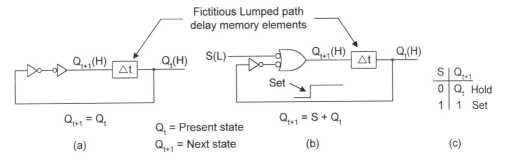

FIGURE 10.10
Development of the concept of Set. (a) Wire loop with a fictitious lumped path delay memory element and two inverters used to restore an imaginary signal. (b) Wire loop with one inverter replaced by a NAND gate used to introduce a Set condition. (c) Truth table obtained from the logic expression for Q_{t+1} in (b) showing the Hold and Set conditions.

The truth table in Fig. 10.10c is constructed by introducing the values {0, 1} for S into this equation and is another means of representing the behavior of the circuit in Fig. 10.10b. The *hold condition* $Q_{t+1} = Q_t$ occurs any time the next state is equal to the present state, and the Set condition occurs any time the next state is a logic 1, i.e., $Q_{t+1} = 1$.

The circuit of Fig. 10.10b has the ability to introduce a Set condition as shown, but no means of introducing a *Reset* ($1 \rightarrow 0$) condition is provided. However, this can be done by replacing the remaining inverter with an NAND gate performing the AND operation as shown in Fig. 10.11a. Then, if $R(L) = 1(L)$ when $S(L) = 0(L)$, a Reset condition is introduced into the circuit. Thus, both a Set and Reset condition can be introduced into the circuit by external means. This basic memory element is called the *set-dominant basic cell* for which the logic circuit in Fig. 10.11a is but one of seven ways to represent its character, as discussed in the following paragraphs.

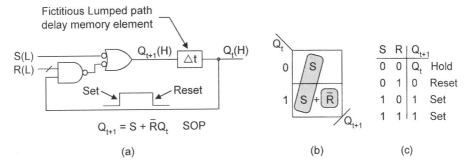

FIGURE 10.11
The set-dominant basic cell represented in different ways. (a) Logic circuit showing the Set and Reset capability, and the Boolean equation for the next state function, Q_{t+1}. (b) EV K-map with minimum cover indicated by shaded loops. (c) Operation table for the set-dominant basic cell showing the Hold, Set, and Reset conditions inherent in the basic memory cell.

430 CHAPTER 10 / INTRODUCTION TO SYNCHRONOUS STATE MACHINE DESIGN

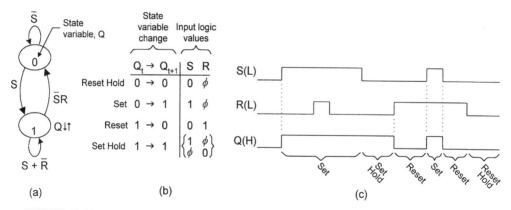

FIGURE 10.12
The set-dominant basic cell (contd.). (a) State diagram derived from the operation table in Fig. 10.11c. (b) Excitation table derived from the state diagram. (c) Timing diagram illustrating the operation of the set-dominant basic cell.

Reading the circuit in Fig. 10.11a yields the following SOP expression for the next state function:

$$Q_{t+1} = S + \bar{R}Q_t. \qquad (10.5)$$

When this expression is plotted in an EV K-map, Fig. 10.11b results, where minimum cover is indicated by shaded loops. From this expression or from the EV K-map, it is clear that a set condition is introduced any time $S = 1$, and that a reset condition results only if $R = 1$ and $S = 0$. However, if both inputs are inactive, that is, if $S = R = 0$, it follows that $Q_{t+1} = Q_t$, which is the hold condition for the basic cell. The Hold, Set, and Reset conditions are easily observed by inspection of the *operation table* for the set-dominant basic cell given in Fig. 10.11c. The basic cell is called *set-dominant* because there are two input conditions, $S\bar{R}$ and SR, that produce the Set condition as indicated by the operation table in Fig. 10.11c. Notice that Fig. 10.11 represents four ways of representing the set-dominant basic cell: logic circuit, NS function, NS K-map, and operation table.

By using the operation table in Fig. 10.11c, the state diagram for the set-dominant basic cell can be constructed as given in Fig. 10.12a. To clarify the nomenclature associated with any fully documented state diagram, the following definitions apply to the state variable changes and will be used throughout this text:

$$\begin{cases} 0 \to 0 = \text{Reset Hold} \\ 0 \to 1 = \text{Set} \\ 1 \to 0 = \text{Reset} \\ 1 \to 1 = \text{Set Hold} \end{cases}. \qquad (10.6)$$

Thus, for the state diagram in Fig. 10.12a, $\bar{S}$ is the Reset Hold branching condition, S is the Set branching condition, $\bar{S}R$ is the Reset branching condition, and $S + \bar{R}$ is the Set Hold branching condition. The output Q is issued (active) only in the Set state (state 1), not in the Reset state (state 0). Notice that for each of the two states the sum rule ($\sum = 1$)

10.4 THE BASIC MEMORY CELLS

holds as it must. But all branching conditions are easily deduced from an inspection of the operation table. For example, the Set condition is the Boolean sum $S\bar{R} + SR = S$, or the Set Hold condition is the sum $\bar{S}\bar{R} + S\bar{R} + SR = S + \bar{R}$, which is simply the complement of the Reset branching condition $\overline{SR} = S + \bar{R}$ in agreement with the sum rule.

From the state diagram of Fig. 10.12a another important table is derived, called the *excitation table*, and is presented in Fig. 10.12b. Notice that a don't care ϕ is placed in either the S or R column of the excitation table for the basic cell to indicate an unspecified input branching condition. For example, the Set branching condition S requires that a 1 be placed in the S column while a ϕ is placed in the R, column indicating that R is not specified in the branching condition for Set. Similarly, for the Set Hold branching path $1 \rightarrow 1$, the branching condition $S + \bar{R}$ requires a 1 and ϕ to be placed in the S and R columns for the S portion of the branching condition, and that a ϕ and 0 to be placed in the S and R columns for the $\bar{R}$ portion, respectively. Thus, the excitation table specifies the input logic values for each of the four corresponding state variable changes in the state diagram as indicated.

As a seventh and final means of representing the behavior of the set-dominant basic cell, a timing diagram can be constructed directly from the operation table in Fig. 10.11c. This timing diagram is given in Fig. 10.12c, where the operating conditions Set, Set Hold, Reset, and Reset Hold are all represented—at this point no account is taken of the path delay through the gates. Notice that the set-dominant character is exhibited by the $S, R = 1, 0$ and $S, R = 1, 1$ input conditions in both the operation table and timing diagram.

10.4.2 The Reset-Dominant Basic Cell

By replacing the two inverters in Fig. 10.10a with NOR gates, there results the logic circuit for the *reset-dominant basic cell* shown in Fig. 10.13a. Now, the Set and Reset inputs are presented active high as $S(H)$ and $R(H)$. Reading the logic circuit yields the POS logic expression for the next state,

$$Q_{t+1} = \bar{R}(S + Q_t), \tag{10.7}$$

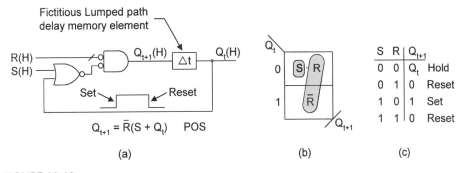

FIGURE 10.13
The reset-dominant basic cell represented in different ways. (a) Logic circuit showing the Set and Reset capability, and the Boolean equation for the next state function, Q_{t+1}. (b) EV K-map with minimum cover indicated by shaded loops. (c) Operation table for the reset-dominant basic cell showing the Hold, Set, and Reset conditions inherent in the basic memory cell.

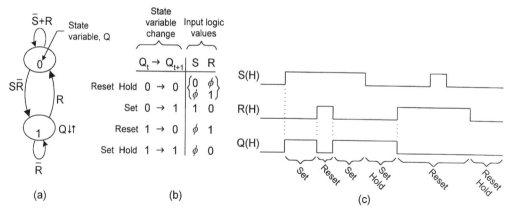

FIGURE 10.14
The reset-dominant basic cell (contd.). (a) State diagram derived from the operation table in Fig. 10.13c. (b) Excitation table derived from the state diagram. (c) Timing diagram illustrating the operation of the reset-dominant basic cell.

which is plotted in the EV K-map in Fig. 10.13b with minimum cover indicated by shaded loops. The operation table for the reset-dominant basic cell is constructed directly from the Boolean expression for Q_{t+1} and is given in Fig. 10.13c, where input conditions for Hold, Reset, and Set are depicted. Notice that the Set condition is introduced only when $S\bar{R}$ is active, whereas the Reset condition occurs any time R is active, the reset-dominant character of this basic memory element.

The state diagram for the reset-dominant basic cell is constructed from the operation table in Fig. 10.13c with the result shown in Fig. 10.14a. Here, the Set condition $S\bar{R}$ is placed on the $0 \rightarrow 1$ branching path. Thus, it follows that the Reset Hold condition is $\bar{S} + R$, which can be read from the operation table as $\bar{S}\bar{R} + \bar{S}R + SR = \bar{S} + R$, or is simply the complement of the Set input condition $S\bar{R} = \bar{S} + R$, a consequence of the sum rule. The remaining two branching conditions follow by similar reasoning.

The excitation table for the reset-dominant basic cell is obtained directly from the state diagram in Fig. 10.14a and is presented in Fig. 10.14b. Again, a don't care ϕ is placed in either the S or R column of the excitation table for the basic cell to indicate an unspecified input branching condition, as was done in the excitation table for the set-dominant basic cell of Fig. 10.12b. The nomenclature presented to the left of the excitation table follows the definitions for state variable change given by Eqs. (10.6).

The seventh and final means of representing the reset-dominant basic cell is the timing diagram constructed in Fig. 10.14c with help of the operation table in Fig. 10.13c. Again, no account is taken at this time of the gate propagation delays. Notice that the reset-dominant character is exhibited by the $S, R = 0, 1$ and $S, R = 1, 1$ input conditions in both the operation table and the timing diagram.

At this point the reader should pause to make a comparison of the results obtained for the set-dominant and reset-dominant basic cells. Observe that there are some similarities, but there are also some basic differences that exist between the two basic memory elements. Perhaps these similarities and differences are best dramatized by observing the timing diagrams in Figs. 10.12c and 10.14c. First, notice that the S and R inputs arrive active low to the set-dominant basic cell but arrive active high to the reset-dominant cell. Next,

10.4 THE BASIC MEMORY CELLS

observe the difference in the $Q(H)$ waveform for these two types of basic cells. Clearly, the set-dominant character is different from the reset-dominant character with regard to the $S, R = 1, 1$ input condition. This difference may be regarded as the single most important difference between these two memory cells and will play a role in the discussion that follows.

10.4.3 Combined Form of the Excitation Table

The excitation table for a memory element has special meaning and utility in state machine design. In subsequent sections it will be shown that the basic memory cell plays a major role in the design and analysis of flip-flops, the memory elements used in synchronous state machine design. Two such excitation tables have been identified so far: one associated with the set-dominant basic cell and the other associated with the reset-dominant cell. For purposes of flip-flop design these two excitation tables are inappropriate because of the different way they behave under the $S, R = 1, 1$ input condition. To overcome this difference, the two may be combined to give the single *generic (combined) excitation table* as shown in Fig. 10.15. Here, common S, R input conditions for the two excitation tables in Figs. 10.15a and 10.15b are identified for each of the four branching paths given and are brought together to form the combined excitation table in Fig. 10.15c. The important characteristic of the combined excitation is that *the $S, R = 1, 1$ condition is absent*. This leads to the following important statements:

- Because the $S, R = 1, 1$ condition is not present in the combined excitation table, it is applicable to either the set-dominant basic cell or the reset-dominant basic cell.
- Throughout the remainder of this text only the combined excitation table will be used in the design of other state machines, including other memory elements called flip-flops.

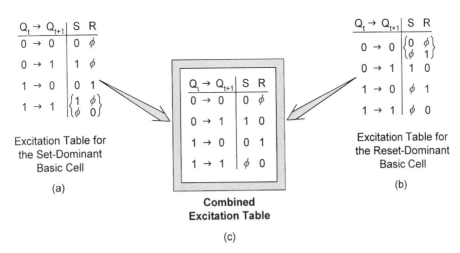

FIGURE 10.15
The excitation table for the basic cell. (a) Excitation table for the set-dominant (NAND-based) basic cell. (b) Excitation table for the reset-dominant (NOR-based) basic cell. (c) Generic (combined) excitation table applicable to either of the basic cells since the $S, R = 1, 1$ condition is absent.

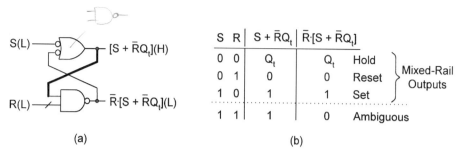

FIGURE 10.16
Mixed-rail outputs of the set-dominant basic cell. (a) Logic circuit showing the mixed-logic output expressions from the two gates. (b) Truth table indicating the input conditions required for mixed-rail outputs.

Thus, the individual excitation tables for the set-dominant and reset-dominant basic cells will be of no further use in the discussions of this text.

10.4.4 Mixed-Rail Outputs of the Basic Cells

There are subtle properties of the basic cells, yet to be identified, that are essential to the design of other memory elements. These properties deal with the output character of the basic cells. Referring to the logic circuit in Fig. 10.11a, only one output is identified for the set-dominant basic cell. However, by removing the fictitious lumped path delay (LPD) memory element Δt and arranging the conjugate NAND gate forms one above the other, there results the well-known "cross-coupled" NAND gate configuration shown in Fig. 10.16a. There is but one feedback path for the basic cell (indicated by the heavy line), though it may appear to the reader as though there are two.

The mixed-logic output expression from each of the two conjugate NAND gate forms in the set-dominant basic cell is read and presented as shown in Fig. 10.16a. Using these two output expressions, the truth table in Fig. 10.16b is constructed. In this table it is observed that all input conditions except $S, R = 1, 1$ generate what are called *mixed-rail outputs* from the two conjugate NAND gate forms. This means that when a $0(H)$ is produced from the OR form, a $0(L)$ appears on the output of the AND form. Or when the former is $1(H)$, the latter is $1(L)$. The $S, R = 1, 1$ input condition, meaning $S(L) = R(L) = 1(L)$, produces outputs that are $1(H)$ and $0(L) = 1(H)$ from the OR and AND forms, respectively, and are not mixed-rail outputs — the NAND gate outputs are ambiguous, since they cannot be labeled as either Set or Reset.

A similar procedure is used in defining the mixed-rail outputs from the reset-dominant basic cell. Shown in Fig. 10.17a are the "cross-coupled" NOR gates where the fictitious LPD memory element Δt has been removed, and outputs from the two conjugate NOR gate forms are given in mixed-logic notation. Notice again that only one feedback path exists as indicated by the heavy line.

The input conditions required to generate mixed-rail outputs from the reset-dominant basic cell are presented in the truth table of Fig. 10.17b. This table is obtained from the logic circuit and mixed-logic expressions in Fig. 10.17a. Notice that all input conditions except the $S, R = 1, 1$ condition generate *mixed-rail outputs* from the two conjugate NOR gate forms, similar to the case of the set-dominant basic cell in Fig. 10.16b. Thus, again, the

10.4 THE BASIC MEMORY CELLS

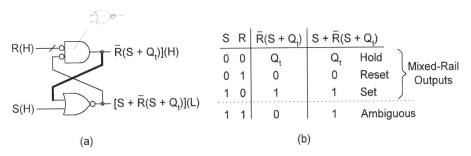

FIGURE 10.17
Mixed-rail outputs of the reset-dominant basic cell. (a) Logic circuit showing the mixed-logic output expressions from the two confugate gate forms. (b) Truth table indicating the input conditions required to produce mixed-rail output conditions.

$S, R = 1, 1$ condition produces an ambiguous output, since the outputs from the conjugate NOR gates are neither a Set nor a Reset.

Clearly, the mixed-rail outputs of the two types of basic memory cells and the combined excitation table representing both basic cells all have something in common. From the results of Figs. 10.15c, 10.16b, and 10.17b, the following important conclusion is drawn:

> *The mixed-rail output character of the set- and reset-dominant basic cells is inherent in the combined excitation table of Fig. 10.15c, since the $S, R = 1, 1$ input condition is absent.*

Use of this fact will be made later in the design of the memory elements, called flip-flops, where the basic cells will serve as the memory. Thus, if the $S, R = 1, 1$ condition is never allowed to happen, mixed-rail output response is ensured. But how is this output response manifested? The answer to this question is given in the following subsection.

10.4.5 Mixed-Rail Output Response of the Basic Cells

From Subsection 10.4.4, one could gather the impression that a mixed-rail output response from the conjugate gate forms of a basic cell occurs simultaneously. Actually, it does not. To dramatize this point, consider the set-dominant basic cell and its mixed-rail output response to nonoverlapping Set and Reset input conditions shown in Fig. 10.18a. It is observed that the active portion of the waveform from the ANDing operation is symmetrically set inside of that from the ORing (NAND gate) operation by an amount τ on each edge. Here, it is assumed that $\tau_1 = \tau_2 = \tau$ is the propagation delay of a two-input NAND gate. Thus, it is evident that the mixed-rail output response of the conjugate gate forms does not occur simultaneously but is delayed by a gate propagation delay following each Set or Reset input condition. The circuit symbol for a set-dominant basic cell operated under mixed-rail output conditions is given in Fig. 10.18b. Should an $S, R = 1, 1$ input condition be presented to the set-dominant basic cell at any time, mixed-rail output response disappears, and the circuit symbol in Fig. 10.18b is no longer valid. That is, the two Q's in the logic symbol assume the existence of mixed-rail output response.

In a similar manner, the mixed-rail output response of the reset-dominant basic cell to nonoverlapping Set and Reset input conditions is illustrated in Fig. 10.18c. Again, it is

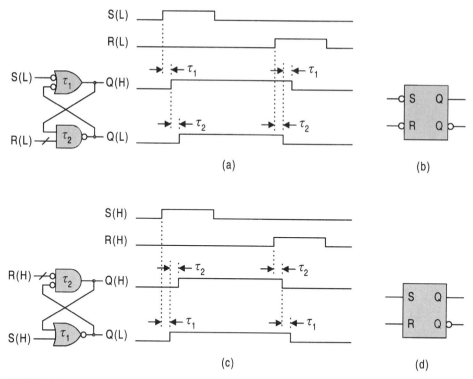

FIGURE 10.18
Mixed-rail output response of the basic cells and circuit symbols. (a) Logic circuit and mixed-rail output response for the set-dominant basic cell. (b) Circuit symbol symbol for the set-dominant basic cell. (c) Logic circuit and mixed-rail output response for the reset-dominant basic cell. (d) Circuit symbol for the reset-dominant basic cell.

observed that the active portion of the waveform from the ANDing (NOR gate) operation is symmetrically set within that of the ORing operation by an amount equal to $\tau = \tau_1 = \tau_2$, the propagation delay of a NOR gate. The circuit symbol for the reset-dominant basic cell operated under mixed-rail output conditions is given in Fig. 10.18d. The difference in circuit symbols for set- and reset-dominant basic cells is indicative of the fact that the former requires active low inputs while the latter requires active high inputs. As is the case for the set-dominant basic cell, an $S, R = 1, 1$ input condition eliminates mixed-rail output response and invalidates the circuit symbol in Fig. 10.18d. The two Q's in the logic symbol assume the existence of mixed-rail output response.

10.5 INTRODUCTION TO FLIP-FLOPS

The basic cell, to which the last section was devoted, is not by itself an adequate memory element for a synchronous sequential machine. It lacks versatility and, more importantly, its operation cannot be synchronized with other parts of a logic circuit or system. Actually, basic cells are asynchronous FSMs without a timing control input but which are essential to

10.5 INTRODUCTION TO FLIP-FLOPS

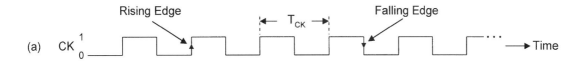

FIGURE 10.19
Clock logic waveforms. (a) Regular clock waveform showing rising and falling edges and a fixed clock period T_{CK}. (b) Irregular clock waveform having no fixed clock period.

the design of *flip-flops*, the memory elements that are used in the design synchronous state machines. A flip-flop may be defined as follows:

> *A flip-flop is an asynchronous one-bit memory element (device) that exhibits sequential behavior controlled exclusively by an enabling input called CLOCK.*

A flip-flop samples a data input of one bit by means of a clock signal, issues an output response, and stores that one bit until it is replaced by another. One flip-flop is required for each state variable in a given state diagram. For example, FSMs that are capable of generating the 3-bit binary sequences shown in Fig. 10.2 each require three flip-flops for their design.

The enabling input, clock, can be applied to the flip-flops as either a regular or irregular waveform. Both types of clock waveforms are represented in Fig. 10.19. The regular clock waveform in Fig. 10.19a is a periodic signal characterized by a clock period T_{CK} and frequency f_{CK} given by

$$f_{CK} = \frac{1}{T_{CK}}, \qquad (10.8)$$

where f_{CK} is given in units of Hz (hertz) when the clock period is specified in seconds. The irregular clock waveform in Fig. 10.19b has no fixed clock period associated with it. However, both regular and irregular clock waveforms must have rising ($0 \to 1$) and falling ($1 \to 0$) edges associated with them, as indicated in Fig. 10.19.

10.5.1 Triggering Mechanisms

In synchronous sequential machines, state-to-state transitions occur as a result of a *triggering mechanism* that is either a rising or falling edge of the enabling clock waveform. Flip-flops and latches that trigger on the rising edge of the clock waveform are said to be *rising edge triggered* (*RET*), and those that trigger on the falling edge of the clock waveform are referred to as *falling edge triggered* (*FET*). These two triggering mechanisms are illustrated in Fig. 10.20, together with the logic symbols used to represent them. The distinction between flip-flops and latches will be made in Section 10.7.

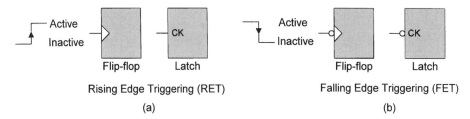

FIGURE 10.20
Flip-flop and latch logic circuit symbology. (a) Rising-edge triggering. (b) Falling-edge triggering.

Mechanisms involving a two-stage triggering combination of RET and FET flip-flops are classified as *master–slave (MS)* triggering mechanisms and flip-flops that employ this two-stage triggering are called, accordingly, *master–slave (MS) flip-flops*. MS flip-flops will be dealt with together with edge-triggered flip-flops in subsequent sections.

10.5.2 Types of Flip-Flops

The designer has a variety of flip-flops and triggering mechanisms from which to choose for a given FSM design. The mechanisms are classified as either edge triggered (ET), meaning RET or FET, or master–slave (MS). The types of flip-flops and the mechanisms by which they operate are normally chosen from following list:

D flip-flops (ET or MS triggered)
T flip-flops (ET or MS triggered)
JK flip-flops (ET or MS triggered)

The generalized definitions of the flip-flop types D, T, and JK are internationally accepted and will be discussed in turn in the sections that follow. There are other flip-flop types (e.g., SR flip-flops) and other triggering mechanism interpretations, and these will be noted where appropriate. It is the intent of this text to concentrate on the major types of flip-flop memory elements.

10.5.3 Hierarchical Flow Chart and Model for Flip-Flop Design

In checking the data books on flip-flops it becomes clear that there exists an interrelationship between the different types suggesting that in many cases there exists a "parent" flip-flop type from which the others are created — a hierarchy for flip-flop design. In fact, it is the D flip-flop (D-FF) that appears to be the basis for the creation of the other types of flip-flops, as indicated in the flow chart of Fig. 10.21. However, it is the JK flip-flop types that are called *universal flip-flops* because they operate in all the modes common to the D, T, and SR type flip-flops. Also, once created, the JK flip-flops are most easily converted to other types of flip-flops (e.g., JKs converted to Ts), as suggested by the flow chart. With few exceptions, flip-flops other than D, JK, and SR types are rarely available commercially. Of the latches, one finds that only the D and SR latches are available commercially.

There are, of course, exceptions to this hierarchy for flip-flop design, but it holds true for most of the flip-flops. The miscellaneous category of flip-flops includes those with special properties for specific applications. The SR flip-flop types fall into this category.

10.5 INTRODUCTION TO FLIP-FLOPS

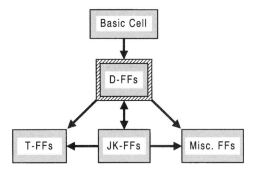

FIGURE 10.21
Flow chart for flip-flop design hierarchy, showing D type flip-flops as central to the design of other flip-flops.

The model that is used for flip-flop design is the basic model given in Fig. 10.3c but adapted specifically to flip-flops. This model, presented in Fig. 10.22a, is applied to a generalized, fictitious RET XY type flip-flop and features one or more basic memory cells as the memory, the next state (NS) forming logic, external inputs including clock (CK), the S and R next state functions, and the present state (PS) feedback paths. Had the fictitious XY-FF been given an FET mechanism, a bubble would appear on the outside of the clock triggering symbol (the triangle). Note that the S and R next state functions would each be represented by dual lines if two basic cells are used as the memory for the XY flip-flop. The logic circuit symbol for the RET XY flip-flop (XY-FF) is given in Fig. 10.22b.

Not all flip-flops to be discussed in the sections that follow have two data inputs and not all have PS feedback paths as in Fig. 10.22. And not all flip-flops will be rising edge triggered as in this fictitious flip-flop. Furthermore, flip-flops classified as master–slave flip-flops do not adhere to the model of Fig. 10.22, since they are two-stage memory elements composed of two memory elements of one type or another. Nevertheless, the model of Fig. 10.22 presents a basis for flip-flop design and will be used in the discussions that follow.

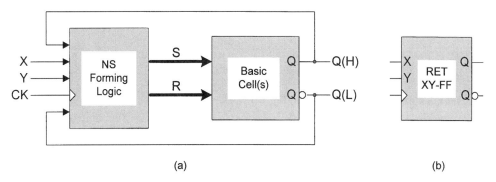

FIGURE 10.22
(a) The basic model adapted to a fictitious RET XY type flip-flop showing the basic cell(s) as memory, the NS forming logic, the S and R next state fucntions, the external data inputs X and Y, the clock (CK) input, and the present state (PS) feedback lines from the mixed-rail outputs Q. (b) Circuit symbol for the RET XY flip-flop.

440 CHAPTER 10 / INTRODUCTION TO SYNCHRONOUS STATE MACHINE DESIGN

10.6 PROCEDURE FOR FSM (FLIP-FLOP) DESIGN AND THE MAPPING ALGORITHM

The following three-step procedure will be used in the design of FSMs including flip-flops:

1. Select the FSM (e.g., a flip-flop type) to be designed and represent this FSM in the form of a state diagram. The output-forming logic can be mapped and obtained at this point.
2. Select the memory element (e.g., a basic cell or flip-flop) to be used in the design of the FSM (e.g., in the design of another flip-flop) and represent this memory element in the form of an excitation table.
3. Obtain the NS function(s) for the FSM in the form of NS K-maps by combining the information represented in the state diagram with that represented in the excitation table for the memory. To accomplish this, apply the following *mapping algorithm*:

> **Mapping Algorithm for FSM Design**
>
> *AND the memory input logic value in the excitation table with the corresponding branching condition (BC) in the state diagram for the FSM to be designed, and enter the result in the appropriate cell of the NS K-map.*

The mapping algorithm is of general applicability. It will be used not only to design and convert flip-flops, but also to design synchronous and asynchronous state machines of any size and complexity. The idea behind the mapping algorithm is that all FSMs, including flip-flops, are characterized by a state diagram and a memory represented in the form of an excitation table. The mapping algorithm provides the means by which these two entities can be brought together in some useful fashion so that the NS functions can be obtained. For now, the means of doing this centers around the NS K-maps. But the procedure is general enough to be computerized for CAD purposes by using a state table in place of the state diagram. Use will be made of this fact in the latter chapters of this text.

10.7 THE D FLIP-FLOPS: GENERAL

Every properly designed and operated D-FF behaves according to a *single* internationally accepted definition that is expressed in any one or all of the three ways. Presented in Fig. 10.23 are the three means of defining the D flip-flop of any type. The first is the operation table for any D-FF given in Fig. 10.23a. It specifies that when D is active Q must be active (Set condition), and when D is inactive Q must be inactive (Reset condition). The state diagram for any D-FF, given in Fig. 10.23b, is best derived from the operation table and expresses the same information about the operation of the D-FF. Thus, state 0 is the Reset state ($Q_t = 0$) when $D = 0$, and state 1 is the Set state ($Q_t = 1$) when $D = 1$.

The excitation table for any D-FF, given in Fig. 10.23c, is the third means of expressing the definition of a D-FF. It is best derived directly from the state diagram in Fig. 10.23b. In this table the $Q_t \rightarrow Q_{t+1}$ column represents the state variable change from PS to NS, and the D column gives the input logic value for the corresponding branching path in the state

10.7 THE D FLIP-FLOPS: GENERAL

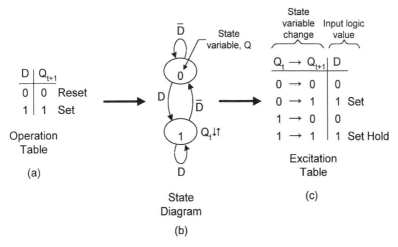

FIGURE 10.23
Generalized D flip-flop definition expressed in terms of the operation table (a), the state diagram (b), and the excitation table (c).

diagram. For example, the Reset hold branching path $0 \to 0$ is assigned $D = 0$ (for $\bar{D}$), and the Set branching path $0 \to 1$ is assigned the $D = 1$ for branching condition D. The excitation table for the D-FF is extremely important to the design of other state machines, including other flip-flops, as will be demonstrated in later sections.

Now that the foundation for flip-flop design has been established, it is necessary to consider specific types of D flip-flops. There are three types to be considered: the D-latch, the edge triggered (ET) D flip-flop, and the master–slave (MS) D flip-flop, all of which adhere to the generalized definition of a D flip-flop expressed in Fig. 10.23. Each of these D-type flip-flops is represented by a unique state diagram containing the enabling input clock (CK) in such a way as to identify the triggering mechanism and character of the D flip-flop type. In each case the memory element used for the design of the D flip-flop is the basic cell (set-dominant or reset-dominant) that is characterized by the combined excitation table given in Fig. 10.15c. The design procedure follows that given in Section 10.6 where use is made of the important *mapping algorithm*.

10.7.1 The D-Latch

A flip-flop whose sequential behavior conforms to the state diagram presented in Fig. 10.24a is called an RET *transparent (high) D latch* or simply *D latch*. Under normal flip-flop action the RET D latch behaves according to the operation table in Fig. 10.23a, but only when enabled by *CK*. The transparency effect occurs when *CK* is active ($CK = 1$). During this time Q goes active when D is active, and Q goes inactive when D is inactive — that is, Q tracks D when $CK = 1$. Under this transparency condition, data (or noise) on the D input is passed directly to the output and normal flip-flop action (regulated by *CK*) does not occur. If the D latch is itself to be used as a memory element in the design of a synchronous FSM, the transparent effect must be avoided. This can be accomplished by using a pulse narrowing circuit of the type discussed later. The idea here is that minimizing the active

442 CHAPTER 10 / INTRODUCTION TO SYNCHRONOUS STATE MACHINE DESIGN

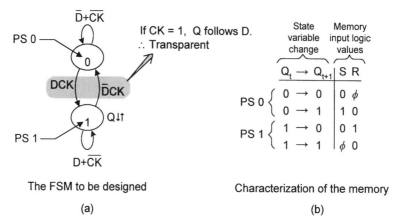

FIGURE 10.24
The RET D latch. (a) State diagram for the D latch showing transparency effect when CK = 1. (b) Excitation table for the basic cell and characterization of the memory.

portions of the clock waveform also minimizes the probability that the transparent effect can occur. Although this is likely to be true, it is generally recommended that the *D* latch not be considered as a viable option when selecting a memory element for FSM design. Of course, if *D* can never go active when *CK* is active, the D latch can be considered as a viable option for memory in FSM design.

The memory element to be used in the design of the D latch is one or the other of two basic cells (Fig. 10.18) characterized by the combined excitation table given in Fig. 10.24b. The plan for design of the D latch is simply to take the information contained in the state diagram of Fig. 10.24a and in the excitation table in Fig. 10.24b, and bring the two kinds of information together in the form of next-state K-maps by using the mapping algorithm given in Section 10.6. When this is done the following information is used for the K-map entries:

For $0 \to 0$ $\begin{cases} \text{place } 0 \cdot (\bar{D} + \overline{CK}) = 0 \text{ in Cell 0 of the } S \text{ K-map} \\ \text{place } \phi \cdot (\bar{D} + \overline{CK}) = \phi(\bar{D} + \overline{CK}) \text{ in Cell 0 of the } R \text{ K-map} \end{cases}$

For $0 \to 1$ $\begin{cases} \text{place } 1 \cdot (DCK) = DCK \text{ in Cell 0 of the } S \text{ K-map} \\ \text{place } 0 \cdot (DCK) = 0 \text{ in Cell 0 of the } R \text{ K-map} \end{cases}$

For $1 \to 0$ $\begin{cases} \text{place } 0 \cdot (\bar{D}CK) = 0 \text{ in Cell 1 of the } S \text{ K-map} \\ \text{place } 1 \cdot (\bar{D}CK) = \bar{D}CK \text{ in Cell 1 of the } R \text{ K-map} \end{cases}$

For $1 \to 1$ $\begin{cases} \text{place } \phi \cdot (D + \overline{CK}) = \phi(D + \overline{CK}) \text{ in Cell 1 of the } S \text{ K-map} \\ \text{place } 0 \cdot (\bar{D}CK) = 0 \text{ in Cell 1 of the } R \text{ K-map} \end{cases}$.

This results in the next state EV K-maps, minimum next state functions for *S* and *R*, and the logic circuit and symbol all shown in Fig. 10.25. The four null (zero) entries are omitted in the EV K-maps, leaving only the two essential and two nonessential (don't care) entries for

10.7 THE D FLIP-FLOPS: GENERAL

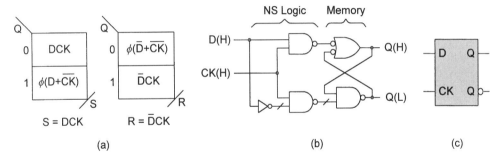

FIGURE 10.25
Design of the RET D latch by using a basic cell as the memory. (a) EV K-maps and minimum Boolean expressions for the S and R next-state functions. (b) Logic circuit showing the NS logic from part (a) and the set-dominant basic cell as the memory. (c) Logic circuit symbol for the RET D latch.

use in extracting minimum cover. Note that DCK (in the S K-map) is contained in ϕD, that $\bar{D}CK$ (in the R K-map) is contained in $\phi \bar{D}$, and that the logic circuit conforms to the model in Fig. 10.22 exclusive of PS feedback. The CK input to the circuit symbol in Fig. 10.25c is consistent with that for a latch as indicated in Fig. 10.20a.

The behavior of the RET D latch is best demonstrated by the timing diagram shown in Fig. 10.26. Here, normal D flip-flop (D-FF) action is indicated for D pulse durations much longer than a CK period. For normal D-FF behavior, Q goes active when CK samples (senses) D active, and Q goes inactive when CK samples D inactive. However, when CK is active and D changes activation level, the transparency effect occurs. This is demonstrated in the timing diagram of Fig. 10.26.

The FET (*transparent low*) D latch is designed in a similar manner to the RET D latch just described. All that is required is to complement CK throughout in the state diagram of Fig. 10.24a, as shown in Fig. 10.27a. Now, the transparency effect occurs when CK is inactive ($CK = 0$). If a set-dominant basic cell is again used as the memory, there results the logic circuit of Fig. 10.27b, where an inverter is the only added feature to the logic circuit shown in Fig. 10.25b. The logic circuit symbol for the FET D latch is given in Fig. 10.27c. Here, the active low indicator bubble on the clock input identifies this as a falling edge

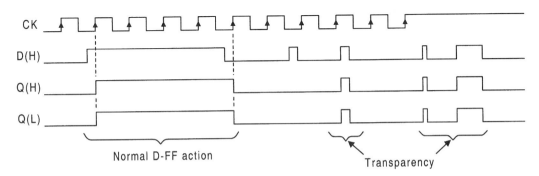

FIGURE 10.26
Timing diagram for an RET D latch showing normal D-FF action and the transparency effect that can occur when CK is active, where no account is taken of gate path delays.

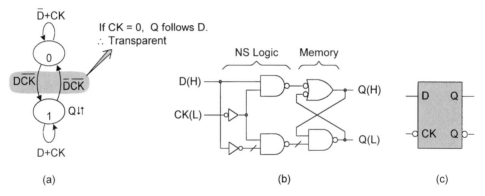

FIGURE 10.27
The FET D latch. (a) State diagram showing condition for transparency. (b) Logic circuit assuming the use of a set-dominant basic cell as the memory for design. (c) Logic circuit symbol.

triggered device consistent with Fig. 10.20(b). A CK(H) or CK(L) simply means RET or FET, respectively.

If either the RET D latch or the FET D latch is to be used as the memory element in the design of a synchronous FSM, extreme care must be taken to ensure that the transparency effect does not occur. Transparency effects in flip-flops result in unrecoverable errors and must be avoided. This can be accomplished by using a pulse narrowing circuit of the type shown in Fig. 10.28a. Here, an inverting delay element of duration Δt is used to produce narrow pulses of the same duration in the output logic waveform as indicated in Fig. 10.28b. The delay element can be one or any odd number of inverters, an inverting buffer, or an inverting Schmitt trigger. In any case, the delay element must be long enough to allow the narrow pulses to reliably cross the switching threshold. If the delay is too long, the possibilty of transparency exists; if it is too short, flip-flop triggering will not occur.

10.7.2 The RET D Flip-Flop

The transparency problem inherent in the D latch, discussed in the previous subsection, places rather severe constraints on the inputs if the latch is to be used as a memory element

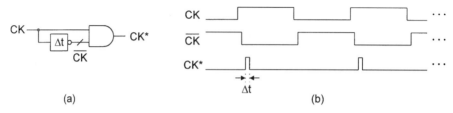

FIGURE 10.28
Pulse narrowing circuit. (a) Logic circuit showing an inverting delay element Δt used to produce narrow pulses from long input pulses. (b) Positive logic timing diagram showing the resulting narrow pulses of duration Δt on the output waveform.

10.7 THE D FLIP-FLOPS: GENERAL

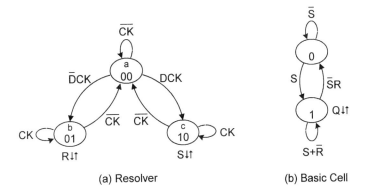

FIGURE 10.29
The RET D flip-flop as represented by state diagrams. (a) Resolver FSM input stage. (b) Set-dominant basic cell output stage.

in the design of a state machine. This problem can be overcome by using an edge triggered D flip-flop that possesses *data lockout character* as discussed in the following paragraph. Shown in Fig. 10.29a is the *resolver* FSM that functions as the input stage of an RET D flip-flop. Here, state a is the sampling (unresolved) state, CK is the sampling (enabling) input, and states b and c are the resolved states. Observe that the outputs of the resolver are the inputs to the basic cell shown in Fig. 10.29b, and that the output of the basic cell is the output of the D flip-flop. Thus, an input FSM (the resolver) drives an output FSM (the basic cell) to produce the D flip-flop which conforms to the general D flip-flop definitions given in Fig. 10.23. Note that both the resolver and basic cell are classified as asynchronous state machines, yet they combine to produce a state machine (flip-flop) that is designed to operate in a synchronous (clock-driven) environment. But the flip-flop itself is an asynchronous FSM!

To understand the function of the RET D flip-flop, it is necessary to move stepwise through the operation of the two FSMs in Fig. 10.29: Initially, let Q be inactive in state a of the resolver. Then, if CK samples D active in state a, the resolver transits $a \rightarrow c$ and issues the output S, which drives the basic cell in Fig. 10.29b to the set state 1 where Q is issued. In state c, the resolver holds on CK, during which time Q remains active; and the data input D can change at any time without altering the logic status of the flip-flop—this is the *data lockout* feature. When CK goes inactive ($\overline{CK}$), the resolver transits back to state a, where the sampling process begins all over again, but where Q remains active. Now, if CK samples D inactive ($\bar{D}$) in state a, the resolver transits $a \rightarrow b$, at which time R is issued. Since the branching condition $\bar{S}R$ is now satisfied, the basic cell is forced to transit to the reset state 0, where Q is deactivated. The resolver holds in state b on active CK. Then when CK goes inactive ($\overline{CK}$), the resolver transits back to the unresolved state a, at which time the sampling process begins all over again, but with Q remaining inactive.

The design of the RET D flip-flop follows the design procedure and mapping algorithm given in Section 10.6. Since the logic circuit for the set-dominant basic cell is known and given in Fig. 10.18a, all that is necessary is to design the resolver circuit. This is done by using what is called the *nested cell model*, which uses the basic cells as the memory elements. Shown in Fig. 10.30 are the state diagram for the resolver (the FSM to be designed), the

446 CHAPTER 10 / INTRODUCTION TO SYNCHRONOUS STATE MACHINE DESIGN

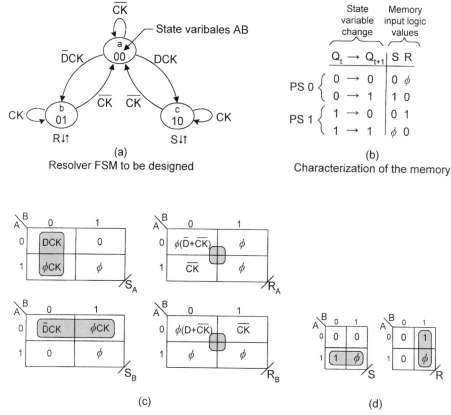

FIGURE 10.30
Resolver design for the RET D flip-flop. (a) State diagram for the resolver. (b) Characterization of the memory. (c) EV K-maps for the next state functions required to drive the two basic cells. (d) Output K-maps for the resolver.

characterization of the memory (combined excitation table for the basic cell), and the EV K-maps for the next state and output functions.

The mapping algorithm requires that the information contained in the state diagram of Fig. 10.30a be combined with the excitation table of Fig. 10.30b to produce the next state EV K-maps. This has been done in Fig. 10.30c by introducing the following information obtained by a step-by-step application of the mapping algorithm:

State 00 (K-map cell 0)

$$\text{Bit A} \begin{cases} 0 \to 1, \text{ place } 1 \cdot (DCK) = DCK \text{ in the } S_A \text{ K-map} \\ 0 \to 0, \text{ place } \phi \cdot (\bar{D} + \overline{CK}) \text{ in the } R_A \text{ K-map} \end{cases}$$

$$\text{Bit B} \begin{cases} 0 \to 1, \text{ place } 1 \cdot (\bar{D}CK) = \bar{D}CK \text{ in the } S_B \text{ K-map} \\ 0 \to 0, \text{ place } \phi \cdot (D + \overline{CK}) \text{ in the } R_B \text{ K-map} \end{cases}$$

10.7 THE D FLIP-FLOPS: GENERAL

State 01 (K-map cell 1)

Bit A $\begin{cases} 0 \to 0, \text{ place } 0 \text{ in the } S_A \text{ K-map,} \\ \quad\quad \text{place } \phi \text{ in the } R_A \text{ K-map} \end{cases}$

Bit B $\begin{cases} 1 \to 0, \text{ place } 1 \cdot (\overline{CK}) = \overline{CK} \text{ in the } R_B \text{ K-map} \\ 1 \to 1, \text{ place } \phi \cdot (CK) = \phi CK \text{ in the } S_B \text{ K-map} \end{cases}$

State 10 (K-map cell 2)

Bit A $\begin{cases} 1 \to 0, \text{ place } 1 \cdot (\overline{CK}) = \overline{CK} \text{ in the } R_A \text{ K-map} \\ 1 \to 1, \text{ place } \phi \cdot (CK) = \phi CK \text{ in the } S_A \text{ K-map} \end{cases}$

Bit B $\begin{cases} 0 \to 0, \text{ place } 0 \text{ in the } S_B \text{ K-map,} \\ \quad\quad \text{place } \phi \text{ in the } R_B \text{ K-map} \end{cases}$.

Notice that for every essential EV entry in a given K-map cell there exists the complement of that entry ANDed with ϕ in the same cell of the other K-map. This leads to the following modification of the mapping algorithm in Section 10.6 as it pertains to S/R mapping:

1. Look for Sets $(0 \to 1)$ and Resets $(1 \to 0)$ and make the entry $1 \cdot$(Appropriate BC) in the proper S_i or R_i K-map, respectively, according to the combined excitation table for the basic cell. (Note: BC = branching condition.)

2. For each Set entry (from [1]) in a given cell of the S_i K-map, enter $\phi \cdot \overline{\text{(Appropriate BC)}}$ in the *same cell* of the corresponding R_i K-map.
   For each Reset entry (from [1]) in a given cell of the R_i K-map, enter $\phi \cdot \overline{\text{(Appropriate BC)}}$ in the *same cell* of the corresponding S_i K-map.

3. For Hold Resets $(0 \to 0)$ and Hold Sets $(1 \to 1)$, enter $(0,\phi)$ and $(\phi,1)$, respectively, in the (S_i, R_i) K-maps in accordance with the combined excitation table for basic cell given in Fig. 10.15c.

Continuing with the design of the RET D flip-flop, the minimum NS and output functions extracted from the EV K-maps in Figs. 10.30c and 10.30d are

$$\begin{cases} S_A = \bar{B}DCK & R_A = \overline{CK} \\ S_B = \bar{A}\bar{D}CK & R_B = \overline{CK} \\ \quad S = A & R = B \end{cases}, \quad (10.9)$$

which are implemented in Fig. 10.31a. Here, the basic cells for bits A and B are highlighted by the shaded areas within the resolver section of the RET D flip-flop. Notice that the requirement of active low inputs to the three set-dominant basic cells is satisfied. For example, in the resolver FSM this requirement is satisfied by $R_A(L) = R_B(L) = \overline{CK}(L) = CK(H)$. The circuit symbol for the RET D flip-flop is given in Fig. 10.31b, where the triangle on the CK input is indicative of an edge triggered flip-flop with data-lockout character and is consistent with Fig. 10.20a.

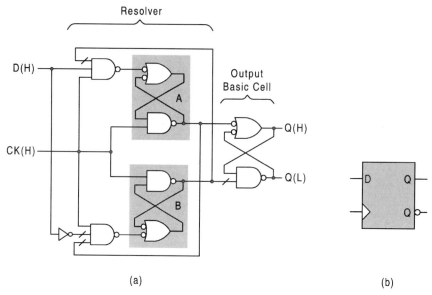

FIGURE 10.31
(a) Logic circuit for the RET D flip-flop as constructed from Eqs. (10.9) showing the resolver and output basic cell stage. (b) Logic circuit symbol.

The operation of the RET D flip-flop is best represented by the timing diagram in Fig. 10.32, where arrows on the rising edge of the clock waveform provide a reminder that this is an RET flip-flop. The edge-triggering feature is made evident by the vertical dashed lines, and the data lockout character is indicated by the absence of a flip-flop output response to narrow data pulses during the active and inactive portions of the clock waveform. For the sake of simplicity, no account is taken of gate propagation delay in Fig. 10.32.

10.7.3 The Master–Slave D Flip-Flop

Another useful type of D flip-flop is the master–slave (MS) D flip-flop defined by the two state diagrams in Fig. 10.33a and that conforms to the general definitions for a D flip-flop given in Fig. 10.23. The MS D flip-flop is a two-stage device consisting of an RET D latch

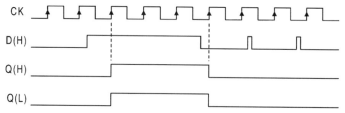

FIGURE 10.32
Timing diagram showing proper operation of the RET D flip-flop.

10.7 THE D FLIP-FLOPS: GENERAL

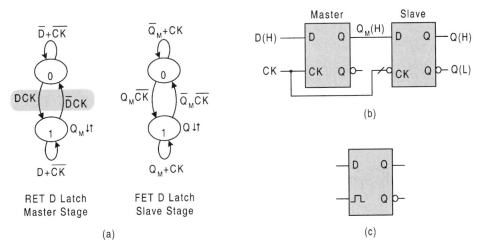

FIGURE 10.33
The master–slave (MS) D flip-flop. (a) State diagram for the master and slave stages. (b) Logic circuit. (c) Circuit symbol.

as the master stage and an FET D latch as the slave stage. The output of the master stage is the input to the slave stage. Thus, the transparency problem of the D latch in Fig. 10.24a has been eliminated by the addition of the slave stage that is triggered *antiphase* to the master. Thus, should signals pass through the master stage when *CK* is active, they would be held up at the slave stage input until *CK* goes inactive.

The design of the MS D flip-flop can be carried out following the same procedure as given in Figs. 10.24, 10.25, and 10.27. However, this is really unnecessary, since the logic circuits for both stages are already known from these earlier designs. The result is the logic circuit given in Fig. 10.33b, where the output of the master RET D latch symbol is the input to the slave FET D latch symbol. The logic circuit symbol is shown in Fig. 10.33c and is identified by the pulse symbol on the clock input.

The operation of the MS D flip-flop is illustrated by the timing diagram in Fig. 10.34, where no account is taken of gate propagation delay. Notice that signals that are passed through the master stage during active *CK* are not passed through the slave stage, which is

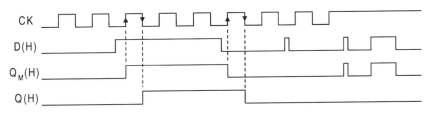

FIGURE 10.34
Timing diagram for the MS D flip-flop showing the output response from master and slave stages, and the absence of complete transparency with no account taken of gate path delays.

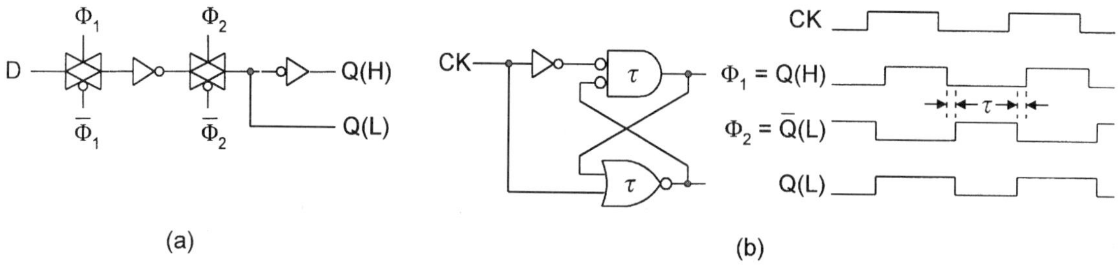

FIGURE 10.35
(a) The MS D flip-flop configured with CMOS transmission gates and inverters and requiring two-phase (2Φ) clocking. (b) The reset-dominant basic cell used to generate 2Φ clocking as indicated by the output logic waveforms.

triggered antiphase to the master. However, there is the possibility of noise transfer, though of low probability. If logic noise should appear at the input to the slave stage just at the instant that CK goes through a falling edge, that noise can be transferred to the output.

One important advantage the MS D flip-flop has over the edge triggered variety is that the MS D flip-flop can be configured with transmission gates and inverters. Such a configuration is shown in Fig. 10.35a, where two CMOS transmission gates are used together with two inverters. To achieve the two-stage effect required by the MS configuration, the CMOS transmission gates must be operated by using two-phase (2Φ) clocking such that the active portions of the clock phases are nonoverlapping. Shown in Fig. 10.35b is a reset-dominant basic cell used to generate the two clock phases (Φ_1 and Φ_2) whose active portions are separated in time by an amount τ, the path delay of a NOR gate. Notice that both phase waveforms (Φ_1 and Φ_2) are given in positive logic, similar to the physical voltage waveforms but without rise and fall times. These clock phase signals must each be supplied to the CMOS transmission gates in complementary form. This means that when Φ_1 is at LV, $\bar{\Phi}_1$ must be at HV and vice versa. The same must be true for Φ_2. Each complementary form is achieved by the use of an inverter with a buffer in the HV path for delay equalization, if necessary.

10.8 FLIP-FLOP CONVERSION: THE T, JK FLIP-FLOPS AND MISCELLANEOUS FLIP-FLOPS

In Fig. 10.21 a hierarchy for flip-flop design is given with the understanding that the D flip-flop is central to such a process. In this text, this is the case, as will be demonstrated by the design of the other important types of flip-flops. First, however, certain information must be understood.

To design one flip-flop from another, it is important to remember the following:

> *The new flip-flop to be designed inherits the triggering mechanism of the old (memory) flip-flop.*

10.8 THE T, JK FLIP-FLOPS AND MISCELLANEOUS FLIP-FLOPS

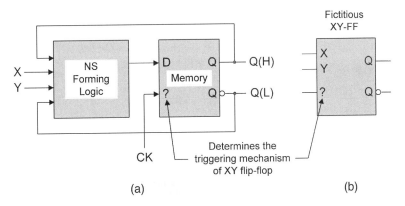

FIGURE 10.36
(a) Model and (b) logic symbol for a fictitious XY flip-flop derived from a D flip-flop having an unspecified triggering mechanism.

This important fact can best be understood by considering the fictitious XY flip-flop shown in Fig. 10.36. This fictitious flip-flop has been derived from a D flip-flop of some arbitrary triggering mechanism indicated by the question mark (?) on the clock input.

The model in Fig. 10.36a can be compared with the basic model in Fig. 10.22 for the same fictitious XY flip-flop, where now a D flip-flop is used as the memory instead of basic cells. In either case the XY flip-flop is designed according to the design procedure and mapping algorithm presented in Section 10.6, but the characterization of memory is different. As will be recalled from Section 10.7, flip-flops designed by using one or more basic cells require that the memory be characterized by the combined excitation table for the basic cell given in Fig. 10.15c. Now, for flip-flop conversion by using a D flip-flop as the memory, the excitation table for the D flip-flop in Fig. 10.23c must be used.

10.8.1 The T Flip-Flops and Their Design from D Flip-Flops

All types of T flip-flops behave according to an internationally accepted definition that is expressed in one or all of three ways. Presented in Fig. 10.37 are three ways of defining the T flip-flop, all expressed in positive logic as was true in the definition of the D flip-flops. Shown in Fig. 10.37a is the operation table for any T flip-flop. It specifies that when T is active, the device must toggle, meaning that $0 \to 1$ and $1 \to 0$ transitions occur as long as $T = 1$. When $T = 0$, the T flip-flop must hold in its present state. The state diagram for T flip-flops in Fig. 10.37b is derived from the operation table and conveys the same information as the operation table. Here, the toggle character of the T flip-flop is easily shown to take place between Set and Reset states when T is active, but holding in these states when T is inactive.

The excitation table presented in Fig. 10.37c is the third means of expressing the definition of T flip-flops. It is easily derived from the state diagram and hence conveys the same information regarding T flip-flop operation. This excitation table will be used to characterize the memory in the design of FSMs that require the use of T flip-flops as the memory elements.

452 CHAPTER 10 / INTRODUCTION TO SYNCHRONOUS STATE MACHINE DESIGN

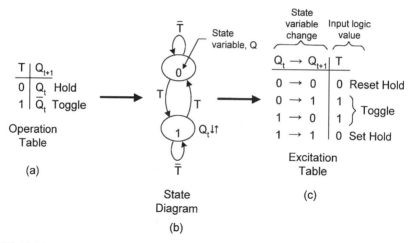

FIGURE 10.37
Generalized T flip-flop definition expressed in terms of the operation table (a), the state diagram (b), and the excitation table (c).

Design of the T Flip-Flops from D Flip-Flops Since T flip-flops are to be designed (converted) from D flip-flops, the excitation table for the D flip-flop must be used to characterize the memory. This excitation table and the state diagram representing for the family of T flip-flops must be brought together by using the mapping algorithm set forth in Section 10.6. This is done in Fig. 10.38, parts (a), (b), and (c), where the next state logic for flip-flop conversion is found to be

$$D = T \oplus Q. \tag{10.10}$$

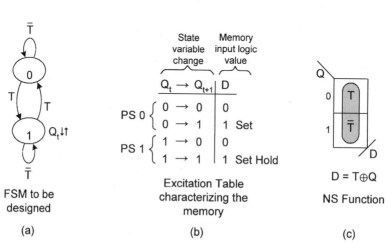

FIGURE 10.38
Design of the T flip-flops. (a) The state diagram for any T flip-flop. (b) Excitation table for the D flip-flop memory. (c) NS K-map and NS function resulting from the mapping algorithm.

10.8 THE T, JK FLIP-FLOPS AND MISCELLANEOUS FLIP-FLOPS

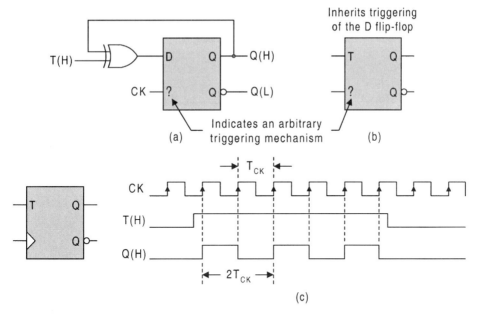

FIGURE 10.39
(a), (b) Implementation of Eq. (10.10) and logic circuit symbol for a T flip-flop of arbitrary triggering mechanism. (c) Logic symbol and timing diagram for an RET T flip-flop showing toggle and hold modes of operation.

Implementation of the NS function given in Eq. (10.10) is shown in Fig. 10.39a together with the symbol for the T flip-flop in Fig. 10.39b, which as yet has not been assigned a triggering mechanism — the designer's choice indicated by the question mark (?) on the clock input. Remember that the new FSM (in this case a T flip-flop) inherits the triggering mechanism of the memory flip-flop (in this case a D flip-flop). Shown in Fig. 10.39c is the logic circuit symbol and timing diagram for an RET T flip-flop, the result of choosing an RET D flip-flop as the memory. The timing diagram clearly indicates the toggle and hold modes of operation of the T flip-flop. For the sake of simplicity no account is taken of the propagation delays through the logic.

Were it desirable to produce an MS T flip-flop, the memory element in Fig. 10.39a would be chosen to be a MS D flip-flop. The timing diagram for an MS T flip-flop would be similar to that of Fig. 10.39c, except the output from the slave stage would be delayed from the master stage by a time period $T_{CK}/2$. This is so because the slave stage picks up the output from the master stage only on the falling edge of CK, that is, the two stages are triggered antiphase to one another.

10.8.2 The JK Flip-Flops and Their Design from D Flip-Flops

The flip-flops considered previously are single data input flip-flops. Now, consideration centers on a type of flip-flop that has two data inputs, J and K. The members of the JK flip-flop family conform to the internationally accepted definition expressed in terms

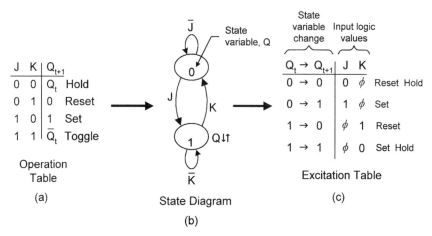

FIGURE 10.40
Generalized JK flip-flop definition expressed in terms of the operation table (a), the state diagram (b), and the excitation table (c).

of an operation table, a state diagram, or an excitation table provided in Fig. 10.40. The operation table in Fig. 10.40a reveals the four modes of JK flip-flop operation: Hold, Reset, Set, and Toggle. Thus, it is seen that the JK type flip-flops operate in all the modes common to SR, T, and D type flip-flops, though SR flip-flops (clocked SR latches) are yet to be discussed. For this reason the JK flip-flops are sometimes referred to as the *universal flip-flops*. The state diagram in Fig. 10.40b is best derived from the operation table. For example, the Set $(0 \rightarrow 1)$ branching condition follows from the Boolean sum $(\text{Set} + \text{Toggle}) = J\bar{K} + JK = J$, and the Reset $(1 \rightarrow 0)$ branching condition results from the sum $(\text{Reset} + \text{Toggle}) = \bar{J}K + JK = K$. The Set–Hold and Reset–Hold conditions result from the sums $J\bar{K} + \bar{J}\bar{K} = \bar{K}$ and $\bar{J}K + \bar{J}\bar{K} = \bar{J}$, respectively. However, given the set and reset branching conditions, the sum rule in Eq. (10.3) can and should be used to obtain the two hold conditions.

The excitation table for the JK flip-flops in Fig. 10.40c is easily derived from the state diagram in (b). For example, the Reset–Hold branching path requires a branching condition $\bar{J}$ that places a 0 and a ϕ in the J and K columns of the excitation table. A ϕ is used for unspecified inputs in branching conditions. Similarly, a 1 and ϕ are placed in the J and K columns for the Set branching condition J. Notice that this excitation table bears some resemblance to that of the combined excitation table for the basic cells in Fig. 10.15c, but with two additional don't cares. The excitation table for the JK flip-flops will be used rather extensively to characterize the memory in the design of FSMs that require JK flip-flops as memory elements.

Design of the JK Flip-Flops from the D Flip-Flops The process used previously in the design of T flip-flops from D flip-flops is now repeated for the case of the JK flip-flops defined in Fig. 10.40 in terms of the operation table, state diagram, and excitation table. Shown in Fig. 10.41a is the state diagram representing the family of JK flip-flops, the FSMs to be designed. Since a D flip-flop is to be used as the memory element in the design, its excitation table must be used to characterize the memory and is provided in Fig. 10.41b.

10.8 THE T, JK FLIP-FLOPS AND MISCELLANEOUS FLIP-FLOPS

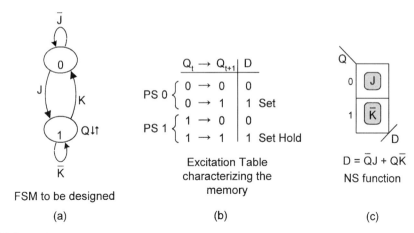

FIGURE 10.41
Design of the JK flip-flops. (a) State diagram for any JK flip-flop. (b) Excitation table for the D flip-flop memory. (c) NS K-map and NS function required for flip-flop conversion.

By using the mapping algorithm in Section 10.6 together with the state diagram for a JK flip-flop and the excitation table for the memory D flip-flop, there results the NS logic K-map and NS forming logic shown in Fig. 10.41c. Notice that only the Set and Set Hold branching paths produce non-null entries in the NS K-map for D, a fact that is always true when applying the mapping algorithm to D flip-flop memory elements.

The minimum NS logic function extracted from the K-map is

$$D = \bar{Q}J + Q\bar{K} \qquad (10.11)$$

and is shown implemented in Fig. 10.42a with a D flip-flop of an arbitrary triggering mechanism as the memory. Its circuit symbol is given in Fig. 10.42b, also with a question mark (?) indicating an arbitrary triggering mechanism determined from the D flip-flop memory element. In Fig. 10.42c is shown the circuit symbol and timing diagram for an FET JK flip-flop that has been derived from an FET D flip-flop. The timing diagram illustrates the four modes of JK flip-flop operation: Hold (Reset or Set), Reset, Set, and Toggle. Notice that once a set condition is sampled by clock, that condition is maintained by the flip-flop until either a reset or toggle condition is sampled by the falling edge of the clock waveform. Similarly, once a reset condition is executed by clock, that condition is maintained until either a set or toggle condition is initiated. As always, the toggle mode results in a divide-by-two of the clock frequency.

Equation (10.11) has application beyond that of converting a D flip-flop to a JK flip-flop. It is also the basis for converting D K-maps to JK K-maps and vice versa. K-map conversion is very useful in FSM design and analysis since it can save time and reduce the probability for error. The subject of K-map conversion will be explored in detail later in this chapter.

10.8.3 Design of T and D Flip-Flops from JK Flip-Flops

The procedures for converting D flip-flops to T and JK flip-flops, used in the preceding subsections, will now be used for other flip-flop conversions. The conversions JK-to-T and

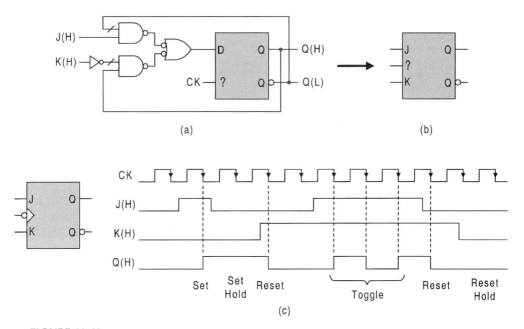

FIGURE 10.42
(a) Implementation of Eq. (10.11), and (b) logic circuit symbol for a JK flip-flop of arbitrary triggering mechanism. (c) Logic symbol and timing diagram for an FET JK flip-flop designed from an FET D flip-flop showing all four modes of operation indicated by the operation table in Fig. 10.40a.

JK-to-D are important because they emphasize the universality of the JK flip-flop types. Presented in Fig. 10.43, for JK-to-T flip-flop conversion, are the state diagram for the T flip-flops (a), the excitation table characterizing the JK memory (b), and the NS K-maps and NS functions for J and K (c). Plotting the NS K-maps follows directly form application of the mapping algorithm given earlier in Section 10.6. Notice that the ϕ's in the NS K-maps result from summing of the branching condition values relative to the branching paths of a particular present state (PS). For example, in PS state 1, a ϕ is placed in cell 1 of the J K-map, since $\phi T + \phi \bar{T} = \phi$ as required by the $1 \to 0$ and $1 \to 1$ branching paths, respectively. By using the don't cares in this manner, the minimum cover for the NS functions is

$$J = K = T. \tag{10.12}$$

Thus, to convert any JK flip-flop to a T flip-flop of the same triggering character, all that is necessary is to connect the J and K input terminals together to become the T input, as indicated by the logic circuit symbols in Fig. 10.43d. Equation (10.12) will also be useful for converting JK K-maps to T K-maps and vice versa.

The conversion of JK flip-flops to D flip-flops follows in a similar manner to that just described for converting JK to T flip-flops. Presented in Fig. 10.44 are the state diagram for the family of D flip-flops (a), the excitation table for the memory JK flip-flop (b), and the NS K-maps and conversion logic extracted from the K-maps (c). The minimum NS

10.8 THE T, JK FLIP-FLOPS AND MISCELLANEOUS FLIP-FLOPS

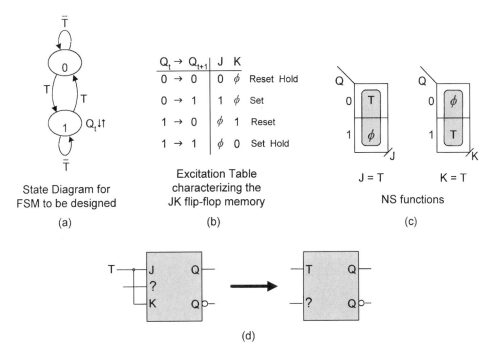

FIGURE 10.43
Design of the T flip-flops from the JK flip-flops. (a) State diagram representing the family of T flip-flops. (b) Excitation table characterizing the JK memory element. (c) NS K-maps and NS functions for the flip-flop conversion. (d) Logic circuit and symbol for a T flip-flop of arbitrary triggering mechanism.

functions, as extracted from the NS K-maps, are given by

$$J = D \quad \text{and} \quad K = \bar{D}. \qquad (10.13)$$

Shown in Fig. 10.44d is the logic circuit and its circuit symbol for D flip-flop conversion from a JK flip-flop of arbitrary triggering mechanism. Clearly, all that is necessary to convert a JK flip-flop to a D flip-flop is to connect D to J and D to K via an inverter.

10.8.4 Review of Excitation Tables

For reference purposes, the excitation tables for the families of D, T, and JK flip-flops, discussed previously, are provided in the table of Fig. 10.45. Also shown in the table is the excitation table for the family of SR flip-flops and all related SR devices which include the basic cells. Notice the similarity between the JK and SR excitation tables, which leads to the conclusion that *J is like S and K is like R*, but not exactly. The only difference is that there are two more don't cares in the JK excitation table than in the SR excitation table. Also observe that the D values are active for Set and Set Hold conditions, and that the T values are active only under toggle $1 \to 0$ and $0 \to 1$ conditions. These facts should serve as a mnemonic means for the reader in remembering these important tables.

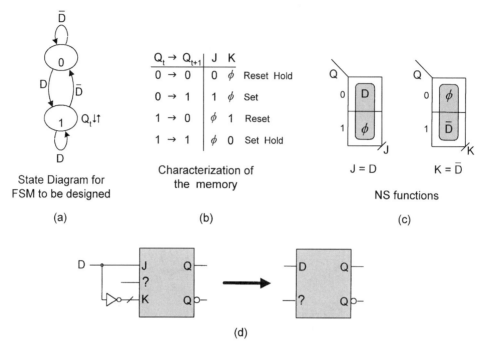

FIGURE 10.44
Design of the D flip-flops from the JK flip-flops. (a) State diagram representing the family of D flip-flops. (b) Excitation table characterizing the JK memory element. (c) NS K-maps and NS functions for the flip-flop conversion. (d) The logic circuit and symbol for a D flip-flop of arbitrary triggering mechanism.

Eventually, construction of the NS K-maps will become so commonplace that specific mention of either the mapping algorithm or the particular excitation table in use will not be necessary.

Any of the excitation tables given in Fig. 10.45 can be used to characterize the flip-flop memory for the purpose of applying the mapping algorithm in Section 10.6 to obtain the NS forming logic for an FSM. In fact, that is their only purpose. For example, if D flip-flops are required as the memory in the design of an FSM, the excitation table for the family of D flip-flops is used. Or if JK flip-flops are to be used as the memory, the excitation table for the JK flip-flops is used for the same purpose, etc.

	$Q_t \rightarrow Q_{t+1}$	D	T	J	K	S	R
Reset Hold	$0 \rightarrow 0$	0	0	0	ϕ	0	ϕ
Set	$0 \rightarrow 1$	1	1	1	ϕ	1	0
Reset	$1 \rightarrow 0$	0	1	ϕ	1	0	1
Set Hold	$1 \rightarrow 1$	1	0	ϕ	0	ϕ	0

FIGURE 10.45
Summary of the excitation tables for the families of D, T, JK, and SR flip-flops.

10.8 THE T, JK FLIP-FLOPS AND MISCELLANEOUS FLIP-FLOPS

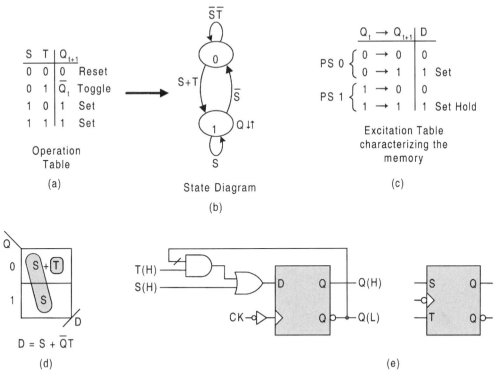

FIGURE 10.46
Design of a special-purpose FET ST flip-flop. (a) Operation table for the family of ST flip-flops. (b) State diagram derived from (a). (c) Characterization of the D flip-flop memory. (d) NS K-map and NS function for flip-flop conversion. (e) Logic circuit and circuit symbol for the FET ST flip-flop.

10.8.5 Design of Special-Purpose Flip-Flops and Latches

To emphasize the applicability and versatility of the design procedure and mapping algorithm given in Section 10.6, other less common or even "nonsense" flip-flops will now be designed. These design examples are intended to further extend the reader's experience in design procedures.

An Unusual (Nonsense) Flip-Flop Suppose it is desirable to design an FET ST (Set/Toggle) flip-flop that is defined according to the operation table in Fig. 10.46a. The state diagram for the family of ST flip-flops, derived from the operation table, is shown in Fig. 10.46b. Also, suppose it is required that this flip-flop is to be designed from an RET D flip-flop. Therefore, the memory must be characterized by the excitation table for the D flip-flop presented in Fig. 10.46c. By using the mapping algorithm, the NS K-map and NS forming logic are obtained and are given in Fig. 10.46d. Implementation of the NS logic with the RET D flip-flop to obtain the FET ST flip-flop is shown in part (e) of the figure. Notice that the external Q feedback is necessary to produce the toggle character required by the operation table and state diagram for the family of ST flip-flops. If it had been required to design an MS ST flip-flop, then an MS D flip-flop would have been used as the memory element while retaining the same NS forming logic.

The external hardware requirememts in the design of the FET ST flip-flop can be minimized by using an RET JK flip-flop as the memory in place of a D flip-flop. If the D excitation table in Fig. 10.46c is replaced by that for the JK flip-flops in Fig. 10.40c, the NS functions become $J = S + T$ and $K = \bar{S}$, a reduction of one gate. It is left to the reader to show the mapping details.

A Special-Purpose Clocked SR Latch As used in this text, the term *latch* refers to gated or clocked memory elements that do not have data lockout character and that exhibit transparency, or that lose their mixed-rail output character under certain input conditions. The D latch in Fig. 10.24 is an example, since it exhibits the transparency effect under the condition $CK(H) = 1(H)$. The family of SR latches also fall into this category. One such SR latch is defined by the operation table in Fig. 10.47a from which the state diagram in Fig. 10.47b is derived. This latch is observed to Set under the $S\bar{R}$ branching condition, Reset under condition $\bar{S}R$, and hold if S,R is either 0,0 or 1,1. Notice that CK is part of the input branching conditions, and that the basic cell is to be used as the memory characterized by the excitation table in Fig. 10.47c. Applying the mapping algorithm yields the NS K-maps and NS-forming logic given in part (d) of the figure. Implementing with a

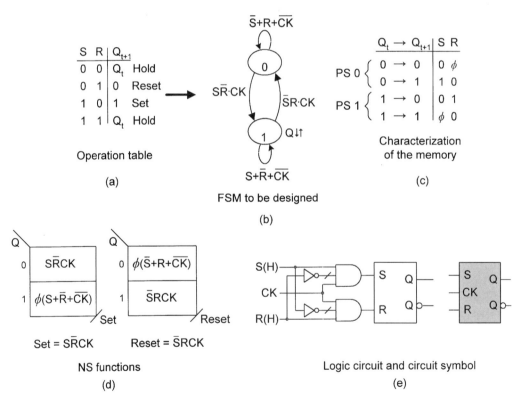

FIGURE 10.47
Design of a special-purpose SR latch. (a) Operation table for this family of SR flip-flops and latches. (b) State diagram for the special SR latch derived from the operation table in (a). (c) Characterization of the basic cell memory. (d) NS K-maps and NS-forming logic. (e) Logic circuit and circuit symbol.

10.9 LATCHES AND FLIP-FLOPS WITH SERIOUS TIMING PROBLEMS   461

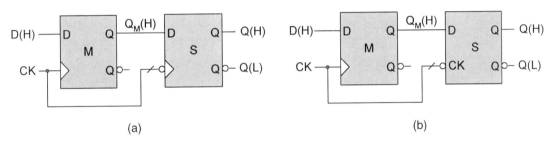

FIGURE 10.48
The D data lockout flip-flop. (a) All edge triggered flip-flop variety. (b) Same as (a) except with an FET D latch as the slave stage.

reset-dominant basic cell yields the logic circuit and circuit symbol shown in Fig. 10.47e. Clearly, an $S,R = 1,1$ condition cannot be delivered to the basic cell output stage. But there is a partial transparency effect. For example, a change $\bar{S}\bar{R} \to S\bar{R}$ while in state 0 with CK active ($CK = 1$) will cause a transition to state 1 where Q is issued. Thus, Q follows S in this case, which is a transparency effect. Similarly, a change $\bar{S}\bar{R} \to \bar{S}R$ while in state 1 when CK is active causes a transition $1 \to 0$ with an accompanying deactivation of Q. Again, this is a transparency effect, since Q tracks R when $CK = 1$.

The Data Lockout MS Flip-Flop   The *data lockout MS flip-flop* is a type of master–slave flip-flop whose two stages are composed of edge-triggered flip-flops or are an edge-triggered/latch combination. Only the master stage must have the data lockout character (hence must be edge triggered). Shown in Fig. 10.48a is a D data lockout flip-flop composed of an RET D flip-flop master stage and an FET D flip-flop slave stage, and in (b) an RET D flip-flop master with an FET D latch as the slave stage. The design in Fig. 10.48b needs less hardware than that in (a) because of the reduced logic requirements of the D latch. Another possibility is to use JK flip-flops in place of the D flip-flops in Fig. 10.48a, thus creating a JK data lockout flip-flop. But the JK flip-flops require more logic than do the D flip-flops, making the JK data lockout flip-flop less attractive. In any case, there is little advantage to using a data lockout flip-flop except when it is necessary to operate peripherals antiphase off of the two stage outputs, Q_M and Q, in Fig. 10.48.

10.9   LATCHES AND FLIP-FLOPS WITH SERIOUS TIMING PROBLEMS: A WARNING

With very few exceptions, two-state flip-flops have serious timing problems that preclude their use as memory elements in synchronous state machines. Presented in Fig. 10.49 are four examples of two-state latches that have timing problems—none have the data lockout feature. The RET D latch (a) becomes transparent to the input data when $CK = 1$, causing flip-flop action to cease. The three remaining exhibit even more severe problems. For example, the FET T latch (b) will oscillate when $T \cdot \overline{CK} = 1$, and the RET JK latch (c) will oscillate when $JK \cdot CK = 1$, requiring that $J = K = CK = 1$, as indicated in the figure. Notice that the branching conditions required to cause any of the latches to oscillate is found simply by ANDing the $0 \to 1$ and $1 \to 0$ branching conditions. Any

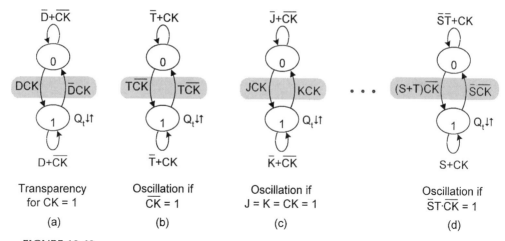

FIGURE 10.49
Timing problems in latches. (a) RET D latch. (b) FET T latch. (c) RET JK latch. (d) FET ST latch.

nonzero result is the branching condition that will cause oscillation. Thus, the FET ST latch in Fig. 10.49d will oscillate under the condition $(S + T)\bar{S} \cdot \overline{CK} = \bar{S}T \cdot \overline{CK} = 1$, that is if $S = CK = 0$ and $T = 1$. The reason for the oscillation in these latches is simply that CK no longer controls the transition between states since the branching condition between the two states is logic 1. These FSMs are asynchronous, as are all flip-flops and latches, and if the transitions are unrestricted by CK, they will oscillate. Thus, none of these two-state latches should ever be considered for use as memory elements in the design of synchronous FSMs. The one exception is the JK latch, which can be used as a memory element providing that J and K are never active at the same time — thus, operating as an SR latch.

There is an MS flip-flop that is particularly susceptible to timing problems. It is the MS JK flip-flop defined by the two state diagrams shown in Fig. 10.50a and implemented in (b).

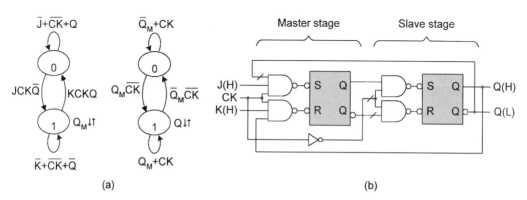

FIGURE 10.50
An MS JK flip-flop that exhibits the error catching problem. (a) State diagrams for the MS JK flip-flop which exhibit a handshake configuration. (b) Logic circuit derived from the state diagrams in (a).

Here, a handshake configuration exists between the master and slave stages. A handshake configuration occurs when the output of one FSM is the input to another and vice versa. This FSM is susceptible to a serious *error catching* problem: In the reset state, if CK is active and a glitch or pulse occurs on the J input to the master stage, the master stage is irreversibly set, passing that set condition on to the slave stage input. Then when CK goes inactive, the output is updated to the set state. This is called *1's catching* and is an unrecoverable error, since the final set state was not regulated by CK. Similarly, in the set state, if CK is active and a glitch or pulse occurs on the K input, the master stage is irreversibly reset, passing that reset condition on to the slave stage input. Then when CK goes inactive, the output is updated to the reset state. This is called *0's catching* and is also an unrecoverable error.

Because of the error catching problem just described, the MS JK flip-flop in Fig. 10.50b, derived from the "handshake" state diagrams in Fig. 10.50a, should never be considered for application as a memory element in a synchronous state machine. If an MS JK flip-flop is needed as the memory element, it is best designed by using Eq. (10.11) and Fig. 10.42a for conversion from an MS D flip-flop that has no error catching problem. Also, because the MS D flip-flop can be implemented by using transmission gates and inverters, as in Fig. 10.35, the conversion to a MS JK can be accomplished with a minimum amount of hardware.

10.10 ASYNCHRONOUS PRESET AND CLEAR OVERRIDES

There are times when the flip-flops in a synchronous FSM must be initialized to a logic 0 or logic 1 state. This is done by using the asynchronous preset and clear override inputs to the flip-flops. To illustrate, a D latch is shown in Figs. 10.51a and 10.51b with both preset and clear overrides. If the flip-flop is to be initialized a logic 0, then a $CL(L) = 1(L)$ is presented to NAND gates 1 and 4, which produces a mixed-rail reset condition, $Q(H) = 0(H)$ and $Q(L) = 0(L)$ while holding $PR(L) = 0(L)$. Or to initialize a logic 1, a $PR(L) = 1(L)$ is presented to NAND gates 2 and 3, which produces a mixed-rail set condition, $Q(H) = 1(H)$ and $Q(L) = 1(L)$, but with $CL(L)$ held at $0(L)$. Remember from Subsection 10.4.4 that $S(L)$ and $R(L)$ cannot both be 1(L) at the same time or else there will be loss of mixed-rail output. Thus, the $CL, PR = 1,1$ input condition is forbidden for this reason. The following relations summarize the various possible preset and clear override input conditions applicable to any flip-flop:

$$\begin{cases} CL(L)=1(L) \\ PR(L)=0(L) \end{cases} \text{Initialize 0} \quad \begin{cases} CL(L)=0(L) \\ PR(L)=1(L) \end{cases} \text{Initialize 1} \\ \begin{cases} CL(L)=0(L) \\ PR(L)=0(L) \end{cases} \text{Normal Operation} \quad \begin{cases} CL(L)=1(L) \\ PR(L)=1(L) \end{cases} \text{Forbidden} \quad . \quad (10.14)$$

The timing diagram in Fig. 10.51c best illustrates the effect of the asynchronous preset and clear overrides. In each case of a PR(L) or CL(L) pulse, normal operation of the latch is interrupted until that pulse disappears and a clock triggering (rising) edge occurs. This asynchronous override behavior is valid for any flip-flop regardless of its type or triggering mechanism, as indicated in Fig. 10.52. For all flip-flops, these asynchronous overrides act directly on the output stage, which is a basic cell.

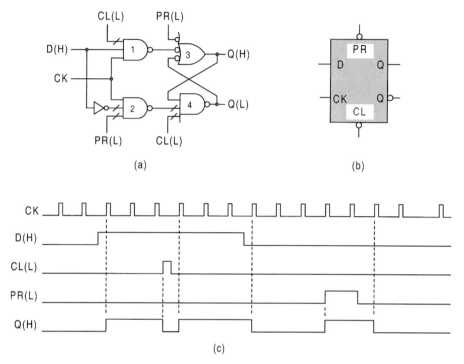

FIGURE 10.51
Asynchronous preset and clear overrides applied to the D latch. (a) Logic circuit for the D latch showing the active low preset and clear connections. (b) Logic circuit symbol with active low preset and clear inputs indicated. (c) Timing diagram showing effects of the asynchronous overrides on the flip-flop output.

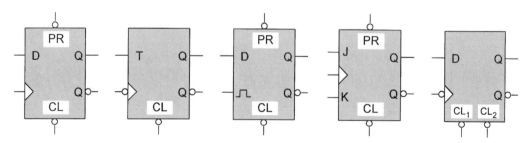

FIGURE 10.52
Examples of flip-flops with asynchronous preset and/or clear overrides.

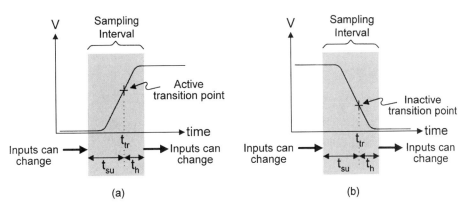

FIGURE 10.53
Clock voltage waveforms showing sampling interval ($t_{su} + t_h$) during which time the data inputs must remain stable at their proper logic levels. (a) Rising edge of the clock waveform. (b) Falling edge of the clock waveform.

10.11 SETUP AND HOLD-TIME REQUIREMENTS OF FLIP-FLOPS

Flip-flops will operate reliably only if the data inputs remain stable at their proper logic levels just before, during, and just after the triggering edge of the clock waveform. To put this in perspective, the data inputs must meet the *setup and hold-time* requirements established by clock, the *sampling variable* for synchronous FSMs. The setup and hold-time requirements for a flip-flop are illustrated by voltage waveforms in Fig. 10.53, where both rising and falling edges of the clock signal are shown. The sampling interval is defined as

$$Sampling\ interval = (t_{su} + t_h), \qquad (10.15)$$

where t_{su} is the *setup time* and t_h is the *hold time*. It is during the sampling interval that the data inputs must remain fixed at their proper logic level if the outcome is to be predictable. This fact is best understood by considering the definitions of setup and hold times:

- Setup time t_{su} is the time interval preceding the active (or inactive) transition point (t_{tr}) of the triggering edge of CK during which all data inputs must remain stable to ensure that the intended transition will be initiated.
- Hold time t_h is the time interval following the active (or inactive) transition point (t_{tr}) of the triggering edge of CK during which all data inputs must remain stable to ensure that the intended transition is successfully completed.

Failure to meet the setup and hold-time requirements of the memory flip-flops in an FSM can cause improper sampling of the data that could, in turn, produce erroneous transitions, or even metastability, as discussed in the next chapter. A change of the data input at the time *CK* is in its sampling interval can produce a *runt pulse*, a pulse that barely reaches the switching threshold. An incompletely sampled runt pulse may cause erroneous FSM behavior. As an example of proper and improper sampling of an input, consider a portion of the resolver state diagram for an RET D flip-flop shown in Fig. 10.54a. Assuming that

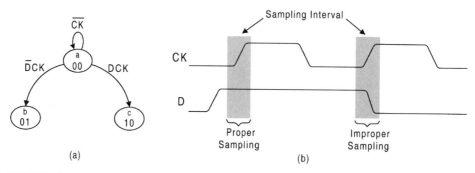

FIGURE 10.54

Examples of proper and improper sampling of the data input. (a) Portion of the resolver state diagram for an RET D flip-flop. (b) Voltage waveforms showing proper and improper sampling of the D waveform during the sampling interval of CK.

the FSM is in state a and that the rising edge of CK is to sample the D input waveform, two sampling possibilities are illustrated by the voltage waveforms for CK and D in Fig. 10.54b. Proper sampling occurs when the data input D is stable at logic level 1 in advance of the rising edge of CK and maintained during the sampling interval. Improper sampling results when D changes during the sampling interval.

The setup and hold-time intervals are important design parameters for which manufacturers will normally provide worst-case data for their flip-flops. Awareness and proper use of this data is vital to good state machine design practice. Ignoring this data may lead to state machine unpredictability or even failure. The means to deal with this problem is discussed later in Section 11.4.

10.12 DESIGN OF SIMPLE SYNCHRONOUS STATE MACHINES WITH EDGE-TRIGGERED FLIP-FLOPS: MAP CONVERSION

Where nearly ideal, high-speed sampling is required, and economic considerations are not a factor, edge-triggered flip-flops may be the memory elements of choice. The setup and hold-time requirements for these flip-flops are the least stringent of all, and they possess none of the problems associated with either the latches or MS flip-flops discussed earlier. In this section two relatively simple FSMs will be designed to demonstrate the methodology to be used. The emphasis will be on the procedure required to obtain the next state and output functions of the FSM. This procedure will involve nothing new. Rather, it will be the continued application of the design procedure and mapping algorithm discussed Section 10.6, and an extension of the flip-flop conversion examples covered in Section 10.8 but now applied to K-map conversion and FSM design.

10.12.1 Design of a Three-Bit Binary Up/Down Counter: D-to-T K-map Conversion

In Fig. 10.2d a bidirectional binary sequence of states is used to represent a Mealy machine. Now, that same binary sequence of states will be completed in the form of a three-bit binary up/down counter as shown by the eight-state state diagram in Fig. 10.55a. It is this counter

10.12 DESIGN OF SIMPLE SYNCHRONOUS STATE MACHINES

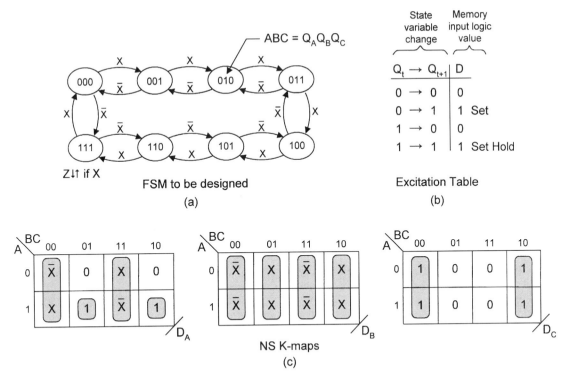

FIGURE 10.55
Design of a three-bit up/down binary counter by using D flip-flops. (a) State diagram for the three-bit up/down counter with a conditional (Mealy) output, Z. (b) Excitation table characterizing the D flip-flop memory. (c) NS K-maps plotted by using the mapping algorithm showing BC domain subfunctions indicated with shaded loops.

that will be designed with D flip-flops. Using the mapping algorithm, the excitation table for D flip-flops in Fig. 10.55b is combined with the state diagram in (a) to yield the entered variable (EV) NS K-maps shown in Fig. 10.55c.

The extraction of gate-minimum cover from the EV K-maps in Fig. 10.55c is sufficiently complex as to warrant some explanation. Shown in Fig. 10.56 are the compressed EV K-maps for NS functions D_A and D_B, which are appropriate for use by the CRMT method, discussed at length in Section 5.7, to extract multilevel gate minimum forms. The second-order K-maps in Fig. 10.56 are obtained by entering the BC subfunction forms shown by the shaded loops in Fig. 10.55c. For D_A, the CRMT coefficients g_i are easily seen to be $g_0 = A \oplus \bar{C}\bar{X}$ and $g_1 = (A \oplus CX) \oplus (A \oplus \bar{C}\bar{X}) = CX \oplus \bar{C}\bar{X}$, as obtained from the first-order K-maps in Fig. 10.56b. Similarly, for D_B the CRMT coefficients are $g_0 = \bar{B} \oplus X$ and $g_1 = 1$. When combined with the f coefficients, the gate minimum becomes

$$\begin{cases} D_A = A \oplus \bar{B}\bar{C}\bar{X} \oplus BCX \\ D_B = \bar{B} \oplus C \oplus X \\ D_C = \bar{C} \\ Z = ABCX \end{cases}, \qquad (10.16)$$

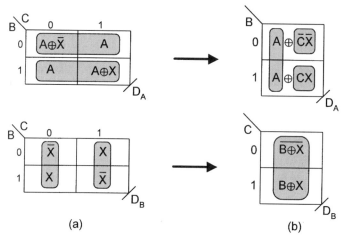

FIGURE 10.56
Compressed EV K-maps required to extract a multilevel logic minimum for NS functions D_A and D_B of Fig. 10.54. (a) Second-order EV K-maps. (b) First-order EV K-maps.

which is a three-level result (due to D_A) with an overall gate/input tally of 7/18, excluding inverters. The next state function for D_C is obtained by inspection of the third-order K-map in Fig. 10.55c, and the output Z is read directly off of the state diagram. Note that the expressions for D_A and D_B in Eqs. (10.16) can be obtained directly from the first-order K-maps in Fig. 10.56b by applying the mapping methods discussed in Section 5.2. The minimum cover is indicated by the shaded loops.

Toggle character is inherent in the binary code. This is evident from an inspection of the state diagram in Fig. 10.55a. State variable C toggles with each transition, state variable B toggles in pairs of states, and state variable A toggles in groups of four states. Thus, it is expected that the T flip-flop design of a binary counter will lead to a logic minimum, and this is the case. Shown in Fig. 10.57 is the design of the binary up/down counter by using T flip-flops as the memory represented by the excitation table in Fig. 10.57b. The NS K-maps, shown in (c) of the figure, are plotted by using the mapping algorithm. Extracting minimum cover from these K-maps (see shaded loops) yields the two-level results

$$\begin{cases} T_A = \bar{B}\bar{C}\bar{X} + BCX \\ T_B = \bar{C}\bar{X} + CX \\ T_C = \bar{X} + X = 1 \\ Z = ABCX \end{cases} \quad (10.17)$$

with an overall gate input tally 7/18 excluding inverters. Although the gate/input tally is the same as that produced by the three-level result given by Eqs. (10.16), the two-level result is expected to be faster and, of course, amenable to implementation by two-level programmable logic devices (e.g., PLAs).

Implementation of Eqs. (10.17) is shown in Fig. 10.58a, where the NS forming logic, memory and output forming logic are indicated. The present state is read from the flip-flop

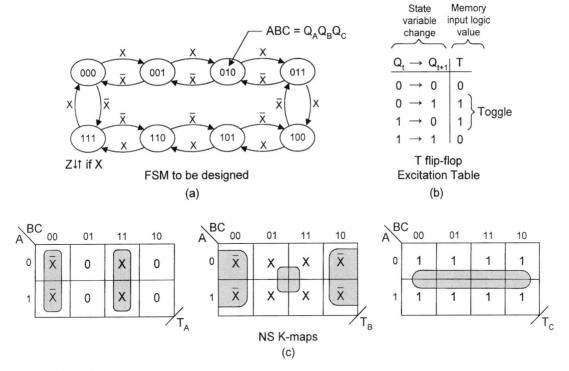

FIGURE 10.57
Design of the three-bit up/down binary counter by using T flip-flops. (a) State diagram for the three-bit up/down counter with a conditional (Mealy) output, Z. (b) Excitation table characterizing the T flip-flop memory. (c) NS K-maps plotted by using the mapping algorithm and showing minimum cover.

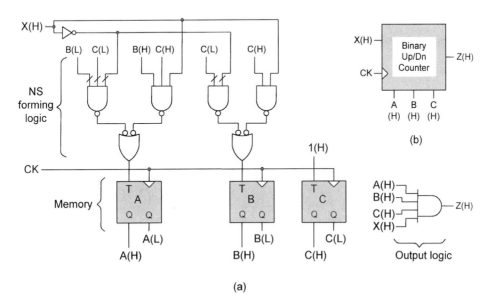

FIGURE 10.58
Implementation of the up/down binary counter represented by Eqs. (10.17). (a) NS-forming logic, T flip-flop memory, and output-forming logic stages. (b) Block diagram for the counter.

469

outputs $A(H)$, $B(H)$, and $C(H)$, where $Q_A = A$, $Q_B = B$, and $Q_C = C$, and the Mealy output Z is issued from the AND gate in state 111 but only when input X is active, i.e., only when the counter is in an up-count mode. The block symbol for this counter is shown in Fig. 10.58b.

D K-map to T K-map Conversion Once the NS D K-maps have been plotted, it is unnecessay to apply the mapping algorithm a second time to obtain the NS T K-maps. All that is necessary is to use the $D \to T$ flip-flop conversion equation, Eq. (10.10), but written as

$$D = Q \oplus T = \bar{Q}T + Q\bar{T}. \tag{10.18}$$

Applied to the individual state variables in a $D \to T$ K-map conversion process, Eq. (10.18) takes on the meaning

$$\begin{cases} D_A = \bar{Q}_A T_A + Q_A \bar{T}_A \\ D_B = \bar{Q}_B T_B + Q_B \bar{T}_B \\ D_C = \bar{Q}_C T_C + Q_C \bar{T}_C \\ \vdots \end{cases} \tag{10.19}$$

In Fig. 10.59 are the D and T K-maps for the three-bit binary up/down counter reproduced from Figs. 10.55 and 10.57. The heavy lines indicate the domain boundaries for the three state variables A, B, and C. An inspection of the K-maps together with Eqs. (10.19) results in the following algorithm for D-to-T K-map conversion:

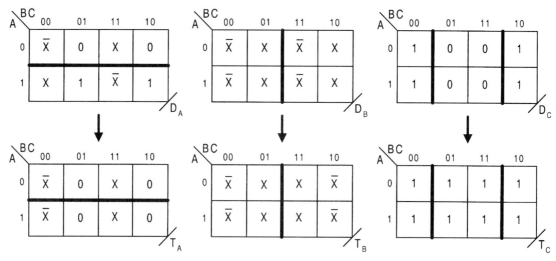

FIGURE 10.59
D and T K-maps for the three-bit binary up/down counter showing the domain boundaries for state variable bits A, B, and C.

10.12 DESIGN OF SIMPLE SYNCHRONOUS STATE MACHINES 471

> **Algorithm 10.1:** $D \to T$ K-map Conversion (Refer to Eq. 10.19)
>
> (1) For all that is NOT A in the D_A K-map, transfer it to the T_A K-map directly ($\bar{A}T_A$).
> (2) For all that is A in the D_A K-map, transfer it to the T_A K-map complemented ($A\bar{T}_A$).
> (3) Repeat steps (1) and (2) for the $D_B \to T_B$ and $D_C \to T_C$, etc., K-map conversions.

Notice that the word "complemented," as used in the map conversion algorithm, refers to the complementation of the contents of each cell in the domain indicated.

10.12.2 Design of a Sequence Recognizer: D-to-JK K-map Conversion

It is required to design a sequence recognizer that will issue an output any time an overlapping sequence ...01101... is detected as indicated in Fig. 10.60a. To do this a choice is made between the Moore or Mealy constructions shown in Figs. 10.60b and 10.60c, respectively. For the purpose of this example, the Mealy construction is chosen. Let the external

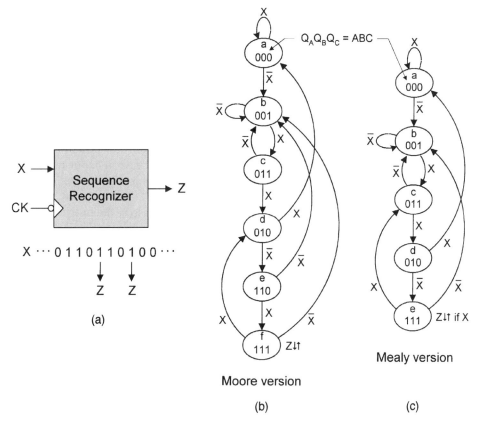

FIGURE 10.60
A simple sequence recognizer for an overlapping sequence ···01101···. (a) Block diagram and sample overlapping sequence. (b) Moore FSM representation. (c) Mealy FSM representation.

472   CHAPTER 10 / INTRODUCTION TO SYNCHRONOUS STATE MACHINE DESIGN

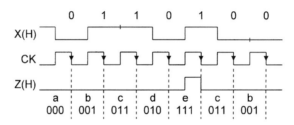

FIGURE 10.61
Timing diagram for the Mealy version of the sequence recognizer in Fig. 10.60c.

input X be synchronized *antiphase* to clock, meaning, for example, that X is synchronized to the rising edge of clock when the memory is FET flip-flops. An *overlapping sequence* is one for which a given sequence can borrow from the latter portions of an immediately preceding sequence as indicated in Fig. 10.60a. The loop $\cdots d \to e \to f \to d \cdots$ in the Moore construction or the loop $\cdots c \to d \to e \to c \cdots$ in the Mealy construction illustrates the overlapping sequence. A *nonoverlapping sequence* requires that each sequence of pulses be separate, i.e., independent of any immediately preceding sequence. Note that the Mealy state diagram is constructed from the Moore version by merging states e and f in Fig. 10.60b, and by changing the unconditional output to a conditional output.

The timing diagram showing the sequence of states leading to the conditional (Mealy) output is presented in Fig. 10.61, where the state identifiers and state code assignments are indicated below the Z waveform. Notice that input X is permitted to change only on the rising edge of the clock waveform and that the arrows indicate a FET flip-flop memory. Thus, when the FSM enters state e on the falling edge of clock, an output is issued when X goes active, and is deactivated when the FSM leaves state e. Any deviation from the sequence $\ldots 01101 \ldots$ would prevent the sequence recognizer from entering state e and no output would be issued. Also, once in state e the overlapping loop $\cdots e \to c \to d \to e \cdots$ would result in repeated issuance of the output Z.

Consider that the Mealy version of the sequence recognizer is to be designed by using D flip-flops. Shown in Fig. 10.62 are the excitation table and the resulting D K-maps obtained by applying the mapping algorithm. The shaded loops reveal the minimum covers for the output and NS functions, which are easily read as

$$\begin{cases} D_A = B\bar{C}\bar{X} \\ D_B = B\bar{C}\bar{X} + CX \\ D_C = \bar{X} + \bar{B}C + A \\ Z = AX \end{cases} \quad (10.20)$$

Notice that the term $B\bar{C}\bar{X}$ is a shared PI since it appears in two of the three NS functions.

D K-map to JK K-map Conversion   Assuming it is desirable to design the sequence recognizer of Fig. 10.60c by using JK flip-flops instead of D flip-flops, the process of

10.12 DESIGN OF SIMPLE SYNCHRONOUS STATE MACHINES

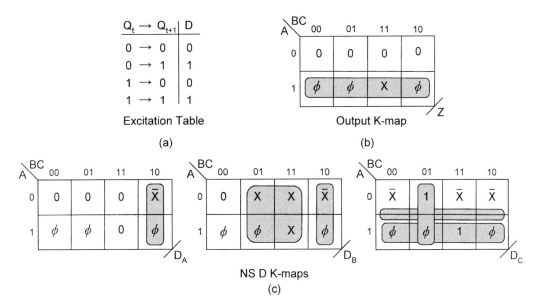

FIGURE 10.62
D K-map construction for the Mealy version of the sequence recognizer in Fig. 10.60c. (a) Excitation table for D flip-flops. (b), (c) Output K-map and NS D K-maps showing minimum cover.

obtaining the NS JK K-maps can be expedited by K-map conversion. It will be recalled from Eq. (10.11) that $D \to JK$ flip-flop conversion logic is given by

$$D = \bar{Q}J + Q\bar{K}.$$

When this equation is applied to the individual state variables in a $D \to JK$ K-map conversion, Eq. (10.11) takes the meaning

$$\begin{cases} D_A = \bar{Q}_A J_A + Q_A \bar{K}_A \\ D_B = \bar{Q}_B J_B + Q_B \bar{K}_B \\ D_C = \bar{Q}_C J_C + Q_C \bar{K}_C \\ \vdots \end{cases}. \quad (10.21)$$

Shown in Fig. 10.63 are the JK K-maps converted from the D K-maps. From these K-maps the minimum cover is easily observed to be

$$\begin{cases} J_A = B\bar{C}\bar{X} & K_A = 1 \\ J_B = CX & K_B = C \oplus X \\ J_C = \bar{X} & K_C = \bar{A}BX \end{cases}, \quad (10.22)$$

which represents a gate/input tally of 4/10 compared to 5/12 for the NS functions in Eq. (10.20), all exclusive of possible inverters.

474 CHAPTER 10 / INTRODUCTION TO SYNCHRONOUS STATE MACHINE DESIGN

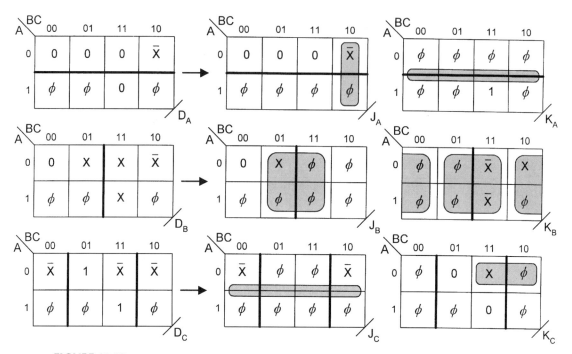

FIGURE 10.63
D-to-JK K-map conversion for the sequence recognizer of Fig. 10.60c, showing domain boundaries for state variables A, B, and C, and minimum cover for the JK K-maps.

Implementation of Eqs. (10.22) is provided in Fig. 10.64, together with the output-forming logic given in Eqs. (10.20). Notice that NOR/XOR/INV logic is used for this purpose and that notation for the present state follows established practice in this text, namely $Q_A = A$, $Q_B = B$, and $Q_C = C$. The clock symbol $CK(L)$ simply indicates FET memory elements.

An inspection of the $D \rightarrow JK$ K-map conversion in Fig. 10.63 together with Eqs. (10.21) evokes the following algorithm:

Algorithm 10.2: $D \rightarrow JK$ K-map Conversion [Refer to Eq. (10.21)]

(1) For all that is NOT A in the D_A K-map, transfer it to the J_A K-map directly ($\bar{A} J_A$).
(2) For all that is A in the D_A K-map, transfer it to the K_A K-map complemented ($A \bar{K}_A$).
(3) Fill in empty cells with don't cares.
(4) Repeat steps (1), (2), and (3) for the $D_B \rightarrow J_B$, K_B and $D_C \rightarrow J_C$, K_C, etc., K-map conversions.

It is important to note that the "fill-in" of the empty cells with don't cares is a result of the don't cares that exist in the excitation table for JK flip-flops. The reader should verify that the JK K-map results in Fig. 10.63 are also obtained by directly applying the JK excitation table and the mapping algorithm to the Mealy form of the sequence recognizer given in Fig. 10.60c. In doing so, it will become apparent that the $D \rightarrow JK$ K-map conversion

10.12 DESIGN OF SIMPLE SYNCHRONOUS STATE MACHINES 475

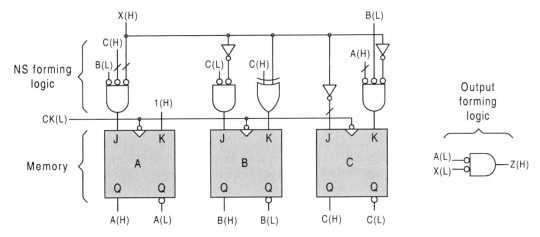

FIGURE 10.64
Implementation of Eqs. (10.22) for the sequence recognizer of Fig. 10.60c showing the NS-forming logic, memory, and output-forming logic.

method is quicker and easier than the direct method by using the excitation table for JK flip-flops. Furthermore, the K-map conversion approach permits a comparison between, say, a D flip-flop design and a JK K-map design, one often producing a more optimum result than the other. For these reasons the K-map conversion approach to design will be emphasized in this text.

Missing-State Analysis To this point no mention has been made of the missing (don't care) states in Fig. 10.60c. Missing are the states 100, 101, and 110, which do exist but are not part of the primary routine expressed by the state diagram in Fig. 10.60c. Each don't care state goes to ($\rightarrow$) a state in the state diagram as indicated in Fig. 10.65. For example, $100 \rightarrow 001$ unconditionally, but $110 \rightarrow 111$ if $\bar{X}$ or $110 \rightarrow 001$ if X, etc. The NS values are determined by substituting the present state values A, B, and C into the NS functions given in Eqs. (10.20).

The missing state analysis gives emphasis to the fact that FSMs, such as the sequence recognizer in Fig. 10.60, must be initialized into a specific state. On power-up, the sequence recognizer of Fig. 10.64 could initialize into any state, including a don't care state. For

Present State A B C	Next State $D_A D_B D_C$	Conclusion			
1 0 0	0 0 1	$100 \rightarrow 001$			
1 0 1	0 X 1	$101 \xrightarrow{\bar{X}} 001$	or	$101 \xrightarrow{X} 011$	
1 1 0	X̄ X̄ 1	$110 \xrightarrow{\bar{X}} 111$	or	$110 \xrightarrow{X} 001$	

FIGURE 10.65
Missing state analysis for the Mealy version of the sequence recognizer given in Fig. 10.60c.

example, if the FSM should power up into don't care state 110 with X inactive ($\bar{X}$), it would transit to state 111 on the next clock triggering edge and would falsely issue an output Z if X goes active. Ideally, on power-up, this FSM should be initialized into state 000 to properly begin the sequence. Section 11.7 discusses the details by which this can be accomplished.

10.13 ANALYSIS OF SIMPLE STATE MACHINES

The purpose of analyzing an FSM is to determine its sequential behavior and to identify any problems it may have. The procedure for FSM analysis is roughly the reverse of the procedure for FSM design given in Section 10.6. Thus, in a general sense, one begins with a logic circuit and ends with a state diagram. There are six principal steps in the analysis process:

1. Given the logic circuit for the FSM to be analyzed, carefully examine it for any potential problems it may have and note the number and character of its flip-flops and its outputs (Mealy or Moore).
2. Obtain the NS and output logic functions by carefully reading the logic circuit.
3. Map the output logic expressions into K-maps, and map the NS logic expressions into K-maps appropriate for the flip-flops used. If the memory elements are other than D flip-flops, use K-map conversion to obtain D K-maps.
4. From the D K-maps, construct the Present State/Inputs/Next State (PS/NS) table. To do this, observe which inputs control the branching, as indicated in each cell, and list these in ascending canonical word form together with the corresponding NS logic values. Ascending canonical word form means the use of minterm code such as $\bar{X}\bar{Y}\bar{Z}$, $\bar{X}\bar{Y}Z$, $\bar{X}Y\bar{Z}$ for branching dependency on inputs X, Y, and Z relative to a given K-map cell.
5. Use the PS/NS table in step 4 and the output K-maps to construct the fully documented state diagram for the FSM.
6. Analyze the state diagram for any obvious problems the FSM may have. These problems may include possible hang (isolated) states, subroutines from which there are no exits, and timing defects (to be discussed in Chapter 11). Thus, a redesign of the FSM may be necessary.

A Simple Example To illustrate the analysis procedure, consider the logic circuit given in Fig. 10.66a, which is seen to have one input X and one output, Z, and to be triggered on the falling edge of the clock waveform. Also, the external input arrives from a negative logic source. Reading the logic circuit yields the NS and output logic expressions

$$J_A = B \odot X, \qquad J_B = AX$$
$$K_A = X, \qquad K_B = A \qquad (10.23)$$
$$Z = \bar{A}\bar{B}X,$$

where A and B are the state variables. These expressions are mapped into JK K-maps and

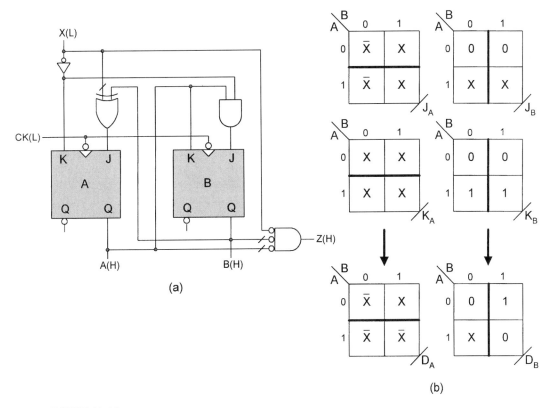

FIGURE 10.66
(a) The logic circuit to be analyzed. (b) The NS JK K-maps for Eqs. (10.23) and their conversion to the D K-maps needed to construct the PS/NS table.

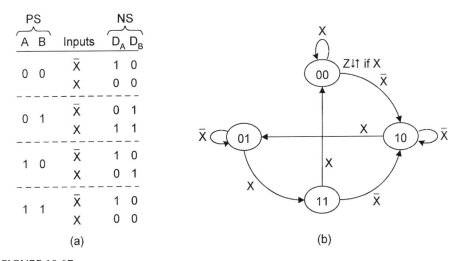

FIGURE 10.67
(a) PS/NS table constructed from the D K-maps in Fig. 10.66 and (b) the resulting state diagram for the FSM of Fig. 10.66a.

converted to D K-maps as shown in Fig. 10.66b. Here, use is made of Algorithm 10.2 for the reverse conversion process, that is, for the JK-to-D K-map conversion. Notice that the domain boundaries are indicated by heavy lines as was done in Fig. 10.63.

Step 4 in the analysis procedure, given previously, requires the construction of the PS/NS table from the D K-maps that are provided in Fig. 10.66b. This is done in Fig. 10.67a, from which the state diagram follows directly as shown in Fig. 10.67b.

There are no serious problems with this FSM other than the potential to produce an output race glitch (ORG) as a result of the transition 10 → 01 under branching condition X. The problem arises because two state variables are required to change during this transition, but do not do so simultaneously. The result is that the FSM must transit from state 10 to 01 via one of two race states, 00 or 11. If the transition is by way of state 00, Z will be issued as a glitch that could cross the switching threshold. A detailed discussion of the detection and elimination of output race glitches is presented in Section 11.2.

A More Complex Example The following NS and output expressions are read from a logic circuit that has five inputs, U, V, W, X, and Y, and two outputs, *LOAD* (*LD*) and *COUNT* (*CNT*):

$$J_A = U + B\bar{W} \qquad J_B = AX + AY$$
$$K_A = B\bar{X} + \bar{X}Y \qquad K_B = A(\bar{X} + VY) \qquad (10.24)$$
$$LD = \bar{A}\bar{B}X \qquad CNT = \bar{A}B\bar{X}Y.$$

Presented in Fig. 10.68 are the JK-to-D K-map conversions for the NS functions given in Eqs. (10.24). As in the previous example, Algorithm 10.2 is used for the reverse conversion from JK to D K-maps. The domain boundaries are again indicated by heavy lines.

The PS/NS table for the NS functions, shown in Fig. 10.69a, is constructed from the D K-maps in Fig. 10.68. Notice that only the input literals indicated in a given cell of the D K-maps are represented in the PS/NS table as required by step 4 of the analysis procedure

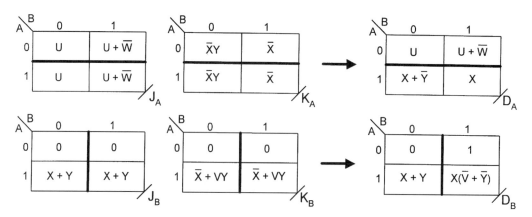

FIGURE 10.68
JK-to-D K-map conversions for the NS functions given in Eqs. (10.24).

10.13 ANALYSIS OF SIMPLE STATE MACHINES

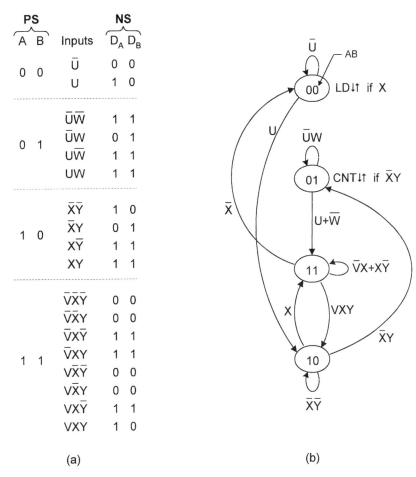

FIGURE 10.69
(a) PS/NS table obtained from the D K-maps in Fig. 10.68 and (b) the resulting state diagram for the FSM represented by Eqs. (10.24).

given previously in this section. Representation of these literals in canonical form ensures that the sum rule is obeyed—all possible branching conditions relative to a given state are taken into account.

The state diagram for the Mealy FSM represented by Eqs. (10.24) is derived from the PS/NS table in Fig. 10.69a and is shown in Fig. 10.69b. Both Mealy outputs are deduced directly from the output expressions in Eqs. (10.24). This FSM has the potential to form an output race glitch (ORG) during the transition from state 11 to state 00 under branching condition $\bar{X}Y$. Thus, if state variable A changes first while in state 11, the FSM could transit to state 00 via race state 01, causing a positive glitch in the output CNT, which is issued conditional on the input condition $\bar{X}Y$. No other potential ORGs exist. A detailed discussion of ORGs together with other kinds of logic noise is provided in Chapter 11.

10.14 VHDL DESCRIPTION OF SIMPLE STATE MACHINES

An introduction to VHDL description of devices is given in Section 6.10. There, certain key words are introduced in bold type and examples are given of the behavioral and structural descriptions of combinational primitives. In Section 8.10, VHDL is used in the description of a full adder to illustrate three levels of abstraction. In this section, the behavioral descriptions of two FSMs (a flip-flop and a simple synchronous state machine) are presented by using the IEEE standard package *std_logic_1164*.

10.14.1 The VHDL Behavorial Description of the RET D Flip-flop

(Note: Figure 10.51a provides the symbol for the RET D flip-flop that is being described here.)

library IEEE
use IEEE.std_logic_1164.all;
entity RETDFF **is**
    generic (SRDEL, CKDEL: **Time**);
    port (PR, CL, D, CK: **in** bit; Q, Qbar: **out** bit)   —PR and CL are active low inputs
end RETDFF
architecture behavioral **of** RETDFF **is**
begin
    process (PR, CL, CK)
begin
    if PR = '1' and CL = '0' **then**   —PR = '1' and CL = '0' is a clear condition
        Q <= '0' **after** SRDEL;   —'0' represents LV
        Qbar <= '1' **after** SRDEL;   —'1' represents HV
    elseif PR = '0' and CL = '1' **then**   —PR = '0' and CL = '1' is a preset condition
        Q <= '1' **after** SRDEL;
        Qbar <= '0' **after** SRDEL;
    elseif CK' event and CK = '1' and PR = '1' and CL = '1' **then**
        Q <= D **after** CKDEL;
        Qbar <= (**not** D) **after** CKDEL;
    end if;
    end process;
end behavioral;

In the example just completed, the reader is reminded that the asynchronous overrides are active low inputs as indicated in Fig. 10.51a. However, VHDL descriptions treat the '1' and '0' as HV and LV, respectively. Therefore, it is necessary to apply Relations (3.1) in Subsection 3.2.1 to properly connect the VHDL description to the physical entity.

10.14 VHDL DESCRIPTION OF SIMPLE STATE MACHINES

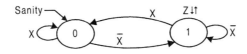

FIGURE 10.70
A simple FSM that is used for a VHDL description.

10.14.2 The VHDL Behavioral Description of a Simple FSM

Shown in Fig. 10.70 is the state diagram for a two-state FSM having one input, X, and one output, Z. It also has a Sanity input for reset purposes.
The following is a VHDL behavioral description of the FSM in Fig. 10.70:

library IEEE;
use IEEE.std_logic_1164.all;
entity FSM **is**
       port (Sanity, CK, X: **in** bit; Z: **out** bit);   —Sanity is an active low reset input
end FSM;
architecture behavioral of FSM **is**
       type statetype **is** (state0, state1);
       signal state, NS : statetype := state0;
begin
sequence_process: **process** (state, X);
begin
       case state **is**
       when state0 =>
              if X = '0' **then**
                 NS <= state1;
                 Z <= '1';
           else NS <= state0;
                 Z <= '0';
           end if;
       when state1 =>
              if X = '1' **then**
                 NS <= state0;
                 Z <= '0';
           else NS <= state1;
                 Z <= '1';
           end if;
       end case;
end process sequence_process;

```
CK_process: process;
begin
wait until (CK'event and CK = '1');
                if Sanity = '0' then    — '0' represents LV
                    state <= state0;
            else state <= NS;
            end if;
        end process CK_process;
end behavorial;
```

In this example the effect of the Sanity input is presented at the end of the behavioral description. But it could have been placed in front of the sequence_process. Also, a keyword not encountered in all previous examples is **type**. This keyword is used to declare a name and a corresponding set of declared values of the type. Usages include scalar types, composite types, file types, and access types. References on the subject of VHDL are cited in Further Reading at the end of Chapter 6.

10.15 FURTHER READING

Nearly all texts on the subject of digital design offer coverage, to one extent or another, of flip-flops and synchronous state machines. However, only a few texts approach these subjects by using fully documented state (FDS) diagrams, sometimes called *mnemonic state diagrams.* The FDS diagram approach is the simplest, most versatile, and most powerful pencil-and-paper means of representing the sequential behavior of an FSM in graphical form. The text by Fletcher is believed to be the first to use the FDS diagram approach to FSM design. Other texts that use FDS diagrams to one degree or another are those of Comer and Shaw. The text by Tinder is the only text to use the FDS diagram approach in the design and analysis of latches, flip-flops, and state machines (synchronous and asynchronous). Also, the text by Tinder appears to be the only one that covers the subject of K-map conversion as it is presented in the present text.

[1] D. J. Comer, *Digital Logic and State Machine Design*, 3rd ed. Saunders College Publishing, Fort Worth, TX, 1995.
[2] W. I. Fletcher, *An Engineering Approach to Digital Design*. Prentice Hall, Englewood Cliffs, NJ, 1980.
[3] A. W. Shaw, *Logic Circuit Design*. Sanders College Publishing, Fort Worth, TX, 1993.
[4] R. F. Tinder, *Digital Engineering Design: A Modern Approach*. Prentice Hall, Englewood Cliffs, NJ, 1991.

The subjects of setup and hold times for flip-flops are adequately treated in the texts by Fletcher (previously cited), Katz, Taub, Tinder (previously cited), Wakerely, and Yarbrough.

[5] R. H. Katz, *Contemporary Logic Design*. Benjamin/Commings Publishing, Redwood City, CA, 1994.

PROBLEMS

[6] H. Taub, *Digital Circuits and Microprocessors*. McGraw-Hill, New York, 1982.
[7] J. F. Wakerly, *Digital Design Principles and Practices*, 2nd ed. Prentice-Hall, Englewood Cliffs, NJ, 1994.
[8] J. M. Yarbrough, *Digital Logic Applications and Design*. West Publishing Co., Minneapolis/St. Paul, MN, 1997.

With the exception of texts by Katz and Taub, all of the previously cited references cover adequately the subject of synchronous machine analysis. The texts by Fletcher, Shaw, and Tinder in particular present the subject in a fashion similar to that of the present text. Other texts that can be recommended for further reading on this subject are those by Dietmeyer and by Nelson *et al.*, the former being more for the mathematically inclined.

[9] D. L. Dietmeyer, *Logic Design of Digital Systems*, 2nd ed. Allyn and Bacon, Inc., Boston, MA, 1978.
[10] V. P. Nelson, H. T. Nagle, B. D. Carroll, and J. D. Irwin, *Digital Logic Circuit Analysis and Design*. Prentice Hall, Englewood Cliffs, NJ, 1995.

For detailed information on the subject of VHDL, the reader is referred to Further Reading at the end of Chapter 6.

PROBLEMS

10.1 (a) Complete the timing diagram in Fig. P10.1 for the set-dominant basic cell shown in Figs. 10.18a and 10.18b. To do this, sketch the resulting waveforms, taking into account the path delay through a NAND gate represented by τ_p.

(b) Test the results of part (a) by simulating the circuit.

10.2 (a) Complete the timing diagram in Fig. P10.2 for the reset-dominant basic cell shown in Figs. 10.18c and d. To do this, sketch the resulting waveforms, taking into account the path delay through a NOR gate represented by τ_p.

(b) Test the results of part (a) by simulating the circuit.

10.3 The set-dominant clocked basic cell (also called a *gated basic cell* or *gated SR latch*) is represented by the expression

$$Q_{t+1} = SCK + Q_t(\overline{RCK}),$$

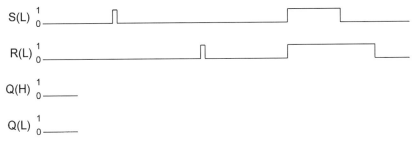

FIGURE P10.1

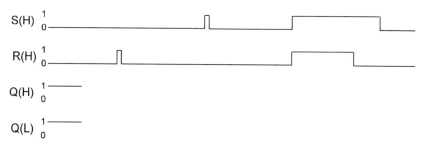

FIGURE P10.2

where Q_{t+1} is the next state, Q_t is the present state, CK is the clock input, and S and R are the set and reset inputs, respectively.

(a) From the preceding expression, plot the first-order EV K-map for this device showing minimum cover. Follow the example in Fig. 10.11b.

(b) From the expression for Q_{t+1}, construct the NAND logic circuit for the gated basic cell. Thus, show that it consists of four two-input NAND gates (nothing else), which includes a set-dominant basic cell represented as two "cross-coupled" NAND gates as in Fig. 10.18a.

(c) By using the logic circuit in part (b), complete the timing diagram shown in Fig. P10.3. Neglect the path delays through the NAND gates, and note that the arrows indicate rising edge triggering by clock. [Hint: The logic waveforms for $Q(H)$ and $Q(L)$ can be deduced qualitatively from the equation for Q_{t+1}.]

(d) Test the results of part (c) by simulating the circuit of part (b) with a logic simulator.

10.4 (Note: This problem should be undertaken only after completing Problem 10.3.) The state diagram for the set-dominant basic cell is shown in Fig. 10.12a.

(a) Add CK to this state diagram in Fig. 10.12a to create the state diagram for the rising edge triggered (RET) set-dominant SR latch of Problem 10.3 in a manner similar to the state diagram given in Fig. 10.47b. (Hint: If this is properly done, Fig. 10.12a will result when $CK = 1$.)

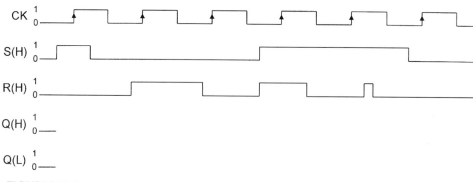

FIGURE P10.3

PROBLEMS

(b) Redesign the gated basic cell of Problem 10.3 by using the set-dominant basic cell as the memory. To do this, follow the examples in Figs. 10.24, 10.25, and 10.47 by plotting EV K-maps for Set and Reset. Thus, it is necessary to combine the information in the state diagram of part (a) with the excitation table in Fig. 10.15c via the mapping algorithm given in Section 10.6.

(c) Construct the NAND/INV logic circuit from the results of part (a). In what way does it differ from that constructed in part (b) of Problem 10.3? What can be said about the $S, R = 1, 1$ condition relative to these two designs? (Hint: Only one inverter is used.)

(d) Read this circuit and write a single expression similar to that given in Problem 10.3. Then construct a first-order EV K-map from this result. Compare the K-map with that in part (a) of Problem 10.3. Are these two K-maps the same? Explain your answer.

(e) Complete the timing diagram in Fig. P10.3 for this design. What do you conclude relative to the $S, R = 1, 1$ condition?

(f) Test the results of part (d) by simulating the circuit of part (b) with a logic simulator.

10.5 (a) By using Eq. (10.5), implement the set-dominant basic cell by using one 2-to-1 MUX and one AND gate (nothing else). [Hint: Plot Eq. (10.5) in a first-order K-map of axis S, and remember that the S and R inputs are introduced active low into the basic cell.]

(b) Construct the logic circuit for the design of part (a). To do this, construct the logic circuit for the 2-to-1 MUX and provide both active high and active low outputs as in Fig. 6.4d. Qualitatively, discuss how the mixed-rail output response of this design compares with that of Fig. 10.18a.

10.6 (a) Convert an RET D flip-flop to a set-dominated RET SR flip-flop. To do this, use minimum external logic and assume that the S and R inputs arrive active high.

(b) Complete the timing diagram in Fig. P10.3 for this flip-flop. Is mixed-rail output response preserved in this flip-flop? Explain your answer.

10.7 Shown in Fig. P10.4 are the operation tables for four unusual (perhaps nonsense) flip-flops.

(1) Construct the two-state state diagram and excitation table for each of these. To do this, follow the example of the JK flip-flop in Fig. 10.40.

N	Q_{t+1}
0	$\bar{Q}_t$
1	1

(a)

L	N	Q_{t+1}
0	0	1
0	1	Q_t
1	0	$\bar{Q}_t$
1	1	0

(b)

S	P	Q_{t+1}
0	0	0
0	1	1
1	0	$\bar{Q}_t$
1	1	1

(c)

A	B	Q_{t+1}
0	0	0
0	1	1
1	0	1
1	1	Q_t

(d)

FIGURE P10.4

$Q_t \to Q_{t+1}$	F	G
0 → 0	ϕ	0
0 → 1	ϕ	1
1 → 0	1	0
1 → 1	$\begin{Bmatrix} 0 \\ \phi \end{Bmatrix}$	$\begin{Bmatrix} \phi \\ 1 \end{Bmatrix}$

(a)

$Q_t \to Q_{t+1}$	P	K
0 → 0	1	0
0 → 1	$\begin{Bmatrix} 0 \\ \phi \end{Bmatrix}$	$\begin{Bmatrix} \phi \\ 1 \end{Bmatrix}$
1 → 0	1	ϕ
1 → 1	0	ϕ

(b)

$Q_t \to Q_{t+1}$	S	F
0 → 0	1	0
0 → 1	$\begin{Bmatrix} 0 \\ \phi \end{Bmatrix}$	$\begin{Bmatrix} \phi \\ 1 \end{Bmatrix}$
1 → 0	$\begin{Bmatrix} 0 \\ 1 \end{Bmatrix}$	$\begin{Bmatrix} 1 \\ 0 \end{Bmatrix}$
1 → 1	$\begin{Bmatrix} 0 \\ 1 \end{Bmatrix}$	$\begin{Bmatrix} 0 \\ 1 \end{Bmatrix}$

(c)

$Q_t \to Q_{t+1}$	R	M
0 → 0	$\begin{Bmatrix} 0 \\ \phi \end{Bmatrix}$	$\begin{Bmatrix} \phi \\ 0 \end{Bmatrix}$
0 → 1	1	1
1 → 0	0	0
1 → 1	$\begin{Bmatrix} 1 \\ \phi \end{Bmatrix}$	$\begin{Bmatrix} \phi \\ 1 \end{Bmatrix}$

(d)

FIGURE P10.5

(2) Find the gate-minimum logic circuit required to convert any type D flip-flop to each of these flip-flops. To do this, follow the model in Fig. 10.36 and assume that all inputs arrive active high.

10.8 Shown in Fig. P10.5 are four excitation tables for unusual (perhaps nonsense) flip-flops.

(1) Construct the two-state state diagram and operation table for each of these.

(2) Find the gate-minimum logic circuit required to convert any type of JK flip-flop to each of these flip-flops. To do this, follow the model in Fig. 10.36 and assume that all inputs arrive active high.

10.9 Find the gate-minimum logic circuit required for the following flip-flop conversions. To do this, use the excitation tables in Fig. P10.5.

(a) Convert an MS D flip-flop to a MS FG flip-flop with inputs active low.

(b) Convert an RET T flip-flop to an FET PK flip-flop with inputs $P(L)$ and $K(H)$.

(c) Convert an RET D flip-flop to an RET RM flip-flop with inputs active high.

10.10 (a) Draw the two-state state diagram that describes the operation of an RET JK flip-flop that has an active low *synchronous* clear input—one that takes effect only on the triggering edge of the clock signal.

(b) Find the gate-minimum logic circuit required to convert an RET D flip-flop to the JK flip-flop of part (a) by using NAND/INV logic only. Assume that the inputs arrive active high.

10.11 Given the circuit in Fig. P10.6, complete the timing diagram to the right and determine the logic function F. (Hint: Construct a truth table for Q_1, Q_2, and Z.)

10.12 The results of testing an FSM indicate that when its clock frequency f_{CK} exceeds 25 MHz the FSM misses data. The tests also yield the following data:

$$\tau_{FF(max)} = 15 \text{ ns}$$

$$\tau_{NS(max)} = 13 \text{ ns},$$

where $\tau_{FF(max)}$ is the maximum observed delay through the memory flip-flops, and $\tau_{NS(max)}$ is the maximum observed delay through the next-state-forming logic, both given in nanoseconds.

PROBLEMS

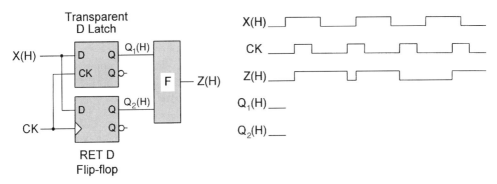

FIGURE P10.6

(a) Calculate the minimum setup time $t_{su(min)}$ from the foregoing information. Note that the hold time plays no significant role here.

(b) On a sketch of the voltage waveform for clock, illustrate the relative values for $\tau_{FF(max)}$, $\tau_{NS(max)}$, $t_{su(min)}$, and T_{CK}.

10.13 Shown in Fig. P10.7 are three relatively simple FSMs. First, check each state diagram for compliance with the sum rule and mutually exclusive requirement. Then, for this problem, design each of these FSMs by using RET D flip-flops as the memory. To do this use a gate-minimum NS and output logic and assume that the inputs and outputs are all active high.

10.14 Repeat Problem 10.13 by using RET JK flip-flops as the memory.

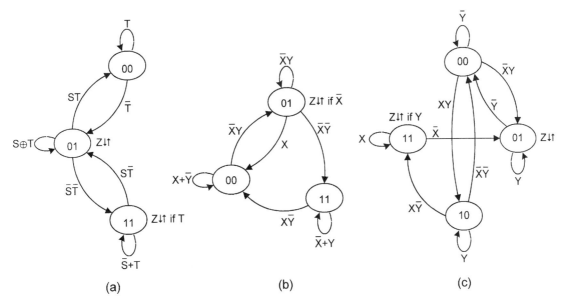

FIGURE P10.7

10.15 Design the FSM in Fig. P10.7b by using PK flip-flops that are characterized by the excitation table in Fig. P10.5b. To do this, find the gate-minimum logic required for the NS- and output-forming logic. Do not implement the result. Thus, the information in Figs. P10.7b and P10.5b must be brought together via the mapping algorithm in Section 10.6. (Hint: The easiest approach to this problem is to obtain the NS K-maps for a D flip-flop design and then apply the conversion logic for D-to-PK K-map conversion. See Subsection 10.12.2 for assistance if needed.)

10.16 (a) Construct a *four-state* state diagram for an FSM that samples (with clock) a continuous stream of data on an input X. The FSM is to issue an output Z any time the sequence ...1001... occurs. Consider that the sequence can be *overlapping* as, for example, ...100100100....

(b) By using two state variables, give this state diagram any valid state code assignment.

10.17 Repeat Problem 10.16 for a nonoverlapping sequence ...0101....

10.18 Construct the state diagram for an FSM that samples (with clock) a continuous stream of data on an input X. The FSM is to issue an output Z any time the sequence ...10110... occurs. Consider that the sequence can be *overlapping* as, for example, ...10101011011010..., where an output is issued twice in this series. The state diagram must conform to the following representations:
(a) A Moore FSM representation with six states.
(b) A Mealy FSM representation with five states.

10.19 (a) Design a *serial 2's complementer* logic circuit by using *two* RET D flip-flops and a gate-minimum NS and output forming logic. To do this, follow Algorithm 2.6 in Subsection 2.6.2 and the ASM chart in Fig. 13.29b. The inputs are *Start* and *Bin* (for binary), and the output is T (for two's complement), all active high. (Hint: There are at least three states and the process is unending.)

(b) Test the design of the serial 2's complementer by simulation and compare the results with the timing diagram in Fig. 13.30.

(c) Repeat part (a), except use two RET JK flip-flops. Which design is more optimum? Explain.

10.20 Shown in Fig. P10.8 is the logic circuit for an FSM that has two inputs, X and Y, and two outputs, P and Q. Analyze this FSM to the extent of constructing a fully documented state diagram. To do this, follow the examples in Section 10.13.

10.21 Presented in Fig. P10.9 is the logic circuit for a two-input/one-output FSM that is to be analyzed. To do this, construct a fully documented state diagram by following the example in Figs. 10.66 and 10.67. Indicate any possible branching problems this FSM may have. Such problems may include states for which there is no entrance as, for example, don't care states.

10.22 In Fig. P10.10 is the logic circuit for a single-input/single-output FSM. Analyze this FSM by constructing a fully documented state diagram. Indicate any possible branching problems this FSM may have. Such problems may include states for which there is no entrance.

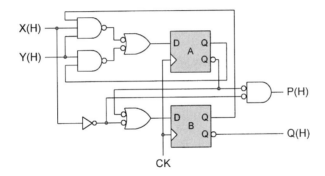

FIGURE P10.8

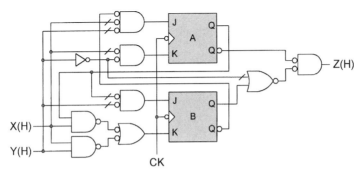

FIGURE P10.9

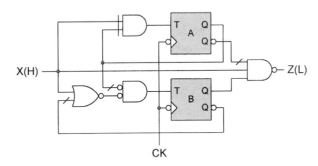

FIGURE P10.10

10.23 (1) Shown in the following are the NS- and output-forming logic expressions for five FSMs. Analyze these FSMs to the extent of constructing a fully documented state diagram for each. To do this follow the examples in Section 10.13. Note that the state variables are, as always, A, B, and C, and the outputs are P and Q. All other literals are inputs. For FSM (d), use the excitation table in Fig. 10.5a to characterize the FG flip-flop memory. (Hint: For FSM (d), convert D K-maps to FG K-maps to obtain the NS logic.)

(2) Indicate any possible branching problems that may exist in each of the FSMs. Such branching problems may include states for which there is no entrance and that might be don't care states.

$J_A = BXY + \bar{X}\bar{Y}$ $\quad T_A = \bar{A}B\bar{X}\bar{Y} + AXY$ $\quad J_A = \bar{B}XY + B\bar{X}Y$
$K_A = \bar{B}X\bar{Y} + B\bar{X}Y$ $\quad T_B = \bar{A}BXY + \bar{B}\bar{Y}$ $\quad K_A = BX + XY\bar{Z}$
$J_B = \bar{A}\bar{X} + \bar{X}\bar{Y}$ $\quad P = AB\bar{X} + \bar{A}B$ $\quad J_B = A(\bar{X} + \bar{Z})$
$K_B = XY$ $\quad\quad\quad\quad\quad\quad\quad\quad\quad\quad\quad\quad\quad\quad K_B = \bar{A}\bar{Y}$
$Q = A\bar{B}X + B\bar{X}$ $\quad\quad\quad\quad\quad\quad\quad\quad\quad\quad\quad Q = \bar{B}X\bar{Y} + A\bar{B}\bar{Y}$

$\quad\quad$(a)$\quad\quad\quad\quad\quad\quad\quad\quad\quad$(b)$\quad\quad\quad\quad\quad\quad\quad\quad\quad$(c)

$F_A = X\bar{Y}$ $\quad\quad\quad\quad\quad\quad D_A = \bar{A}B\bar{C}\bar{W} + A\bar{C}\bar{N}$
$G_A = B\bar{X}\bar{Y} + AY$ $\quad\quad D_B = \bar{A}CX\bar{Y} + \bar{A}B + B\bar{C}$
$F_B = X\bar{Y} + \bar{A}X$ $\quad\quad\quad D_C = \bar{A}\bar{B}\bar{C}S + \bar{A}\bar{B}C\bar{Y} + BCH + ABN$
$G_B = \bar{X}Y + B\bar{X} + AY$ $\quad\quad\quad\quad\quad + ABT + ABC$
$P = B\bar{X} + A$ $\quad\quad\quad\quad\quad P = \bar{B}CSY + ABC$
$\quad\quad\quad\quad\quad\quad\quad\quad\quad\quad Q = A\bar{C}N$

$\quad\quad$(d)$\quad\quad\quad\quad\quad\quad\quad\quad\quad$(e)

10.24 (a) Prove that if Eq. (10.4) is satisfied, Eq. (10.3) is also satisfied.

(b) Prove that the mutually exclusive requirement is uniquely satisfied in Fig. 10.8 if

$$f_{i \leftarrow j} \cdot f_{k \leftarrow j} = 0$$

for all i and k, where $i \neq k$. Here i, j, and k are integers with values 0, 1, 2, 3, ...

CHAPTER 11

Synchronous FSM Design Considerations and Applications

11.1 INTRODUCTION

A number of design considerations and problem areas were purposely avoided in the previous chapter. This was done to focus attention on the basic concepts of design and analysis. These design considerations and problem areas include logic noise in the output signals; problems associated with asynchronous inputs, metastability, and clock distribution; and the initialization and reset of the FSM. It is the purpose of this chapter to discuss these and other subject areas in sufficient detail so as to develop good, reliable design practices.

11.2 DETECTION AND ELIMINATION OF OUTPUT RACE GLITCHES

Improper design of an FSM can lead to the presence of *logic noise* in output signals, and this noise can cause the erroneous triggering of a next stage switching device to which the FSM is attached. So it may be important that FSMs be designed to issue signals free of unwanted logic transients (noise) called *glitches*.

There are two main sources of output logic noise in an FSM:

- Glitches produced by state variable race conditions
- Glitches produced by static hazards in the output logic

In this and the following section, both types of logic noise will be considered, with emphasis on their removal by proper design methods.

A glitch that occurs as a result of two or more state variable changes during a state-to-state transition is called an *output race glitch* or simply *ORG*. Thus, an ORG may be regarded as an *internally initiated function hazard* (see Section 9.4), since two or more state variables try to change simultaneously but cannot. A glitch is an unwanted transient in an otherwise steady state signal and may appear as either a logic 0-1-0 (positive glitch) or as a logic 1-0-1 (negative) glitch, as indicated by the following:

As a voltage transient, an ORG may not develop to an extent that it crosses the switching threshold of a next-stage device. Even so, the wise designer must expect that the ORG might cross the switching threshold and take corrective measures to eliminate it.

By definition, a *race condition* is any state-to-state transition involving a change in two or more state variables. Thus, race conditions do not exist between logically adjacent states. The fact is that there are $n!$ possible (alternative) race paths for state-to-state transitions involving a change in n state variables. For example, a change in two state variables requires two alternative race paths, while a change in three state variables allows for $3! = 6$ alternative race paths. But since the specific, real-time alternative race path that an FSM may take during a state-to-state transition is not usually predictable, all possible alternative race paths must be analyzed if corrective action is to be taken. It is on this premise that the following discussion is based.

An ORG produced by a transition involving the change of two state variables is illustrated in Fig. 11.1a by a portion of a state diagram. Here, two race paths, associated with the transition from state 011 to state 110, are indicated by dashed lines. One race path is by way of race state 010 and the other via race state 111, a don't care state. Notice that the

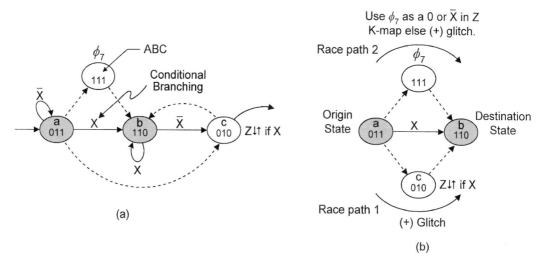

FIGURE 11.1
Illustration of an ORG caused by a transition from state 011 to state 110 involving a change in two state variables. (a) Portion of a state diagram showing the two alternative race paths associated with the transition. (b) Simplified diagram showing the origin and destination states, the race paths and possible ORGs.

11.2 DETECTION AND ELIMINATION OF OUTPUT RACE GLITCHES

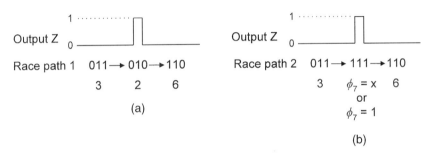

FIGURE 11.2
Logic sketches of the ORGs that can be produced by the state diagram segment in Fig. 11.1. (a) Positive (+) glitch produced in state 010 via race path 1. (b) Conditional positive (+) glitch produced in don't-care state 111 via race path 2.

branching from state a to state b is conditional on X and that this satisfies the conditional output Z if X in race state 010. The race paths are best shown by the simplified diagram in Fig. 11.1b, where all nonessential details are removed. Note that neither the *origin state* nor the *destination state* are output states — an important observation in ORG analysis.

Should state variable C change first, the transition from state 011 to state 110 will take place via race path 1 (race state 010) and the output Z would glitch, as indicated in Fig. 11.1b. On the other hand, if state variable A should change first, the transition will occur via race path 2 through don't-care state ϕ_7. In this case the question of whether or not an ORG will be produced depends on how ϕ_7 is used in extracting the cover for Z in the output K-map. If ϕ_7 is used as a 1, state 111 becomes an unconditional output state and an ORG will be produced by race path 2. Similarly, if ϕ_7 is used as an X, state 111 becomes a conditional (Mealy) output state that will produce an ORG via race path 2. As indicated in Fig. 11.1b, a choice of $\phi_7 = 0$ or $\phi_7 = \bar{X}$ eliminates the possibility of an ORG in state 111 via race path 2. That an ORG is an *output discontinuity* is illustrated by the logic sketches in Fig. 11.2, which represents the ORGs produced by the state diagram segment in Fig. 11.1. Notice that in both cases, the ORGs are (+) glitch discontinuities in the Z output, which should have been maintained as a steady-state logic 0 signal during the transition 011 → 110. The ORG shown in Fig. 11.2b is said to be conditional because its production depends on how ϕ_7 is used in the K-map for Z, as discussed previously.

Given that ORGs are present in the FSM segment of Fig. 11.1, corrective action must be taken to eliminate them, assuming it is necessary to do so. The easiest corrective action involves changing the state code assignments to eliminate the race condition that caused the ORGs. When state codes 110 and 010 in Fig. 11.1a are interchanged, the race condition disappears together with the ORGs. A simple alteration of the state code assignments is not always possible, but should be one of the first corrective measures considered.

Another example is presented that demonstrates the use of other methods for eliminating a possible ORG. Shown in Fig. 11.3a is a three-state FSM that has one input X, two outputs, Y and Z, and a possible ORG produced during the transition 00 → 11. To help understand the ORG analysis of this FSM, only the essential features of Fig. 11.3a are presented in Fig. 11.3b. Here, it is easily seen that if state variable A is caused to change first, the transition from state 00 to state 11 will take place via race path 1 through race state 10

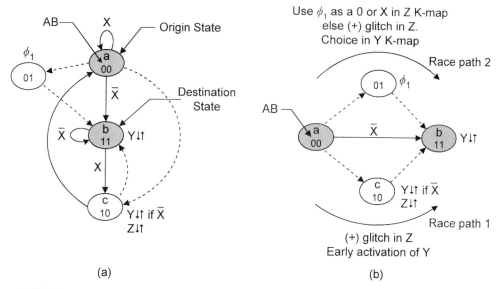

FIGURE 11.3
Example of an ORG in a simple three-state FSM with two outputs, Y and Z, and one input X. (a) State diagram showing the two alternative race paths for a transition from state a to state b. (b) Simplified state diagram segment showing only the essential details of the race paths and ORG possibilities.

causing an ORG in output Z, but not in output Y. This can be understood by noting that the output Z is not issued in the *origin state* 00 at the time the transition takes place. Nor is Z issued in the *destination state* 11 at the time of arrival. But Z is issued unconditionally in race state 10, thereby causing a discontinuity in the Z signal via race path 1 — an ORG. Output Y, on the other hand, is issued conditionally on $\bar{X}$ in race state 10 and unconditionally in the destination state 11. As a result, only early activation of Y is possible by way of race path 1 — there is no ORG in Y.

The don't-care state 01 can potentially produce an ORG in output Z via race path 2 if ϕ_1 is used as a 1 or as an $\bar{X}$ in the K-map for output Z. This ORG can be avoided by using ϕ_1 as a 0 or X in extracting cover from the Z K-map. If X is used, for example, then the output Z in state 01 is conditional on X. That is, the output is Z ↑↓ if X, which does not satisfy the branching condition $\bar{X}$ for the transition $a \rightarrow b$, and no ORG in Z is possible. The choice $\phi_1 = 0$ in the Z K-map clearly makes state 01 a nonoutput state for Z. In the case of output Y, don't care ϕ_1 can take on any value in the Y K-map, since only early activation of Y is possible should Y be caused to be issued in state 01.

The NS and output K-maps and the resulting minimum covers for the simple FSM of Fig. 11.3a are shown in Figs. 11.4a and 11.4b. (Remember that the NS K-maps are plotted by using the mapping algorithm in Section 10.6). Notice that ϕ_1 is taken to be 0 in the Z K-map, in agreement with the requirements indicated in Fig. 11.3b. Thus, an ORG in Z is not possible via race path 2. But ϕ_1 is used as a 1 in the Y K-map, which is permissible since, by race path 2, only early activation of output Y is possible. Early activation or late deactivation of an output is not a problem and, in some cases, may even be preferred as a means of optimizing output logic.

11.2 DETECTION AND ELIMINATION OF OUTPUT RACE GLITCHES

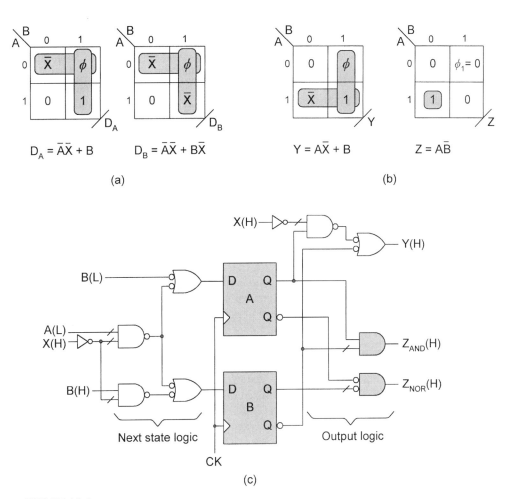

FIGURE 11.4
Implementation of the FSM in Fig. 11.3. EV K-maps for the NS and output functions. (b) Logic circuit showing two means of generating output Z.

Output Z is shown to be generated from both an AND gate and from a NOR gate in Fig. 11.4c. This has been done to demonstrate that the choice of hardware can influence the outcome of ORGs. An examination of the mixed-rail outputs from basic cells in Fig. 10.18 will help the reader understand the following discussion and how choice of hardware can be used as a means of eliminating ORGs. If the output is taken from the AND gate, $Z_{AND}(H)$, and if the D flip-flops are NAND based, state variable A will change before B by as much as a gate delay, causing the ORG to be formed. However, if the NOR gate is used to generate Z, $Z_{NOR}(H)$, and the D flip-flops are NAND based, no ORG will result. This is true since state variable B will change before state variable A, forcing the FSM to take race path 2 during the transition $a \to b$. But the use of NOR-based D flip-flops has the opposite effect for each of the two cases just stated. In this case, $Z_{AND}(H)$ should be used for a glitch-free output.

11.2.1 ORG Analysis Procedure Involving Two Race Paths

Certain facts emerge in the ORG discussions that have been presented so far. For reference purposes these facts are presented in the following procedure:

- Notice whether or not the *origin and destination states* in a given state-to-state transition have the same or different output action relative to a given output. If the origin and destination states have the same output action relative to a given output (that is, both are output states or both are not output states), then check to see if a potential ORG exists via the race states. If the origin and destination states have different output actions (that is, one is an output state but the other is not), *no* ORG is possible.
- If a potential ORG exists, corrective action should be taken to eliminate it by one of several means — an ORG may erroneously cross the switching threshold and trigger a next stage.

When more than two state variables are required to change during a given state-to-state transition, the analysis procedure becomes much more difficult. Recall that for a change of n state variables during a transition there are $n!$ possible race paths.

11.2.2 Elimination of ORGs

Six methods for eliminating an ORGs are cited here for reference purposes. Three of these methods, the first three listed below, have been discussed previously. These six methods may be viewed as listed in decreasing order of importance or desirability:

1. If possible, for a don't-care state that lies in a race path, make *proper* choice of its value in the output K-map to eliminate the ORG.
2. Change the state code assignment to either remove the race condition that caused the ORG, or move the race condition elsewhere in the state diagram where an ORG cannot form.
3. If possible, and with reference to Fig. 10.18 and the flip-flop technology, choose the output hardware necessary to eliminate the ORG.
4. Filter the output logic signal containing the ORG.
5. Use a buffer (fly) state to remove the race condition that caused the ORG.
6. Increase the number of state variables to eliminate the race condition that caused the ORG.

An inspection of the state diagram in Fig. 11.3a indicates that a simple change in the state code assignment can eliminate the potentially active ORG in output Z just discussed. This is demonstrated in Fig. 11.5a and in the state diagram segment of Fig. 11.5b. By using the state code assignment shown, all ORGs are eliminated. Now, the conditional branching $a \rightarrow b$ in Fig. 11.5a is logically adjacent and cannot cause an ORG. The unconditional branching from 10 to 00 ($c \rightarrow a$) is also logically adjacent and will cause no ORG. The only race condition that exists is the conditional branching from state 01 to 10 ($b \rightarrow c$) and for this no ORG is possible, as indicated in Fig. 11.5b. Branching $b \rightarrow c$ via race path 1

11.2 DETECTION AND ELIMINATION OF OUTPUT RACE GLITCHES 497

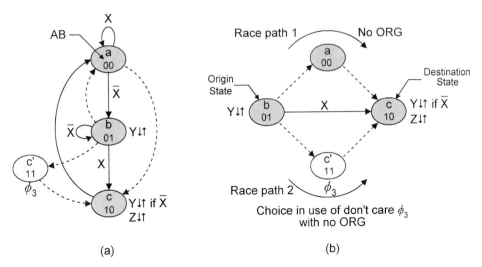

FIGURE 11.5
Use of a change in state code assignment to eliminate an ORG. (a) State diagram of Fig. 11.3a, showing change in state code assignment and new race paths for transition $b \to c$ (dashed lines). (b) Diagram segment for (a) showing elimination of all ORGs.

results in normal deactivation of Y and normal activation of Z. Assigning ϕ_3 a logic 1 or X in the Y K-map merely results in a late deactivation of output Y via race path 2. Or, using ϕ_3 as a logic 1 in the Z K-map results in an early activation of output Z via race path 2. Early or late activation or deactivation of an output is of no concern in most cases. In fact, it is only under the most stringent of timing conditions that such late or early activation or deactivation may become an important consideration. Use of ϕ_3 as a logic 0 results in the normal output response in either case. The following paragraphs offer two simpler alternatives for the elimination of all ORGs in the FSM of Fig. 11.3, alternatives that may or may not provide the best solution.

Shown in Fig. 11.6 are two examples of how a potential ORG can be eliminated by using a buffer (fly) state to eliminate the race condition causing the ORG. In Fig. 11.6a, don't-care state 01 is used to eliminate the ORG in output Z by removing the race condition between state a and b in Fig. 11.3a. In doing this, an additional clock period is introduced for the transition from a to b. The use of the buffer state in Fig. 11.6b removes the potential ORG in output Y but creates an additional clock period delay for the transition from state c to state a. These additional clock period delays caused by the presence of a buffer state may or may not be acceptable, depending on the design requirements for the FSM. Clearly, the best solution to this ORG problem is that indicated in Fig. 11.5.

Another approach to eliminating ORGs is to filter them. This is easily accomplished since ORGs, like all forms of logic noise, occur immediately following the triggering edge of the clock waveform. Shown in Fig. 11.7a is an acceptable filtering scheme involving an edge-triggered flip-flop that is triggered *antiphase* to the FSM memory flip-flops. Thus, if the memory flip-flops of the FSM are triggered on the rising edge of the clock waveform (RET), then the D flip-flop filter must be triggered on the falling edge of the clock waveform (FET) or vice versa. The timing diagram in Fig. 11.7b illustrates the filtering action of the

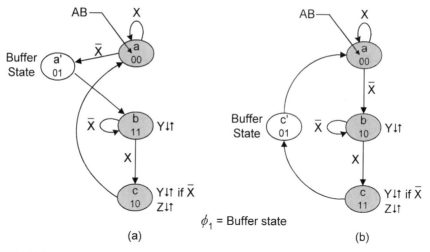

ϕ_1 = Buffer state

FIGURE 11.6
Elimination of possible ORGs by using buffer states to eliminate race conditions. (a) Use of don't-care state 01 as a buffer state to eliminate the ORG in Z shown in Fig. 11.3b. (b) Use of don't-care state 01 as a buffer state to eliminate a possible ORG in Y after interchanging the state codes 11 and 10 in Fig. 11.3a.

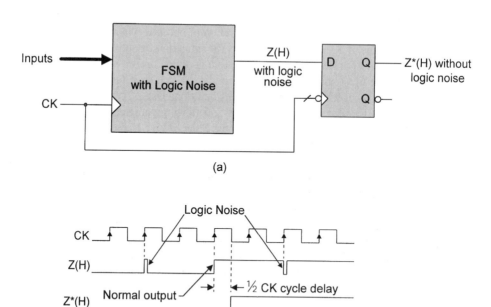

FIGURE 11.7
Filtering method for eliminating logic noise. (a) Logic circuit showing the use of an FET D flip-flop, triggered antiphase to the FSM memory, to filter logic noise. (b) Timing diagram showing filtering action of the FET D flip-flop.

D flip-flop. Notice that one-half of a clock cycle is lost because of the action of the filter. If the D flip-flop is triggered *in phase* with the FSM memory flip-flops, an entire clock cycle will be lost in the filtering process. To help understand how the filtering process eliminates logic noise such as ORGs, the following is presented for reference purposes:

> **Remember:** All forms of *logic noise* (glitches), including ORGs, occur immediately following the triggering edge of the clock waveform, and the duration of any logic noise pulse will always be much less than one-half the clock period.

Because logic noise occurs immediately following the triggering edge (RET or FET) of the clock waveform, it is very easily filtered. Another type of noise, called *analog noise*, is more difficult to filter since it is usually produced randomly from sources outside the logic system. The filtering of analog noise will not be discussed in this text.

Which Methods Should Be Used to Eliminate a Potential ORG?

If possible, make the proper choice of don't-care values in output K-maps or change the state code assignments as needed to eliminate ORGs. These two reliable methods are least likely to increase hardware requirements and slow down FSM operation. Choosing the output hardware in accordance with Fig. 10.18 is somewhat "iffy," since this method may depend on a single gate delay to force branching along a particular non-ORG race path. Unlike methods 1 and 2, method 3 does not offer assurance that a given ORG will not form.

Methods 4 and 5 both involve delays in the performance of the FSM and in most cases increase the hardware requirements. The filter method (4) is the most desirable of the two, since only a half CK cycle (plus the path delay through the D flip-flop) is involved. The filter method also has another advantage. By using a bank of such flip-flops (called an *output holding register*) to filter multiple outputs from an FSM, the outputs can be delivered synchronously to the next stage. Use of a buffer state (5) to eliminate a race condition in a branching path (one that caused the ORG) introduces an additional clock cycle in that branching path, and this may not be an acceptable option.

Least desirable, usually, is method 6. Although increasing the number of state variables may not alter the performance appreciably, this method does require an additional flip-flop and additional feedback paths for each state variable that is added. Thus, method 6 usually requires additional NS logic and may even require additional output-forming logic. The *one-hot code* method, discussed in Section 13.5, offers some advantages over conventional coding methods, but at the expense of requiring as many state variables (hence, also flip-flops) as there are states in the state diagram (see Table 2.11).

11.3 DETECTION AND ELIMINATION OF STATIC HAZARDS IN THE OUTPUT LOGIC

A detailed treatment of static hazards in combinational logic circuits is provided in Sections 9.1 and 9.2 and forms the basis for discussion of hazards in the output logic of FSMs presented in this section. It is recommended that the reader review these sections before continuing on in this section. Unique to state machines is the fact that the static hazards can be either *externally initiated* (as in combinational logic) or *internally initiated* because of

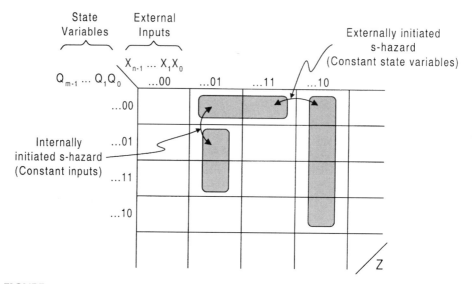

FIGURE 11.8
Output table for fictitious output Z showing externally and internally initiated static hazard transitions for either static 1-hazards or static 0-hazards.

a change in a state variable. The basic difference between these two types of static hazards (s-hazards) is illustrated by the generalized *output table* for output Z shown in Fig. 11.8. Here, the double arrows represent hazardous transitions for either static 1-hazards or static 0-hazards. Notice that an externally initiated s-hazard is created by a change in a *single* external input while all state variables are held constant, meaning that an externally initiated s-hazard takes place under a *hold condition*. Conversely, an internally initiated s-hazard is created by a change in a *single* state variable with all external inputs held constant. But in this latter case it will take an input change to initiate the transition that produces the s-hazard. The following discussions will consider both externally and internally initiated s-hazards.

11.3.1 Externally Initiated Static Hazards in the Output Logic

Externally initiated static hazards can occur only in Mealy FSMs. A simple example is presented in Fig. 11.9a, which is a resolver state machine configuration similar to that shown in Fig. 10.29a for the RET D flip-flop. This Mealy machine has two inputs, X and Y, and one output, Z. It is the function of this FSM that its output Z remain constant unless input Y changes while the FSM is in a resolved state, 01 or 10. Thus, a change in Y while the FSM is in a resolved state deactivates the output Z. Note that the FSM never enters state 11, a don't-care state.

The minimum SOP cover is shown in Fig. 11.9b. Notice that the coupled variable is identified as the external input Y. Also, observe that the state variables in the two coupled terms, $\bar{A}\bar{Y}$ and $\bar{B}Y$, are read in minterm code as $\bar{A}\bar{B} = 00$ to indicate that the hazard is produced by a change $Y \to \bar{Y}$ in state 00 under the holding condition $\bar{X}$. When the SOP

11.3 DETECTION AND ELIMINATION OF STATIC HAZARDS

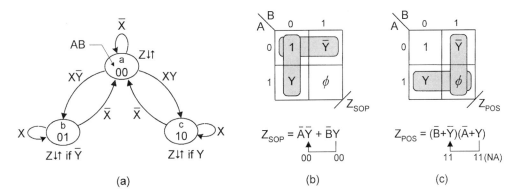

FIGURE 11.9
(a) Resolver FSM configuration with two inputs, X and Y, and one output, Z. (b) Minimum SOP cover for Z showing a $1 \rightarrow 0$ change in coupled variable Y while in state 00. (c) Minimum POS cover for Z indicating a $0 \rightarrow 1$ change in Y while in state 11, a don't-care state, which is not applicable (NA).

consensus law in Eqs. (3.14) is applied, the ANDed residue of the coupled terms is simply $\bar{A}\bar{B}$, which is the hazard cover (see Section 9.2). Thus, adding the hazard cover to the minimum SOP expression yields

$$Z_{SOP} = \bar{A}\bar{Y} + \bar{B}Y + \underbrace{\bar{A}\bar{B}}_{\text{Hazard cover}}, \tag{11.1}$$

which ensures that the static 1-hazard will not form under any set of circumstances.

The NAND/INV logic and timing diagrams for the Z_{SOP} function with and without hazard cover are shown in Fig. 11.10. Figure 11.10a illustrates the formation of the static 1-hazard resulting from a $1 \rightarrow 0$ change in external input Y, hence an externally initiated s-hazard. Implementation of Eq. (11.1), shown in Figure 11.10b, illustrates the removal of the s-hazard as a result of adding hazard cover $\bar{A}\bar{B}$ to the minimum SOP cover. In this latter case the hazard is removed regardless of the activation level of input Y, (H) or (L), and regardless of the delay imposed by the inverter.

To reinforce what has been said in the foregoing discussion, the function Z_{SOP} is represented in the output table of Fig. 11.11. The hazardous transition, indicated by an arrow, shows that the static 1-hazard can be produced only by a $1 \rightarrow 0$ change in Y assuming that input Y arrives active high. The hazard is eliminated by the hazard cover, which must cover the hazardous transition as indicated in the figure. Notice that the hazardous transition occurs in state 00 under holding condition $\bar{X}$ as required by Figs. 11.9a and 11.9b.

The minimum POS cover is indicated in Fig. 11.9c. Reading the coupled terms in maxterm code indicates that a $0 \rightarrow 1$ change in Y must occur in state 11, which is a don't-care state. Since this FSM never enters state 11, the static 0-hazard never occurs and, accordingly, hazard cover is said to be not applicable (NA). The gate/input tally for the POS expression is 3/6 compared to 4/9 for Eq. (11.1), both exclusive of inverter count. Thus, hardware-wise, a discrete logic POS implementation of the output Z is favored over the SOP expression in Eq. (11.1).

502   CHAPTER 11 / SYNCHRONOUS FSM DESIGN CONSIDERATIONS

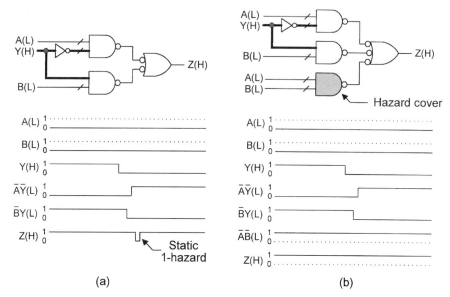

FIGURE 11.10
Formation and removal of the static 1-hazard in the Mealy FSM of Fig. 11.9(a) NAND/INV logic for output Z in Fig. 11.9b and timing diagram showing the formation of the s-hazard. (b) Implementation of Eq. (11.1) and timing diagram with hazard cover showing removal of the s-hazard.

11.3.2 Internally Initiated Static Hazards in the Output of Mealy and Moore FSMs

The following discussion can best be understood by a reexamination of the mixed-rail output responses of the set- and reset-dominant basic cells given in Fig. 10.18. Notice that in both cases the active portion of the waveform from the ANDing operation is symmetrically

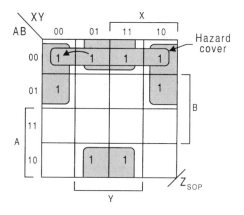

FIGURE 11.11
Output K-map showing hazardous transition (arrow) and hazard cover for Z_{SOP} in Eq. (11.1) and Fig. 11.9b.

11.3 DETECTION AND ELIMINATION OF STATIC HAZARDS

placed within that from the ORing operation by an amount equal to the path delay through a gate. From this information the following conditions for s-hazard can be stated:

- Conditions for static 1-hazard (SOP hazard) formation:
    1. A 1 → 0 change in the Q output of the flip-flop when $Q(H)$ leads $Q(L)$
    2. A 0 → 1 change in the Q output of the flip-flop when $Q(L)$ leads $Q(H)$
- Conditions for static 0-hazard (POS hazard) formation:
    3. A 0 → 1 change in the Q output of the flip-flop when $Q(H)$ leads $Q(L)$
    4. A 1 → 0 change in the Q output of the flip-flop when $Q(L)$ leads $Q(H)$

Note that these four conditions are similar to those for combinational logic when considering the activation level of the initiating (coupled-variable) input. In Figs. 9.2 and 9.3, for example, the coupled-variable inputs to the SOP logic circuits are the external inputs $A(H)$ and $A(L)$, respectively. For internally initiated s-hazard formation, the coupled-variable input is assumed to be the output from a basic cell, $Q(H)$ and $Q(L)$. This is a valid assumption since the output stage of the most common flip-flops is a basic cell with mixed-rail outputs.

By relating the mixed-rail output response of the basic cells to the conditions for s-hazard formation just stated, two useful conclusions can be drawn. To understand how these conclusions come about, it is necessary to revisit Fig. 10.18. Presented in Fig. 11.12 are the mixed-rail output responses of the basic cells together with the conditions for s-hazard formation for each, as deduced from conditions 1 through 4 previously stated. An inspection of Fig. 11.12a reveals that the mixed-rail output response for the set-dominant (SOP) basic cell generates the conditions for POS hazard (static 0-hazard) formation. In dual fashion, the mixed-rail output response for the reset-dominant (POS) basic cell in Fig. 11.12b generates the conditions for SOP hazard (static 1-hazard) formation. That the set- and reset-dominant basic cells can be called SOP and POS circuits, respectively, is easily deduced from an inspection of Figs. 10.11a and 10.13a.

From Fig. 11.12 and the forgoing discussion, the following two conclusions can be drawn, subject to the assumption that the coupled variables in an output expressions are state variables produced from the mixed-rail outputs of the flip-flops:

- For flip-flops with NAND-centered (SOP) basic cells, s-hazards produced by either a 1 → 0 or a 0 → 1 change in the coupled variable are *not* possible if SOP output logic is used.
- For flip-flops with NOR-centered (POS) basic cells, s-hazards produced by either a 0 → 1 or a 1 → 0 change in the coupled variable are *not* possible if POS output logic is used.

The ramifications of the forgoing conclusions are important in FSM design. If the output logic and that for the basic cell output stage of the flip-flop are matched, that is, both SOP or both POS logic, internally initiated s-hazards are not possible in the logic domain. For example, if the output logic of an FSM is to be implemented by using an SOP device, such as a PLA, the choice of NAND-centered flip-flops avoids the possibility of internally initiated s-hazard formation in the output logic of that FSM. On the other hand, if the form of the output logic and that for the basic cell of the flip-flops are different, one SOP and the other POS, s-hazards are possible on either a 1 → 0 or a 0 → 1 change of the coupled variable (state

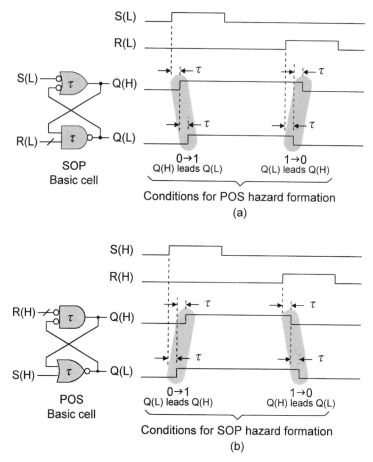

FIGURE 11.12
Mixed-rail output response of the basic cells and conditions for internally initiated s-hazard formation. (a) Logic circuit and mixed-rail output response for the set-dominant (SOP) basic cell showing conditions for POS hazard formation. (b) Logic circuit and mixed-rail output response for the reset-dominant basic cell showing conditions for SOP hazard formation.

variable). If such an s-hazard is formed, it will most likely cross the switching threshold since the delay τ in Fig. 11.12 represents an entire gate delay.

There is another means by which mixed-rail outputs can be produced from a flip-flop. Shown in Fig. 11.13 is an RET D flip-flop whose output response is $Q(H)$ from the flip-flop, but now $Q(L)$ is taken from an inverter, not from the flip-flop. As a consequence, the active portion of the $Q(L)$ waveform is skewed in time relative to that from $Q(H)$. Though the delay difference is now only that of an inverter, it does make possible the formation of a static 1-hazard on a $1 \rightarrow 0$ change of the coupled variable, or the formation of a static 0-hazard on a $0 \rightarrow 1$ change of the coupled variable. Matching the output-forming logic to the logic of the flip-flops can no longer be used as a possible means of eliminating internally initiated s-hazards in the output logic.

11.3 DETECTION AND ELIMINATION OF STATIC HAZARDS

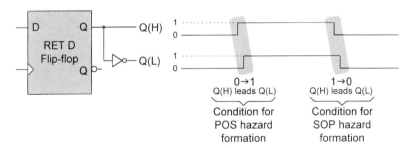

FIGURE 11.13
Conditions for s-hazard formation from the mixed-rail output response of a flip-flop where $Q(L)$ is produced by an inverter.

When Is It Necessary to Run a Static Hazard Analysis on an FSM? If the output logic of an FSM appears to have the potential for s-hazard formation, there arises the question of when an s-hazard analysis should be run on an FSM. There are specific guidelines one can use in deciding this issue. These guidelines are stated next, but not in any particular order of importance. A static hazard analysis should be carried out:

1. If it is known that s-hazard production in an output can cause a problem in the next stage.
2. Always following an ORG analysis and any corrective action that may result.
3. If it is determined that the output in question is not to be filtered.
4. If there is no match of the output logic character with that of the flip-flop output stages.
5. If the logic character of the flip-flops is unknown.

Generally, all five guidelines should be considered, but particular notice should be paid to guidelines 2 and 3. Hazard analyses should always be carried out following any corrective action required by an ORG analysis. If, for example, an ORG analysis requires a change in the state code assignments or requires the particular use of a specific don't care in an output K-map, the output logic is certain to change. It is useless to run a hazard analysis before the final output logic is known. It is also useless to run either an ORG or a hazard analysis if it is known that the output is to be filtered.

A Simple Example Consider the state diagram for the Mealy FSM presented in Fig. 11.14a. It is shown to have two inputs, S and T, and one output, Y. The ORG analysis, which must be run before the hazard analysis, is shown in Fig. 11.14b. No ORG is possible in this FSM, and ϕ_4 can be used in any way to extract minimum cover for output Y. Notice that late deactivation of Y is possible via race state 001 if the branching condition from $c \rightarrow a$ is $\bar{S}T$, but normal deactivation if $\bar{S}\bar{T}$.

Hazard analyses for the FSM in Fig. 11.14 are carried out in Fig. 11.15. The hazard analysis in Fig. 11.15a indicates that a static 1-hazard is possible if NOR-based flip-flops are used with SOP output logic, and that the hazard cover required in that case is $C\bar{S}T$, the

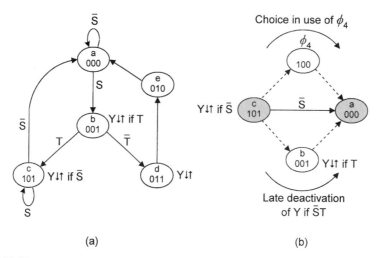

FIGURE 11.14
A Mealy FSM having two inputs, S and T, and one output, Y. (a) State diagram. (b) ORG analysis showing no ORG is possible.

ANDed residue of the coupled terms. The final result, after adding hazard cover, is given by the following expression:

$$Y_{SOP} = \bar{A}CT + A\bar{S} + BC + \underbrace{C\bar{S}T}_{\text{Hazard cover}}.$$
(11.2)

Note that if NAND-based flip-flops are used with the SOP output logic (Y_{SOP}), no s-hazard is possible in the logic domain (see Fig. 11.12a) and no hazard cover should be added.

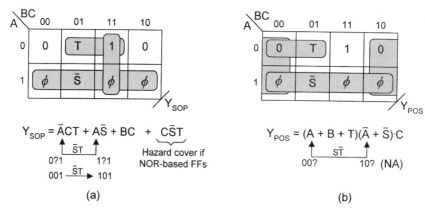

FIGURE 11.15
S-hazard analysis for the FSM in Fig. 11.14. (a) Static 1-hazard analysis showing hazard cover necessary only if NOR-based flip-flops are used with the SOP output logic. (b) Static 0-hazard analysis showing that the transition b-to-c has no s-hazard associated with it since the input requirements indicated are not met. Thus, POS hazard cover is not applicable (NA).

11.3 DETECTION AND ELIMINATION OF STATIC HAZARDS

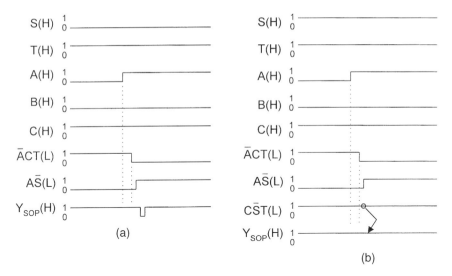

FIGURE 11.16
Timing diagrams for the static 1-hazard (SOP hazard) analysis shown in Fig. 11.15a. (a) Timing diagram for Y_{SOP} showing a static 1-hazard formed during the transition b-to-c assuming the use of NOR-based flip-flops for the memory. (b) Timing diagram for Y_{SOP} showing the removal of the static 1-hazard by addition of hazard cover.

The timing diagrams for the hazard analysis in Fig. 11.15a are presented in Fig. 11.16. In Fig. 11.16a a static 1-hazard in Y_{SOP} is indicated for the transition $001 \rightarrow 101$, assuming that NOR-centered flip-flops are used for the FSM memory. If NAND-centered flip-flops are used instead, no static 1-hazard (SOP hazard) will occur in Y_{SOP}. Remember that when the logic character of the flip-flops matches that of the output logic, internally initiated s-hazards are not possible in the logic domain. In Fig. 11.16b the static 1-hazard is shown removed because of the presence of static hazard cover $C\bar{S}T$. In fact, the hazard is removed regardless of the magnitude of any asymmetrically located delays associated with the $b \rightarrow c$ transition in the logic or physical domain. Notice that account is taken of the gate path delays in the timing diagrams of Fig. 11.16.

The memory and output logic for the FSM of Fig. 11.14 is shown in Fig. 11.17, assuming the use of NOR-based flip-flops and SOP output logic. The external inputs are assumed to arrive active high, and the shaded NAND gate is the hazard cover, $C\bar{S}T$.

The POS hazard analysis in Fig. 11.15b indicates that a static 0-hazard is not possible under any circumstances. The reason: only the transition $00? \rightarrow 10? = 001 \rightarrow 101$ is possible and that must take place under the branching condition $\bar{S} + T = \overline{S\bar{T}}$, which does not meet the branching requirements for the $b \rightarrow c$ transition shown in Fig. 11.14a. In any POS analysis it must be remembered that the ORed branching condition, as deduced from the coupled terms, must be complemented for comparison with the state diagram. This is so because the state diagram, like any K-map, is a minterm-code-based graphic. In extracting minimum cover from the Y_{POS} K-map, the domains for the state variables (A, B, and C) are complemented, but the entered variable inputs are not.

The output-forming logic for Y_{POS} is provided in Fig. 11.18a where it is assumed that the external inputs, S and T, arrive active high as before. The timing diagram for Y_{POS}, given

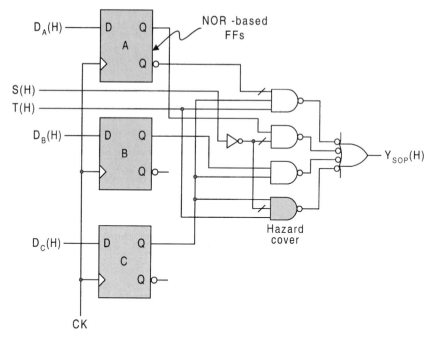

FIGURE 11.17
Memory flip-flops (assumed to be NOR-based) and SOP output logic with hazard cover for the FSM of Fig. 11.14a.

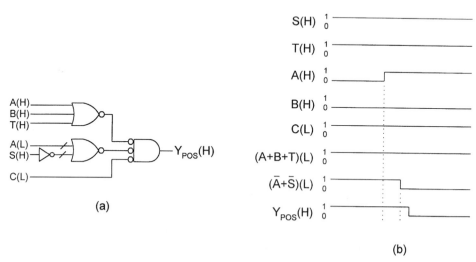

FIGURE 11.18
(a) Output logic circuit for Y_{POS} in Fig. 11.15b. (b) Timing diagram for Y_{POS} showing absence of a static 0-hazard during the b-to-c transition assuming the use of NAND-based flip-flops for the memory.

in Fig. 11.18b, clearly shows that a static 0-hazard is not possible for the transition $001 \rightarrow 101$. This is so because the $b \rightarrow c$ branching condition in Fig. 11.14a requires that input T be active, which is contrary to the requirements indicated by the coupled terms, $(S\bar{T})$. Thus, independent of whether or not there is a match between the logic character of the flip-flops and that of the output logic, no s-hazard is possible in Y_{POS}. Therefore, the addition of hazard cover is not applicable (NA) as indicated in Fig. 11.15b. Notice that the gate/input tallies for Y_{POS} and Y_{SOP} are 3/8 and 4/10, respectively, exclusive of inverters and hazard cover. If hardware cost is the only consideration, the best choice for output logic would be Y_{POS}, as given in Fig. 11.15b.

11.3.3 Perspective on Static Hazards in the Output Logic of FSMs

Static hazards in the *next-state-forming logic* are never a problem in synchronous FSMs simply because the memory flip-flops act as a filtering stage for such logic noise. However, in the case of s-hazards in the *output-forming-logic*, the situation is much different. As has been discussed, a static hazard in the output function of an FSM can cause malfunction of a next-stage logic device to which the output function is an input. But not every s-hazard may cross the switching threshold of that next stage device. The problem is that the designer cannot afford to gamble on that, and instead should take corrective measures such as adding hazard cover or filtering the output to eliminate the hazard.

Externally initiated s-hazards pose a special dilemma for the designer, since the asymmetric delay is usually caused by an inverter. The previous discussion suggests that if the coupled terms require a branching direction opposite to that actually present in the state diagram, hazard cover may be ignored. In fact, the s-hazard may still be formed if a delay in the alternative path (not through the inverter) is larger than the inverter. Thus, it may be desirable to apply the "shotgun approach" to all externally initiated s-hazards in the output functions. This means that hazard cover would be assigned to all externally coupled terms regardless or whether they represent a $1 \rightarrow 0$ change or a $0 \rightarrow 1$ change of the coupled external variable as indicated by the state diagram. This action would certainly make computer-aided corrective action simpler for externally initiated s-hazards.

There is the remote possibility that internally initiated s-hazards may form even if the logic character of the flip-flops matches that of the output-forming logic. For an s-hazard to be produced under this condition, a delay larger than that of a basic cell gate would have to exist in an alternative path so as to effectively reverse the symmetrical inset of the waveforms in Fig. 10.18. Though the probability that this may happen is low, it is something of which the designer should be aware.

The following set of guidelines are offered to help eliminate any confusion the foregoing discussion may have caused and to help establish safe and reliable design practices:

- Add hazard cover for all externally initiated s-hazards in the output logic expressions as required by the coupled terms. There is one exception to this rule: If the state in which the externally initiated hazard exists is an extraneous state (one that neither exists in the state diagram nor serves as a race state), as was the case in Fig. 11.9c, no hazard cover is needed and none should be added.
- If internally initiated s-hazards are present and the goal is to achieve an optimum design, match the logic character of the flip-flops with that of the output-forming

logic and take no corrective action on these s-hazards. Then, in configuring a circuit layout try to minimize parasitic effects by minimizing lead lengths between the flip-flops and output logic. If the logic character of the flip-flops is unknown, *always* add hazard cover for all internally coupled terms in the output logic for which a valid hazardous transition exists. Note that Fig. 11.13 applies to any PLD in which the internal flip-flops lack $Q(L)$ outputs.

- If the outputs of FSM A are the inputs to another FSM B, take caution in assuming that logic noise (e.g., s-hazards) from FSM A will be filtered by the memory of FSM B. Whether or not such logic noise will be filtered by the memory of FSM B depends on many factors, including the type of input conditioning circuits that exist, the nature of the NS-forming logic, and the character of FSM B itself. If this information is unknown or questionable, the safest action is to provide clean output signals from FSM A by using the methods described previously.

11.4 ASYNCHRONOUS INPUTS: RULES AND CAVEATS

A synchronous input is one that is synchronized with clock to the extent that it cannot change its logic level during a sampling interval (see Fig. 10.53). Any input that does not meet this requirement is said to be an *asynchronous input*, defined as follows:

> *An asynchronous input is one that can change logic levels at any time, particularly during the sampling interval established by the sampling variable, CK.*

As was pointed out in Section 10.11, an input to a synchronous FSM must meet the setup and hold-time requirements established by clock (the sampling variable) or proper transitions cannot generally be guaranteed. Simply stated, a synchronous FSM may not function properly if more than one asynchronous input is present. Remember that clock is, by definition, an asynchronous input. Therefore, CK should be considered to be the only permissible asynchronous input controlling the branching from a given state.

11.4.1 Rules Associated with Asynchronous Inputs

To reduce the probability for FSM malfunction due to the presence of asynchronous inputs, the following two rules should be observed:

> *Rule 1 (Branching Dependency Rule): Avoid branching dependency on more than one asynchronous input.*
> *Rule 2 (Conditional Output Rule): Do not attempt to generate an output conditional on an asynchronous input.*

These two rules are easily justified by discussing the consequences of their violation. For example, if more than one asynchronous input controls the branching from a given state, the sequential behavior can become unpredictable, resulting in the malfunction of the FSM. Furthermore, an output that is conditional on an asynchronous input can, under certain conditions, be no more than an underdeveloped (runt) pulse that may cause problems in

11.4 ASYNCHRONOUS INPUTS: RULES AND CAVEATS

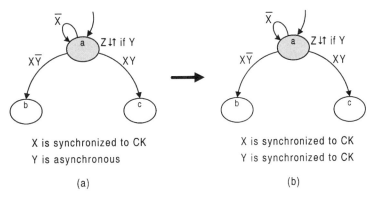

FIGURE 11.19
(a) Improper branching and output generation conditional on two asynchronous inputs, CK and Y.
(b) Proper branching and output generation conditional on one asynchronous input, CK.

the next stage to which it is an input. These problems and their proper solutions are now considered in more detail.

Shown in Fig. 11.19 is a portion of a common resolver configuration that is used here to illustrate the problems associated with asynchronous inputs and violation of rules 1 and 2. There are three inputs to the resolver, X, Y, and CK, that control the branching from state a, where CK is understood to be the sampling variable and is not included in the state diagram. In Fig. 11.19a both the branching from state a and the output, Z, are conditional on two asynchronous inputs, Y and CK, which is a violation of both rule 1 and rule 2. Should input Y change during the sampling interval established by CK, the branching and output are not predictable. Worse yet, a runt pulse can be produced in the memory flip-flops forcing the FSM into a metastable condition (discussed in Subsection 11.4.4) or possibly causing an error transition in the FSM. Furthermore, output, Z, could be generated as a runt pulse that could cause problems in another FSM to which it is an input. This is so because the conditional output can be in its development stage at the time the flip-flops trigger. Remember, it takes longer for the flip-flops to execute a transition than it does to generate a conditional output by combinational logic from a given state. An output should always be presented as a reliably detectable signal to the next stage and never as a pulse of unpredictable duration.

In short, the proper solution to the problems implied by Fig. 11.19a is to synchronize all external inputs to the CK waveform, as indicated in Fig. 11.19b. Now, input Y will be stable at its proper logic level at the time CK goes through its sampling interval; the sampling variable, CK, remains the only permissible asynchronous input. Even though output Z is issued on an exiting condition in state a, it will nonetheless be generated well in advance of the transition so as to be a reliably detectable pulse by the next stage. The important issue of synchronizing inputs is discussed in the following subsection.

11.4.2 Synchronizing the Input

A reliable approach to dealing with the problem of asynchronous inputs is to synchronize each asynchronous input to the clock waveform before it is introduced into the next state

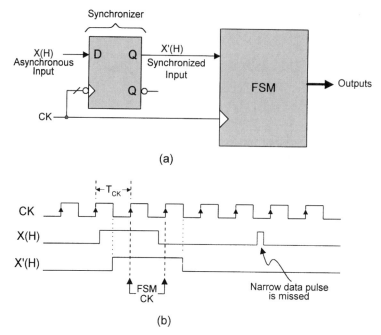

FIGURE 11.20
(a) Synchronizing scheme for an asynchronous input X showing synchronizer triggered antiphase to the FSM. (b) Timing diagram indicating results of synchronizing the input.

logic section of the FSM. This is accomplished by using a synchronizer D flip-flop triggered antiphase to the FSM, as shown in Fig. 11.20a. Here, $X(H)$ is an asynchronous input to the FET synchronizer that issues a synchronized output, $X'(H)$, to the RET FSM.

The timing diagram, presented in Figure 11.20(b), illustrates the action of the synchronizer. Notice that the FSM can pick up the input X (as the synchronized X' signal) after a delay ranging approximately from 1/2 to 3/2 of a clock period depending on when the signal $X(H)$ changes relative to CK, and assuming the setup and hold times are met and are much smaller than a clock period. This pickup delay is the price that must be paid to present a reliably readable data signal to the FSM. Notice that the arrows on the clock waveform represent the rising edge triggering of the FSM. Also shown in Fig. 11.20b is a data pulse too narrow to be picked up by the synchronizer. The means by which a narrow data pulse can be read by the FSM is considered next.

11.4.3 Stretching and Synchronizing the Input

If it is known that the data can arrive as asynchronous pulses of duration less than that of the clock period, a means must be sought to stretch as well as synchronize the data signals. An effective scheme for accomplishing this is presented in Fig. 11.21a. The narrow asynchronous pulse is first stretched by the set-dominant basic cell (stretcher), then synchronized by the synchronizer. The active low output of the synchronizer is fed back to the $R(L)$ input of the stretcher to reset it in readiness for the next narrow pulse. Notice that the output of

11.4 ASYNCHRONOUS INPUTS: RULES AND CAVEATS

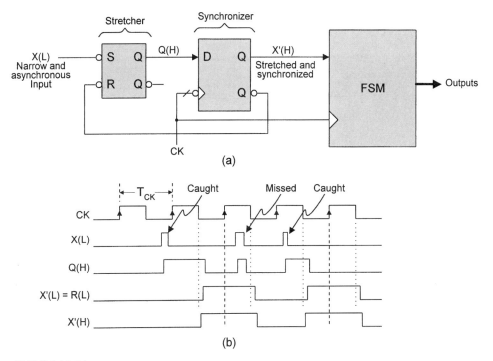

FIGURE 11.21
Stretching and synchronizing the input. (a) Logic circuit showing stretcher and synchronizer stages. (b) Timing diagram illustrating the action of the stretcher cell and synchronizer, and showing caught and missed narrow pulses.

the synchronizer, $X'(H)$, is both stretched and synchronized, thereby providing a reliably detectable signal to the FSM regardless of the pulse duration. If X' must be presented active low to the FSM, the $Q(L)$ output from the synchronizer can be used. Also, if the data is presented to the stretcher as $X(H)$, an inverter can be used on the line to the stretcher's active low input. Alternatively, a double complementation can be used somewhere between the $X(H)$ input to the stretcher and the input to the FSM, meaning (H) to (L) and (L) to (H). For example, $Q(L)$ from the stretcher can be used as the input to the synchronizer. Note that a reset-dominant basic cell cannot be used as the stretcher cell for positive pulse trains, since sustained positive data pulses would be reset by the feedback from the synchronizer leading to false data input to the FSM.

The actions of the stretcher and synchronizer are illustrated in Fig. 11.21b. Here, it is observed that not all narrow pulses can be caught by the synchronizer and presented to the FSM. If a second pulse appears before the stretcher cell is reset, it cannot be picked up by the stretcher as a discrete data pulse. Consequently, a second narrow pulse having a leading edge separated by less than $2T_{CK}$ from the leading edge of the first pulse cannot be guaranteed to be caught by the FSM, and a second leading pulse edge separated by less than T_{CK} from the first can never be caught. These limiting conditions are based on the assumption that the setup and hold times are negligibly small compared to the clock period, usually a valid assumption. Again, observe that the arrows on the clock waveform represent

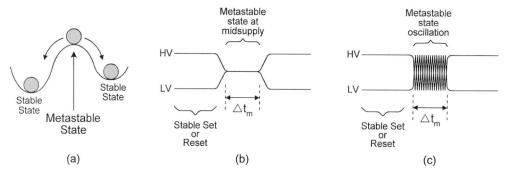

FIGURE 11.22
Qualitative representations of metastability. (a) Mechanical analogue. (b) FSM metastability manifested as a midsupply state. (c) FSM metastability manifested as an oscillatory state. Δt_m = Metastable exit time.

the rising edge triggering of the FSM. Thus, for each narrow $X(L)$ pulse that is caught, a stretched $X'(H)$ pulse is picked up by the RET FSM well into the active portion of the pulse.

11.4.4 Metastability and the Synchronizer

An important function of the synchronizer is to protect an FSM from the effects of *metastability* if caused by an input change during the sampling interval of the clock waveform. The problem is that the synchronizer is itself subject to the effects of metastability caused by data input changes occurring during its sampling interval. Metastability is a very low-probability event, but it can happen and can be a potential problem in any system with feedback. Just as the second law of thermodynamics cannot be violated in attempting to invent a perpetual motion machine, no "fix-it" scheme exists that will reduce to zero the probability that metastability will occur in a given FSM. But there are synchronizing schemes that can come close! Before proceeding with the means by which this can be accomplished, it will be helpful to define metastability in qualitative terms.

Three qualitative representations of metastability are depicted in Fig. 11.22. First is the mechanical analogue, shown in Fig. 11.22a, featuring a ball or round disk metastably situated atop a convex surface such that any slight perturbation would send it to one stable state or another. More appropriate to the needs of FSM design is the electrical representation of the metastable state that lies somewhere between a set and a reset condition, say at midsupply, as illustrated in Fig. 11.22b. Here, the time that the FSM spends in the metastable state, denoted by Δt_m, is called the *metastable exit time*. This is a statistical period of time that cannot be predicted. The two double-line regions preceding and following the metastable state represent a stable set or reset condition, one or the other. However, it cannot be predicted which logic level (set or reset) will emerge following exit from the metastable state. The oscillatory metastable state illustrated in Fig. 11.22c is also a possibility in some FSMs, which if exists, could pose a more serious problem for the FSM than a simple midsupply "hangup." Here again, the logic level (set or reset) following exit from the metastable state cannot be predicted. But this is probably a moot point, since an oscillatory condition can potentiallly cause far more serious problems than an unpredictable outcome following exit from that state.

The foregoing discussion applies to any FSM, including flip-flops. As an example, the resolver section of a D flip-flop shown in Fig. 10.31a can go metastable and cause both the flip-flop and the FSM in which it is operating to malfunction. Thus, the synchronizer in Fig. 11.20a is subject to the metastable condition and can pass that metastable state on

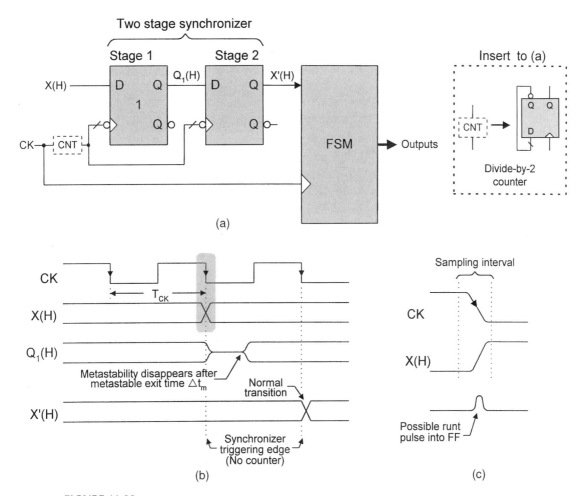

FIGURE 11.23
A two stage synchronizer scheme to greatly reduce the probability of a metastable state occurrence. (a) Logic circuit showing two synchronizer FET D flip-flops in series, and use of a divide-by-2 counter to increase T_{CK} of the synchronizers relative to the FSM. (b) Timing diagram for the two stage synchronizer scheme with no counter showing a possible metastable state developed in stage 1. (c) Blow up of shaded area in (b) showing a possible runt pulse formation as a result of data X changing during the sampling interval of CK.

to the FSM it is supposed to protect. A practical solution is illustrated by the two-stage synchronizer configuration shown in Fig. 11.23a. The idea depicted here is that in the event synchronizer 1 should go metastable, it would emerge from that metastable state long before synchronizer 2 is triggered, as illustrated in Fig. 11.23b. This, in turn, greatly reduces the probability that synchronizer 2 will become metastable and cause malfunction of the FSM. Of course, it is assumed that the metastable exit time, Δt_m, will always be less than T_{CK}, an assumption that may or may not be valid.

The blown-up region in Fig. 11.23c illustrates one means by which a metastable state can be introduced into stage 1 of the synchronizer. If asynchronous input $X(H)$ changes during the sampling interval of clock, a *runt pulse* could form and be introduced into the D flip-flop as neither a set nor a reset condition, and this could initiate the metastable state. Such a runt pulse could cross the switching threshold but lack the "strength" or duration needed to resolve the flip-flop into a set (or reset) condition, and a metastable condition could result.

Experimentally, it is found that the *mean time between failures* (MTBF) of the single D flip-flop synchronizer in Fig. 11.20a is determined by the equation

$$\text{MTBF} = \left\{ \frac{e^{(T_{CK}-t_{su})/\tau}}{T_0 \cdot f_{CK} \cdot f_D} \right\} \text{ in seconds,} \tag{11.3}$$

where T_{CK} is the clock period in nanoseconds (ns); f_{CK} is the clock frequency in hertz (Hz); t_{su} is the setup time in ns (see Fig. 10.53); f_D is the average number of asynchronous data input changes per second (data frequency) in Hz; and τ (in ns) and T_0 (in seconds, s) are empirical constants, provided by the flip-flop manufacturers, that depend on the electrical characteristics of the flip-flop and on the physical conditions under which the flip-flop is operated. For most applications, it is reasonable to assume that $f_{CK} \gg f_D$. Note that MTBF refers to probabilistic failure caused by a metastable condition when $\Delta t_m > T_{CK} - t_{su}$ in a single D flip-flop synchronizer.

Clearly, the larger the MTBF, the better is the action of the synchronizer flip-flop and vice versa. Ideally, an infinite value for MTBF would be the most desirable, albeit impossible to achieve with the synchronizing scheme of Fig. 11.23a. A value of $10^{10} s = 317$ years might be achievable, but under what conditions? An important feature of Eq. (11.3) is the sensitive inverse dependence of the MTBF on flip-flop clock frequency $f_{CK} = 1/T_{CK}$ and on the empirical τ constants T_0:

$$\text{MTBF} \propto \frac{e^{(f_{CK} \cdot \tau)^{-1}}}{T_0 \cdot f_{CK}}.$$

Thus, for a high MTBF, it desirable to have a low f_{CK} (high T_{CK}) and low values for τ and T_0. To achieve reasonably high values of the MTBF in a single D flip-flop, it is necessary to use D flip-flops from a fast technology such as the 74HCnn series or, better yet, the 74Fnnn or 74ASnn series (see Subsection 6.1.4 for an explanation of part numbers). For these D flip-flops τ can be as low as 0.3 ns with values of T_0 down in the microsecond ($\mu s = 10^{-6} s$) range. When operated at relatively low frequencies MTBF values of $10^{10} s$ may be possible, but only for small t_{su}.

Still, at the high frequencies required by modern technology, a single D flip-flop synchronizer is not sufficient, and use must be made of the two-stage synchronizer shown in Fig. 11.23 together with counters on the clock inputs to the two stages. By using fast D

11.5 CLOCK SKEW 517

flip-flop technology and by creating a large T_{CK} for the synchronizers relative to the FSM, large values of the MTBF can be achieved even with high frequencies. Note that use of a delay circuit in place of the counter would be worse than having no delay at all. A divide-by-2 counter doubles the clock period (see Subsection 12.3.1). Now, the clock period for the two synchronizers is at least double that of the FSM, greatly improving chances for $2T_{CK} > \Delta t_m$. The divide-by-2 counter should be the slow 74SL74 with a $Q(L) \to D(H)$ feedback as indicated in the insert to Fig. 11.23a. If this is not sufficient, there are other alternatives. One alternative is to replace the divide-by-2 counter in Fig. 11.23a by a divide-by-4 *ripple counter* (see Section 12.5 for details). As another alternative, a *multiple-stage synchronizer* scheme can be used with or without a counter on the clock inputs to the stages as in Fig. 11.23. Also, Schmitt triggers can be used on the data lines between stages for additional discrimination of a metastable signal.

All of the synchronizing schemes just mentioned are used at the expense of system throughput, the price that must be paid to introduce reliably readable data to the protected FSM. Also, it must be remembered that because metastability is a statistical phenomenon and is unpredictable, no synchronizer "fix-it" scheme can be devised that will eliminate entirely the possible occurrence of the metastable state. All that can be done is to reduce the probability for metastability occurrence to acceptable levels for a given application. In Chapter 16 an externally asynchronous/internally clocked (EAIC) system will be discussed that will deal with the problem in a different and more effective manner. EAIC configurations are pausable systems capable of yielding an infinite MTBF value with no required external synchronizing logic of the type shown in Fig. 11.23.

11.5 CLOCK SKEW

In synchronous sequential machines the triggering edge of the clock waveform is assumed to reach each flip-flop of the memory at approximately the same time. Sometimes, however, this does not happen because of the presence of asymmetric path delays caused mainly by resistance and parasitic capacitance effects in the clock leads to the memory devices or by poor clock buffering methods. When such delays become large enough to cause a shift in the triggering edge of one flip-flop relative to another, *clock skew* is said to exist. Clock skew can become a serious problem in digital systems, particularly in complex systems operated at very high frequencies.

Illustrated in Fig. 11.24 is one type of problem that can occur as a result of clock skew. Shown in Fig. 11.24a are two RET D flip-flops configured in series with delays Δt_1 and Δt_2 indicated on the clock inputs to flip-flops 1 and 2, respectively. If the delays are equal, $\Delta t_2 = \Delta t_1 = 0$, no clock skew exists and proper flip-flop output response to a change in data input $X(H)$ results, as indicated in Fig. 11.24b. Observe that $X(H)$ is synchronized to the falling edge of the CK = CK_1 waveform. On the other hand, the condition $\Delta t_2 > \Delta t_1$ can result in an erroneous output, as indicated in Fig. 11.24c. Such an error will occur in output $Q_2(H)$ if $\Delta t_2 - \Delta t_1 > \tau_{ff}$, where τ_{ff} is the flip-flop propagation delay. Timing anomalies of this type can lead to unrecoverable errors in the operation of shift registers and other devices. The reverse skew, $\Delta t_1 > \Delta t_2$, on the other hand, will not cause an output error in these devices, but will delay the issuance of the outputs by the amount of the skew $\Delta t_1 > \Delta t_2$. The subject of shift registers will be discussed in detail in Section 12.2. Finally, note that if the configuration indicated in Fig. 11.24 is used as a two-stage

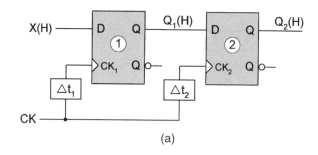

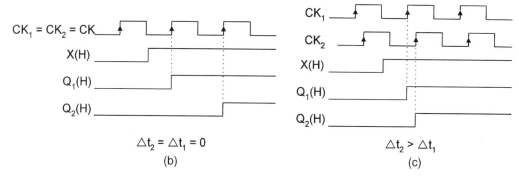

FIGURE 11.24

Clock skew illustrated by using two D flip-flops in series. (a) Logic circuit showing delays on the clock inputs to the two flip-flops. (b) Timing diagram showing correct response of the flip-flop outputs to a data input change if $\Delta t_1 = \Delta t_2 = 0$. (c) Timing diagram showing erroneous output response of flip-flops due to a data change when $\Delta t_2 > \Delta t_1$.

synchronizer under the condition $\Delta t_2 - \Delta t_1 > \tau_{ff}$, little or no protection is provided against metastability.

Another type of clock skew problem can occur when the data are presented in parallel, as depicted in Fig. 11.25. Shown in Fig. 11.25a are two D flip-flops configured in parallel with individual delays to the clock inputs and with identical data inputs $X(H)$ synchronized in phase with the CK input. For the sake of simplicity the inputs are made identical. Under the condition that the delays are equal, $\Delta t_2 = \Delta t_1$, correct output response results, as indicated by the timing diagram in Fig. 11.25b. Here, since $X(H)$ is synchronized to the rising edge of the CK waveform, both outputs, $Q_1(H)$ and $Q_2(H)$, must change simultaneously, one clock period following the data input change. This, however, may not happen if $\Delta t_2 > \Delta t_1$, as indicated by the timing diagram in Fig. 11.25c. In the event that $\Delta t_2 - \Delta t_1 > \tau_{ff}$, $X(H)$ can be picked up by flip-flop 2 one clock period in advance of flip-flop 1 as illustrated. This is a serious and unrecoverable error in the output signals. Note that because the flip-flops are configured in parallel, the preceding discussion applies equally well to the reverse skew $\Delta t_1 > \Delta t_2$.

The clock skew problem demonstrated in Fig. 11.25c supports the need to synchronize the data input $X(H)$ antiphase to the clock triggering edge. In this case the system issues

11.5 CLOCK SKEW

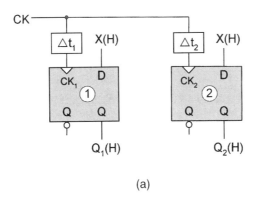

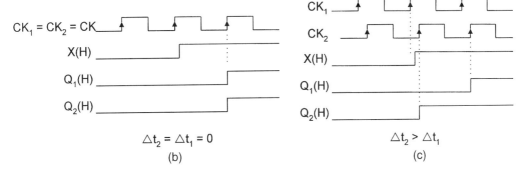

FIGURE 11.25
Clock skew illustrated by using two D flip-flops in parallel. (a) Logic circuit showing delays on the clock inputs to the two flip-flops. (b) Timing diagram showing correct response of the flip-flop outputs to a data input change if $\Delta t_1 = \Delta t_2 = 0$. (c) Timing diagram showing an erroneous output response of flip-flops due to a data change when $\Delta t_2 > \Delta t_1$.

outputs separated by no more than the skew $\Delta t_2 - \Delta t_1$ or $\Delta t_1 - \Delta t_2$. This, of course, can become a problem only if the skew exceeds the tolerable limits permitted by the design specifications.

The elimination of clock skew in simple synchronous FSMs, e.g., in shift registers, is not usually a difficult task. Providing that the clock skew is stable (that is, not time or temperature dependent), one simply balances the delays by using inverter pairs, noninverting drivers, and the like. For high-frequency systems, transmission line delays on leads can be substantial, and this can cause the balancing procedure to become more difficult. In any case, the elimination of clock skew problems can be ensured only if all clock lead delays are symmetric or nearly so.

Clock skew problems are more difficult to diagnose and deal with in very complex systems operated by a system clock that must drive many independent devices at high frequency. Modern VLSI circuits, WSI circuits, and ASICs are good examples. Other examples include the use of FPGAs discussed in Subsection 7.7.3. The best advice that can be given to the designer of such systems is to "think symmetrically" when laying out a circuit or programming the routing paths in FPGAs. Try to avoid obvious sources of asymmetric path delays, particularly those associated with the system clock leads. Often, a conscious effort in this regard can save much time and expense.

11.6 CLOCK SOURCES AND CLOCK SIGNAL SPECIFICATIONS

Various timing problems relative to the clock waveform have been discussed, but no mention has been made of the clock signal source and specifications. How, in fact, is a high-frequency, highly precise clock waveform produced, and how must it be specified so as to perform predictably as the system clock to a synchronous FSM? The answer is not a simple one, but it can be dealt with in semiquantitative terms. First, there must be a reference frequency source, one that has the following desirable characteristics:

High-frequency capability
Frequency stability
Starting reliability
Duty cycle control
Reasonable square-wave output capability

11.6.1 Clock-Generating Circuitry

Shown in Fig. 11.26 are two oscillator circuits that possess characteristics suitable for rather different applications. Figure 11.26a presents an inexpensive self-starting oscillator circuit that is limited to relatively low frequencies that are somewhat adjustable by the RC time constant. This particular oscillator circuit has little or no useful application in modern sophisticated state machine design. The oscillator circuit in Fig. 11.26b is considerably more expensive, but has all of the desirable characteristics mentioned previously except duty cycle control. There are oscillator circuits more and less sophisticated than that shown in Fig. 11.26b. However, all oscillators capable of delivering a stable high frequency within

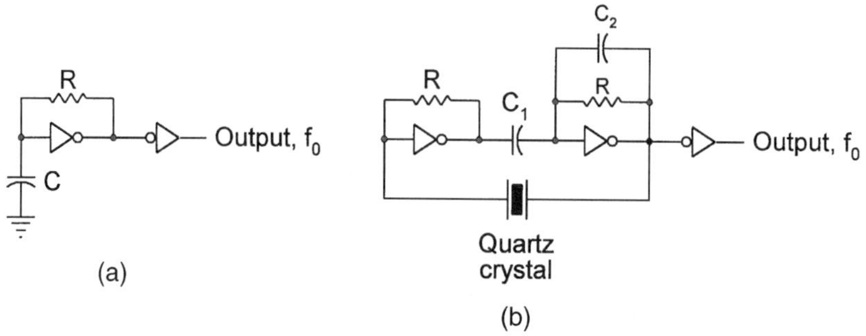

FIGURE 11.26
Example of clock oscillator circuits. (a) A simple, inexpensive, self-starting oscillator circuit that is frequency limited. (b) A high-frequency, crystal-controlled oscillator with good starting capability and frequency stability.

11.6 CLOCK SOURCES AND CLOCK SIGNAL SPECIFICATIONS

0.1% precision will be crystal controlled. Quartz crystals, which can be cut (dimensioned) to oscillate at a specific frequency to a great precision, are an ideal choice for use in a crystal-controlled oscillator. Such specific frequencies can be in the megahertz range.

The *duty cycle* of a clock waveform is defined by the relation

$$Duty\ cycle \equiv \frac{T_{Active}}{T_{CK}} \times 100\ \text{(in percent)}, \qquad (11.4)$$

where T_{Active} is the active portion of a clock cycle and T_{CK} is the clock period, both given in seconds. Thus, a 50% duty cycle means that the active and inactive portions of the clock waveforms are equal. Duty cycle control by an oscillator circuit is important but requires additional circuit elements and raises the cost of the device.

An oscillator, such as that in Fig. 11.26b, provides the reference frequency f_0 that may have to become some multiple of f_0 to achieve the high frequencies required by modern sequential machines. Dividing frequency is easily accomplished by using a counter, as explained later in Section 12.3. However, obtaining an integer multiple of the reference frequency, nf_0, is a much more complex operation. One means of accomplishing this is to use a *phase-locked loop* with a programmable divider in the feedback called a *frequency synthesizer*, the details of which are beyond the scope of this text. Properly designed, the frequency synthesizer will provide all of the desirable oscillator characteristics previously mentioned. Information on this and related subjects can be obtained in references cited in Further Reading at the end of this chapter.

11.6.2 Clock Signal Specifications

At some point in the design of a synchronous FSM, the designer must fashion the digital electronics of the FSM to a given clock frequency or, perhaps less likely, the reverse. In either case, it is necessary to know on what parameters an optimum clock frequency depends. A view of Figs. 10.58 and 10.64, which are typical logic circuits for synchronous FSMs, indicates that the clock period cannot be less than the propagation delay through the flip-flop (including the setup time) plus the delay through the next state-forming logic. In mathematical terms, the minimum clock period is usually evaluated from the maximum system cycle time

$$T_{CK} > \tau_{ff_{max}} + \tau_{ns_{max}} + t_{su_{max}} \qquad (11.5)$$

or

$$T_{CK} = \tau_{ff_{max}} + \tau_{ns_{max}} + t_{su_{max}} + \Delta t_{fs}$$
$$= 1/f_{CK}, \qquad (11.6)$$

where $\tau_{ff_{max}}$ is the maximum flip-flop propagation delay, $\tau_{ns_{max}}$ is the maximum propagation delay through the NS forming logic, $t_{su_{max}}$ is the maximum setup time (defined in Section 10.11), and Δt_{fs} is a *factor of safety*. The factor of safety allows for some variance in the values used for the other parameters and for the possibility of clock skew on clock lines to the flip-flops. The maximum flip-flop propagation delay is determined from the t_{phl} and t_{plh} parameters, as illustrated for an RET D flip-flop in Fig. 11.27. Thus, the average value for τ_{ff} is obtained by introducing the data from Fig. 11.27 into Eq. (6.1), but it is clear that $\tau_{ff_{max}} = \tau_{phl}$ in this case. Normally, the manufactures of the flip-flop devices will provide

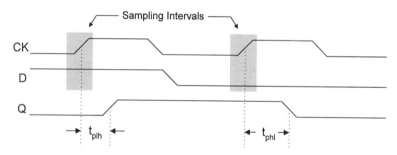

FIGURE 11.27
Propagation delay data for an RET D flip-flop from which the maximum propagation delay can be obtained.

sufficient information to evaluate Eq. (11.6), providing that acceptable values for $\tau_{ns_{max}}$ and Δt_{fs} are used. The value for $\tau_{ns_{max}}$ must be obtained with knowledge of the NS logic technology, which is usually available from the manufacturer. An acceptable value for Δt_{fs} might be 20% of $(\tau_{ff} + \tau_{ns} + t_{su})_{max}$, giving $T_{CK} = 1.2(\tau_{ff} + \tau_{ns} + t_{su})_{max}$ as a safe minimum clock period. Then from this, a safe maximum clock frequency $f_{CK} = 1/T_{CK}$ can be obtained.

11.6.3 Buffering and Gating the Clock

There are other considerations regarding the use of a clock waveform signal in a synchronous system. Normally, the system clock signal from the clock-generating circuitry must be buffered for fan-out reasons. As used in this text, the word *buffer* is synonymous with *line driver*. In large systems where the clock signals must be supplied to a large number of flip-flops, there may be insufficient fan-out to drive the flip-flops. In this case buffers must be used but in a way that does not cause clock skew. The best way to buffer the clock signals is to use packaged IC buffers (as opposed to individual buffers or inverters off chip) and to do so "symmetrically" to minimize clock skew.

Also, if it is necessary to gate the clock signals in addition to buffering them, the best choice may be to use tri-state drivers (see Fig. 3.8 for CMOS tri-state drivers). If it is necessary to gate some clock signals but not others, an asymmetric delay may result that can produce clock skew problems. The solution to this potential problem is to place a delay on each nongated clock line that is equal to the delay of the tri-state driver — again, think symmetrically. Generally, it is a bad idea to gate the clock signals by using discrete gates. To do this invites clock skew problems. If logic gates must be used for the gating action, use ICs and make certain that all delays on clock lines are equal or nearly so — once again, think symmetrically.

11.7 INITIALIZATION AND RESET OF THE FSM: SANITY CIRCUITS

An important part of the operation of any sequential machine is that it be initialized (on power-up) into a specific state, or that it be reset into a specific state once in operation. If initialization and reset of the FSMs were not possible, one can imagine the chaos that could

11.7 INITIALIZATION AND RESET OF THE FSM: SANITY CIRCUITS

result. Take, for example, the cruise control of an automobile. Failure of it to initialize or reset into a startup state could be disastrous. Imagine not being able to initialize or reset the controller of one's computer. Equally important, no FSM should ever be designed such that it can initialize or reset into a "hang" state or subroutine that is not part of the intended sequence. Whether the FSM is the controller for an elevator or traffic light system, or the controller for a robotics or audio playback system, it should be obvious that initialization and reset capabilities are vitally important.

11.7.1 Sanity Circuits

What is needed for initialization and reset of the FSM is a signal that can be used to drive an FSM momentarily into a specific starting or reference state whenever it is necessary to do so — that is, during power-up to initialize the FSM or during a reset operation. Shown in Fig. 11.28 is a *sanity* circuit as it is used to power up or reset a three-bit D flip-flop memory into the 001 state. It is called a sanity circuit because it adds "sanity" to a situation that could otherwise be chaotic (insane) for the designer. The need for initialization and reset was established in Subsection 10.12.2 by the missing state analysis in Fig. 10.65 following the design of a sequence recognizer.

In Fig. 11.28 notice that *Sanity(L)* is connected to an active low asynchronous clear (CL) override to initialize or reset a logic 0, but is connected to an active low asynchronous preset (PR) override to initialize or reset a logic 1. It is important to observe that only one active low asynchronous override per flip-flop can be connected to a sanity line and that all others must be connected to 0(L) for normal operation of the flip-flop. The reader should review the subject of asynchronous preset and clear overrides in Section 10.10 before proceeding further on this subject.

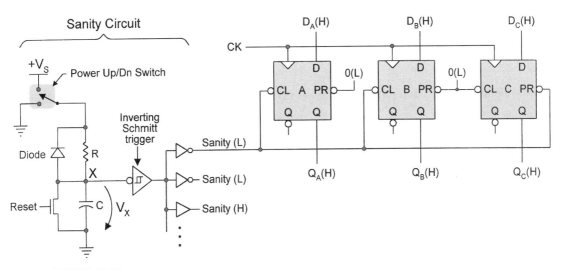

FIGURE 11.28
Sanity circuit and proper connections required to initialize or reset a three-bit memory into the 001 state.

To understand how the sanity circuit works, it is necessary to first focus attention on the R–C component of the circuit in Fig. 11.28. Students in electrical engineering will recognize this as a first-order R-C circuit, where RC is called the *time constant* of the circuit. With Reset at LV and neglecting the influence of the Schmitt trigger in Fig. 11.28, the following approximations for V_X (voltage at node X) result:

On *Power Up* at time $t = 0$ (power Up/Dn switch connected to the supply voltage $+V_S$ with Reset set to LV):

$$V_X(t) \cong V_S\{1 - e^{-t/RC}\} \tag{11.7}$$

$$@t = 0^+ \quad V_X \cong 0V, \text{ therefore } X = 0(H) = 1(L) = Sanity(L)$$

$$@t \cong 5RC \quad V_X \cong V_S, \text{ therefore } X = 1(H) = 0(L) = Sanity(L)$$

On *Power Down or Reset* at time $t = 0$ (power Up/Dn switch connected to ground or Reset is set to HV, one or the other):

$$@t = 0^+ \quad V_X \cong V_S, \text{ therefore } X = 1(H) = 0(L) = Sanity(L)$$

$$@t > 0 \quad V_X \cong 0V, \text{ therefore } X = 0(H) = 1(L) = Sanity(L)$$

On power up with the Reset input to the NMOS at LV, the capacitor is charged through the resistance R since the diode is nonconducting in reverse bias. (Recall that a diode conducts in forward bias only, the direction of the "arrow" in its symbol.) The result is Eq. (11.7) for the approximate time dependent rise of voltage at node X. In effect, it is these voltage values at node X that are presented to the preset and clear overrides of the flip-flops during initialization or reset of the memory flip-flops, as in Figure 11.28. In the logic domain this means that $X(L) = Sanity(L)$. Notice that power down or reset is abrupt with no significant exponential decay in voltage at node X. This is so because the capacitor is discharged to ground either through the diode on power down or through the NMOS switch at reset. In either case the discharge of the capacitor is extremely rapid. An abrupt power down or reset is important during short power interruptions so as to ensure that proper initialization of the flip-flops occurs during the power recovery event.

Thus, for a short period of time, say $<3RC$, each flip-flop is initialized to either a logic 0 or a logic 1 via the $Sanity(L)$ input to its active low asynchronous clear or preset override. Then, beyond a period of about $5RC$ all flip-flops are free to function normally since their active low preset and clear overrides are now at 0(L). Typical values for RC may range from the millisecond to the microsecond range by adjusting the values for R (in ohms) and C (in farads). Values of the time constant that are too short may fail to properly initialize the flip-flops in the memory, and values too large may cause unnecessary delays in the initialization process. Therefore, it is worthwhile for the designer to match the RC time constant to the logic family of the flip-flop memory. Note that $Sanity(H)$ signals (see Fig. 11.28) are useful in initializing asynchronous FSMs as described later in Section 14.11.

The results of the foregoing discussion are illustrated in Fig. 11.29. Here, V_{pu} is the power-up switching threshold of the Schmitt trigger, and V_{pd} is the power-down switching threshold. Unlike an inverter whose upward and downward bound switching thresholds are

11.7 INITIALIZATION AND RESET OF THE FSM: SANITY CIRCUITS 525

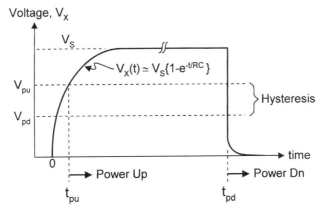

FIGURE 11.29
$V-t$ characteristic at node X for the sanity circuit in Fig. 11.28 showing power-up (V_{pu}) and power-down (V_{pd}) switching thresholds and hysteresis effect of the Schmitt trigger.

the same at about mid-supply, the Schmitt trigger exhibits a hysterisis effect illustrated in Fig. 11.29 and discussed in the following paragraph. Clearly, power up can occur before V_X reaches the supply level, V_S, and that is permissible provided that t_{pu} is sufficient time for all flip-flops in the memory to be initialized. Proper choice of the RC time constant would satisfy this requirement.

The Schmitt trigger has three important characteristics that make it an ideal choice for use in a sanity circuit. It has good fan-out capability, abrupt triggering, and the ability to reject unwanted signals, a feature called *noise immunity*. These characteristics are best understood by an inspection of its CMOS implementation and its I/O voltage waveforms shown in Fig. 11.30. The configuration of the CMOS circuit in Fig. 11.30a is that of an inverter with double NMOS and PMOS transistors for improved fan-out (compared to a simple inverter), and for feedback purposes. The transistors M_P and M_N supply the feedback voltages V_{FP} and V_{FN} necessary to cause the output from the Schmitt trigger to change abruptly following triggering at the dual thresholds, V_{pu} and V_{pd}, respectively. The I/O voltage waveforms in Fig. 11.30b illustrate these characteristics. The input waveform for V_{Xin} shows that slow changing voltage ramps become abrupt changes in the output waveform V_{Xout}. The hysteresis effect shown in Fig. 11.30b corresponds to that in Fig. 11.29 and is expressed as the difference $V_{pu} - V_{pd}$. Both the abruptly changing output waveform and the hysteresis are due mainly to the internal feedback. Note that input line noise of amplitude less than that of the hysteresis is rejected in the output signal, a feature that can produce clean, noise-free outputs from the Schmitt trigger.

There are variations on the theme for implementing a sanity circuit. For example, the Schmitt trigger in Fig. 11.28 can be replaced by an odd number of inverters. The problem with this arrangement is that the inverters, which have a hysteresis of approximately zero, have virtually no noise immunity and they do not switch abruptly. Another variation of the sanity circuit is to replace the electronic NMOS Reset switch with a mechanical switch or eliminate the reset feature altogether. Or alternatively, one or more external Master Reset lines can be introduced to node X in lieu of or in parallel to the Reset switch, electronic or

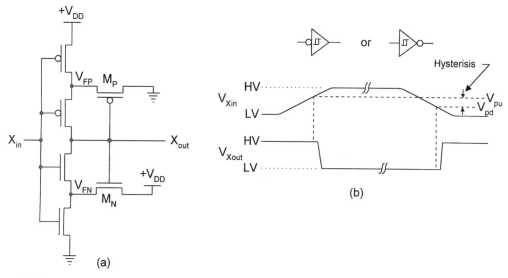

FIGURE 11.30
The inverting Schmitt trigger. (a) CMOS implementation showing feedback voltages. (b) Logic symbols, and input and output voltage waveforms showing hysteresis.

mechanical. In any case, the bank of parallel inverters and/or buffers shown in Fig. 11.28 must be used to meet fan-out needs.

11.8 SWITCH DEBOUNCING CIRCUITS

A common problem in digital system design is to provide human interface to the system. The use of push-button switches is a typical example. Asynchronous input signals from push-button switches often produce a phenomenon called *switch bounce* that derives from the mechanical structure of the switch and the physical nature of the contact surfaces. Multiple open/close transitions may occur immediately following the depression or release of a button switch, or any mechanical switch for that matter. Serious problems can result in an FSM if a high-frequency clock catches the bounce signals produced by a mechanical switch. This is equivalent to the introduction of false data.

11.8.1 The Single-Pole/Single-Throw Switch

Shown in Fig. 11.31a is a simple normally closed single-pole/single-throw (SPST) mechanical switch, and in Fig. 11.31b the contact noise (bounce) that occurs as a result of opening or closing the switch. Unfortunately, there is no solution to the problem of debouncing a SPST switch other than to provide a delay greater than the bounce periods, Δt_B. This can be accomplished by using an RC circuit of the type shown in Fig. 11.31c. Here, the delays are determined by the choice of the R and C components together with the hysteresis effect of the inverting Schmitt trigger as in Figs. 11.29 and 11.30. The voltage $V_X(t)$ across the

11.8 SWITCH DEBOUNCING CIRCUITS

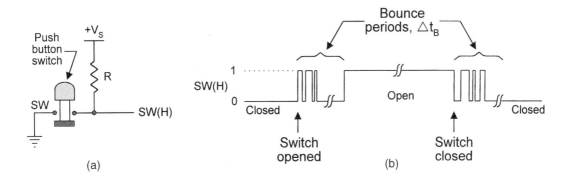

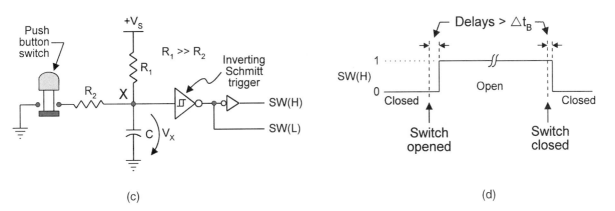

FIGURE 11.31
Debouncing the normally closed single-pole/single-throw (SPST) mechanical switch. (a) A non-debounced SPST switch. (b) Timing diagram showing logic bounce periods for the switch in (a). (c) A possible debouncing circuit for the SPST switch. (d) Bounce-free timing diagram for the debounced SPST switch.

capacitor depends on whether the switch is opened or closed and is given approximately by

$$\text{Switch opened} \quad @t = 0 \quad V_X(t) \cong V_S\{1 - e^{-t/R_1 C}\} \quad (11.8)$$

$$\text{Switch Closed} \quad @t = 0 \quad V_X(t) \cong V_S\{e^{-t/R_2 C}\}, \quad (11.9)$$

where Eq. (11.9) expresses an exponential decay that depends on an $R_2 C$ time constant. The value of R_2 is chosen such that $R_2 \ll R_1$, a requirement for rapid discharge of the capacitor following switch closure.

The result of the debouncing action is illustrated in Fig. 11.31d, where the delay periods are indicated to be greater than the bounce periods Δt_B, as they must be. The circuit delay on switch closure will be less than that for opening the switch since $R_2 \ll R_1$. Depending on how the SPST switch is to be used, the $R_1 C$ and $R_2 C$ time constants must be chosen so that the circuit delays are always greater that the worst-case bounce periods for opening

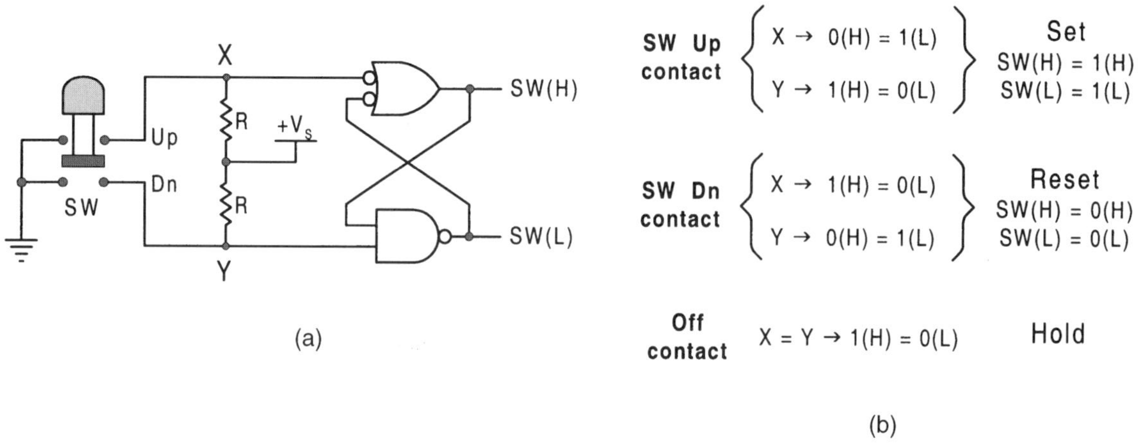

FIGURE 11.32
Debouncing the single-pole/double-throw (SPDT) switch by using a set-dominant basic cell. (a) Logic circuit. (b) Logic values for Up-, Dn-, and Off-contact positions of switch, SW.

and closing the switch, respectively. Thus, it is important that R_2 not be chosen too small — certainly not zero if closing the switch can affect the behavior of an FSM. On the other hand, if closure of the SPST switch can have no effect on the behavior of an FSM, R_2 can be set to zero.

11.8.2 The Single-Pole/Double-Throw Switch

Unlike the single-pole/single-throw switch just discussed, the single-pole/double-throw (SPDT) switch can be debounced very easily and precisely by using a basic cell. Shown in Fig. 11.32a is the debouncing circuit for a SPDT switch. Notice that when the switch button is in the up contact position the basic cell is set, and when it is in the down contact position it is reset as indicated in Fig. 11.32b. Furthermore, in an off-contact position the basic cell is forced to hold the previous mixed-rail output (see Fig. 10.16). What this means is that the first contact bounce to cross the switching threshold of the basic cell on an Up or Dn position of the switch will set or reset the basic cell, respectively. All subsequent bounces are ignored. That is, any contact bounce that is produced following the first can do nothing but hold the basic cell in either a set or reset condition. The set-dominant basic cell in Fig. 11.32a can be replaced by a reset-dominant basic cell if the $+V_S$ and ground terminals are interchanged. The interchange is necessary to maintain the off-contact hold requirement.

The debouncing arrangement in Fig. 11.32 can be used with most any CMOS family, but there is a relatively high price tag for this type of circuit. For low-budget needs, a simpler configuration can suffice under certain conditions. Shown in Fig. 11.33 is a simple debouncing circuit for the SPDT switch consisting of two cross-coupled inverters and buffers. The circuit functions somewhat the same as that in Fig. 11.32 with one major difference. Upon switching from the Up contact position to the Dn contact position, or vice

11.8 SWITCH DEBOUNCING CIRCUITS

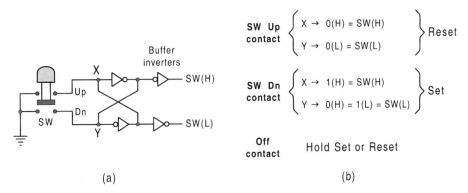

FIGURE 11.33
A simple, low-budget SPDT switch debouncing circuit for low-speed CMOS. (a) Logic circuit with buffer inverters. (b) Logic values for Up-, Dn-, and Off-positions of SW.

versa, there is a short time (approximately the path delay of the feedback loop consisting of the two inverters) during which HV is shorted to ground. This can give rise to switching transients that can cause problems in the FSM to which the debouncing circuit is connected. Furthermore, the relatively high power drain during these periods may or may not be acceptable. For best results the cross-coupled inverters should be implemented with CMOS that will *not* source high current in the active high state. The 74SL04 CMOS inverter appears to be a good choice for this purpose.

As a final thought, not all mechanical switch inputs need to be debounced. Switched inputs that are set prior to the initiation of a sequential process need not be debounced provided that the resulting switch signal is stable at the time the sequential process is to begin. Examples are the so-called *DIP switches* in computers that are preset when the computer is not in operation. The design of the one- to three-pulse generator in Section 11.9 illustrates the difference in dealing with preset switches as opposed to those that are not.

11.8.3 The Rotary Selector Switch

A variation on the theme of Fig. 11.32 can be applied to the debouncing of a four-post rotary selector switch shown in Fig. 11.34. Here, each NAND gate receives a feedback line from each of the other three NAND gates but not from itself, and each set-dominant basic cell serves basically the same purpose as in the debouncing of the SPDT switch. Together, the basic cells and the feedback inputs to the NAND gates permit the output logic levels for all switches to be maintained during an off-contact bounce. The first selector–post contact that crosses the switching threshold of the basic cell sets that switch and resets the other switches via the feedback paths. All subsequent bounces cause the inputs to the basic cell of that switch to fluctuate between the set and hold conditions. The resetting of the other basic cells occurs after about one gate delay following the first threshold contact.

There are alternative means of debouncing a rotary switch that may or may not be recommendable depending on how the rotary switch is to be used. One alternative for

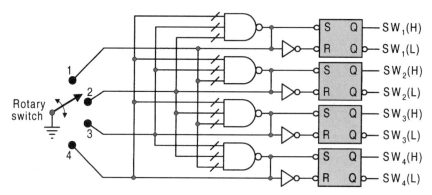

FIGURE 11.34
Debouncing a rotary select switch with four posts.

low-budget needs is to replace each basic cell in Fig. 11.34 with cross-coupled inverters as in Fig. 11.33a. Another alternative applies to the case of a large number of posts where fast throughput of the input signal is important and where fan-in limitations become a problem when using CMOS NAND gates. Here, the debouncing circuit in Fig. 11.34 can best be implemented with NOR gates, reset-dominant basic cells and with the rotary switch ground replaced by the supply voltage, $+V_S$. In this case the NOR gates can be configured as in Fig. 8.46 with no fan-in limitations. Note that it is not recommended that the basic cells be removed in any of these debouncing circuits. To do this would allow bounce transients to occur over a period of at least two gate delays before the circuit stabalizes — there are no RC components present that can produce delays to outlast the bounce periods.

11.9 APPLICATIONS TO THE DESIGN OF MORE COMPLEX STATE MACHINES

The design of FSMs in Section 10.12 was limited to relatively simple state machines for which few problems existed. However, the design of the sequence recognizer in Subsection 10.12.2 did point to the need for initialization, one of several design considerations covered in this chapter. Now, it is necessary to move on to more complex FSMs so as to apply some of these design considerations.

11.9.1 Design Procedure

For reference purposes, a seven-part design procedure is presented here. Although not every design consideration is included, the procedure is complete enough to serve as a guideline for most FSM designs. This procedure is intended to be an augmentation of the three-step procedure given in Section 10.6 and should be used in a manner dictated by the nature and complexity of the design project. For example, only portions of this procedure need be used for the design of relatively simple FSMs. On the other hand, very complex design problems might require going beyond the coverage of this procedure. The reader should

11.9 APPLICATIONS TO THE DESIGN OF MORE COMPLEX STATE MACHINES

review the contents Section 10.6, in particular the *mapping algorithm*, before continuing in this section.

Part I. Understand the Problem

1. Develop a thorough understanding of the functional requirements and I/O specifications of the FSM to be designed. The construction of block diagrams can be helpful in this regard.
2. Note any specific timing constraints that must be met. Not all information regarding timing constraints and timing problems may be apparent initially and may have to be gathered as the design proceeds.

Part II. Construct a State Diagram

1. Choose a model (e.g., a Moore or Mealy model) and construct a fully documented state diagram that meets the requirements of the algorithm and timing constraints of the FSM. Use flowcharts and timing diagrams if necessary. Several attempts at constructing a state diagram may be necessary in obtaining the one best suited to the design.

   The use of algorithmic state machine (ASM) charts and state tables can be very useful in arriving at a suitable state diagram. Section 11.10 discusses ASM chart nomenclature and the use of state tables together with their relationship to the state diagram and to a hardware description language such as VHDL.

2. If asynchronous inputs are present, make certain that the branching dependency and conditional output rules, given in Subsection 11.4.1, are obeyed. Decide at this point if any or all of the asynchronous inputs are to be synchronized—usually, they will have to be synchronized.

Part III. Obtain the Output Functions

1. Choose the NS and output logic hardware and memory devices to be used and then obtain the output functions. Knowing how the output functions are to be implemented and the character of the flip-flops to be used can influence the design strategy with regard to static hazards in the output, as discussed in Section 11.3.
2. If logic noise is determined to be a problem in the output signals of the FSM, corrective action must be taken.
   (a) If output race glitches (ORGs) are present, eliminate them by using one or more of the methods considered in Subsection 11.2.2.
   (b) If static hazards exist in the output functions, eliminate them by adding hazard cover to the output functions as discussed in Section 11.3, or use the filtering method illustrated in Fig. 11.7. If one or more of the s-hazards are of the internally initiated type, a proper choice of flip-flops can be used to eliminate them, as indicated in Subsection 11.3.2.

Part IV. Obtain the Next-State Functions Plot the NS K-maps by using the *mapping algorithm* given in Section 10.6 and then extract minimum or reduced cover for the NS functions. Implement these results by using discrete logic as in Section 10.12, or by using a PLD as discussed later in Section 13.2. If a ROM is to be used to implement the NS and output logic, program the ROM directly from the state diagram (also discussed in Section 13.2). If a shift register or counter is used as the memory, obtain the NS logic according to the procedure discussed later in Sections 13.3 and 13.4. For one-hot designs, the NS and output logic function are read directly from the state diagram or ASM chart (see Section 13.5).

Part V. Select the Circuits Necessary for I/O Conditioning and Initialization/Reset

1. Select the appropriate input debouncing and synchronizing/stretching circuits and the output filtering hardware to be used, if any. Refer to Sections 11.2, 11.4, and 11.8 for assistance in making the proper choice. Make certain that all timing requirements are met.
2. Select the initialization/reset hardware (sanity circuit) appropriate for the design as discussed in Section 11.7.

Part VI. Construct the Logic Circuit Construct a complete logic circuit of the FSM, preferably in mixed-logic notation, and make any necessary comments for future reference. Avoid the use of unusual logic symbols unless accompanied by appropriate labels.

Part VII. Test the Logic Circuit Simulate the logic circuit to ascertain whether or not it operates correctly in the logic domain. Use both high (gate) level and low (SPICE) level simulations, in that order, if possible. The final test, of course, is that in real time by using testing equipment such as a waveform analyzer.

11.9.2 Design Example: The One- to Three-Pulse Generator

The problem is to design a pulse generator that will issue one, two, or three clean, discrete pulses or no pulses depending on the settings of two switches, SW_1 and SW_0. A general description of the pulse generator is provided by the block diagram in Fig. 11.35a and by the operation table in Fig. 11.35b. It is required that each pulse issued by this FSM be of the same active duration as clock, and that the two switches SW_1 and SW_0 be preset well in advance of the start switch S command. Thus, SW_1 and SW_0 need not be debounced or synchronized. It is also required that these two preset switches remain fixed at their proper logic level for a period of time exceeding that required for a pulse generating sequence. It is further required that the pulse generator be initialized into a non-output state, and that the start signal S be returned to the inactive condition following a pulse-generating sequence and before initiating another pulse sequence.

The switch inputs S, SW_1, and SW_0 are asynchronous inputs. However, only the start switch S is required to be debounced and synchronized. It is best to synchronize S antiphase to the FSM memory, which is arbitrarily chosen to be FET flip-flops, as indicated in Fig. 11.35a. The timing diagram in Fig. 11.35c illustrates the operation of the pulse generator by showing one- and three-pulse generations in agreement with the operation table in Fig. 11.35b. Notice that the first pulse of a sequence is issued with the next active portion of the clock following an active Start command and the sampling of the preset switch logic levels.

11.9 APPLICATIONS TO THE DESIGN OF MORE COMPLEX STATE MACHINES   533

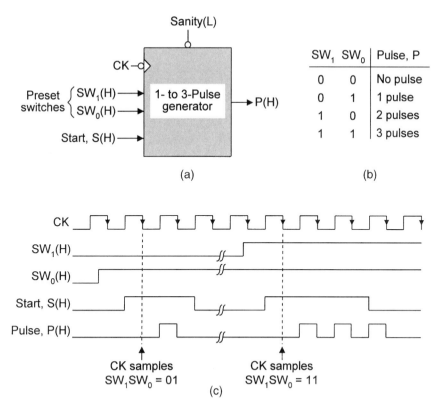

FIGURE 11.35
Description and operation of the one- to three-pulse generator. (a) Block diagram. (b) Operation table. (c) Timing diagram showing one- and three-pulse generations.

The flowchart that satisfies the algorithm and timing requirements of the one- to three-pulse generator is provided in Fig. 11.36a. The flow chart is a "thinking tool" that is used in connection with the operation table to assist in the construction of the state diagram shown in Fig. 11.36b. The shaded action squares in the flow chart are the same as the oval state symbols in a state diagram. Notice how much more vividly the sequential behavior is represented by the state diagram than by the flow chart. The state diagram has five states that require the use of three state variables named $Q_A Q_B Q_C = ABC$. Each state is seen to satisfy the sum rule given by Eq. (10.3) and illustrated in Fig. 10.7.

The next step in the design of the one- to three-pulse generator is to run an ORG analysis followed by a hazard analysis. This is done in Fig. 11.37, where it is seen that no ORG exists if ϕ_2 is taken to be either 0 or CK in the K-map for P. Keep in mind that none of the outputs P if CK are issued immediately on entrance into a given state, since each state-to-state transition occurs on the falling edge of CK, that is, on $\overline{CK}$. Note that if $\phi_2 = 1$ in the P K-map, an ORG is possible via the $000 \rightarrow 010 \rightarrow 011$ race path. Finally, since there are no coupled terms in the output function $P = A(CK) + B(\overline{CK})$, there are no static hazards possible. From these results it is concluded that there are no restrictions placed on

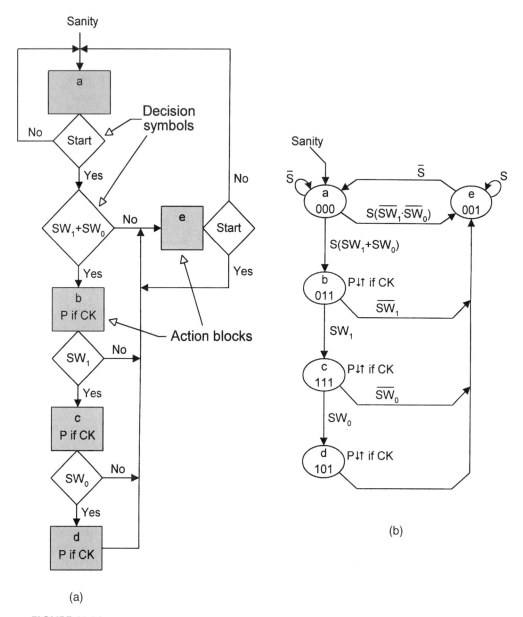

FIGURE 11.36
Sequential description of the one- to three-pulse generator. (a) Flow chart. (b) State diagram derived from the flow chart showing a suitable state code assignment.

the technology of the memory flip-flops, as discussed in Section 11.3, and that no filtering of the output signal is necessary.

Having run the ORG and hazard analyses, all that remains is to map the NS functions, extract minimum cover, and construct the logic circuit with the appropriate input conditioning

11.9 APPLICATIONS TO THE DESIGN OF MORE COMPLEX STATE MACHINES 535

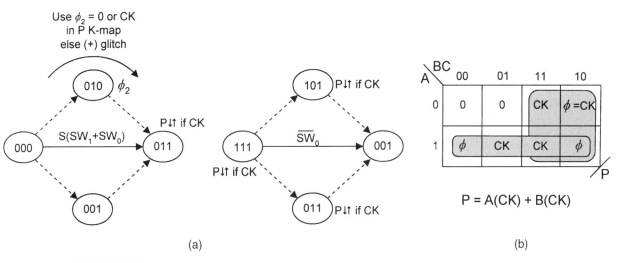

FIGURE 11.37
(a) Output race glitch (ORG) analysis showing conditions for a race-glitch-free output. (b) The output K-map and the minimum hazard-free output function.

and initialization circuits. Presented in Fig. 11.38 are the NS K-maps for one- to three-pulse generator of Fig. 11.36 assuming the use of D flip-flops. The resulting minimum NS and output functions are easily seen to be

$$\begin{cases} D_A = \bar{A}B(SW_1) + AB(SW_0) \\ D_B = \bar{A}B(SW_1) + \bar{C}S(SW_1) + \bar{C}S(SW_0) \\ D_C = S + A + B \\ P = A(CK) + B(CK) \end{cases}, \qquad (11.10)$$

which represent a total gate/input tally of 10/26 in two-level logic. Notice the shared PI, $\bar{A}B(SW_1)$, in the expressions for D_A and D_B.

Implementation of Eqs. (11.10) by using discrete logic is shown in Fig. 11.39. Also shown are the debouncing, synchronizing and initialization (sanity) circuits. Notice that

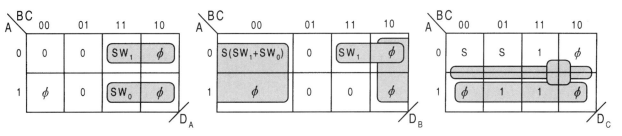

FIGURE 11.38
Next state K-maps showing minimum cover for the one- to three-pulse generator of Fig. 11.36b.

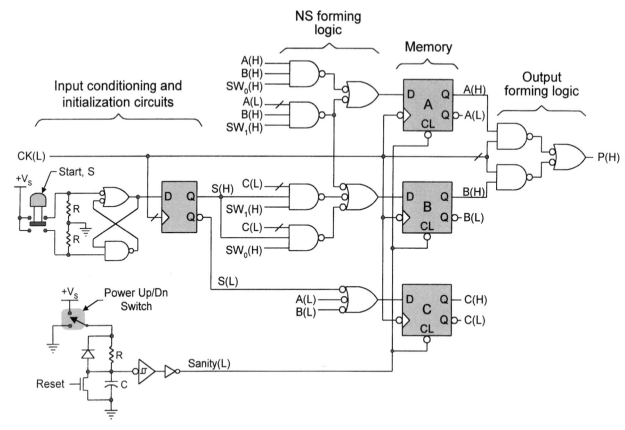

FIGURE 11.39
Implementation of Eqs. (11.10) for the one- to three-pulse generator showing the debouncing, synchronizing, and initialization circuits.

Sanity(L) is connected to the active low asynchronous clear overrides of the flip-flops for initialization or reset into the 000 state. Also note that only one RET D flip-flop is used for the synchronizing stage of the start switch input, S. A more robust synchronizing stage would be one such as that illustrated in Fig. 11.23a.

Although discrete two-level logic is used in Fig. 11.39 for the NS- and output-forming logic, there exists a variety of other alternatives. These include the use of MUXs, decoders, and array logic devices. Also, the memory can be made up of other devices such as shift registers and counters. These and other alternative approaches to FSM design are considered in Chapter 13.

11.10 ALGORITHMIC STATE MACHINE CHARTS AND STATE TABLES

In Subsection 11.9.1 a design procedure is laid out followed by the design of a relatively simple FSM, the one- to three-pulse generator. In this design procedure, it is implied that the final goal in describing the sequential behavior of a state machine is to arrive at a suitable

11.10 ALGORITHMIC STATE MACHINE CHARTS AND STATE TABLES

state diagram from which the design can be completed following the indicated procedure. In this text, the fully documented state diagram is the easiest to work with in carrying out the design of relatively simple FSMs. There are, however, other useful means of expressing the sequential behavior of state machines. These other means include the use of *algorithmic state machine (ASM)* charts and *state tables*. The use of these as an aid in constructing the state diagram will now be explored.

11.10.1 ASM Charts

Just as the flowchart functions as a useful thinking tool in the construction of the state diagram, so also does the ASM chart serve as a useful thinking tool. In fact, the two are very similar, the ASM chart being the more useful in creating VHDL FSM descriptions. Shown in Fig. 11.40 are the symbols used in the construction of ASM charts. The state block symbol in Fig. 11.40a is used to give the state identifier, the state code assignment (if known), and a listing of all unconditional (Moore) outputs associated with that state. The decision symbol in Fig. 11.40b contains the input conditions on which depend the branching from a given state. To assist in creating a VHDL description of the FSM, a separate symbol is provided for conditional (Mealy) outputs, as indicated in Fig. 11.40c. A conventional flow chart representation would combine all outputs, unconditional and conditional, into the state block symbol, as was done in Fig. 11.36a, where only Mealy outputs exist. The entry path to the conditional output symbol of Fig. 11.40c is always from a decision symbol, but its exit path can be either to a state block symbol or to another decision symbol. Notice that in comparing flowchart and ASM chart notation, the following interchangeability of symbols applies:

$$1 \longleftrightarrow \text{True} \longleftrightarrow \text{Yes}$$
$$0 \longleftrightarrow \text{False} \longleftrightarrow \text{No}$$

All of these are based on positive logic, as is true for any logic graphic, including state diagrams.

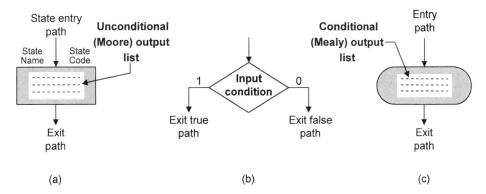

FIGURE 11.40
Traditional ASM chart symbology. (a) State block symbol and list of unconditional (Moore) outputs. (b) Decision symbol showing true and false exit condition paths. (c) Conditional output symbol and list of Mealy outputs.

538 CHAPTER 11 / SYNCHRONOUS FSM DESIGN CONSIDERATIONS

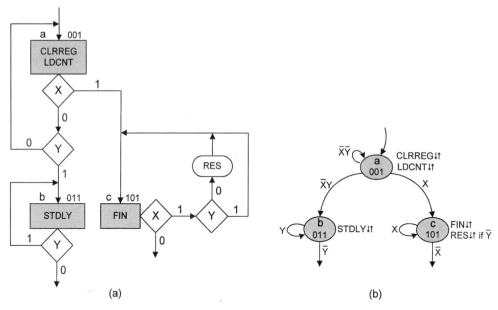

FIGURE 11.41
ASM chart and state diagram for a resolver configuration having two inputs and having both conditional and unconditional outputs. (a) ASM chart by using symbols in Fig. 11.40. (b) The equivalent fully documented state diagram for the ASM chart in (a).

An example of the application of ASM chart notation is presented in Fig. 11.41a, together with its state diagram equivalent in Fig. 11.41b. This is seen to be a three-state resolver segment having two inputs, X and Y, and both conditional (Mealy) and unconditional (Moore) outputs. Notice the manner in which the conditional output *RES* must be represented in the ASM chart. A conventional flowchart representation would have combined the conditional output *RES*, indicated as *RES* if $\bar{Y}$, with the unconditional output *FIN* in state c. Notice also how much easier it is to read the state diagram than the ASM chart. Imagine how difficult it would be to obtain the NS-forming logic by using the ASM chart. Clearly, the fully documented state diagram is much more suitable for this purpose. This leads to the following guidelines regarding the use of ASM charts vs the use of state diagrams:

> For state machines of up to moderate complexity, the ASM chart, like the flow chart, should be used as a "thinking tool" in the construction of a fully documented state diagram. Extracting the NS- and output-forming logic from fully documented state diagrams is much simpler for such FSMs than the use of ASM charts. It is rare in modern times that ASM charts are used in the design of state machines. Rather, it is more likely that experienced designers will design modern complex state machines by using a high-level description language such as VHDL or Verilog. The ASM chart or flow chart can be useful in obtaining a VHDL or Verilog description of a given state machine, but will not likely be used to design it. The one notable exeption to this is the use of the one-hot code method in state machine design, as described later in Section 13.5. There, the ASM chart is shown to be useful in writing the NS and output functions directly from the ASM chart without the need for K-maps or minimization algorithms.

11.10 ALGORITHMIC STATE MACHINE CHARTS AND STATE TABLES 539

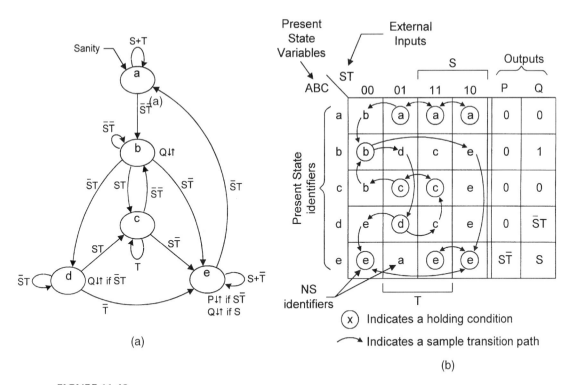

FIGURE 11.42
Representation of a fictitious five-state FSM having two external inputs and two outputs. (a) State diagram representation. (b) The equivalent state table for the FSM in (a).

11.10.2 State Tables and State Assignment Rules

The tabular representation of the state diagram is called the *state table*, or *next state table* if output data is excluded. Shown in Fig. 11.42 are two representations for a Mealy FSM having two inputs S and T, and two outputs P and Q. The state diagram for this FSM, lacking only a suitable state code assignment, is given in Fig. 11.42a, and its equivalent state table representation is presented in Fig. 11.42b. In both representations, literals (a, b, c, d, e) are used for state identification. On the vertical axis of the state table they represent the present state (PS), and within the state table they represent the next state (NS). The encircled state identifiers indicate a holding condition for which PS = NS. Thus, in state a the FSM must hold on input condition $S + T$, so the identifier a is encircled in row a for ST input values 01, 11, and 10, meaning $\bar{S}T + ST + S\bar{T} = S + T$. The state identifiers that are not encircled in the state table represent unstable conditions. For instance, in state a under holding condition $\bar{S}T$, a transition to state b takes place if input T changes $1 \to 0$, as indicated by the two transition paths. Or in state b, holding on $\bar{S}\bar{T}$, a transition to state e will occur if input S changes $0 \to 1$. The FSM cannot transit from state b to state c without changing both inputs simultaneously, a condition that should be avoided if possible. Clearly, the state table presents all features of the state diagram and is, therefore, the tabular equivalent of the state diagram or ASM graphic representation. But the sequential behavior of the FSM is much more easily grasped from the state diagram than from the state table. Furthermore, given a suitable state code assignment, it should be obvious that the state diagram is far easier to use for a "pencil-and-paper" design of an FSM than is the state table. There are, however, several important usages of state tables, among them being their use for CAD purposes explored in Section 11.11.

The state table provides a relatively simple means of obtaining the state code assignments required for the optimum or near-optimum NS and output logic of an FSM by using D flip-flops as the memory. There are three state assignment rules by which this can achieved, listed in *descending order* of priority:

Rule 1 (The "into rule"): Make logically adjacent assignments to present states that branch "into" a common next state, provided that their input conditions are the same.

Rule 2 (The "from rule"): Make logically adjacent assignments to states that are the next states "from" a common present state, provided that their input conditions are logically adjacent.

Rule 3 (The output rule): Make logically adjacent assignments to states having the same outputs. Rule 3 is relatively unimportant except where large numbers of outputs are involved.

In Fig. 11.43a use is made of the next state table in applying rules 1 and 2 to the FSM of Fig. 11.42. Here, rule 1 has the highest priority and is applied to state adjacency sets in columns under constant input conditions. Thus, by rule 1, states within the set $\{abc\}$ should be made logically adjacent, and those within set $\{de\}$ should be made adjacent, both sets being under the same input condition $I_0 = \bar{S}\bar{T}$. Similarly, states within sets $\{ae\}$ and $\{bd\}$, under input condition $I_1 = \bar{S}T$, should be made logically adjacent, etc. Rule 2, of lesser priority, is applied to the rows of the state table as indicated in Fig. 11.43a. Now the input conditions must be logically adjacent. For example, in present state d, states with sets $\{de\}$, $\{cd\}$, and $\{ce\}$ should be made logically adjacent. State sets that appear in both rule 1 and rule 2 are given the highest priority and are indicated in dashed boxes. These are followed in priority by those that appear only in rule 1. Those of least priority appear only in rule 2. Notice that not all sets appearing in rules 1 and 2 can be accommodated, hence the reason to prioritize, as just discussed. For example, it is not possible to include set $\{ce\}$ together with the higher priority sets.

By incorporating rules 1 and 2, as indicated in Fig. 11.43a, there results the following three-bit state assignments:

$$a = 000, \quad b = 001, \quad c = 011, \quad d = 101, \quad \text{and} \quad e = 100.$$

These assignments are used in the state diagram of Fig. 11.43b and will generate an optimum or near-optimum set of next-state functions, but only in three bits. It is possible, albeit unlikely, that a four-bit set could result in a more optimum set of next-state functions. However, no attempt will be made to explore this possibility. Note that ORGs are possible in both P and Q.

The NS K-maps are plotted from the state diagram in Fig. 11.43b, assuming the use of D flip-flops, and are given in Fig. 11.44 together with the output K-maps. Also shown are the minimum covers for the K-maps that yield the following NS and output

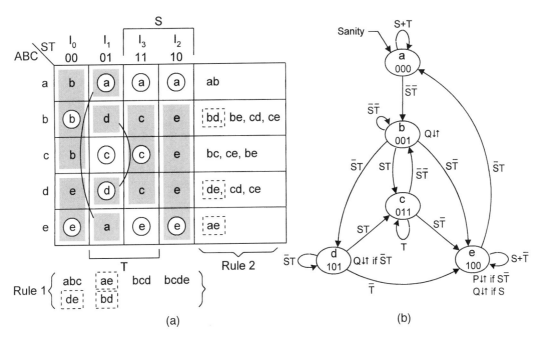

FIGURE 11.43
Application of state assignment rules 1 and 2 to the FSM of Fig. 11.42a. (a) Next state table showing grouping of states (shaded areas) that satisfy rule 1, and the results of rule 2. (b) The fully documented state diagram showing an optimum or near optimum set of state code assignments resulting from application of rules 1 and 2 in (a).

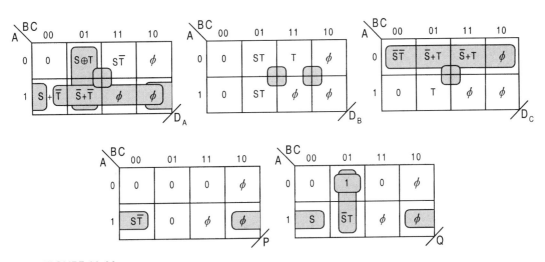

FIGURE 11.44
Next-state and output K-maps plotted from the state diagram in Fig. 11.43b assuming the use of D flip-flops and showing minimum cover.

functions:

$$\begin{cases} D_A = C\bar{S}\bar{T} + \bar{B}C\bar{S}T + A\bar{C}S + A\bar{T} \\ D_B = CST + BT \\ D_C = \bar{A}\bar{S}\bar{T} + CT \\ P = A\bar{C}S\bar{T} \\ Q = A\bar{C}S + \bar{B}C\bar{S}T + \bar{A}\bar{B}C \end{cases}. \tag{11.11}$$

These results represent a total gate/input tally of 14/40, excluding possible inverters. Eqs. (11.11) will be compared with the results generated by using the array algebraic approach to design discussed next in Section 11.11.

11.11 ARRAY ALGEBRAIC APPROACH TO LOGIC DESIGN

Results similar to those of Eqs. (11.11) can be obtained by using what is called the *array algebraic approach* to state machine design. This approach is applicable to any FSM for which each state-to-state transition ends in a holding condition, and each state obeys the sum rule. Thus, the FSM in Fig. 11.36 would not be suitable for this method since there are states without holding conditions.

The array algebraic approach can be used for the computer automated design (CAD) of either synchronous or asynchronous FSMs, and without the need to use either state diagrams or K-maps. Furthermore, the array algebra that is used bears a close resemblance to matrix algebra, but there are some important differences. To properly launch this subject and to minimize the difficulty index, the various matrix arrays and equations will be given using the FSM in Fig. 11.43 as an example. In this way, the reader can follow the operations with little difficulty.

Given the state code assignments that are generated by using the next-state table in Fig. 11.43a,

$$a = 000, \quad b = 001, \quad c = 011, \quad d = 101, \quad \text{and} \quad e = 100,$$

the state matrix **S** is defined as

$$\mathbf{S} = \begin{matrix} a \\ b \\ c \\ d \\ e \end{matrix} \begin{bmatrix} 0 & 0 & 0 \\ 0 & 0 & 1 \\ 0 & 1 & 1 \\ 1 & 0 & 1 \\ 1 & 0 & 0 \end{bmatrix} = \text{State matrix}.$$

Also obtained from the next-state table in Fig. 11.43a is the destination matrix **D**, given by

$$\mathbf{D} = \begin{matrix} & I_0 & I_1 & I_3 & I_2 \\ a \\ b \\ c \\ d \\ e \end{matrix} \begin{bmatrix} 0 & ae & a & a \\ abc & 0 & 0 & 0 \\ 0 & c & bcd & 0 \\ 0 & bd & 0 & 0 \\ de & 0 & e & bcde \end{bmatrix} = \text{Destination matrix}.$$

11.11 ARRAY ALGEBRAIC APPROACH TO LOGIC DESIGN

The **D** matrix is formed by combining all states in a given column that are associated with a holding condition. For example, state set $\{abc\}$ is associated with the holding condition b in the I_0 column. Similarly, state set $\{de\}$ is associated with holding condition e in the same column. Notice that all state set entries involving two or more states in the **D** matrix are an expression of rule 1, as indicated in Fig. 11.43a. Single literals appear when a present state identifier is associated exclusively with a holding condition, as in row 1 columns 3 and 4 or in row 5 column 3 of Fig. 11.43a. A zero appears when there is no next state associated with the present state.

By taking the transpose of the **S** matrix ($\mathbf{S}^t$) and multiplying it with the **D** matrix there results the function matrix $\mathbf{F_{NS}}$ given by

$$\mathbf{F_{NS}} = \mathbf{S}^t\mathbf{D} = \begin{bmatrix} 0 & 0 & 0 & 1 & 1 \\ 0 & 0 & 1 & 0 & 0 \\ 0 & 1 & 1 & 1 & 0 \end{bmatrix} \begin{bmatrix} 0 & ae & a & a \\ abc & 0 & 0 & 0 \\ 0 & c & bcd & 0 \\ 0 & bd & 0 & 0 \\ de & 0 & e & bcde \end{bmatrix} \qquad (11.12)$$

$$= \begin{bmatrix} de & bd & e & bcde \\ 0 & c & bcd & 0 \\ abc & bcd & bcd & 0 \end{bmatrix} = \text{Function matrix}.$$

Notice how sets combine in array algebra. For example, bcd in column 2, row 3 of the function matrix results from an implied matrix operation $c + bd$. This is one of the peculiarities of array algebra. Thus, bcd results from the Boolean product $c \cdot bd$.

Now it is necessary to evaluate the function matrix **F** in terms of the state variables. This can be accomplished in either of two ways. For an automated design approach, the tabular representation of the state assignment in Fig. 11.45a can be used in connection with a minimization algorithm such as that of Quine–McCluskey (Q-M) discussed in Subsection 4.8.1. Alternatively, a K-map representation of the state assignments, as in Fig. 11.45b, can be used. In either case, if all state identifiers are present in a given state adjacency set, that set becomes, logic 1. State identifiers not part of a set are assigned a logic 0. Remember that in the case of the Q-M algorithm, don't cares are treated as minterms and therefore

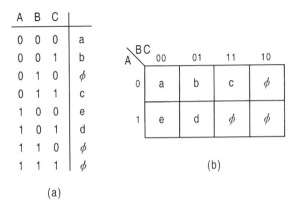

FIGURE 11.45
State assignment representation for the state machine in Fig. 11.43b.

take logic 1. For K-map evaluations, the don't cares are treated as don't cares. However evaluated, the results for the six state adjacency sets in the function matrix are

$$de = \sum m(4,5,6,7) = A, \quad bd = \sum m(1,5) = \bar{B}C, \quad e = \sum m(4,6) = A\bar{C},$$

$$c = \sum m(2,3,6,7) = B$$

$$abc = \sum m(0,1,2,3) = \bar{A}, \quad bcd = \sum m(1,2,3,5,6,7) = C,$$

$$bcde = \sum m(1,3,4,5,6,7) = A + C,$$

all of which must be substituted into Eq. (11.12) before proceeding. Again, to use the Q-M algorithm in evaluating the six adjacency sets just presented, care must be taken to include the three don't cares $\phi(2,6,7)$ as minterms $m(2,6,7)$ in each set, if they are not already included. Thus, for use with the Q-M algorithm, the adjacency sets become

$$de = \sum m(2,4,5,6,7), \quad bd = \sum m(1,2,5,6,7), \quad e = \sum m(2,4,6,7),$$

$$c = \sum m(2,3,6,7)$$

$$abc = \sum m(0,1,2,3,6,7), \quad bcd = \sum m(1,2,3,5,6,7),$$

$$bcde = \sum m(1,2,3,4,5,6,7),$$

which will yield the same results as given previously. Because of the simplicity of the adjacency set minimization process, the Q-M algorithm is quite suitable for CAD purposes even for relatively complex state machines. However, most any minimization algorithm is suitable for this purpose.

After making the appropriate substitutions into Eq. (11.12), the next-state functions can be evaluated. This is done by multiplying the function matrix $\mathbf{F_{NS}}$ by the input matrix $\mathbf{I}$ to yield the following next-state function matrix $\mathbf{NS}$:

$$\mathbf{NS} = \mathbf{F_{NS}I} = \begin{bmatrix} A & \bar{B}C & A\bar{C} & (A+C) \\ 0 & B & C & 0 \\ \bar{A} & C & C & 0 \end{bmatrix} \begin{bmatrix} I_0 \\ I_1 \\ I_3 \\ I_2 \end{bmatrix} = \begin{bmatrix} D_A \\ D_B \\ D_C \end{bmatrix} \quad (11.13)$$

By carrying out the indicated matrix multiplication, there result the NS equations

$$D_A = AI_0 + \bar{B}CI_1 + A\bar{C}I_3 + (A+C)I_2$$
$$= A\bar{S}\bar{T} + \bar{B}C\bar{S}T + A\bar{C}ST + AS\bar{T} + CS\bar{T}$$
$$D_B = BI_1 + CI_3$$
$$= B\bar{S}T + CST$$
$$D_C = \bar{A}I_0 + CI_1 + CI_3$$
$$= \bar{A}\bar{S}\bar{T} + CT,$$

11.11 ARRAY ALGEBRAIC APPROACH TO LOGIC DESIGN

which compare closely with the NS functions in Eqs. (11.11). Notice that $A\bar{S}\bar{T}+AS\bar{T} = A\bar{T}$ in D_A.

The output functions can be obtained by following the same procedure. Now, however, the state matrices for the outputs are those obtained directly from the state table in Figure 11.42(b). When this is done, the following results are obtained for outputs P and Q:

$$\mathbf{F_P} = \mathbf{P^t D} = \begin{bmatrix} 0 & 0 & 0 & 0 & S\bar{T} \end{bmatrix} \begin{bmatrix} 0 & ae & a & a \\ abc & 0 & 0 & 0 \\ 0 & c & bcd & 0 \\ 0 & bd & 0 & 0 \\ de & 0 & e & bcde \end{bmatrix}$$

$$= [de \quad 0 \quad e \quad bcde]S\bar{T},$$

or

$$\mathbf{F_P} = \mathbf{P^t D} = [A\bar{S}\bar{T} \quad 0 \quad A\bar{C}S\bar{T} \quad (A+C)S\bar{T}]$$

Then, by multiplying $\mathbf{F_P}$ by the input matrix $\mathbf{I}$, the output P is found to be

$$P = \mathbf{F_P I} = [A\bar{S}\bar{T} \quad 0 \quad A\bar{C}S\bar{T} \quad (A+C)S\bar{T}] \begin{bmatrix} I_0 \\ I_1 \\ I_3 \\ I_2 \end{bmatrix}$$

$$= AS\bar{T} + CS\bar{T},$$

where $I_0 = \bar{A}\bar{B}$, $I_1 = \bar{A}B$, $I_3 = AB$, and $I_2 = A\bar{B}$. Similarly, for output Q there results

$$\mathbf{F_Q} = \mathbf{Q^t D} = [0 \quad 1 \quad 0 \quad \bar{S}T \quad S]\mathbf{D}$$
$$= [S \quad bd\bar{S}T \quad eS \quad bcdeS]$$
$$= [S \quad \bar{B}C\bar{S}T \quad A\bar{C}S \quad (A+C)S]$$

and

$$Q = \mathbf{F_Q I} = \bar{B}C\bar{S}T + A\bar{C}ST + AS\bar{T} + CS\bar{T}$$

Altogether the NS and output functions generated from the array algebraic approach are

$$\begin{cases} D_A = A\bar{S}\bar{T} + \bar{B}C\bar{S}T + A\bar{C}ST + AS\bar{T} + CS\bar{T} \\ D_B = B\bar{S}T + CST \\ D_C = \bar{A}\bar{S}\bar{T} + CT \\ P = AS\bar{T} + CS\bar{T} \\ Q = \bar{B}C\bar{S}T + A\bar{C}ST + AS\bar{T} + CS\bar{T} \end{cases}, \qquad (11.14)$$

which represent a total gate/input tally of 14/43 compared to 14/40 for the standard K-map approach of Eqs. (11.11), all excluding possible inverters. Observe that all p-terms in the output expressions of Eqs. (11.14) are covered by the D_A expression. In fact, it is characteristic of the array algebraic approach that most, if not all, of the p-terms in the output functions will be shared PIs. This is so because the output functions are obtained by using the same form of the function equation $\mathbf{F} = \mathbf{Z}^t \cdot \mathbf{D}$, where $\mathbf{Z}$ is any output function matrix and $\mathbf{D}$ is the destination matrix used in obtaining the NS functions. Notice that reducing the expression for D_A by factoring the terms $A\bar{S}\bar{T} + AS\bar{T} = A\bar{T}$ results in a gate/input tally of 14/41, only a very minor improvement. Generally, in using the array algebraic approach to design, significant savings in hardware can result by considering all of the factoring/reduction possibilities that exist, particularly within the NS logic expressions. However, account must also be taken of the shared PIs that might be lost in the factoring/reduction process.

An inspection of the Q output function in Eqs. (11.14) reveals an externally initiated static hazard in the coupled terms $A\bar{C}ST + AS\bar{T}$. This s-hazard can occur on a $1 \rightarrow 0$ change in input T while in state 100 under holding condition S (see Fig. 11.43b). As indicated in Section 11.3, this hazard can be eliminated either by adding the hazard cover $A\bar{C}S$ or by filtering. Note that the p-term $A\bar{C}ST$ can be replaced by $A\bar{C}S$ in the expression for Q, thereby requiring no hazard cover. This results from the simplification $A\bar{C}ST + AS\bar{T} = A\bar{C}S + AS\bar{T}$ after applying the absorptive law. Note that no hazards exist in the output functions P or Q of Eqs. (11.11). In any case, since ORGs are possible in both outputs, they should be filtered thereby eliminating all logic noise—*hence no hazard analysis is needed.*

In attempting to automate the design of state machines by the array algebraic method, the most difficult part, the "bottleneck," is to obtain the state adjacency sets of function matrix $\mathbf{F}$ in terms of the state variables. Fortunately, these problems break up into single-output minimization problems, as is indicated by the example given earlier, and often can be easily handled by tabular minimization algorithms such as that of Q-M. But a given minimization problem can be cyclical in the sense that more than one minimum is possible. Petrick's algorithm (see Further Reading) can be used to solve such problems, and it is easily implemented on a computer, but for only simple to moderately complex problems. On the other hand, an optimum solution may not be necessary, and one of the minimum solutions for an adjacency set can be arbitrarily chosen on the basis of some criterion built into the CAD algorithm. Full-blown heuristic-type minimization algorithms are usually not required for this purpose. However, if needed, none are better for very large minimization problems than the Espresso-II algorithm briefly discussed in Section 4.8. In any case, after the $\mathbf{F}$ matrix has been expressed in terms of the state variables, the array algebraic process can continue to fruition.

Before we leaving this subject, one final thought is worth mentioning. The array algebraic approach is perfectly general. It can be applied to any FSM, synchronous or asynchronous that meet the minimum requirements mentioned at the beginning, and to any set of state code assignments that is used. The results may or may not be optimum, but will be at least near optimum depending, of course, on the choice of state code assignments. In this section, the array algebraic method is used to design a synchronous state machine, of moderate complexity, whose state assignments are obtained by applying state assignment rules 1 and 2 given previously. However, applications of rules 1 and 2 do not eliminate ORGs.

In Section 14.12 the array algebraic method is again used, but to design the fastest asynchronous FSMs possible, called *single-transition-time (STT) machines*. For these FSMs

11.12 STATE MINIMIZATION   547

the state code assignments are chosen by using special partitioning methods that avoid ORGs and other serious timing problems. These partitioning methods, not involving state assignment rules 1 and 2, are used to construct the state table from which the **S** matrix is derived. These methods are also applicable to synchronous D flip-flop designs, but with an increased number of state variables required — the price of avoiding ORGs.

11.12 STATE MINIMIZATION

Formal state minimization procedures are available that involve state tables, implication charts, merger graphs, and the like. Further Reading at the end of this chapter cites references on this subject. However, such procedures are rarely used in modern state machine design. For state machines of up to moderate complexity, a minimum or near minimum number of states can be obtained simply by visual inspection of the state diagram or state table. In fact, it may not be desirable to obtain a minimum number of states for a particular FSM. There are occasions where a nonminimum number of states may lead to a more optimum set of NS and output functions for a state machine. Furthermore, if the state machine is relatively complex and if it is to be implemented, say, with an FPGA or PAL, it really doesn't matter whether or not a minimum number of states exist. In cases where hardware capability far exceeds the state machine requirements, it is only necessary to make certain that the FSM performs its tasks properly — hardware limitation is not a factor.

In this section, a visual method is used to demonstrate how states can be merged to produce a more optimum design of relatively simple state machines. Consider the requirements for the pulse width adjuster (PWA) in Fig. 11.46, which has a single input X and a single output P. It is required that X be synchronized in phase with the RET D flip-flops of the memory. The PWA is to function according to the operation table in Fig. 11.46a, where the pulse widths are adjusted to one, two, or three clock periods, T_{CK}, as indicated. The timing diagram in Fig. 11.46b illustrates the pulse width relationship between the input pulse waveform X and the output waveform P relative to seven states $a, b, c, d, e, f,$ and g.

The state diagram that corresponds to the requirements of the PWA set forth in Fig. 11.46 is shown in Fig. 11.47a. An inspection of the seven states in the state diagram and in the state table of Fig. 11.47b indicates that two merging operations are possible. If states c and d are merged to form state c', and if states $e, f,$ and g are merged to form state d', there results the much simplified state diagram of Fig. 11.47c. This four-state PWA, functions the same as the seven-state PWA, but at a significant reduction in hardware cost. There are other advantages to state reductions. For example, in the case of the four-state PWA in Fig. 11.47c, it can be coded in Gray code so as to eliminate any possibility of ORGs occurring in the output. However, state reductions alone may or may not eliminate static hazards in the output.

Notice how easy it is to recognize the merging patterns of the states in Fig. 11.47. Such visual approaches to the state-merging process can be carried out even on much more complex FSMs. Usually it is not necessary to apply formal techniques to this process. The point is that, if the FSM is to be implemented with an array logic device, such as a ROM, PLA, FPGA, or PAL, it may not matter whether or not a state-minimum design exists. What is more important is the correct operation of the FSM. Of course, if the FSM is to be

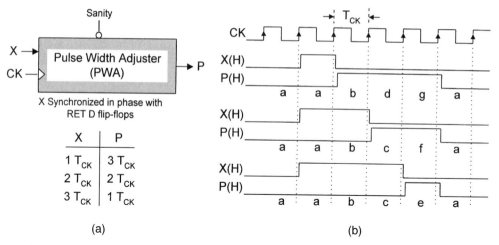

FIGURE 11.46
The pulse width adjuster (PWA) FSM. (a) Block diagram and operation table. (b) Timing diagrams showing the three operations of the PWA relative to seven states a, b, c, d, e, f, and g.

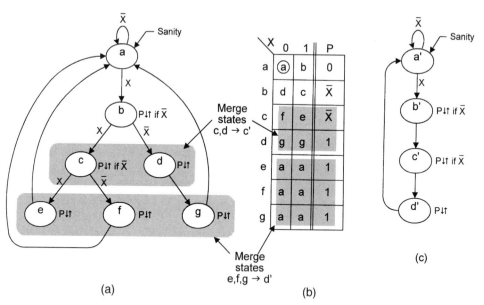

FIGURE 11.47
Use of a pulse width adjuster (PWA) to demonstrate state reduction by merging. (a), (b) State diagram and state table showing merging of states c, d to state c' and merging of states e, f, g to state d'. (c) The resulting state diagram for the PWA.

designed on chip to be manufactured by the millions, an optimum design may be necessary (hardware-wise and/or speed-wise), but with or without a minimum number of states.

FURTHER READING

Few texts cover the subject of output race glitches (ORGs). The known sources on this subject are the texts of Fletcher, Shaw, and Tinder, and of these the last is by far the most comprehensive. It is equally difficult to find further reading on the subject of static hazards in the outputs of synchronous FSMs. The reason for this is not exactly clear. Again the best source appears to be the text by Tinder.

[1] W. I. Fletcher, *An Engineering Approach to Digital Design*. Prentice Hall, Englewood Cliffs, NJ, 1980.
[2] A. W. Shaw, *Logic Circuit Design*. Sanders College Publishing, Fort Worth, TX, 1993.
[3] R. F. Tinder, *Digital Engineering Design: A Modern Approach*. Prentice Hall, Englewood Cliffs, NJ, 1991.

The subjects of asynchronous inputs, synchronizers and their failure, and metastability appear to be covered to one extent or another by most texts in the field and in many journal articles. Perhaps the best coverage for further reading on these subjects is found in the text by Wakerly, with others by Fletcher, Tinder (both previously cited), Daniels, Katz, McCluskey, and Unger all being a distant second choice.

The texts of Wakerly and Daniels cover the subject of mean time between failure (MTBF) of synchronizer flip-flops and are recommended for further reading on this subject.

[4] J. D. Daniels, *Digital Design from Zero to One*. John Wiley & Sons, New York, 1996.
[5] R. H. Katz, *Contemporary Logic Design*. Benjamin/Cummings Publishing, Redwood City, CA, 1994.
[6] E. J. McCluskey, *Logic Design Principles*. Prentice Hall, Englewood Cliffs, NJ, 1986.
[7] S. H. Unger, *The Essence of Logic Circuits*. Prentice Hall, Englewood Cliffs, NJ, 1989.
[8] J. F. Wakerly, *Digital Design Principles and Practices*, 2nd. ed. Prentice-Hall, Englewood Cliffs, NJ, 1994.

Of the journal articles on metastability and the synchronizer, none are more important than those by Chaney, who has over many years established himself as a leading authority on the metastability problem in synchronizers. In Chaney's article will be found measured data on the MTBF of a variety of common flip-flops. Also, there are the earlier works of Chaney *et al.*, Stoll, and Veedrick that are worth reading for a more complete grasp of the synchronizer problem. The advanced reader may find the theoretical work of Kleeman and Cantoni more contributive to an understanding of the problem.

[9] T. J. Chaney, "Measured Flip-Flop Responses to Marginal Triggering," *IEEE Trans. Comput.* **C-32**(12), 1207–1209 (1983).
[10] T. J. Chaney, S. M. Ornstein, and W. M. Littlefield, "Beware the Synchronizer," Dig. COMPCON, San Francisco, Sept. 1972, pp. 317–319.
[11] L. Kleeman and A. Cantoni, "On the Unavoidability of Metastable Behavior in Digital Systems," *IEEE Trans. on Comput.* **C-36**(1), 109–112 (1987).

[12] P. A. Stoll, "How to Avoid Synchronization Problems," *VLSI Design*, Nov.–Dec., pp. 56–59 (1982).
[13] H. J. M. Veedrick, "The Behavior of Flip-Flops Used as Synchronizers and Prediction of their Failure Rate," *IEEE Journal of Solid State Circuits* **SC-15**(2), 169–176 (1980).

Adequate treatments of clock skew are found in the texts of Fletcher, McCluskey, Tinder, and Wakerly, all previously cited. Excellent coverage of clock generating circuitry is provided in the text by Fletcher. Discussions on clock signal specifications, buffering, and gating can be found in the text by Wakerly. For the advanced reader needing information on the techniques for generating high-frequency clock waveforms from frequency synthesizers, the texts by Best, Egan, and Rhode are recommended.

[14] R. G. Best, *Phase-Locked Loops — Theory, Design and Applications*. McGraw-Hill, New York, 1984.
[15] W. F. Egan, *Frequency Synthesis by Phase Lock*. Wiley Interscience, New York, 1981.
[16] U. L. Rhode, *Digital PLL Frequency Synthesizers Theory and Design*. Prentice Hall, Englewood Cliffs, NJ, 1983.

Further reading on the subject of initialization (sanity) circuits is best found in the text by Langdon and that by Tinder (previously cited). On the subject of debouncing circuits the texts by Langdon, Tinder, and Wakerly are recommended, although the subject is to one degree or another covered in other texts such as those by Daniels, Katz, and Unger, all previously cited.

[17] B. G. Langdon, Jr., *Computer Design*. Computeach Press, Inc., San Jose, CA, 1982.

References covering the uses of ASMs, state tables, and state assignment rules in state machine design are numerous. Good examples of all three of these subjects are found in the texts by Hayes, Nelson *et al.*, Roth, Wakerly (previously cited), and Yarbrough. The text by Comer uses a unique graphical representation of sequential machines that appears to draw from a combination of ASM chart notation and state diagram notation. Of the journal articles on optimal state assignments, that by De Micheli *et al.* is perhaps the most authoritative available.

[18] D. J. Comer, *Digital Logic and State Machine Design*, 3rd ed. Saunders College Publishing, Fort Worth, TX, 1995.
[19] G. De Micheli, R. Brayton, and A. Sangiovanni-Vincentelli, "Optimal State Assignment for Finite State Machines," *IEEE Trans. on CAD/ICAS* **CAD-4**(3), 269–284 (1985).
[20] J. P. Hayes, *Introduction to Digital Design*. Addison-Wesley, Reading, MA, 1993.
[21] V. P. Nelson, H. T. Nagle, B. D. Carroll, and J. D. Irwin, *Digital Logic Circuit Analysis and Design*. Prentice Hall, Englewood Cliffs, NJ, 1995.
[22] C. H. Roth, *Fundamentals of Logic Design*, 4th ed. West, St. Paul, MN, 1992.
[23] J. M. Yarbrough, *Digital Logic Applications and Design*, West, Minneapolis/St. Paul, MN, 1997.

The formal approach to state reduction is nicely covered by numerous texts, including those of Hayes, Katz, McCluskey, Nelson *et al.*, and Yarbrough, all previously cited.

BIBLIOGRAPHY   551

For the more theoretically inclined, the texts by Dietmeyer, De Micheli, and Kohavi, are recommended.

[24] D. L. Dietmeyer, *Logic Design of Digital Systems*, 2nd ed. Allyn and Bacon, Inc., Boston, MA, 1978.
[25] G. De Micheli, *Synthesis and Optimization of Digital Circuits*. McGraw-Hill, New York, 1994.
[26] Z. Kohavi, *Switching and Finite Automata Theory*. McGraw-Hill, New York, 1978.

There are no known simple references on the subject of the array algebraic approach to logic design of synchronous state machines. The advanced reader may find the treatment by Dietmeyer (previously cited) helpful, but some background in array Boolean algebra notation is needed. For references covering Petrick's algorithm and related subjects, the reader is referred to the texts by Hayes, Nelson *et al.*, and Roth, all previously cited.

PROBLEMS

11.1 Inspect all three state diagrams in Figure P10.7 for possible output race glitches (ORGs) and static hazards. If any exist, indicate their origin and type following the examples in Sections 11.2 and 11.3.

11.2 Shown in Fig. P11.1 are the state diagrams for two fictitious FSMs.

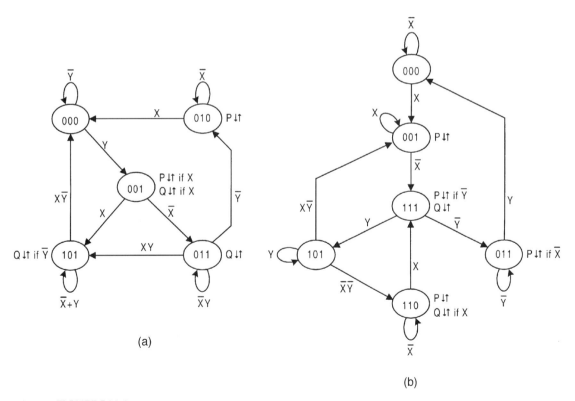

FIGURE P11.1

(1) Run a *complete output race glitch (ORG) analysis* on each FSM. To do this, follow the examples in Section 11.2. Thus, if ORGs exist, indicate their origin and type (+ or −). Do not alter the state diagram in any way.

(2) In consideration of part (1), run a *complete static hazard analysis* on each of these FSMs. To do this, follow the examples in Section 11.3. Assume that each FSM is to be implemented with NAND-based flip-flops, and indicate whether an existing static hazard is externally or internally initiated. Consider both SOP and POS output-forming logic and give the gate/input tally for each, including static hazard cover (if any). Do not alter the state code assignment and do not construct a logic circuit for the FSM.

11.3 Shown in Fig. P11.2 are the state diagrams for two fictitious FSMs.

(1) Run a *complete output race glitch (ORG) analysis* on each FSM. To do this, follow the examples in Section 11.2. Thus, if ORGs exist, indicate their origin and type (+ or −). Do not alter the state diagram in any way.

(2) In consideration of part (1), run a *complete static hazard analysis* on each of these FSMs. To do this, follow the examples in Section 11.3. Assume that the FSM is to be implemented with NOR-based flip-flops, and indicate whether an existing static hazard is externally or internally initiated. Consider both SOP and POS output-forming logic and give the gate/input tally for each, including

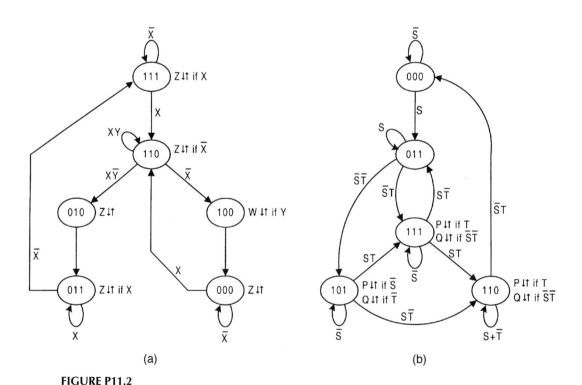

FIGURE P11.2

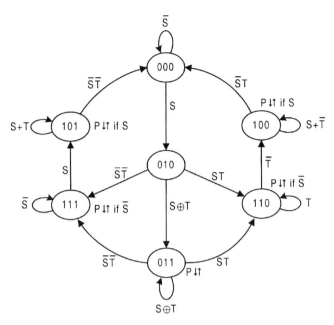

FIGURE P11.3

static hazard cover (if any). Do not alter the state code assignment and do not construct a logic circuit for the FSM.

11.4 The FSM in Figure P11.3 has two inputs, S and T and one output, P.
(a) Run a *complete output race glitch (ORG) analysis* on this FSM. To do this, follow the examples in Section 11.2. Thus, if ORGs exist, indicate their origin and type (+ or −). Do not alter the state diagram in any way.
(b) In consideration of part (a), run a *complete static hazard analysis* on this FSM. To do this, follow the examples in Section 11.3. However, it is not known whether to use NAND- or NOR-based flip-flops for its design. Consider both SOP and POS output-forming logic and give the gate/input tally for each (including any static hazard cover). Based on this information, make a selection as to the type of flip-flop (NAND- or NOR-based) that will yield the most optimum design. Do not alter the state code assignment and do not construct a logic circuit for the FSM.

11.5 Carry out complete ORG and static hazard analyses on the FSM in Fig. 11.43b. To do this, use may be made of the NS and output expressions in Eqs. (11.11). If any of these timing defects exist in the output signals, indicate the best means of eliminating them. (Hint: See Subsection 11.2.2.)

11.6 Suppose it is desirable to estimate the mean time between failures (MTBF) for a synchronizing system that is required to protect a hypothetical FSM operated at 200 MHz when the asynchronous data change at an average rate of 10 kHz. By experiment, the average setup time t_{su} is 1 nanosecond (ns) for the high-speed

D flip-flops to be used for the synchronizer. Consider that at 200 MHz, it is necessary to use a synchronizing scheme of the type shown in Fig. 11.23a for which a divide-by-4 counter is used in the diagram.
(a) Calculate the MTBF (in seconds, days and years) if $\tau = 0.5$ ns and $T_0 = 1 \times 10^{-6}$ seconds. [Hint: Use Eq. (11.3) and take into account the cumulative effects of both stages.]
(b) Repeat part (a) if a divide-by-two counter is used in the diagram. Calculate the MTBF in seconds for comparison with the result in part a.

11.7 It is desired to find a safe operating clock frequency for a given FSM. The following data is collected relative to the operation of the FSM:

Flip-flop parameters $f_{plh} = 6$ ns; $f_{phl} = 9$ ns
Maximum delay through the NS logic $\tau_{NS} = 7$ ns
Maximum flip-flop setup time $t_{su} = 3$ ns

Calculate a safe operating clock frequency f_{CK} based on a 15% safety factor.

11.8 Derive the expressions for $V_X(t)$ in Eqs. (11.8) and (11.9) relative to Fig. 11.31c. Assume that $R_1 \gg R_2$ and that the switch is opened (or closed) at time $t = 0$ only after steady-state conditions are reached. State any simplifying assumptions that are made relative to the Schmitt trigger and inverter. (Note: This exercise involves solving a first-order RC circuit.)

11.9 Presented in Fig. P11.4 is the state diagram for a one-bit serial adder. The operand bits, a and b, are introduced serially and are synchronized antiphase to the clock triggering edge. The outputs are S (sum) and C_o (carry-out). Assume that the FSM is initialized (reset) after each addition operation.
(a) Complete the state diagram by giving it a state code assignment that is free of ORGs.

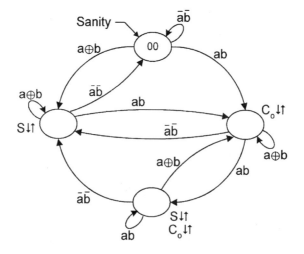

FIGURE P11.4

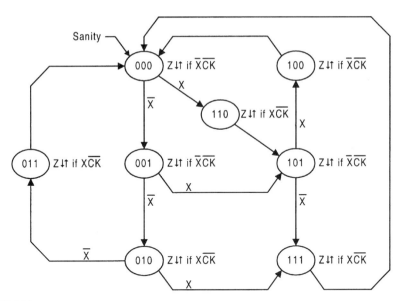

FIGURE P11.5

(b) Design this serial adder by using two RET D flip-flops and a gate-minimum NS and output logic. To do this, use XOR patterns where appropriate. Initialize into the 00 state and show the sanity circuit and connections. Assume that all inputs and outputs are active high.

(c) Construct the timing diagram for the operation of the one-bit serial adder. To do this, include all four addition operations as indicated in the truth table of Fig. 8.1c.

11.10 Shown in Fig. P11.5 is the state diagram representing a serial BCD-to-XS3 converter. A synchronous BCD waveform is presented on the X input, and a synchronous XS3 waveform is issued on the Z output. Note that all output signals are issued on an exciting condition, and that the BCD code arrives serially LSB first.

(a) Use RET D flip-flops for an *optimum* logic circuit design of this converter. To do this, use XOR patterns for the output function. Assume that input X is synchronized to the falling edge of the clock waveform and that both X and Z are active high. Initialize the FSM into the 000 state and show the sanity circuit and its connections to the converter.

(b) Determine if ORGs or static hazards are present in the output. If they exist, then take the necessary steps to eliminate them and alter the logic circuit accordingly. Otherwise, do nothing. In any case, do not alter the state diagram.

(c) Construct the timing diagram for the BCD-to-XS3 converter by introducing a BCD waveform equivalent to decimal 2 followed by decimal 7 (both introduced LSB first). Thus, include waveforms for $X(H)$, CK, $A(H)$, $B(H)$, $C(H)$, and $Z(H)$. Use a clock waveform with a 50% duty cycle. Explain the difference in active durations of the input and output pulses. (Hint: On the timing diagram, the four BCD code bits will appear in reverse order since they are introduced LSB first.)

(d) Verify the timing diagram of part (c) by simulating the logic circuit of part (a).

11.11 A 3-bit serial odd-parity detector is to be designed that will issue an active output pulse $P_{Odd\,Det}(L)$ any time a series of three clock periods samples an odd number of active pulses (one or three, in any order) on an input pulse string X. The output must be issued only when clock is active.

(a) Construct an *optimum state diagram and state table* for this detector. To do this, make effective use of the "from rule" discussed in Subsection 11.10.2 and initialize into the 000 state to begin the process. Remember that the FSM must issue an output on the active portion of the clock waveform. (Hint: This is a Mealy machine of five, six, or seven states depending on the design.)

(b) Design the logic circuit for this detector by using three FET JK flip-flops and a gate-minimum NS and output logic. To do this consider using XOR patterns where appropriate. Assume that X arrives active high from a mechanical switch, and that it must be *debounced and synchronized* antiphase to the clock triggering edge. Show all input conditioning circuitry and their connections to the FSM. Plan to use a SPDT debouncing circuit of the type shown in Fig. 11.32a. [Hint: If Part (a) is done correctly, two to four gates will be required for the NS and output logic.]

11.12 An FSM is to be designed that will issue an output according to the following requirements:

If clock samples S *active* with both X and Y *inactive*, then Z is issued on Y following $X\bar{Y}$ or X following $\bar{X}Y$, provided that these events are spaced one clock period apart. If these conditions are not met (an EQV condition), then Z will not be issued, and the FSM must wait for S to be sampled *inactive* before the FSM can return to the initial state and start the process over again. The output Z must be issued for only one clock period, after which the FSM must return unconditionally to the initial state.

Construct a *state diagram and state table* for this FSM and give it a glitch-free state code assignment. Plan to initialize the FSM into the 000 state. (Hint: Properly done, the state diagram will have only six states.)

11.13 Shown in Fig. P11.6 is a state diagram for an FSM that has two inputs, X and Y, and one output, Z.

(a) Given the state code assignment indicated, use the *array algebraic approach* to obtain the NS expressions for this FSM. To do this, first construct the state table to obtain the state matrix **S** and destination matrix **D**. Then find the function matrix $\mathbf{F_{NS}}$ and the next matrix **NS** by following the example in Section 11.11. End with an optimum set of logic equations for D_A and D_B.

(b) Repeat the array algebraic approach to obtain the output function for Z.

PROBLEMS

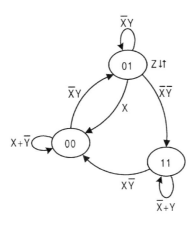

FIGURE P11.6

(c) From the results of parts (a) and (b), analyze this FSM by constructing the revised state diagram. To do this, follow the examples in Section 10.13. Are ORGs now possible? Are they possible in the state diagram of Fig. P11.6?

11.14 Shown in Fig. P11.7 is a state diagram for an FSM that has two inputs, X and Y, and three outputs, P, Q, and R.

(a) Given the state code assignment indicated, use the *array algebraic approach* to obtain the NS expressions for this FSM. To do this, first construct the *state table* to obtain state matrix **S** and the destination matrix **D**. Then find the function matrix $\mathbf{F_{NS}}$ and the next matrix **NS** by following the example in Section 11.11. End with an optimum set of logic equations for D_A, D_B, and D_C.

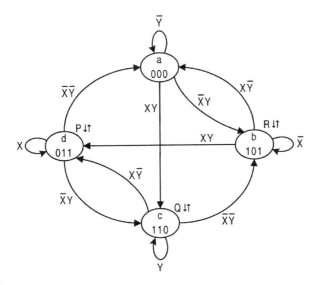

FIGURE P11.7

(b) Repeat the algebraic approach to obtain the output functions for P, Q, and R. Are static 1-hazards present? If so, indicate whether they are internally initiated or externally initiated, and give the hazard cover required to eliminate them.

(c) From the results of parts (a) and (b), analyze this FSM by constructing the revised state diagram. To do this, follow the examples in Section 10.13. Are ORGs possible? Are they possible in the original state diagram of Fig. P11.7?

(d) Noticing that none of the principal states are used as race states, obtain each output function in terms of the three variables of the state in which the output is issued. Now comment on the presence or absence of ORGs and static hazards. Is this a valid set of output function expressions and, if so, is this a special case? Which is best, the results of (b) or those of (d)?

11.15 Presented in Fig. P11.8 is the state table for an FSM having two inputs, X and Y, and two outputs, P and Q. Notice that it follows the format given in Fig. 11.43a and that the best compliance possible is made of the state assignment rules for three state variables. (See state assignment rules 1 and 2 in Subsection 11.10.2.)

(a) Given the state code assignment indicated, use the *array algebraic approach* to obtain the NS expressions for this FSM. To do this, first obtain the state matrix **S** and destination matrix **D**. Then find the function matrix $\mathbf{F_{NS}}$ and the next matrix **NS** by following the example in Section 11.11. End with an optimum set of logic equations for D_A, D_B, and D_C. Thus, some function minimization is necessary.

(b) Repeat the algebraic approach to obtain the output functions for P and Q. Are static 1-hazards present? If so, indicate whether they are internally initiated or externally initiated, and give the hazard cover required to eliminate them.

(c) It will be observed that the array algebraic approach eliminates ORGs but typically creates redundant output states. Use the results of part (b) to find an optimum set of output functions that will still eliminate ORGs but that will no longer require hazard cover. Keep in mind that the array algebraic approach tends to maximize the number of shared PIs in the output functions, but often at the expense of creating static 1-hazards.

ABC \ XY	I_0 00	I_1 01	I_3 11	I_2 10	P	Q
111 → a	(a)	(a)	b	(a)	0	0
110 → b	a	d	(b)	c	0	0
010 → c	(c)	e	e	(c)	0	1
100 → d	f	(d)	(d)	f	1	0
011 → e	a	(e)	(e)	(e)	1	0
000 → f	(f)	(f)	b	(f)	0	0

FIGURE P11.8

PROBLEMS

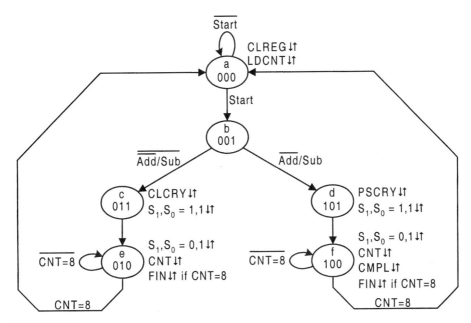

FIGURE P11.9

(d) Prove that the results of parts (a) and (c) are valid sets of NS and output expressions by constructing a state diagram for this FSM. To do this, it will be necessary to create a PS/NS table from the EV K-maps that derive from parts (a) and (b). Compare this state diagram with that generated directly from the state table in Fig. P11.8. Are ORGs possible in the original FSM?

11.16 Collapse the redundant state diagram (for a serial adder/subtractor) in Fig. P11.9 into one of *three* states. It is required that outputs CLCRY and PSCRY accompany the USR mode control outputs $S_1, S_0 = 1, 1$, and that each of the two sets of mode control outputs shown in the figure be assigned to separate states. It is further required that the *Start* signal be active for a period greater than one clock period and that it must go inactive before *CNT* and *FIN* signals can be issued. Assume that any ORGs that occur after the three-state process is complete have no effect on the proper operation of the FSM.

CHAPTER 12

Module and Bit-Slice Devices

12.1 INTRODUCTION

In Chapter 10 use was made of both the basic cell and the flip-flop as the memory in the design of relatively simple state machines such as other flip-flops and a sequence recognizer. In Section 11.9 use was again made of flip-flops as memory devices in the design of a more complex FSM, the one- to three-pulse generator. In this chapter, devices such as *shift registers* and *counters* are considered. Registers and counters constitute two very important classes of FSMs that are functionally different, and that are commonly used in the following ways:

- As stand-alone devices
- As data path devices in a controlled system
- As memory devices in controller design

As will become evident, there is a variety of different types of shift registers and even a greater variety of counters, some relatively simple and some relatively complex. Where applicable, use will be made of the *modular approach* to register and counter design, meaning that the modules can be cascaded into larger units. After completing this chapter the reader will be familiar with the design and operation of almost any shift register or counter.

12.2 REGISTERS

For reference purposes, there are four modes of bitwise register operation:

- True Hold a logic 0 or logic 1
- Shift Right a logic 0 or logic 1
- Shift Left a logic 0 or logic 1
- Parallel Load a logic 0 or logic 1

Not all shift registers are designed to operate in all four modes. The simplest register, one that can neither shift nor true hold, is called the *storage (holding) register*. The condition whereby a device can sustain any set of logic output values over any number of clock cycles independent of its input logic status is called *true hold*. The most complex shift register and one that is designed to operate in all four modes just listed is called the *universal shift register*. These and other shift registers will be considered in some detail during the discussions that follow. Here, the *modular approach* to register design is emphasized, featuring the design of a 1-bit slice, the Jth stage, which can be cascaded to form a register of any size. After completing this section, the reader will be familiar with most any shift register available commercially.

Registers are used in a wide variety of digital systems. They are used in the temporary storage of binary data, in data transmission, in arithmetic operations, in counter design, in accumulators, and in a host of other specialized applications. Registers are even used as memory elements in FSM design.

12.2.1 The Storage (Holding) Register

A register whose only function is to store information is called a *storage register* and is sometimes referred to as a *parallel-in/parallel-out* or *PIPO* register. It is, of course, the simplest of all registers, since it consists of nothing more than an array of synchronously triggered D flip-flops with independent data inputs. Shown in Figs. 12.1a and 12.1b are the state diagram and NS K-map for the Jth stage of a storage register. Notice that the NS function D_J obtained from the K-map is trivial since it can be easily deduced from the state diagram.

Storage (holding) registers are commonly used in the output stage of FSMs to filter out logic noise. A one-bit holding register is featured in Fig. 11.7, where it is used to filter the logic noise in output Z from the FSM. Storage registers are also used to provide ordered delivery of parallel data. For example, combinational adders, subtractors, multipliers, dividers, and arithmetic logic units all require the data to be introduced in an ordered and parallel fashion, an operation that is easily accomplished by using storage registers. A four-bit adder/subtractor of the type shown in Fig. 8.9 would require two four-bit PIPO registers, one for word A and the other for word B.

12.2.2 The Right Shift Register with Synchronous Parallel Load

The operation table for the Jth stage of a unidirectional shift register that can operate in only two modes, right shift and parallel load, is given in Fig. 12.2a. It is the function of this shift register that when the mode control S is inactive ($S = 0$) the register must shift right one bit on each triggering edge of clock, and when S is active ($S = 1$) it must parallel load synchronously. Synchronous parallel load means that the load values appearing on the input to the register will be loaded into the register's flip-flops by the action of clock, not via the flip-flop's asynchronous preset and clear overrides.

Shown in Fig. 12.2b is the state diagram for the shift register as derived from the operation table in Fig. 12.2a. Notice that the branching condition f_{ab} is obtained by ANDing the mode control logic (in the S column of the operation table) with the corresponding next state action parameter (in the NS$_J$ column) for each operation that can introduce a set condition ($0 \rightarrow 1$) into the register, and then ORing the results. Thus, both a right shift Q_{J+1} or a parallel load P_J operation can produce a set condition, hence $f_{ab} = \bar{S}Q_{J+1} + SP_J$. The set hold branching condition, f_b, must contain all that is in f_{ab}, hence $f_b = f_{ab}$, as is true for the D flip-flop in Fig. 10.23b. The other two branching conditions follow from the sum rule but need not be specified in plotting the K-map for D_J, given in Fig. 12.2c. The NS logic for

12.2 REGISTERS

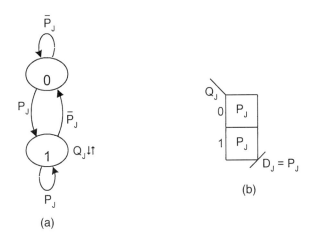

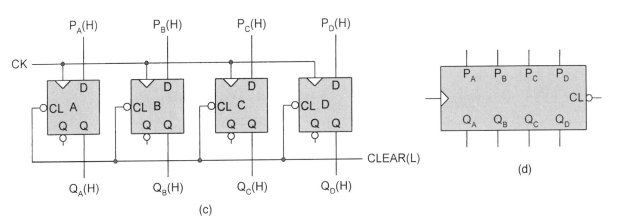

FIGURE 12.1
Design of the storage (PIPO) register. (a) State diagram and (b) NS K-map for the Jth stage. (c) Circuit diagram and (d) circuit symbol for a four-bit storage register.

the Jth module is easily seen to be

$$D_J = \bar{S} Q_{J+1} + S P_J, \tag{12.1}$$

which can be read directly from either the operation table or state diagram.

The next-state logic, as given by Eq. (12.1), can be implemented by using discrete logic or by using a 2-to-1 MUX—the logic is the same. For this design the latter is chosen. Shown in Figs. 12.3a and 12.3b are the MUX K-map for the Jth stage, and the connections for the $(J+1)$th, Jth, and $(J-1)$th stages. Here, S is the mode control and R is the serial input (SI) for right shifting. The serial out (SO) is taken from the least significant Q output bit, which in this case is Q_{j-1}. Notice that this shift register has PIPO, parallel-in/serial-out (PISO), serial-in/parallel-out (SIPO), and serial-in/serial-out (SISO) capability. But it can only shift right and parallel load. It does not have true hold, meaning that it cannot hold information over any number of clock periods independent of the parallel

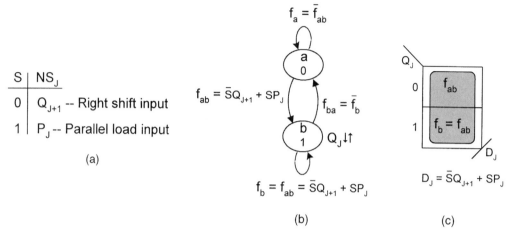

FIGURE 12.2
Design of a 1-bit slice parallel loadable right shift register. (a) Operation table showing mode control, S, and NS action for the Jth stage. (b) State diagram derived from (a). (c) NS K-map plotted from (b) assuming the use of D flip-flops.

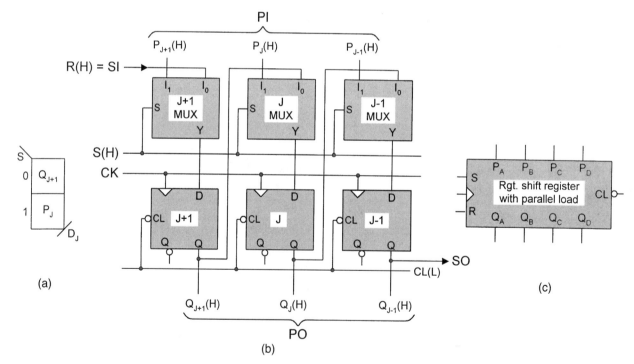

FIGURE 12.3
Modular design of a right shift register with synchronous parallel load and asynchronous clear capability. (a) Two-to-one MUX K-map for the Jth stage. (b) Modular connections for three bits. (c) Block diagram symbol for the 4-bit parallel loadable right shift register.

12.2 REGISTERS

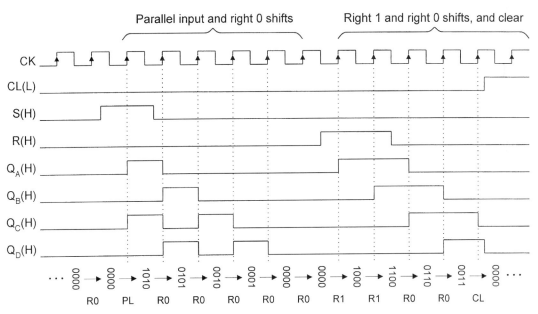

FIGURE 12.4
Timing diagram for the parallel loadable right shift register of Fig. 12.3c showing a parallel load of 1010 and subsequent right shifts for R values of 0, 1 and 0, and asynchronous clear.

load logic values. For true hold to exist, each module would have to feed its output back to itself on command of the mode control, which cannot happen in the shift register of Fig. 12.3.

Presented in Fig. 12.4 is the timing diagram for the four-bit, parallel loadable, right shift register represented by the block symbol in Fig. 12.3c. As indicated, a parallel load of $P_A P_B P_C P_D = 1010$ is introduced followed by right shifts for serial inputs set at 0, 1 and 0, and ending with an asynchronous clear $CL(L) = 1(L)$. For the sake of simplicity, no account is taken of the propagation delay through the logic.

Variations of the shift register in Fig. 12.3 are possible. By connecting the Q output of each flip-flop to the I_0 MUX input of the *next most significant bit* (MSB) stage, a parallel loadable left-shift register results. The 8-bit version of this shift register is equivalent to the commercial 74xx166 shift register. Or by eliminating the MUX of each module in Fig. 12.3b and by connecting each flip-flop output to the D input of the next MSB or next LSB stage, a simple left or right shift register results but, of course, without the parallel load feature. In the subsection that follows, a shift register having all these features and more is discussed in detail.

12.2.3 Universal Shift Registers with Synchronous Parallel Load

A shift register that possesses all four bitwise modes of operation, given at the beginning of this section, is called the *universal shift register* (USR). Its operation table in Fig. 12.5a indicates that the USR requires two mode control inputs, S_1 and S_0, for the four modes of

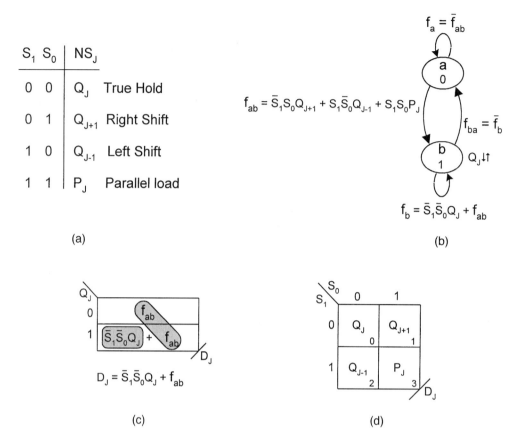

FIGURE 12.5
Design of a 1-bit slice universal shift register (USR). (a) Operation table for the Jth stage. (b) State diagram for the Jth stage. (c) NS logic K-map plotted from (b) assuming the use of D flip-flops. (d) MUX K-map for D_J.

operation. The state diagram for the Jth stage, shown in Fig. 12.5b, is obtained directly from the operation table. For example, the branching condition f_{ab} is the Boolean sum of all set producing conditions, each formed by ANDing the mode controls with its corresponding NS action parameter. Thus, since a set condition can be introduced by a shift right operation, the term $\bar{S}_1 S_0 \cdot Q_{J+1}$ must be included in the expression for f_{ab}. Because a left shift or a parallel load can also introduce a set condition into the register, two more terms are added for a total of three ANDed terms in the expression f_{ab} as indicated. Similarly, the set hold condition f_b must include $\bar{S}_1 \bar{S}_0 \cdot Q_J$ as well as all the set terms in f_{ab}. The two remaining branching conditions, f_{ba} and f_a, can be found from the sum rule, but are irrelevant for a D flip-flop design.

The minimum NS logic for the Jth stage is obtained from the K-map in Fig. 12.5c, which is plotted from the state diagram in Fig. 12.5b, assuming the use of D flip-flops. The logic expression for D_J, as read from the K-map, is

$$D_J = \bar{S}_1 \bar{S}_0 Q_J + \bar{S}_1 S_0 Q_{J+1} + S_1 \bar{S}_0 Q_{J-1} + S_1 S_0 P_J, \qquad (12.2)$$

12.2 REGISTERS

which is just the set hold condition in the state diagram. Equation (12.2) can be implemented either with discrete logic or by using a 4-to-1 MUX. For this example, the latter is chosen and its MUX K-map representation is given in Fig. 12.5d. From this K-map an n-bit USR can be configured. Shown in Fig. 12.6a is a 3-bit slice USR given for stages $J+1$, J, and $J-1$, all deduced from the MUX K-map in Fig. 12.5d. Notice that for this 3-bit USR, the serial input for right shifting, R, replaces the MUX input Q_{j+2} from the next MSB stage. Similarly, the serial input for left shifting, L, replaces the MUX input Q_{j-2} from the next LSB stage. The nature of the true hold mode is evident by observing that the output of each stage is fed

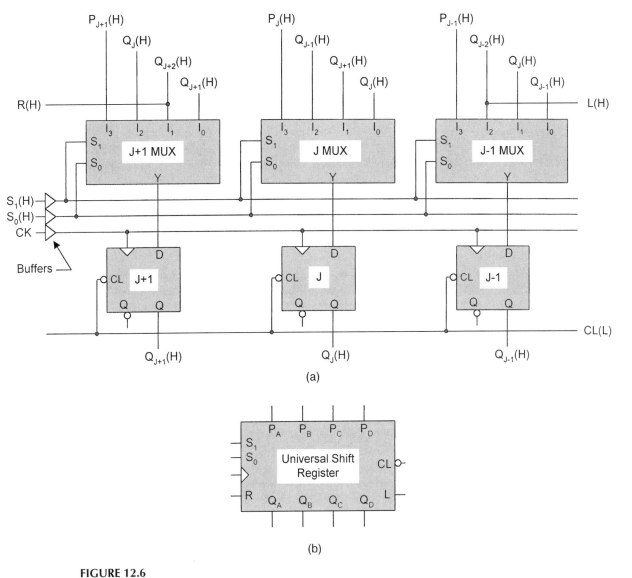

FIGURE 12.6
(a) MUX implementation of a 3-bit slice universal shift register (USR). (b) Block diagram symbol for a 4-bit USR triggered with RET D flip-flops showing buffered inputs.

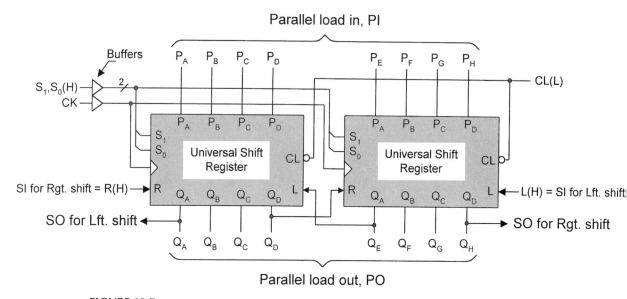

FIGURE 12.7
Block diagram symbols showing two cascaded 4-bit USRs to form an 8-bit USR.

back to its own I_0 MUX input, a requirement of cell 0 in the MUX K-map of Fig. 12.5d. The parallel load inputs for each stage, P_{J+1}, P_J, and P_{J-1} are the MUX inputs I_3, as required by cell 3 of the MUX K-map. The block diagram symbol for a 4-bit USR is given in Fig. 12.6b.

The USR stages in Fig. 12.6 can be cascaded to form a USR of any size. Shown in Fig. 12.7 is an 8-bit USR formed by using two 4-bit USRs. The external serial inputs, R and L, for right and left shifting, are the MSB stage MUX I_1 input and the LSB stage MUX I_2 input, respectively, of the cascaded system. The SO outputs are taken from the Q outputs at the extreme opposite ends of the two cascaded USRs as shown. Also, note that mode control inputs, S_1 and S_0, and the clock input CK must be buffered for fan-out purposes, as indicated in Figs. 12.6 and 12.7. Proper buffering of such signals is important to avoid the introduction of degraded signals to the various components of the USR (see Section 3.5 and Subsection 11.6.3). Individually and in cascade, but with FET D flip-flops, the USRs in Fig. 12.7 are equivalent to the 4-bit 74xx194 and to the 8-bit 74xx299 commercial USRs, respectively.

12.2.4 Universal Shift Registers with Asynchronous Parallel Load

The universal shift registers in Figs. 12.6 and 12.7 are parallel loaded synchronously as required by the operation table given in Fig. 12.5a. USRs can be parallel loaded *asynchronously* by removing the parallel load mode of operation from the operation table and by implementing it via the asynchronous preset and clear overrides of the flip-flops. When this is done, the new operation table, state diagram, NS K-map, and NS function D_J for the Jth module become those shown in Figs. 12.8a, b, and c, respectively. The resulting logic

12.2 REGISTERS

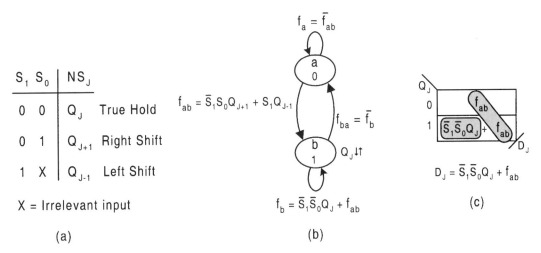

FIGURE 12.8
Design of a 1-bit slice universal shift register (USR) with asynchronous parallel load capability. (a) Operation table for the Jth stage. (b) State diagram for the Jth stage. (c) NS logic K-map plotted from (b), and minimum cover for NS function assuming the use of D flip-flops.

expression for D_J is

$$D_J = \bar{S}_1 \bar{S}_0 Q_J + \bar{S}_1 S_0 Q_{J+1} + S_1 Q_{J-1}, \tag{12.3}$$

which can be implemented by using either discrete logic or an SSI device such as a 4-to-1 MUX. Clearly, use of a MUX would not be the most efficient use of the logic, since there are only three terms, not four as in Eq. (12.2). Recall from Section 6.2 that full use of a 2^n-to-1 MUX as a function generator requires that 2^n unique functions be generated by the use of n data select inputs. This is not the case in Eq. (12.3). However, if optimized use of hardware is not required, use of an off-the-shelf MUX to implement Eq. (12.3) can suffice quite nicely.

The advantage of asynchronous parallel loading is that the load values can be introduced directly into the register's memory via the preset and clear overrides of the flip-flops and that shifting can occur on the rising edge of the clock waveform immediately following the release of the load command. In comparison, synchronous parallel loading can occur only on the triggering edge of the clock waveform, but the external load inputs should be synchronized to the clock signal. The load inputs for asynchronous parallel loading do not have to be synchronized.

A combinational logic truth table must be constructed to provide the external logic necessary for the *asynchronous parallel load* capability. This truth table is given in Fig. 12.9, together with the K-maps and minimum cover for the asynchronous *PRE* and *CLR* override inputs of the flip-flops. The minimum cover yields the following expressions for *PRE* and *CLR*:

$$\begin{cases} PRE = \overline{CL} \cdot LD \cdot P_J \\ CLR = LD \cdot \bar{P}_J + CL \\ = \overline{PRE} \cdot LD + CL \end{cases} \tag{12.4}$$

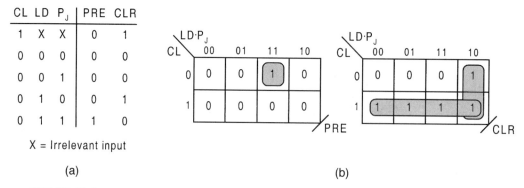

FIGURE 12.9
Combinational logic required for asynchronous parallel load. (a) Truth table. (b) K-maps and minimum cover for the preset and clear overrides of a flip-flop.

The second expression for *CLR* is obtained by complementing the K-map for *PRE*, ANDing with *LD* and ORing with *CL*, as the terms suggest. This alternative expression is the one used in this example.

Shown in Fig. 12.10a is the logic circuit for the Jth stage of a USR with asynchronous parallel load and asynchronous clear capability, where use has been made of Eqs. (12.4). Here, the NS-forming logic for D_J is implemented with discrete logic, and an FET D flip-flop is used as the memory. Notice that the load control input is made active low, *LD(L)*, which is commonly done for such devices. The Q_{J+1} and Q_{J-1} inputs are taken from the next MSB and next LSB stages, respectively, as was done in Fig. 12.6a. Proper buffering of input signals is indicated in Fig. 12.10a.

Cascading four identical stages results in a 4-bit USR having the block circuit symbol shown in Fig. 12.10b. Observe that it differs from that in Fig. 12.6b only by the presence of the *LD(L)* input required for asynchronous parallel load capability. An 8-bit USR is produced by cascading two 4-bit modules as was done in Fig. 12.8, but with the added *LD(L)* input buffered and connected to both 4-bit USRs. Functionally, the 4-bit and 8-bit USRs are equivalent to the commercial 74xx194 and 74xx299 USRs, respectively, but with asynchronously parallel loaded data inputs. A perspective on synchronous vs asynchronous parallel loading of data is given later in Subsection 12.3.6 following a detailed discussion of counters.

12.2.5 Branching Action of a 4-Bit USR

In Section 13.3 the USR is used as the memory as a form of alternative architecture in the design of state machines. To program the USR in such applications requires that its branching action be labeled as illustrated in Fig. 12.11 for a fictitious state machine. Here, it is assumed that shifting action has priority over parallel load. Notice that for the branching action of the USR there are six possibilities:

$$H, SL0, SL1, SR0, SR1, \text{ and } PL,$$

representing hold, shift left 0 or 1, shift right 0 or 1, and parallel load, respectively.

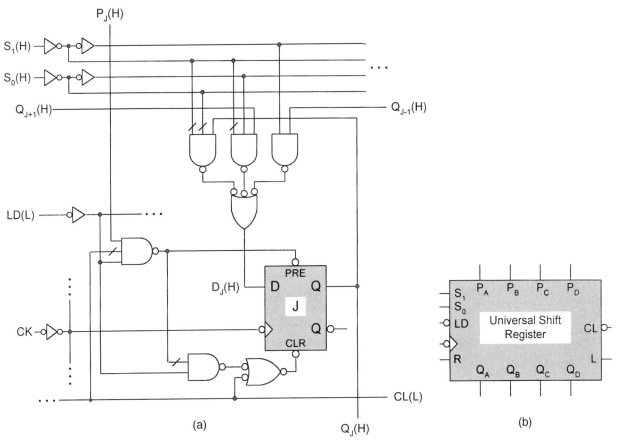

FIGURE 12.10
(a) The Jth stage for a USR with asynchronous parallel load capability. (b) Block circuit symbol for a 4-bit USR with asynchronous parallel load capability.

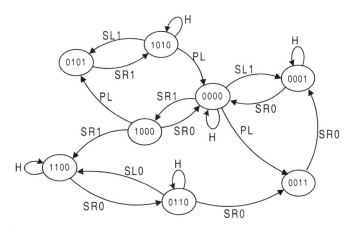

FIGURE 12.11
Illustration of the branching action of a USR used as the memory for a fictitious state machine.

12.3 SYNCHRONOUS BINARY COUNTERS

Synchronous counters form a class of FSMs for which each state code assignment of its state diagram is taken to be a number in a count sequence. Most simple synchronous counters are degenerate Moore machines that obey the basic model of Fig. 10.3c, since their only outputs are the state variables. Other synchronous binary counters are those that have control inputs and unconditional or conditional outputs, and that adhere to either the Moore or Mealy model (Fig. 10.4 or 10.5). In any case, such binary counters are classified as *modulo-N counters* or as *divide-by-N counters*, where N is the number of states of the sequence. The divide-by-N designation results from the fact that the clock frequency is divided by $N(f_{CK}/N)$ if taken from the MSB output of the counter. The up/down binary counter of Figs. 10.57 and 10.58 is classified as a modulo-8 (divide-by-8) *bidirectional counter*. But as will soon become evident, it is also a divide-by-8, divide-by-4, or divide-by-2 binary counter depending on from which output A, B, or C the count is taken, respectively. Any of these counters can be designed with synchronous or asynchronous parallel load capability, which means that these counters can begin the synchronous count from the parallel load state.

The state sequence of a synchronous counter need not conform to a regular binary count, up or down. Synchronous counters can be designed to count in any of the codes defined in Section 2.10, and in any direction. The most common of these for use in counter design are the decimal codes, specifically the BCD code. A BCD counter has 10 states and is accordingly called a *decade* or *divide-by-10* counter. Still, the count sequence does not have to be binary. Counters can be designed to count in a unit distance code sequence of the type given in Table 2.12. The most common of these is the Gray code counter that sequences through states shown in column (2) of Table 2.12, assuming it to be of four bits.

Counters discussed so far are classified as *synchronous counters* because their flip-flops are all triggered simultaneously by the clock signal. Counters whose flip-flops are each triggered by the output of the next LSB stage flip-flop are called *ripple counters* or *asynchronous counters*. Thus, the triggering action of the flip-flops ripples from the LSB stage flip-flop, where the external clock enters the counter, to the MSB stage flip-flop. Ripple counters can be designed to up count, or down count, or both. These counters will be discussed in detail in Section 12.5.

Finally, there is a broad class of synchronous counters that can be designed by using shift registers of the type discussed in Section 12.2. One such counter, called a *ring counter*, sequences through a series of one-hot code states as in column (c) of Table 2.11. Another counter in this class of counters is called the *twisted ring counter* (also called the *Johnson* or *Mobius counter*), which sequences through a series of creeping code states as in column (7) of Table 2.10. Still other counters can be configured with D flip-flops and XOR gates to form what are called *autonomous linear feedback shift register counters* or simply *ALFSR counters*. ALFSR counters are useful in generating pseudo-random sequences of n-bit binary numbers, among other uses.

For future reference, the following lists several members of the rather diverse family of synchronous counters:

Code counters
    Binary divide-by-2^n counters
    Decimal counters (e.g., BCD, XS3 counters)
    Gray code counters

12.3 SYNCHRONOUS BINARY COUNTERS

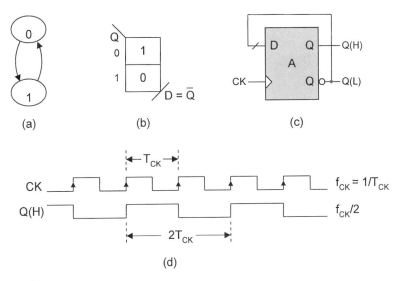

FIGURE 12.12
D flip-flop design of the divide-by-2 counter. (a) State diagram. (b) NS K-map. (c) Logic circuit. (d) Timing diagram.

Bidirectional (up/down) counters
Multisequence counters (e.g., binary/Gray code counters)
Shift register counters
    Standard ring counters
    Twisted ring (Johnson or Mobius) counters
    Linear feedback shift register (LFSR) counters

12.3.1 Simple Divide-by-N Binary Counters

Although these counters represent some of the simplest state machines discussed thus far, their coverage is important to an understanding of some of the basic concepts involved.

The Divide-by-2 Counter Shown in Fig. 12.12 are the state diagram, K-map, logic circuit, and timing diagram for a divide-by-2 ($\div 2^1$) binary counter that has been implemented by using an RET D flip-flop. Because it exhibits only toggle character, it is also called a *toggle module*. The toggle module is used in the design of ripple counters (Section 12.5), in the design of data-triggered counters (Subsection 13.6.2), and as a memory element for pulse-mode state machine design (Chapter 15). Of course, as a divide-by-2 counter, it performs the simple function of dividing the clock frequency by 2, as indicated in Fig. 12.12d.

The Divide-by-3 Counter The divide-by-3 counter has just three states, and therefore is not a divide-by-2^n-counter — it does not complete the 2^2 count, resulting in some interesting consequences. Shown in Fig. 12.13a is the state diagram for a divide-by-3 counter where the sequence is binary $\cdots 00 \rightarrow 01 \rightarrow 10 \rightarrow 00 \cdots$. The NS K-maps are given in Fig. 12.13b, assuming the use of D flip-flops, and the timing diagram is presented in Fig. 12.13c. Notice that each of the two outputs from the flip-flops divides the clock frequency by 3 ($f_{CK}/3$)

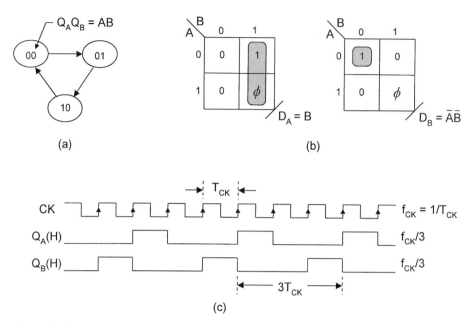

FIGURE 12.13
Design of the divide-by-3 binary counter with D flip-flops. (a) State diagram. (b) NS K-maps. (c) Timing diagram.

and has a $33^1/_3\%$ duty cycle, independent of the clock duty cycle. For the sake of brevity, no logic circuit is given for this example.

The Divide-by-4 Counter The divide-by-4 ($\div 2^2$) counter completes the $2^n = 2^2$ count so that advantage can be taken of the *toggle character* inherent in a divide-by-2^n counter. This means that the use of T flip-flops can be used advantageously in the design of such counters. Shown in Fig. 12.14 are the state diagram, NS K-maps, logic circuit for a T flip-flop design, and the timing diagram for this modulo-4 counter. The toggle character is obvious by an inspection of the state diagram: bit A toggles every other bit and bit B toggles on each bit. It is for this reason that the T flip-flop design generates the simplest NS logic. Keep in mind, however, that extra logic is required to convert a D flip-flop to a T flip-flop, as indicated in Fig. 10.39; T flip-flops are not normally available commercially. Notice that the outputs each exhibit a duty cycle of 50% independent of the duty cycle of the regular clock waveform. Also, observe that the output from the MSB flip-flop (A) divides the frequency by 4, whereas the output from the LSB flip-flop (B) divides it by 2. This fact will be expanded upon in the discussion that follows.

Perspective on Divide-by-N Counters Before moving on to examples of more complex counters, it is worth while to pay attention to some important characteristics of the divide-by-N binary counters. These counters can be divided into two categories: those that are divide-by-2^n (modulo 2^n) counters and those that are not ($N \neq 2^n$). The outputs from a divide-by-2^n binary counter are always of a 50% duty cycle and have frequency division in descending

12.3 SYNCHRONOUS BINARY COUNTERS

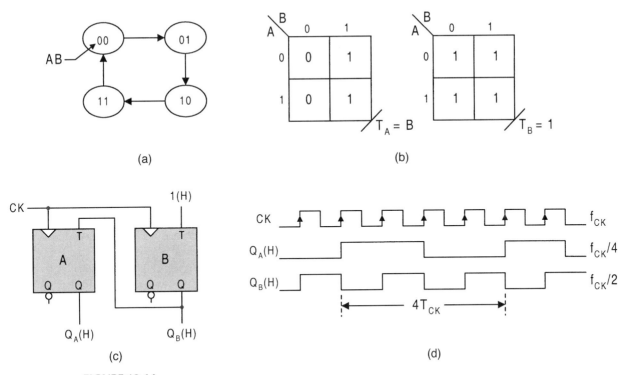

FIGURE 12.14
Design of the divide-by-4 counter by using RET T flip-flops. (a) State diagram. (b) NS K-maps and minimum cover. (c) Logic circuit. (d) Timing diagram showing divide-by-2 and divide-by-4 outputs.

orders of 2^n beginning with the MSB flip-flop and ending with the LSB flip-flop, where $n = 1, 2, 3, 4 \cdots$. Thus, for a 4-bit binary counter Q_A is a $f_{CK} \div 2^4$ output, Q_B is a $f_{CK} \div 2^3$ output, Q_C a $f_{CK} \div 2^2$ output, Q_D a $f_{CK} \div 2^1$, and are all independent of count direction, up or down. In contrast, divide-by-N counters, for which $N \neq 2^n$, do not have frequency division in descending orders of 2^n and do not always have outputs of the same duty cycle.

There is one further and important distinction between these two categories of counters. The divide-by-2^n binary counters have complete toggle character for which the use of T flip-flops yields minimum NS logic. Divide-by-N counters, for which $N \neq 2^n$, do not complete the 2^n count and consequently do not have complete toggle character. For these counters the use of JK flip-flops will most likely yield NS logic of least cost hardware-wise. The following example of a BCD counter is evidence of this latter fact.

12.3.2 Cascadable BCD Up-Counters

The BCD counter is designed to sequence states 0000 through 1001 in binary, after which it must start over. A review of the BCD code is provided in Subsection 2.4.1. In order for the BCD counter to be useful in representing a range of weighted digits ($\cdots$ 100, 10, 1, 0, 0.1, 0.01 $\cdots$), it is necessary to design cascading capability into the counter. Shown in Fig. 12.15 is the state diagram for a BCD (decade) up-counter that can be cascaded to represent multiple

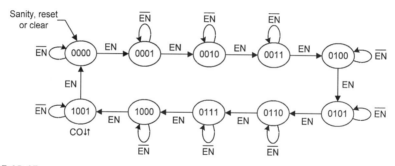

FIGURE 12.15
State diagram for a cascadable BCD up-counter.

decades. It has 10 states and has an enable (EN) input and an unconditional carryout (CO) output for cascading purposes.

The NS K-maps are derived directly from the state diagram in Fig. 12.15 by using the mapping algorithm given in Section 10.6 as applied to T flip-flops. The results are given in Fig. 12.16 together with the output K-map for CO. Also shown are the NS K-maps for a JK flip-flop design, which are obtained by K-map conversion. The T-to-JK K-map conversion algorithm is easily deduced from Eq. (10.12) and by observing the domain partitions in the K-maps indicated with heavy lines. The algorithm is stated as follows:

Algorithm 12.1: $T \rightarrow JK$ K-map Conversion [Refer to Eq. (10.12)]

(1) For all that is NOT A in the T_A K-map, transfer it to the J_A K-map directly.
(2) For all that is A in the T_A K-map, transfer it to the K_A K-map directly.
(3) Fill in the empty cells with don't cares.
(4) Repeat steps (1), (2), and (3) for the $T_B \rightarrow J_B$, K_B and $T_C \rightarrow J_C$, K_C, etc., K-map conversions, always by observing the domain partitions.

Note that Algorithm 12.1 can be applied in reverse — that is, for $JK \rightarrow T$ K-map conversion. Thus, for domain A, all that is NOT A in the J_A K-map is transferred directly to the T_A K-map, and all that is A in the K_A K-map is transferred directly to the T_A K-map, etc.

An inspection of the NS K-maps in Fig. 12.16 indicates that the JK NS logic is simpler than that for the T NS logic. This results from the don't cares that are inherent in the excitation table for the family of JK flip-flops given in Fig. 10.45. Because of the simpler logic, the cascadable BCD counter is implemented with FET JK flip-flops, as shown in Fig. 12.17a. The D flip-flop implementation of this counter is not considered here, but would involve considerably more NS logic.

The timing diagram for the 4-bit BCD counter is given in Fig. 12.18 together with frequency division and duty cycle information for the four state variable outputs. Notice that all but the $D(H)$ output are divide-by-10 outputs and that the duty cycles vary from 20% for the $A(H)$ output to 50% for $D(H)$. Output $C(H)$ has a split duty cycle. Such information can be important for some applications.

The four-bit BCD counter module in Fig. 12.17a can be cascaded to form any number of weighted digits. For example, cascading two 4-bit counters permits a 0 to 99 count or

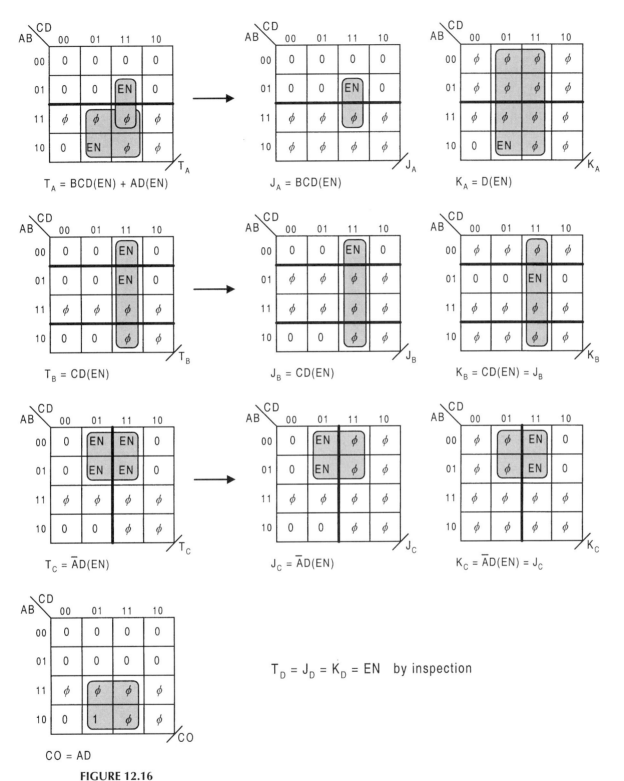

FIGURE 12.16
NS and output K-maps for the T or JK flip-flop design of the cascadable BCD counter in Fig. 12.15.

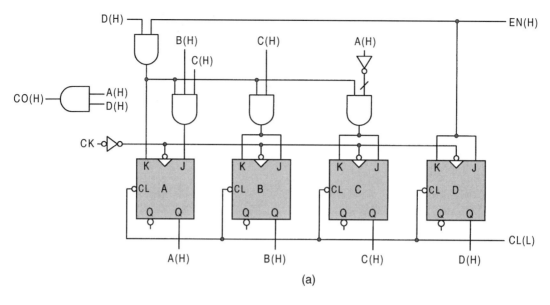

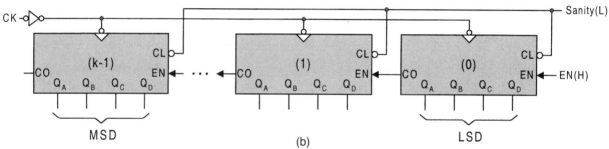

FIGURE 12.17
(a) Implementation of the 4-bit cascadable BCD up-counter by using FET JK flip-flops. (b) Cascaded 4-bit modules to produce a k-digit BCD counter.

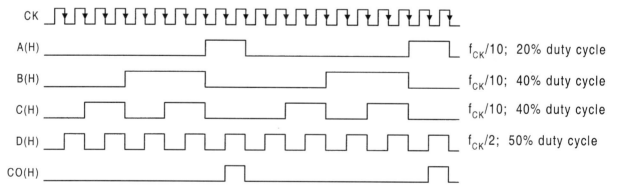

FIGURE 12.18
Timing diagram for the BCD up-counter of Fig. 12.17 showing the frequency division and duty cycle percentages for the state variables, and the output CO in state 1001.

578

12.3 SYNCHRONOUS BINARY COUNTERS

0.1 to 9 count, etc., depending on how one views the count. Cascading three such modules gives a 0 to 999 count. Or generally, cascading k of these modules, as in Fig. 12.17b, forms a k-digit BCD counter with a 0 to $10^k - 1$ count, where k is an integer ($k = 0, 1, 2, 3, \ldots$). The EN input to the LSD stage is, of course, a counter enable control. If $EN(H) = 1(H)$, the counter is enabled. But if $EN(H) = 0(H)$, the counter is disabled and is caused to hold in whatever state it is in at the time. The EN input can be replaced by an ANDing operation permitting two signals to control the operation of the counter: a count enable input and an inhibit input, both performing basically the same function.

The manner in which a cascaded BCD counter operates is as follows: Each full count of the LSD (0) stage sends a CO signal to the next MSD (1) stage which is properly enabled on the next rising edge of the clock pulse. Thus, for each full count (0-to-9) of stage (0), stage (1) is bumped up 1. So after 10 such full counts of stage (0), stage (1) completes its full count (0-to-9) and enables the next MSB stage, which is bumped up one on the next rising edge of clock. Any output race glitches (ORGs) that occur are of no consequence, since the single-output CO can enable the next stage only if is issued for a complete clock period. ORG glitches occur immediately following the triggering edge of the clock waveform and damp out long before the output CO can be picked up by the next MSB stage.

12.3.3 Cascadable Up/Down Binary Counters with Asynchronous Parallel Load

For the most part the design details for this counter have already been established. Equations (10.17) in Subsection 10.12.1 give the NS and output logic for a 3-bit (divide-by-2^3) up/down binary counter, assuming the use of T flip-flops. There, a single input, X, controls the direction of the count such that if $X = 1$ the count is Up or if $X = 0$ the count is Down. The state diagram for a 4-bit (divide-by-2^4) up/down counter is given in Fig. 12.19, where two direction controls are used, Up and Dn (down). Clearly, these two direction controls can never be active at the same time. Should this happen, the FSM would not know how to respond and would malfunction. Notice that all holding conditions are omitted in the state diagram of Fig. 12.19. This is permissible since a T flip-flop design is anticipated — *a given state cannot toggle to itself.*

Equations (10.17) can easily be extended and applied to the 4-bit up/down counter of Fig. 12.19 by noting the trend in the equations and by taking $X = Up$ and $\bar{X} = Dn$. When this is done, there results the following NS and output equations for the cascadable divide-by-2^4

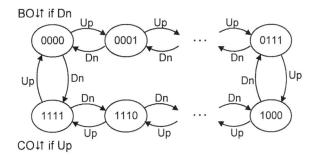

FIGURE 12.19
State diagram for a cascadable up/down binary counter.

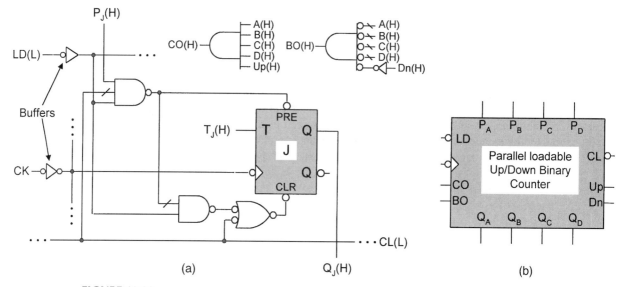

FIGURE 12.20
Implementation of the 4-bit up/down binary counter of Fig. 12.19 with asynchronous parallel load. (a) Logic circuit for the Jth stage with CO and BO logic. (b) Block circuit symbol.

binary up/down counter:

$$\left\{\begin{array}{l} T_A = BCD \cdot Up + \bar{B}\bar{C}\bar{D} \cdot Dn \\ T_B = CD \cdot Up + \bar{C}\bar{D} \cdot Dn \\ T_C = D \cdot Up + \bar{D} \cdot Dn \\ T_D = Up + Dn \\ CO = ABCD \cdot Up \\ BO = \bar{A}\bar{B}\bar{C}\bar{D} \cdot Dn \end{array}\right\} . \qquad (12.5)$$

The trend in the NS and output logic having been established, the equations for any size bidirectional divide-by-2^n binary counter can be written directly without the need for K-maps or minimization algorithms.

All that remains to complete the design of this counter is to obtain the external logic required for the asynchronous parallel load capability. This has already been done in the form of Eqs. (12.4) and Fig. 12.10a for the USR with asynchronous parallel load capability (see Subsection 12.2.4). Presented in Fig. 12.20a is the logic circuit for the jth stage. Here, it is understood that T_J represents the four different NS functions (T_A, T_B, T_C, and T_D) in Eqs. (12.5) taken in turn, and that $LD(L)$ is the command to parallel load the value $P_J(H)$. Observe that when parallel loading, the asynchronous overrides PRE and CLR are never active at the same time.

The block circuit symbol for the 4-bit parallel loadable up/down counter is provided in Fig. 12.20b. Here, each stage is that of Figure 12.20(a), but with the appropriate NS function taken from Eqs. (12.5). Shown in Fig. 12.21 is a $4k$-bit parallel loadable up/down binary counter consisting of k 4-bit counters in cascade. The output logic, CO and BO, for each 4-bit counter is given in Fig. 12.20a. Notice that the input $LD(L)$ is common to all k modules

12.3 SYNCHRONOUS BINARY COUNTERS

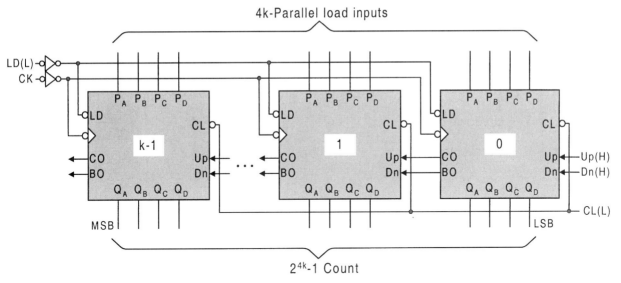

FIGURE 12.21
Cascading of 4-bit parallel loadable up/down counters to form a $4k$-bit counter.

for purposes of introducing the $4k$ parallel load values to the counter asynchronously via the preset and clear overrides of its $4k$ T flip-flops.

The operation of the k-stage up/down binary counter in Fig. 12.21 is straightforward. Any time $LD(L)$ is active the counter is parallel loaded asynchronously and the counting operation is interrupted. When $LD(L)$ becomes inactive, counting is resumed up or down from that parallel load value. The direction of count is determined by which input, $Up(H)$ or $Dn(H)$, is active. Obviously, both cannot be active at the same time. When one of the k stages has completed its full count, an output (CO or BO) is issued to the next stage enabling it on the next triggering edge of clock. If the count is up, CO is issued; if the count is down, BO is issued. The maximum number of states through which *this* k-stage counter can sequence is $2^{4k} - 1$. As is true for the counter in Fig. 12.17, any ORG that exists in the CO or BO output signal disappears long before the next stage is triggered.

The advantage of an asynchronous parallel load feature is that the load values are introduced directly into the memory without having to be clocked in or synchronized, as in the synchronous parallel load arrangement. An asynchronous parallel load capability can be added to any counter that has flip-flops with preset and clear overrides. For example, the BCD counters in Fig. 12.17 can be designed with this feature if the flip-flops are given both preset and clear input overrides. Considered next are the cascadable up/down binary counters with synchronous parallel load and true hold capability.

12.3.4 Binary Up/Down Counters with Synchronous Parallel Load and True Hold Capability

The design of this counter creates a special dilemma. The parallel load must be introduced to the counter synchronously, but T flip-flops lack the capability to do this. One approach would be to design this counter by using D flip-flops with D NS logic. However, the NS

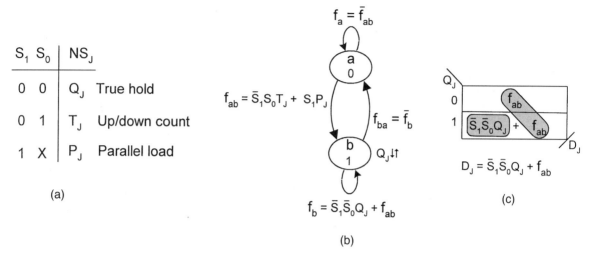

FIGURE 12.22
Design of a 4-bit slice up/down counter with synchronous parallel load and true hold capability. (a) Operation table for the Jth stage. (b) State diagram for the Jth stage. (c) NS logic K-map and minimum cover obtained from (b) assuming the use of D flip-flops.

logic would be too costly (hardware-wise) to justify a design by this means. A much simpler approach is to use D flip-flops for the parallel load and true hold capability but convert them to T flip-flops for the up/down count, all on command of two mode-control inputs. This is the method that is used in this example.

Shown in Fig. 12.22 are the operation table, state diagram, NS K-map, and MUX K-map for the Jth stage of an up/down counter with synchronous parallel load and true hold capability. The NS K-map yields the NS function for the Jth stage D flip-flop as

$$D_J = \bar{S}_1 \bar{S}_0 Q_J + \bar{S}_1 S_0 T_J + S_1 P_J \rightarrow \bar{S}_1 \bar{S}_0 Q_J + \bar{S}_1 S_0 (T_J \oplus Q_J) + S_1 P_J, \quad (12.6)$$

where S_1 and S_0 are the mode controls for this counter. The NS functions and output functions for the up/down count are given by Eqs. (12.5) and are reproduced here for the convenience of the reader:

$$\begin{cases} T_A = BCD \cdot Up + \bar{B}\bar{C}\bar{D} \cdot Dn \\ T_B = CD \cdot Up + \bar{C}\bar{D} \cdot Dn \\ T_C = D \cdot Up + \bar{D} \cdot Dn \\ T_D = Up + Dn \\ CO = ABCD \cdot Up \\ BO = \bar{A}\bar{B}\bar{C}\bar{D} \cdot Dn \end{cases}. \quad (12.7)$$

Thus, T_J in the operation table and state diagram of Fig. 12.22, and each T_J in Eq. (12.6) and in Eqs. (12.7), becomes $T_J \oplus Q_J$ to permit conversion between T and D flip-flops, as explained in the next paragraph.

12.3 SYNCHRONOUS BINARY COUNTERS

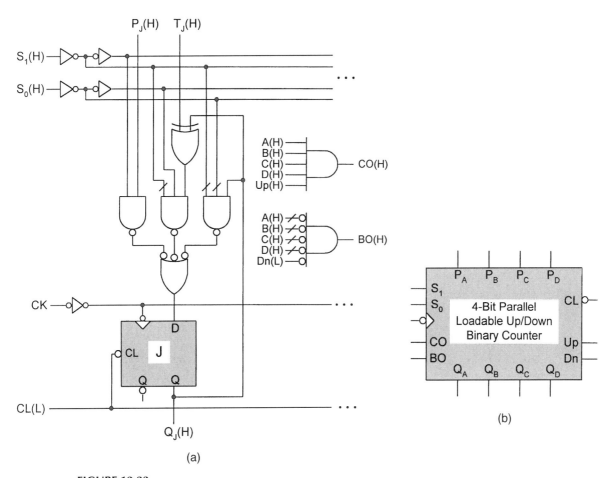

FIGURE 12.23
Implementation of the cascadable up/down binary counter with synchronous parallel load and true hold capability represented in Fig. 12.22 and by Eqs. (12.6) and (12.7). (a) The Jth stage showing D-to-T flip-flop conversion logic and output logic. (b) Block diagram symbol for the 4-bit counter.

Implementation of Eqs. (12.6) and (12.7) is given in Fig. 12.23a for the Jth stage of a 4-bit cascadable up/down binary counter with synchronous parallel load and true hold capability. The block circuit symbol for a 4-bit version of this counter is provided in Fig. 12.23b. This design requires the use of FET D flip-flops, which permit the counter to parallel loaded or hold in a particular state for any number of CK cycles, all depending on the mode controls as indicated in Fig. 12.22a. But to count up or down, the D flip-flops are converted to T flip-flops by Eq. (10.10) or, in this case, by $D_J = T_J \oplus Q_J$ for the Jth stage. This approach provides the best of both worlds: D flip-flops for parallel load and true hold capability, and T flip-flops for an efficient means of dealing with the toggle character inherent in the binary up or down count. Counter designs by this means are especially attractive for implementation by registered PLD devices. These devices include V-type PALs (see Section 7.4), or the Xilinx FPGAs discussed in Subsection 7.7.3, all of which have edge-triggered D flip-flops with asynchronous preset and clear overrides built into them. Implementation of this counter by using FPGA-type devices will require special programming software.

The cascadable up/down binary counter in Figs. 12.22 and 12.23 can be designed with asynchronous parallel load capability while retaining the true hold feature. To do this simply requires that the synchronous parallel load feature be removed from the operation table in Fig. 12.22a, and then applied as an asynchronous parallel load via the *PRE* and *CLR* overrides of each D flip-flop, as is done in Fig. 12.20. Now, only one mode control remains (*S*), permitting the use of a 2-to-1 MUX to implement the new D_J function. With a few changes, the bidirectional 4-bit counter in Fig. 12.23 is equivalent to the commercial 74xx169 counter.

12.3.5 One-Bit Modular Design of Parallel Loadable Up/Down Counters with True Hold

A one-bit modular approach will now be used to design a binary counter that can count up or down, that can be parallel loaded synchronously, and that has *true hold* capability. The operation table for a 1-bit slice counter of this type is presented in Fig. 12.24a. The *LD* and *EN* inputs are the mode control inputs that determine whether the counter will hold, count or parallel load synchronously. The count function *CNT* represents an up- or down-count depending on the count direction parameter $D/\bar{U}$ introduced in the following discussion. The state diagram, shown in Fig. 12.24b, is constructed directly from the operation table in Fig. 12.24a. For example, the branching condition f_{ab} is the Boolean sum of all set producing conditions, each formed by ANDing the mode control inputs on the left with the corresponding NS action parameter on the right of the operation table. Thus, count and parallel load are the set-producing modes of operation that constitute f_{ab}. The set hold condition must contain the true hold condition as well as f_{ab}. The remaining two branching conditions can be obtained from the sum rule, but are of no consequence when designing for D flip-flops.

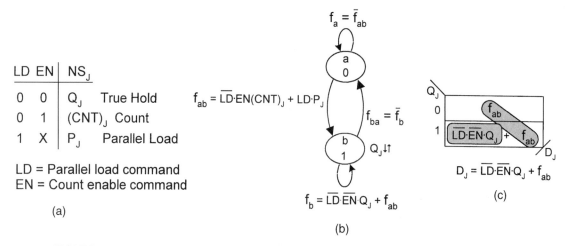

FIGURE 12.24
Design of a 1-bit slice up/down counter with synchronous parallel load and true hold capability. (a) Operation table. (b) State diagram for the *J*th stage. (c) K-map and minimum cover for the *J*th stage assuming the use of D flip-flops.

12.3 SYNCHRONOUS BINARY COUNTERS

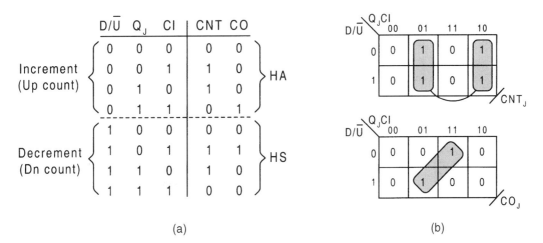

FIGURE 12.25
Truth table representing the increment and decrement operations for the Jth stage of the counter. (b) K-maps for CNT (sum or difference) and CO (carry-out or borrow-out).

The NS K-map, shown in Fig. 12.24c, is obtained from the state diagram in (b) by applying the mapping algorithm assuming the use of D flip-flops. The resulting NS equation for the Jth 1-bit slice is given by

$$D_J = \overline{LD} \cdot \overline{EN} \cdot Q_J + \overline{LD} \cdot EN \cdot (CNT)_J + LD \cdot P_J, \qquad (12.8)$$

which can be implemented with discrete logic or by using an SSI device such as a 4-to-1 MUX. Use of a MUX, however, would not be an efficient use of the device since not all four function terms are present in the expression. For this example, discrete logic will be used to implement Eq. (12.8).

All that remains to be done is to find the logic for the *CNT* parameter representing either a count up or a count down. This is accomplished by constructing a combinational logic truth table for the Jth stage, as shown in Fig. 12.25a. Notice that the first four rows of the truth table correspond to that of a half adder (HA) for up count while the latter four rows correspond to that of a half subtractor (HS) for down count. Here, a new parameter $D/\bar{U}$ is introduced to indicate count direction. The carry-out output *CO* serves as both the carry-out for increments and borrow-out for decrements. Inputs Q_j and *CI* can be thought of as $A \pm B$, where $A = Q_J$ and $B = CI$. For a review of adders and subtractors the reader is referred to Sections 8.2 and 8.3.

The CNT and CO outputs for the Jth stage are mapped in Fig. 12.25b and minimum cover is extracted by using XOR patterns to yield the results

$$\begin{cases} CNT_J = Q_J \oplus CI \\ CO_J = CI \cdot (D/\bar{U} \oplus Q_J) \end{cases}. \qquad (12.9)$$

Implementing these equations together with Eq. (12.8) gives the logic circuit in Fig. 12.26a for the Jth 1-bit slice. This module can be cascaded to form a counter of any number of bits.

The block circuit symbol for a 4-bit counter of this type is shown in Fig. 12.26b, where it is required that the CI of the LSB stage be set at 1(H) so as to enable the AND gate for CO from

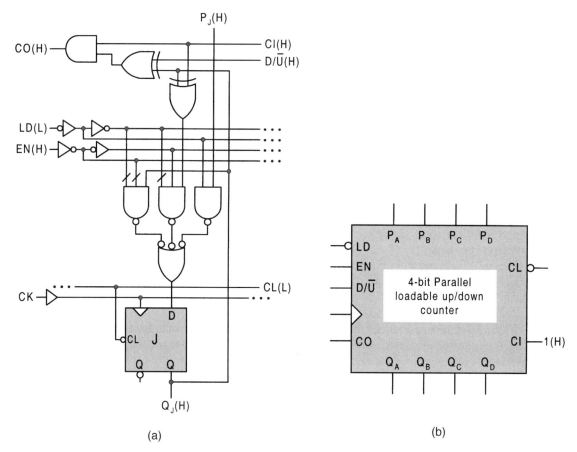

FIGURE 12.26

Implementation of Eqs. (12.8) and (12.9) for the Jth 1-bit counter module with synchronous parallel load and true hold. (b) Block circuit symbol for the 4-bit counter implemented by cascading four 1-bit slices as in (a).

this stage. The gate/input tally for the NS forming logic to this 4-bit synchronously parallel loadable counter is 28/68, excluding possible inverters. For comparison, the gate/input tally for the 4-bit synchronously parallel loadable up/down counter of Fig. 12.23b is 32/86, also excluding possible inverters. Thus, by gate tally alone, the half-adder/half-subtractor approach in Fig. 12.26 is favored over that of Fig. 12.23. Both designs have three-level NS forming logic for each stage, depending on how one views the implementation of the XOR gates. A PAL or FPGA design of a parallel loadable up/down counter would probable favor the design of Fig. 12.23b since one fewer XOR gate is involved. Remember that these array logic devices are registered with only D flip-flops and that the XOR gate must likely be implemented by using the two-level SOP logic of Eq. (3.4) in Section 3.9. However, array logic devices such as the Xilinx FPGAs having arithmetic units might be more amenable to the half-adder/half-subtractor approach of Fig. 12.26.

The 4-bit counter in Fig. 12.26b or that in Fig. 12.23b can be cascaded to form k 4-bit stages that can sequence through $2^{4k} - 1 = 16^k - 1$ states, hence a divide-by-2^{4k} counter.

12.3 SYNCHRONOUS BINARY COUNTERS

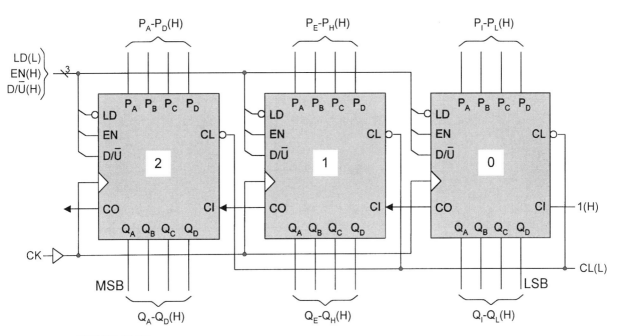

FIGURE 12.27
A divide-by-2^{12}(16^3) up/down binary counter with synchronous parallel load and true hold capability formed by cascading three 4-bit counters of the type given in Fig. 12.26.

Shown in Fig. 12.27 is a three-stage counter of the former type that can sequence through $16^3 - 1 = 4095$ states. At any point in the operation of the counter, it can be given the command $LD(L) = 1(L)$ to parallel load a binary word of 12 bits. Then when $LD(L) = 0(L)$, the counter can hold that number if EN is also inactive $EN(H) = 0(H)$, or it can count up or down from that number if $EN(H) = 1(H)$. The direction of count, of course, depends on the setting of the direction control $D/\bar{U}$: $D/\bar{U}(H) = 0(H)$ for Up, and $D/\bar{U}(H) = 1(H)$ for down. Thus, by parallel loading any number between 0 and 4095, any count sequence or frequency division in that range can be obtained. For example, by parallel loading $000100100111 (= 295_{10})$, the counter can count up from 295 to 4095, a frequency division of $f_{CK} \div 3800$. Or if the counter is set to count down from that parallel load, a frequency division of $f_{CK} \div 295$ would result. One application of the frequency division aspect of counter operation is the production of relatively long periods of time. Thus, by parallel loading 295_{10}, a time $T = 3800T_{CK}$ can be produced by an up count, assuming it completes the count from 295 to 4095. Alternatively, a down count from this value results in time period of $295T_{CK}$, assuming that it is the final CO signal that is sensed. Remember that the 12 outputs of the counter can be tapped for frequency divisions within the range of the complete count, thus allowing for a wide range of time periods.

The counter of Fig. 12.26 can also be designed with asynchronous parallel load capability. To do this requires only that the parallel load feature be removed from the operation table in Fig. 12.24a and implemented by using Eqs. (12.4) together with the *PRE* and *CLR* overrides of the D flip-flops. Shown in Fig. 12.28 are the operation table, MUX K-map for D flip-flops, and logic circuit for the Jth 1-bit stage of such a counter. This 1-bit slice can be cascaded

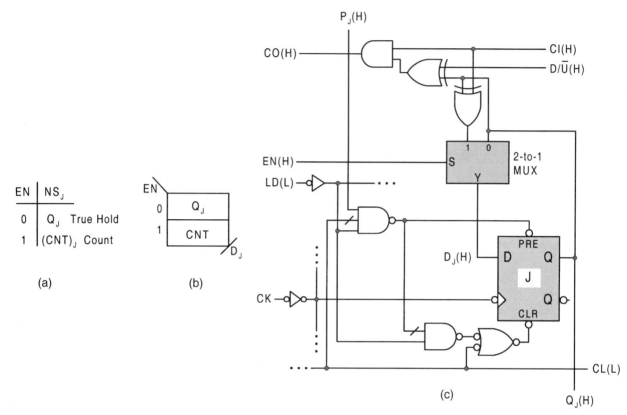

FIGURE 12.28
Design of an up/down counter with asynchronous parallel load and true hold capability. (a) Operation table. (b) MUX K-map assuming the use of D flip-flops. (c) Logic circuit for the Jth 1-bit slice.

as in Fig. 12.27, the only difference being that the loading is produced asynchronously and that FET D flip-flops are used.

12.3.6 Perspective on Parallel Loading of Counters and Registers: Asynchronous vs Synchronous

As has been mentioned or implied at various points in previous discussions, there are two advantages to asynchronously parallel loading data into the memory flip-flops of registers and counters. These advantages are as follows:

1. Introducing the parallel data directly into the memory via the preset and clear overrides of the flip-flops permits the device to change modes of operation before the next triggering edge of clock. This can save time and speed up the processing of data — time can be wasted waiting for the data to be clocked input the memory as is required in synchronously loaded data.

2. Data inputs that are asynchronously loaded never have to be synchronized with clock since the loading process interrupts the operation of the flip-flops by temporarily introducing a clear or preset condition into the flip-flops. Data inputs that are synchronously loaded should be synchronized with clock for reasons discussed in Subsection 11.4.4.

Generally, the cost in hardware of asynchronous parallel loading will be somewhat greater than that for synchronous parallel loading. Comparing Figs. 12.26a and 12.28c is indicative of a small difference in the external logic to the D flip-flop, two gates per stage including the 2-to-1 MUX. Speedwise there is little difference between the two means of parallel loading. The choice of flip-flop type (e.g., T or D or JK) can be a more significant factor in both hardware and speed. However, these factors may be unimportant if the register or counter is implemented by using an array logic device such as a V-type PAL or Xilinx 4000 series FPGA. These devices have built-in D flip-flops, SSI devices, and a host of other features of which use can be made. But such devices will usually require the use of proprietary software to program them, as discussed in Section 7.8.

12.3.7 Branching Action of a 4-Bit Parallel Loadable Up/Down Counter

In Section 13.4 a parallel loadable up/down counter is used as the memory in state machine design — an alternative architecture. To program the counter in such applications, it is necessary to specify the branching action of the counter for each state-to-state transition in the state diagram. The specification of this action is illustrated in Fig. 12.29 for a fictitious FSM. Notice that there are just four possibilities for the branching action of a parallel loadable up/down counter:

$$(H), (I), (D), \text{ and } (PL),$$

representing hold, increment, decrement, and parallel load, respectively. Here, highest priority must be given to the count action if efficient use is to be made of the counter as the memory in the design of a state machine. If only PL branching were used, the design would revert to the use of discrete flip-flops as was the case in the designs of Chapters 10 and 11.

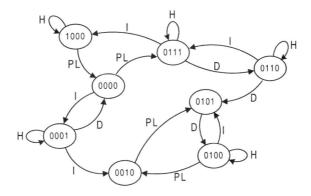

FIGURE 12.29
Illustration of the branching action of a 4-bit parallel loadable up/down counter used as the memory for a fictitious state machine.

12.4 SHIFT-REGISTER COUNTERS

Shift registers are normally designed to be operated in one or more of the noncyclic "data" modes of operation given at the beginning of Section 12.2. However, the flip-flops of a register can be configured to operate in a "nondata" cyclic fashion with or without external logic depending on the desired effect. Registers that are configured to operate in a cyclic fashion are called *shift-register counters*. These counters cycle through a sequence of states that generally conform to one or three types of codes: 1-hot code (Table 2.11), creeping code (Table 2.10), and a pseudo-random code. For future reference, the names of these counters are

Ring counters
Twisted ring (Johnson) counters
Linear feedback shift register (LFSR) counters

An introduction to these counters is provided in the next few subsections.

12.4.1 Ring Counters

A counter that consists of n states and n state variable outputs, such that each output corresponds to the integers (decimal values) 0 to $n-1$ in *1-hot code*, is called a *ring counter*—the simplest of the shift-register counters. This, of course, assumes that the counter is initially loaded with a binary word having a single "1." A 10-state 1-hot code is given in column (c) of Table 2.11 in Subsection 2.10.2. This means that a ring counter of 10 states would sequence through this 1-hot code in cyclical fashion, but would require 10 flip-flops to accomplish this. In comparison a binary counter having 10 flip-flops would sequence through $2^{10} = 1024$ states.

Shown in Fig. 12.30a is the state diagram for a simple 4-bit ring counter that will sequence through a 1-hot code of 4 bits. The present-state/next-state (PS/NS) table for this counter is given in Fig. 12.30b. A brief inspection of the columns in this table yields the NS logic expressions, given in Fig. 12.30c, without the need to use K-maps. The nature of the NS functions requires that the single 1 be circulated around the counter in cyclic fashion.

The logic circuit for this ring counter is shown in Fig. 12.30d. Once initialized into the 0001 state, the "1" will be circulated as illustrated in the state diagram of Fig. 12.30a. Actually, any bit pattern can be circulated in this fashion. For example, if two 1's are initialized into the counter to form an even-parity code, that bit pattern would be circulated according to the NS functions in Fig. 12.30c.

The ring counter of Fig. 12.30 must be initialized into one of the 1-hot code states. If it is not initialized, it could power up into any one of the five extraneous subroutines, including two "hang" states. Even if the counter is properly initialized, there is the possibility that one of the extraneous subroutines could be entered because of noise or power fluctuation. To avoid this problem the ring counter can be made self-correcting. To accomplish this, a missing-state analysis must be made of the 12 don't-care states, as in Fig. 12.31a, where five extraneous subroutines are discernible by close inspection of the table. The two hang states are easily seen to be states 0000 and 1111, since these states branch to themselves. Correction follows in Fig. 12.31b if it is recognized that all present states must be shifted left with a 0 except for state 0000, which must be shifted left with a 1. Notice that all 12 extraneous states eventually transit to a 1-hot state, though over varying numbers of clock

12.4 SHIFT-REGISTER COUNTERS

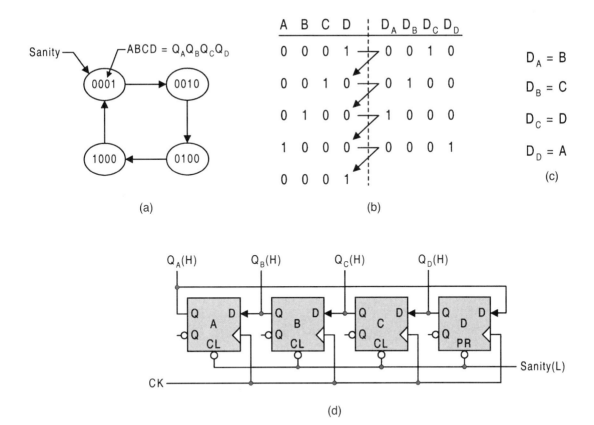

FIGURE 12.30
Design of a 4-bit ring counter. (a) State diagram. (b) PS/NS table. (c) NS functions. (d) Logic circuit configured for left shifting and initialized into the 0001 state.

cycles. As an example, state 9 transits directly to state 2 (a 1-hot state) after one clock cycle. But state 13 must transit $13 \rightarrow 10 \rightarrow 4$ over two clock cycles and state 7 must transit $7 \rightarrow 14 \rightarrow 12 \rightarrow 8$ over three clock cycles. Thus, while this counter is self-correcting, it may take up to three clock cycles for it to recover to a 1-hot state. This fact justifies initialization into the 0000 state, which must enter state 0001 on the next triggering edge of clock. If no initialization occurs, the counter can power up into any state, including a state such as 7 that requires three clock pulses for recovery.

Inspection of PS/NS tables in Figs. 12.30b and 12.31b indicates that only states 0000 and 1000 must be shifted left with a logic 1. The remaining 14 states are shifted left with a logic 0. Shown in Fig. 12.32a is the K-map for left shifting giving the result $L = \bar{B}\bar{C}\bar{D}$, which makes self-correcting any 4-bit 1-hot ring counter. Shown in Fig. 12.32b is a 4-bit USR that has been configured to produce a self-corrected ring counter that is initialized into the 0000 state. Notice that it is wired for left shifting according to the operation table for a USR given in Fig. 12.5a.

Ring counters can be expanded to include k states each with k state variables and can circulate any binary pattern once parallel loaded into the shift register. As one example, the

								Non-self-correcting				Self-correcting		
PS				NS					Decimal			Decimal		
A	B	C	D	D_A	D_B	D_C	D_D	RL	PS		NS	PS		NS
0	0	0	0	0	0	0	0	RL	0	→	0 Hang state	0	→	1 SL1
0	0	1	1	0	1	1	0	RL	3	→	6	3	→	6 SL0
0	1	0	1	1	0	1	0	RL	5	→	10	5	→	10 SL0
0	1	1	0	1	1	0	0	RL	6	→	12	6	→	12 SL0
0	1	1	1	1	1	1	0	RL	7	→	14	7	→	14 SL0
1	0	0	1	0	0	1	1	RL	9	→	3	9	→	2 SL0
1	0	1	0	0	1	0	1	RL	10	→	5	10	→	4 SL0
1	0	1	1	0	1	1	1	RL	11	→	7	11	→	6 SL0
1	1	0	0	1	0	0	1	RL	12	→	9	12	→	8 SL0
1	1	0	1	1	0	1	1	RL	13	→	11	13	→	10 SL0
1	1	1	0	1	1	0	1	RL	14	→	13	14	→	12 SL0
1	1	1	1	1	1	1	1	RL	15	→	15 Hang state	15	→	14 SL0

RL = Rotate left   SL = Shift left

(a)   (b)

FIGURE 12.31
Missing-state analysis and correction of the 4-bit ring counter in Fig. 12.30 (a) PS/NS table for the non-self-correcting counter. (b) PS/NS table for the self-correcting ring counter showing the 1-hot code states shaded.

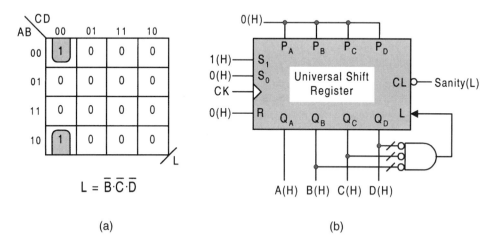

$L = \overline{B} \cdot \overline{C} \cdot \overline{D}$

(a)   (b)

FIGURE 12.32
Initialized and self-correcting 4-bit ring counter designed with the USR of Fig. 12.6. (a) K-map for the left serial input to the USR. (b) The USR wired for left shifting with external logic required for self-correction.

12.4 SHIFT-REGISTER COUNTERS

ring counter of Fig. 12.32b can be expanded to an 8-bit, 1-hot ring counter by cascading two USRs as in Fig. 12.7. But without parallel loading a 1-hot state, the self-correction logic becomes $L = \bar{B}\bar{C}\bar{D}\bar{E}\bar{F}\bar{G}\bar{H}$, a fan-in of 7 for a single gate. Thus, self-correction takes place within seven clock pulses. Generally, an n-bit ring counter will self-correct within $n - 1$ clock pulses. For such counters and where permitted, the CMOS NOR gate, shown in Figure 8.46, is preferred since it operates free of fan-in limitations. Alternatively, any bit pattern can be parallel loaded into an n-bit ring counter and circulated with self-correction.

The advantage of the ring counter is that it provides glitch-free decoded outputs directly from the flip-flops. This means that one and only one flip-flop is active for each state of the 1-hot sequence. This feature can be very useful for timing sequence generation in control applications. A down side to the ring counter is that it does not encode its states as efficiently as binary counters — one flip-flop must be used for each state. Considered next is a type of counter that can generate twice as many states as the ring counter but with only a minor increase in overall hardware.

12.4.2 Twisted Ring Counters

A counter that circulates a *creeping code*, such as that in Column (7) of Table 2.10 in Subsection 2.10.1, is called a *twisted ring counter*, or sometimes called a *Johnson counter*. The "twist" aspect of this counter is created simply by interposing an inverter in the feedback line of a standard ring counter or by tapping the feedback line off of the active low output of the flip-flop. Shown in Fig. 12.33a is the state diagram for a 4-bit twisted ring counter together with the required branching action for a USR design of this counter. The PS/NS table for this counter, given in Fig. 12.33b, indicates that a left shift of $\bar{A}$ generates the next state for each of the eight states in the sequence. This is shown more vividly in the K-map of Fig. 12.33c for L, the left-shift serial input of a USR. Notice that all eight extraneous states are assigned a don't-care symbol.

A USR design of this twisted-ring counter is given in Fig. 12.33d. Notice that it is initialized into the 0000 state, which is one of eight states of the creeping code sequence. Thus, once initialized the counter will cycle through the creeping code states — that is, unless the unexpected occurs and the counter is caused to enter an extraneous state. To avoid this potential problem, the counter can be made self-correcting. But to do this requires additional logic, as was the case for the ring counter of Fig. 12.32b.

The twisted ring counter of Fig. 12.33d can be made self-correcting by making use of the shift left and parallel load capability of the USR. Shown in Figs. 12.34a and 12.34b are the K-maps for S_0 and L of the USR. Here, left-shifting of the eight creeping code states is the same as in Fig. 12.33c, except that state 0 along with states 2, 4, and 6 are parallel loaded into the 0001 state. Also, the remaining five extraneous states (0101, 1001, 1010, 1011, and 1101) are shifted left an $\bar{A}$ eventually to states 2, 4, or 6, where they are subsequently parallel loaded into state 0001. Up to $n - 1 = 3$ clock pulses are necessary for the self-correction of this counter shown configured with the USR in Fig. 12.34c.

Twisted ring counters of any size can be designed. By cascading k 4-bit USRs, a twisted ring counter of $4k$ bits results that will sequence through $8k$ creeping code states. For these counters the external logic maintains the same form, namely $L = \bar{A}$ and $S_0 = \bar{Q}_{MSP} \cdot \bar{Q}_{LSB}$, for self-correction. If self-correction is neglected, then $S_0 = 0$ as in Fig. 12.33d. The advantage of the twisted ring counter over its cousin, the standard ring counter, is that $2n$ states are generated for n flip-flops as opposed to n states for the ring counter.

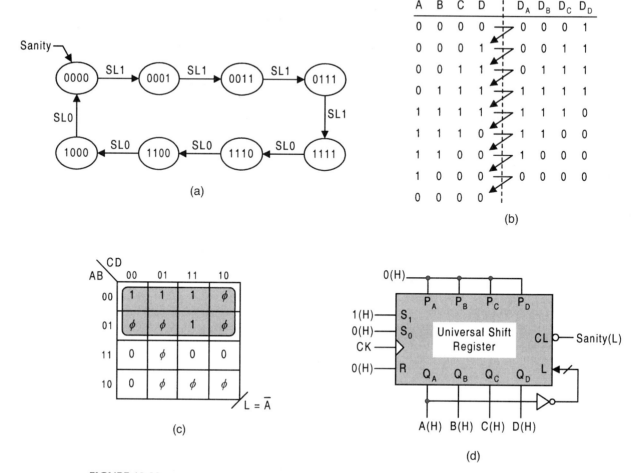

FIGURE 12.33
Design of a 4-bit twisted ring counter by using a USR. (a) State diagram showing USR branching action. (b) PS/NS table. (c) K-map and minimum cover for serial input L of the USR. (d) Initialized USR configured as the twisted ring counter.

Furthermore, the creeping code, like the Gray code, is unit distance, meaning that each state in the sequence is surrounded by states that differ by no more than one bit. Because of their simplicity, twisted ring counters can be easily used to produce time delays.

12.4.3 Linear Feedback Shift Register Counters

A series connection of D flip-flops with feedback paths via XOR gates but with no external inputs is called an *autonomous linear feedback shift register* (ALFSR) counter. If external inputs are involved, the counter is usually referred to simply as an LFSR counter. The ALFSR counter generates pseudorandom test patterns (vectors) that are useful in testing both combinational and sequential machines. Shown in Fig. 12.35a is an ALFSR counter

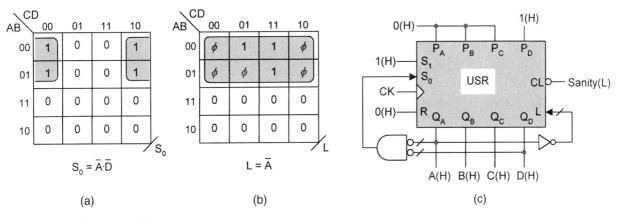

FIGURE 12.34
External logic necessary for self-correction of the twisted ring counter designed with a USR. (a) K-map and minimum cover for the S_0 mode control. (b) K-map and minimum cover for the left-shift serial input. (c) Logic diagram showing external logic and a parallel load of 0001 required for self-correction.

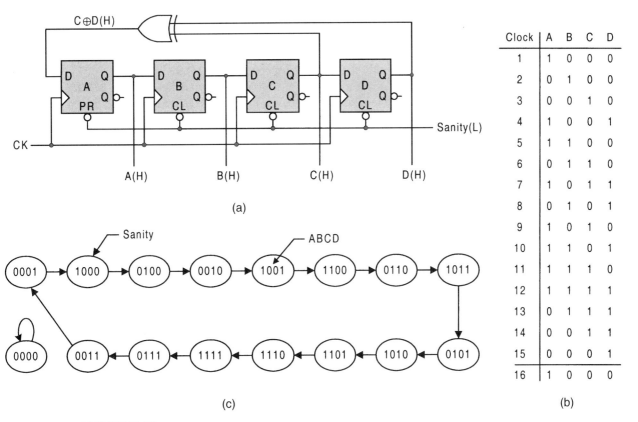

FIGURE 12.35
Analysis of a 4-bit near-maximum length ALFSR counter. (a) Logic diagram. (b) Truth table showing clock pulses and sequence of states. (c) State diagram for the 16 states.

that is to be analyzed. The D flip-flops are connected in a series for right shifting with feedback $C \oplus D$ to the MSB stage. Compare this with the ring counter in Fig. 12.30d. The sequence of states is easily generated from knowledge of the feedback function, $C \oplus D$, and is given in Fig. 12.35b together with the corresponding clock pulse number. Notice that this ALFSR counter is initialized into the 1000 state and that it will sequence through 15 of the 16 states in pseudorandom fashion. The ALFSR counter is not allowed to enter the 0000 state since that state is a hang state, as indicated in the state diagram of Fig. 12.35c. Once in the 0000 state it must remain there indefinitely unless reinitialized.

The ALFSR counter just described is termed a *near-maximum-length* ALFSR counter since it can sequence through $2^n - 1$ states, one short of the maximum of 2^n, where n is the number of state variables (or flip-flops). Other feedback functions used for the ALFSR counter of Fig. 12.35 may not generate a near-maximum-length sequence of pseudorandom states. For example, it is easily shown that the $B \oplus C$ feedback function can generate only eight unique pseudo-random states, and $A \oplus B$ only five, each being initialized into the 1000 state. But other initialization states may be used with similar results. If, for example, the ALFSR counter of Fig. 12.35 is initialized into the 1111 state, 8 unique states are generated with feedback function $B \oplus C$, 15 unique states result if $C \oplus D$ is used, and 5 unique states result if $A \oplus B$ is used, all as before. Note that for right shifting, all XOR combinations not containing D must be avoided if initializing into the 0001. The reason is simply that on the second clock pulse the ALFSR counter is caused to enter the hang state 0000, from which there is no exit. Thus, a valid feedback function must contain the LSB state variable.

If a packaged shift register (e.g., a USR) is used to externally configure an ALFSR counter, a means must be found to initialize the counter into the 0000 state with the ability to cycle through all 2^n pseudorandom states. Shown in Fig. 12.36a is the new state diagram with all 16 states represented, and the K-map for D_A is plotted in Fig. 12.36b. Since it is known that XOR functions are involved, there is an opportunity to use the Reed–Muller (R-M) transformation forms as discussed in Section 5.7. Following the examples given there, the following R-M g coefficients become

$$g_0 = g_1 = g_4 = g_6 = g_8 = g_{10} = g_{12} = g_{14} = 1$$
$$g_2 = g_3 = g_5 = g_7 = g_9 = g_{11} = g_{13} = g_{15} = 0.$$

When these coefficients are introduced into Eq. (5.17) the NS function D_A is found to be

$$\begin{aligned}D_A &= 1 \oplus D \oplus B \oplus BC \oplus A \oplus AC \oplus AB \oplus ABC \\ &= 1 \oplus D \oplus B\bar{C} \oplus A\bar{C} \oplus AB\bar{C} \\ &= 1 \oplus D \oplus B\bar{C} \oplus A\bar{B}\bar{C} \\ &= 1 \oplus D \oplus \bar{C}(B + A\bar{B}) \\ &= 1 \oplus D \oplus [\bar{C}(A + B)],\end{aligned}$$

which yields the gate minimum result

$$D_A = D \oplus (\bar{A}\bar{B} + C). \tag{12.10}$$

12.4 SHIFT-REGISTER COUNTERS

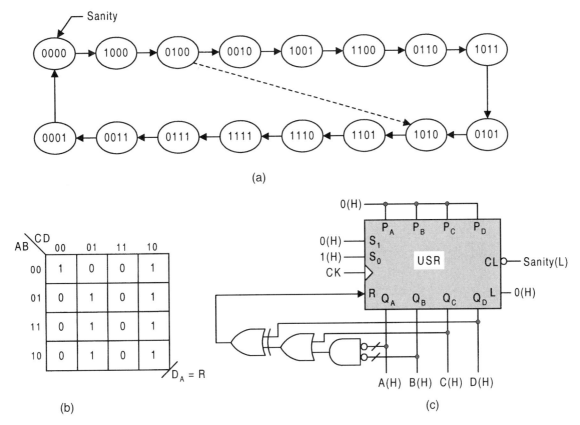

FIGURE 12.36
USR implementation of a ALFSR counter that will sequence through all 2^n states. (a) State diagram showing sequence of 16 states. (b) K-map for R plotted from the state diagram in (a). (b) Logic circuit showing the external logic given by Eq. (12.10).

Here, use has be made of the identities in XOR algebra given in Subsection 3.11.2. The other NS functions remain the same: that is, $D_B = A$, $D_C = B$, and $D_D = C$. Figure 12.36c shows a USR configured with the corrected feedback function of Eq. (12.10). This ALFSR counter will initialize into the 0000 state and will sequence through all $2^4 = 16$ maximum-length states in cyclic fashion. Without making the correction expressed by Eq. (12.10), it would not be possible to initialize or parallel load into the 0000 state and then sequence through all 2^n pseudorandom states.

By altering the logic expressed by Eq. (12.10), it is possible to selectively reduce the number of unique pseudorandom states from the maximum length of 16 shown in Fig. 12.36a. To do this the following procedure can be applied:

(1) Select the number of states, $S < (2^n - 1)$, that is desired.

(2) Find a pair of states separated by $\{(2^n - 1) - S\}$ other states such that the smaller is an even digit and 1 less than the larger.

(3) Advance 1 state from the larger and draw an arrow. The result is a modified state diagram of S states from which the new feedback logic for a USR can be obtained.

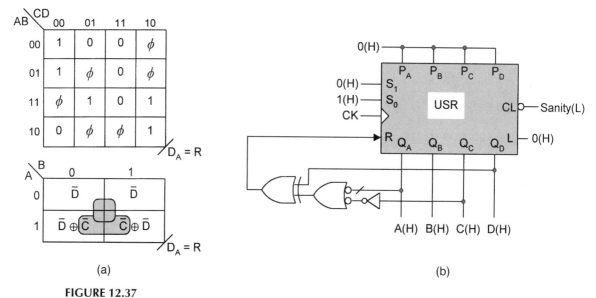

FIGURE 12.37
Design of the decade ALFSR counter in Fig. 12.36a. (a) Conventional and EV K-maps for the NS logic, D_A. (b) A USR configured with the logic of Eq. (12.11).

As an example of the application of this procedure, suppose it is desired to design a decade ALFSR counter from the pseudo-random states in Fig. 12.36a. A pair of states separated by $16 - 1 - 10 = 5$ must be found such that the smaller is even and 1 less than the larger. An inspection of the state diagram in Fig. 12.36a indicates that states 4 and 5 satisfy this requirement. By advancing by 1 state from state 5, an arrow is drawn from state 4 to state 10, as indicated by the dashed arrow. This is the 10-state state diagram for which a new D_A must be found, the other three NS functions remaining the same. Presented in Fig. 12.37a is the NS logic for D_A as determined from the new state diagram in Fig. 12.36a. Here, the conventional (1's and 0's) K-map is converted to an EV K-map, which permits the use of XOR patterns to extract gate-minimum logic as discussed in Section 5.2. The result is expressed as

$$D_A = \bar{D} \oplus (A\bar{C}) = D \oplus (\bar{A} + C), \qquad (12.11)$$

where use has been made of Eqs. (3.27) and (3.15). The remainder of the NS functions remain the same as $D_B = A$, $D_C = B$, and $D_D = C$. A USR, configured with Eq. (12.11), is shown in Fig. 12.37b. This counter will initialize into the 0000 state and thereafter sequence in iterative fashion through all 10 pseudorandom states shown to the left of the dashed arrow in Fig. 12.36a.

The foregoing discussion has dealt with ALFSR counter variations based on a single set of feedback XOR taps, namely $C \oplus D$. However, there are numerous possible XOR taps that can be used for the feedback function of a 4-bit ALFSR counter, but not all will generate a maximum length ALFSR counter. This is also true for ALFSR counters having a larger number of bits. For example, the feedback function $D \oplus E \oplus F \oplus H$ applies

12.4 SHIFT-REGISTER COUNTERS

Table 12.1 Examples of Feedback Functions for Near-Maximum-Length ALFSR Counters

SR size, n-bits	Feedback function $f(Q)$	Feedback function (literal notation)	Near maximum length (In nos. of states)
4	$Q_1 \oplus Q_0$	$C \oplus D$	15
5	$Q_2 \oplus Q_0$	$C \oplus E$	31
8	$Q_4 \oplus Q_3 \oplus Q_2 \oplus Q_0$	$D \oplus E \oplus F \oplus H$	255
12	$Q_6 \oplus Q_4 \oplus Q_1 \oplus Q_0$	$F \oplus H \oplus K \oplus L$	4,095
16	$Q_5 \oplus Q_4 \oplus Q_3 \oplus Q_0$	$K \oplus L \oplus M \oplus P$	65,535
24	$Q_7 \oplus Q_2 \oplus Q_1 \oplus Q_0$	$Q \oplus V \oplus W \oplus X$	16,777,215
32	$Q_{22} \oplus Q_2 \oplus Q_1 \oplus Q_0$	—	4,294,967,295

to an 8-bit near-maximum-length ALFSR counter in literal notation. Or for 12- and 16-bit near-maximum-length ALFSR counters, the feedback functions $F \oplus H \oplus K \oplus L$ and $K \oplus L \oplus M \oplus P$ apply, respectively. Shown in Table 12.1 are a few feedback functions that apply to right-shifted, near-maximum-length ALFSR counters. Note that for the numeral notation Q_0 is always the LSB, and that for the literal notation $Q_A = A$ is always the MSB of the counter.

As has been pointed out earlier, ALFSR counters are very useful in generating pseudorandom test vectors suitable for testing a variety of machines, combinational and sequential. Take, for example, a 16-bit ALFSR counter. It can sequence through $2^{16} - 1 = 65,535$ unique pseudorandom states in iterative fashion if the all-zero state is forbidden, or through 65,536 states if corrected to include the all-zero state. If a 32-bit ALFSR counter is used for testing, a total of 4,294,967,296 unique pseudorandom states are available with correction to include the all-zero state. Some large state machines are designed with ALFSR counter elements in them to provide a *built-in-self-test* (BIST) capability. BIST capability facilitates and automates testing of these machines without need for an external testing facility.

Correction for inclusion of the all-zero state in the general case for maximum-length ALFSR counters is not trivial, but it is not difficult either. Consider that upon initializing into the all-zero state $00000 \cdots 00$ the next transition must be into the $10000 \cdots 00$ state to begin the pseudorandom sequence. Then, at the end of the 2^n sequence, in the $00000 \cdots 01$ state, the ALFSR counter must return to the all-zero state. For all of this to happen, a correction function must be found and XORed with the feedback function. Noting that all feedback functions in Table 12.1 end with Q_0, it follows that the correction function must be the ANDed complements of all ALFSR counter outputs except Q_0, that is, $\bar{Q}_{n-1} \cdot \bar{Q}_{n-2} \cdot \ldots \cdot \bar{Q}_2 \cdot \bar{Q}_1$. Here, Q_{n-1} is the MSB and Q_0 is the LSB. Therefore, the corrected feedback function is given by

$$f(corrected) = (\bar{Q}_{n-1} \cdot \bar{Q}_{n-2} \cdot \ldots \cdot \bar{Q}_2 \cdot \bar{Q}_1) \oplus f(Q), \qquad (12.12)$$

where $f(Q)$ is the numeral feedback function in column 2 of Table 12.1. Thus, it follows that the corrected feedback functions for 4-bit, 5-bit, and 8-bit ALFSR counters are $(\bar{Q}_3 \bar{Q}_2 \bar{Q}_1) \oplus Q_1 \oplus Q_0$, $(\bar{Q}_4 \bar{Q}_3 \bar{Q}_2 \bar{Q}_1) \oplus Q_2 \oplus Q_0$, and $(\bar{Q}_7 \bar{Q}_6 \bar{Q}_5 \bar{Q}_4 \bar{Q}_3 \bar{Q}_2 \bar{Q}_1) \oplus Q_4 \oplus Q_3 \oplus Q_2 \oplus Q_0$, respectively. Applying Eq. (12.12) to the 16-state ALFSR counter in Fig. 12.36 yields

Eq. (12.10),

$$D_A = (\bar{A}\,\bar{B}\,\bar{C}) \oplus C \oplus D = D \oplus (\bar{A}\bar{B} + C),$$

in literal form, which can be proved by applying the laws of XOR algebra in Section 3.11.

A cursory inspection of Eq. (12.12) and Table 12.1 indicates that the maximum number of XOR operations is three, independent of ALFSR counter size, but only after simplifying by application of Eqs. (3.25) and (3.13). Fortunately, three XOR operations can be handled very easily by a standard four-input, even-parity generator module such as that in Fig. 6.32c. The problem is, of course, in dealing with the large number of ANDed complements present in the correction functions. If CMOS logic is permitted, the multiple input NOR gate in Figure 8.46 can be used to great advantage. It has no fan-in limitations.

Table 12.1 and Eq. (12.12) apply to right-shifted ALFSR counters that sequence through all 2^n pseudo-random states and that are initialized into the all-zero state. Table 12.1 and Eq. (12.12) can also be applied to *left-shifted ALFSR counters* if the Q_i outputs are interpreted in "reverse" fashion such that Q_0 is the MSB and Q_{n-1} the LSB. The corrected feedback function now becomes $(\bar{Q}_1 \cdot \bar{Q}_2 \cdots \bar{Q}_{n-2} \cdot \bar{Q}_{n-1}) \oplus f(Q')$ by omitting Q_0, where $f(Q')$ is the feedback function in Table 12.1 interpreted in reverse order. For example, $Q_0 \oplus Q_2 \oplus Q_3 \oplus Q_4 = A \oplus C \oplus D \oplus E$ for $n = 8$.

12.5 ASYNCHRONOUS (RIPPLE) COUNTERS

All counters discussed in Sections 12.3 and 12.4 are classified as synchronous counters because the flip-flops, of which the counters are constructed, are all triggered simultaneously or very nearly so. Counters composed of T flip-flops that are triggered in series are called *ripple counters*. Each T flip-flop is triggered off of the output from the next LSB flip-flop. For this reason, they are classified as asynchronous counters even though the LSB flip-flop is triggered by the external CK signal.

Shown in Fig. 12.38a is a general divide-by-2^n ripple counter composed of *toggle modules* of the type shown in Fig. 12.12c, a toggle module being nothing more than a divide-by-2 counter. Notice that the $Q(H)$ output of each toggle module is the input to the FET clock of the next MSB stage. In Fig. 12.38b is the timing diagram for the three LSB stages of this ripple counter. The count, taken from outputs $Q_{n-1} \cdots Q_2 Q_1 Q_0$ is shown to be in ascending binary, an up-count, and that frequency is divided beginning with $f_{CK} \div 2$ for Q_0 and ending with $f_{CK} \div 2^n$ for Q_{n-1}. A sanity input permits the counter to be initialized into the $0 \cdots 000$ state.

The direction of the count ($Dn = 1$ for down count and $Dn = 0$ for up count) of any ripple counter can be altered by making any odd number of changes in the expression

$$Dn = RET \oplus Q_{CK} \oplus Q_{OUT}. \qquad (12.13)$$

In this equation, $RET = 1$ for RET flip-flops or $RET = 0$ for FET flip-flops, $Q_{CK} = 1$ if triggering is from $Q(H)$ of the next LSB stage or $Q_{CK} = 0$ if triggering is from $Q(L)$, and $Q_{OUT} = 1$ if the count is read from $Q(H)$ or $Q_{OUT} = 0$ if read from $Q(L)$. Thus, any odd number of changes (parameters or operators) in Eq. (12.13) changes the count direction. For example, if RET toggle modules are used in Fig. 12.38a, a down-count occurs. Thus,

12.5 ASYNCHRONOUS (RIPPLE) COUNTERS

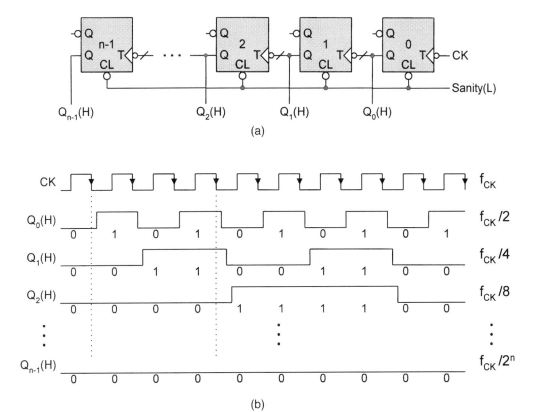

FIGURE 12.38
An *n*-bit ripple up-counter. (a) Logic circuit implemented with toggle modules as the memory elements. (b) Timing diagram showing frequency division and transition delays due to clock ripple effect.

Eq. (12.13) becomes

$$Dn = 1 \oplus 1 \oplus 1 = 1 = Down-count.$$

This is easily verified in Fig. 12.38b by shifting in turn each of the outputs Q_0, Q_1, and Q_2 to the left such that each output change occurs on the rising edge of the next LSB output, the output Q_0 being shifted to the left by one half of a CK period. The same result could have been achieved by triggering the FET toggle modules in Fig. 12.38a off of $Q(L)$ from the next LSB stage. Now $Q_{CK} = 0$ so that $Dn = 0 \oplus 0 \oplus 1 = 1 = Down-count$. But applying both changes ($RET = 1$ and $Q_{CK} = 0$) given above would leave the count unaltered, that is, $Dn = 1 \oplus 0 \oplus 1 = 0 = Up-count$.

Any memory element capable of the toggle mode is suitable for use in a ripple counter. For reference purposes, three types of flip-flops are shown in Fig. 12.39, all configured to operate in the toggle mode. They are (a) an FET JK flip-flop, (b) an FET T flip-flop, and (c) an RET D flip-flop wired as a toggle module (divide-by-2 counter). The toggle module

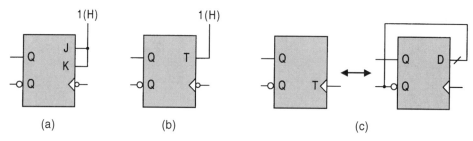

FIGURE 12.39
Three examples of memory elements suitable for use in a ripple counter. (a) FET JK flip-flop in toggle mode. (b) FET T flip-flop in toggle mode. (c) RET D flip-flop wired as a toggle module.

in Fig. 12.39c is the least costly (hardware-wise) of the three and is the one featured in the ripple counter of Fig. 12.38a, but as an FET toggle module.

Take Care in Using the Ripple Counter There are two major problems that can arise in using of ripple counters. The problems are stated as follows together with some suggestions for proper use:

- All ripple counters suffer from a progressive noise (glitch) generation problem if any attempt is made to decode their outputs. An inspection of the timing diagram in Fig. 12.38b shows the transition delays that result from the series (ripple) triggering of the flip-flops. If decoding of the outputs is not necessary or if decoding is used but glitch production can cause no problem, ripple counters can be used advantageously — they require no external logic for their operation. If these conditions cannot be met, *no* attempt should be made to use ripple counters. Instead, use should be made of synchronous binary counters whose output transitions are synchronous or very nearly so.
- Ripple counters are inherently slow compared to synchronous counters. This is so because the output changes must propagate through the counter one flip-flop at a time. For an n-bit ripple counter this propagation delay may be expressed as

$$\tau_{Ripple\ Counter} = n \times \tau_{ff}, \qquad (12.14)$$

where τ_{ff} is the delay through a single flip-flop in the ripple counter. This counter delay would be required for completion of a 2^n binary count. In comparison, the delay of a divide-by-2^n synchronous binary counter required to complete the count is

$$\tau_{Synch\ Counter} = (\tau_{ff} + \tau_{NS}) \qquad (12.15)$$

and is not progressive. In Eq. (12.15) τ_{NS} is the propagation delay through the next-state-forming logic required in the design of the synchronous counter.

Therefore, if counter speed is not a consideration and if the outputs are not decoded in any way, use of ripple counters can be recommended. In fact, if these conditions apply,

12.5 ASYNCHRONOUS (RIPPLE) COUNTERS

ripple counters can be cascaded to any size simply by connecting the appropriate output from the MSB stage of one counter to the LSB clock input another, etc. For example, the n-bit ripple counter of Fig. 12.38a can be cascaded to produce a $2n$-bit up-counter by simply connecting its Q_{n-1} output to the LSB FET clock input of the other n-bit ripple counter. But remember, that ripple counter delay increases in proportion to the number of flip-flops in the counter.

It is also possible to design a ripple counter that will count through $N < 2^n$ states. This is demonstrated in Fig. 12.40 by the design of a decade (divide-by-10) ripple counter that is initialized into the 0000 state. The truth table in Fig. 12.40a gives the values of the asynchronous preset (PR) and clear (CL) overrides to the flip-flops required to force a series of asynchronous transitions from the "jump" state 1010 to the origin state 0000. Notice that the PR and CL override values must be 0 for proper counter operation in the 10-state count sequence. All states between state 1010 and 0000 are don't-care states.

The state diagram for this decade ripple counter is given in Fig. 12.40b. It illustrates the fact that when the counter attempts to enter the jump state 1010 (the 11th state) the counter will be forced to transition asynchronously through some path to the origin state 0000 (see looped arrow). The correction logic required to do this is found from the conventional K-map in Fig. 12.40c, which yields $CL_A = CL_C = A \cdot C$, all other override values being

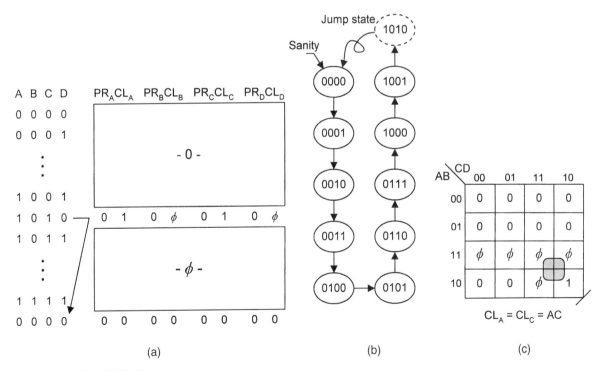

FIGURE 12.40
Design of decade ripple up-counter. (a) Truth table showing the values of the PR and CL asynchronous override inputs to the flip-flops. (b) State diagram showing jump state 1010. (c) K-map and minimum cover for the CL inputs, CL_A and CL_C.

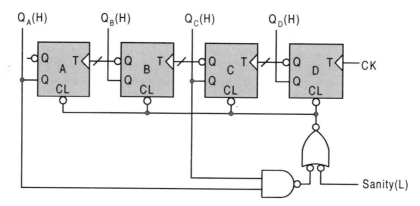

FIGURE 12.41
Implementation of the decade ripple counter of Fig. 12.40 showing correction and sanity logic.

logic 0. The result is the logic circuit in Fig. 12.41, which shows both the correction logic and that required to initialize the counter into the 0000 state. Here, this logic is configured in a manner that simply clears all memory elements. Thus, when the counter is initialized or when it attempts to enter the jump state 1010, it will be forced asynchronously into state 0000 via the external logic and the four *CL* overrides of the toggle modules. Notice that the asynchronous *PR* overrides need not be present since they are all set at logic 0.

The decade ripple counter in Fig. 12.41 suffers the same problems described earlier. That is, it will exhibit progressive logic noise generation if the outputs are decoded, and it is inherently slow. In addition, this counter breaks the 2^n count, thereby requiring it to undergo an asynchronous transition from the 1010 state to the origin state 0000 via the asynchronous *CL* overrides. But this can occur only after a delay through the external and internal logic associated with *CL*. Consequently, additional timing problems can be created if an attempt is made to decode the output signals. Therefore, as a rule of thumb, it is advisable to use ripple counters in the absence of any output decoding logic. This rule is especially important if the decoded signals are used as inputs to other switching devices. The few exceptions to this rule were noted earlier.

Synchronous unidirectional binary counters, such as the cascadable BCD up-counter in Fig. 12.17a, can be made bidirectional by reading the outputs from 2-to-1 MUXs placed on the mixed-rail outputs of its flip-flops. Reading from the $Q(L)$ outputs of the BCD counter in Fig. 12.17a has the effect of complementing the output waveforms in Fig. 12.18 — a down count. Ripple counters can also be made bidirectional by reading their outputs from 2-to-1 MUXs placed on the mixed-rail outputs of the toggle modules and by applying Eq. (12.13). For example, the MUX output for the *J*th stage of the ripple counter in Fig. 12.38a would be

$$Y_J = XQ_J + \bar{X}\bar{Q}_J, \qquad (12.16)$$

where X is the direction control, and Q_J and $\bar{Q}_J$ are the active high and active low outputs from the *J*th stage toggle module, respectively. Thus, the count is up if $X = 1$ or down if $X = 0$, but only for FET toggle modules. Reading the outputs from $\bar{Q}_J$ instead of Q_J has the effect of complementing the waveforms in Fig. 12.38b to give a down count. However, to do this does not allow the CK waveform to be used to add an additional bit in the count, as it can be in Fig. 12.38b. Note that for RET toggle modules, X and $\bar{X}$ must be interchanged in Eq. (12.16).

One important precaution must be recognized when using MUXs together with Eq. (12.13) for bidirectional ripple counter designs. It is *not* a good idea to use the 2-to-1 MUXs for purposes of altering the triggering activation level of the toggle modules by placing the MUXs between modules. To do so makes it possible for the counter to change count simply by changing the direction control X while the external clock signal is idle.

FURTHER READING

Nearly every modern text in digital design covers the subject of shift registers to one extent or another. Perhaps the best sources for further reading are the texts by McCluskey, Nelson *et al.*, Shaw, Tinder, Wakerly, and Yarbrough. Of these, the texts of Nelson *et al.*, Wakerly, and Yarbrough provide the best coverage of commercial MSI registers and their applications.

[1] E. J. McCluskey, *Logic Design Principles*. Prentice Hall, Englewood Cliffs, NJ, 1986.
[2] V. P. Nelson, H. T. Nagle, B. D. Carroll, and J. D. Irwin, *Digital Logic Circuit Analysis and Design*. Prentice Hall, Englewood Cliffs, NJ, 1995.
[3] A. W. Shaw, *Logic Circuit Design*. Sanders College Publishing, Fort Worth, TX, 1993.
[4] R. F. Tinder, *Digital Engineering Design: A Modern Approach*. Prentice Hall, Englewood Cliffs, NJ, 1991.
[5] J. F. Wakerly, *Digital Design Principles and Practices*, 2nd ed. Prentice-Hall, Englewood Cliffs, NJ, 1994.
[6] J. M. Yarbrough, *Digital Logic Applications and Design*. West Publishing Co., Minneapolis/St. Paul, MN, 1997.

Again, almost every text will provide some information regarding synchronous binary counter design and application. The references just cited regarding registers are good examples of this. The texts by Nelson *et al.*, Wakerly, and Yarbrough seem particularly strong in their treatment of commercial MSI counters and their applications. Tinder's text is the only one that covers the one-bit modular design of counters by using half adders and half subtractors. The text by Katz and the lesser-known text by Taub are also worth reading on this subject. For the advanced reader the older text by Dietmeyer can be worthwhile.

[7] D. L Dietmeyer, *Logic Design of Digital Systems*, 2nd ed. Allyn and Bacon, Boston, MA, 1971.
[8] R. H. Katz, *Contemporary Logic Design*. Benjamin/Cummings Publishing, Redwood City, CA, 1994.
[9] H. Taub, *Digital Circuits and Microprocessors*, McGraw-Hill, New York, 1982.

For further reading on the subjects of ring and twisted ring (Johnson) counters, the texts by McCluskey, Nelson *et al.*, and Wakerly are recommended. Of these three, the text by Nelson *et al.* appears to be the most thorough.

The subject of linear feedback shift register (LFSR) counters is somewhat esoteric, with recommended further reading limited to a few sources. The most important of these devices are the autonomous LFSR counters or ALFSR counters. The best treatment on these devices appears to be found in the texts by McCluskey, Nelson *et al.*, and Wakerly, all previously cited. The feedback functions listed in Table 12.1 of this text are generated by *primitive polynomials* that can be found in texts by McClusky, Wakerly, and Golumb.

[10] S. W. Golumb, *Shift Register Sequences*. Aegean Park Press, Laguna Hills, CA, 1982.

The subject of ripple (or asynchronous) counters is somewhat special, and useful information may be more difficult to find. With the exception of texts by Wakerly and Yarbrough, all of those previously cited cover this subject adequately. For the advanced reader, the text by Dietmeyer (previously cited) is recommended.

PROBLEMS

12.1 The shift registers that are featured in this chapter are all built around edge-triggered D flip-flops. Suppose one decided to design a shift register with transparent D latches instead of edge-triggered D flip-flops. What are the negative consequences (if any) of this design? If this poses a problem, are there any conditions under which such a design would be acceptable? Explain.

12.2 Problem 10.6 in Chapter 10 features the conversion of an RET D flip-flop to an RET SR flip-flop. What would be the advantage or disadvantage of using RET SR flip-flops in place of RET D flip-flops in the design of a shift register?

12.3 A four-bit storage (PIPO) register is featured in Fig. 12.1. Reconfigure this register so that it is a *tri-state register*. To do this, use tri-state drivers so that a 1(L) on either of two enable inputs, EN_1 or EN_2, enables the active high outputs, and a 0(L) on both of the two enable inputs disables the active high outputs. (Hint: See Figure 3.8.)

12.4 (a) Use the four-bit right shift register in Fig. 12.3c and a single OR gate (nothing else) to generate the waveform shown in Figure P12.1 from any one of its four outputs Q_A, Q_B, Q_C, and Q_D.

(b) Run a missing state analysis on the resulting FSM and determine whether or not it is necessary to initialize it into one of its states.

12.5 A *cascadable left-shift register* is to be designed. It is to have true hold and asynchronous parallel load capability. Also, it is to have asynchronous clear (reset) capability.

(a) Give the operation table and state diagram for the *J*th 1-bit slice for this register.

(b) Construct the logic circuit for the *J*th stage. To do this, use a 2-to-1 MUX and an RET D flip-flop together with the external logic required for the asynchronous

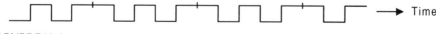

FIGURE P12.1

PROBLEMS

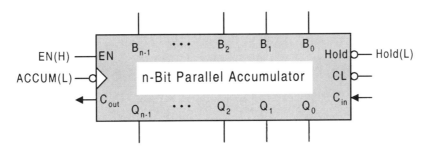

FIGURE P12.2

   parallel load capability. Include all truth tables, K-maps, and logic expressions that are used in constructing the logic circuit.

12.6 Shown in Fig. P12.2 is the block diagram for a cascadable *n*-bit *parallel accumulator*. It is the function of this accumulator to add and store (accumulate) with each triggering edge of clock the numerical data that appears on the *B* word lines. For example, if *ACCUM(L)* first samples ... 0001 on the first triggering edge, a ... 0001 will be stored in the flip-flops and delivered to their outputs. Then, if on the second triggering edge, *ACCUM(L)* samples a ... 0111 on the *B* word lines, a ... 1000 will be stored in the flip-flops and delivered to the outputs.

   (a) Design this accumulator by using *n* full adders and *n* FET D flip-flops. The accumulator is to have asynchronous clear, and tri-state output capability such that the outputs are enabled only if $EN(H) = 1(H)$. For this part, disregard the *Hold(L)* input.

   (b) Add true hold capability to this accumulator, meaning that at any time the current accumulated sum can be stored and delivered to the outputs for any number of clock (ACCUM) cycles. Assume that this occurs under the condition $Hold(L) = 1(L)$.

12.7 The waveforms for the divide-by-3 counter in Fig. 12.13c show a frequency of $f_{CK}/3$ and a duty cycle of $33\frac{1}{3}\%$ for each of the two outputs. What would be the frequency and duty cycle for a divide-by-3 counter if the state code assignment were changed to Gray code, $00 - 01 - 11 - 00 - \ldots$?

12.8 (a) Construct the waveforms for a divide-by-5 counter by using a binary count beginning with 000. From these waveforms determine the frequency and duty cycle for each of the three outputs.

   (b) Repeat part (a) for a Gray code count beginning with 000.

   (c) Repeat part (a) for the pseudo-random count $011 - 001 - 111 - 101 - 010 - 011 - \ldots$.

12.9 Design a 2-bit bidirectional binary/Gray code counter that will operate according to the mode control and count requirements given in Fig. P12.3. To do this, use two 4-to-1 MUXs, RET D flip-flops as the memory, a 2-to-4 decoder for the outputs, and a gate-minimum NS logic. Assume that the mode control inputs, *X* and *Y*, are asynchronous and must be synchronized antiphase to clock, and that all inputs and outputs are active high.

X	Y	Count	Outputs
0	0	Up Gray	UPGRY
0	1	Dn Gray	DNGRY
1	0	Up Binary	UPBIN
1	1	Dn Binary	DNBIN

FIGURE P12.3

12.10 A counter is to be designed that will count through the following sequence of states in three-bit code:

$$\text{Sequence I} \quad \cdots 0 \to 1 \to 3 \to 2 \to 0 \cdots \text{ If } x$$
$$\text{Sequence II} \quad \cdots 7 \leftarrow 6 \leftarrow 4 \leftarrow 5 \leftarrow 7 \cdots \text{ If } \bar{x}$$

It is required that the counter change sequence at any time beginning with the complement of the state in the previous sequence. For example, if $x \to \bar{x}$ while in state 2 of Sequence I, then Sequence II will begin with state 5, that is, 010 → 101 and so on.
(a) Construct the state diagram and state table for this counter.
(b) Design this counter with RET T flip-flops and a gate-minimum NS forming logic. Assume that the input x arrives asynchronously and is active high. (Hint: Use XOR patterns.)

12.11 Design a 1-bit slice (Jth stage) for a *cascadable parallel loadable up-counter* by using the hardware given below (nothing else). The counter is to have asynchronous parallel load and asynchronous clear capability. End with an optimum logic circuit showing all inputs and outputs. Block symbols may be used where appropriate for the hardware parts listed.

Allowable Hardware
One half adder
One RET D flip-flop
One 2-to-1 MUX
Gates as needed for the asynchronous parallel load

12.12 By using the simplest means possible, convert the 4-bit binary counter of Fig. 12.23 to the following counters such that each will count *continuously* via a count command $CNT(H) = 1(H)$. To do this, use $LD(L)$ as the command to parallel load and set the $CL(L)$ input to the counter to $0(L)$.
(a) Divide-by-8 (modulo 8) up-counter beginning with state 0000.
(b) Divide-by-10 (BCD) down-counter beginning with state 1010.
(c) Divide-by-11 (modulo 11) down-counter beginning with state 1111.
(d) Divide-by-10 (XS3) up-counter beginning with state 0011.

12.13 A psychology student requires a special timer for a research experiment that is being performed. Design a timer that will deliver a single pulse after a 45-second period

PROBLEMS 609

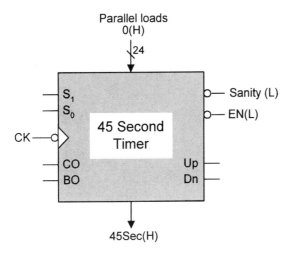

FIGURE P12.4

from a 100-kHz clock on command of an enable pulse $EN(L)$ that is less than 45 seconds. However, if the enable signal is fixed at $EN(L) = 1(L)$, the timer will issue pulses at 45-second intervals. The 45-second period must be delivered with an accuracy of $\pm 0.5\%$. Also, the system is to be initialized into the all-zero state from which the count will begin when enabled. Fig. P12.4 gives the block symbol for the timer and the permitted hardware to be used. Note that the 8-bit counters are constructed of counters shown in Fig. 12.23b. [Hint: Use the NAND gate on the most significant counter stage to generate the 45 second pulse and a clear, and use the $R(L)$ input to the basic cell for the enable.]

12.14 A design project requires the use of both a universal shift register (USR) and a parallel loadable bidirectional counter, devices that must be operated interchangeably with the same set of mode controls. It is also required that the carry and borrow outputs be disabled except during counter operation. The problem is that limited space requires a compact system. With this information in mind, do the following:

(a) Construct the operation table, state diagram, and MUX K-map for the Jth stage of the USR/counter device. Base your decisions on the hardware requirements given in Fig. P12.5a.

(b) From the information in part (a), first design the Jth stage for the USR/counter. It must be cascadable and bi-directional with true hold, asynchronous parallel load, and clear capability. Then, show how that stage can be cascaded to form the 4-bit device shown in Fig. P12.5b. (Hint: The counter design should be a combination of the counter design examples featured in Subsections 12.3.3 and 12.3.4.) Also, no *CO* or *BO* signal is permitted during a true hold or shift operation.

12.15 (a) Construct the complete state diagram for the self-correcting twisted ring counter featured in Fig. 12.34. In doing so, demonstrate that all extraneous states eventually end up in state 0001.

Required Hardware for the Jth stage

One FET D flip-flop with PR and Cl overrides
One 4-to-1 MUX
One XOR gate
Necessary gates for the NS and output logic

(a)

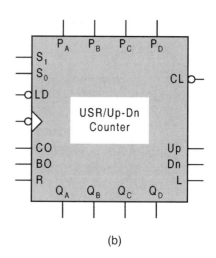

(b)

FIGURE P12.5

(b) Are there any states other than state 0001 that can be used for parallel loading in Fig. 12.34? If so, name them. Can state 0000 be used for parallel loading? Explain.

12.16 (a) Construct a table to indicate the pseudorandom states through which a right shifting 5-bit ALFSR counter would sequence. To do this, use Table 12.1 and assume that the ALFSR counter has been corrected to include the all-zero state.

(b) Construct the logic circuit for the 5-bit ALFSR counter of part (a). Include the gate-minimum correction logic and plan to initialize this counter into the all-zero state.

12.17 Repeat Problem 12.16 for a 4-bit ALFSR counter that is left shifted.

12.18 Design a 4-bit maximum length ALFSR counter that will right shift or left shift as determined by a mode control X. Plan to initialize it into the all-zero state. Thus, assume that it possesses the required correction logic for the right or left shift of a universal shift register (USR).

12.19 Design a *ripple down-counter* that will sequence through the following states:

$$111 \rightarrow 110 \rightarrow 101 \rightarrow 100 \rightarrow 011 \rightarrow 010 \rightarrow 111 \cdots.$$

To do this, use FET T flip-flops and take the count from the $Q(H)$ outputs of the flip-flops. Initialize the counter into the 111 state.

12.20 Shown in Fig. P12.6a is the block symbol for a 4-bit ($\div 2^4$) ripple Up/Down counter. The count direction is determined by the following:

$$Up \text{ if } X(H) = 1(H)$$
$$Dn \text{ if } X(H) = 0(H)$$

PROBLEMS

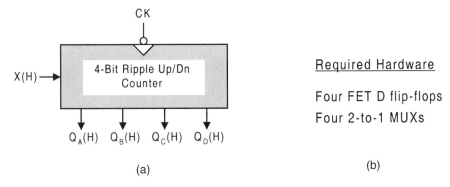

FIGURE P12.6

(a) Construct the state diagram and state table for this counter.

(b) Design this counter by using the hardware indicated in Fig. P12.6b (nothing else), and end with a logic diagram. Plan to initialize into the 0000 state and take the count from the $Q(H)$ outputs. (Hint: Review Fig. 12.39 and read the discussion at the end of Section 12.5.)

(c) Alter the design slightly to provide cascading capability for this counter.

CHAPTER 13

Alternative Synchronous FSM Architectures and System-Level Design

13.1 INTRODUCTION

It is in this chapter that an attempt will be made to bring the subjects of Chapters 10 through 12 together in some meaningful fashion so that useful controller and system-level designs can be created. This is, to state it mildly, no simple task, since an almost endless number of alternatives are available to the designer. Accordingly, and without apology, the treatment will be limited to a few select topics that are representative of some of the more popular and constructive approaches to state machine and system-level design. The "creativity" aspect of the design task is highly valued and should be exercised by the skilled designer whenever it is profitable to do so. To a reasonable extent this creativity ethic will be used in this chapter, but only if it serves to edify the reader's experience in design without unnecessary effort. Cute or novel designs that add little or nothing to an understanding of design fundamentals will be left to the reader's imagination.

13.1.1 Choice of Components to be Considered

The first thing that must done before proceeding is to list the various devices that should be considered for use in a given design architecture. The various components available to the designer are divided into the following five categories:

1. **Next state and output-forming logic.** Choose from the following:

    Discrete logic (gates mainly)
    MUXs
    Decoders
    ROMs
    PLAs
    Basic I/O PALs

2. **Memory.** Choose from the following:

   Discrete flip-flops (D, JK, or T that are edge-triggered or master/slave)
   Shift registers
   Counters

3. **Registered PLDs for total state machine design.** Choose from the following:

   R- and V-type PALs
   FPGAs (e.g., Actel and Xilinx)
   GALs, EPLDs, PLSs, etc. (see Subsection 7.7.4 for definitions)

4. **Input and output conditioning circuits.** Choose from the following:

   Synchronizers
   Synchronizer/stretchers
   Debouncing circuits
   Output holding (storage) registers for filtering

5. **Initialization and reset circuits.** Choose from the following:

   Sanity circuits

The preceding list of components may not be exhaustive, but it covers most of the components that are commonly used in modern state machine and system-level design. Clearly, the choice of components depends on various considerations, including intended use, physical realization, programmability, and a host of other factors. For example, if it is the intent of the designer to place the state machine *on chip*, the choice is somewhat limited. In this case, a proper choice might include the use of a PLA and discrete flip-flops chosen from categories 1 and 2, together with the appropriate input and output conditioning circuits and initialization circuit. On the other hand, if the choice is *off chip*, many more alternatives are available, mainly because of the vast numbers of possibilities contained in categories 1, 2, and 3, perhaps limited only by the imagination of the designer. In the following sections, several exemplary design architectures are demonstrated in the design of various FSMs and controlled systems. Before proceeding it is recommended that the reader review the design procedure laid out in Subsection 11.9.1.

13.2 ARCHITECTURES CENTERED AROUND NONREGISTERED PLDs

The model used for designs centered around nonregistered PLDs is shown in Fig. 13.1, together with block symbols representing possible input and output conditioning circuits. This model is sometimes referred to as the *Huffman model*. The PLD represents a ROM, a PLA, or a basic I/O PAL and is used to generate *both* the NS- and output-forming logic. The memory can be any of the devices listed previously in category 2, namely discrete flip-flops

13.2 ARCHITECTURES CENTERED AROUND NONREGISTERED PLDs

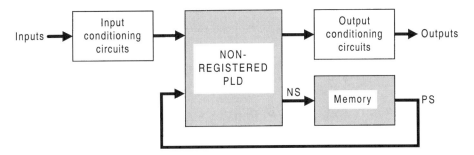

FIGURE 13.1
Model used for architectures centered around a nonregistered PLD showing block symbols representing input and output conditioning circuits for the model.

of some type, a shift register, or a counter. If a shift register or counter is used, the highest priority is given to shifts or counts in assigning state codes so as to make the most efficient use of that particular memory device. Such devices are assumed to be of the off-the-shelf type and should be used in the model shown in Fig. 13.1 only if the sequential nature of the FSM is amenable to their use. For some FSMs, it may be advisable to use discrete flip-flops as the memory elements of choice — a designer's option. The input and output conditioning circuit elements are taken from category 4 in the previous section.

13.2.1 Design of the One- to Three-Pulse Generator by Using a PLA

For purposes of comparison and for a simple first example, consider the design of the one- to three-pulse generator in Fig. 13.2 by using a PLA to generate the NS and output forming logic. Figure 13.2a is a reproduction of that in Fig. 11.36b discussed in Subsection 11.9.2. For a review of PLAs and the actual programming of MOS-oriented PLAs, refer to Section 7.3.

Shown in Fig. 13.2b is the p-term table for Eqs. (11.10), which are obtained from the K-maps in Fig. 11.38 and which are provided as follows for the convenience of the reader:

$$\begin{cases} D_A = \bar{A}B(SW_1) + AB(SW_0) \\ D_B = \bar{A}B(SW_1) + \bar{C}S(SW_1) + \bar{C}S(SW_0) \\ D_C = S + A + B \\ P = A(CK) + B(CK) \end{cases}. \qquad (13.1)$$

Notice that the p-terms are listed in the order of those for D_A, D_B, D_C, and output P. It is a good idea to organize the p-term table in such manner for ease of future reading. Also, note that the p-term $\bar{A}B(SW_1)$ is a shared PI for next state functions D_A and D_B and is given only once in the p-term list. For the AND plane (the decoder portion of the PLA), an existing input is represented either as a logic 1 if uncomplemented or as a logic 0 if complemented in the p-term. A dash is used to indicate the absence of an input in the p-term to the left.

Some explanation of the CK input to the PLA is necessary. In Section 7.5 the subject of active low inputs and outputs relative to PLAs and ROMs is discussed. However, the periodic CK signal (waveform) is really an *"apolar" input* to a state machine and is treated

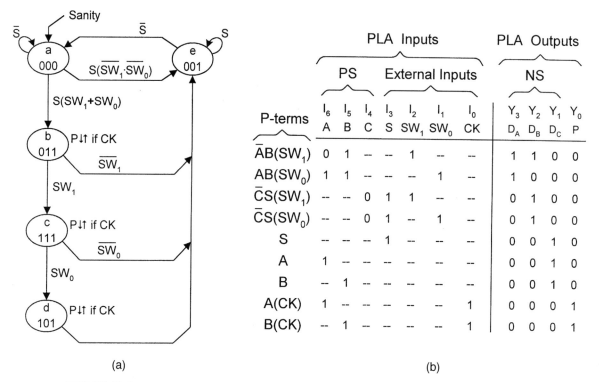

FIGURE 13.2
Design of the one- to three-pulse generator centered around a PLA. (a) State diagram. (b) P-term table suitable for programming a 7 × 9 × 4 or larger PLA.

as such in this text. Thus, a CK waveform need not have an activation level, high or low, associated with it. In a few cases, CK may be assigned an activation level indicator for emphasis or clarification, particular if I/O conditioning circuits are involved. Figure 11.39 is an example.

Presented in Fig. 13.3 is an $n \times p \times m = 8 \times 16 \times 4$ FPLA that is programmed to generate the NS and output forming logic for the one- to three-pulse generator of Fig. 13.2. Here, the symbolism represents the bit position patterns illustrated in Fig. 7.6. The tri-state drivers serve to enable the FPLA if $EN(L) = 1(L)$ or to disable the FPLA if $EN(L) = 0(L)$. Notice that all nine p-terms in Fig. 13.2b are represented and that one, $\bar{A}B(SW_1)$, contributes to both the D_A and D_B NS functions — hence a shared PI, as pointed out previously. Observe also that the array of square dots and circles in the OR plane of Fig. 13.3 is the same as the PLA output array of 1's and 0's in the p-term table of Fig. 13.2b. This will always be so for the symbolic representations of nonsequential PLDs. Note that the square dots and circles store a 1(L) and 0(L), respectively, in agreement with Fig. 7.6.

Implementation of the programmed FPLA in Fig. 13.3 is illustrated in Fig. 13.4. The one- to three-pulse generator is unique in the sense that CK is an input (like any other input) to the PLA. This, of course, is required if the output P, shown in Fig. 13.2a, is to be conditional on CK. Recall, in Subsection 11.9.2, Fig. 11.35, that the pulses are required to be issued only when CK is active. Actually, it is possible to remove the CK input to the

13.2 ARCHITECTURES CENTERED AROUND NONREGISTERED PLDs 617

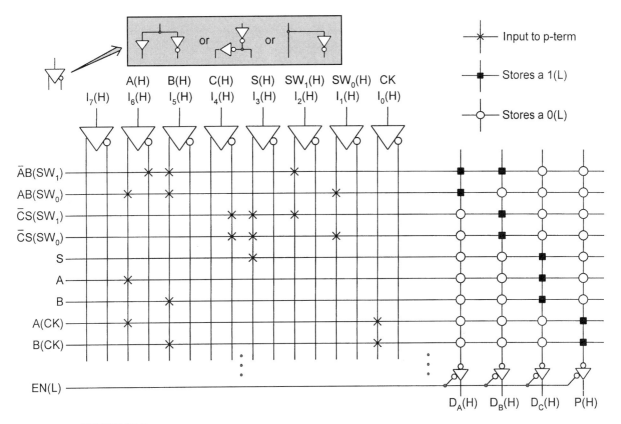

FIGURE 13.3
Symbolic representation of the fusible bit position patterns for an 8 × 16 × 4 FPLA that is programmed to generate the NS and output forming logic required by the one- to three-pulse generator in Figure 13.2.

PLA provided that the output $P(H)$ is ANDed with CK externally. This would satisfy the requirement just mentioned while requiring one less input to the PLA. Also, notice that the actual debouncing, synchronizing, and initialization circuits are not shown in Fig. 13.4 since they are exactly the same as those provided in Fig. 11.39.

13.2.2 Design of the One- to Three-Pulse Generator by Using a PAL

Unlike the PLA, a PAL device can be programmed only in the AND plane. The OR plane has a fixed number of inputs for each output and is, therefore, nonprogrammable. It is for this reason that all p-terms must be programmed separately into the PAL device — shared PIs cannot be used, as in the case of a PLA. Shown in Fig. 13.5 is the symbolic representation of the fusible bit position patterns for an 8 × 16 × 4 basic I/O PAL that is programmed to generate the NS and output logic required by the one- to three-pulse generator in Fig. 13.2. Notice that all 10 p-terms in Eqs. (13.1) are programmed into the AND plane and that the p-term $\bar{A}B(SW_1)$ is listed twice and not shared as in the FPLA of Fig. 13.3. In the nonprogrammable OR plane, three p-term connections [filled squares each storing 1(L)] are provided for each output. If fewer than three p-term connections are needed, the unused

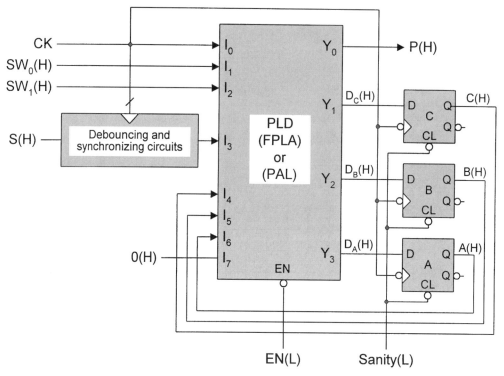

FIGURE 13.4

Implementation of the one- to three-pulse generator with a PLD such as an FPLA or basic I/O PAL, where the debouncing, synchronizing, and initialization circuits are given in Fig. 11.39.

connections are not programmed. This is the case for outputs $D_A(H)$ and $P(H)$, each with one unused p-term. On the other hand, if more ORed connections are needed than are provided by the fixed OR plane of the PAL, the outputs must be ORed external to a basic I/O PAL. However, an L-type PAL has feedback paths that could be used for that purpose. The tri-state drivers serve the same purpose as in Fig. 13.3.

The basic I/O PAL in Fig. 13.5 is nonregistered, meaning that it lacks the capability to be used to implement a state machine without using external memory elements (flip-flops). The R- and V-type PALs, discussed in Section 7.4, are much more versatile devices since they can be programmed to implement state machines without the need for external memory — they have built-in flip-flops and feedback paths. Erasable PALs are also available, a feature that makes them even more attractive to the designer. The acronym PAL is a registered trademark of Advanced Micro Devices, Inc. Therefore, use of this acronym acknowledges AMD's right of trademark for all PAL-type devices.

13.2.3 Design of the One- to Three-Pulse Generator by Using a ROM

Whereas the PAL is programmed in the AND plane, the ROM is programmed only in the OR plane. But programming the ROM is simpler in the sense that the canonical ROM program

13.2 ARCHITECTURES CENTERED AROUND NONREGISTERED PLDs 619

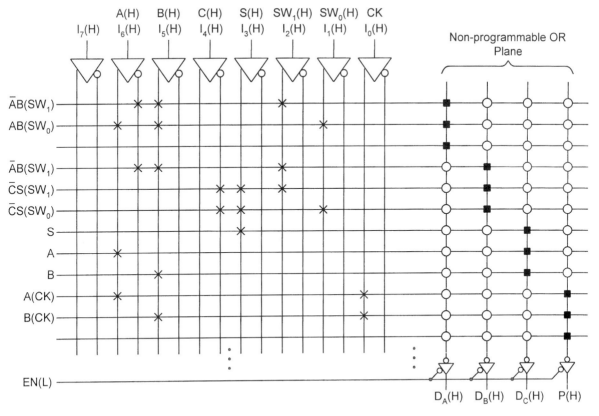

FIGURE 13.5
Symbolic representation of the fusible bit position patterns for an 8 × 16 × 4 basic I/O PAL that is programmed to generate the NS- and output-forming logic required by the one- to three-pulse generator in Fig. 13.2. PAL is a registered trademark of AMD, Inc.

table is obtained directly from either the state diagram or from the state table. Shown in Fig. 13.6a is the state table for the one- to three-pulse generator. An inspection of the state table shows it to be the same as the state diagram in Fig. 13.2a with one major exception. The outputs P are no longer conditional on CK. This has been done to reduce the number of inputs to the ROM, PROM in this case. Recall that ROM size increases by a factor of 2 for each additional ROM input. Now, however, another means must be found to produce a pulse output conditional on CK. This is done by ANDing P with CK as is illustrated later.

The ROM program table is given in Fig. 13.6b. In this case, it is constructed directly from the state table in Fig. 13.6a with unconditional P outputs. Notice that the program table is canonical (1's and 0's only, as it must be) and that the irrelevant input symbol X is used to collapse it to only 11 rows. The fully expanded truth table would require $2^6 = 64$ rows of I/O data, which is not necessary to program the ROM. The missing states in the program table are all assigned X's on the input side and don't-cares on the output side of the table. Remember that an irrelevant input, like a don't care, can be assigned a logic 1 or

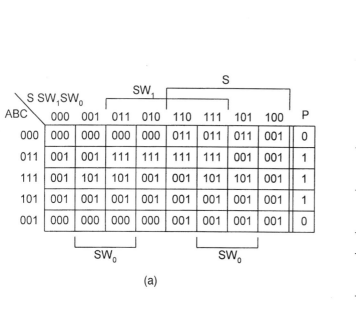

FIGURE 13.6
ROM design of the one- to three-pulse generator in Fig. 13.2, except with CK removed as an input to the ROM. (a) State table showing an unconditional output P for states b, c, and d in Fig. 13.2a. (b) Collapsed program table for a PROM implementation.

a logic 0; it doesn't matter. This, of course, is true only if the FSM is properly initialized, in this case into the 000 state. Finally, remember that all outputs in a ROM program table are indicated relative to the present state (PS), never the NS.

Presented in Fig. 13.7 is a $2^n \times m = 2^6 \times 4$ PROM that is programmed to generate the NS- and output-forming logic for the one- to three-pulse generator represented in Fig. 13.6b. Here, the filled square dots and circles represent the fusible bit position patterns for storage of 1's and 0's shown in the generalized PROM structure of Fig. 7.2. As before, the tri-state driver outputs permit the PROM to be enabled if $EN(L) = 1(L)$ or disabled if $EN(L) = 0(L)$, according to Fig. 3.8d, assuming CMOS logic.

Implementation of the one- to three-pulse generator by using a PROM is shown in Fig. 13.8. Here, the required dependence of output P on CK is removed from the PROM and placed external to it by using an AND gate. Thus, output P' cannot be issued except in states 011, 111, and 101 of the state table, and then only when CK is active, as required

13.2 ARCHITECTURES CENTERED AROUND NONREGISTERED PLDs

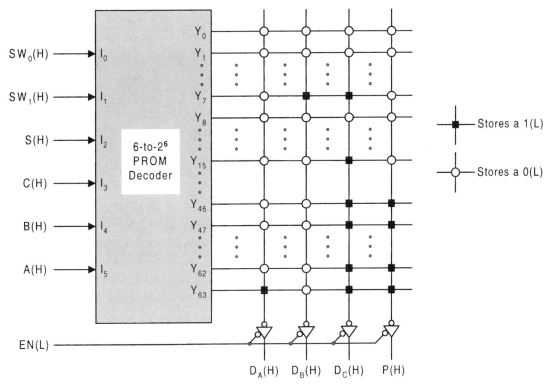

FIGURE 13.7
PROM decoder and symbolic representation of the fusible bit patterns required to program the one-to three-pulse generator represented in Fig. 13.6.

by the design specifications. Note that the input debouncing and synchronizing circuits are the same as those shown in Fig. 11.39, which was also true for the FPLA design in Fig. 13.4.

There still remains the question of output race glitches (ORGs) and static hazards in the PROM implementation of the one- to three-pulse generator. First, according to Fig. 11.37, there are no ORGs possible in this FSM. Second, the PROM generates minterms for output P, as illustrated in Fig. 13.9a. The expression for P, obtained from the K-map in 13.9a or from the state table in Fig. 13.6a, is given in Fig. 13.9b. It indicates the possibility of two internally initiated static 1-hazards. This can be verified by comparing the coupled terms in the logic expression for P with the state table. The possible hazardous transitions are $011 \rightarrow 111$ and $111 \rightarrow 101$. Assuming that the flip-flops are NOR-based, static 1-hazards will be produced in output $P(H)$ shown in Fig. 13.8. However, these hazards cannot possibly appear in the output $P'(H)$, since they are filtered out by the AND gate. Remember that all logic noise is produced immediately following the triggering edge of the CK waveform. Since FET flip-flops are used for the memory and since the pulses P are coincident with CK active, the output is filtered by the ANDing operation permitting clean pulses to be issued.

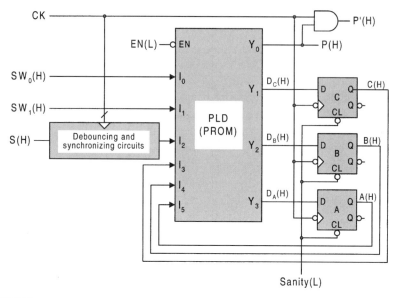

FIGURE 13.8
Implementation of the one- to three-pulse generator with a six-input, four-output PROM showing the external logic required to generate pulses conditional on CK.

13.2.4 Design of a More Complex FSM by Using a ROM as the PLD

As a second and more complex example of ROM-centered implementation, consider the state diagram for a fictitious FSM in Fig. 13.10a. This state machine features four synchronous inputs, one of which is active low, and four outputs, one of which is also active low. This machine is interesting because it possesses up to three-way branching where branching is dependent on all four inputs, and has both conditional and unconditional outputs. Thus, the ROM program table will be somewhat more complex than that of Fig. 13.6b. Though this FSM has only seven states, it is as complex (branching-wise) as one is likely to encounter in the field.

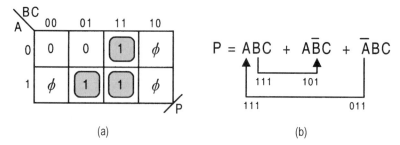

FIGURE 13.9
Static hazard analysis of the PROM implementation of output P taken from the state table in Fig. 13.6a. (a) K-map showing cover for P required by the PROM. (b) Expression for P showing coupled p-terms and internally initiated hazard transitions.

13.2 ARCHITECTURES CENTERED AROUND NONREGISTERED PLDs

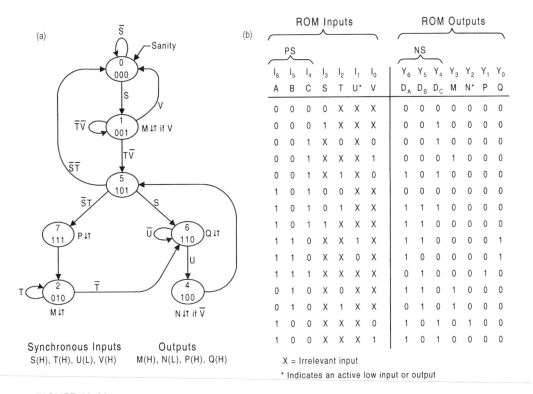

FIGURE 13.10
Design of the NS and output logic for a fictitious FSM by using a ROM. (a) State diagram. (b) Collapsed ROM program table constructed directly from the state diagram in (a).

The ROM program table is constructed directly from the state diagram and is given in Fig. 13.10b. As can be seen, this table is a collapsed canonical truth table involving only 15 rows. Fully expanded, this table would require $2^7 = 128$ rows, which is unnecessary for programming purposes. Remember that the irrelevant input X is to inputs as the don't care ϕ is to outputs. Thus, all input data for state 011 (not shown) take X's and all output data relative to this state take don't cares.

There are other features of this ROM program table that are noteworthy. Active low inputs to a ROM can be dealt with by complementing the logic values in columns of those active low inputs, or by placing inverters on these inputs to the ROM and not complementing the columns. Similarly, active low outputs can be handled by either complementing their columns or by placing an inverter on these outputs, but not both actions. The input $U(L)$ and the output $N(L)$ are represented by using an asterisk in the ROM I/O table to indicate that one of the two actions just stated is necessary to accommodate their active low logic level. For this example, only the U input column is complemented in the ROM program table of Fig. 13.10b. This eliminates the need to use an inverter on the input to produce $U(L)$. The active low output $N(L)$ will be issued from an output holding register as discussed later in this section. Section 7.5 provides a review of this subject.

One other feature of the ROM program table, mentioned earlier, is important to remember: *The outputs are always given relative to the present state, never the next state.* The reader can verify this by comparing the ROM program tables in Figs. 13.6b and 13.10b with their respective state table and state diagram in Figs. 13.6a and 13.10a.

The ROM program table in Fig. 13.10b could have been constructed from a state table. However, this approach would have been more difficult, or at least more tedious, mainly because of the size of the state table needed. Nevertheless, a state table can be constructed by using state identifiers, thereby permitting the construction of the ROM program table to proceed with little effort. Use of a state table for this purpose is the method of choice if a CAD approach is used.

Shown in Fig. 13.11 is an 8-to-2^8 EPROM decoder and symbolic representation of the fusible bit position patterns in the OR plane required to generate the NS- and output-forming logic for the FSM in Fig. 13.10. The meaning of the filled squares and circles and use of

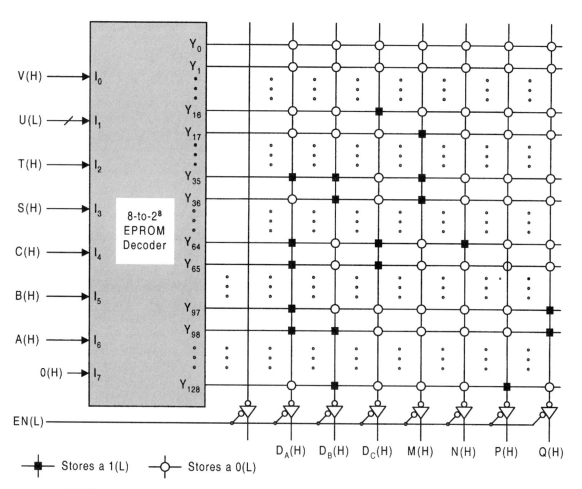

FIGURE 13.11
EPROM decoder and symbolic representation of the fusible bit position patterns in the OR plane required to program the fictitious FSM in Fig. 13.10.

13.2 ARCHITECTURES CENTERED AROUND NONREGISTERED PLDs

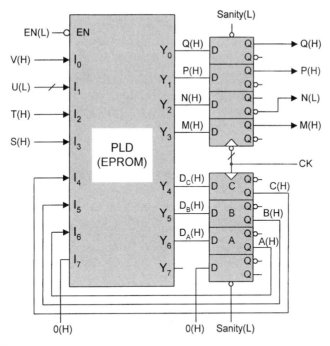

FIGURE 13.12
Implementation of the fictitious FSM in Figs. 13.10 and 13.11 by using an 8-input/8-output EPROM, and two 4-bit storage registers.

the tri-state drivers was discussed previously relative to Fig. 13.7. Notice that the array of filled squares and circles in Fig. 13.11 is exactly the same as the array of 1's and 0's in the ROM program table of Fig. 13.10b. Note that the last 127 minterms are not used.

Implementation of the fictitious FSM in Fig. 13.10 is shown in Fig. 13.12. It follows closely that of the one- to three-pulse generator in Figure 13.8, but with some significant changes in ROM size and in external logic. In this case use is made of a $2^8 \times 8$ EPROM, in agreement with Fig. 13.11, though only a $2^7 \times 7$ EPROM is necessary. Also different is the use of two 4-bit storage registers, one used for the 3-bit memory and the other used as a holding register to filter out the ORGs that can occur in the four outputs. Notice that the output holding (filtering) register is triggered antiphase to the memory register, a necessary feature for filtering logic noise, as discussed in Subsection 11.2.2. Static hazards in the output logic are not possible in this FSM; but if they were possible, they also would be filtered out. The output holding register serves one other function. It can also be used to deliver the four outputs synchronously to the next stage independent of any logic and routing delay differences that can occur within the EPROM.

Finally, notice that the active low input $U(L)$ and active low output $N(L)$ are properly dealt with in Fig. 13.12. Since the input column for U is to be complemented before the EPROM is programmed, no inverter on this input is necessary. However, since the output column for N in the ROM program table is not complemented prior to programming, $N(L)$ must be delivered by the register as indicated in Fig. 13.12. Thus,

626   CHAPTER 13 / ALTERNATIVE SYNCHRONOUS FSM ARCHITECTURES

$N(L)$ must be issued from the $Q(L)$ output of the register's flip-flop, or by an inverter on $Q(H)$.

13.3 STATE MACHINE DESIGNS CENTERED AROUND A SHIFT REGISTER

There are times when the designer might like to consider using an off-the-shelf universal shift register (USR) in the design of a state machine, one that is amenable to the shifting character of the USR. Remember, it makes little sense to use a USR for this purpose if most of the FSM's state-to-state transitions are parallel load actions. For such an FSM, it would be best to use discrete flip-flops as has been done in all examples up to this point. In making the state code assignments for an FSM, shifting operations must be given the highest priority if the most efficient use is to be made of the shift register.

Shown in Fig. 13.13a is the state diagram for an FSM that would be considered amenable to the use of a USR as the memory. Notice that it has what could be termed a linear array

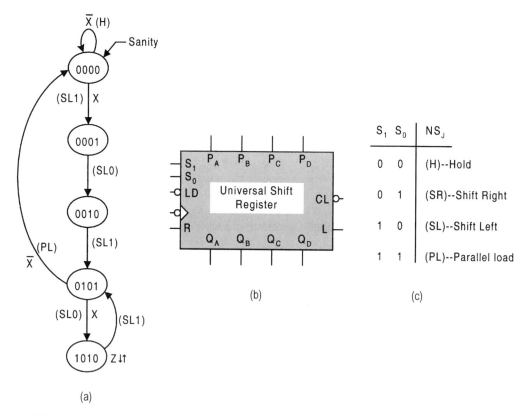

FIGURE 13.13
State machine design by using the USR as the memory. (a) State diagram for a fictitious FSM showing state-to-state branching actions of the USR. (b) Block diagram symbol for the USR. (c) Operation table and branching action for the USR.

13.3 STATE MACHINE DESIGNS CENTERED AROUND A SHIFT REGISTER

of states — hence, a *linear state machine*. The branching action required by the USR is given for each state-to-state transition. Notice that only left shifting and parallel loading are required of the USR.

In Figs. 13.13b and (c) are the block circuit symbol and operation table for the USR. A view of the logic circuit symbol indicates that logic must be found for inputs S_1, S_0, R, L, and the four parallel load inputs P_A, P_B, P_C, and P_D before the USR can be used in the design of this FSM. However, the external logic required to drive the USR through the sequence of states in Fig. 13.13a turns out to be quite simple. Shown in Fig. 13.14a are the K-maps for the two mode controls and the serial input for left shifting. The minimum cover,

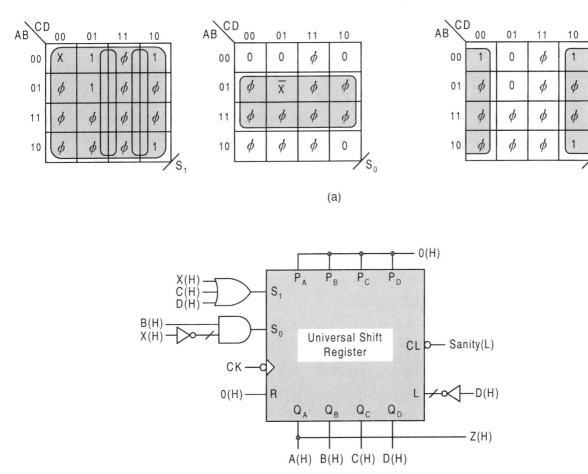

FIGURE 13.14
Implementation of the fictitious FSM in Fig. 13.13 by using a USR as the memory. (a) Mode control and serial input K-maps showing minimum cover. (b) The USR and external logic derive from (a) and from the state diagram.

indicated by shaded loops, yields the expressions:

$$\begin{cases} S_1 = X + C + D \\ S_0 = B\bar{X} \\ L = \bar{D} \end{cases}, \qquad (13.2)$$

where $R = 0$ or 1 and $P_A = P_B = P_C = P_D = 0$ by inspection of the state diagram in Fig. 13.13a. The logic for L can also be deduced from an inspection of the state diagram since state variable D toggles in complementary fashion with respect to the left shifting of 1's and 0's. Observe that state pairs, 0101 and 1010, are the only two four-bit patterns that can be cycled exclusively with either a left shift or a right shift. Knowledge of this fact can be useful in state machine designs centered around a shift register, as in this case. Implementation of Eqs. (13.2) is shown in Fig. 13.14b, where the Moore output is simply $Z(H) = A(H)$, as deduced from the state diagram in Figure 13.13a.

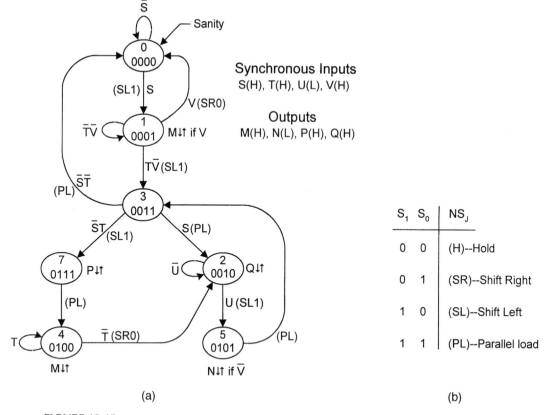

FIGURE 13.15
Design of the fictitious FSM of Fig. 13.10a by using a USR as the memory. (a) State diagram with a state code assignment amenable to a USR design showing state-to-state branching actions in parentheses. (b) Operation table for a USR.

13.3 STATE MACHINE DESIGNS CENTERED AROUND A SHIFT REGISTER

A More Complex Example of State Machine Design Centered around a USR It has just been demonstrated that a "linear state machine" can be well suited to the use of a shift register as the memory. However, if this approach is applied to an FSM design where multiple branchings are involved, use of a USR as the memory element loses some of its appeal. Consider the state machine in Fig. 13.15a, which is the FSM of Fig. 13.10a but coded in such a way as to take better advantage of the shift character of the USR. The branching actions of the USR, defined in Fig. 13.15b, are indicated in parentheses for each state-to-state transition. As in the previous example, this is very helpful in obtaining the required logic external to the USR. Notice that the MSB state variable A is left inactive so as to minimize the external logic commitment — its use is not needed in this case. Deactivation of a state variable in shift register designs can be done only if care is taken to ensure that the shifting and parallel load actions do not create problems at this bit position.

The third-order K-maps for the mode control and the parallel load inputs are provided in Figs. 13.16a and 13.16b. Because the MSB state variable is inactive, only the remaining state variables, B, C, and D, need be used in K-map construction. No minimum cover is indicated in the mode control K-maps because MUXs are to be used to implement S_1 and S_0 — a designer's call. Note that a K-map for P_A is not necessary since, by inspection of the state diagram, it is evident that $P_A = 0$. K-maps for serial inputs L and R are also unnecessary since, by inspection of the state diagram, $L = 1$ and $R = 0$. That is, all indicated shift-left operations are $SL1$ and all indicated shift-right operations are $SR0$; all others are,

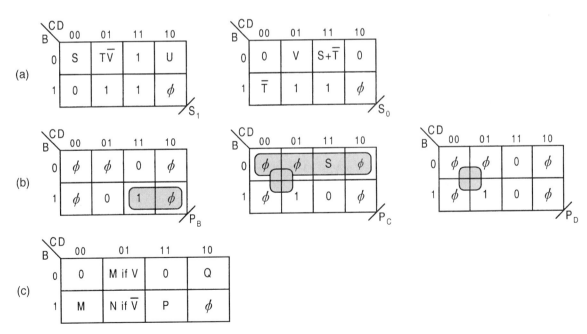

FIGURE 13.16
K-maps for the fictitious FSM of Figure 13.15(a). (a) Mode control K-maps appropriate for MUX implementation. (b) Parallel load input K-maps and minimum cover. (c) Composite K-map for the four outputs.

of course, don't cares. The composite output K-map in Fig. 13.16c is useful since a state decoder is to be used to generate all Moore outputs and to reduce the logic necessary for the Mealy outputs.

Gathering the results so far, the expressions for the serial and parallel load inputs are

$$\begin{cases} L = 1 \\ R = 0 \\ P_A = 0 \\ P_B = BC = \text{State 7} \\ P_C = \bar{B}S + \bar{C} = (\text{State 3}) \cdot S + \text{State 5} \\ P_D = \bar{C} = \text{State 5} \end{cases}, \quad (13.3)$$

and for the four outputs

$$\begin{cases} M = (\text{State 1}) \cdot V + \text{State 4} \\ N = (\text{State 5}) \cdot \bar{V} \\ P = \text{State 7} \\ Q = \text{State 2} \end{cases}, \quad (13.4)$$

where it is understood that the mode control inputs, S_1 and S_0, are to be generated by 8-to-1 MUXs and that a state decoder is to be used to produce the State values given in Eqs. (13.3) and (13.4). It is important for the reader to realize that the *State* values in the parallel load and output equations can be read directly from the state diagram. For example, $P_B = \text{State 7}$ since the only parallel load involving state variable B is the branching 0111 → 0100. Similarly, $P_C = (\text{State 3}) \cdot S + \text{State 5}$ is due to the parallel load 0011 → 0010 under branching condition S and the unconditional parallel load transition 0101 → 0011. Or, in the case of an output, $N = (\text{State}) \cdot \bar{V}$ results since N is conditional on $\bar{V}$ in state 0101. Thus, the use of a state decoder can save time and reduce the number of external gates required for implementation, which, of course, comes at the cost of adding a state decoder.

Shown in Fig. 13.17 is the FSM of Fig. 13.15a centered around a USR and state decoder with 8-to-1 MUXs used to generate the mode controls, S_1 and S_0. Here, the external logic to the MUXs is the logic contained in the cells of the mode control K-maps, and the parallel load and output logic follow Eqs. (13.3) and (13.4). An output holding register, triggered anti-phase to the memory, is necessary since ORGs abound, as can be seen by an inspection of the state diagram in Fig. 13.15a. Observe that the shifting and parallel load action required by this FSM presents no problem at the inactive MSB position, A, since that position accepts a logic 0 in all cases.

The use of the state decoder in Fig. 13.17 is to be considered a design convenience, and so its presence is arbitrary. A state decoder helps to minimize the parallel load and output-forming logic and reduces the overall effort in obtaining this logic. In the absence of a state decoder, one can expect a significant increase in the number of gates required to implement the parallel load and output logic. For example, without the state decoder, the

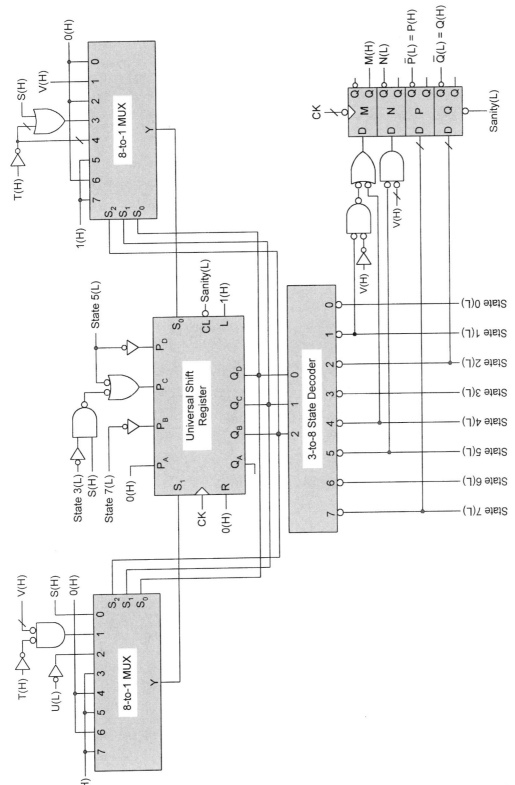

FIGURE 13.17
Implementation of the FSM in Fig. 13.15a centered around a USR with application of Eqs. (13.3) and (13.4) and 8-to-1 MUXs for the external logic.

631

output-forming logic in Eqs. (13.4) becomes

$$\begin{cases} M = \bar{B}\bar{C}DV + B\bar{D} \\ N = B\bar{C}D\bar{V} \\ P = BC \\ Q = C\bar{D} \end{cases}, \qquad (13.5)$$

as read from the composite K-map in Fig. 13.16c. This represents an increase of three gates over that required by Eqs. (13.4). But again, the price to be paid for convenience and for the reduction in external gate logic is the added state decoder hardware.

13.4 STATE MACHINE DESIGNS CENTERED AROUND A PARALLEL LOADABLE UP/DOWN COUNTER

For purposes of comparison, it will be interesting to design the same FSMs as in Section 13.3 but now centered around a parallel loadable up/down counter instead of a USR. Shown in Fig. 13.18a is the "linear state machine" of Fig. 13.13a, but now state coded in a count sequence. Notice that the MSB state variable is inactive and that only one parallel load transition exists, 0011 → 0000.

The counter to be used for this design is that featured in Fig. 12.23. This is a binary up/down counter with synchronous parallel load and true hold capability. The logic symbol and operation table for this counter are reproduced from Subsection 12.3.4 and are presented in Figs. 13.18b and 13.18c for convenience of the reader. An inspection of the logic circuit symbol indicates that external logic must be found for inputs S_1, S_0, Up, Dn, and the four parallel load inputs P_A, P_B, P_C, and P_D. But, as it turns out, this external logic is quite simple. From the K-maps and minimum cover for the mode control and Up/Dn inputs given in Fig. 13.19a, there results the external logic given by

$$\begin{cases} S_1 = CD\bar{X} \\ S_0 = X + B + C + D \\ Up = \bar{B} \\ Dn = B \end{cases}, \qquad (13.6)$$

where it is understood that $P_A = P_B = P_C = P_D = 0$ by inspection of the state diagram in Fig. 13.18a. That is, the parallel load inputs are necessarily all zero because the only parallel load branching is from state 0011 to state 0000. Notice that the logic for Up and Dn could also have been deduced from the state diagram.

The resulting logic circuit for the FSM in Fig. 13.18a is shown in Fig. 13.19b. Here, it is easily seen from the state diagram that the single Moore output is simply $Z(H) = B(H)$. Comparing Fig. 13.19b with Fig. 13.14b indicates that both a USR and parallel loadable up/down counter design of this linear state machine result in only minimal external logic. Remember that to accomplish these designs it is necessary that the USR and counter have both parallel load and true hold capability.

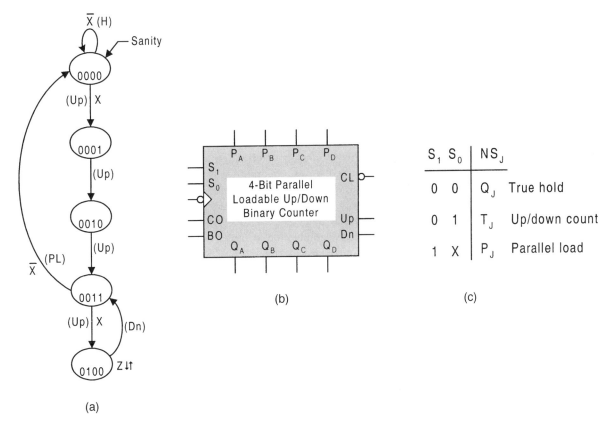

FIGURE 13.18
State machine design by using the parallel loadable up/down counter of Fig. 12.23 as the memory. (a) State diagram for the FSM showing state-to-state branching actions of the counter. (b) Block diagram symbol for the parallel loadable up/down counter. (c) Operation table for the parallel loadable up/down counter.

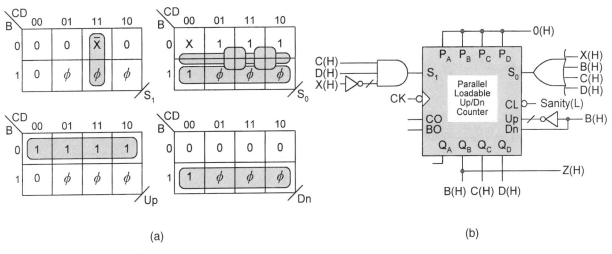

FIGURE 13.19
Implementation of the state machine in Fig. 13.18a by using an up/down counter with parallel load and true hold capability. (a) K-maps and minimum cover for the mode control and *Up* and *Dn* inputs. (b) Logic circuit for the state machine.

A More Complex Example of FSM Design Centered around a Parallel Loadable Up/Down Counter The previous example demonstrated that the a linear state machine could be favorably designed by using a parallel loadable up/down counter as was the case in using a USR for the memory. But what would be the consequence of using such a counter-based design for a more complex FSM? The comparison between the various approaches to FSM design now continues with the design of the FSM in Figs. 13.10 and 13.15 by using as the memory the 4-bit parallel loadable up/down counter with true hold capability featured in Fig. 12.26. Shown in Fig. 13.20a is the state diagram for the FSM with a state code assignment suitable to counter design. Clearly, the number and character of its inputs and outputs, together with up to three-way branching, is much more complex than the linear FSM of Fig. 13.18a. Consequently, it is predictable that the external logic required for a counter design of this FSM will be considerably more complex than that in Fig. 13.19b. In fact, an architecture similar to that for the USR design in Fig. 13.17 is to be expected.

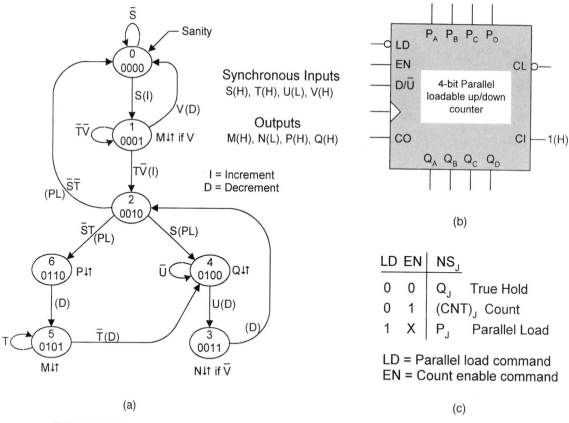

FIGURE 13.20
Design of the fictitious FSM of Fig. 13.10a by using a parallel loadable up/down counter with true hold capability. (a) State diagram with state code assignments suitable for counter design showing counter branching actions in parentheses. (b) Logic circuit symbol for the counter of Fig. 12.26. (c) Operation table for the counter.

13.4 STATE MACHINE DESIGNS

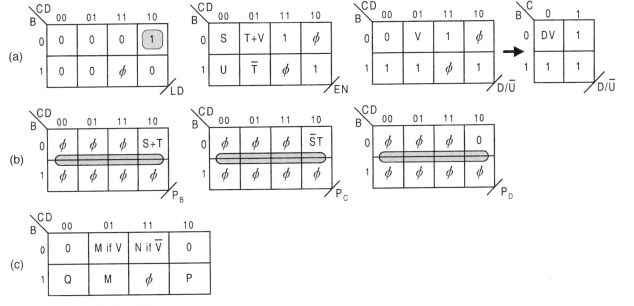

FIGURE 13.21
K-maps for the fictitious FSM of Fig. 13.20a consistent with the operation table in Fig. 13.20c. (a) Mode control and count direction EV K-maps. (b) Parallel load input K-maps and minimum cover. (c) Composite K-map for the four outputs.

An inspection of the logic circuit symbol in Fig. 13.20b indicates that to drive this counter through the state sequence in Fig. 13.20a, external logic must be found for inputs LD, EN, $D/\bar{U}$ and the four parallel load inputs P_A, P_B, P_C, and P_D. The EV K-maps for all external inputs except P_A are shown in Figs. 13.21a and 13.21b, together with the minimum cover for LD and the three parallel load inputs. It follows that $P_A = 0$ since the MSB state variable A is inactive as can be seen in the state diagram of Fig. 13.20a. The outputs are represented by the composite K-map in Fig. 13.21c. From these K-maps there results the following expressions for the inputs:

$$\left\{\begin{array}{l} LD = \bar{B}C\bar{D} = State\ 2 \\ P_B = S + T \\ P_C = \bar{S}T \\ P_D = 0 \end{array}\right\}. \tag{13.7}$$

For the four outputs, assuming the use of a state decoder,

$$\left\{\begin{array}{l} M = (State\ 1) \cdot V + State\ 5 \\ N = (State\ 3) \cdot \bar{V} \\ P = State\ 6 \\ Q = State\ 4 \end{array}\right\}. \tag{13.8}$$

Inputs EN and $D/\bar{U}$ are not included in Eqs. (13.7) since the choice is made to use MUXs to implement these parameters. In the case of $D/\bar{U}$, the option of using either an 8-to-1 MUX or a 4-to-1 MUX is indicated by the compressed EV K-maps in Fig. 13.21a. For this example, the 8-to-1 MUX will be used. Predictably, there is similarity between Eqs. (13.8) and (13.4).

Shown in Fig. 13.22 is the implementation of the FSM in Fig. 13.20a centered around the parallel loadable up/down counter of Fig. 12.26. A state decoder is used primarily to reduce the external gate logic required to generate the four outputs. The choice is made to implement the count enable and direction controls, EN and $D/\bar{U}$, by using 8-to-1 MUXs although, in the latter case, discrete logic or a 4-to-1 MUX would make more efficient use of hardware. An additional gate would be necessary to produce the p-term DV if a 4-to-1 MUX is used, as indicated in Fig. 13.21a. Notice that the external logic to the MUXs is exactly that contained in the cells of the EN and $D/\bar{U}$ K-maps. As in the USR design of this fictitious FSM shown in Fig. 13.17, an output holding register is used to filter the several ORGs that are produced in the operation of this FSM.

The state decoder in Fig. 13.22 can be eliminated, but only at the expense of additional external logic. From the K-maps in Fig. 13.21, the change in the external gate commitment would be

$$LD = \bar{B}C\bar{D}, \quad M = \bar{B}\bar{C}DV + BD, \quad N = CD\bar{V}, \quad P = BC, \quad Q = B\bar{C}\bar{D},$$

which is an increase of four gates over that required with a state decoder. Notice that use of a 16-to-1 MUX to generate EN would eliminate the need for the OR gate shown in Fig. 13.22.

13.5 THE ONE-HOT DESIGN METHOD

As evident from the previous examples, designing a state machine to have a minimum number of state variables (hence a minimum number of flip-flops) involves a considerable effort. Functions often must be mapped and minimized before the design process can be completed. Furthermore, for such designs, no direct relation exists between states of the FSM and the NS and output functions that result.

An alternative design architecture exists that greatly reduces the design effort and ends with a direct relationship between the states of the FSM and the NS and output logic that results. This method is aptly dubbed the *one-hot method* for state machine design — a single "1" per state. But the advantages provided by this method come at a price: one flip-flop per state each with NS-forming logic. A 10-bit one-hot code is given in Column (c) of Table 2.11 in Subsection 2.10.2.

A big advantage of the one-hot method is that the NS and output functions are generated directly from either the state diagram, state table or from an ASM chart — *no specific state code assignments are needed!* Shown in Fig. 13.23a is a state diagram segment for the jth reference state that serves as the model for the one-hot method. Here, it is understood that any branching condition $f_{j \leftarrow j}$ represents the holding condition for the jth state, where j is an integer $j = 0, 1, 2, \ldots, (m-1)$. Since only one logic 1 is permitted in each state code, the use of D flip-flops make it necessary to know only the branching conditions for states that transition *into* a given reference state. The result is the generalized NS (D_j) and output (Z_l) forming logic for m states and r total outputs presented in Fig. 13.23b. These functions are expressed succinctly by

$$D_j = \sum_{k=0}^{m-1} Q_k \cdot f_{j \leftarrow k} \quad \text{and} \quad Z_l = \sum_{j=0}^{m-1} Q_j \cdot f_{j,l}(X), \tag{13.9}$$

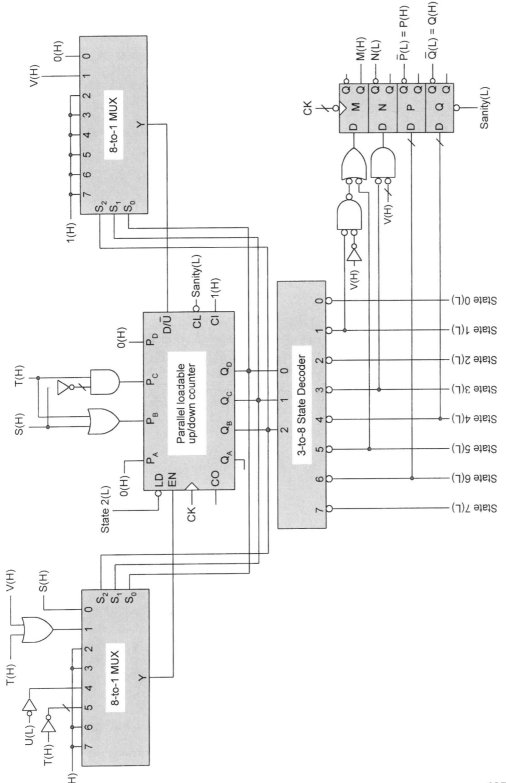

FIGURE 13.22
Implementation of the FSM in Fig. 13.20a centered around the parallel loadable up/down counter of Fig. 12.26 with application of Eqs. (13.7) and (13.8) and with 8-to-1 MUXs for count enable and direction controls.

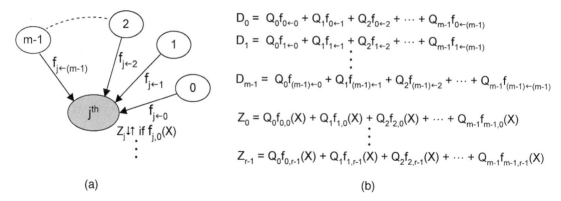

(a) (b)

FIGURE 13.23
Model for the one-hot method expressed by Eqs. (13.9). (a) State diagram segment showing "into" branching conditions and Mealy outputs for the jth reference state. Here, any branching condition $f_{j \leftarrow j}$ is understood to represent the holding condition for the jth state. (b) Generalized one-hot NS- and output-forming logic for D flip-flop designs by application of Eqs. (13.9) to m states and r total outputs.

where $f_{j,l}(X)$ represents the jth function of external inputs X for the lth output, the Q's are the state variables, and the integer $l = 0, 1, 2, \ldots, (r-1)$. Notice that Eqs. (13.9) give the minimum NS- and output-forming logic for a D flip-flop design by the one-hot method — but without the use of K-maps!

To illustrate the application of Eqs. (13.9), consider the state diagram and state table in Figs. 13.24a and b, which represent the FSM in Fig. 13.13a but with only state identifiers

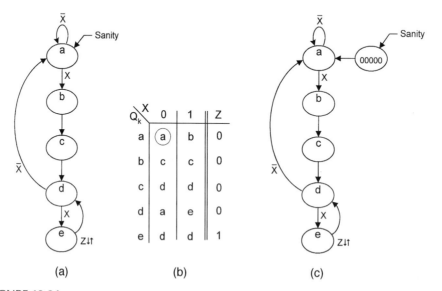

(a) (b) (c)

FIGURE 13.24
State machine design by using the one-hot method. (a) State diagram for a fictitious FSM. (b) State table for the FSM in (a). (c) State diagram of part (a) suitable for initialization into state 00000 by using the one-hot-plus-zero approach.

13.5 THE ONE-HOT DESIGN METHOD

indicated for the states. Applying Eqs. (13.9) directly to either the state diagram or state table, there results the following NS and output functions:

$$\begin{cases} D_a = a\bar{X} + d\bar{X} \\ D_b = aX \\ D_c = b \\ D_d = c + e \\ D_e = dX \\ Z = e \end{cases}, \quad (13.10)$$

where the assignment of specific one-hot codes is *not* necessary. If one were to make one-hot state code assignments for this FSM, the specific code words would be chosen from the set {00001, 00010, 00100, 01000, 10000} in any order. But to do this is an apparent waste of the designer's time and can even be misleading. All that is important to know is that Eqs. (13.10) can be read directly from either the state diagram or state table without the assistance of K-maps, and that no specific one-hot state code assignments are required or even desired. These are the salient features of the one-hot method that set it apart from the alternative approaches. But, of course, the advantages afforded by the one-hot method come at the price of an increased hardware commitment.

One potential problem with the one-hot method for state machine design is the initialization into a one-hot state as in Fig. 13.24a. To do this requires that the D flip-flops have both preset (*PR*) and clear (*CL*) asynchronous overrides, or that one flip-flop have a PR override while the other four have *CL* overrides. However, many MSI devices, such as storage registers, come with only *CL* asynchronous overrides. To overcome this limitation on the use of the one-hot method, a *one-hot-plus-zero* approach can be used, as indicated in Fig. 13.24c. Now the FSM can be initialized into the 00000 state with flip-flops having only CL asynchronous overrides. But the cost of this convenience is the extra logic required for the D_a function given by $D_a = a\bar{X} + d\bar{X} + \bar{a}\bar{b}\bar{c}\bar{d}\bar{e}$. Shown in Figs. 13.25a and 13.25b are the logic circuits for the one-hot and one-hot-plus-zero approaches, respectively, based on Eqs. (13.10). To avoid fan-in limitations by the one-hot-plus-zero method, the correction for generalized "0" state initialization $\bar{a}\bar{b}\bar{c}\bar{d}\bar{e}\cdots$ is best implemented by using the CMOS NOR gate shown in Fig. 8.46.

A More Complex Example of the One-Hot Design Method To further illustrate the use of Eqs. (13.9), consider the state diagram and state table for a fictitious FSM in Fig. 11.42 that is reproduced in Fig. 13.26 for the convenience of the reader. Reading directly from the state diagram or state table, Eqs. (13.9) become

$$\begin{cases} D_a = aS + aT + e\bar{S}T + \bar{a}\bar{b}\bar{c}\bar{d}\bar{e} \\ D_b = a\bar{S}\bar{T} + b\bar{S}\bar{T} + c\bar{S}\bar{T} \\ D_c = bST + cT + dST \\ D_d = b\bar{S}T + d\bar{S}T \\ D_e = bS\bar{T} + cS\bar{T} + d\bar{T} + eS + e\bar{T} \\ P = eS\bar{T} \\ Q = d\bar{S}T + eS + b \end{cases}, \quad (13.11)$$

where it is understood that $a = Q_a$, $b = Q_b$, $c = Q_c$, $d = Q_d$, and $e = Q_e$. To initialize this FSM into the 00000 state instead of state *a*, in agreement with Fig. 13.26a, D_a must

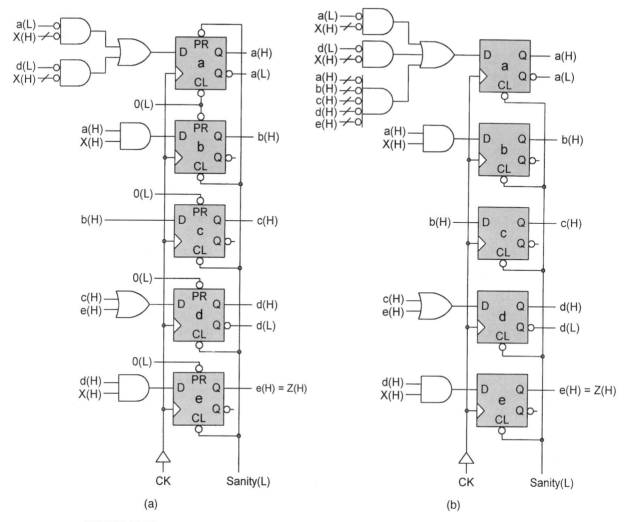

FIGURE 13.25
Implementation of the FSM in Fig. 13.13a by using the one-hot functions given by Eqs. (13.10). (a) External logic required if FSM is initialized into one-hot state a. (b) External logic required if FSM is initialized into the 00000 state by using the one-hot-plus-zero method implied by Fig. 13.24c.

include the term $\bar{a}\bar{b}\bar{c}\bar{d}\bar{e}$, as indicated in Eqs. (13.11). This irreversibly directs the FSM into state a on the next clock triggering edge following initialization. Note the increased hardware required by Eqs. (13.11) compared to that required by Eqs. (11.11) in Subsection 11.10.2, the extra cost for use of the one-hot method. But ORGs and s-hazards in the output logic are not possible, as explained later in Subsection 13.5.4.

13.5.1 Use of ASMs in One-Hot Designs

The one-hot method holds some unique advantages over other approaches to state machine design. Because there is a direct relation between each state of the FSM and the NS and

13.5 THE ONE-HOT DESIGN METHOD 641

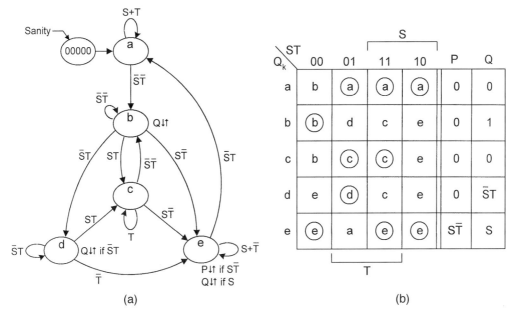

FIGURE 13.26
Reproductions of the FSM in Figure 11.42 for use in the one-hot design method. (a) Fully documented state diagram representation showing state identifiers. (b) Equivalent state table representation.

output functions that result in the one-hot method, a registered PLD can be programmed directly from the state table, the ASM chart, or the state diagram. In fact, the ASM chart can thought of as a graphical representation of the one-hot NS and output equations from which the logic circuit is constructed.

Consider the state diagram and equivalent ASM segments given in Figs. 13.27a and 13.27b. From either of these, the one-hot NS and output equations are read directly as

$$\begin{cases} D_a = (\ ) \\ D_b = a + b\bar{X} \\ D_c = bX \\ M = bX \\ P = b \\ R = c \end{cases}, \quad (13.12)$$

with the resulting logic circuit shown in Fig. 13.27c. Notice how the ASM chart or the state equations translate directly to the logic circuit.

As a second and more complex example of the use of ASMs in logic circuit construction by the one-hot method, consider the resolver configuration in Fig. 11.41. Reproduced in Figs. 13.28a and 13.28b are the state diagram and ASM chart for this resolver, from which the following NS and output functions are derived by application

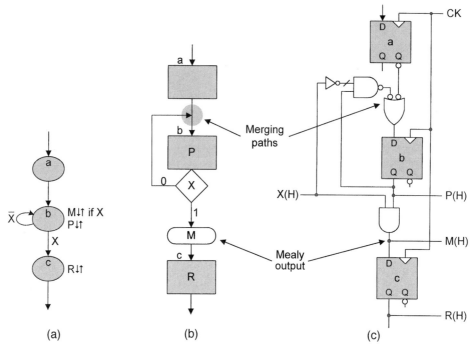

FIGURE 13.27
One-hot state machine configuration derived from a state diagram, an ASM chart, or from state equations. (a) State diagram segment. (b) ASM segment equivalent to (a). (c) One-hot logic circuit derived from the state diagram segment, ASM segment, or from Eqs. (13.12).

of Eqs. (13.9):

$$\left\{\begin{array}{l} D_a = a\bar{X}\bar{Y} + (\ ) \\ D_b = a\bar{X}Y + bY \\ D_c = aX + cX \end{array}\right\} \quad \text{and} \quad \left\{\begin{array}{l} CLRREG = LDCNT = a \\ STDLY = b \\ FIN = c \\ RES = c\bar{Y} \end{array}\right\}. \quad (13.13)$$

Again, notice the ease with which the one-hot Eqs. (13.13) are generated from the state diagram.

The logic circuit for the resolver configuration, shown in Fig. 13.28c, is easily produced from the ASM chart in Fig. 13.28b. But the logic circuit is also easily constructed either from the NS and output functions in Eqs. (13.13) or from the state diagram. In fact, any one of these (the state diagram, the NS and output function, or the ASM chart) can be used with equal ease in constructing the one-hot logic circuit. The fully documented state diagram can replace the ASM chart for this purpose if it is recognized that a holding condition is a merging path that contributes to the NS function according to Eqs. (13.9). The reader can confirm this be comparing the NS functions in Eqs. (13.13) with the state diagram in Fig. 13.28a.

13.5 THE ONE-HOT DESIGN METHOD

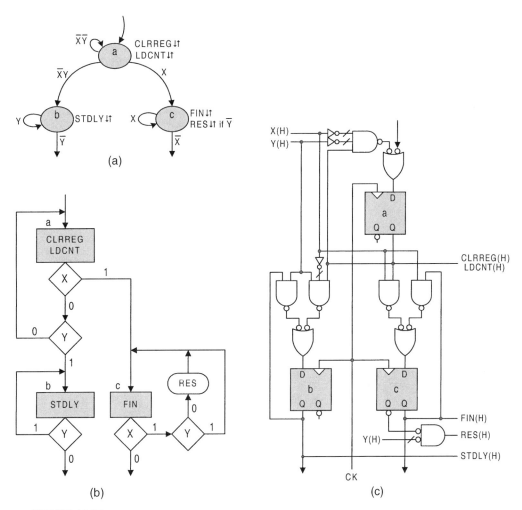

FIGURE 13.28
Resolver configuration of Fig. 11.41 implemented by the one-hot design method. (a) Fully documented state diagram for the resolver. (b) Equivalent ASM chart. (c) One-hot logic circuit constructed directly from the ASM chart, state diagram, or from the NS and output functions in Eqs. (13.13).

13.5.2 Application of the One-Hot Method to a Serial 2's Complementer

Algorithm 2.6 in Section 2.6 presented a simple "pencil-and-paper" method of obtaining the 2's complement of a binary number. As a simple example of the application of the one-hot method in state machine design, Algorithm 2.6 will now be implemented. It is recommended that the reader review and fully understand this algorithm before continuing in this subsection.

Shown in Fig. 13.29a is the block diagram symbol for the serial 2's complementer indicating that the binary input (*Bin*) is introduced LSB first to the complementer and that the 2's complement output (*T*) is issued LSB first. The ASM chart and state diagram

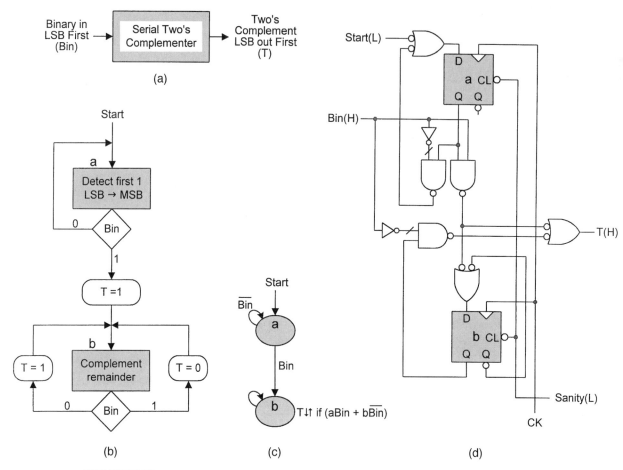

FIGURE 13.29
Design of the serial 2's complementer by using Algorithm 2.6 and the one-hot method. (a) Block symbol of the complementer. (b) ASM chart and (c) state diagram representations of Algorithm 2.6. (d) Logic circuit for the serial 2's complementer derived directly from either the ASM chart or the state diagram.

expressions of Algorithm 2.6 are given in Figs. 13.29b and 13.29c. Notice how much simpler it is to read the state diagram representation than it is to read the ASM chart. From either the state diagram or ASM chart there results the following one-hot NS and output expressions:

$$\begin{cases} D_a = a\overline{Bin} + (Start) \\ D_b = aBin + b \\ T = aBin + b\overline{Bin} \end{cases}, \tag{13.14}$$

where it follows that $D_b = aBin + bBin + b\overline{Bin} = aBin + b$, the result obtained from the state diagram.

13.5 THE ONE-HOT DESIGN METHOD   645

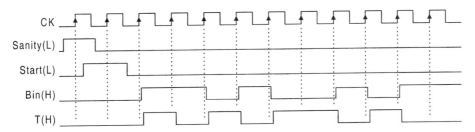

FIGURE 13.30
Timing diagram for the serial 2's complementer in Fig. 13.29c, showing the binary input Bin and the 2's complement output T together with the initialization and start signals.

Equations (13.14) are implemented with the one-hot logic circuit shown in Fig. 13.29d. Here, it is observed that the FSM is initialized into the 00 state following which a Start signal must be applied over at least one clock cycle to begin the process. In effect, the Start signal irreversibly forces the FSM into a one-hot state from the 00 state following deactivation of the sanity input. Notice also that the sequence is open-ended in the sense that it never returns to the initial state a. Thus, the process will continue ad infinitum, or until the circuit is reset by the sanity input.

The results of a logic simulation of the serial 2's complementer is given in Fig. 13.30. Here, the serial input *Bin* is shown synchronized in phase with clock, and the circuit is forced into state a by *Sanity(L)* following initialization. Notice that the Start signal is sampled by the triggering edge of the clock waveform immediately following release of the Sanity initialization signal. This is necessary to permit the process to begin.

13.5.3 One-Hot Design of a Parallel-to-Serial Adder/Subtractor Controller

For this example, consider that two 8-bit USRs, one for word A and the other for word B, shift each bit into a single Full Adder (FA) LSB first. The sum is then issued serially from the FA LSB first. One bit, say bit B, is introduced to the FA via a controlled inverter (XOR gate) for purposes of adding bit B to or subtracting (in 2's complement) bit B from bit A. A D flip-flop is used to supply the carry-out of one operation to the carry-in of the next bitwise serial operation. The D flip-flop must also have PRE and CLR overrides to preset the carry-in (PSCRY) to the FA for the subtraction operation, as required by Eq. (2.14) in Subsection 2.6.2, or to clear the carry-in (CLCRY) if addition. An n-bit binary counter is used to indicate when the 8-bit addition/subtraction process is complete so that the system can be reset for the next 8-bit series of bit-wise operations.

Shown in Fig. 13.31a is the state diagram representing the sequence of events that must take place during the process of serially adding or subtracting two 8-bit operands. Thus, this state diagram represents the controller for the process. Notice that use is made of the one-hot-plus-zero approach allowing the FSM to be initialized into the 000 state. The process begins in state a by loading the counter (*LDCNT*) in preparation for counting, by clearing the registers (*CLREG*), and by pushing the start button (*Start*) to begin the process. In state b, the external D flip-flop is initialized for either subtraction or addition (*PSCRY* or *CLCRY*), and the mode controls to the USRs are set to parallel load the 8-bit operands ($S_1 = 1, S_0 = 1$). Finally, in state c the mode control S_1 goes inactive for right shifting

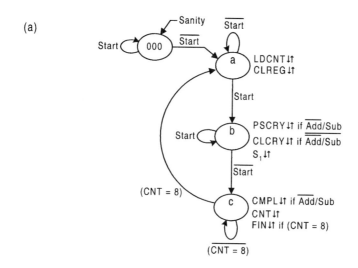

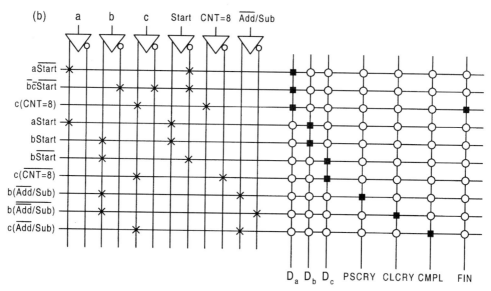

FIGURE 13.31
Design of the parallel-to-series adder/subtractor controller by the one-hot method. (a) State diagram for the controller initialized into the 000 state by using the one-hot-plus-zero approach. (b) Symbolic representation of the fusible bit position patterns for an FPLA programmed to generate the NS and Mealy output logic in Eqs. (13.15).

($S_1 = 0$); the XOR gate is set to complement (*CMPL*) operand *B* if subtraction or not if addition, hence *CMPL* if ($\overline{Add/Sub}$); counting is begun (*CNT*); and a completion signal (*FIN*) is issued at the end of 8 counts, *FIN* if (*CNT* = 8). Notice that the mode control S_0 is set to logic 1 throughout the process.

13.5 THE ONE-HOT DESIGN METHOD

From the state diagram in Fig. 13.31a, the one-hot NS and output functions are read directly by applying Eqs. (13.9), and the results are

$$\begin{cases} D_a = a\overline{Start} + \bar{b}\bar{c}\overline{Start} + c(CNT=8) \\ D_b = a\overline{Start} + bStart \\ D_c = b\overline{Start} + c\overline{(CNT=8)} \end{cases} \quad \text{and} \quad \begin{cases} LDCNT = CLREG = a \\ PSCRY = b(\overline{Add/Sub}) \\ CLCRY = b\overline{(Add/Sub)} \\ S_1 = b \\ CMPL = c(\overline{Add/Sub}) \\ CNT = c \\ FIN = c(CNT=8) \end{cases}, \quad (13.15)$$

where if follows that $D_a = a\overline{Start} + \bar{a}\bar{b}\bar{c}\overline{Start} + c(CNT=8) = a\overline{Start} + \bar{b}\bar{c}\overline{Start} + c(CNT=8)$. In the state diagram and in Eqs. (13.15) it is understood that the start signal (*Start*) must be active for a period of time greater than the clock period and that it must be debounced. It is not necessary to synchronize *Start* because of the GO/NO-GO configurations that exist relative to states a and b. Finally, the exact nature of the counter is not highly relevant at this time since its only function is to issue the signal $CNT = 8$ at the end of the process. Thus, $CNT = 8$ is necessarily a synchronous output from the counter.

The one-hot implementation of the parallel-to-series adder/subtractor controller is illustrated in Fig. 13.31b, where an FPLA is programmed to generate the NS and Mealy output functions of Eqs. (13.15). The Moore outputs in Eqs. (13.15) are not included because they are generated by the outputs from the flip-flops, an important characteristic of the one-hot method. Note that with a little care, it is possible to program the FPLA directly from the state diagram by application of Eqs. (13.9). For more complex FSMs, however, it is still a good idea to construct a p-term table from the NS and output equations to help reduce programming errors and to establish a record for future use.

The logic circuit for the adder/subtractor controller is shown in Fig. 13.32 where an FPLA and a 4-bit storage register are used for the implementation. Three individual FET D flip-flops could be used in place of the 4-bit storage register, but the 4-bit storage register is conveniently available as the 74xx175 MSI chip. Notice that all four of the Moore outputs are issued directly from the flip-flop outputs. The $6 \times 10 \times 7$ FPLA indicated is the minimum size required. The actual size of the FPLA may be larger, its choice being left to the discretion of the designer. The debouncing circuit is chosen from those discussed in Section 11.8.

13.5.4 Perspective on the Use of the One-Hot Method: Logic Noise and Use of Registered PLDs

The subject of logic noise in the output of one-hot FSMs is conspicuously absent in all previous discussions. The reason: No logic noise is possible in the FSMs considered! Since the output functions never involve coupled state variables, internally initiated static hazards are not possible. Externally initiated static hazards are also not possible since a properly designed one-hot FSM cannot hold in a two-one's race state. Furthermore, if care is taken in the use of two-one's race states as output states, ORGs will not be generated (see

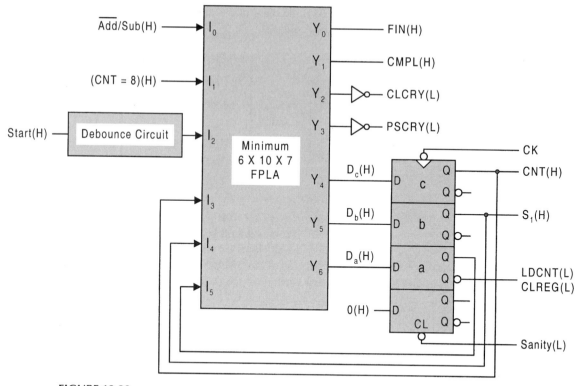

FIGURE 13.32

Implementation of the parallel-to-serial adder/subtractor controller by using a minimum $6 \times 10 \times 7$ FPLA and a 4-bit storage register to implement the one-hot NS and output functions in Eqs. (13.15).

Section 11.2 for a discussion of ORGs). Remember that each state-to-state transition is forced through a unique state having two ones—never through the all zeroes state. The reason for this is that the action of the flip-flop in a given state-to-state transition holds the "1" of the origin state active until the transition to the destination state is complete. Consequently, the use of the one-hot-plus-zero approach presents no problem even if the all-zero state is used as an output state.

Registered PLDs, such as the R- and V-type PALs discussed in Section 7.4, are a natural choice for the one-hot implementations of relatively small FSMs with only Moore (unconditional) outputs. As has been pointed out, Moore outputs are generated directly from the flip-flop outputs in one-hot designs. The problem encountered in dealing with Mealy (conditional) outputs in one-hot designs is that each Mealy output requires an ANDing operation between an external input and a one-hot state variable (flip-flop output). But PALs with R- or V-type macrocells lack the capability of generating Mealy outputs directly from internal ANDing operations. Therefore, each Mealy output must be generated by an ANDing operation external to the PAL. Alternatively, the one-hot state variable can be fed back into an unused macrocell and ANDed with the external input. But this uses up a macrocell and delays that Mealy output by a clock cycle. Remember that in the one-hot method, each state

13.6 SYSTEM-LEVEL DESIGN   649

requires a macrocell and if each Mealy output must also use a macrocell, the capability of the PAL can be quickly used up for all but relatively small FSMs. Therefore, as a rule, it is best to use registered PAL devices for one-hot designs of relatively small FSMs with only Moore outputs. Used in this manner, registered PAL designs by the one-hot method offer quick, convenient and reliable results, and without the need for K-maps or programming software. See subsection 16.4.4 for information regarding synchronous one-hot programmable sequencers.

If registered PLDs are to be used to implement large Mealy state machines by the one-hot method, FPGAs are the best choice. A good example is the use of the 4000 series Xilinx FPGAs. As explained in Subsection 7.7.3, these devices are extremely versatile and have the capacity to handle very large one-hot state machine designs with both Moore and Mealy outputs. The one drawback in the use of these FPGAs is that they require dedicated software to program them. For all but the experienced user of Xilinx FPGAs, this requirement is an impediment to design and may even preclude their use. Xilinx FPGAs accept VHDL descriptions of state machines from which the FSM can be synthesized automatically by synthesis tools such as *AutoLogic VHDL* by Mentor Graphics. More information on this and related subjects can be obtained from references cited in Further Reading at the end of this chapter.

There still remains the question of initializing registered PLDs for one-hot designs. R-type PAL devices apparently lack initialization capability and are not recommended for use in most one-hot designs. The macrocells of V-type PAL devices contain D flip-flops with both PRE and CLR asynchronous overrides. Thus, V-type PALs can be initialized directly into a one-hot state but are otherwise limited in their use in one-hot applications as explained earlier. The configurable logic blocks (CLBs) of all Xlinx FPGAs contain D flip-flops with both PRE and CLR overrides and consequently are suitable for one-hot state initialization. Generally, registered PLDs having D flip-flops with only CLR overrides can be used, but only for the one-hot-plus-zero approach as indicated by previous examples. Alternatively, for some FSM designs, it may be possible to initialize into a "zero" state and then force the FSM into a one-hot state by using a start input, as in Fig. 13.29 for the series 2's complementer.

13.6 SYSTEM-LEVEL DESIGN: CONTROLLER, DATA PATH, AND FUNCTIONAL PARTITION

One very common view of a digital system is the use of an FSM as the *controller* for a set of components parts that comprise the *controlled system* called the *data path*. This view is expressed in Fig. 13.33, where all input and output (I/O) conditioning logic has been omitted to focus attention on the main features of this architecture. Here, it is understood that the data path devices generally consist of a mixture of both sequential and combinational logic machines. Typical among these are registers, counters, ALUs, PLDs of various types, decoders, MUXs, shifters, comparators, digital-to-analog (D/A) converters, and the like. The architecture represented in Figure 13.33 is the one emphasized in this text.

All sections in this chapter up to this point have been devoted to various architectures that can and should be considered in controller design. Chapters 10, 11, and 12 supply the necessary background information needed to build reliable controllers as well as those

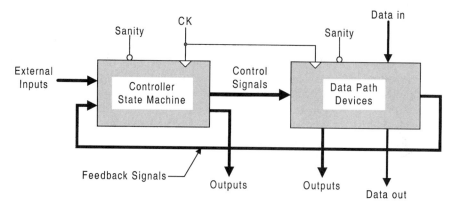

FIGURE 13.33
Controller/data-path architecture for digital system design.

FSMs that comprise the data path. Chapters 2 through 9 provide the necessary background for the design of a wide range of combinational logic devices, many of which are widely used in the data path of digital systems. In short, this section may be considered as the culmination of all developments necessary to build reliable digital systems.

The controller for a digital system is an FSM, perhaps like the one in Fig. 13.31a. But it is also the "brains" of the system. Its function is to coordinate precisely the operation of the various components of the data path so as to perform the specific tasks required by the system. Thus, the controller must issue instructions (control signals) to the *data path unit* (DPU) based on the external inputs it receives and on the feedback information received from the DPU. A configuration such as this, where the outputs of one unit are the inputs to another, and vice versa, is called a *handshake interface*. Feedback from the DPU is not a requirement for all systems, but is common in most. Note that both the controller and data path devices may receive signals from and issue signals to the outside world.

Designing a complex digital system requires a "divide-and-conquer" approach. The system must be divided into subsystems that in turn must be broken down into well-defined parts that can be implemented with available hardware. The detailed block diagram that conveys this information is appropriately called the *functional partition* of the system. Thus, the functional partition contains a block representation of the controller, all the peripheral devices that constitute the DPU, all inputs from and outputs to the outside world, and the I/O conditioning circuits. Consequently, the functional partition contains all the information needed for "hookup" and operation of the system given the details of the controller design, which must be treated as an integral part of the design process.

The functional partition and a detailed flowchart or ASM chart for the controller of a digital system are usually interdependent and must be developed together. For a complex digital system this development process may require two or more attempts at representing the functional partition and flowchart or ASM chart before satisfactory representations can be found. Simple block diagrams are often useful in this process, since they can provide a physical picture of the overall system. The use of timing diagrams is usually a necessary part of the development stages of the design process — in some designs timing considerations are of paramount importance. Finally, remember that a flowchart or ASM chart is considered

13.6 SYSTEM-LEVEL DESIGN

to be only a "thinking tool" for the construction of the *state diagram* or *state table* from which the controller is designed.

There may be more than one good design for a given digital system. This is particularly true for complex digital systems. The success of the design will usually depend on the engineering creativity, intuition, and generally the experience of the digital designer. But the manner in which a digital system is to operate in a particular environment can also be an important factor. For example, suppose a stepping motor control system is to be designed to move a certain mass from one fixed position to another in a smooth, nonjerky fashion. Clearly, the design considerations for the stepping motor controller, based on mass, time, and distance constraints, are different for the operation of a small robotic arm than for the operation of an elevator. The point is that important detailed information regarding timing and functional constraints must be factored into the design process from the beginning stages if successful designs are to result.

13.6.1 Design of a Parallel-to-Serial Adder/Subtractor Control System

A brief description of the parallel-to-serial adder/subtractor system was given in Subsection 13.5.3. There, the one-hot-plus-zero approach was used to design the system controller shown in Figs. 13.31 and 13.32. Now, it is necessary to construct the functional partition for this system. This is done in Fig. 13.34, where block circuit symbols are used to represent the controller and data path devices. The data path unit (DPU) consists of two 8-bit USRs, a full adder (FA), an RET D flip-flop, a controlled inverter (XOR gate), and a 4-bit parallel loadable up/down counter of the type shown in Fig. 12.20. Of course, there are many "variations on the theme" in the design of the DPU. For example, right shift registers with parallel load capability can replace the USRs, a simple 3-bit binary up counter with asynchronous CL can replace the 4-bit parallel loadable up/down counter, and a transparent D latch with asynchronous PR and CL overrides (Fig. 10.51) can replace the edge-triggered D flip-flop. If operands larger than 8-bits are to be added or subtracted, larger registers must be used. Thus, two 8-bit registers can be cascaded in series to accommodate 16-bit operands, or four 8-bit registers can be cascaded to accommodate 32-bit operands, etc.

Presented in Fig. 13.35a is a reconstruction of the state diagram for the parallel-to-serial adder/subtractor controller in Fig. 13.31a, but now with a state code assignment suitable for a conventional design. In Fig. 13.35b is shown the timing diagram for an 8-bit serial subtraction operation by the adder/subtractor system. Notice that the sequence of events indicated in the timing diagram are the same as those in the state diagram and that they, together with the functional partition in Fig. 13.34, provide a complete stepwise description of this system: Following initialization of the adder/subtractor in state a, the controller loads a 0000 into the counter and clears the USRs. After the start button *Start* is pressed (for a period of time greater than a clock period) the controller transits from state a to state b. In state b the RET D flip-flop and mode control S_1 are set to logic 1 by the controller in preparation for subtraction. The carry-in CI to the FA is now initialized to logic 1, as required for subtraction by 2's complement. After the release of the *Start* switch button (hence $\overline{Start}$), the controller transits to state c, where counting by the counter is begun. During this time, the two 8-bit USRs deliver the operands serially LSB first to the full adder (FA) via a controlled inverter on the B line, which is now set to complement B $[CMPL(H) = 1(H)]$ as required for subtraction. With each clock triggering edge, bitwise

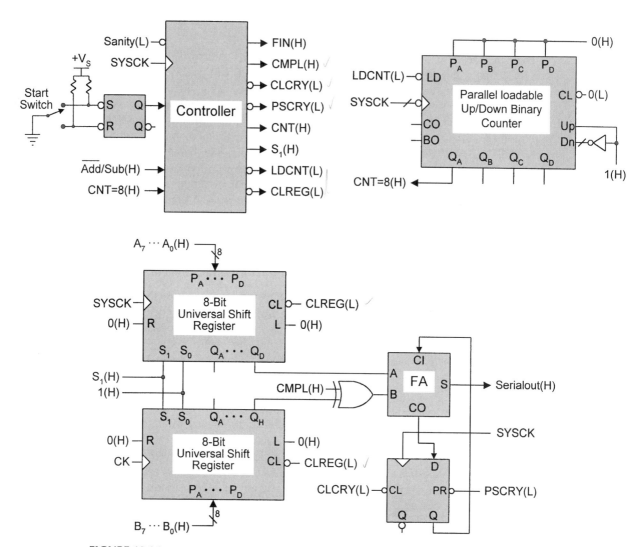

FIGURE 13.34
Functional partition for the parallel-to-serial adder/subtractor system showing block symbols for the controller and data path devices.

subtraction of operand B from A occurs and continues until all eight operand bits of A and B have been processed. At this time the controller receives the signal $CNT = 8$, a completion signal *FIN* is issued, and the controller returns to state a in readiness for the next 8-bit addition/subtraction process.

All that remains is to find an acceptable logic design for the controller. In contrast to the one-hot approach represented by Eqs. (13.15) and implemented in Figs. 13.31 and 13.32, a minimum result will now be found. Shown in Fig. 13.36 are the K-maps and minimum covers for the NS- and output-forming logic as plotted from the state diagram in Fig. 13.35a, assuming the use of JK flip-flops. The resulting NS and output functions are

13.6 SYSTEM-LEVEL DESIGN

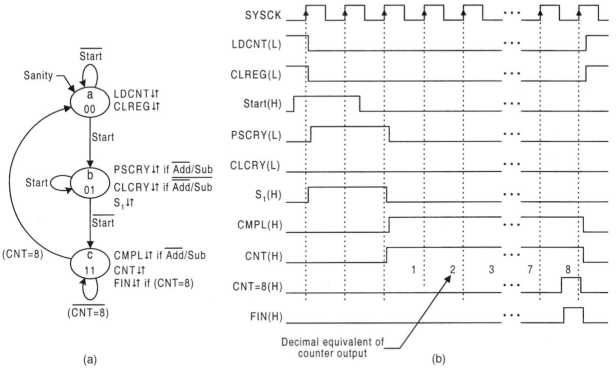

FIGURE 13.35
Design and timing considerations for the parallel-to-serial adder/subtractor system. (a) State diagram suitable for a conventional controller design. (b) Timing diagram for an 8-bit serial subtraction operation showing input signals to and output signals from the controller.

given by

$$\left\{\begin{array}{l} J_A = B\overline{Start} \\ K_A = (CNT=8) \\ J_B = Start \\ K_B = A(CNT=8) \end{array}\right\} \quad \text{and} \quad \left\{\begin{array}{l} LDCNT = CLREG = \bar{B} \\ PSCRY = \bar{A}B(\overline{Add/Sub}) \\ CLCRY = \bar{A}B(\overline{\overline{Add/Sub}}) \\ S_1 = \bar{A}B \\ CMPL = A(\overline{Add/Sub}) \\ CNT = A \\ FIN = A(CNT=8) \end{array}\right\}, \quad (13.16)$$

which contain one shared PI, $A(CNT=8)$. Notice that there are four Mealy outputs and four Moore outputs, none of which have static hazards associated with them.

Equations (13.16) are implemented in Fig. 13.37 by using a minimum number of gates external to the flip-flops for a total gate/input tally of 6/12 excluding the single inverter. This may be compared with the one-hot-plus-zero design given by Eqs. (13.15) which represent a total gate/input tally of 13/28 excluding inverters and taking account of the one shared,

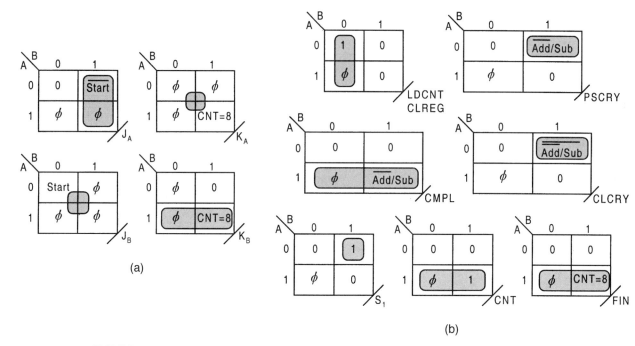

FIGURE 13.36
Next-state- and output-forming logic for the adder/subtractor controller as obtained from the state diagram in Fig. 13.35a. (a) NS K-maps and minimum cover for a JK flip-flop design. (b) Output forming logic showing minimum cover.

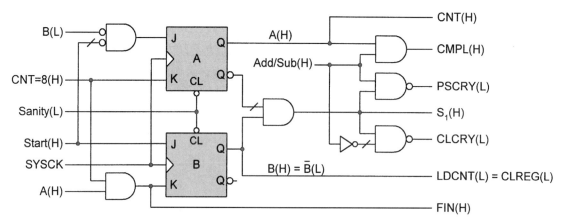

FIGURE 13.37
Logic minimum design of the parallel-to-serial adder/subtractor controller of Fig. 13.35a by using JK flip-flops.

13.6 SYSTEM-LEVEL DESIGN

PI $c(CNT = 8)$. Thus, roughly twice as much external hardware and one extra flip-flop are needed by the one-hot approach for the convenience of reading and implementing the NS- and output-forming logic directly from the state diagram without the use for K-maps. If D flip-flops are used instead of JK flip-flops, it is easily shown by map conversion that the NS functions become

$$\begin{cases} D_A = \bar{A}B\overline{Start} + A\overline{(CNT = 8)} \\ D_B = \bar{A}Start + A\overline{(CNT = 8)} + \bar{A}B \end{cases}, \qquad (13.17)$$

the output logic remaining the same. This would bring the total gate/input tally for the D flip-flop design to 10/22 exclusive of inverters. Thus, the gap narrows between a conventional D flip-flop design and that for the one-hot method. Also shown, by map conversion, is the T flip-flop design that falls in between the JK and D designs, yielding

$$\begin{cases} T_A = \bar{A}B\overline{Start} + A(CNT = 8) \\ T_B = \bar{B}Start + A(CNT = 8) \end{cases} \qquad (13.18)$$

for the NS functions, giving a total gate/input tally of 8/19 excluding inverters.

There still remains the question of ORGs in the design of this FSM. The transition from state 11 to state 00 can result in the production of ORGs if the race path is via the 01 state. In Fig. 13.36b it is evident that ϕ_2 is not used in the K-maps for *PSCRY*, *CLCRY*, or S_1 and, consequently, ORGs are not possible by the 10-race state path. But this discussion is made moot by the fact that the $11 \to 00$ transition completes the process and the FSM is brought to an initialized condition in state 00. Therefore, it does not matter that ORGs are produced during this transition — no logic noise problems exist. This fact can be useful in the design of other system controllers.

13.6.2 Design of a Stepping Motor Control System

Stepping motors convert a series of pulses into angular motion that permits very accurate positioning of the motor's rotor without feedback control. Also, stepping motors are useful in systems where there is space only for a small motor to drive a relatively massive part. Linear angular accelerations and decelerations of the motor can prevent slippage, chattering, or jerky motion that could lead to mechanical failure or adversely affect mechanical operation. Stepping motors exhibit zero steady-state error positioning and can develop torque up to 15 Nm (Newton-meters). They are used in robotics to accurately operate mechanical parts in some manner, in fluid control systems for precise adjustment of fluid control valves, in wire-wrap processing of circuit boards, and in a variety of other applications too numerous to mention here.

Stepping motors will accept pulse strings in the range of 1500 to 2500 pulses per second. The design of the control system required to generate these pulse strings is the subject of this subsection. The nature and design of the stepping motor to which the control system is attached fall outside the scope of this text and will not be discussed further (see Further Reading for information on this subject).

The overall operational characteristics for the stepping motor control system are provided in Fig. 13.38. In Fig. 13.38a are shown the angular velocity/time requirements of the motor. The GO command causes a linear angular acceleration of the motor while a HALT command

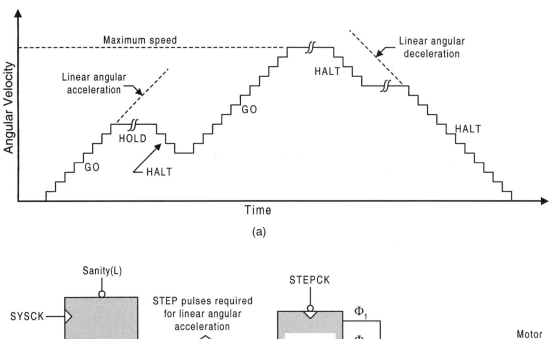
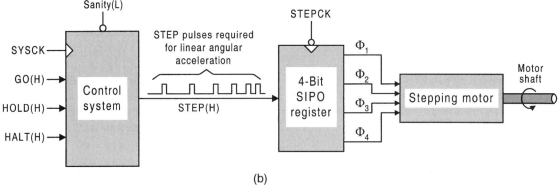

FIGURE 13.38
Overall operational characteristics of the stepping motor control system. (a) Angular velocity vs time requirements of the control system. (b) Physical picture showing input controls, STEP pulse train required for linear angular acceleration, and register outputs to stepping motor.

produces a linear angular deceleration. The HOLD command causes the motor to maintain the angular velocity that is reached at the time the HOLD command is given. The motor must operate between zero speed and a maximum angular velocity that is set by the number of steps in the speed/time characteristic, 16 in the case of Fig. 13.38a.

The physical picture for the overall system is presented in Fig. 13.38b. Here, the control system receives one of the three (nonoverlapping) asynchronous input signals, GO, HOLD, or HALT, and issues a series of STEP pulses in response to that input signal. In the physical picture, a GO signal is implied, resulting in a STEP pulse series required to cause a linear angular acceleration of the motor. Each STEP pulse is received by the SIPO register, which, in turn, delivers a set of four phase pulses (Φ_1, Φ_2, Φ_3, Φ_4) to the power transistors of the stepping motor, causing the motor to rotate by a certain amount. The SIPO register is triggered by the STEPCK waveform, which is exactly twice the frequency of SYSCK, the waveform used to trigger the control system.

13.6 SYSTEM-LEVEL DESIGN   657

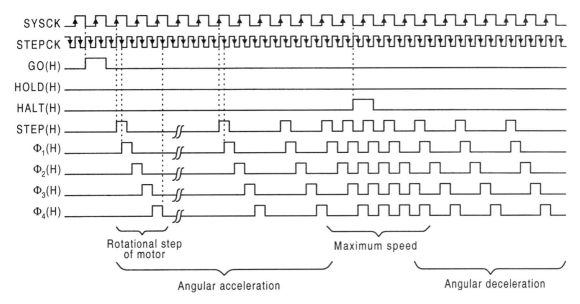

FIGURE 13.39
Acceptable timing relationships between synchronized external inputs and STEP pulse signals to the stepping motor.

An acceptable timing relationship between external inputs, the STEP pulse, and the phase pulse signals to the stepping motor is given in Fig. 13.39. Each STEP pulse width is specified to be one period of the STEPCK waveform and to be active coincidentally with the active portion of the SYSCK waveform. When a STEP pulse is received by the register, that pulse is shifted from the LSB stage toward the MSB stage on each falling edge of the STEPCK pulse. Thus, a set of four time-shifted pulses is generated from the shift register outputs by each STEP pulse as indicated in Figure 13.39. The maximum rotational velocity is set by the frequency of the STEPCK waveform, which is assumed to be low enough to match the inertial characteristics of the motor. The maximum rotational velocity (speed) is illustrated midway through the timing diagram in Figure 13.39 followed by an angular deceleration mode as indicated. Note that the SYSCK waveform can be generated from the STEPCK waveform simply by using a divide-by-two counter. Such a counter is shown in Fig. 12.12c.

The functional partition of the stepping motor control system is shown in Fig. 13.40. Synchronous, nonoverlapping inputs GO, HOLD, and HALT are introduced to the controller from input conditioning circuits. The data path (DPU) devices consist of a parallel-loadable right shift register, as in Fig. 12.3, but triggered by FET flip-flops; a special parallel-loadable, up/down *data-triggered counter*; and a parallel-loadable up/down counter, of the type shown in Fig. 12.20, set for up-count only and hereafter called the "up-counter." The special data-triggered counter is similar to that in Fig. 12.20, except that the NS functions in Eqs. (12.5) are the clock inputs to the FET T flip-flops — hence, data triggered. This counter is triggered off of the falling edge of the Up (*DECDLY*) or Dn (*INCDLY*) input pulse as indicated by its design shown at the end of this subsection in Fig. 13.46. The up-counter is triggered on the falling edge of the SYSCK waveform and issues a CO (*CNT*) signal at the end of count

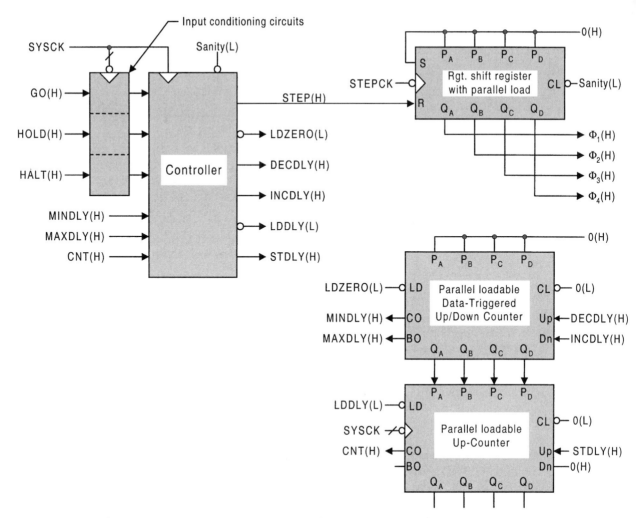

FIGURE 13.40
Functional partition for the stepping motor control system showing controller and data path components, and an input conditioning circuit block.

15 that is picked up by the controller on the next rising edge of SYSCK. To ease timing restrictions, both counters are designed to be parallel loaded asynchronously via the PRE and CLR overrides on their flip-flops.

Constructed in coordination with the functional partition is the ASM chart shown in Fig. 13.41. It is to be used as a thinking tool in the construction of the sate diagram from which the controller will be designed. The chart expresses the basic algorithm involved that is physically carried out by the functional partition in Fig. 13.40. Briefly, this algorithm requires that in the GO mode, and with each successive STEP pulse issued, the up count is decreased from a maximum of 15 to a minimum of 0 SYSCK cycles via the parallel load of the up/down data-triggered counter. In the HALT mode this processed is reversed with each successive issue of the STEP pulse. And in either case, the acceleration or deceleration

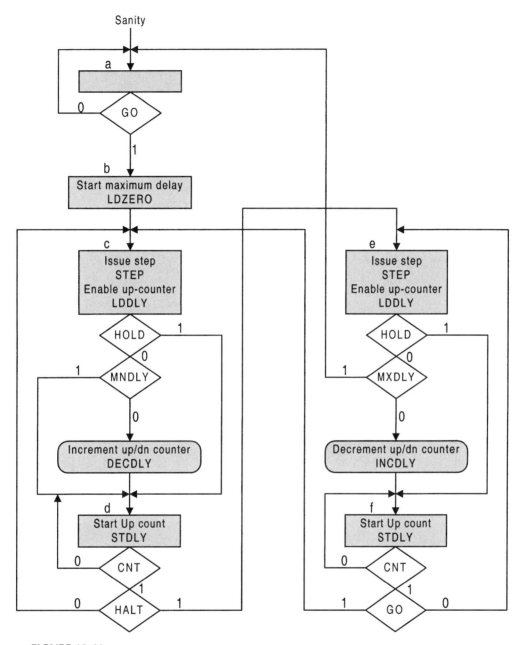

FIGURE 13.41
ASM chart for the stepping motor controller used as a thinking tool for the construction of the state diagram.

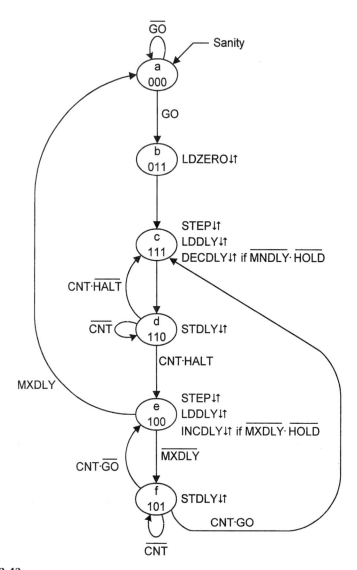

FIGURE 13.42
Fully documented state diagram for the stepping motor controller as derived from the ASM chart in Fig. 13.41.

process can be bypassed by a HOLD command. Notice that the ASM chart is that of a Mealy machine.

The state diagram that is derived from the ASM chart is shown in Fig. 13.42. A state code assignment has been given that yields no output race glitches (ORGs), provided that the don't cares ϕ_1 and ϕ_2 are not used for minimizing the output logic. The output *LDZERO* is an exception to this requirement, permitting these don't-care states to be used as needed for its minimum cover (see transition 000 → 011 in the state diagram). Notice how much more

13.6 SYSTEM-LEVEL DESIGN

vividly the sequential behavior of this FSM is portrayed by the state diagram than by the ASM chart. But the ASM chart serves as a better thinking tool than the state diagram when used to represent the algorithmic behavior of the FSM while constructing the functional partition. Of course, a flowchart can serve the same purpose as the ASM chart in this regard.

At this point it is decided that the controller is to be designed by using a PLA together with RET D flip-flops. To this end the K-maps for the NS and output functions are plotted from the state diagram in Fig. 13.42 and are presented in Fig. 13.43. Here, the minimum cover is shown by shaded loops, as is customary in this text, and the results are given as follows:

$$\begin{cases} D_A = A \cdot \overline{MXDLY} + B + C \\ D_B = C \cdot CNT \cdot GO + \bar{A} \cdot GO + B \cdot \overline{CNT} + B \cdot \overline{HALT} + BC \\ D_C = B\bar{C} \cdot CNT \cdot \overline{HALT} + A\bar{B}\bar{C} \cdot \overline{MXDLY} + \bar{B}C \cdot \overline{CNT} \\ \qquad + \bar{B}C \cdot GO + \bar{A} \cdot GO + \bar{A}C \\ \qquad LDZERO = \bar{A}C \\ \qquad STEP = LDDLY = A\bar{B}\bar{C} + ABC \\ \qquad DECDLY = ABC \cdot \overline{MNDLY} \cdot \overline{HOLD} \\ \qquad INCDLY = A\bar{B}\bar{C} \cdot \overline{MXDLY} \cdot \overline{HOLD} \\ \qquad STDLY = A\bar{B}C + AB\bar{C} \end{cases} \quad (13.19)$$

Notice that there are two shared PIs, $\bar{A} \cdot GO$ and $\bar{A}C$, bringing the p-term count to 19 for the combined NS and output functions. The requirement that ϕ_1 and ϕ_2 not be used for the output functions, other than *LDZERO*, is evident in the K-map for *STDLY*.

The p-term table based on the results given by Eqs. (13.19) is presented in Fig. 13.44. Here, two inputs are each marked with a single asterisk and two outputs are each marked with a double asterisk to indicate that they are active low. Active low inputs to and active low outputs from PLAs are discussed in Section 7.5. Recall from that section that the active low inputs can be accommodated by either complementing their columns in the p-term table or by using an inverter on their input lines, but not both. Acive low outputs from a PLA-type device require the use of inverters. Notice that there are 9 inputs, 8 outputs, and 19 p-terms indicated in the p-term table. Thus, the minimum size PLA required for this controller has dimensions $9 \times 19 \times 8$, but any larger PLA device can suffice. The use of a ROM to implement this FSM would be an inefficient application (an overkill) of the device, since only a small fraction of the $2^9 = 512$ minterm capability of the ROM would be utilized. For a review of array logic devices and their uses, the reader is referred to Sections 7.2 through 7.6.

Having completed the functional partition and the p-term table for the controller, all that remains is an overview of the controller architecture. This is done in Fig. 13.45 where a $9 \times 20 \times 8$ FPLA is used to generate the NS and output functions and a 4-bit storage register is used as the memory. To satisfy the requirement that *STEP* be issued coincidentally with *SYSCK*, an AND gate is used to AND the *SYSCK* waveform with the *STEP* signal issued by the FPLA. The input conditioning circuits and *SYSCK* generating circuits are provided for completeness. Notice that it is a divide-by-2 counter that generates SYSCK from a STEPCK input. The input conditioning circuits each consist of debouncing and synchronizing stages. If it is known that the input signals are of duration less

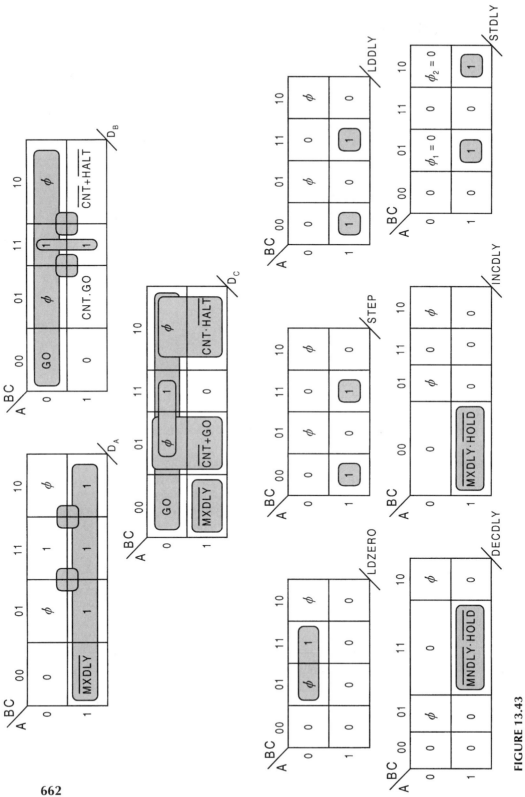

FIGURE 13.43
NS and output K-maps plotted from the state diagram in Fig. 13.42 showing minimum cover for a glitch-free design of the stepping motor controller.

	PLA Inputs									PLA Outputs							
	A	B	C	MNDLY*	MXDLY*	CNT	GO	HALT	HOLD	D_A Y_7	D_B Y_6	D_C Y_5	LDZERO** Y_4	STEP LDDLY** Y_3	DECDLY Y_2	INCDLY Y_1	STDLY Y_0
P-terms	I_8	I_7	I_6	I_5	I_4	I_3	I_2	I_1	I_0								
A·$\overline{\text{MAXDLY}}$	1	–	–	–	0	–	–	–	–	1	0	0	0	0	0	0	0
B	–	1	–	–	–	–	–	–	–	1	0	0	0	0	0	0	0
C	–	–	1	–	–	–	–	–	–	1	0	0	0	0	0	0	0
C·CNT·GO	–	–	1	–	–	1	1	–	–	0	1	0	0	0	0	0	0
$\overline{A}$·GO	0	–	–	–	–	–	1	–	–	0	1	1	0	0	0	0	0
B·$\overline{\text{CNT}}$	–	1	–	–	–	0	–	–	–	0	1	0	0	0	0	0	0
B·$\overline{\text{HALT}}$	–	1	–	–	–	–	–	0	–	0	1	0	0	0	0	0	0
BC	–	1	1	–	–	–	–	–	–	0	1	0	0	0	0	0	0
B$\overline{C}$·CNT·$\overline{\text{HALT}}$	–	1	0	–	–	1	–	0	–	0	0	1	0	0	0	0	0
A$\overline{B}\overline{C}$·$\overline{\text{MXDLY}}$	1	0	0	–	0	–	–	–	–	0	0	1	0	0	0	0	0
$\overline{B}$C·$\overline{\text{CNT}}$	–	0	1	–	–	0	–	–	–	0	0	1	0	0	0	0	0
$\overline{B}$C·GO	–	0	1	–	–	–	1	–	–	0	0	1	0	0	0	0	0
$\overline{A}$C	0	–	1	–	–	–	–	–	–	0	0	0	1	0	0	0	0
A$\overline{B}\overline{C}$	1	0	0	–	–	–	–	–	–	0	0	0	0	1	0	0	0
ABC	1	1	1	–	–	–	–	–	–	0	0	0	0	1	0	0	0
ABC·$\overline{\text{MNDLY}}$·$\overline{\text{HOLD}}$	1	1	1	0	–	–	–	–	0	0	0	0	0	0	1	0	0
A$\overline{B}\overline{C}$·$\overline{\text{MXDLY}}$·$\overline{\text{HOLD}}$	1	0	0	–	0	–	–	–	0	0	0	0	0	0	0	1	0
A$\overline{B}$C	1	0	1	–	–	–	–	–	–	0	0	0	0	0	0	0	1
AB$\overline{C}$	1	1	0	–	–	–	–	–	–	0	0	0	0	0	0	0	1

* Indicates an active low input—complement column or use an inverter on the input.
** Indicates an active low output—must use an inverter on the output.

FIGURE 13.44

P-term table for the PLA implementation of the NS and output functions of the stepping motor controller expressed by Eqs. (13.19).

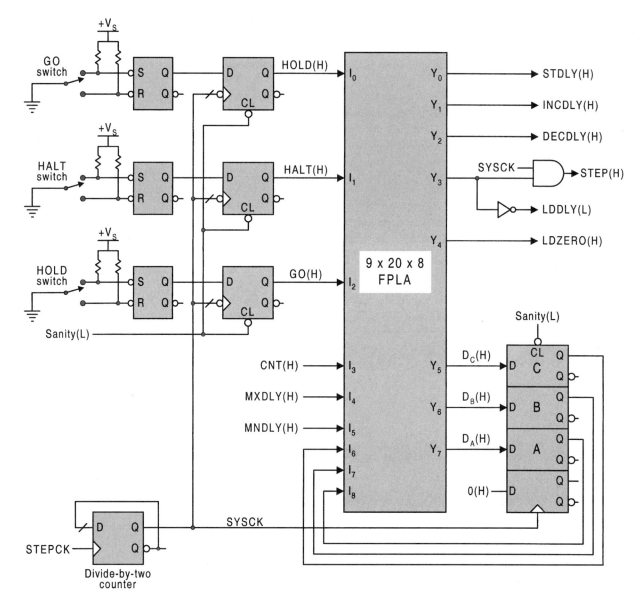

FIGURE 13.45
Architecture for the stepping motor controller centered around an FPLA and showing input conditioning and clock generation circuitry.

than one clock period, a stretcher stage (see Fig. 11.21) must be added to each of these circuits.

The data-triggered up/down counter indicated in the functional partition of Fig. 13.40 is somewhat different from any counters discussed previously. This counter is triggered on the falling edge of either the *Up* pulse or the *Dn* input pulse with the strict requirement that these pulses *never* be overlapping. This requirement is necessarily met by the controller. An

13.6 SYSTEM-LEVEL DESIGN

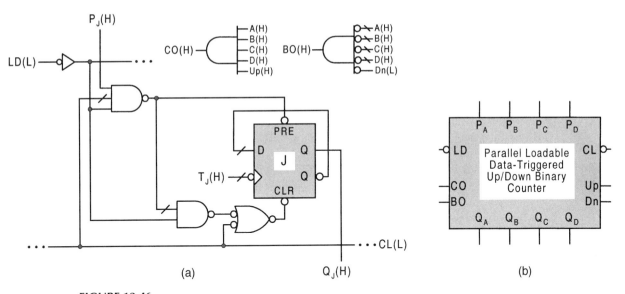

FIGURE 13.46
Implementation of the 4-bit data-triggered up/down binary counter with asynchronous parallel load. (a) Logic circuit for the Jth stage showing the CO and BO output logic, where the NS functions $T_J(H)$ are given by Eqs. (12.5). (b) Block circuit symbol.

inspection of the state diagram in Fig. 13.42 indicates that *DECDLY* and *INCDLY*, the Up and Dn inputs to the counter, can never be active at the same time — they are issued in separate states of the state diagram. Another unique feature of this data-triggered counter is that the memory of each stage is an FET D flip-flop configured as an FET *toggle module* (a divide-by-2 counter). Thus, the only data input to each memory element is by way of the clock input, where T_J in Fig. 13.46a represents T_A, T_B, T_C, and T_D in Eqs. (12.5). The FET feature of the toggle module ensures that triggering will occur on the falling edge of the data pulse, a necessary feature of this type of counter. Were triggering to occur on the rising edge of the pulse, the nonoverlapping requirement of the input pulses could not generally be assured in applications other than the present one. In effect, the data-triggered up/down counter in Fig. 13.46 is an asynchronous state machine since it operates independent of a clock signal. It is said to operate in the *pulse mode*. An in-depth coverage of this subject and related matters is provided in Chapter 15, which deals with the pulse mode design of state machines.

Just as is true for the first system-level design in Subsection 13.6.1, there are many acceptable variations possible in implementing the stepping motor control system. Some of these variations may depend on the type and character of the stepping motor itself. But aside from that possibility, there exists other suitable variations. Take the DPU, for example. The parallel-loadable right shift register can be replaced by a USR set for right shifting, and both counters can be the parallel loadable up/down type featured in Subsection 12.3.5 with the appropriate changes in the functional partition. If this change in counters is made, then the up/down (upper) counter shown in the functional partition must be triggered on the rising edge of the *SYSCK* waveform while the lower up/down counter is set for up count and triggered on the falling edge of the *SYSCK* waveform. In this case the parallel loading is best accomplished asynchronously by using the counter design presented in Fig. 12.28.

13.6.3 Perspective on System-Level Design in This Text

System-level designs can connote a different meaning to different designers. To some, system-level design might refer strictly to a combinational system. Or to others, it might mean the design of a microprocessor or computer. In the sense used in this text, a system-level design will always imply the presence of a controller unit (CU) and a controlled system called the data path unit (DPU). The CU will always be an FSM, which must not be confused with the CPU or *central processing unit* of a microprocessor or computer. Commonly, the CPU contains both a CU and a DPU; the DPU (or *execution unit*, EU) typically consists of registers, shifters, and an ALU. The design of microprocessors and computers will not be covered in this text. It is the philosophy of the author that digital design fundamentals and the design of microprocessors and computers cannot be treated effectively within a single text. Further Reading at the end of this chapter cites references on the subject of microprocessor and computer design for the reader wishing to develop in that direction.

In system-level designs, the CU and DPU take on an entirely different identity and functionality and may differ greatly in their individual hardware requirements. In one case the DPU may be far more complex in its hardware makeup than the CU, while in other cases the reverse may be true. Two illustrative examples of system-level design have been presented in Subsections 13.6.1 and 13.6.2. Both are examples of the case where the DPU is more complex hardware-wise than the CU. There are, of course, many more examples of system-level design that could be offered in this section, and each could be used to illustrate specific facets of the design process and involve system designs both larger and smaller than those previously presented. This, however, is not practical given the space limitation of a text, and would take up space at the expense of other important subject matter. Learning how to design at the system level requires practice, practice, and more practice — there is no substitute for practical experience in this field. Threrefore, as an alternative, other illustrative system-level design problems are provided in the problem section to this chapter. And to help the reader in the decision-making process for these problems, a few suggestions are offered regarding hardware, input conditioning, and so on. Again, it is emphasized that these problems are all open-ended in the sense that they have no single best solution. Consequently, the reader's design skills and engineering intuition can be exercised within the limits provided by the description of the problem. But, the instructor can also permit greater latitude in arriving at an acceptable solution — all under the heading of "variations on the theme." This attitude toward design can be quite rewarding to both student and instructor alike.

13.7 DEALING WITH UNUSUALLY LARGE CONTROLLER AND SYSTEM-LEVEL DESIGNS

In using the model given in Figure 13.1, it is assumed that both the NS- and output-forming logic functions for a controller FSM can be handled by a single nonregistered PLD. In the event that this is not the case and the controller requirements exceed the limitations of a single PLD, separate PLDs of the same or different type can be used to implement the NS and output forming logic. The idea here is to invoke the concept of "divide and conquer." Such a scheme is shown in Fig. 13.47 for a Mealy FSM together with input and

13.7 DEALING WITH UNUSUALLY LARGE CONTROLLER 667

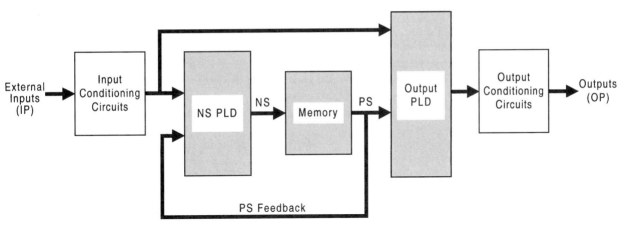

FIGURE 13.47
Separate PLD approach to the implementation of unusually large Mealy controller FSMs showing input and output conditioning circuit blocks.

output conditioning circuit blocks. Furthermore, this scheme can be used for very large FSM design even if the PLDs are individually insufficient for the task. Section 7.6 explains how multiple PLD schemes can be used to augment input and output capability, but only when decoders are used with PLDs having tri-state enables. There may be times when these arrangements are both expedient and advantageous to the designer. However, for unusually large controller and system-level designs, there are better options available to the designer as explained in the following paragraphs.

For the complete design of very large controller FSMs by using registered PLDs, excellent choices are the Xilinx FPGAs. The Xilinx XC4000E (0.5 micron–5 volt) series, for example, offers a wide variation in FPGA capability and features ranging from 100 (10×10 array) configurable logic blocks (CLBs) and 360 flip-flops for the XC4003E to 1024 (32×32 array) CLBs and 2560 flip-flops for the XC4025E, and operating up to 66 MHz. The gate equivalence for the XC4000E series ranges from 2000 to 45,000. At the 0.25-micron and 2.5-volt end, Xilinx offers the XC4000XV series. These devices range from 4624 (68×68 array) CLBs for the XC40125 to 8454 (92×92 array) CLBs for the XC40250 family with a gate equivalency ranging from 80,000 to 500,000. This series will soon be extended to the 2,000,000 gate-equivalency level. Xilinx claims that the XC4000XV series FPGAs can operate at over 100 MHz with minimum power consumption by today's standards. Quite clearly, these devices lie in the VLSI range and are large and versatile enough to be used for an entire system-level design—both combinational and sequential. Their use is leading away from on-chip designs for many applications and may even replace on-chip microprocessor design for specialized, low-volume applications. The XC40250XV has more than 100,000,000 transistors, compared to the 7,500,000 transistors used in the Pentium II microprocessor.

Whether or not it is desirable to use an FPGA, say, for the implementation of the DPU devices, is a matter left to the discretion of the designer. Also, the reader must understand that to design with these FPGAs requires the use of sophisticated software to cover all aspects of the design. The software, provided by Xilinx Corp., can be used for schematic capture, simulation, and the automatic block placement and routing of interconnects. Obviously, considerable knowledge of the use and interpretation of this software is needed before reliable designs can result. But even with that knowledge, the designer must still deal with a variety of timing problems. In some system-level designs timing is everything and improper routing delays can cause malfunction of the system. Fortionately, Xilinx Corp. has taken this into account and has provided generous routing resources in their XC4000E and XC4000XV series FPGAs and have made them reprogrammable an unlimited number of times. The section on Futher Reading at the end of this chapter cites relevant sources of information on this subject.

If it is the designer's intent to use a so-called programmable logic sequencer (PLS) for total system design, be aware of the limitations of such a device. Although many of these devices conform to the model in Fig. 13.47, the number of flip-flops they provide may be quite limiting. For example, the Signetics PLS155 provides the equivalent of a $16 \times 45 \times 12$ PLA but is equipped with only four edge-triggered flip-flops on chip. Of course, such devices can be combined to accommodate larger designs, but compared to what FPGAs can offer, it may not seem worthwhile. This is not to say that individual PLSs cannot be useful in simple controller designs. Even the Signetics PLS155 can be useful in the design of FSMs having four or fewer state variables. Remember that FSMs up to 16 states can be designed by using four flip-flops as the memory. But for very large controller- and system-level designs, it is advisable to look elsewhere for a suitable PLD. In particular, FPGAs should be considered as the ideal choice for such FSMs provided that the appropriate software is available for programming.

FURTHER READING

To one extent or another, every text on digital design contributes something to the subject of alternative architectures in synchronous controller design and, perhaps to a lesser extent, to system-level design. Useful sources for further reading on the subject of alternative controller designs of state machines can best be found in texts by Fletcher and Tinder, and to a lesser extent in the texts by Katz and Roth. The texts by Fletcher and Tinder provide extensive coverage of counter- and register-based controller design. The use of MUXs and state decoders is also covered in these two references.

[1] W. I. Fletcher, *An Engineering Approach to Digital Design*. Prentice Hall, Englewood Cliffs, NJ, 1980.
[2] R. H. Katz, *Contemporary Logic Design*. Benjamin/Cummings Publishing, Redwood City, CA, 1994.
[3] C. H. Roth, *Fundamentals of Logic Design*, 4th ed. West Publishing Co., St. Paul, MN, 1992.
[4] R. F. Tinder, *Digital Engineering Design: A Modern Approach*, Prentice Hall, Englewood Cliffs, NJ, 1991.

Further reading on the subject of controller design centered around nonregistered PLDs, mainly ROMs and PLAs, can be found in the four previously cited references. In addition, the text of Nelson *et al.* provides useful further reading on this subject.

[5] V. P. Nelson, H. T. Nagle, B. D. Carroll, and J. D. Irwin, *Digital Logic Circuit Analysis and Design*. Prentice Hall, Englewood Cliffs, NJ, 1995.

FURTHER READING 669

Good coverage of the use of registered PLDs in digital system design can be found in the texts by Bolton, Carter, Katz (previously cited), Lala, Pellerin and Holley, and Wakerly, and these are recommended for further reading. For the automatic logic design of digital systems, the book by Edwards is recommended.

[6] M. Bolton, *Digital Systems Design with Programmable Logic*. Addison-Wesley, Reading, MA, 1990.
[7] J. W. Carter, *Digital Designing with Programmable Logic Devices*. Prentice Hall, Englewood Cliffs, NJ, 1997.
[8] T. K. Edwards, *Automatic Logic Synthesis for Digital Systems*. McGraw-Hill, New York, 1992.
[9] P. K. Lala, *Digital System Design Using Programmable Logic Devices*. Prentice Hall, Englewood Cliffs, NJ, 1990.
[10] D. Pellerin and M. Holley, *Practical Design Using Programmable Logic*. Prentice Hall, Englewood Cliffs, NJ, 1991.
[11] J. F. Wakerly, *Digital Design Principles and Practices*, 2nd ed. Prentice-Hall, Englewood Cliffs, NJ, 1994.

The one-hot method in state machine design is apparently offered for significant further reading in only two texts, those by Hayes and by Nelson *et al.* (previously cited). Both contribute something different to the subject and are recommended. To a lesser extent this subject is covered in the text by Comer.

[12] D. J. Comer, *Digital Logic and State Machine Design*, 3rd ed. Saunders College Publishing, Fort Worth, TX, 1995.
[13] J. P. Hayes, *Introduction to Digital Design*. Addison-Wesley, Reading, MA, 1993.

Other sources for further reading on the subject of system-level design where examples are provided are found in the texts by Fletcher (previously cited) and Shaw.

[14] A. W. Shaw, *Logic Circuit Design*. Sanders College Publishing, Fort Worth, TX, 1993.

For the reader who wishes to have more information on stepping motors, mentioned in this chapter in connection with a stepping motor controller design, the book by Kenjo is recommended.

[15] T. Kenjo, *Stepping Motors and Their Microprocessor Controls*. Oxford University Press, 1984.

Finally, it should be noted that for logic system design by using registered PLDs, PLSs, and FPGAs, there may be no better sources than the data books published by Advanced Micro Devices, Signetics, Xilinx, Actel, and Altera. GAL devices are covered by Lattice Semiconductor's data book. For EPLD component specifications and applications, the reader will find Intel's data book useful.

[16] *ACT Family Field Programmable Gate Array Databook*. Actel Corp., Sunnyvale, CA, 1991.
[17] *Altera Data Book*. Altera Corp., San Jose, CA, 1995.
[18] *GAL Data Book*. Lattice Semiconductor, Hillsboro, OR, 1992.
[19] *PAL Device Data Book*. Advanced Micro Devices, Inc., Sunnyvale, CA, 1992.
[20] *Programmable Gate Array Data Book*. Xilinx, Inc., San Jose, CA, 1995.

[21] *Programmable Logic Data Book*, Intel Corp., Santa Clara, CA, 1994.
[22] *Programmable Logic Devices Data Handbook*. Signetics Co., Sunnyvale, CA, 1992.
[23] *The Programmable Logic Data Book*. Xilinx, Inc., San Jose, CA, 1996.
[24] *XACT, Logic Cell Array Macro Library*. Xilinx, Inc., San Jose, CA, 1992.

Most texts in digital design do not attempt to cover digital design fundamentals together with the organization and design of microprocessors (or microcontrollers) and computers. Of those that do attempt this and for the reader who is interested in microprocessor and computer design but who has had no previous experience in the field, the texts by Hayes and Katz (both previously cited) and that by Shaw are given a qualified recommendation. Usually the subject of computer organization and design is a challenge to develop in a single dedicated text. So one might expect the treatment to be somewhat on the thin side in the three texts cited above. Better sources for the beginning reader can be found in the text by Mano and Kime, and in that by Pollard. In these last two references the reader will find much more detailed information on computer organization and design. However, the reader should expect to find only token coverage of digital design fundamentals in these texts.

[25] A. W. Shaw, *Logic Circuit Design*. Saunders College Publishing, Fort Worth, TX, 1991.
[26] M. M. Mano and C. R. Kime, *Logic and Computer Design Fundamentals*. Prentice-Hall, Englewood Cliffs, NJ, 1997.
[27] L. H. Pollard, *Computer Design and Architeture*. Prentice-Hall, Englewood Cliffs, NJ, 1990.

PROBLEMS

13.1 Shown in Fig. P13.1 is an FSM that has two inputs, X and Y, and two outputs, P and Q. It is to be designed by using RET D flip-flops as the memory, and an FPLA for the NS- and output-forming logic.

(a) Run both output race glitch and static hazard analyses on this FSM and determine the requirements for glitch-free outputs. In doing this, select the type of

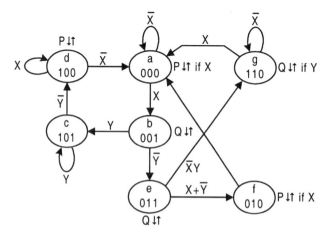

FIGURE P13.1

flip-flop (NAND-based or NOR-based) that should be used. Remember that a PLA is an SOP device.

(b) By using third-order K-maps, obtain an optimal set of expressions for the NS functions.

(c) Construct the p-term table and block diagram for an FPLA design of this FSM. To do this, use a 5 × 13 × 5 FPLA (as a block symbol) to design an optimum glitch-free logic circuit for this FSM. Plan to filter the output signals only if necessary. Do not alter the state diagram. Assume that input X is asynchronous from a mechanical switch (switch Down in Fig. 11.32a is the active state), that input Y is synchronous, and that both arrive active low. Take the outputs as active high.

13.2 A counter is to be designed that will drive the seven-segment display in Fig. 6.22a directly from its seven state variables, that is, from the flip-flop outputs, A, B, C, D, E, F, and G. When the counter is connected to the common cathode LED array in Fig. 6.23b, seven-segment Arabic numerals will appear. The FSM must have a count-up enable control X and must be cascadable so that numerals greater than 9 can be displayed. Thus, two such counters in cascade will count with each clock cycle, $\cdots$ 0–1–2–3–$\cdots$–90–0$\cdots$, but only if enabled.

(a) Construct the state diagram for this FSM. Plan to initialize it into the *decimal* zero state.

(b) Assuming the use of D flip-flops, map the state diagram directly into seven fourth-order EV K-maps and extract minimum or near minimum cover for each of the seven NS functions by using a logic minimizer such as BOOZER. (Suggestions: The simplest approach is to use the map format $AB/CD \parallel E/FG$ by following the example in Fig. 5.7 as an array of third-order K-maps. Each cell of a given fourth-order NS K-map will represent a third-order submap with axes E/FG and one entered variable, X. Thus, each NS K-map represents a fourth-order compression. It will be helpful to divide each state code assignment of the state diagram into two parts, the most significant four bits for the K-map axes AB/CD and the least significant three bits for the submaps. Note that the use of submaps is necessary only for cells 6, 14 and 15.)

(c) Use an 8 × 32 × 8 FPLA to implement the NS- and output-forming logic. Assume that the inputs and outputs are active high. To do this, construct the p-term table and block diagram for this FSM.

13.3 The state diagram in Fig. P13.2a represents the controller for a candy-bar vending machine. The controller has six inputs and four outputs, all of which are defined in Fig. P13.2b.

(a) Construct a minimum size p-term table for implementation of the NS and output functions by using an FPLA. To do this, assume that D flip-flops are to be used as the memory, and note that only one of the inputs LT, GT, or EQ can be active at any given time — they are the outputs from a comparator. Furthermore, assume that all inputs and outputs are active high.

(b) From the results of part (a), construct the logic circuit for the vending machine controller. Plan to use RET D flip-flops and to initialize into the 000 state. If ORGs are present, take the necessary steps to eliminate them, but do not change the state code assignment that is given. Use a block symbol for the PLA and

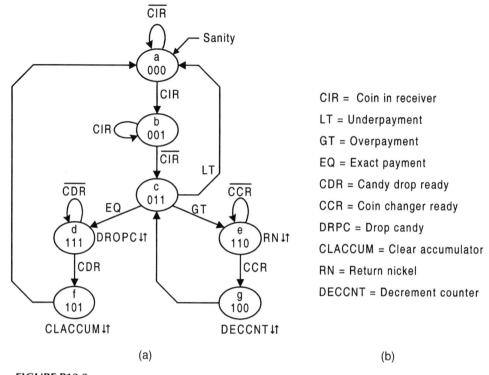

FIGURE P13.2

note that synchronizing of the inputs is not necessary for GO/NO-GO branching actions. Also, assume that the inputs from the comparator are synchronized by the *CIR* signal.

13.4 (a) Without altering the state diagram in Fig. P13.2a, construct the ROM program table for the candy bar vending machine controller directly from the state diagram in Fig. P13.2a. Again, note that only one of the inputs *LT*, *GT*, or *EQ* can be active at any given time — they are the outputs from a comparator.

(b) Repeat part (b) of Problem 13.3. Also, assume that all inputs and outputs are active high.

13.5 Shown in Fig. P13.3 is the p-term table for an FSM that has five inputs, U, W, X, Y, and Z, and four outputs, P, Q, R, and S.

(a) Obtain the state diagram for this FSM. (Hint: First, construct the NS and output K-maps.)

(b) Run complete ORG and hazard analyses on this FSM. If these timing defects exist in any of the outputs, indicate what corrective action is most appropriate to eliminate them.

13.6 The state diagrams for two FSMs are presented in Fig. P13.4.

(1) Construct the collapsed ROM program table for each of these FSMs directly from the state diagram. List the present state, ABC, in ascending binary

Inputs								Outputs						
A	B	C	U	W	X	Y	Z	D_A	D_B	D_C	P	Q	R	S
0	0	1	0	-	-	-	-	1	0	0	0	0	0	0
1	-	0	-	0	-	-	-	1	0	0	0	0	0	0
1	-	1	-	-	-	0	-	1	0	0	0	0	0	0
-	1	0	-	-	-	-	-	1	1	0	0	0	0	0
1	1	-	-	-	-	-	-	1	0	0	0	0	0	0
0	1	-	-	-	-	-	1	0	1	0	0	0	0	0
1	-	-	-	0	-	-	-	0	1	0	0	0	0	0
1	0	-	-	-	-	-	-	0	1	0	0	0	0	0
0	0	0	-	-	1	-	-	0	0	1	0	0	0	0
1	0	0	-	0	-	-	-	0	0	1	0	0	0	0

Inputs								Outputs						
A	B	C	U	W	X	Y	Z	D_A	D_B	D_C	P	Q	R	S
1	-	1	-	-	-	1	-	0	0	1	0	0	0	0
-	1	1	-	-	-	-	0	0	0	1	0	0	0	0
1	1	-	-	-	1	-	-	0	0	1	0	0	0	0
1	1	1	-	-	-	-	-	0	0	1	0	0	0	0
1	-	0	-	-	-	-	-	0	0	0	1	0	0	0
0	1	0	-	1	-	-	-	0	0	0	0	1	0	0
1	0	1	-	-	-	-	-	0	0	0	0	1	0	0
-	1	1	-	-	-	-	-	0	0	0	0	0	1	0
0	0	1	-	-	-	-	0	0	0	0	0	0	0	1
1	1	1	-	-	1	-	-	0	0	0	0	0	0	1

FIGURE P13.3

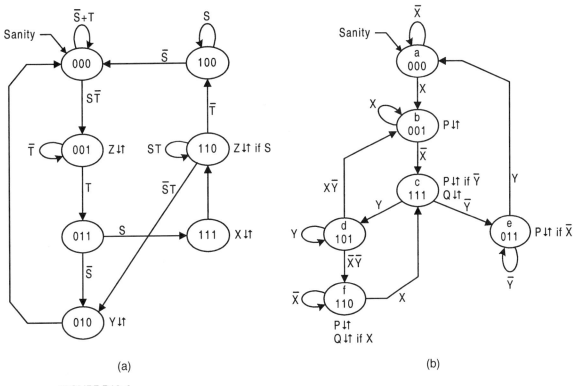

(a)   (b)

FIGURE P13.4

order, and assume that the activation levels for the inputs and outputs are as follows:

For Fig. P13.4a — $S(L)$ and $T(H)$; $X(H)$, $Y(H)$ and $Z(H)$

For Fig. P13.4b — $X(H)$ and $Y(H)$; $P(L)$ and $Q(L)$

Note that inverters cannot be used in dealing with an active low input or output. For the FSM in Fig. P13.4b, make a clear distinction between the input X and the irrelevant input symbol, X, used in the collapsed ROM program table.

(2) Construct the logic circuit for the ROM implementation of each of these FSMs. Use a block symbol for the ROM and assume the use of FET D flip-flops. Consider that S and T are synchronous inputs and bounce-free. However, in Fig. P13.4b both X and Y are asynchronous inputs, and input X arrives from a mechanical switch. Thus, include any input or output conditioning circuits that are necessary for a reliable glitch-free operation of the FSM. Initialize as indicated in the state diagrams.

13.7 Construct the collapsed ROM program table for the FSM in Fig. P11.3. List the present state, ABC, in ascending binary order. List any assumptions made.

13.8 The ROM program table in Fig. P13.5 represents an FSM having two inputs, S and T, and two outputs, P and Q.
(a) Construct the state diagram for this FSM directly from the program table. Indicate which, if any, are don't-care states.
(b) Point out any problems or potential problems this FSM may have.

13.9 Shown in Fig. P13.6 is the state diagram for a sequence recognizer. This FSM is the same as that in Fig. 10.60c, but with a state code assignment best suited for a shift register design. Design this FSM by using a universal shift register (USR) following the example in Fig. 13.14. Assume that both the input and output are

A	B	C	S	T	D_A	D_B	D_C	P	Q		A	B	C	S	T	D_A	D_B	D_C	P	Q
0	0	0	0	0	0	0	1	0	0		1	0	0	0	X	1	0	0	0	1
0	0	0	0	1	0	0	0	0	0		1	0	0	1	X	0	0	0	0	1
0	0	0	1	X	0	0	0	1	0		1	0	1	1	1	1	0	1	0	0
0	0	1	0	0	0	0	1	0	0		1	0	1	0	X	1	0	0	0	0
0	0	1	0	1	0	0	1	0	1		1	0	1	1	0	1	1	1	0	0
0	0	1	1	0	1	0	1	0	0		1	1	0	X	0	1	0	0	0	0
0	0	1	1	1	1	0	1	1	0		1	1	0	X	1	1	0	0	1	0
0	1	0	X	X	0	1	1	0	1		1	1	1	0	X	1	1	0	0	1
0	1	1	X	X	0	1	0	1	1		1	1	1	1	X	1	1	1	0	1

X = Irrelevant input

FIGURE P13.5

PROBLEMS

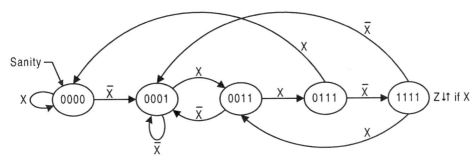

FIGURE P13.6

active high. Use a gate-minimum external logic and plan to initialize the FSM into the 0000 state. (Hint: Look for XOR patterns in the S_1 and S_0 K-maps.)

13.10 A candy bar vending machine is described in Problem 13.3 and is represented by the state diagram in Fig. P13.2. Change the state code assignment in the state diagram as follows:

$$a \to 0000 \quad b \to 0001 \quad c \to 0011 \quad d \to 0111$$
$$e \to 0010 \quad f \to 0110 \quad g \to 0101$$

Now design this FSM by using a universal shift register (USR) and a state decoder. To do this, follow the architecture used for the example in Fig. 13.17. Is an output holding register necessary? Explain your answer.

13.11 In Fig. P13.7 is given the logic circuit for an FSM that is built around a universal shift register. This architecture is similar to that used for the example in Fig. 13.17, but with neither a state decoder nor an output holding register.

(a) From the logic diagram, obtain the state diagram for this FSM. Indicate which, if any, of the states are don't care states. (Hint: Construct the PS/NS table from the K-maps associated with the USR.)

(b) Analyze the FSM for any possible problems.

13.12 Shown in Fig. P13.8 is the state diagram for the sequence recognizer in Fig. P13.6, but with a state code assignment that is best suited for a design centered around a counter. Design this FSM by using a parallel loadable up/down counter following the example in Fig. 13.19. Assume that both the input and output are active high. Find a gate-minimum external logic and plan to initialize the FSM into the 0000 state. (Hint: Look for XOR patterns.)

13.13 The candy bar vending machine is described in Problem 13.3 and is represented by the state diagram in Fig. P13.2. Alter the state code assignment as indicated below and design this FSM by using a parallel loadable up/down counter and a state decoder. To do this, follow the architecture used for the example in Fig. 13.22. Is an output holding register necessary? Explain your answer.

$$a \to 0000 \quad b \to 0001 \quad c \to 0010 \quad d \to 0101$$
$$e \to 0100 \quad f \to 0110 \quad g \to 0011$$

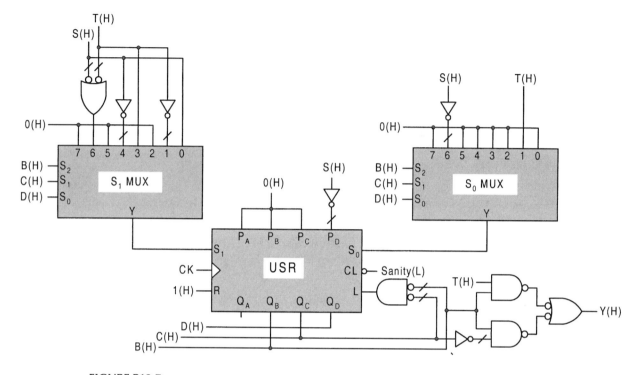

FIGURE P13.7

13.14 Shown in Fig. P13.9 is the logic circuit for an FSM that is built around a parallel loadable up/down counter. This architecture is similar to that used for the example in Fig. 13.22, but with neither a state decoder nor an output holding register.

(a) Obtain the state diagram for this FSM. Indicate which, if any, of the states are don't care states. (Hint: Construct the PS/NS table from the K-maps associated with the counter.)

(b) Analyze the FSM for any possible problems.

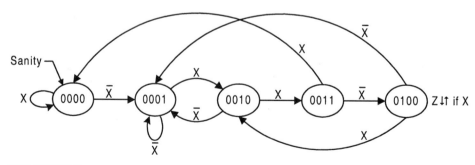

FIGURE P13.8

PROBLEMS

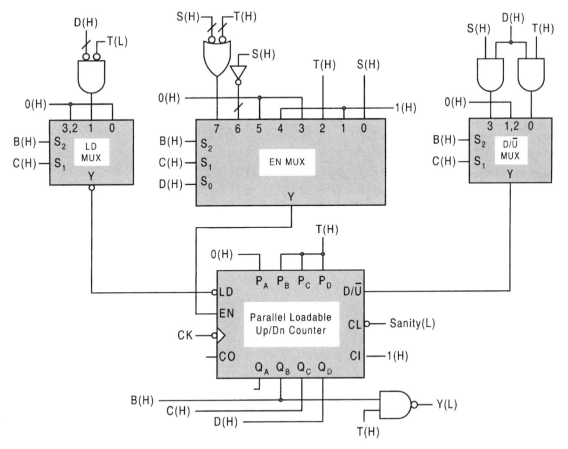

FIGURE P13.9

13.15 The NS and output logic functions for a *one-hot FSM* are as follows:

$$D_a = a\bar{X}Y + c\bar{X}Y$$
$$D_b = a\bar{X}\bar{Y} + b\bar{X} + bY$$
$$D_c = aX + bX\bar{Y} + cX + c\bar{Y}$$
$$Z = bY$$

Here, a, b, and c are the state identifiers, X and Y are the external inputs, and Z is the output.

(a) Construct the state diagram directly from the NS and output functions given above.

(b) Show how the NS function D_a must be altered to initialize the FSM into the 000 state, but thereafter be driven into the one-hot state a. Implement the logic for D_a and connect it to an RET D flip-flop symbol together with all other required

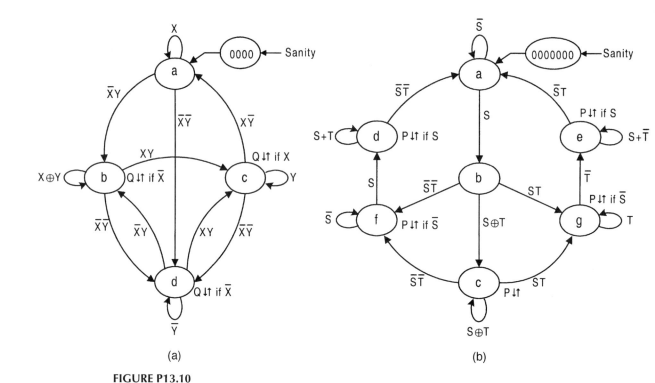

FIGURE P13.10

inputs to and outputs from that flip-flop. Do not implement the logic for flip-flops *b* and *c*.

13.16 Presented in Figure P13.10 are the state diagrams for two FSMs, each with two inputs and one output.

  (1) Given the indicated state identifiers, use the *one-hot approach* to obtain the NS and output expressions for each of these FSMs directly from the state diagrams. Assume that RET D flip-flops are used as the memory elements.

  (2) Obtain the logic necessary to initialize each FSM into the all-zero state after which each must be forced to transit irreversibly into one-hot state *a* — the one-hot-plus-zero approach.

  (3) Construct the logic circuit, including the initialization logic into the all-zero state for each one-hot design.

  (4) Comment on the presence or absence of ORGs and s-hazards in the outputs. If ORGs are possible, indicate where they exist and by what race path. If they are not possible, explain your reasoning.

13.17 (a) The state diagram for a candy bar vending machine controller is presented in Figure P13.2. Given the state identifiers indicated for this FSM, write the one-hot NS and output expressions directly from the state diagram. Plan to initialize directly into the one-hot state *a*.

PLA Inputs							PLA Outputs					PLA Inputs							PLA Outputs				
a	b	c	d	Q	K	M	D_a	D_b	D_c	D_d	P	a	b	c	d	Q	K	M	D_a	D_b	D_c	D_d	P
1	–	–	–	0	–	–	1	0	0	0	0	–	1	–	–	1	–	1	0	1	0	0	0
1	–	–	–	–	0	–	1	0	0	0	0	–	–	1	–	1	1	1	0	1	0	0	0
–	1	–	–	0	–	–	1	0	0	0	0	–	–	1	–	–	0	–	0	0	1	0	1
–	1	–	–	–	0	0	1	0	0	0	0	–	–	–	1	–	0	–	0	0	1	0	1
–	–	1	–	0	1	–	1	0	0	0	0	–	–	–	1	1	1	–	0	0	0	1	0
–	–	–	1	0	1	–	1	0	0	0	0	1	–	–	–	1	1	–	0	0	0	1	0
0	0	0	0	–	–	–	1	0	0	0	0	–	1	–	–	1	1	0	0	0	0	1	0
												–	–	1	–	1	1	0	0	0	0	1	0

FIGURE P13.11

(b) Implement the NS and output functions by using a FPLA and RET D flip-flops. To do this construct the p-term table together with a logic circuit and the necessary connections for initialization. Assume that the all inputs and outputs are active high. Are ORGs possible in this design? Explain your reasoning.

(c) Is a ROM implementation of the NS and output functions for this one-hot FSM a wise choice? Are there FSMs for which the ROM implementation of a one-hot FSM has an advantage over a PLA or PAL implementation? Explain your answers to these questions.

13.18 Shown in Fig. P13.11 is the p-term table for the *one-hot* design of an FSM that has three inputs, Q, K, and M, and one output P. Here, the state identifiers are a, b, c, and d.

(a) Construct the state diagram directly from the p-term table. Pay particular attention to how the FSM is to be initialized.

(b) Analyze this FSM for possible ORGs and static hazards.

THE FOLLOWING PROBLEMS ARE TO BE CARRIED OUT AT THE SYSTEM LEVEL.

(Note that typically there is more than one correct solution for each system-level design.)

13.19 (a) Design a multiple pulse generator that will issue, on the *Pulse* output, 0 to 99 clean (glitch-free), evenly spaced pulses with an active duration the same as that for the system clock. To do this, it is necessary to design a controller and two interconnected BCD down-counters with an active low borrow-out (*BO*). Use RET D flip-flops for the counter design and FET D flip-flops for the controller, both with Preset and Clear overrides. Thus, state-to-state transitions of the controller are made on the falling edge of the system clock.

A *Start* signal is required to *load* the counters and initiate the process. Assume that the count settings are made by individual switches and are loaded asynchronously into the counters via the Preset and Clear overrides prior to the *Start* signal. The count begins at the particular setting of the switches and ends when the count reaches zero. The pulses are to be generated with active clock by the *Pulse* output from the controller, and the counters are to count with inactive clock on each rising edge of the Count command *CNT* from the controller. An *END(L)* signal from the counters ends the count process when zero has been reached. Make certain that the counters are loaded at least one clock period before the *CNT* and *Pulse* signals are issued by the controller. Plan to use four states for the controller design, and make certain that only one series of pulses can be issued on a start command.

To design the BCD down counters, follow the example in Subsection 12.3.2, but for a down count, and with asynchronous preset and clear override capability as in Fig. 12.20. Let *CNT(H)* be the enabling input to the MSD counter, and connect the two counters in series by connecting the *BO(L)* of the MSD counter to the *EN(H)* of the LSD counter. Note that a pulse is never issued in the 0000 end state, and that any false data setting (1010 to 1111) must *not* result in pulse generation.

(b) Construct a timing diagram of the results of part a assuming a count of 03. To do this, include the waveforms for *CK, START(H), LD(L), CNT(H), PULSE(H), END(L)*, and present states *A(H)* and *B(H)*.

13.20 An election between two competing candidates for mayor is to be held in a small community of 752 registered voters. Design a voter booth tabulation system that will tally the vote count on each of two competing candidates. The booth will show an "Enter" light when not occupied. When the voter enters the booth and closes the door, a Voter-in (*VI*) signal is sent to the controller, the "Enter" light is turned off, and an "Occupied" light is turned on. This is accomplished by a motion detector working in coordination with door and light switches (matters of no concern to this design problem). Once in the booth with the door closed, the voter pushes one of two switches for the candidate of his or her choice. When either button is pressed a corresponding counter is incremented, and the door is automatically opened for the voter to exit. The current count of each counter is stored in a register as a BCD number ready to be presented later as a seven-segment display. If both buttons are pressed simultaneously, neither counter is incremented and the door is opened. It is not possible for a voter to vote twice while in the booth. Once the door is opened and the voter exists the booth ($\overline{VI}$), the process is ready to begin again. Assume that the entrence to the voter booth is minitored in some way so as to prevent an individual from voting more than once.

The block diagram for the controller is provided in Fig. P13.12(a) and the input and output symbology is defined in Fig. P13.12(b). Take all inputs and outputs as active high and note that the switch inputs Ba and Bb are presented to the controller asynchronously from mechanical switches.

Design the controller for this system by using the *one-hot-plus-zero approach*, and construct the functional partition for its operation. To do this, use RET D flip-flops for the controller FSM, and RET cascaded BCD up-counters for the count. In addition to the controller, plan to initialize the counters, registers and the appropriate input conditioning circuits. Carefully consider how best to trigger the registers relative to the counters and controller. Assume that the lights and door opening mechanisms are available to the designer. No acknowledge signal following an increment is necessary from the counters. Finally, make certain that all required input conditioning circuits are included.

13.21 A traffic light control system is to be designed that will operate traffic lights at the intersection of a main highway and an infrequently used farm road. Traffic sensors are placed on both the highway and the farm road to indicate when traffic is present. If no traffic is sensed on the farm road, traffic on the highway is allowed to flow. But when a vehicle activates the sensor on the farm road, the highway light signals are activated immediately if the traffic sensor on the highway is not active. Otherwise, the vehicle on the farm road must wait 30 seconds or until the highway is clear, whichever occurs first, before the highway signals are reactivated. Once the farm road is clear, the system must activate the farm and highway lights so as to permit highway traffic to flow, but only after a 30-second time interval to allow the farm road to clear.

In designing the control system, two interval timers (counters) must be designed, one for the 30 second time interval and the other for the 5-second yellow light time interval. These timers accept an input to signal the start of the time interval and return an output to indicate the end of the time interval. Upon receiving the count enable input signal, the timer begins timing. At the end of the specified time, the output signal is activated and remains active until the count enable input signal is deactivated.

Construct a suitable *controller state diagram and functional partition* for the traffic light control system. Make any reasonable state code assignment for the controller and use an architecture centered around a PLA, as in Figure 13.1. Construct the p-term table for programming the PLA device and provide a block diagram for the controller. Assume that all inputs and outputs are active high. Use the abbreviations given next and assume that F and H are asynchronous inputs.

Controller Inputs: $F =$ Farm road active; $H =$ Highway active; $30 = 30$ seconds complete; $5 = 5$ seconds complete.

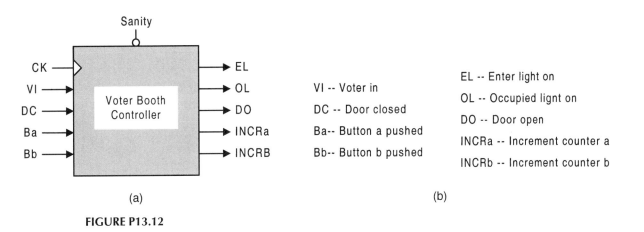

FIGURE P13.12

Controller Outputs: *FR* = Farm Red; *FG* = Farm Green; *FY* = Farm Yellow; *HR* = Highway Red; *HG* = Highway Green; *HY* = Highway Yellow; *S30* = Start 30 seconds; *S5* = Start 5 seconds; *LDCNT* = Load counters

Use a minimum number of RET D flip-flops and an optimum NS and output logic for the controller. Pay particular attention to timer requirements as they pertain to the handshake interface, counter design, and clock frequency. To do this, construct timing diagrams if necessary. Initialize the system properly and deal with any asynchronous input requirements. Assume that the clock frequency is 13.1 kHz.

Hints and suggestions:

(1) Six or seven states are adequate for the state diagram.
(2) Plan to synchronize inputs and filter outputs as needed.
(2) Use divide-by-16^4 and divide by 16 parallel loadable binary counters to generate the 5 second and 30 second time intervals. Counters will need to be initialized.
(3) Counter design should follow that in Figs. 12.19, 12.20, and 12.21.
(4) By law, a green light never changes directly to red, but must first change to yellow.
(5) Assume that mechanisms for light generation exist and that they are unaffected by logic noise.
(6) The output *LDCNT* must be free of logic noise.

CHAPTER 14

Asynchronous State Machine Design and Analysis: Basic Concepts

14.1 INTRODUCTION

In Chapters 10 through 13 the emphasis was directed toward synchronous sequential machine design. These chapters developed a rather thorough understanding of the concepts necessary for the meaningful and reliable design of these machines. Now, it is necessary to move on to another type of sequential machine — the asynchronous FSM. In Fig. 14.1 is presented an overview of the various types of digital machines. Observe that combinational machines are classified as asynchronous because they operate in the absence of a clock signal, but they do not have feedback. Combination logic machines were the subjects of Chapters 6 through 8. As is indicated in Fig. 14.1, all sequential machines must have feedback, but they can be divided into two categories, synchronous and asynchronous.

The major aim of this chapter is, of course, to develop a working-level understanding of asynchronous FSMs, their design and analysis, and to design state machines that operate at speeds exceeding those possible for their synchronous FSM counterparts. But the mission of this chapter is really broader than that. In the course of the various discussions, the reader will develop a better understanding of those concepts involved in synchronous machine design and analysis. In fact, an understanding of asynchronous sequential machines is required before synchronous sequential machines can be fully understood.

So why has the subject of asynchronous machine concepts and methodologies been delayed to this point? The answer is simple. The study of asynchronous FSMs forces one to deal with the complexities of sequential machines in greater depth than was required for the simpler synchronous machines. Putting it another way, the study of synchronous FSMs permitted the reader to develop capabilities sufficient to design and analyze large systems without having to deal with the intricacies of asynchronous machine design. Remember that *all* digital machines can eventually be broken down into their component asynchronous parts. For example, the synchronous FSMs, studied in Chapters 10 through 13, use memory elements (flip-flops) that are themselves asynchronous machines but that are designed to operate in a clock-driven environment.

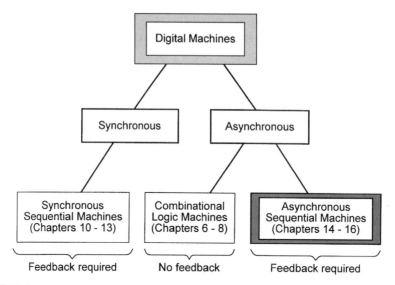

FIGURE 14.1
Breakdown of the various classes of digital machines showing chapters in this text pertinent to each class with emphasis on asynchronous sequential machines, the subject of this and subsequent chapters.

14.1.1 Features of Asynchronous FSMs

All sequential machines have certain characteristics in common. However, there are features owned more or less exclusively by asynchronous FSMs:

- The presence of memory in the absence of the familiar clocked flip-flop
- The appearance of the asynchronous machine as a combinational logic circuit with feedback

Other more subtle features distinguish asynchronous FSMs from those that are synchronous. These features include the possible existence of certain timing defects such as endless cycles (oscillations), critical races (races that can produce error transitions), static hazards in both the NS and output logic, and essential hazards. Static hazards that are generated in the NS-forming logic of asynchronous FSMs can cause the malfunction of these machines. Static hazards that form in the NS logic of synchronous FSMs are of no consequence since they are filtered out by the action of clock in the memory flip-flops. Also, in synchronous FSMs static hazards that are produced in the output-forming logic can be filtered out by using an output holding register. This advantage is also not shared by asynchronous FSMs, since there is no system clock with which to trigger a register. In short, the benefits of clock, which are taken for granted in synchronous machine design, do not exist in asynchronous FSMs. The reality is that endless cycles, critical races, and hazards can occur in asynchronous FSMs and, if present, can and do cause the machines to malfunction. A detailed study of these and other timing defects and the actions required to eliminate them constitute a significant portion of this chapter.

14.1.2 Need for Asynchronous FSMs

It is perhaps natural to believe that the data processing in and passage through a sequential machine must be regulated by some periodic sampling (enabling) function, the system clock. This, of course, is a requirement of the synchronous sequential machine. But one never questions the absence of a clock in the combinational logic circuits covered in Chapters 6, 7, and 8, yet these circuits are asynchronous machines of a type — those without feedback (i.e., nonsequential). Why then the concern about the need for a clock to regulate synchronous sequential operations? And when is it advantageous, if ever, to perform sequential operations asynchronously? The complete answers to these questions will be forthcoming, but only after most of the contents of this chapter has been considered. For now let it suffice to say that it may be desirable to use asynchronous designs for the following reasons:

- The speed requirements of the system may exceed the capability of synchronous machines. Properly designed, a synchronous FSM can only approach (not equal) the *speed* of a properly designed asynchronous FSM performing the same sequential operation(s). There are exceptions to this rule.
- Use of a system clock to synchronize a given sequential machine may not be possible or even desirable. Clock distribution problems (*clock skew*) may seriously limit the use of synchronous designs, particularly in complex digital systems operated at very high frequencies.
- Since flip-flops and clock oscillator circuits are absent, an asynchronous design may occupy less real estate on an IC chip and use less power than an equivalent synchronous design. However, this statement may not be true for complex asynchronous FSMs, the components of which must communicate through handshake configurations.
- Just as there are some designs that should be carried out synchronously, there are other designs that lend themselves quite naturally to asynchronous design. This statement may be even more relevant in integrated systems, systems containing both synchronous and asynchronous state machines, where maximum speed is required.

Clearly, there is potential for use of asynchronous machines. In fact, it is predictable that designers will become more familiar with this type of machine, that asynchronous design techniques will improve, and that asynchronous FSM methods will play an important role in the design of future superhigh-speed microprocessors and computers. It is the judgment of many digital designers that the continued upward climb of system size and speed will require more integration of asynchronous FSMs into "conventional" system-level designs.

14.2 THE LUMPED PATH DELAY MODELS FOR ASYNCHRONOUS FSMs

In synchronous FSMs the memory function is formed by using flip-flops. But if asynchronous FSMs are characterized by the absence of such devices as flip-flops, how, then, does

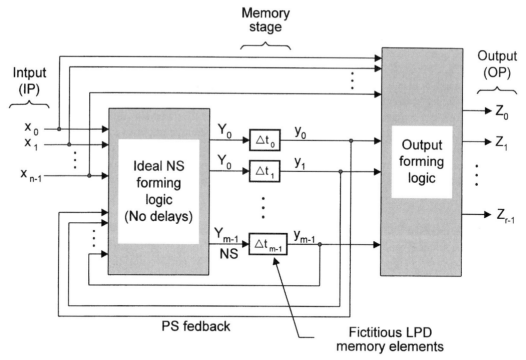

FIGURE 14.2
Lumped path delay (LPD) model for an asynchronous Mealy FSM operated in the fundamental mode.

memory manifest itself in these machines? The answer to this question lies in the fact that data transport through an FSM is not instantaneous. Propagation time delays are an inherent part of any circuit, and it is these path delays that constitute the memory stage of an asynchronous FSM. Recall that this is precisely the basis for the heuristic development of the basic cells presented in Section 10.4. It is this heuristic development that provides the basis for the generalized and more formal treatment that follows.

Consider the Mealy model for the asynchronous FSMs shown in Fig. 14.2. This model is called the generalized *lumped path delay (LPD) model* and is applicable to FSMs that are operated in the *fundamental mode*.

> *Operation in the fundamental mode requires that no external input to an FSM may change until all internal signals are stabilized, and that only one input can change at a time.*

The LPD model is characterized by an NS-forming logic that is treated as ideal (free of path delays) for which the propagation time delays are separated out into a minimum number of distinct lumped memory elements $\Delta t_0, \Delta t_1, \Delta t_2, \ldots \Delta t_{m-1}$, each delay element being associated with a state variable. It is these fictitious lumped memory elements, taken in toto, that constitute the memory stage for an asynchronous FSM.

The model of Fig. 14.2 is the most degenerate (fundamental) form of the Mealy model depicted in Fig. 10.5. This model can be broken down into the more rudimentary forms similar in appearance to those in the development of the basic model in Fig. 10.3 or to the Moore model in Fig. 10.4, but always with a memory stage composed of fictitious LPD memory elements. Regarding the memory stage, it will be recalled that the memory for the basic model in Fig. 10.3(c) is interpreted as basic cells in Fig. 10.22 or as a flip-flop in the case of Fig. 10.36. In fact, the more general models given in Figs. 13.1 and 13.47 can be used in synchronous systems where the memory is interpreted as discrete flip-flops, or flip-flops in a register or counter. Now, the reader should consider that *memory* in all these models can be interpreted as any one of the following forms given in the order of increasing degeneracy:

$$\{\text{Flip-flops} \rightarrow \text{BasicCells} \rightarrow \text{LPD Memory Elements}\}$$

Thus, if the memory is composed of flip-flops and clocked, the FSM is called synchronous. But if the more degenerate forms are used for the memory (e.g., basic cells or LPD memory elements) the FSM becomes asynchronous. In this text, the *nested cell model* is characterized by the use of basic cells as the memory elements, while the *LPD model* is characterized by the use of fictitious LPD memory elements.

14.3 FUNCTIONAL RELATIONSHIPS AND THE STABILITY CRITERIA

The parameters used in Fig. 14.2 are defined by

$$\begin{aligned}
x_i &= x_{n-1}, \ldots x_2, x_1, x_0 = \text{Input State (IP)} \\
Y_k &= Y_{m-1}, \ldots Y_2, Y_1, Y_0 = \text{Next State (NS)} \\
y_j &= y_{m-1}, \ldots y_2, y_1, y_0 = \text{Present State (PS)} \\
Z_l &= Z_{r-1}, \ldots Z_2, Z_1, Z_0 = \text{Output State (OP)},
\end{aligned} \qquad (14.1)$$

all of which have been arranged in positionally weighted form to represent binary words. These parameters are functionally related to each other and to the inputs and outputs by the following set of logic equations written in subscript notation:

$$\begin{aligned}
Y_j(t) &= y_j(t + \Delta t) \\
Y_k(t) &= Y_k[x_i(t), y_j(t + \Delta t)] \\
Z_l(t) &= Z_l[x_i(t), y_j(t)],
\end{aligned} \qquad (14.2)$$

or simply

$$Y = f(\text{IP}, \text{PS})$$
$$Z = f'(\text{IP}, \text{PS}).$$

The subscripts in Eqs. (14.2) are assigned the ranges of values

$$i = 0, 1, 2, \ldots, n - 1$$
$$j = 0, 1, 2, \ldots, m - 1$$
$$k = 0, 1, 2, \ldots, m - 1$$
$$l = 0, 1, 2, \cdots, r - 1$$

The fact that the inputs, x_i, can be multivariable functions implies that one asynchronous FSM may be controlled by another asynchronous FSM.

Inspection of Eqs. (14.2) indicates that corresponding NS and PS variables are separated in time by distinct lumped delay memory elements, Δt_j. This leads directly to the important *stability criteria* for asynchronous FSMs operated in the fundamental mode:

Stability Criteria

If the PS is logically equal to the NS at some point in time, then

$$Y_j(t) = y_j(t) \quad \text{(for } all \text{ } j\text{)}, \tag{14.3}$$

and the asynchronous FSM is stable in that state.

If the PS and NS are not logically equal at any point in time, then

$$Y_j(t) \neq y_j(t) \quad \text{(for } any \text{ } j\text{)}, \tag{14.4}$$

and the asynchronous FSM is unstable in that state and must transit to another state.

Here, the presence of a lumped memory element for each feedback loop ensures that all path delays within the NS forming logic are represented. A much less attractive alternative is the *distributed path delay model*, which requires a memory element for each gate and as many state variables. The LPD model has the decided advantage of simplicity — it requires a minimum of lumped memory elements and hence a minimum number of state variables. Use of the distributed path delay model would be prohibitively difficult for all but the simplest state machines.

14.4 THE EXCITATION TABLE FOR THE LPD MODEL

The excitation table for the LPD model of Fig. 14.2 and all of its degenerate forms is derived directly from the stability criteria given by Eqs. (14.3) and (14.4). The results are shown in Figs. 14.3a and 14.3b, where a stable condition exists for $y_t = Y_t$, and an unstable condition exists if $y_t \neq Y_t$. Here, $y_t \to y_{t+1}$ represents a transition from the PS to the NS, implying that $y_{t+1} = Y_t = $ NS. It is important to notice the similarity between the excitation table in Fig. 14.3b and that for the D flip-flop in Fig. 14.3c. Thus, it is expected that some LPD design methods apply to synchronous D flip-flop designs and vice versa. The excitation table for the LPD model is essential to the design of asynchronous FSMs to be operated in the fundamental mode and will be used extensively throughout the remainder of this text.

14.5 STATE DIAGRAMS, K-MAPS, AND STATE TABLES FOR ASYNCHRONOUS FSMs

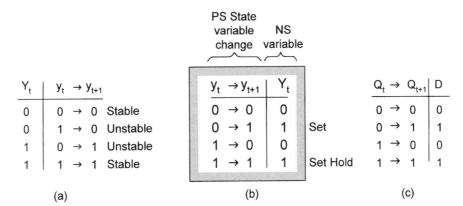

FIGURE 14.3
(a) Excitation table for the LPD model as derived from Eqs. (14.3) and (14.4). (b) The excitation table of (a) arranged in the form familiar for flip-flops. (c) Excitation table for the D flip-flop shown for comparison.

14.5 STATE DIAGRAMS, K-MAPS, AND STATE TABLES FOR ASYNCHRONOUS FSMs

This section deals with subject matter that has been covered in Chapters 10 and 11, but now applied to asynchronous FSMs. Thus, the concepts involved here are basically the same as in synchronous FSM design. Therefore, the reader who is familiar with this subject matter may wish to simply browse through this short section for a sufficient understanding of its contents.

14.5.1 The Fully Documented State Diagram

The sequential behavior of any FSM (synchronous or asynchronous) is revealed most effectively by a fully documented state diagram representing the sequential behavior of the FSM. However, the state diagram itself does not indicate whether the machine is synchronous or asynchronous. For example, the state diagram in Fig. 11.42 could be interpreted as that for either an synchronous or asynchronous FSM. But once the FSM is declared to be an asynchronous FSM and to be operated in the fundamental mode, then the design process can begin by applying the model and excitation table of Figs. 14.2 and 14.3b to the state diagram.

Shown in Fig. 14.4 is a section of a generalized, fully documented state diagram applicable to any FSM, in particular to an asynchronous FSM. The features are the same as those in Fig. 10.6, except that the PS variables are specifically identified as $y_{m-1} \cdots y_2 y_1 y_0$ to distinguished them from those for a synchronous FSM $Q_A Q_B Q_C Q_D \cdots = ABCD \cdots$, as used in this text. The branching conditions are given in subscript notation where, for example, $f_{ab}(x_i)$ represents conditional branching on inputs x_i from state a to state b, and $f_b(x_i)$ is the holding condition in state b, again a function of inputs x_i. Also, the output in state c is conditional on some function of inputs x_i.

Sum Rule and Mutually Exclusive Requirement The *sum rule* and *mutually exclusive requirement* for state diagrams representing asynchronous FSMs are given by Eqs. (10.3) and (10.4); the conditions under which they can be violated are discussed in Section 10.3.

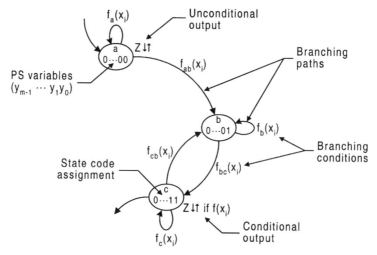

FIGURE 14.4
Fully documented state diagram as interpreted for an asynchronous FSM.

Thus, there is no difference in the applications of these rules to either synchronous or asynchronous FSMs. See Problem 10.24 for more information on the relationship between these two rules, and a more direct means of testing for mutually exclusivity of branching conditions.

ASM Charts and Flowcharts Flowcharts were used in Subsection 11.9.2 and ASM charts were used in Subsections 11.10.1, 13.5.2, and 13.6.2 as *thinking tools* in the construction of the fully documented state diagrams for synchronous FSM design. So also can these thinking tools be used for the purpose of constructing state diagrams for asynchronous FSM design. Furthermore, it may be recalled that the ASM chart is used effectively to design a one-hot state machine in Subsection 13.5.2, but the fully documented state diagram is shown to be equally effective for such a purpose. The point to be made here is that the ASM chart or flowchart should be considered only as a thinking tool in the construction of the fully documented state diagrams or state tables. In this text, it is the fully documented state diagram or state table that is considered to be the simplest and most effective means of representing the sequential behavior of an FSM (synchronous or asynchronous) in preparation for its design.

14.5.2 Next-State and Output K-maps

When using the LPD model for asynchronous FSMs, the entered variable (EV) K-maps for the NS variables are easily constructed by applying the mapping algorithm in Section 10.6 to the state diagram and the excitation table for the LPD model in Fig. 14.3b. Thus, this NS mapping process for the LPD model is very similar to that used in the D flip-flop designs of synchronous FSMs, the only difference being the nomenclature for the PS and NS parameters. Shown in Fig. 14.5 are the generalized EV K-maps for the NS and output functions as applied to the LPD model of an asynchronous FSM. The K-maps are of the mth order with state variables for their axes and inputs as the EVs. It is important for the

14.5 STATE DIAGRAMS, K-MAPS, AND STATE TABLES FOR ASYNCHRONOUS FSMs 691

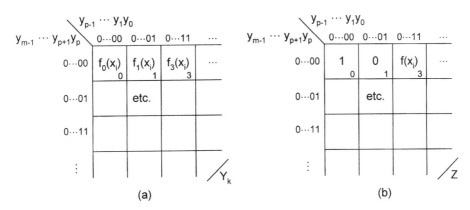

FIGURE 14.5
K-maps for asynchronous FSMs. (a) EV K-map for the kth NS state variable. (b) EV K-map for the output Z.

reader to understand that the state variables must always be the K-map axes variables, never the K-map EVs. Therefore, for state variables numbering between four and nine, K-map formats of the types shown in Fig. 4.38 of Section 4.7 and in Fig. 5.7 of Section 5.9 are recommended. For larger numbers of state variables, computer-aided design should be considered as the only reasonable alternative.

14.5.3 State Tables

State tables and NS tables were used previously in connection with the use of state assignment rules in Subsection 11.10.2 and in the array algebraic approach to synchronous FSM design discussed in Section 11.11. State tables are, of course, the tabular equivalent of a state diagram. In this chapter state tables will be used in the design of asynchronous single-transition-time (STT) state machines by using the array algebraic approach. STT machines are the fastest state machines possible but require special state coding procedures that were not needed in Section 11.11. Shown in Fig. 14.6 are the state diagram and state table for the FSM in Figs. 11.42 and 11.43 but interpreted as an asynchronous FSM to be operated in the fundamental mode. Notice that each cell entry in Fig. 14.6b is a state identifier representing the specific state code assignment shown on the vertical axis of the state table and in agreement with those in the state diagram of Fig. 14.6a. State variables should not be used as cell entries in state tables.

Recall from the discussion in Subsection 11.10.2 that the encircled state identifiers in state tables indicate a holding condition. But a holding condition in an asynchronous FSM means that Eq. (14.3) of the *stability criteria* is satisfied and that the FSM is stable in that state. So it follows, for example, that in state $a = 000$ the FSM is stable in that state under input conditions $\bar{S}T + ST + S\bar{T} = S + T$. Conversely, if the FSM is unstable in a given state according to Eq. (14.4), it must transit to another state. Thus, should the input conditions change to $\bar{S}\bar{T}$ while in state a, the FSM must transit to state b as indicated by the vertical down arrow in the $\bar{S}\bar{T}$ column of Fig. 11.6b. To summarize, the encircled state identifiers in a state table indicate FSM stability in agreement with Eq. (14.3) while the vertical arrows

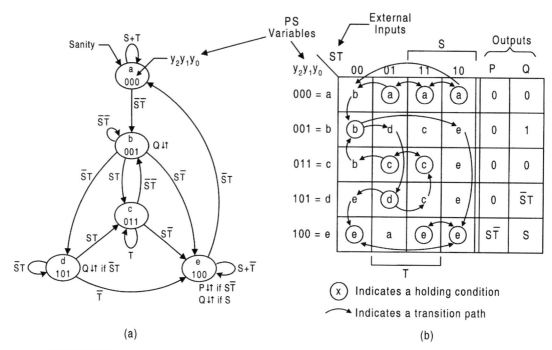

FIGURE 14.6
The FSM in Figs. 11.42 and 11.43 interpreted as an asynchronous FSM to be operated in the fundamental mode. (a) State diagram. (b) State table.

indicate FSM instability according to Eq. (14.4). Perhaps the most important use of state tables is in the designs of FSMs without the use of K-maps. This will be demonstrated later in connection with the design of STT machines.

14.6 DESIGN OF THE BASIC CELLS BY USING THE LPD MODEL

The reader who is familiar with Section 10.4 will recall that the set-dominant basic cell and the reset-dominant basic cell were developed by a heuristic approach. This was done to avoid having to use asynchronous design methods which, at that time, would have caused unnecessary delays in the primary goals of that chapter. Now, it is appropriate that the basic memory cells be designed from first principles by using the LPD model. The basic cells represent two of the simplest asynchronous FSMs that are operated in the fundamental mode.

14.6.1 The Set-Dominant Basic Cell

The state diagram for the set-dominant basic cell is shown in Fig. 14.7a and is a reproduction of that given in Fig. 10.12a. The NS logic for this basic cell is easily found by using the mapping algorithm, given in Section 10.6, to combine information contained in the state diagram with that contained in the excitation table for the LPD model in Fig. 14.7b. The result

14.6 DESIGN OF THE BASIC CELLS BY USING THE LPD MODEL

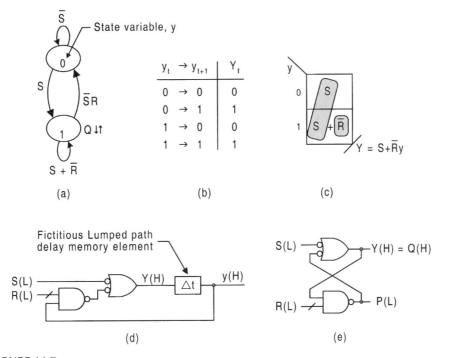

FIGURE 14.7
Design of the set-dominant basic memory cell by using the LPD model. (a) State diagram. (b) Excitation table for the LPD model. (c) NS K-map and minimum cover. (d) Logic circuit showing fictitious LPD memory element. (e) Final logic circuit with fictitious memory element removed.

is the NS K-map and minimum cover shown in Fig. 14.7c. From this there results the NS logic function in LPD notation given by

$$Y = S + \bar{R}y. \tag{14.5}$$

Except for the difference in PS and NS nomenclature, Eqs. (14.5) and (10.5) are identical, as they must be. The logic circuit that results from Eq. (14.5) is presented in Fig. 14.7d and is seen to be identical (again except for PS and NS nomenclature) with that given in Fig. 10.11a. Since the lumped path delay element Δt is fictitious, it may be removed, resulting in the familiar "cross-coupled NAND gate" circuit shown in Fig. 14.7e. Recall that in Fig. 10.18a this latter circuit was analyzed as to its mixed-rail output response. But more information remains on this deceptively simple machine, as discussed in the following paragraph.

If the inputs $S(L)$ and $R(L)$ should undergo a simultaneous $1(L) \to 0(L)$ change, the basic cell may become metastable and either "hang up" in a state that is neither a set nor reset, or oscillate. This condition is illustrated in Fig. 14.8, which represents a logic (ideal) simulation of the basic cell. The oscillation occurs because the identical cross-coupled NAND gates drive each other in antiphase fashion to produce an oscillation of period $2\tau_p$, where τ_p is the propagation delay through a NAND gate. Under ideal conditions, oscillatory behavior of this type is predictable and indicative of a possible metastable condition.

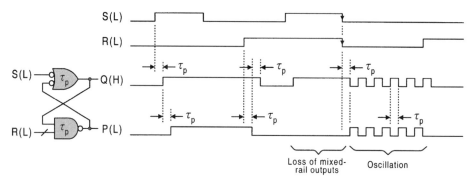

FIGURE 14.8
Timing diagram for the set-dominant basic cell showing loss of mixed-rail outputs for the $S, R = 1, 1$ condition, and the oscillatory behavior that results when S and R change $1 \to 0$ simultaneously.

However, an actual physical test of a basic cell will most likely not yield these same results, since metastability is a low-probability condition. But it can occur! In fact, in real-time tests of closely matched NAND gates, the basic cell is likely to show short-duration instability when subjected to simultaneous $1(L) \to 0(L)$ of the $S(L)$ and $R(L)$ inputs. It is because of loss of mixed-rail output character and the possibility of metastable behavior that the $S, R = 1, 1$ condition is normally avoided in using basic cells for FSM design. Remember that it is only for mixed-rail conditions that $P(L) = Q(L)$. Subsection 10.4.4 discusses the importance of mixed-rail character of the basic cell.

14.6.2 The Reset-Dominant Basic Cell

The design of the reset-dominant basic cell follows closely that of the set-dominant basic cell in the previous section. Shown in Fig. 14.9 are the state diagram, excitation table for the LPD model, the NS K-map and minimum cover, and the logic circuits with and without the fictitious LPD memory element. The NS function read in maxterm code from the NS K-map is given by

$$Y = \bar{R}(S + y) \tag{14.6}$$

and is seen to be identical with that of Eq. (10.7), except for the change in PS and NS notation.

Simultaneous $1(H) \to 0(H)$ changes of the inputs $S(H)$ and $R(H)$ to the reset-dominant basic cell can cause timing problems similar to those that can occur in the set-dominant basic cell. Shown in Fig. 14.10 is a timing diagram for the reset-dominant basic cell similar to that in Fig. 14.8. As indicated for ideal cross-coupled NOR gates, loss of mixed-rail output conditions can lead to oscillatory behavior under simultaneous $1 \to 0$ changes in the inputs. This again supports the need to avoid the $S, R = 1, 1$ condition when using basic cells for design purposes. Although real cross-coupled NOR gates may not oscillate as in Fig. 14.10, they may go logically unstable or may go metastable for a short period of time if simultaneous $1 \to 0$ input changes are permitted.

14.7 DESIGN OF THE RENDEZVOUS MODULES

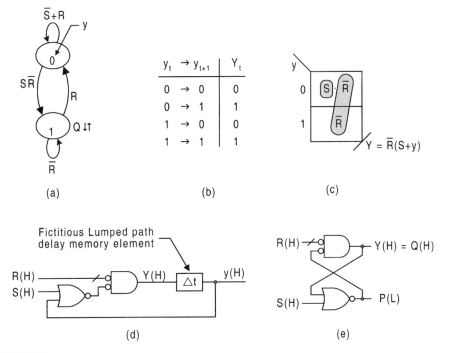

FIGURE 14.9
Design of the reset-dominant basic memory cell by using the LPD model. (a) State diagram. (b) Excitation table for the LPD model. (c) NS K-map and minimum cover. (d) Logic circuit showing fictitious LPD memory element. (e) Final logic circuit with fictitious memory element removed.

14.7 DESIGN OF THE RENDEZVOUS MODULES BY USING THE NESTED CELL MODEL

A *rendezvous module* (*RMOD*) is an asynchronous state machine whose output is issued active when all external inputs become active and is issued inactive when all external inputs

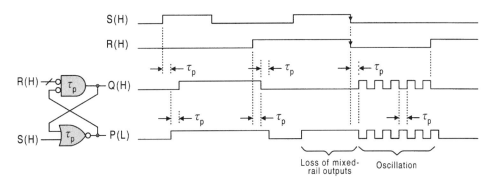

FIGURE 14.10
Timing diagram for the reset-dominant basic cell showing loss of mixed-rail outputs for the $S, R = 1, 1$ condition, and the oscillatory behavior that results when S and R change $1 \to 0$ simultaneously.

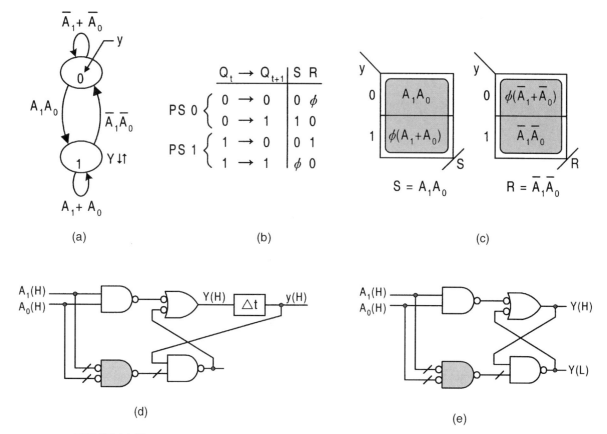

FIGURE 14.11
Design of the two-input rendezvous module (RMOD) by using the nested cell model. (a) State diagram. (b) Excitation table for the basic cell. (c) NS K-maps and minimum NS functions. (d), (e) Logic circuits with and without the fictitious LPD memory element for the basic cell.

become inactive. The RMOD, also known as a *majority gate* or *Muller C (concurrency) module*, is used in the design of other useful asynchronous circuits. The name *C-module* has also been used for this device. In effect, the RMOD acts like an AND gate in issuing an active output but acts like an OR gate in issuing an inactive output. Thus, the external inputs must all *rendezvous* with the same activation level before a change in the output is possible. Since the acronym RMOD is easy to remember and seems more descriptive of the device's function, it will be retained in this text.

Shown in Fig. 14.11a is the state diagram for a two-input RMOD. Clearly, the transition from the inactive state to the active state is possible only if both inputs become active — an AND function. Then, while in the active state, a transition to the inactive state is possible only if both inputs become inactive — an OR function. Applying the mapping algorithm in Section 10.6 to the state diagram together with the excitation table of Fig. 10.15c gives the NS K-maps and minimum covers for the *nested cell* design of the RMOD in Fig. 14.11c. From the K-maps there results the NS S and R functions

$$S = A_1 A_0 \quad \text{and} \quad R = \bar{A}_1 \bar{A}_0 \tag{14.7}$$

14.7 DESIGN OF THE RENDEZVOUS MODULES

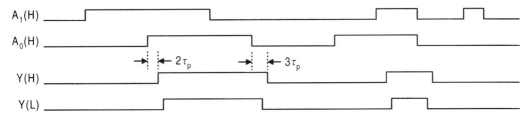

FIGURE 14.12
Timing diagram for the two-input RMOD showing input conditions for active and inactive mixed-rail outputs from the basic cell.

for external inputs A_1 and A_0. Implementation of Eq. (14.7) yields the NAND/OR logic circuits with and without the fictitious LPD memory element, Δt, shown in Figs. 14.11d and 14.11e. In the strict sense, the RMOD operates out of the fundamental mode since it is no longer required that one input "settle in" before another input changes. Notice that the OR gates are shaded in these figures.

The timing diagram for the two-input RMOD is given in Fig. 14.12. The input conditions are shown for active and inactive mixed-rail outputs. Also shown is a $2\tau_p$ delay following input active level conjunction and $3\tau_p$ following inactive level conjunction, where τ_p is the propagation delay through a NAND gate or OR gate, the two types of gates being treated the same in this case.

Multiple input RMODs can be designed. Presented in Fig. 14.13a is the state diagram for an n-input RMOD where the similarity between it and the two-input RMOD is evident. Thus, the NS functions for multiple inputs follow in similar fashion to those given in Eq. (14.7) and are

$$S = A_{n-1} \cdot \cdots \cdot A_1 A_0 \quad \text{and} \quad R = \bar{A}_{n-1} \cdot \cdots \cdot \bar{A}_1 \bar{A}, \tag{14.8}$$

which leads to the NOR/INV logic circuit given in Fig. 14.13b. Here, the multiple-input NOR gates (shown shaded) can be configured as in the specially designed CMOS NOR gate

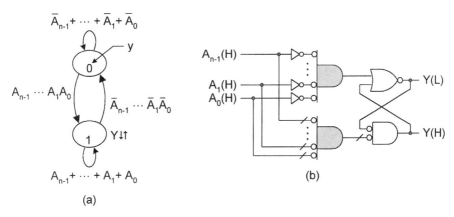

FIGURE 14.13
The multiple input RMOD. (a) State diagram. (b) Logic circuit based on the nested cell model.

698 CHAPTER 14 / ASYNCHRONOUS STATE MACHINE DESIGN AND ANALYSIS

of Fig. 8.46 so as to avoid possible fan-in problems. Recall that propagation delay increases with increasing number of gate inputs. Notice that a reset-dominant basic cell is used as the memory element in this case.

14.8 DESIGN OF THE RET D FLIP-FLOP BY USING THE LPD MODEL

The RET D flip-flop was previously designed in Subsection 10.7.2 by using the basic cell as the memory. In this section the same flip-flop will be designed by using the LPD model. Shown in Fig. 14.14 are the state diagrams for the resolver and set-dominant basic cell FSMs, both reproduced from Fig. 10.29 for the convenience of the reader. Note the change in the resolver state code assignment.

Since the set-dominant basic cell has previously been designed in Fig. 14.7, all that remains is to design the resolver for the D flip-flop by using the LPD model and then connect the two. In Fig. 14.15a is the resolver state diagram reproduced from Fig. 14.14(a), and in Figs. 14.15(b) and (c) are the NS and output K-maps with minimum covers indicated by shaded loops. The NS K-maps are constructed by combining the information in the state diagram with the excitation table in Fig. 14.3b via the mapping algorithm in Section 10.6. Reading the minimum cover in the K-maps yields the following results for the NS and output functions:

$$Y_1 = y_0 DCK + y_1 CK$$
$$= (y_0 D + y_1)CK$$
$$Y_0 = y_0 D + y_1 + \overline{CK} \qquad (14.9)$$
$$S = y_1$$
$$R = \bar{y}_0,$$

where factorization has been use so that the term $(y_0 D + y_1)$ appears in both NS functions, Y_1 and Y_0 for optimization purposes.

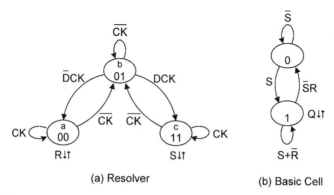

FIGURE 14.14
The RET D flip-flop as represented by state diagrams. (a) Resolver FSM input stage. (b) Set-dominant basic cell output stage.

14.8 DESIGN OF THE RET D FLIP-FLOP BY USING THE LPD MODEL

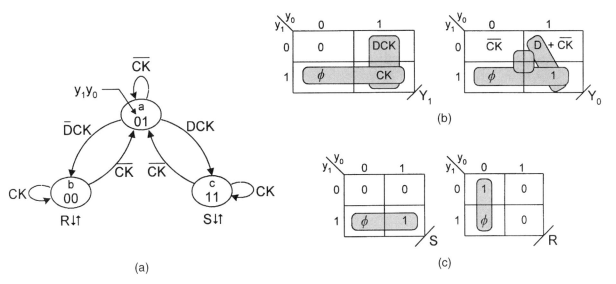

FIGURE 14.15
Design of the RET D flip-flop resolver by using the LPD model. (a) State diagram. (b) NS K-maps and minimum cover. (c) Output K-maps and minimum cover.

To connect the resolver to a set-dominant basic cell it is necessary that the resolver outputs be active low, that is, $S(L) = y_1(L)$ and $R(L) = \bar{y}_0(L) = y_0(H)$. Therefore, by applying the NS functions in Eqs. (14.9), there results the logic circuit for the RET D flip-flop in Fig. 14.16a, which is shown with the fictitious LPD memory elements indicated in their proper positions. Notice that only four NAND gates are necessary to implement the resolver circuit, whereas six NAND gates are required in the earlier design shown in Fig. 10.31. The logic circuit in Fig. 14.16b is the same as that in Fig. 14.16a but with the fictitious memory elements removed and showing the asynchronous PR and CL override connections (dashed lines). It is equivalent to the 74LS74 RET D flip-flop. An FET D flip-flop results by adding an inverter to the CK input.

The fictitious LPD memory elements are removed in Fig. 14.16b and asynchronous PR and CL override connections are added for completeness. Notice that all gates are now three-input NAND gates. An explanation of the override connections follows closely that for the transparent D latch in Fig. 10.51. A review of the discussion in Section 10.10 will help with an understanding of the reasoning behind these connections. The introduction of a CL override signal is straightforward. With $PR(L) = 0(L)$, a, $CL(L) = 1(L)$, forces gates 2 and 6 to issue a $0(L)$, which, in turn, forces gate 5 to issue a $0(H)$, thereby completing the mixed-rail clear output of the flip-flop. Remember that the asynchronous overrides PR and CL can never be active at the same time.

The introduction of a PR override signal is a little more involved but still easily explained. With $CL(L) = 0(L) = 1(H)$, introducing a $PR(L) = 1(L)$ forces gates 1 and 5 to issue a $1(H)$, which is now the input to gate 2. Then for gate 6 to issue a $1(L)$, as required for a mixed-rail set output, gate 3 must issue a $1(H)$. This is made possible because a $1(H)$ output from gate 3 results directly or indirectly from the CK input. Thus, if $CK(H) = 1(H) = 0(L)$, the input to gate 3 from gate 2 is $1(L)$, forcing a $1(H)$ from gate 3. Or, if $CK(H) = 0(H) = 1(L)$,

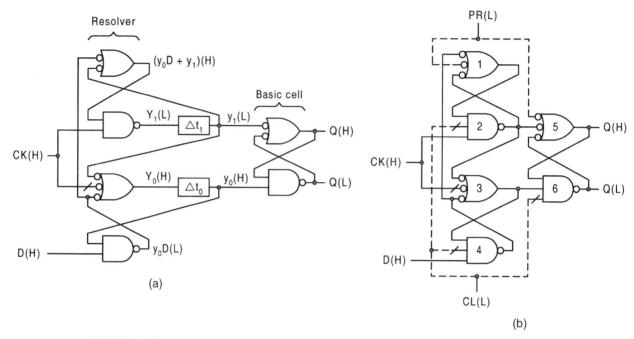

FIGURE 14.16
Implementation of the RET D flip-flop. (a) Implementation showing the intermediate functions, and indicating the proper position of the fictitious LPD memory elements. (b) The same circuit as in (a) but with the fictitious LPD memory elements removed and showing the asynchronous PR and CL override connections (dashed lines).

the output from gate 3 is again $1(H)$. Since in either case all three inputs to gate 6 are now $1(H)$, its output is $1(L)$, thereby completing the mixed-rail set output from the flip-flop. In short, it is the *CK* input that makes possible a mixed-rail set output from this RET D flip-flop.

14.9 DESIGN OF THE RET JK FLIP-FLOP BY FLIP-FLOP CONVERSION

The conversion of a D flip-flop to a JK flip-flop is illustrated in Fig. 10.42a by using Eq. (10.11) for the conversion. So, if the D flip-flop is of the design in Fig. 14.16b, nine gates would be required, three for the conversion logic and six for the resolver FSM. The conversion can be optimized by introducing Eq. (10.11) into the expression $y_0 D(L)$ (given in Fig. 14.16a) to obtain the following result:

$$y_0 D(L) = y_0(\bar{Q}J + Q\bar{K})(L)$$
$$= (y_0\bar{Q}J + y_0 Q\bar{K})(L) \qquad (14.10)$$

In fact, the resolver for the RET JK flip-flop can be constructed simply by introducing Eq. (10.11) into the D flip-flop resolver in Fig. 14.15a. If gate 4 in Fig. 14.16b is replaced

14.10 DETECTION AND ELIMINATION OF TIMING DEFECTS

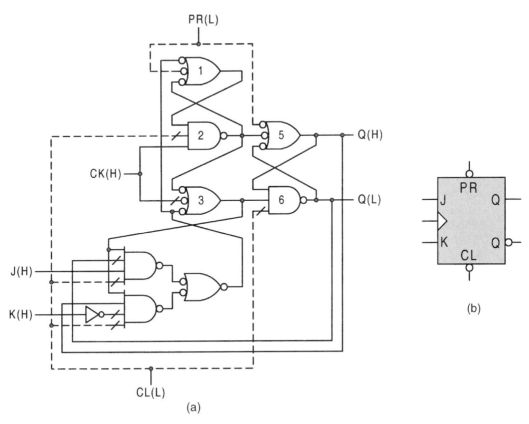

FIGURE 14.17
(a) Optimized conversion of the RET D flip-flop in Fig. 14.16b to an RET JK flip-flop. (b) Logic circuit symbol.

by the logic of Eq. (14.10), there results the optimized RET JK flip-flop and logic circuit symbol shown in Figs. 14.17a and 14.17b. This logic circuit is equivalent to the 74LS109 JK flip-flop but with the added *PR* override. Note that an FET JK flip-flop results simply by adding an inverter to the *CK* input.

14.10 DETECTION AND ELIMINATION OF TIMING DEFECTS IN ASYNCHRONOUS FSMs

The preceding sections are intended to be only an introduction to asynchronous FSM design. Much more must be known regarding the complexities of asynchronous sequential machines before meaningful designs are possible. The subject matter of this section is not only essential to the development of good design practices for asynchronous FSMs but should improve the reader's understanding of synchronous FSMs as well.

In Subsection 14.1.1, it was indicated that certain timing defects such as endless cycles (oscillations), critical races, static hazards, and essential hazards, can exist in asynchronous

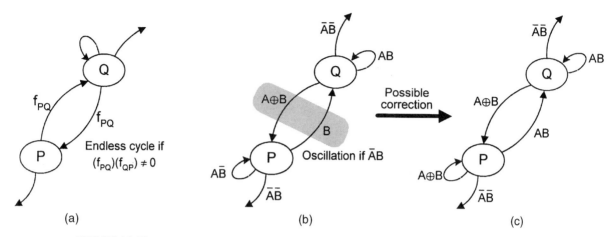

FIGURE 14.18
Endless cycles in asynchronous FSMs. (a) A segment of a state diagram used as a model for endless cycle analysis. (b) Example of an endless cycle. (c) Elimination of the endless cycle in (b).

FSMs and can cause the FSMs to malfunction. In Section 10.9, oscillations were shown to exist in some two-state flip-flops making them useless for most any application. Now it is necessary to learn how to detect and eliminate these timing defects, so that reliable asynchronous FSM designs can result.

14.10.1 Endless Cycles

The transition of an asynchronous FSM from one stable state to another stable state through one or more unstable states is called a *cycle*. When an asynchronous FSM enters a cycle for which there is no stable state, an *endless cycle* or *oscillation* is said to exist. Although cycles are necessary to the operation of some asynchronous FSMs, endless cycles must *always* be avoided.

Shown in Fig. 14.18a is a segment of a state diagram for which the branching conditions are f_{PQ} and f_{QP} between two states P and Q. The condition under which an endless cycle can exist is expressed by

$$f_{PQ} \cdot f_{QP} \neq 0, \tag{14.11}$$

meaning that any residue of this Boolean product $f_{PQ} \cdot f_{QP}$ is the branching condition for which an endless cycle exists. A typical example is presented in Fig. 14.18b. Here, an endless cycle is caused to occur under the branching condition $(A \oplus B) \cdot B = \bar{A}B$. If the algorithm for this fictitious FSM permits, the endless cycle can be eliminated by making the appropriate changes in the branching conditions associated with state P as indicated in Fig. 14.18c. Of course, if the branching condition $\bar{A}B$ can never exist, no correction of this state diagram segment is necessary. Endless cycles, as in Fig. 14.18, need not be limited to two states. Although less likely, multiple-state configurations can also support endless cycles. For example, suppose an asynchronous FSM exists having a sequence of

14.10 DETECTION AND ELIMINATION OF TIMING DEFECTS

interconnected states P, Q, R, S, P. In this case an endless cycle exists if

$$f_{PQ} \cdot f_{QR} \cdot f_{RS} \cdot f_{SP} \neq 0. \tag{14.12}$$

14.10.2 Races and Critical Races

The set of alternative cycle paths that lead to the same state is called a race. Recall that this subject was discussed in Section 11.2 relative to output race glitches (ORGs) in synchronous FSMs. In an asynchronous FSM a race results when the FSM undergoes a transition to a next state that differs from the present state by two or more bits. There are $n!$ race paths for a race condition involving the change of n state variables. Since no two feedback variables can change precisely at the same time, one variable will always change before another, even though the time span between the two events may be very small. Thus, the alternative race path taken by the FSM will depend on which feedback variable changes first to meet the stability criterion of Eq. (14.3), and this is not usually predictable.

The generalized state diagram segment in Fig. 14.19a serves as a model for detection of race and critical race conditions associated with the transition from state P to state Q under branching condition f_{PQ}. The noncritical race conditions, given in Fig. 14.19b, indicate that a proper transition from the *origin state* P to the *destination state* Q requires the following conditions: that input condition I_{PQ} be contained in the branching condition f_{PQ}, that I_{PQ} be contained in either f_Q or f_{QX}, and that a valid branching path be available from the race state R and S to state Q. This last condition requires that I_{PQ} be contained in both branching conditions f_{RQ} and f_{SQ}. Note that the symbol $\subseteq$ is standard algebraic notation for "is contained in" and a slash through it signifies its negation.

If, on the other hand, I_{PQ} is contained in either f_R or f_S, a *critical race* exists as indicated in Fig. 14.19c. Now the FSM can be stuck in either state R or S in attempting

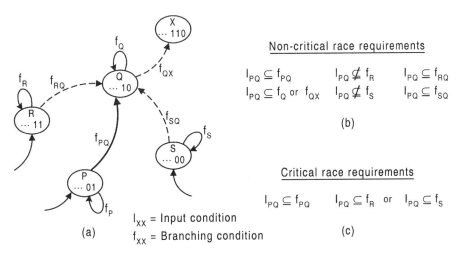

FIGURE 14.19
Races and critical races in asynchronous FSMs. (a) Generalized state diagram segment used as a model for detection of races and critical races. (b) Requirements for noncritical races. (c) Requirements for critical races.

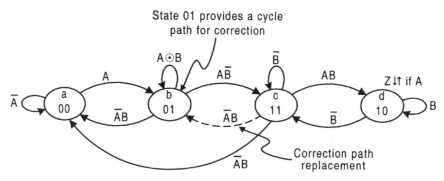

FIGURE 14.20
State diagram of an FSM showing a critical race for the 11 → 00 transition and its elimination by using a correction path.

the $P \rightarrow Q$ transition. Basically, the requirement for a critical race is that the holding condition of any race state contain the input condition for transition from the origin state to the destination state. Where only two state variable changes are involved in a given state-to-state transition, detection of a critical race is easily accomplished by a cursory inspection of the state diagram. Thus, the following procedure should be followed without exception:

- Look for race conditions in the state diagram. If the holding condition for any race state contains I_{PQ}, a critical race exists.
- Make certain that *valid* branching paths exist between the race states and the destination state.
- Eliminate the critical race by any one of several means discussed later in this subsection. Critical races must never be permitted to exist in any asynchronous FSM designed to operate in the fundamental mode.

As an example, consider the state diagram for an FSM in Fig. 14.20. Shown is a critical race during the 11 → 00 transition under branching condition $\bar{A}B$. Thus, if the race path is via state 10 during 11 → 00 transition, the FSM will reside stably and improperly in state 10 under holding condition B. Thus, $I_{ca} \subseteq f_d$ which is the requirement for a critical race according to Fig. 14.19c. On the other hand, if the race path is via the 01 state, the FSM will cycle correctly to state 00. The problem is, of course, that it cannot be predicted by which race path the FSM will transit. The critical race is eliminated by replacing the 11 → 00 branching path with the 11 → 01 path indicated by the dashed arrow, but under the same branching condition $\bar{A}B$. Thus, the basic algorithm has not been altered in making this correction. State 01 now provides an cycle path from state 11 → 01 → 00 under the same branching condition $\bar{A}B$.

Methods for Eliminating Critical Races The methods for eliminating critical races are straightforward and similar to those discussed in Subsection 11.2.2 relative to ORGs in synchronous FSMs. The causal race condition can be eliminated by one of the following actions given in descending order of importance or desirability:

14.10 DETECTION AND ELIMINATION OF TIMING DEFECTS   705

1. Altering the branching path without changing the basic algorithm of the FSM.
2. Changing the state code assignment to remove all race conditions or move the race condition elsewhere in the state diagram without creating any new timing defects.
3. Adding a buffer (fly) state to remove the race condition without violating critical timing constraints.
4. Adding additional state variables.

The action of removing the critical race by moving the causal race condition elsewhere in the state diagram is complicated by the possible formation of ORGs. Not only must the change be scrutinized as to the formation of another critical race, but the formation of static hazards and ORGs (not originally present) must be considered. The issue of static hazards in the NS-forming logic will be considered next. Obviously, the safest course of action in removing a critical race is to change the state code assignment so as to eliminate all race conditions in the FSM. Doing so, eliminates all race-related timing defects automatically.

14.10.3 Static Hazards in the NS and Output Functions

Before proceeding with this subsection, the reader should review Sections 9.2 and 11.3 dealing with static hazards (s-hazards) in combinational circuits and in the output of synchronous FSMs, respectively. In Section 11.3, the treatment of s-hazards in the NS-forming logic was not an issue since all such timing defects are filtered out by the memory flip-flops. Of course, asynchronous FSMs have no such filtering mechanism and are therefore subject to the problems that hazards can create in the NS-forming logic. Whereas s-hazards in the output-forming logic of asynchronous FSMs cannot cause the parent FSM to malfunction, s-hazards in the NS-forming logic can and do cause FSMs to malfunction. This may be viewed as yet another complicating timing defect that distinguishes the asynchronous FSM from its cousin, the synchronous state machine.

Static hazards in *both* the NS and output forming logic of asynchronous FSMs fall into two general categories: *externally initiated* and *internally initiated* static hazards, as illustrated in Fig. 11.8 for s-hazards in the output logic of synchronous FSMs. In fact, there is little difference in the methods used for detection and elimination of s-hazards in asynchronous and those used for s-hazards in the output functions of synchronous FSMs. It is important for the reader to remember the following:

> *Any suspect hazardous transition found by analyzing the NS and output functions of an asynchronous FSM must be verified by inspection of the state diagram* — this is standard operating procedure for s-hazard analysis in such state machines.

Since s-hazards in the output-forming logic cannot cause malfunction of the asynchronous FSM itself, attention in this subsection will be devoted to these timing defects in the NS forming logic. The analysis of static hazards in the output logic of asynchronous FSMs follows closely developments in Section 9.2 and in Subsection 11.3.1.

As the first and simple example, the transparent D latch, discussed in Subsection 10.7.1, will be designed by using the LPD model and then analyzed for an s-hazard timing problem. Shown in Fig. 14.21a is the state diagram for the D latch reproduced from Fig. 10.24a. By using the mapping algorithm in Section 10.6 to combine the excitation table for the

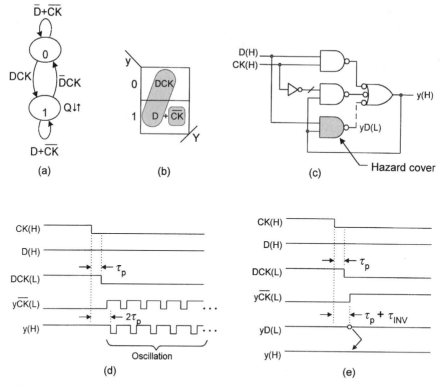

FIGURE 14.21
Example of an externally initiated s-hazard in the RET D latch. (a) State diagram. (b) NS K-map and minimum cover based on the LPD model. (c) Logic circuit derived from Eq. (14.13). (d) Timing diagram for the D latch without hazard cover showing oscillation effect of s-hazard. (e) Timing diagram with hazard cover showing elimination of the hazard.

LPD model with the state diagram, the NS K-map and minimum cover result as shown in Fig. 14.21b. The NS function is easily read to be

$$Y = DCK + y\overline{CK} + \underbrace{yD}_{\text{Hazard cover}}, \qquad (14.13)$$

which includes hazard cover, the ANDed residue of the coupled terms. The notation in Eq. (14.13) is intended to indicate that a static 1-hazard occurs on a $1 \rightarrow 0$ change in CK in state 1 when input D is active — hence, an *externally initiated* s-hazard. The logic circuit for the LPD design of the D latch is given in Fig. 14.21c, where the s-hazard cover yD is indicated by the shaded NAND gate. This circuit should be compared with that in Fig. 10.25b, where the basic cell is used as the memory—the nested cell design.

From Eq. (14.13), the timing diagrams are constructed without and with hazard cover as shown in Figs. 14.21d and 14.21e, respectively. Here, τ_p is the path delay for any two- or

14.10 DETECTION AND ELIMINATION OF TIMING DEFECTS   707

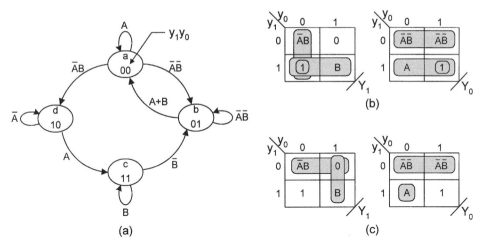

FIGURE 14.22
Hazard analysis of a four-state FSM. (a) State diagram. (b) SOP NS K-maps showing minimum cover. (c) POS NS K-maps and minimum cover.

three-input NAND gate and $\tau_{INV} = \frac{3}{5}\tau_p$ is the path delay for an inverter. The oscillation, shown in Fig. 14.21d, occurs as a result of the s-hazard formation. When the s-hazard is removed by adding hazard cover the D latch functions normally as indicated in Fig. 14.21e. It is true that a real-time test of this hazardous transition in the D latch may not show logic instability in the absence of hazard cover. That is, the asymmetric path delay imposed by an inverter may not be sufficient to cause the formation of the s-hazard. However, a proper design of this FSM would make no such presupposition and would include hazard cover. Adding the hazard cover yD means that the FSM cannot malfunction due to an s-hazard even if the inverter creates an enormous delay. In fact, with hazard cover, an s-hazard cannot be formed as a result of an asymmetric delay of any magnitude on either path of CK to the output ORing stage.

As a second and more complex example, consider the state diagram for the four-state FSM in Fig. 14.22a. The NS K-maps and minimum cover are given in Figs. 14.22b and 14.22c for SOP and POS logic, respectively. Remember, it is the mapping algorithm of Section 10.6 that is used to bring together the information in the state diagram with that of the LPD excitation table to construct the NS K-maps. The SOP NS-forming logic is read from the minimum cover in Fig. 14.22b to give the following results:

$$Y_1 = \bar{y}_0 \bar{A} B + y_1 B + y_1 \bar{y}_0, \quad Y_0 = \bar{y}_1 \bar{A}\bar{B} + y_1 A + y_1 y_0 + \underbrace{y_0 \bar{A}\bar{B}}_{\substack{\text{Hazard} \\ \text{cover}}} \quad (14.14)$$

$$\underset{01}{\uparrow} \qquad \underset{\bar{A}\bar{B}}{} \qquad \underset{11}{\uparrow}$$

Equations (14.14) also includes the hazard analysis following the procedure established earlier in Section 9.2 for combinational logic circuits. From these results, it is clear that an *internally initiated* static 1-hazard (an SOP hazard) may exist in the Y_0 function, that the coupled terms are $y_1 y_0$ and $\bar{y}_1 \bar{A}\bar{B}$, and that the hazardous transition is from state 11 to state

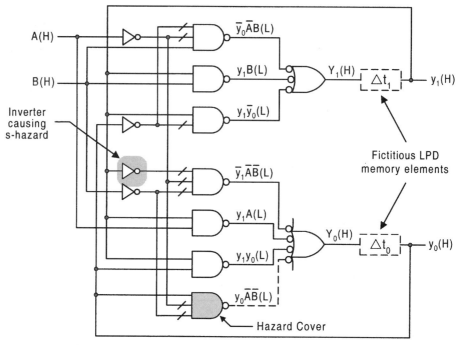

FIGURE 14.23
LPD logic circuit diagram derived from Eqs. (14.14) and the state diagram in Fig. 14.22

01 (a change of $1 \rightarrow 0$ in y_1) under constant-input conditions $\bar{A}\bar{B}$. An inspection of the state diagram indicates that this can occur, thereby validating the existence of the s-hazard. As indicated in Eqs. (14.14), the s-hazard is eliminated by adding the hazard cover $y_0\bar{A}\bar{B}$. The removal of the hazard is verified in Fig. 14.24b by the presence of the hazard cover. As is always true, the removal of the s-hazard by adding hazard cover eliminates any possibility that this hazard will form no matter how large the delay is through the inverter.

The LPD logic circuit is constructed from the NS functions in Eqs. (14.14) and is presented in Fig. 14.23. The LPD memory elements, indicated in dashed boxes, are included only as a reminder that a fictitious memory stage exists. Hereafter, these fictitious LPD memory elements will be excluded in, but implied by, the logic circuit. The shaded gate in Fig. 14.23 is the hazard cover that eliminates the s-hazard indicated in Eqs. (14.14).

Timing diagram verification of the existence of the s-hazard indicated in Eqs. (14.14), is given in Fig. 14.24. Here, attention is focused on the $11 \rightarrow 01$ transition (see Fig. 14.22a). To simplify the timing diagram, certain terms have been omitted purposely because they are logic 0 for this transition and, hence, do not contribute to the hazard analysis. Thus, according to Eqs. (14.14), $y_0 = 1$, $A = 0$, and $B(1 \rightarrow 0)$. Therefore, it follows that $\bar{y}_0\bar{A}B = y_1\bar{y}_0 = y_0 A = 0$ and need not be included in the timing analysis. These results lead to the following simplified NS functions: $Y_1 = y_1 B$ and $Y_0 = \bar{y}_1\bar{B} + y_1 y_0$ when no hazard cover is added. Notice in Fig. 14.24 that the delays are given in terms of τ_p and τ_{INV}, the path delay of a NAND gate (two or three input) and that of an inverter, respectively. This is done to help the reader trace through the sequence of steps leading to the formation of the s-hazard. For

14.10 DETECTION AND ELIMINATION OF TIMING DEFECTS   709

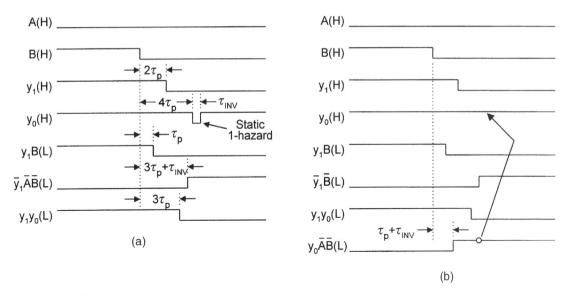

FIGURE 14.24
Timing diagrams for the FSM in Fig. 14.23, showing (a) formation of the static 1-hazard in y_0 and (b) its elimination by adding hazard cover (see arrow).

this timing analysis, as in the previous example, the relative delay values are expressed by $\tau_{INV} = \frac{3}{5}\tau_p$.

The formation of the static 1-hazard, shown in Fig. 14.24a, occurs as a result of the asymmetric path delay imposed by the highlighted inverter shown in Fig. 14.23. Thus, there are two alternative paths of the coupled (feedback) variable y_1 to output y_0: one through gate $y_1 y_0$ and the other through gate $\bar{y}_1 \bar{A} \bar{B}$ via the highlighted inverter. The reader can follow the sequence of events that lead up to this s-hazard formation shown in Fig. 14.24a by noting that the term $y_1 B$ is the first to change after one NAND gate path delay following the change in input B. This is followed by a change in the state variable y_1 after an additional NAND gate path delay. The sequence of events continues as indicated in Fig. 14.24a until the static 1-hazard is formed after four NAND gate path delays.

The s-hazard in Fig. 14.24 is eliminated by applying the SOP form of the consensus law, given in Eqs. (3.14), to the coupled terms $\bar{y}_1 \bar{A} \bar{B}$ and $y_1 y_0$. When this is done the result is the hazard cover term $y_0 \bar{A} \bar{B}$, which eliminates the s-hazard after a delay of one gate delay plus an inverter delay following the change in B well in advance of the hazard, as indicated in Fig. 14.24b. This hazard is eliminated regardless of the magnitude of the asymmetric delay on either of the alternative paths of y_1 to y_0. Notice that the waveforms in Fig. 14.24b are identical to those of Fig. 14.24a except for the presence of hazard cover and the absence of the s-hazard in y_0.

The static 1-hazard shown in Fig. 14.24a is nondisruptive in the sense that the FSM finally resides in the proper 01 state immediately following a brief improper transition to state 00. However, there is a short delay in achieving stability in the 01 state, and this could be highly disruptive to any next-stage FSM to which y_0 is attached if the hazard is sufficiently well developed. Also, if it is required that state 00 issue an output signal, an ORG will result that could be disruptive depending, of course, on how that output signal is used.

A static 0-hazard (POS hazard) also exists in the FSM of Fig. 14.22. From the K-maps in Fig. 14.22c, the following POS NS functions are read in maxterm code:

$$Y_1 = (y_1 + \bar{A})(y_1 + \bar{B})(\bar{y}_0 + B)(y_1 + \bar{y}_0)$$

$$Y_0 = (y_1 + \bar{A})(y_1 + \bar{B})\underbrace{(\bar{y}_1 + y_0 + A)}_{00 \quad \bar{A}B \quad 10} \cdot \underbrace{(y_0 + A + \bar{B})}_{\text{Hazard cover}}. \qquad (14.15)$$

From these NS expressions, it is clear that an internally initiated static 0-hazard exists in function Y_0 and that it occurs on a $00 \to 10$ due to a $0 \to 1$ change in the state variable y_1 under constant branching conditions $\overline{(A + \bar{B})} = \bar{A}B$. Remember that the coupled terms are read in maxterm code and that this requires the input conditions, as read from the coupled terms, to be complemented before a comparison can be made with the state diagram, which is a minterm-code based graphic. Applying the POS consensus law in Eqs. (3.14) means that the hazard cover is the ORed residue of the coupled terms $(y_1 + \bar{B})$ and $(\bar{y}_1 + y_0 + A)$ given by $(y_0 + A + \bar{B})$, as indicated in Eqs. (14.15). The hazard is eliminated by adding this hazard cover, as indicated by the arrow in Fig. 14.25b. The addition of hazard cover $(y_0 + A + \bar{B})$ ensures that this hazard can never form regardless of the size of the asymmetric delay associated with either alternative path of y_1 to the output y_0. This is true for the elimination of any s-hazard after adding hazard cover.

Unlike the static 1-hazard in Fig. 14.24a, the static 0-hazard in Fig. 14.25a is potentially disruptive to the FSM itself. Any time an s-hazard can cause an FSM to go logically unstable as in Fig. 14.25a, the potential is there for malfunction. Of course, it is understood that the s-hazard must develop to the extent that it is picked up by the NS-forming logic. Since

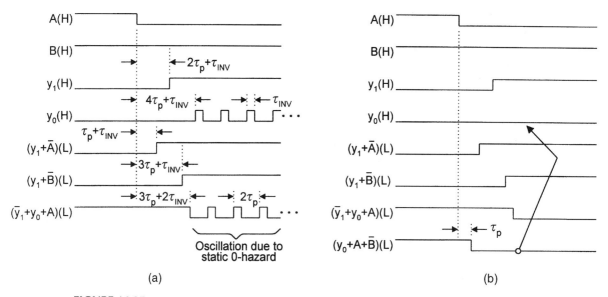

FIGURE 14.25
Timing diagrams for the POS NS functions in Eqs. (14.15) showing (a) formation of the static 0-hazard in y_0 and the resulting oscillation, and (b) elimination of the s-hazard by adding hazard cover (see arrow).

it is only the delay through an inverter that is the causal effect, there is the possibility that the FSM will function properly even without hazard cover. But since this cannot be assured, hazard cover must be added. Again, this should be considered as standard operating procedure in dealing with static hazards in the NS logic as well as the output logic.

14.10.4 Essential Hazards in Asynchronous FSMs

Elimination of all endless cycles, critical races and static hazards from an asynchronous FSM operated in the fundamental mode does not ensure proper operation of the FSM. Certain noncombinational hazards produced by explicitly located asymmetric path delays in gates and/or on leads are guaranteed to cause such FSMs to malfunction. These hazards, called *essential hazards* (*E-hazards*), are steady-state sequential hazards in the sense that they involve the change of two or more state variables in otherwise steady-state output signals. The term "essential" does not imply "needed" or "necessary," but rather, refers to the fundamental mode of FSM operation. Without exception, E-hazards cannot be eliminated by adding redundant cover as can s-hazards.

General Requirements for E-hazard Formation The general requirements that must be met before an E-hazard can form are as follows:

1. The asynchronous FSM must operate in the fundamental mode.
2. There must be at least two state (feedback) variables — hence, at least three states — and at least one external input, designated as the *initiator* input.
3. There must be at least two paths of propagation of the initiator to the *first invariant* state variable: one path directly to the first invariant and at least one other indirect path to the first invariant via the *second invariant* state variable. Both the initiator and second invariant must meet at a specific gate called the *race gate*.
4. An asymmetric path delay must be explicitly located in the direct path of the initiator to the first invariant state variable and must be at least of the minimum magnitude to cause the E-hazard to form.

The process of E-hazard formation involves a "critical" race (to the race gate) between the initiator and the second invariant state variable. If the race is won by the second invariant, an E-hazard is formed. An explicitly located path delay of sufficient duration will ensure that the race is won by the second invariant state variable and, consequently, cause the E-hazard to form.

The path delay requirements for the formation of a *first-order E-hazard* in a two-level NS logic system are given in Fig. 14.26. Here, two race gate (RG) types are identified. For the case of the first-level race gate in Fig. 14.26a, the path delay requirement for E-hazard formation is given by

$$\Delta t_E > \tau_1 + \tau_2, \tag{14.16}$$

and for the second-level race gate in Fig. 14.26b by

$$(\Delta t_E + \tau_1) > \tau_2 + \tau_3 + \tau_4. \tag{14.17}$$

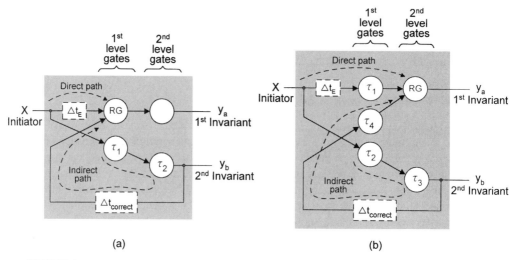

FIGURE 14.26

Illustrations of the path delay requirements for E-hazard formation in two-level logic showing causal delays Δt_E, initiator input X, first and second invariants, gate delays τ_i, race gates (RG), and correction delays to eliminate the E-hazard. (a) First-level race gate. (b) Second-level race gate.

In both Eqs. (14.16) and (14.17) the quantity Δt_E is the asymmetric path delay, shown in Fig. 14.26, that is required to cause the E-hazard to form (y_b wins the race); τ_i are the path delays associated with the gates (including any inverters) and leads. In these equations the correction delay $\Delta t_{correct}$ is assumed to be zero. If a counteracting delay $\Delta t_{correct}$ is added in the indicated feedback path of the 2nd invariant, then the requirements for eliminating the E-hazard are given by

$$\Delta t_E < (\tau_1 + \tau_2 + \Delta t_{correct}), \qquad (14.18)$$

for the first-level race gate, and

$$(\Delta t_E + \tau_1) < (\tau_2 + \tau_3 + \tau_4 + \Delta t_{correct}), \qquad (14.19)$$

for the second-level race gate. Thus, if Δt_E is of sufficient magnitude to cause an E-hazard to become active according to the requirements of Eqs. (14.16) and (14.17), the second invariant y_b wins the race and the FSM is guaranteed to malfunction. However, if a counteracting (correcting) delay is added in the feedback path of the 2nd y-variable invariant, the inequality is reversed, as in Eqs. (14.18) and (14.19). Under this condition, the initiator X wins the race and the E-hazard is eliminated.

The minimum requirements for E-hazard formation are summarized in Fig. 14.27. The state diagram segment, shown in Fig. 14.27a, specifies the first- and second-level race gate SOP terms that must be contained in the first invariant function Y_i before an E-hazard is possible. Notice that the first invariant is the second y-variable to change while the second invariant is the first to change. The minimum requirements for E-hazard formation are continued in Fig. 14.27b, where now another type of E-hazard is identified, the *d-trio*. The

14.10 DETECTION AND ELIMINATION OF TIMING DEFECTS   713

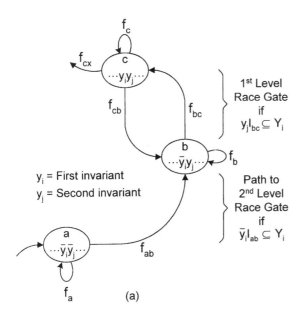

FIGURE 14.27
Minimum requirements for first-order E-hazard and d-trio formation in two-level SOP logic. (a) State diagram segment showing first- and second-level race gate requirements, only one of which will be met in the first-invariant function Y_i. (b) Minimum requirements for E-hazard and d-trio formation indicating assumed input conditions for I_{ab} and I_{bc}.

d-trio (delay-trio) is a special E-hazard that returns the FSM to the intended state but only following a second (error) transition to another state. Thus, the transition path for a d-trio is $a \to b \to c \to b$, while that for a E-hazard is $a \to b \to c$ or $a \to b \to c \to x$, where state x lies beyond state c in Fig. 14.27a. The latter E-hazard path is possible if the input conditions are such that $I_{ab} \subseteq f_{cx}$ in addition to those indicated in Fig. 14.27b. Clearly, the minimum requirements are the same for the E-hazard and d-trio formation, except the E-hazard does not return the FSM to the intended next state. Another important minimum requirement for E-hazard and d-trio formation is that the initiator x_i is permitted to have only one change in f_{ab} and f_{bc} while holding x_j and all other inputs constant.

To summarize, an E-hazard or d-trio can form iff an unintended asymmetric delay Δt_E of sufficient magnitude is explicitly located as shown in Fig. 14.26, and if the minimum requirements indicated in Fig. 14.27 are met. A cursory check of the state diagram is all that is necessary to show whether or not the minimum requirements for E-hazard (and d-trio) formation are met. If they are not met, these potential defects cannot form and no further analysis is necessary. If the minimum requirements are met, the second stage of the analysis is to determine the requirements for the indirect path — that is, the requirements to allow the second y-variable invariant to win the race at the race gate.

Only first-order E-hazards are considered in this text. The reason is that second and higher order E-hazards are far less likely to be activated than first-order E-hazards. A second-order E-hazard, for example, requires two successive invariants in the indirect path (IP), which greatly increases the minimum path delay requirement for activation of the E-hazard.

Indirect Path (IP) Requirements for E-hazard and D-trio Formation The first-order IP requirements are as follows:

1. The IP must *not* be inconsistent with the conditions of the initiating state a in Fig. 14.27a, including its state variables and all input conditions other than the initiator input.
2. The IP must contain the initiator as x_i or $\bar{x}_i$.
3. The IP must follow a path to the RG that is unobstructed. Thus, IP terms in the second invariant function Y_j must not be inconsistent with any input held constant.

With reference to Fig. 14.27a, the IP must not be inconsistent with $\cdots \bar{y}_i, \bar{y}_j \cdots, x_j$ and must contain x_i or $\bar{x}_i$ in Y_j.

A SIMPLE EXAMPLE. Consider the state diagram for the simple two-input FSM shown in Fig. 14.28a. Here, two paths are shown, one for an E-hazard and the other for a d-trio. The shaded states indicate the origin states for the potential defect in question. Thus, the E-hazard path is $c \to b \to a$ while that for the d-trio is $a \to d \to c \to d$ as indicated by the dashed arrows. Notice that there are no endless cycles or critical races present in this FSM.

So that the reader can follow the reasoning process involved in analyzing these potential defects, the NS functions, read from the K-maps in Fig. 14.28b, are provided in Eqs. (14.20) and (14.21) and are used for E-hazard and d-trio analysis, respectively. In these equations RG represents a race gate or a path to a race gate, and IP represents an indirect path term. The

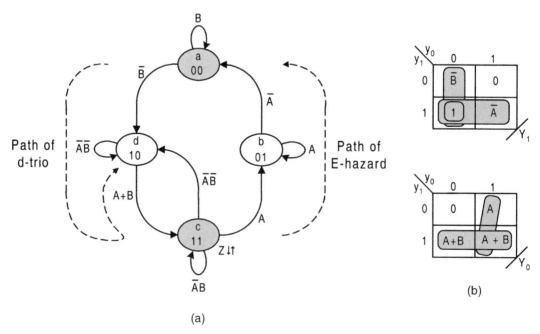

FIGURE 14.28
E-hazard and d-trio analysis for a simple FSM having two inputs and one unconditional output. (a) State diagram showing paths for an E-hazard and for a d-trio. (b) NS K-maps and minimum cover.

14.10 DETECTION AND ELIMINATION OF TIMING DEFECTS

subscripts E and D refer to the E-hazard and d-trio, respectively. Following the procedure given in Subsection 14.10.3, a brief inspection of these NS functions indicates that they are free of static hazards.

$$Y_1 = \bar{y}_0 \bar{B} + \underbrace{y_1 \bar{A}}_{IP_E} + y_1 y_0 \qquad Y_0 = \underbrace{y_0 A + y_1 A + y_1 B}_{\text{ORing RG}_E} \overset{\text{Path to RG}_E}{} \qquad (14.20)$$

$$Y_1 = \underbrace{\bar{y}_0 \bar{B}}_{IP_D} + y_1 \bar{A} + y_1 y_0 \qquad Y_0 = y_0 A + y_1 A + \underbrace{y_1 B}_{\text{ANDing RG}_D} \qquad (14.21)$$

Nearly complete information regarding E-hazard and d-trio formation can be gleaned from Fig. 14.28(a) together with the NS functions given by Eqs. (14.20) and (14.21). The logic circuit, constructed from these NS functions, is presented in Fig. 14.29. Here, ΔT_E and ΔT_D are the unwanted explicitly located path delays (occuring separately) that will cause the formation of the E-hazard and d-trio according to Eqs. (14.22) and (14.23). The race gates, RG_E and RG_D, are shown shaded.

E-hazard Analysis With reference to the state diagram in Fig. 14.28a and to Eqs. (14.20), the following constitute the *minimum* requirements for E-hazard formation as set forth in

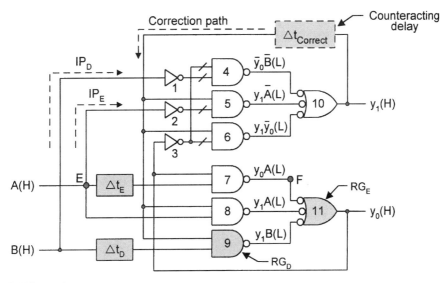

FIGURE 14.29
Logic circuit constructed from the NS functions of Eqs. (14.20) or (14.21) showing causal delays required to form the E-hazard or d-trio, their respective race gates, and the position of the counteracting delay required to eliminate the E-hazard or d-trio.

Fig. 14.27:

$$\bar{A}B \to AB, \quad AB = I_{cb} \subseteq f_b = A, \quad \bar{A}B = I_{ba} \subseteq f_c = \bar{A}B, \quad I_{cb} \nsubseteq f_{ab}.$$

Only a single change in the initiator is indicated in $c \to b$ and $b \to a$ with B held constant. Here, state 11 (state c) is the origin state, A is the initiator input, the intended path is $11 \to 01$ ($c \to b$), the E-hazard (error) path is $11 \to 01 \to 00$, y_0 is the first invariant, and y_1 is the second invariant.

The remaining requirements for E-hazard formation in the FSM of Fig. 14.28a are obtained from Eqs. (14.20) and Fig. 14.29 in accordance with Fig. 14.27 and the indirect path (IP) requirements given previously.

1. A delay Δt_E placed on the initiator A input to the first invariant y_0 causes a critical race to the race gate RG_E between the initiator A and the second invariant y_1. If Δt_E exceeds the minimum path delay requirements, y_1 wins the race and the E-hazard is formed. If Δt_E is not of sufficient magnitude, the initiator input A wins the race and no E-hazard will form.

2. The path to the ORing (2nd level) race gate (RG_E) is indicated by the term $y_0 A$ in Y_0, as shown in Eqs. (14.20). No ANDing RG is possible according to Figs. 14.27a and 14.28a.

3. The indirect path (IP_E) must not be inconsistent with B, y_1, y_0 in Y_1 and must contain y_1 or $\bar{y}_1$ and A or $\bar{A}$ in Y_0. Therefore, the IP is via the term $y_1 \bar{A}$ in Y_1, and either $y_1 A$ or $y_1 B$ in Y_0.

4. Based on the foregoing and with reference to Fig. 14.29, the theoretical minimum path delay requirements for E-hazard formation is given by the inequalities

$$(\Delta t_E + \tau_7) > (\tau_2 + \tau_5 + \tau_{10} + \tau_8) = (\tau_{INV} + 3\tau_p)$$

or

$$\Delta t_E > (\tau_{INV} + 2\tau_p), \quad (14.22)$$

where τ_p is the path delay through a gate (e.g., a two- or three-input NAND gate), and $\tau_{INV} = \tau_2$ is the path delay through an inverter. Thus, Eq. (14.22) does not take into account the gate path delay dependence on fan-in.

D-trio Analysis With reference to the FSM in Fig. 14.28a and Eqs. (14.21), the following constitute the *minimum* requirements for d-trio formation as set forth in Fig. 14.27:

$$\bar{A}B \to \bar{A}\bar{B}, \quad \bar{A}\bar{B} = I_{ad} \subseteq f_d = \bar{A}\bar{B}, \quad \bar{A}B = I_{dc} \subseteq f_a = B, \quad I_{ad} \subseteq f_{cd} = \bar{A}\bar{B},$$

and only a single change of the initiator is indicted in $a \to b$ and $b \to c$ with $\bar{A}$ constant. In this case state 00 (state a) is the origin state, B is the initiator, the intended path is $00 \to 10$ ($a \to b$), the d-trio (error) path is $00 \to 10 \to 11 \to 10$, y_0 is the first invariant, and y_1 is the second invariant.

The remaining requirements for d-trio formation in the FSM of Fig. 14.28a are obtained from Eqs. (14.21), from Figs. 14.27 and 14.29, and from the IP path requirements given previously.

14.10 DETECTION AND ELIMINATION OF TIMING DEFECTS

1. A delay Δt_D placed on the initiator B input to the first invariant y_0 causes a critical race to the race gate between the initiator B and the second invariant y_1. If Δt_D exceeds the minimum path delay requirements, y_1 wins the race and the d-trio will be formed. If Δt_D is not of sufficient magnitude, the initiator input B wins the race and no d-trio will form. Should the d-trio be formed, an output Z will be issued for a duration equal to the difference between Δt_D and the minimum path delay requirements for d-trio formation.
2. An ANDing race (RG_D) is indicated by the term $y_1 B$ in y_0, as indicated in Eqs. (14.21). No ORing RG is possible according to Figs. 14.27a and 14.28a.
3. The indirect path (IP_D) must not be inconsistent with $\bar{A}$, $\bar{y}_1$, $\bar{y}_0$ in Y_1 and must contain either B or $\bar{B}$ in Y_1. Therefore, the IP_D is by way of the term $\bar{y}_0 \bar{B}$ in Y_1.
4. Based on this information and with reference to Fig. 14.29, the theoretical minimum path delay requirement for d-trio formation is given by the inequality

$$(\Delta t_D) > (\tau_1 + \tau_4 + \tau_{10}) = (\tau_{INV} + 2\tau_p), \tag{14.23}$$

where, as previously, τ_p is the path delay through a gate (e.g., a two- or three-input NAND gate), and $\tau_{INV} = \tau$ is the path delay through an inverter. Accordingly, Eq. (14.23) does not take into account the gate path delay dependence on fan-in.

The corrective action required to prevent the E-hazard or d-trio from forming, is indicated in Fig. 14.29 by a counteracting delay in the feedback path of the second-invariant state variable y_1. Thus, the theoretical corrective action required to eliminate these defects is given by the inequalities

$$\Delta t_E < (\tau_{INV} + 2\tau_p + \Delta t_{Correct}) \tag{14.24}$$

and

$$\Delta t_D < (\tau_{INV} + 2\tau_p + \Delta t_{Correct}). \tag{14.25}$$

Notice that if $\Delta t_{Correct} = \Delta t_E$, the inequalities of Eqs. (14.24) and (14.25) are easily satisfied. Also, observe that the delay Δt_E is effective in causing the E-hazard to form at any point along the path E to F (see the large nodes in Fig. 14.29), including the intervening two-input NAND gate. This is characteristic of any ORing race gate, a feature not shared with the ANDing race gate.

Further verification of the results presented so far is provided by the timing diagrams in Fig. 14.30. Presented in Fig. 14.30a is the result of E-hazard formation indicating an error transition $11 \rightarrow 01 \rightarrow 00$ due to a delay $\Delta t_E = 5\tau_p$ positioned anywhere along the path between large nodes E and F (including the intervening NAND gate) shown in Fig. 14.29. Recall that under the input change $\bar{A} B \rightarrow A B$ from state 11 the correct transition should be $11 \rightarrow 01$, but because of the unwanted path delay Δt_E an error transition is forced to occur. A path delay of $\Delta t_E = 5\tau_p$ clearly exceeds the minimum path delay requirements for E-hazard formation given by Eq. (14.22).

The formation of the d-trio is illustrated by the timing diagram in Fig. 14.30b. Here, a delay of $\Delta t_D = 5\tau_p$, positioned as shown in Fig. 14.29, causes an error pulse in state variable y_0 and output Z of duration $5\tau_p - (\tau_{INV} + 2\tau_p) = 3\tau_p - \tau_{INV}$. The d-trio has the appearance

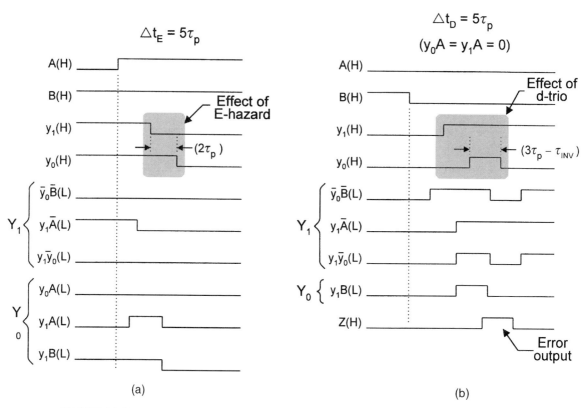

FIGURE 14.30
Timing diagrams derived from simulator tracings showing error transitions caused by delays of $5\tau_p$ located at the positions Δt_E and Δt_D indicated in Fig. 14.29. (a) Timing diagram showing development of the E-hazard consistent with Eq. (14.22). (b) Timing diagram showing development of the d-trio consistent with Eq. (14.23).

of a static 0-hazard but with a pulse width proportional to the difference between the delay Δt_D and the minimum path delay requirements for d-trio formation given by Eq. (14.23). The proper transition from state 00 under input change $\bar{A}B \rightarrow \bar{A}\bar{B}$ should be $00 \rightarrow 10$, but because of the explicitly located delay Δt_D the d-trio transition $00 \rightarrow 10 \rightarrow 11 \rightarrow 10$ is forced to occur.

The E-hazard and d-trio featured in this example can be removed simply by adding a counteracting delay of sufficient magnitude in the feedback path of the second invariant state variable as indicated in Fig. 14.29. When this is done the requirements of Eqs. (14.24) and (14.25) are met. A safe magnitude for the counteracting delay is usually the magnitude of the minimum theoretical delay causing the E-hazard or d-trio to form. If the latter magnitude is not known, then a delay of $2\tau_p$ will usually suffice.

14.10.5 Perspective on Static Hazards and E-hazards in Asynchronous FSMs

Static hazard analyses should always be run and corrective active taken (if needed) prior to carrying out an E-hazard analysis. The reason for this is simple: Static hazard cover may,

14.11 INITIALIZATION AND RESET OF ASYNCHRONOUS FSMs   719

in some cases, actually provide an indirect path for E-hazard formation, thereby making E-hazard formation possible whereas otherwise it would not be. E-hazards are potential defects in the sense that the FSM logic is not "born" with these defects as can be the case for static hazards. E-hazards require explicitly located path delays of magnitude exceeding the minimum requirements before they can form. However, an active E-hazard is guaranteed to cause malfunction of the FSM, whereas active static hazards in the NS logic may or may not be disruptive to the operation of the FSM. Before a static hazard in the NS-forming logic can cause malfunction, it must be "strong" enough to cross the switching threshold, but even then the hazard may not cause malfunction. However, the designer must assume that the static hazard has the potential to cause malfunction of the FSM and must add hazard cover. In fact, some designers find it worthwhile to take the "shotgun approach," which means adding hazard cover to any pair of coupled terms appearing in the NS logic functions.

The d-trio is a special case of an E-hazard that causes the FSM to undergo an error transition before residing in the intended state. Sometimes this has the effect of only delaying the transition from the origin state to the intended state. However, at other times an output can be activated erroneously as in Fig. 14.30b. Such an erroneous output can be just as disruptive as an active E-hazard would be. For this reason, active E-hazards and d-trios are considered equally capable of causing malfunction of an asynchronous FSM and corrective action should be taken where warranted. This action usually amounts to nothing more than adding a delay in the feedback path of the second invariant state variable, a delay equal to about the minimum path delay requirement for E-hazard formation.

Corrective action to prevent the formation of E-hazards can take the form of carefully choosing routing paths in a circuit layout so as to avoid excessive path delays at certain critical locations in the circuit. Thus, an E-hazard analysis is of value in this regard, since knowledge of the position and magnitude of a causal delay can offer the designer the information needed to make an engineering judgment as to possible corrective action. Again, it must be remembered that a strongly active E-hazard is guaranteed to cause malfunction of the FSM. If the minimum path delay requirements are just barely exceeded, a weakly active E-hazard may cause the FSM to become logically unstable or may even permit the FSM to operate properly. But the designer should not take a chance except for the case where a large minimum path delay requirement is indicated. The E-hazard and d-trio effects given in Fig. 14.30 are those of a strongly active E-hazard, since the delay of $5\tau_p$ that is used exceeds the minimum requirements by about a factor of 2. A causal delay that just exceeds the minimum path delay requirements for an E-hazard, as indicated in Eqs. (14.22), will cause the FSM to oscillate when simulated. The same reduction in the causal delay for the d-trio only narrows the error pulse. Real circuits, on the other hand, may require causal dalays considerably in excess of the theoretical minimum.

14.11 INITIALIZATION AND RESET OF ASYNCHRONOUS FSMs

Like synchronous FSMs, most asynchronous FSMs must also be initialized or reset. But unlike synchronous FSMs that can be initialized or reset via sanity circuit inputs to PR and CL overrides of the flip-flops, asynchronous FSMs must be initialized or reset by using sanity circuit inputs to the gates of which the NS logic is configured. The sanity circuit shown in Fig. 11.28 presented in connection with synchronous FSMs applies here also. Figure 14.31 illustrates the means by which an asynchronous FSM must be initialized or reset. Figure 14.31a applies to an active low output from the sanity circuit while Fig. 14.31b

720 CHAPTER 14 / ASYNCHRONOUS STATE MACHINE DESIGN AND ANALYSIS

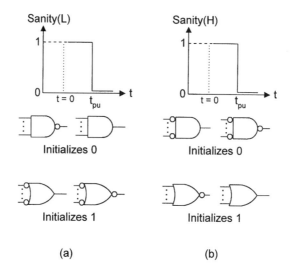

FIGURE 14.31
Gate requirements for initializing a logic 0 or a logic 1. (a) Active low output from the sanity circuit. (b) Active high output from the sanity circuit.

applies to an active high output from the sanity circuit. Generally, an ANDing operation is required to initialize a logic 0, and an ORing operation is required to initialize a logic 1. For example, a $Sanity(L) = 1(L) = 0(H)$ initializes a logic 0 if it is the input to an AND symbol without input active low indicator bubbles, but initializes a logic 1 if it is the input to an OR symbol with input active low indicator bubbles, as in Fig. 14.31a. Conversely, a $Sanity(H) = 1(H) = 0(L)$ initializes a logic 0 if it is the input to an AND symbol with input bubbles, but initializes a logic 1 if it is the input to a OR symbol without the input bubbles, as in Fig. 14.31b. The time t_{pu} is the power-up point beyond which the system can be operated.

As an example of the initialization methods just discussed, consider the two generalized two-level SOP circuits shown in Fig. 14.32. In Fig. 14.32a the NAND circuit can be initialized a logic 0, whereas in Fig. 14.32b the NAND circuit can be initialized a logic 1, both with an input $Sanity(L) = 1(L) = 0(H)$ required for initialization. The difference is, of course, in the way the sanity input is introduced into the circuit. For the former case it is introduced in the input ANDing stage and in the latter case it is introduced into the output ORing stage. Note that if the NAND logic in Fig. 14.32b is replaced by AND/OR logic, the sanity input must be changed to $Sanity(H)$, but no change would be necessary in Fig. 14.32a for AND/OR logic. A dual arrangement results in the case of two-level NOR-based POS logic for which $Sanity(H)$ is the initializing input.

14.12 SINGLE-TRANSITION-TIME MACHINES AND THE ARRAY ALGEBRAIC APPROACH

This important section describes a class of asynchronous FSMs that undergo the fastest state-to-state transition times possible and that avoid all race-associated problems, namely critical races and ORGs. This class of fundamental mode FSMs is commonly called *single transition*

14.12 SINGLE-TRANSITION-TIME MACHINES   721

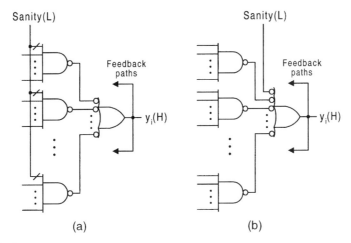

FIGURE 14.32
Initializing two-level NAND SOP logic with a sanity(L) input. (a) *Sanity*(L) = 1(L) = 0(H) used to initialize a logic 0. (b) *Sanity*(L) = 1(L) used to initialize a logic 1.

time (STT) state machines. The array algebraic approach used here is precisely the same as that used in Section 11.11 for synchronous FSM design by using D flip-flops. The reason why the same array algebraic approach can be used lies in the fact that the excitation table for the LPD model is exactly the same as that for D flip-flop designs. Now, however, state code assignments must be found that will eliminate critical races and ORGs and yet yield NS functions that represent the fastest transition times possible. As will be demonstrated in this and the following section, STT FSMs can be designed by using either the LPD model or the nested cell model, both models complying with the requirements of the fundamental mode.

Procedure for Obtaining an STT State Code Assignment

1. Construct a state table *free* of cycles and buffer states, which are strictly forbidden, and assign a state identifier to each state. The state table can be constructed very easily from the state diagram or ASM chart for the FSM. Note that violation of the sum rule can cause critical races.
2. Identify the state that is to be initialized and assume that it will be an all-zero state ($\cdots$ 000) or an all-one state ($\cdots$ 111). This is done to simplify the initialization process (see Section 14.11).
3. Partition the state transitions into groups or sets that eliminate critical races and ORGs. These partitions are called π (partial) *partitions*. The π-partitions result from an extension of the "into rule" (rule 1) used to obtain optimum state code assignments for D flip-flop designs of synchronous FSMs as discussed in Subsection 11.10.2. If present, the state identifier for the initialization state together with all other state identifiers associated with that initialization state must be positioned on the left side of the π-partitions. This is done so as to organize the π-partitions into a form that can be used to obtain a valid STT state code assignment by following the remaining steps of this procedure.

4. Collect the π-partitions into partitions that include *all* states identifiers. These partitions are called τ (total) *partitions*. If this is properly done, all τ-partitions will begin with the state identifier for the initialization state on the left side of the partition.
5. Find a minimum set of τ-partitions that "cover" all π-partitions. The number of τ-partitions is equal to the number of state variables for the FSM. There may be more than one minimum set of τ-partitions. If more than one minimum set of τ-partitions exist, the choice of any one of the minimum sets will lead to an optimum or near optimum STT design — there is usually little difference in their use. A nonminimum set of τ-partitions will usually *not* yield an optimum STT design, but it can happen.
6. Select a valid state code assignment from a minimum set of τ-partitions. Choose the initialization state to be either a $\cdots 000$ state or a $\cdots 111$ state, not a mixture. See Section 14.11 for rules governing the initialization of asynchronous FSMs. Note that for FSMs lacking cross branching the partitioning method defaults to unit distance coding of states as in Fig. 14.22.

At this point the array algebraic approach, discussed in Section 11.11, can be used to obtain the NS and output functions for the STT state machine. The array algebraic approach discussed here is actually an extension of the LPD model, since the lumped path delays in the NS functions are implied.

As an example, consider the FSM represented by the state diagram and state table in Fig. 14.33, presented here for purposes of designing it as an STT state machine. This figure is a reproduction of that presented in Fig. 14.6, exclusive of state code assignments at this point. From the state table in Fig. 14.33b, there result the *seed sets* given by Eqs. (14.26).

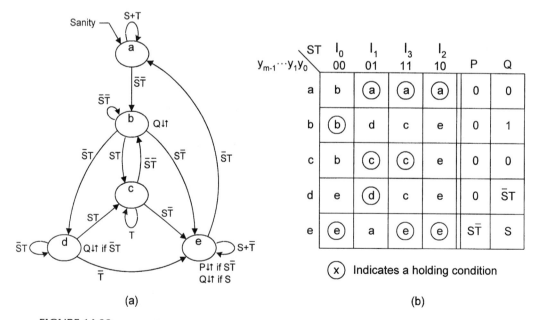

FIGURE 14.33

Reproduction of state machine in Fig. 14.6 for purposes of designing it as an STT machine. (a) State diagram representation. (b) The equivalent state table for the FSM in (a).

14.12 SINGLE-TRANSITION-TIME MACHINES

Seed sets are useful as a aid in establishing the π-partitions and may be disregarded for simple FSMs. Notice that the branching paths within a given seed set contain just one holding condition state identifier and that all branching paths within the set share a common branching condition. This is easily seen by comparing each seed set with the state diagram in Fig. 14.33a. Normally a single state identifier representing a holding condition will not appear singly within a seed set unless it is not otherwise associated with another state identifier within the same seed set.

$$\left.\begin{cases} \text{Seed Set } I_0 = \{ab, bc, de\} \\ \text{Seed Set } I_1 = \{ae, bd, c\} \\ \text{Seed Set } I_3 = \{a, bc, cd, e\} \\ \text{Seed Set } I_2 = \{a, be, ce, de\} \end{cases}\right\} \text{Seed sets} \qquad (14.26)$$

$$\left.\begin{cases} \text{Seed Set } I_0 \to \pi_1 = abc, de \\ \text{Seed Set } I_1 \to \pi_2 = ae, bd \\ \text{Seed Set } I_1 \to \pi_3 = ae, c \\ \text{Seed Set } I_1 \to \pi_4 = bd, c \\ \text{Seed Set } I_3 \to \pi_5 = a, bcd \\ \text{Seed Set } I_3 \to \pi_6 = a, e \\ \text{Seed Set } I_3 \to \pi_7 = bcd, e \\ \text{Seed Set } I_2 \to \pi_8 = a, bcde \end{cases}\right\} \pi\text{-partitions.} \qquad (14.27)$$

The π-partitions are derived from the seed sets in Eqs. (14.26) and are given by Eqs. (14.27), where state a is taken to be the initialization state in agreement with the state diagram in Fig. 14.33a. Observe that when present in a given π-partition, state a always appears on the left side of the partition (the comma). If it is decided to assign $\cdots 000$ to state a, then all state identifiers grouped with a on the left side of the partition must also be assigned logic 0. Accordingly, this requires that all state identifiers on the right side of the partition be assigned a logic 1. For example, from seed set I_0, the π-partition is $\pi_1 = abc, de$ for which state identifiers a, b and c all take logic 0 while state identifiers d and e take logic 1. Notice in particular that the partitions are formed in such a manner that no state variable appears on both sides of the partition, a requirement for discreteness of the partition.

Having completed step 3 of the procedure given previously, it is now required by step 4 that the π-partition be collected into τ-partitions, each of which must contain all the state identifiers. This is done and the results are presented in Eqs. (14.28). Observe that there are five τ-partitions of which only four are necessary to cover all eight π-partitions. The choice of the first four τ-partitions is made, which constitutes a minimum set thereby completing step 5. Hence, four state variables are required.

$$\left.\begin{cases} \tau_1 = abc, de = \pi(1, 6) \\ \tau_2 = ae, bcd = \pi(2, 3, 5) \\ \tau_3 = ace, bd = \pi(2, 4) \\ \tau_4 = a, bcde = \pi(5, 6, 7, 8) \\ \tau_5 = abde, c = \pi(3, 4) \end{cases}\right\} \pi\text{-partitions.} \qquad (14.28)$$

The state matrix **S** can now be established according to step 6 assuming that the initialization state a is assigned all zeros, 0000. If an ascending order of τ-partitions is chosen, the **S** matrix becomes

$$S = \begin{array}{c} \\ a \\ b \\ c \\ d \\ e \end{array} \begin{array}{c} \tau_1 \; \tau_2 \; \tau_3 \; \tau_4 \\ \left[\begin{array}{cccc} 0 & 0 & 0 & 0 \\ 0 & 1 & 1 & 1 \\ 0 & 1 & 0 & 1 \\ 1 & 1 & 1 & 1 \\ 1 & 0 & 0 & 1 \end{array} \right] \end{array} = \text{State matrix}, \qquad (14.29)$$

where Hamming distances of 1, 2, and 3 are required for the STT state-to-state transitions. Note that if the initialization of state a is chosen to be 0000, there are $4! = 24$ ways the columns in the state matrix in Eq. (14.29) can be commuted. Therefore, there are 24 possible state code assignments for which state a is assigned 0000. If state a can be initialized as 1111 in addition to 0000, then there are $2 \times 4! = 48$ possible state code assignments. Generally, for n τ-partitions there are $n!$ **S** arrays possible assuming initialization into either a $\cdots$ 000 state or a $\cdots$ 111 state. Or, if no restrictions are placed on the initialization state code, the number of **S** arrays is expressed as $SA = (2^n - 1)!/(2^n - r)!(n!)$, where n is the number of state variables (τ-partitions) and r is the number of states. In the present case, this would amount to $SA = 1365$ possible state code assignments.

By continuing to follow the procedure described in Section 11.11, the destination matrix becomes

$$D = \begin{array}{c} \\ a \\ b \\ c \\ d \\ e \end{array} \begin{array}{c} I_0 \quad\; I_1 \quad\; I_3 \quad\; I_2 \\ \left[\begin{array}{cccc} 0 & ae & a & a \\ abc & 0 & 0 & 0 \\ 0 & c & bcd & 0 \\ 0 & bd & 0 & 0 \\ de & 0 & e & bcde \end{array} \right] \end{array} = \text{Destination matrix}, \qquad (14.30)$$

which is exactly the same as that given in Eq. (11.12). Then by taking the transpose of the **S** matrix (S^t) and by multiplying it with the destination matrix **D**, there results the NS function matrix F_{NS} given by

$$F_{NS} = S^t D = \begin{bmatrix} 0 & 0 & 0 & 1 & 1 \\ 0 & 1 & 1 & 1 & 0 \\ 0 & 1 & 0 & 1 & 0 \\ 0 & 1 & 1 & 1 & 1 \end{bmatrix} \begin{bmatrix} 0 & ae & a & a \\ abc & 0 & 0 & 0 \\ 0 & c & bcd & 0 \\ 0 & bd & 0 & 0 \\ de & 0 & e & bcde \end{bmatrix}$$

$$= \begin{bmatrix} de & bd & e & bcde \\ abc & bcd & bcd & 0 \\ abc & bd & 0 & 0 \\ 1 & bcd & bcde & bcde \end{bmatrix}, \qquad (14.31)$$

where the entries in the **F** matrix are called the *state adjacency sets*.

The next step is to express the NS function matrix F_{NS} in terms of the state variables $y_3, y_2, y_1,$ and y_0. This can be done by inspection of the state assignment map shown in Fig. 14.34a. Noting that all empty cells of this map are don't cares, the state adjacency sets

14.12 SINGLE-TRANSITION-TIME MACHINES 725

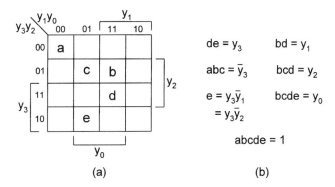

FIGURE 14.34
Evaluation of the state adjacency sets in the **F** matrix of Eq. (14.31). (a) State assignment map for the state matrix of Eq. (14.29). (b) State adjacency sets in terms of the state variables as evaluated by inspection of (a).

are easily expressed in terms of the y-variables as shown in Fig. 14.34b. For example, y_3 covers all states adjacent to states d and e in the y_3 domain. Similarly, $\bar{y}_3$ encompasses all state adjacencies relative to states a, b, and c in the $\bar{y}_3$ domain. If automated designs are required to express the state adjacency sets in terms of the y-variables, tabular methods such as that of Quine–McCluskey can be used as discussed in Section 11.11. However, very large, complex FSMs may require the use of a minimization algorithm such as Espresso-II to accomplish this task.

After the appropriate substitutions are made into Eq. (14.31), the NS functions can be evaluated. This is accomplished by multiplying the function matrix $\mathbf{F}_{NS}$ by the input matrix $\mathbf{I}$ to obtain the following NS function matrix **NS**:

$$\mathbf{NS} = \mathbf{F}_{NS} \cdot \mathbf{I} = \begin{bmatrix} y_3 & y_1 & y_3\bar{y}_1 & y_0 \\ \bar{y}_3 & y_2 & y_2 & 0 \\ \bar{y}_3 & y_1 & 0 & 0 \\ 1 & y_2 & y_0 & y_0 \end{bmatrix} \begin{bmatrix} I_0 \\ I_1 \\ I_3 \\ I_2 \end{bmatrix} = \begin{bmatrix} Y_3 \\ Y_2 \\ Y_1 \\ Y_0 \end{bmatrix}. \qquad (14.32)$$

By carrying out the indicated matrix multiplication, there results the NS equations

$$\begin{bmatrix} Y_3 \\ Y_2 \\ Y_1 \\ Y_0 \end{bmatrix} = \begin{bmatrix} y_3 I_0 + y_1 I_1 + y_3 \bar{y}_1 I_3 + y_0 I_2 \\ \bar{y}_3 I_0 + y_2(I_1 + I_3) \\ \bar{y}_3 I_0 + y_1 I_1 \\ I_0 + y_2 I_1 + y_0(I_3 + I_2) \end{bmatrix}$$

or

$$\begin{cases} Y_3 = y_3 \bar{S}\bar{T} + y_1 \bar{S}T + y_3 \bar{y}_1 ST + y_0 S\bar{T} \\ Y_2 = \bar{y}_3 \bar{S}\bar{T} + y_2 T \\ Y_1 = \bar{y}_3 \bar{S}\bar{T} + y_1 \bar{S}T \\ Y_0 = \bar{S}\bar{T} + y_2 \bar{S}T + y_0 S \end{cases}. \qquad (14.33)$$

The output functions are obtained by using the same procedure. As was indicated in Section 11.11, the state matrices for outputs P and Q are obtained directly from the state table in Fig. 14.33b. By multiplying the transpose of the P state matrix in Fig. 14.33b by the $\mathbf{D}$ matrix and by substituting the appropriate y-variables for the state adjacency sets, the P function matrix is found to be

$$\mathbf{F_P} = \mathbf{P^t D} = \begin{bmatrix} 0 & 0 & 0 & 0 & S\bar{T} \end{bmatrix} \begin{bmatrix} 0 & ae & a & a \\ abc & 0 & 0 & 0 \\ 0 & c & bcd & 0 \\ 0 & bd & 0 & 0 \\ de & 0 & e & bcde \end{bmatrix}$$

$$= [de \ 0 \ e \ bcde]S\bar{T}$$

$$= [y_3 \ 0 \ y_3\bar{y}_1 \ y_0]S\bar{T}$$

or

$$P = \mathbf{F_P I} = [y_3 S\bar{T} \ 0 \ y_3\bar{y}_1 S\bar{T} \ y_0 S\bar{T}] \begin{bmatrix} I_0 \\ I_1 \\ I_3 \\ I_2 \end{bmatrix}$$

$$= y_0 S\bar{T}.$$

The results for output Q follows in similar fashion. By multiplying the transpose of the state matrix for output Q with the $\mathbf{D}$ matrix and by substituting the appropriate y-variables for the state adjacency sets, the Q function matrix becomes

$$\mathbf{F_Q} = \mathbf{Q^t D} = \begin{bmatrix} 0 & 1 & 0 & \bar{S}T & S \end{bmatrix} \begin{bmatrix} 0 & ae & a & a \\ abc & 0 & 0 & 0 \\ 0 & c & bcd & 0 \\ 0 & bd & 0 & 0 \\ de & 0 & e & bcde \end{bmatrix}$$

$$= [(abc + deS) \ bd\bar{S}T \ eS \ bcdeS]$$

or

$$Q = \mathbf{F_Q I} = [(\bar{y}_3 + y_3 S) \ y_1 \bar{S}T \ y_3 \bar{y}_1 S \ y_0 S] \begin{bmatrix} I_0 \\ I_1 \\ I_3 \\ I_2 \end{bmatrix}$$

$$= \bar{y}_3 \bar{S}\bar{T} + y_1 \bar{S}T + y_3 \bar{y}_1 ST + y_0 S\bar{T},$$

where $y_3 S \cdot I_0 = y_3 S \cdot \bar{S}\bar{T} = 0$ has been eliminated.

14.12 SINGLE-TRANSITION-TIME MACHINES

Collectively, the NS and output functions generated from the array algebraic approach are

$$\left\{\begin{array}{l} Y_3 = y_3\bar{S}\bar{T} + y_1\bar{S}T + y_3\bar{y}_1ST + y_0S\bar{T} \\ Y_2 = \bar{y}_3\bar{S}\bar{T} + y_2T \\ Y_1 = \bar{y}_3\bar{S}\bar{T} + y_1\bar{S}T \\ Y_0 = \bar{S}\bar{T} + y_2\bar{S}T + y_0S \\ P = y_0S\bar{T} \\ Q = \bar{y}_3\bar{S}\bar{T} + y_1\bar{S}T + y_3\bar{y}_1ST + y_0S\bar{T} \end{array}\right\}, \qquad (14.34)$$

which represents a gate/input tally of 14/40 taking into account four shared PIs and excluding possible inverters. Notice that *all* five output p-terms are covered by p-terms in the NS logic functions owing to the four shared PIs. This is characteristic of the array algebraic approach to FSM design since the same form of the function matrix $\mathbf{F} = \mathbf{Z}^t\mathbf{D}$ is used for the output functions as for the NS functions.

The NS and output functions in Eqs. (14.34) are guaranteed to be free of critical races and ORGs. This is a result of using the array algebraic approach on the state assignment matrix of Eq. (14.29), the combination of which is inherently exclusionary of all race related problems. However, the result is not expected to be an optimal result. The array algebraic approach to FSM design used in this section is attractive from another point of view: It offers a method for obtaining the NS and output functions of STT state machines that is amenable to computer aided design (CAD).

Logic minimization methods should *rarely*, if ever, be used to obtain the NS and output functions directly from the state diagrams of STT state machines. The reason is that *an STT state code assignment is, by itself, not sufficient to ensure a critical race-free and ORG-free design*. For example, if an optimal K-map minimization approach is used directly on the state diagram in Fig. 14.33a with the STT state code assignments given by Eq. (14.29), a result is obtained that cannot be guaranteed to be free of critical races and ORGs. The NS and output functions must be "looped out" correctly to avoid race-related problems — a task performed automatically by the array algebraic approach.

There remains the question of static hazards in Eqs. (14.34). A static hazard analysis of Eqs. (14.34) indicates that there are seven active static 1-hazards in the NS functions, three in function Y_3, one in Y_2, and three in function Y_0, but all are externally activated. Shown in Eqs. (14.35) are the NS functions with hazard cover included for these seven s-hazards. Also shown is the hazard cover for the singular externally initiated static 1-hazard in the output Q. When this hazard cover is included, the gate/input tally for Eqs. (14.35) becomes 21/68. However, there is one potentially active static 1-hazard in function Y_1 if a delay greater than that of an inverter is placed on the noninverter path, that is, on the T line to gate $y_1\bar{S}T$. There are also two such potentially active s-hazards in output function Q, one between coupled terms $\bar{y}_3\bar{S}\bar{T}$, $y_1\bar{S}T$ and the other between coupled terms $\bar{y}_3\bar{S}\bar{T}$, $y_0S\bar{T}$. In each of these latter two cases an s-hazard could form if a delay exceeding that of an inverter is positioned on the noninverter line of the coupled variable to the coupled term gate. The hazard covers for these hazards are not shown in Eqs. (14.35) since they are not likely to form, although some designers may include them to ensure proper operation of the FSM. The term "potentially active" applies to any hazard that must be activated by an unintended delay that is explicitly located along some path in the circuit — a delay that cannot always be predicted from an analysis of the logic circuit.

$$\begin{cases} Y_3 = y_3\bar{S}\bar{T} + y_1\bar{S}T + y_3\bar{y}_1ST + y_0S\bar{T} + \underbrace{y_3y_1\bar{S} + y_3y_0\bar{T} + y_3\bar{y}_1y_0S}_{\text{Hazard cover}} \\ Y_2 = \bar{y}_3\bar{S}\bar{T} + y_2T + \underbrace{\bar{y}_3y_2\bar{S}}_{\text{Hazard cover}} \\ Y_1 = \bar{y}_3\bar{S}\bar{T} + y_1\bar{S}T \\ Y_0 = \bar{S}\bar{T} + y_2\bar{S}T + y_0S + \underbrace{y_2\bar{S} + y_0\bar{T} + y_2y_0T}_{\text{Hazard cover}} \\ P = y_0S\bar{T} \\ Q = \bar{y}_3\bar{S}\bar{T} + y_1\bar{S}T + y_3\bar{y}_1ST + y_0S\bar{T} + \underbrace{y_3\bar{y}_1y_0S}_{\text{Hazard cover}} \end{cases} \quad (14.35)$$

To assist the reader in understanding the analysis by which the hazard cover in Eqs. (14.35) is obtained, the hazardous $1 \rightarrow 0$ transitions for function Y_3 and the states within which they occur are indicated in Eq. (14.36). Like all the other s-hazards in the NS functions, these are externally initiated static 1-hazards that will occur only under the holding condition of a given state. The static hazard cover is obtained by consensus, that is, by ANDing the residues of the coupled terms involved in the particular hazardous transition. Thus, for this s-hazard occurring in state d, the consensus term is $y_3y_1\bar{S}$, which is the ANDed residue of the coupled terms $y_3\bar{S}\bar{T}$ and $y_1\bar{S}T$. This hazard occurs when the FSM begins the transition from state 1111 to 1001 ($d \rightarrow e$) under input change $\bar{S}T \rightarrow \bar{S}\bar{T}$ during which Y_3 should remain active but instead is forced to undergo a negative glitch caused by the static 1-hazard. For a review of hazard analyses in two-level combinational logic and in synchronous FSMs, see Sections 9.2 and 11.3.

$$Y_3 = y_3\bar{S}\bar{T} + y_1\bar{S}T + y_3\bar{y}_1ST + y_0S\bar{T} \quad (14.36)$$

The two-level NAND/INV logic circuit for Eqs. (14.35) is presented in Fig. 14.35 together with the shared PIs A, B, C, D, and E that are used to implement the outputs Q and P. Also shown are the covers for the seven hazards indicated by shaded gates. Notice that the p-term $y_3\bar{y}_1y_0S$ serves as hazard cover for both the externally initiated s-hazard in NS function Y_3 and the internally initiated s-hazard in output Q. Static hazard cover for an output function is frequently (but not always) found in an NS logic function, including its hazard cover — that is, an output s-hazard cover is frequently a shared PI. Sanity connections are omitted for simplicity but can easily be added following the discussion and figures in Section 14.11.

The results of a logic simulation of the circuit in Fig. 14.35 is shown in Fig. 14.36, where the single transition times are clearly indicated by vertical dotted lines. Thus, the simultaneous change of the state variables is what is meant by single transition time (STT). Clearly, it is easy to understand why race-related timing problems do not exist in such FSMs, even in real time. Though not indicated, the time elapsing between a change in an external input (S or T) and the resulting simultaneous change in y-variables varies from $2\tau_p$ to $(2\tau_p + \tau_{INV})$ for the state-to-state transitions shown. Here, as before, τ_p is the path

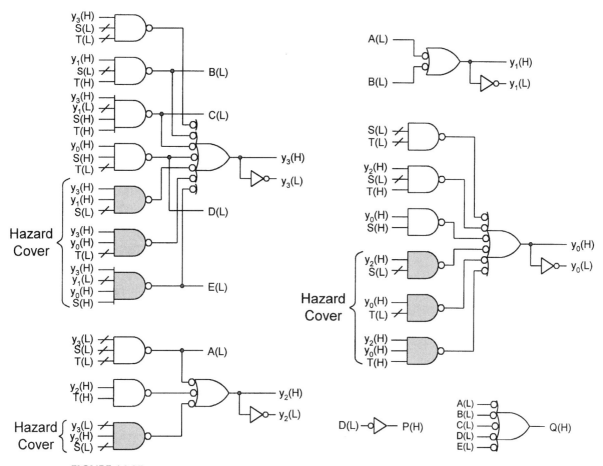

FIGURE 14.35
Two-level implementation of the NS and output functions in Eqs. (14.35) showing hazard cover for the eight s-hazards (shaded gates) and use of the four shared PIs A, B, C, D, and E to implement outputs P and Q.

delay through a gate and τ_{INV} is the path delay through an inverter, where the relative delay values are taken to be $\tau_{INV} = \frac{3}{5}\tau_p$. The changes in Q, for the most part, occur simultaneously with changes in the y-variables, while changes in P occur in the range of τ_p to $(\tau_p + \tau_{INV})$ following an external input change. Changes in Q may precede changes in the y-variables by τ_{INV}, but that is the exception rather than the rule. To assist the reader in following these events, the state codes and state identifiers are provided in Fig. 14.36 following each state-to-state transition.

Hazard analyses, such as that required to arrive at Eqs. (14.35), are not easily carried out and can lead to serious problems if performed incorrectly. Furthermore, as is evident in Eqs. (14.35), there is a significant increase in the hardware commitment required to eliminate the static hazards in complex FSMs such as STT machines. However, there exists a means by which all s-hazards can be eliminated from any fundamental mode state machine (meaning also STT FSMs) without having to add hazard cover. This is the subject of the following section in which the basic memory cell is used as the memory element.

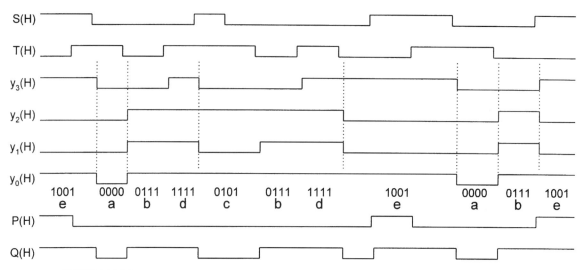

FIGURE 14.36
Results of a simulation for the logic circuit in Fig. 14.35 showing the single transition times of the state variables (dashed lines) that are characteristic of STT machines, the state codes and state identifiers following each transition, and the output fucntion P and Q.

14.13 HAZARD-FREE DESIGN OF FUNDAMENTAL MODE STATE MACHINES BY USING THE NESTED CELL APPROACH

Fundamental mode designs of FSMs are fraught with special problems, not the least of which is that dealing with static hazards in the NS logic. If an s-hazard exits in a NS logic function, it can, under the right conditions, cause the FSM to malfunction. Furthermore, the process of identifying and eliminating these hazards is no trivial task, as has been made evident from discussions in Subsection 14.10.3 and in Section 14.12. No where is this point better illustrated than in Eqs. (14.35), where seven active static 1-hazards are identified in three of four NS functions. But the processes of identifying and eliminating such hazards can be circumvented by using the *nested cell model* as is done in Fig. 10.30 for the RET D flip-flop. The following is one such approach in the use of this model.

Consider the STT NS (LPD) logic in Eqs. (14.34), exclusive of hazard cover, which is to be converted to nested cell form. Next consider that the *LPD-to-SR conversion* follows exactly that of Eq. (10.11) for converting a D flip-flop to a JK flip-flop and is given by

$$Y = \bar{y}S + y\bar{R}. \tag{14.37}$$

Here, J is replaced by S and K is replaced by R, while D and Q are replaced by Y and y, respectively. The reason why these substitutions are permissible is that the combined excitation table for SR in Fig. 10.15c is, in fact, the same as that for the JK flip-flop in Fig. 10.40c as they are applied to design by the mapping algorithm in Section 10.6 — albeit no mention of this has been made previously.

With Eq. (14.37) in mind, a theoretical procedure (applicable to CAD with the appropriate search algorithm) can be devised to execute the conversions directly from the NS Y_i functions. Consider the K-maps in Fig. 10.30c for the nested cell design of the resolver

14.13 HAZARD-FREE DESIGN OF FUNDAMENTAL MODE STATE MACHINES 731

FSM. Essential entries in a given domain excludes that domain character from the NS logic function being extracted. Consequently, the following $Y_i \to S_i$, R_i conversions can be made directly from the Y_i functions in Eqs. (14.34):

$$\left\{\begin{aligned}
Y_3 &= y_3\bar{S}\bar{T} + y_1\bar{S}T + y_3\bar{y}_1 ST + y_0 S\bar{T} \\
S_3 &= \bar{y}_3\bar{y}_1\bar{y}_0(0) + \bar{y}_3\bar{y}_1 y_0(S\bar{T}) + \bar{y}_3 y_1 \bar{y}_0(\bar{S}T) + \bar{y}_3 y_1 y_0(S \oplus T) + y_3 XX\phi \\
&= \bar{y}_1 y_0(S\bar{T}) + y_1 \bar{y}_0(\bar{S}T) + y_1 y_0(S \oplus T) \\
&= y_0 S\bar{T} + y_1 \bar{S}T \\
R_3 &= y_3 XX\phi + y_3\bar{y}_1\bar{y}_0(\overline{\bar{S}\bar{T} + ST}) + y_3\bar{y}_1 y_0(\overline{\bar{S}\bar{T} + ST + S\bar{T}}) \\
&\quad + y_3 y_1 \bar{y}_0(\bar{S}) + y_3 y_1 y_0(\bar{S}\bar{T} + \bar{S}T + S\bar{T}) \\
&= \bar{y}_1\bar{y}_0(S \oplus T) + \bar{y}_1 y_0(\bar{S}T) + y_1\bar{y}_0(S) + y_1 y_0(ST) \\
&= \bar{y}_1 \bar{S}T + y_1 ST + \bar{y}_0 S\bar{T}
\end{aligned}\right\}. \quad (14.38)$$

Here, the results for S_3 and R_3 are precisely those that would be obtained by a K-map conversion of Eqs. (14.34) given the use of Eq. (14.37). Note that XX appearing in the terms $y_3XX\phi$ and $\bar{y}_3XX\phi$ represents all canonical ANDed forms of y_1 and y_0, that is, $\bar{y}_1\bar{y}_0$, $\bar{y}_1 y_0$, $y_1\bar{y}_0$, $y_1 y_0$ (y_2 is absent in Y_3). Similarly, the don't-care symbol ϕ represents all canonical ANDed forms of S and T ($\bar{S}\bar{T}$, $\bar{S}T$, $S\bar{T}$, ST). Thus, $y_3XX\phi$ in S_3 eliminates $\bar{y}_3$ in all p-terms of that function. Similar reasoning is applied to the expression for R_3 where $\bar{y}_3 XX\phi$ eliminates y_3 from all terms in that function.

Continuing this procedure yields the following results for the remaining three functions:

$$\left\{\begin{aligned}
Y_2 &= \bar{y}_3 \bar{S}\bar{T} + y_2 T \\
S_2 &= \bar{y}_2\bar{y}_3(\bar{S}\bar{T}) + \bar{y}_2 y_3(0) + y_2 X\phi \\
&= \bar{y}_3 \bar{S}\bar{T} \\
R_2 &= \bar{y}_2 X\phi + y_2 \bar{y}_3(\bar{S} + T) + y_2 y_3(\bar{T}) \\
&= \bar{y}_3 S\bar{T} + y_3 \bar{T} \\
&= S\bar{T} + y_3 \bar{T}
\end{aligned}\right\}$$

$$\left\{\begin{aligned}
Y_1 &= \bar{y}_3 \bar{S}\bar{T} + y_1 \bar{S}T \\
S_1 &= \bar{y}_1\bar{y}_3(\bar{S}\bar{T}) + \bar{y}_1 y_3(0) + y_1 X\phi \\
&= \bar{y}_3 \bar{S}\bar{T} \\
R_1 &= \bar{y}_1 X\phi + y_1\bar{y}_3(\bar{\bar{S}}) + y_1 y_3(\overline{\bar{S}T}) \\
&= \bar{y}_3 S + y_3(S + \bar{T}) \\
&= S + y_3 \bar{T}
\end{aligned}\right\} \quad (14.38)$$

$$\left\{\begin{aligned}
Y_0 &= \bar{S}\bar{T} + y_2 \bar{S}T + y_0 S \\
S_0 &= \bar{y}_0\bar{y}_2(\bar{S}\bar{T}) + \bar{y}_0 y_2(\bar{S}) + y_0 X\phi \\
&= \bar{S}\bar{T} + y_2 \bar{S} \\
R_0 &= \bar{y}_0 X\phi + y_0\bar{y}_2(S + T) + y_0 y_2(\overline{\bar{S}\bar{T} + \bar{S}T + S}) \\
&= \bar{y}_2 ST + y_2(0) \\
&= \bar{y}_2 ST
\end{aligned}\right\}.$$

Although the NS functions have been converted from Y_i LPD form to hazard-free S,R form in Eqs. (14.38), the output functions must remain as given in Eqs. (14.35), including the hazard cover term in Q. If the output functions were not retained, ORGs would be generated during several of the transitions. With this in mind, the results are given collectively by Eqs. (14.39). The NS and output functions in Eqs. (14.39) represent a gate/input tally of 29/69, including the four basic cells and three shared PIs, but not including possible inverters. This may be compared with a gate/input tally of 21/68 for Eqs. (14.35), including the five shared PIs, but again not including possible inverters. Note that an invalid set of STT NS logic functions would result if the *nested cell model* were applied directly to the state diagram in Fig. 14.33 with the STT state assignments of Eq. (14.29). However, $Y_i \to S_i$, R_i K-map conversions of Eqs. (14.34) will minimize to the NS functions of Eqs. (14.39).

$$\left\{ \begin{array}{l} S_3 = y_0 S\bar{T} + y_1 \bar{S} T \\ R_3 = \bar{y}_1 \bar{S} T + y_1 S T + \bar{y}_0 S \bar{T} \\ S_2 = \bar{y}_3 \bar{S} \bar{T} \\ R_2 = S\bar{T} + y_3 \bar{T} \\ S_1 = \bar{y}_3 \bar{S} \bar{T} \\ R_1 = S + y_3 \bar{T} \\ S_0 = \bar{S}\bar{T} + y_2 \bar{S} \\ R_0 = \bar{y}_2 \bar{S} T \\ P = y_0 S \bar{T} \\ Q = \bar{y}_3 \bar{S} \bar{T} + y_1 \bar{S} T + y_3 \bar{y}_1 S T + y_0 S \bar{T} + \underbrace{y_3 \bar{y}_1 y_0 S}_{\text{Hazard cover}} \end{array} \right.$$ (14.39)

The logic circuit representing Eqs. (14.39) is given in Fig. 14.37, where reset-dominant basic cells are used as memory elements. Here, all sanity inputs have been omitted for simplicity. Initialization of this FSM into the 0000 state requires that all R's be initialized a logic 1 while all S's be initialized an logic 0 (see Section 14.11). Notice the relative simplicity of the NS logic for this "nested cell" model compared to that required by the LPD model in Fig. 14.35. The maximum fan-in for this nested-cell implementation is 4, compared to 7 for Fig. 14.35, all exclusive of sanity inputs.

What has not been discussed here is the relative speed of the two types of implementations, that resulting from Eqs. (14.35) and that from Eqs. (14.39). Predictably, the nested-cell design will be somewhat slower than the LPD design. This is so because the nested-cell design can behave as a three-level implementation whereas the LPD design of Eqs. (14.35) represents a two-level implementation provided that it is not necessary to "tree" any of the NS functions because of fan-in restrictions. Both designs offer smooth fast operation free of critical races and ORGs. But because of the action of the basic cells in the nested cell design, no hazard cover is necessary. Shown in Fig. 14.38 are the simulation results for the nested-cell logic circuit in Fig. 14.37. An examination of this simulation clearly indicates that the y-variable transitions do not necessarily change at the same time as they did in Fig. 14.36. In the case of Fig. 14.38, the time elapsing between a change in an external input and the first y-variable to change varies from $2\tau_p$ to $(3\tau_p + \tau_{INV})$ and the time between y-variable changes for a single transition varies from τ_{INV} to $2\tau_p$. Also, the outputs may precede the first y-variable to change by as much as $2\tau_p$. Note that critical races and ORGs are still precluded from occurring since the NS logic functions of Eqs. (14.34) are used to generate those of Eqs. (14.39) with the output logic and hazard cover remaining the same. Here, as before, τ_p is a gate path delay and $\tau_{INV} = \frac{3}{5}\tau_p$.

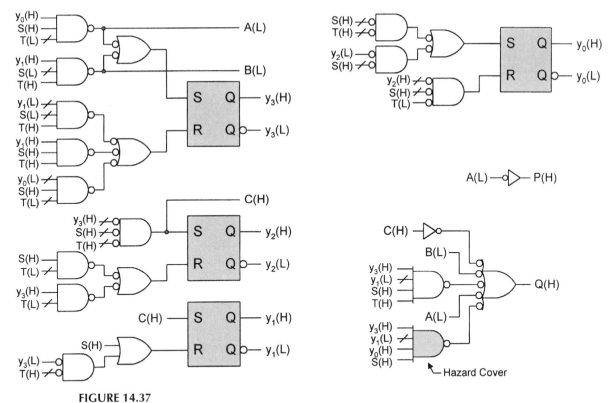

FIGURE 14.37
Implementation of Eqs. (14.39) by using reset-dominant basic cells as memory elements.

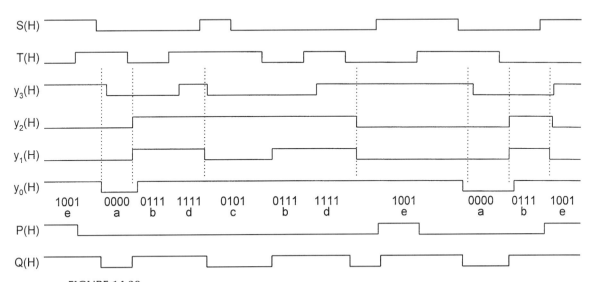

FIGURE 14.38
Results of a simulation for the nested cell logic circuit in Fig. 14.37 showing some variation in the single transition times of the state variables (compare by using dashed lines), the state codes, and state identifiers following each transition, and showing the output functions P and Q.

733

Which method should be used for STT FSM design, the nested cell approach or the LPD approach? Both the nested-cell and the LPD approaches to asynchronous FSM design are generally applicable to any fundamental mode FSM. However, for the nested cell designs of an STT FSM, special methods must be used. The NS functions must be converted from a valid set of Y_i forms that are the result of the array algebraic approach. The conversion process can be accomplished by either $Y_i \rightarrow S_i$, R_i K-map conversions from the Y_i forms or by algebraic means as was demonstrated in this section. Remember that to avoid possible critical races and ORGs in an STT FSM design, it is necessary to use both the partitioning methods and array algebraic approach that were discussed in Section 14.12. Thus, the nested cell design of an STT FSM requires the extra step of converting the NS LPD functions to *S-R* form. Once this is understood, the decision as to which approach to use reduces to the following: An LPD STT approach generally results in a faster logic circuit compared to a comparable nested cell design. However, an LPD design must deal with the static hazard problem in the NS logic, the analysis of which can be complex but can also be automated with some effort. For this approach, fan-in may become an important consideration, particularly if discrete logic is to be used. In contrast, externally initiated s-hazards in the NS logic of a nested cell design cannot affect the operation of the FSM, as demonstrated later in Fig. 14.50. And by applying the requirements of Subsection 11.3.2 to the basic cell, internally initiated s-hazards are also avoided. Finally, all ORGs are easily avoided by using the STT algebraic approach to design, which can be converted to nested cell form if desired.

14.14 ONE-HOT DESIGN OF ASYNCHRONOUS STATE MACHINES

The one-hot design of synchronous FSMs is discussed at length in Section 13.5. Table 2.11 gives a 10-bit one-hot code, a code consisting of a single "1" per state. A model is presented in Fig. 13.23 and by Eqs. (13.9) that applies to the one-hot designs of synchronous FSMs by using D flip-flops. Since the excitation tables for the LPD model and that for D flip-flop designs are the same (see Fig. 14.3), it follows that Eqs. (13.9) also apply to the one-hot design of asynchronous FSMs if the notation changes $D \rightarrow Y$ and $Q \rightarrow y$ are made. The following paragraphs demonstrate this.

In order to apply Eqs. (13.9) to the design of asynchronous one-hot FSMs, however, it is necessary to add another term to the NS function equation. In a synchronous FSM, the single active state variable in the origin state remains active until the transition to the destination state is complete. This happens as a result of the action of the enabling input, clock (CK). But because there is no enabling input such as *CK* controlling the transitions in a fundamental mode FSM, some means must be found to maintain the active state variable in the origin state constant (active) until the transition is complete to the destination state. This is done by altering the NS functions in Eqs. (13.9) in the following way:

$$Y_j = \underbrace{\sum_{k=0}^{m-1} y_k \cdot f_{j \leftarrow k}}_{\text{"Into" terms}} + \underbrace{y_j \cdot \bar{F}_j}_{\text{"Out of" terms}} . \qquad (14.40)$$

Here, F_j is the Boolean sum of all active *y*-variables in states to which the *j*th state transits.

14.14 ONE-HOT DESIGN OF ASYNCHRONOUS STATE MACHINES

Thus, the first part ("into" terms) of Eq. (14.40) represents m minimum NS functions as derived, say, from K-map cover or from a minimization algorithm. This part is identical to that appearing in Eq. (13.9) but with the appropriate symbol changes for the present and next state variables. The second part ("out of" terms) of Eq. (14.40) functions to maintain the state variable of the origin state active until the transition to the destination state is complete. This forces the FSM to transit through a state with two 1's, a state consisting of 1's from the origin and destination states. The r output functions summed over m states are similar to those given by Eqs. (13.9) and are represented by

$$Z_l = \sum_{j=0}^{m-1} y_j \cdot f_{j,l}(X), \qquad (14.41)$$

where $f_{j,l}(X)$ represents the jth function of external inputs X for the lth output with $l = 0, 1, 2, \ldots, (r-1)$.

Application of Eqs. (14.40) and (14.41) is remarkably simple since, as was pointed out in Section 13.5, the NS and output functions can be read directly from the state diagram, from an ASM chart or from a state table — *and without the need for a state code assignment or the use of K-maps*. However, there are a few guidelines that must be followed in state diagram (or state table) construction and initialization of a one-hot state machine:

1. Eliminate all buffer ("fly") states — there is no need for them in a one-hot design.

2. Cycles may be permitted, but cause successive transitions between states with two 1's. This produces an overlap in the two 1's states, resulting in a state variable pulse of short duration, which may or may not be acceptable. Also an intermediate state in a cycle transition should not be an output state, since it would create an output glitch. In short, *it is always best to avoid cycles in one-hot designs*. There is no need for them.

3. If a static hazard exists in the NS-forming logic, it is formed between the "out of" term and an "into" term and is always an internally initiated static 1-hazard in SOP logic. Hazard cover is provided by the "into" holding condition term, which is usually a reduced consensus term. Recall that the consensus term is the ANDed residue of the coupled terms, as discussed in Sections 9.2 and 11.3 and in Subsection 14.10.3. An active static hazard in the NS logic of a one-hot FSM can cause it to malfunction and, therefore, must be eliminated as just described.

4. Initialization of one-hot designs must be accomplished according to Fig. 14.32a together with a term that meets the requirements of the one-hot-plus-zero approach discussed in Section 13.5. Thus, the idea here is to first set all y-variables to logic 0 and then force the FSM into a one-hot state where thereafter it can transition normally from one one-hot state to another. No attempt should be made to initialize according to Fig. 14.32b, because to do so will usually result in the activation of more than one state on startup. Entrance into the intended one-hot routine of the FSM may not take place until the inputs change in some favorable manner.

As an example, and for reasons of comparison, consider the state diagram and state table in Fig. 14.39, which are reproduced from Figs. 13.26 and 14.33 for the convenience of the

736 CHAPTER 14 / ASYNCHRONOUS STATE MACHINE DESIGN AND ANALYSIS

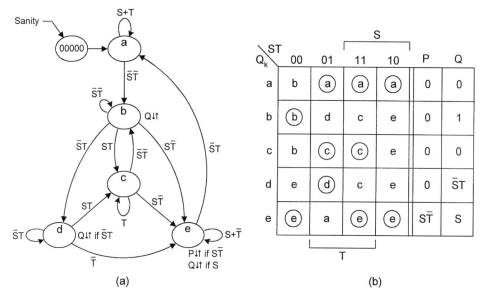

FIGURE 14.39
Reproductions of the FSM in Fig. 13.26 for use by the asynchronous one-hot-plus-zero FSM design method. (a) Fully documented state diagram representation showing only branching conditions and state identifiers. (b) Equivalent state table representation.

reader. For this example, Eqs. (13.11) apply, but with the added "out of" terms required by Eq. (14.41). The result is the set of two-level NS and output functions expressed as

$$\begin{cases} Y_a = aS + aT + e\bar{S}T + a\bar{b} + \bar{a}\bar{b}\bar{c}\bar{d}\bar{e} \\ Y_b = a\bar{S}\bar{T} + b\bar{S}\bar{T} + c\bar{S}\bar{T} + b\bar{c}\bar{d}\bar{e} \\ Y_c = bST + cT + dST + c\bar{b}\bar{e} \\ Y_d = b\bar{S}T + d\bar{S}T + d\bar{c}\bar{e} \\ Y_e = b\bar{S}\bar{T} + c\bar{S}\bar{T} + d\bar{T} + eS + e\bar{T} + e\bar{a} \\ P = e\bar{S}\bar{T} \\ Q = d\bar{S}T + eS + b \end{cases}, \qquad (14.42)$$

where it is understood that $a = y_a$, $b = y_b$, $c = y_c$, $d = y_d$ and $e = y_e$. The $\bar{a}\bar{b}\bar{c}\bar{d}\bar{e}$ term is added to Y_a for initialization purposes — the one-hot-plus-zero approach. Notice the simplicity of the output expressions compared to those of the STT design expressed by Eqs. (14.34). This simplicity derives from the fact that each NS and output function is associated with a specific state.

The NS functions in Eqs. (14.42) are free of critical races, ORGs, and static hazards due to the nature of Eqs. (14.40) and (14.41). The two static 1-hazards that would have been active in the NS functions are each covered by the "into" holding condition term of the state for which the NS function applies. A static hazard in the NS logic of a one-hot design, if present, is always an internally initiated static 1-hazard that is formed between the "out of" term and an "into" term. One internally initiated s-hazard is formed in function Y_b between

14.14 ONE-HOT DESIGN OF ASYNCHRONOUS STATE MACHINES 737

coupled terms $c\bar{S}\bar{T}$ and $bc\bar{d}\bar{e}$, indicating ($c \to b$) under branching conditions $\bar{S}\bar{T}$, but is covered by the holding condition "into" term $b\bar{S}\bar{T}$ for state b. The other s-hazard exists in function Y_c and is produced between coupled terms bST and $c\bar{b}\bar{e}$, meaning $b \to c$ under branching conditions ST, with cover provided by the "into" holding condition term cT for state c. Thus, hazard cover in the NS logic expression of one-hot designs is provided be a reduced consensus term, which turns out to be the "into" holding condition term of the state for which the NS function applies. If left active, s-hazards in the NS logic can cause malfunction of the FSM. No s-hazard is possible in the Q output function of Eqs. (14.42), since the coupled terms eS and $d\bar{S}T$ indicate an externally initiated static 1-hazard that must occur in a two 1's state under a holding condition T that is clearly not possible in a one-hot design.

The logic circuit for the one-hot FSM represented by Eqs. (14.42) is given in Fig. 14.40. Here, it is understood that $a = y_a$, $b = y_b$, $c = y_c$, $d = y_d$, and $e = y_e$, as indicated earlier. Observe that this one-hot FSM is initialized into the 00000 state and that two shared PIs are used in the output function Q. Unlike the case of STT FSMs, which make maximum use of shared PIs in the output expressions, one-hot designs may have few if any shared PIs in their output functions.

The circuit in Fig. 14.40 initializes into state a by first setting *all* state variables to zero and then forcing the FSM into state a by using the one-hot-plus-zero approach described in Section 13.5 and applied here as follows: The initialization process begins with a $Sanity(L) = 1(L) = 0(H)$ input to each of the NAND gates in Fig. 14.40 by following the initializing scheme shown in Fig. 14.32a. Then, when $Sanity$ goes low, that is, when $Sanity(L) = 0(L) = 1(H)$, all inputs to the shaded NAND gate in Fig. 14.40 are set to $1(H)$, which introduces a $1(L)$ into the ORing NAND gate for state a and initializes the FSM into that state. Because use is made of the "all-zero" state in the initialization process, this state should normally not be chosen as an output state. Furthermore, no attempt should be made to initialize a logic one directly into state a by using the scheme shown in Fig. 14.32b. This approach usually results in the activation of more than one state on startup. For a large number of states, it is recommended that the CMOS NOR gate configuration in Fig. 8.46 together with an inverter be used in place of the shaded NAND gate in Fig. 14.40, but with complementary changes in the activation levels of the inputs.

The results of a simulation of the logic circuit in Fig. 14.40 is shown in Fig. 14.41. Vertical dashed lines are placed at specific changes in an external input to emphasize the overlap effect of the "out of" terms in Eqs. (14.42). These terms serve to maintain the y-variable of the origin state active until the transition to the destination state is complete. This, in turn, requires the FSM always to transit through a state of two 1's, one from the origin state and the other from the destination state — the FSM can never transit through the all-zero state. This is easily seen from an inspection of the timing diagram in Fig. 14.41. An analysis of this simulation reveals that the time elapsing between an external variable change and the first y-variable to change is $2\tau_p$ and that the overlap of the y-variables amounts to $(2\tau_p + \tau_{INV})$ in all cases. Here, as in all cases previously, no account is taken of fan-in effects. The relative delay values are $\tau_{INV} = \frac{3}{5}\tau_p$, where τ_p is the delay through any NAND gate. Changes in the outputs P and Q follow a change in the external input by amounts ranging from τ_p to $(2\tau_p + \tau_{INV})$ but fall within the overlap of the y-variable. Thus, the speed of the one-hot design is comparable to that of the LPD STT design in Fig. 14.36.

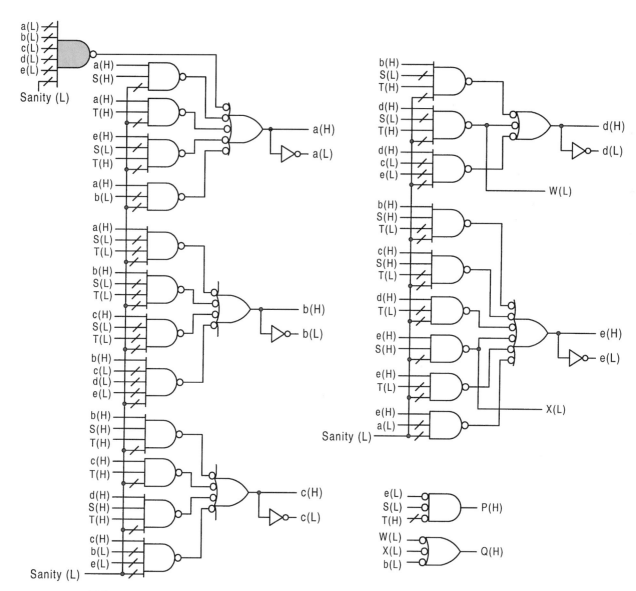

FIGURE 14.40
Implementation of the one-hot FSM represented by Eqs. (14.42) with one-hot-plus-zero initialization circuitry showing shared PIs W and X.

14.15 PERSPECTIVE ON STATE CODE ASSIGNMENTS OF FUNDAMENTAL MODE FSMs

Before the subject of state code assignments can be properly considered, it is necessary to clear up any confusion the reader may have regarding the types of asynchronous FSMs that have been considered. All asynchronous FSMs considered to this point have been those

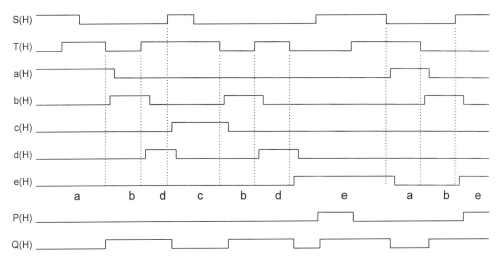

FIGURE 14.41
Results of a simulation for the one-hot SSM of Fig. 14.40 showing effect of the "out of" terms in Eqs. (14.42), which hold each origin state y-variable active until the transition is complete (see dashed lines), and showing the outputs P and Q.

that are said to operate in the fundamental mode. Any fundamental mode FSM requires that no external input to the FSM may change until all internal signals have been stabilized within the FSM and that only one input can change at a time. This requirement holds for STT and one-hot FSMs, both of which can be considered as obeying the LPD model, since fictitious memory elements are always implied. Even the nested cell design of STT FSMs, or the design of flip-flops for that matter, results in state machines that must operate in the fundamental mode. Recall that the basic cell is itself a fundamental mode FSM.

So what really distinguishes one fundamental mode FSM from another? The answer to this question is quite simple. It is the state code assignment as it affects the manner in which the state-to-state transitions occur that ultimately distinguishes one type of fundamental mode FSM from another. Consider that an STT state code assignment is so chosen that critical races and ORGs are eliminated. To do this, the state code assignment may involve multiple y-variable changes during any given state-to-state transition. This introduces the concept of *distance*, i.e., *Hamming distance*. Two adjacent states are said to be *unit-distance coded*, or to have a Hamming distance of 1. An STT design will default to unit-distance coding for FSMs lacking *cross branching*. The FSM in Fig. 14.22a is one such example, since the same unit distance code assignment could have been derived by using the partitioning methods described in Section 14.12. Furthermore, a change of n y-variables during a state-to-state transition involves a Hamming distance of n. The FSM discussed in Section 14.12 possesses several cross branchings and must undergo Hamming distance transitions ranging from 1 to 3 as indicated by the timing diagram in Fig. 14.36. The designer of STT state machines must take extra care to ensure that the sum rule holds for each state in the state diagram or state table and that no cycle paths are present. Failure to meet the sum rule requirement will cause critical races to form.

In comparison, the one-hot approach fixes the Hamming distance at 2, as shown by the timing diagram in Fig. 14.41. Both the STT and the one-hot techniques accomplish the same thing: they both eliminate critical races, but by entirely different approaches to state coding. For the STT design the goal is to arrive at an FSM whose transitions take place simultaneously or nearly so. However, the one-hot design method (for synchronous or asynchronous machines) forces the FSM to cycle through states having exactly two 1's, one from the origin and the other from the destination state. The one-hot approach has the added advantage that static hazards are automatically covered by the holding conditions. Thus, cycle paths must be avoided and the sum rule must always hold in the state diagrams for both STT and one-hot FSMs. Critical races and ORGs are automatically eliminated in STT designs and can easily be avoided in one-hot designs. Associating each output exclusively with its host state automatically eliminates ORGs. If minimization methods are used in one-hot designs, great care must be exercised in using the two-1's race states as output states to avoid ORGs. Finally, recall that static hazards are also eliminated by the one-hot approach to design, an advantage not shared by the STT method.

The alternative to STT or the one-hot approach to state code assignments is to "eyeball" a state code assignment that will eliminate all critical races and ORGs. This usually means making all state-to-state transitions logically adjacent (unit distance coded) by using buffer states where needed to accomplish the task, but all of this is at the expense of speed and the inability to use either the STT or the one-hot method. Dealing with FSMs having complex cross branchings often becomes too arduous and dangerous a task to warrant the use of any method other than an STT or one-hot approach. It is for this reason that these techniques are covered at length in this chapter.

14.16 DESIGN OF FUNDAMENTAL MODE FSMs BY USING PLDs

The rules pertaining to implementation of fundamental mode state machines by using programmable logic devices (PLDs) are not much different than those for synchronous FSMs. However, there are a few important, if not cardinal, rules that must be followed when implementing an asynchronous FSM by using certain types of PLDs. These rules apply to *all* fundamental-mode FSMs, including STT and one-hot designs. The rules are as follows:

1. ROMs should never be used to implement the NS and output logic. They are "noisy," and there is no compelling reason to use them. The logic noise that ROMs can generate in the NS logic of fundamental mode FSMs can cause them to malfunction. This was not the case for synchronous FSMs where the memory flip-flops served as a filters. Of course, it is possible to attach capacitors to the outputs of the NS logic functions from ROMs to filter out the logic noise. But this distorts the signal, which can cause other undesirable effects. There are much better alternatives than to use ROMs!
2. PLAs and PALs are appropriate choices to implement the NS and output functions of fundamental mode FSMs. However, it must be remembered that PALs cannot accept shared PIs and are limited to a fixed number of p-terms within any given Y function. No such restriction is placed on the use of PLAs. PALs (registered trademark of AMD, Inc.) with L-type macrocells are attractive choices because they can come equipped with feedback paths suitable for asynchronous designs. Registered PALs with internal flip-flops such as the R- and V-types should be avoided for fundamental mode FSM design unless the designer is very knowledgeable in their use for such purposes.

3. Both PLA and PAL implementations can be initialized into an all-zero state by adding a sanity input to each p-term as shown in Fig. 14.32a. If it is necessary to initialize a PLA or PAL into an all one state, introduce each y-variable as a separate p-term and connect $Sanity(H)$ to it. Obviously, it is easier to initialize 1's than 0's in a NAND-centered PLD. The reverse is true for a NOR-centered PLD.

4. Whereas FPGAs are attractive PLDs for synchronous FSM design, they can be a source of almost limitless consternation to the designer if used carelessly for fundamental mode FSM design. The reason for this lies in the fact that routing delays can seriously alter the timing behavior of asynchronous state machines (see Subsection 7.7.3). While endless cycles, critical races and static hazards may be designed out of a given FSM, routing delays can cause essential hazard formation that will most certainly cause malfunction. It is recommended that only the most skilled user of FPGAs attempt to use them to implement asynchronous state machines.

As an example, consider the PLA implementation of the NS and output functions given by Eqs. (14.35) and representing the STT FSM in Fig. 14.33a. The $Sanity(L)$ input initializes or resets the FSM into the 0000 state as required by Fig. 14.33a. Notice that all terms in Eqs. (14.35) must be accounted for in the p-term table, including hazard cover. Also, observe that the outputs P and Q have been initialized via the shared PIs. This is really not necessary, but it is convenient. Avoiding initializing the outputs, in this case, would require that the p-terms that make up the output functions be listed separately with 0's for these terms appearing in the Sanity column. A PLA of minimum dimensions $7 \times 16 \times 6$ is required by the p-term table in Fig. 14.42.

Notice that the p-terms for Y_1, P, and Q are not listed separately in the p-term table of Fig. 14.42. However, they are there. Because these particular p-terms are covered by functions Y_3 and Y_2 (they are shared PIs), they need not be listed separately. This is the advantage of PLA implementation over that with a PAL. It may be recalled that because PALs are programmed only in the AND plane they cannot accept shared PIs as is done in PLA p-term table of Fig. 14.42. It is important for the reader to remember this distinction.

14.17 ANALYSIS OF FUNDAMENTAL MODE STATE MACHINES

The procedure for asynchronous FSM analysis is essentially the reverse of that for design. The following summarizes the five-step procedure to be used in analyzing fundamental mode machines:

1. Given the circuit to be analyzed, read the circuit to obtain the NS and output logic expressions.

2. Map the NS and output logic functions into EV K-maps that have as their coordinates the present state variables, y_i. If the asynchronous FSM has been designed by using the nested cell model, the S_i and R_i state variables must be converted to Y-variable form by using the conversion relation given by Eq. (14.37). For state variables exceeding four in number use K-map formats of the type shown in Figs. 4.37, 5.6, and 5.7 all with external inputs as the only EVs.

3. From the Y K-maps, construct the Present State/Inputs/Next State (PS/NS) table with the inputs represented in canonical SOP form. Inclusion of the output data in the PS/NS table is necessary only if the output-forming logic is complex enough to warrant it.

		PLA Inputs							PLA Outputs					
		y_3	y_2	y_1	y_0	S	T	Sanity*	y_3	y_2	y_1	y_0	P	Q
	P-terms	I_6	I_5	I_4	I_3	I_2	I_1	I_0	O_5	O_4	O_3	O_2	O_1	O_0
Y_3	$y_3\bar{S}\bar{T}$	1	-	-	-	0	0	1	1	0	0	0	0	0
	$y_1\bar{S}T$	-	-	1	-	0	1	1	1	0	1	0	0	1
	$y_3\bar{y}_1 ST$	1	-	0	-	1	1	1	1	0	0	0	0	1
	$y_0 S\bar{T}$	-	-	-	1	1	0	1	1	0	0	0	1	1
	⋮	⋮							⋮					
	$y_3\bar{y}_1 y_0 S$	1	-	0	1	1	-	1	1	0	0	0	0	1
Y_2	$\bar{y}_3\bar{S}\bar{T}$	0	-	-	-	0	0	1	0	1	1	0	0	1
	⋮								⋮					
	$\bar{y}_3 y_2 \bar{S}$	0	1	-	-	0	-	1	0	1	0	0	0	0
Y_0	$\bar{S}\bar{T}$	-	-	-	-	0	0	1	0	0	0	1	0	0
	$y_2\bar{S}T$	-	1	-	-	0	1	1	0	0	0	1	0	0
	$y_0 S$	-	-	-	1	1	-	1	0	0	0	1	0	0
	⋮				⋮				⋮					
	$y_2 y_0 T$	-	1	-	1	-	1	1	0	0	0	1	0	0

* Indicates a sanity(L) input.

FIGURE 14.42
P-term table for the PLA implementation of the NS and output functions of the STT FSM expressed by Eqs. (14.35) showing a *Sanity*(L) input as required to initialize into the 0000 state.

4. Construct a fully documented state diagram from the PS/NS table. This diagram should be of the general form illustrated in Fig. 14.4.
5. Analyze the state diagram, together with the NS and output functions, for state construction problems and possible timing defects. The state construction problems may include violations of the sum rule, the mutually exclusive rule, and the initialization requirements. The timing defects include endless cycles, critical races, and static

14.17 ANALYSIS OF FUNDAMENTAL MODE STATE MACHINES

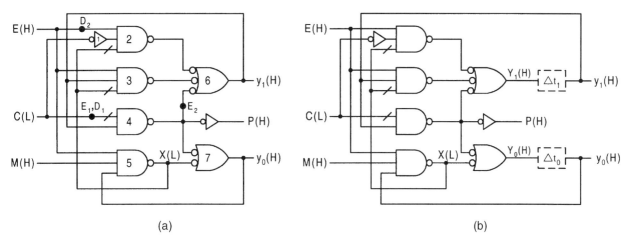

FIGURE 14.43
Logic circuits for the pulse synchronizer module (PSM) used as an analysis example. (a) Mixed-logic circuit without the fictitious LPD memory elements. (b) Logic circuit showing the two fictitious LPD memory elements.

hazards in both the NS and output functions. A complete timing analysis should include essential hazards, though these timing defects are only potentially active depending on the existence of certain unintended path delays at specific locations in the logic circuit. Although this was not mentioned earlier, the nature of the external inputs must also be considered. *Signals from mechanical switches must usually be debounced.* Nowhere can bounce periods be more disruptive to the operation of a sequential FSM than in asynchronous state machine operation. These bounce periods may last into the millisecond range with amplitudes that may cross the switching thresholds tens to thousands of times. Finally, make certain the initialization circuitry is functionally correct.

AN EXAMPLE Consider the logic circuit in Fig. 14.43a representing an FSM called the *pulse synchronizer module* or *PSM*. The PSM has three inputs E (for pulse enable), C (for clock), and M (for mode), and one output P (for pulse). It is the goal of this analysis to determine how the PSM functions and to identify any problems or potential problems it may have. First, it is required to obtain the NS and output functions. To do this, the circuit in Fig. 14.43b, which includes the fictitious LPD memory elements, is read as follows: Let $X = y_0 E M$ and $\bar{X} = \bar{y}_0 + \bar{E} + \bar{M}$. Then

$$\begin{cases} Y_1 = EC\bar{X} + y_1 E\bar{X} + y_1 \bar{C} \\ \quad = \bar{y}_0 EC + EC\bar{M} + y_1 \bar{y}_0 E + y_1 E\bar{M} + y_1 \bar{C} \\ Y_0 = y_1 \bar{C} + y_0 E M \\ \qquad\qquad P = y_1 \bar{C} \end{cases}, \qquad (14.43)$$

where a single shared PI, $y_1\bar{C}$, is seen to exist in both NS functions and in the output function. A hazard analysis of the NS and output functions in Eqs. (14.43) indicates that

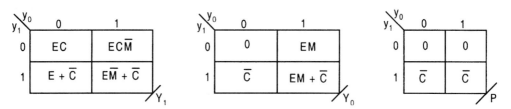

FIGURE 14.44
Next state and output K-maps as plotted from Eqs. (14.43).

the two externally initiated static 1-hazards that would have been active in function Y_1 are covered by terms $y_1 \bar{y}_0 E$ and $y_1 E \bar{M}$. Thus, there are no active static hazards present in the PSM. The positions E_1, D_1, E_2 and D_2 shown in Fig. 14.43a are used later in connection with E-hazard analyses.

The next step is to map the NS and output functions of Eqs. (14.43) as shown in Fig. 14.44. Notice that up to three EVs control the branching of the PSM, and that the shared PI $y_1 \bar{C}$ is readily discernible in the y_1 domain of each of the three K-maps.

The final step is to construct the PS/NS table from the K-maps in Fig. 14.44 and then use the PS/NS table to construct the state diagram as is done in Figs. 14.45a and 14.45b. This, of

PS $y_1 y_0$	Inputs	NS $Y_1 Y_0$	PS $y_1 y_0$	Inputs	NS $Y_1 Y_0$
0 0	$\bar{E}\bar{C}$	0 0	1 0	$\bar{E}\bar{C}$	1 1
	$\bar{E}C$	0 0		$\bar{E}C$	0 0
	$E\bar{C}$	0 0		$E\bar{C}$	1 1
	EC	1 0		EC	1 0
0 1	$\bar{E}\bar{C}\bar{M}$	0 0	1 1	$\bar{E}\bar{C}\bar{M}$	1 1
	$\bar{E}\bar{C}M$	0 0		$\bar{E}\bar{C}M$	1 1
	$\bar{E}C\bar{M}$	0 0		$\bar{E}C\bar{M}$	0 0
	$\bar{E}CM$	0 0		$\bar{E}CM$	0 0
	$E\bar{C}\bar{M}$	0 0		$E\bar{C}\bar{M}$	1 1
	$E\bar{C}M$	0 1		$E\bar{C}M$	1 1
	$EC\bar{M}$	1 0		$EC\bar{M}$	1 0
	ECM	0 1		ECM	0 1

(a)

FIGURE 14.45
(a) PS/NS table obtained from the Y K-maps in Fig. 14.44 and (b) the state diagram for the FSM represented by the logic circuit in Fig. 14.43 as derived from the PS/NS table in part (a) and the output K-map in Fig. 14.44.

14.17 ANALYSIS OF FUNDAMENTAL MODE STATE MACHINES

course, follows the same procedure as used in the analysis of synchronous FSMs discussed in Section 10.13. Remember that the PS/NS table is, in reality, a tabular representation of the state diagram, one that can be read by a computer.

An inspection of the state diagram in Fig. 14.45b indicates that no endless cycles or critical races exist in the PSM. Notice that the race conditions from $01 \to 10$ and from $11 \to 00$ are properly dealt with in the state diagram. In each case the requirements for noncritical race conditions are satisfied in agreement with Figs. 14.19a and 14.19b. Furthermore, a cursory inspection of the state diagram reveals that no ORGs are present. Thus, the PSM is free of any apparent timing problems, including any active static hazards.

There still remains the problem of determining the function of the pulse synchronizer module. Again, an inspection of the state diagram provides the information needed. As can be seen from the state diagram, an output P can be generated only in state 10 and 11 and then only under the input condition $\bar{C}$. A transition $10 \to 11$ must produce an output since the condition $\bar{C}$ is satisfied. However, the output in state 10 during this transition is of little or no consequence since $\bar{C}$ is an exiting condition from this state. If the FSM enters state 11 and then exits from that state to state 00 on $\bar{E}C$ or to state 01 on ECM, only one pulse is issued. If, on the other hand, the PSM is caused to cycle with the C waveform between states 10 and 11 under condition $E\bar{M}$, then a pulse is issued with each falling edge of the C input. Thus, multiple pulses are possible only under the cyclical condition $E\bar{C}\bar{M} \leftrightarrow EC\bar{M}$, whereas a single pulse is issued from state 11 under input conditions $E\bar{C} \to ECM$ or $E\bar{C} \to \bar{E}C$. This assumes that C and E are never permitted to change at the same time in exiting either from state 10 or from state 11. This information is confirmed by the timing diagram shown in Fig. 14.46, which is the result of a logic simulation of the logic circuit in Fig. 14.43a.

As has been previously stated, there are no active timing defects present in the PSM — this FSM will operate as predicted. However, essential hazards can become active if unintended asymmetric delays, exceeding certain magnitudes, occur at specific locations in the logic circuit. Following the minimum requirements for E-hazard and d-trio formation given

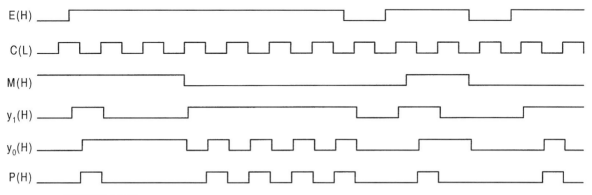

FIGURE 14.46
Simulation results of the PSM logic circuit in Fig. 14.43(a) verifying the pulse P dependence on inputs E, C and M, and showing no critical races, ORGs, or static hazards, all as predicted from Eqs. (14.43) and Fig. 14.45b.

746   CHAPTER 14 / ASYNCHRONOUS STATE MACHINE DESIGN AND ANALYSIS

in Fig. 14.27 and considering the indirect path requirements listed in Subsection 14.10.4, the possible E-hazards and d-trios can be easily determined. However, before continuing in this section, the reader should review the E-hazard analysis of FSMs as discussed in connection with Figs. 14.28, 14.29, and 14.30. With reference to the various gate numbers and delay positions (E_1, D_1, E_2, D_2) in Fig. 14.43a, the following summarizes the ideal requirements for activation of the E-hazards and d-trios in the PSM:

1. The E_1-hazard path is $00 \rightarrow 10 \rightarrow 11 \rightarrow 01$ for input conditions $E\bar{C}M \rightarrow ECM$ while in origin state 00 with a theoretical activation delay of $\Delta t_{E_1} > (\tau_1 + \tau_2 + \tau_6)$ on the C input to ANDing race gate 4 ($y_1\bar{C}$) in y_0 at position E_1, as indicated in Fig. 14.43a. Note that the indirect path is indicated by the inequality which includes an inverter $\tau_1 = \tau_{INV}$. Here, the indirect path must contain the initiator as either C or $\bar{C}$, must not be inconsistent with E, M, and must not be inconsistent with state $a = 00$ meaning $\bar{y}_1\bar{y}_0$. This requires that the indirect path be via gate $EC\bar{X}$ (gate 2) representing $\bar{y}_0EC$ in Y_1.

2. The D_1-trio path is $00 \rightarrow 10 \rightarrow 11 \rightarrow 10$ for input conditions $E\bar{C}\bar{M} \rightarrow EC\bar{M}$ while in origin state 00 with a theoretical delay of $\Delta t_{D_1} > (\tau_1 + \tau_2 + \tau_6)$ on the C line to ANDing race gate 4 at position D_1 shown in Fig. 14.43a. This d-trio causes a glitch in P. The indirect path for the D_1-trio must contain C or $\bar{C}$ and must not be inconsistent with E, $\bar{M}$ (they are constant) or with state 00. Thus, the indirect path and minimum path delay required for activation of the D_1-trio are the same as that for the E_1-hazard just discussed.

3. The E_2-hazard path is $10 \rightarrow 11 \rightarrow 01$ for input conditions $ECM \rightarrow E\bar{C}M$ while in origin state 10 with a theoretical activation delay $\Delta t_{E_2} > (\tau_7 + \tau_5 + \tau_3)$ on the C input to gate 4 ($y_1\bar{C}$) at position E_2, the path to ORing race gate 6 in y_1. Here, the indirect path must contain C or $\bar{C}$ and must not be inconsistent with E, M (which are constant) or with the initiating state 10 meaning $y_1\bar{y}_0$. It follows that the indirect path must be via y_0 through gates 7, 5, and 3.

4. The D_2-trio path is $01 \rightarrow 00 \rightarrow 10 \rightarrow 00$ for input conditions $ECM \rightarrow \bar{E}CM$ while in origin state 01 with a theoretical activation delay of $\Delta t_{D_1} > \tau_5$ on the E input to ANDing race gate 2 ($EC\bar{X}$ contains $\bar{y}_0EC$) at position D_2 to y_1. In this case the indirect path must contain E or $\bar{E}$ and must not be inconsistent with constant inputs C, M or with the initiating state 01 meaning $\bar{y}_1y_0$. Note that this d-trio causes a glitch in y_1 but *not* in P. However, the transition 01 to 00 is delayed by the d-trio path.

Again it is emphasized that E-hazards and d-trios are potential timing defects that can occur only if the minimum requirements are met, which includes an explicity located path delay that exceeds the minimum required to produce the defect. Thus, potential defects 1 and 2 require delays exceeding $2\tau_p + \tau_{INV}$ to activate them, whereas potential defect 3 requires a delay exceeding $3\tau_p$. Therefore, all three are very unlikely to be activated. However, defect 4 requires only a delay exceeding τ_p and, consequently, is more easily activated. Defects 1, 2, and 3 are guaranteed to cause malfunction of the PSM if activated, but defect 4 will cause only a delay in the $01 \rightarrow 00$ transition if activated. A counteracting delay of $2\tau_p + \tau_{INV}$ on the y_1 feedback line essentially eliminates any possibility of defects 1 and 2 from occurring. A counteracting delay of $3\tau_p$ on the y_0 feedback line safely eliminates defect 3 from occurring. To virtually eliminate the D_2-trio, a counteracting delay of magnitude τ_p should be placed

14.17 ANALYSIS OF FUNDAMENTAL MODE STATE MACHINES

on the X feedback line to ANDing race gate 2. As in previous analyses, τ_p is used here to represent the path delay through any gate in Fig. 14.43 irrespective of the number of inputs.

High-level (ideal) simulation results (not shown) verify the theoretical minimum path delay requirements for activation of the E-hazard and d-trio defects 1 through 4 previously discussed. For purposes of simulation, the inverter delay is set at $\tau_{INV} = \frac{3}{5}\tau_P$, which is the value used throughout this chapter. It is left as an exercise for the reader to run simulations on the PSM and verify again its operation and the requirements for E-hazard and d-trio formation.

A SECOND EXAMPLE. Now consider the nested cell logic circuit shown in Fig. 14.47. Notice that this FSM has three state variables y_2, y_1, and y_0, two external inputs A and B, and three Moore outputs W, X and Z, all active high. The basic cells are all of the set-dominant type, and use is made of the mixed-rail outputs from these basic cells to generate the state variable feedback signals. Also, the mixed logic external inputs $A(H)$, $A(L)$, $B(H)$, and $B(L)$ are generated by the use of inverters at the appropriate places in the circuit (but not shown) and are assumed to be bounce-free. Note that the circuit has the sanity connections required for initialization into the 000 state, all according to Fig. 14.32.

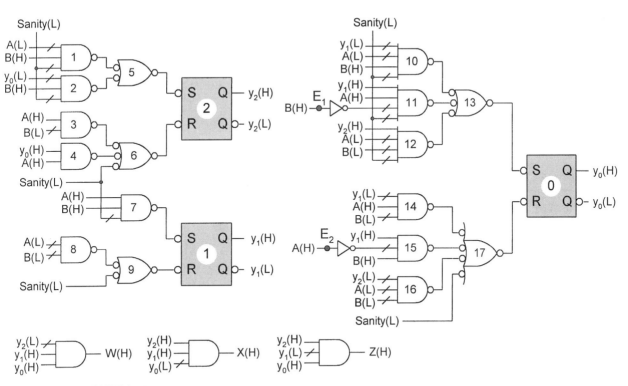

FIGURE 14.47
NS and output logic for a nested cell circuit to be analyzed, showing sanity connections required for initialization into the 000 state.

Reading the circuit in Fig. 14.47 results in the following NS and output functions exclusive of sanity inputs:

$$\begin{cases} S_2 = \bar{A}B + \bar{y}_0 B \\ R_2 = A\bar{B} + y_0 A \\ S_1 = AB \\ R_1 = \bar{A}\bar{B} \\ S_0 = \bar{y}_1 \bar{A}B + y_1 A\bar{B} + y_2 \bar{A}\bar{B} \\ R_0 = \bar{y}_1 A\bar{B} + y_1 \bar{A}B + \bar{y}_2 \bar{A}\bar{B} \\ W = \bar{y}_2 y_1 y_0 \\ X = y_2 y_1 \bar{y}_0 \\ Z = y_2 \bar{y}_1 y_0 \end{cases} \qquad (14.44)$$

In order to analyze this FSM with little difficulty, it is necessary to convert the nested cell NS functions in Eqs. (14.44) to LPD form. This is accomplished by reversing LPD-to-SR conversion expressed by Eq. (14.37), that is, by $Y = \bar{y}S + y\bar{R}$. As it is applied to design by the mapping algorithm in Section 10.6, the LPD-to-SR K-map conversion expressed by this equation is exactly the same as that for D-to-JK K-map conversion given by Algorithm 10.2 in Subsection 10.12.2. Thus, J and K are replaced by S and R, D and Q are replaced by Y and y, and subscripts A, B, and C are replaced by 3, 2, and 1, respectively. With these changes, the S, R functions in Eqs. (14.44) are mapped and converted to LPD K-map form as shown in Fig. 14.48.

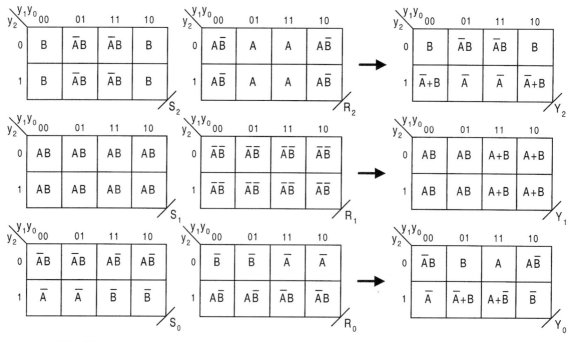

FIGURE 14.48
SR-to-LPD K-map conversion for the NS functions in Eqs. (14.44).

14.17 ANALYSIS OF FUNDAMENTAL MODE STATE MACHINES

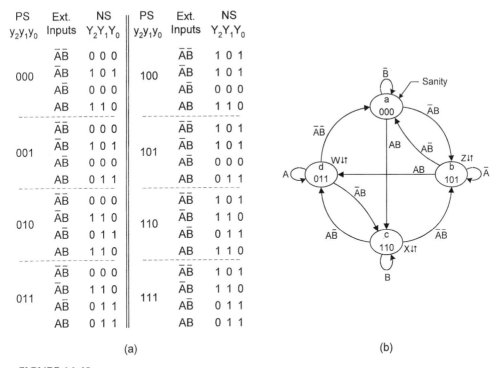

FIGURE 14.49
(a) PS/NS table derived from the K-maps of Fig. 14.48. (b) State diagram for the FSM in Fig. 14.47 as derived from the PS/NS table in (a), but excluding the four don't-care states.

In Fig. 14.49 the PS/NS table is constructed directly from the three Y K-maps in Fig. 14.48. This follows exactly the same procedure as was used for the analyses of synchronous FSMs in Section 10.13. Recall that the excitation table for the LPD model is the same as that for D flip-flops, thereby permitting the analyses for both the synchronous and asynchronous FSMs to proceed in exactly the same fashion. Observe that four of the states in the PS/NS table (001, 010, 100, and 111) have no entrance from any other state and are therefore don't-care states. The remaining states are those that contribute to the sequential behavior as indicated by the state diagram in Fig. 14.49b, which is derived directly from the PS/NS table in Fig. 14.49a and from the output functions in Eqs. (14.44). Notice also that each state-to-state transition involves a Hamming distance of 2, meaning that two state variables must change during the transition.

Having constructed the state diagram representing the FSM in Fig. 14.47, it can now be analyzed for the existence of timing defects and certain other problems it may have. A cursory inspection of the state diagram clearly indicates that no endless cycles exist. An inspection of the state diagram and PS/NS table indicates that none of the four states of the state diagram are used as race states, that each state-to-state transition has a valid branching path to the destination state according to Fig. 14.19, and that each of the three outputs is associated with a specific state in the state diagram. Therefore, no race-related problems exist, meaning that critical races and ORGs do not form. This indicates that this FSM is an STT state machine for which there are six different state code assignments possible by commuting columns of the state matrix **S** defined in Section 14.12. Note that the race states are the four don't-care states mentioned in the previous paragraph.

By analyzing the NS functions in Eqs. (14.44) together with the state diagram in Fig. 14.49b, it is found that one externally initiated static 1-hazard exists in each of the functions, S_0 and R_0. However, these s-hazards are of no consequence since basic cell 0 in Fig. 14.47 effectively filters them out — an important advantage of the nested cell approach to asynchronous FSM design. This is demonstrated by the simulation results shown in Fig. 14.50. Notice that the static 1-hazards that are formed in the S_0 and R_0 NS functions have no effect on the output y_0 because of the filtering action of the basic cell. From the simulation results it is found that the first y-variable to change in response to an external input change varies from $2\tau_p + \tau_{INV}$ to $4\tau_p$, and that the second y-variable change for a given transition may be delayed by as much as $2\tau_p$ relative to the lower limit of the first y-variable change. This means that the STT feature of the LPD design, as illustrated by the timing diagrams in Fig. 14.36, is lost when the nested cell design is used — a conclusion arrived at in Section 14.13 by comparing the simulation results in Fig. 14.38 with those in Fig. 14.36. Nevertheless, the nested cell design of the FSM in Fig. 14.47 is hazard-free and operates reliably with only minor delays in the y-transitions. As with other simulation results given in this chapter, $\tau_{INV} = \frac{3}{5}\tau_p$, where τ_p is the path delay through any gate in Fig. 14.47, including the NAND gates in the basic cells.

Were they present in the nested cell design just described, internally initiated s-hazards would not form since the basic cells are of the set-dominant (NAND) type used with SOP output-forming logic, which in this case is the same as POS logic. The reader may verify these statements by reviewing the subject matter in Subsection 11.3.2. Although Subsection 11.3.2 deals with synchronous FSMs, the conclusions arrived at here are, nevertheless, valid for asynchronous nested cell designs.

Although there are no active timing problems associated with static hazards in the FSM of Fig. 14.47, there is the potential for FSM malfunction due to the formation of essential hazards (E-hazards), as will now be discussed. However, before beginning the E-hazard analysis of this FSM, the reader should review the contents of Subsection 14.10.4.

By following the minimum requirements for E-hazard formation given in Fig. 14.27 and by noting the requirements for indirect E-hazard paths listed in Subsection 14.10.4, one can easily determine the minimum path delays required to activate any potential E-hazard that may exist in this FSM. Thus, an inspection of the state diagram in Fig. 14.49b together with Eqs. (14.44) and Fig. 14.47 indicate that two E-hazards can be activated under the following conditions:

1. The E_1-hazard path is $a \to c \to d$ for input conditions $A\bar{B} \to AB$ while in state a with a theoretical minimum activation delay of $(\Delta t_{E_1} + \tau_{INV}) > (\tau_7 + \tau_{Cell_1})$ on the B line to ANDing race gate 11 ($y_1 A \bar{B}$) in y_0 at position E_1 indicated by an enlarged node in Fig. 14.47. Here, the indirect path must be via state variable y_1, must not be inconsistent with origin state $a = 000$ (meaning $\bar{y}_2\bar{y}_1\bar{y}_0$) or with input A (which is constant), and must contain the initiator as either B or $\bar{B}$. Thus, the indirect path must be via gate 7 (AB) and basic cell 1 in y_1. Note that $\tau_{Cell} = 2\tau_p$, where τ_p is the path delay through a gate, a two-input NAND gate in this case. From this information one deduces that the theoretical minimum delay required to active

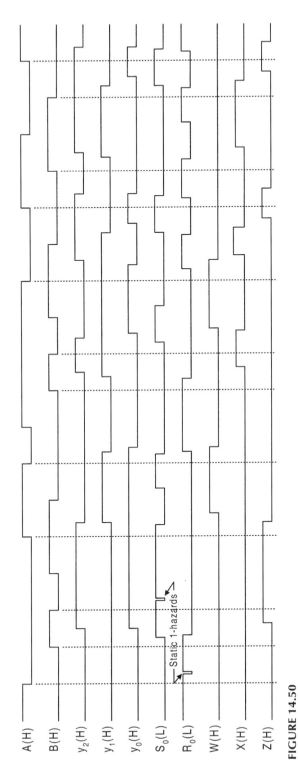

FIGURE 14.50
Simulation results of the logic circuit in Fig. 14.47 showing the NS and output response to input change; also shown are the presence of two externally initiated static 1-hazards in the NS functions, which are filtered out by the basic cell and never affect the state variables or outputs.

the E$_1$-hazard is $\Delta t_{E_1} > (3\tau_p - \tau_{INV}) = \frac{12}{5}\tau$, where $\tau_{INV} = \frac{3}{5}\tau_p$ as in previous examples. Note that for E$_1$-hazard the first and second y-invariants are and y_0 and y_1, respectively.

2. The predicted E$_2$-hazard path is $b \to d \to c$ for input conditions $\bar{A}B \to AB$ while in state b with a theoretical activation delay of $(\Delta t_{E_2} + \tau_{INV}) > (\tau_7 + \tau_{Cell_1})$ on the A line to ANDing race gate 15 ($y_1 \bar{A} B$) in y_0 at position E$_2$ indicated by the enlarged node in Fig. 14.47. Again, the indirect path must be via state variable y_1, but now must contain the initiator as A or $\bar{A}$ and must not be inconsistent with the requirements of state $b = 101$ (meaning $y_2 \bar{y}_1 y_0$) or with input B, which is constant. Thus, the indirect path must again be via gate AB (7). Therefore, the minimum path delay required to activate the E$_2$-hazard is exactly the same as that required to active the E$_1$-hazard, and the first and second y-invariants are y_0 and y_1 as before.

From the results of a high-level (ideal) simulation on the E-hazard problem, the theoretical minimum activation delays for the two E-hazards and their corresponding error transition paths are as predicted in the forgoing discussion. A counteracting delay of $\Delta t_{Correct} \geq (3\tau_p - \tau_{INV}) = \frac{12}{5}\tau$ placed on the y_1 feedback line reduces the probability for E-hazard formation to near zero. This delay is a conservative, usually safe value. However, delays less than Δt_{E_1} can also be effective as long as they meet the requirements of Eq. (14.18). It is left as an exercise for the reader to verify these results by simulation.

The important point to be made here is that E-hazards can form in any FSM of three or more states operated in the fundamental mode. This includes FSMs designed by using either the LPD model or the nested cell model. Since both STT and one-hot FSMs fall into this category, they are also subject to E-hazard formation. The following third example illustrates the point.

A THIRD EXAMPLE. As a third and final example, the logic circuit in Fig. 14.51 is to be analyzed. It is a one-hot FSM having four state variables (y_3, y_2, y_1 and y_0), two external inputs (A and B), and three outputs (W, X, and Z). It is basically the same FSM as in Fig. 14.47, except designed to operate as a one-hot FSM. This is done to compare performance and E-hazard formation between the two design methodologies. Notice that this one-hot FSM is initialized into the 0001 state via the one-hot-plus-zero circuitry, as discussed in Section 13.5 and used in the design an asynchronous FSM in Section 14.14.

Reading the logic circuit in Fig. 14.51 yields the NS and output logic given in Eqs. (14.45). Here, each NS function is separated into the "into" terms and one "out of" term following Eq. (14.40). The "out of" term, it will be recalled, is necessary to maintain the state variable of the origin state active until the transition to the destination state is complete. Each set of "into" terms includes a holding condition term that functions as the hazard cover for the internally initiated s-hazard that is formed between the "out of" term and an "into" term as indicated in Eqs. (14.45). For example, an s-hazard in Y_3 is formed between coupled terms $y_3 \bar{y}_2 \bar{y}_0$ and $y_2 A \bar{B}$ for which the holding condition term $y_3 A$ is the minimum hazard cover, since it contains the consensus term $y_3 \bar{y}_0 A \bar{B}$. Similarly, the s-hazard in Y_1 is formed between the "out of" term $y_1 \bar{y}_3 \bar{y}_0$ and the "into" term $y_0 \bar{A} B$ where the holding condition

14.17 ANALYSIS OF FUNDAMENTAL MODE STATE MACHINES

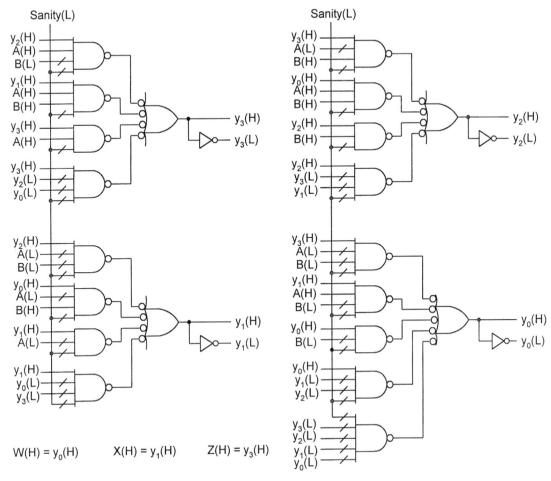

FIGURE 14.51
NS and output logic for a one-hot asynchronous FSM to be analyzed, where the polarized external inputs, A and B, are assumed to be produced by the use of inverters.

term $y_1 \bar{A}$ serves as the hazard cover. Thus, no s-hazards exist in this FSM.

$$\left\{ \begin{array}{l} Y_3 = \overbrace{y_2 A \bar{B} + y_1 A B + y_3 A}^{\text{``Into'' terms}} + \overbrace{y_3 \bar{y}_2 \bar{y}_0}^{\text{``Out of'' terms}} \\ Y_2 = y_3 \bar{A} B + y_0 A B + y_2 B + y_2 \bar{y}_3 \bar{y}_1 \\ Y_1 = y_2 \bar{A} \bar{B} + y_0 \bar{A} B + y_1 \bar{A} + y_1 \bar{y}_3 \bar{y}_0 \\ Y_0 = y_3 \bar{A} \bar{B} + y_1 A \bar{B} + \underbrace{y_0 \bar{B}}_{\substack{\text{Haz.} \\ \text{cov.} \\ \text{terms}}} + y_0 \bar{y}_2 \bar{y}_1 \\ \\ W = y_0 \\ X = y_1 \\ Z = y_3 \end{array} \right\} \quad (14.45)$$

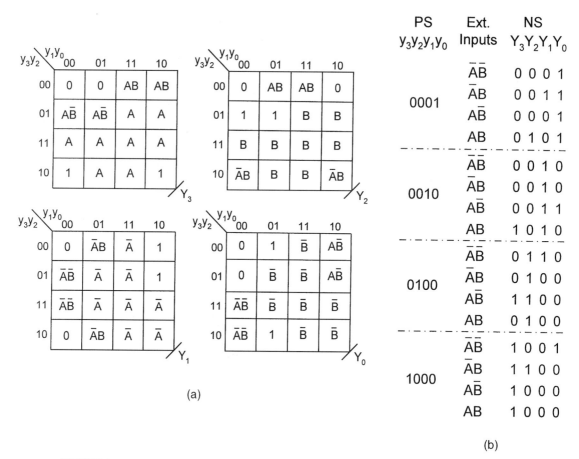

FIGURE 14.52
Analysis of the logic circuit in Fig. 14.51. (a) NS K-maps plotted from Eqs. (14.45). (b) PS/NS table derived from the the K-maps in (a) showing only the necessary logic for the four one-hot states (1, 2, 4 and 8).

The NS K-maps are easily plotted from Eqs. (14.45) and are shown in Fig. 14.52a. Notice that each K-map contains two 0's and two 1's and that a zero always appears in state 0000. Thus, state 0000 is never used as a race state, since a properly designed one-hot FSM is forced to cycle through a state having two 1's, one "1" from the origin state and the other from the destination state.

The PS/NS table for this FSM can be constructed from the K-maps in Fig. 14.52a. This is done in Fig. 14.52b, where only the one-hot states are represented. It is not necessary to represent the cycle states (those with two 1's), since they are easily deduced from the PS/NS table knowing the present and next state and the nature of one-hot FSM operation. All other states are irrelevant. The one-hot-plus-zero path is also excluded from the PS/NS table, although it is easily deduced from the logic circuit, where it is clear that initialization takes place into the 0001 state as discussed in Section 14.14.

Finally, the state diagram is constructed from the PS/NS table and is shown in Fig. 14.53a, where only the four one-hot states are represented. As expected, this state diagram is identical to that in Fig. 14.49b with one exception. The cycle states (those with two 1's) are

14.17 ANALYSIS OF FUNDAMENTAL MODE STATE MACHINES 755

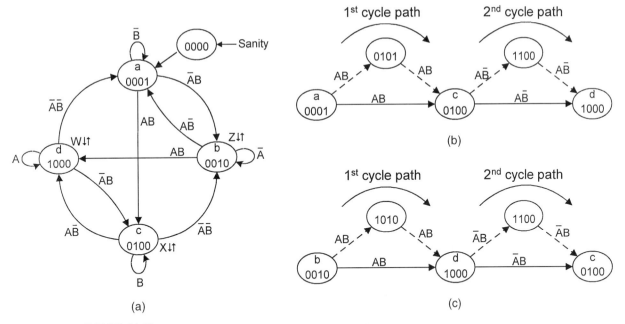

FIGURE 14.53
Analysis of the logic circuit in Fig. 14.51 (contd.). (a) State diagram derived from the PS/NS table in Fig. 14.52b, including no cycle paths but showing the 0000 state as required by the one-hot-plus-zero initialization method. (b) Cycle paths for transitions a-to-c and c-to-d. (c) Cycle paths for transitions b-to-d and d-to-c.

deliberately omitted even though they are an integral part of the state diagram — remember, the FSM is force to cycle through these states in transit from one one-hot state to another. Shown in Fig. 14.53b is a set of two such cycle paths, one for the $a \to c$ transition and the other for the $c \to d$ transition. A similar set for transitions $b \to d$ and $d \to c$ is given in Fig. 14.53c. These two sets of cycle paths will later be used for the E-hazard analysis of this FSM, but also will be useful for comparing timing performance with the previous nested cell design of this FSM given in Fig. 14.50. Notice that the state diagram includes the 0000 state required by the one-hot-plus-zero initialization method applied earlier in Section 14.14. Also, note that no endless cycles, critical races, ORGs, or s-hazards exist in this FSM.

The timing performance of the one-hot FSM is best represented by using timing diagrams taken from simulation results. This is done in Fig. 14.54, where the NS and output response to input change begins following initialization into the 0001 state. From an inspection of these waveforms the cycle paths are easily established. In each case, the destination state overlaps the origin state meaning the FSM is forced to cycle through a state with two 1's. This is accomplished via an "out of" term together with feedback. Thus, the transition 0001-to-0010 requires that state variable y_1 go active via the $y_0 \bar{A} B$ term, which in turn causes the "out of" term $y_0 \bar{y}_1 \bar{y}_2$ to go inactive, followed by the state variable y_0, thereby completing the transition. The active response of state variable $y_1(L)$ to a $0 \to 1$ change in B takes a theoretical time $2\tau_p + \tau_{INV}$, as can be seen from the logic circuit in Fig. 14.51. To deactivate state variable y_0 requires an additional period of $2\tau_p$. Therefore, a given transition

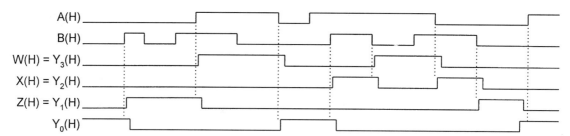

FIGURE 14.54
Simulation results of the logic circuit in Fig. 14.51 showing the NS and output response to input change for comparison with Fig. 14.50.

can be completed in no less time than $4\tau_p + \tau_{INV}$. The vertical dashed lines in Fig. 14.54 are placed for the convenience of the reader to show semiquantitatively these response times and to make it easy to observe the cycle states required for each transition. As in all previous examples, τ_p is the path delay through any gate regardless of its type or number of inputs.

A performance comparison can now be made between the one-hot and nested cell design of this FSM. From the simulation results it is concluded that the nested cell design of this FSM is at best only slightly faster on the average than the one-hot design. The nested cell design will complete a given transition in the theoretical time range of $(2\tau_p + \tau_{INV})$ to $(4\tau_p + \tau_{INV})$, whereas the one-hot design will complete a transition in no less time than $(4\tau_p + \tau_{INV})$. The outputs for the one-hot design, on the other hand, change concurrently with the state variables, as indicated in the timing diagram of Fig. 14.54. In contrast, the output response to input change for the nested cell design falls in the theoretical range of $(3\tau_p + \tau_{INV})$ to $(5\tau_p + \tau_{INV})$ which, on the average, is no faster than the output response for the one-hot design.

Both designs require NS logic for initialization purposes in accordance with Fig. 14.32. However, there is one exception. The nested cell design can be implemented with gated basic cells equipped with PR and CL overrides, the use of which permits initialization of a logic 1 or logic 0. A gated basic cell is nothing more than a basic cell with the S and R inputs introduced into a basic cell via ANDing gates with PR and CL overrides inputs connected as in Fig. 10.51a. Clearly, no CK input is necessary or desired. If gated basic cells are available in chip form, the CK input can be set active, which makes the ANDing gates transparent to the S and R inputs.

The E-hazard analysis of this one-hot FSM is carried out in accordance with Subsection 14.10.4 and Figs. 14.26 and 14.27. Thus, two E-hazard paths are identified, one $a \to c \to d$ and the other $b \to d \to c$, both of which satisfy the minimum requirements for E-hazard formation given in Fig. 14.27. As expected, these E-hazard paths are identical to those for the nested cell design since the state diagrams in Figs. 14.49b and 14.53a are the same. Now however, the cycle states must be taken into account in determining the race gate and indirect path. The following summarizes the conditions under which these two E-hazards can be activated:

1. The predicted E_1-hazard path is $a \to c \to d$ for input conditions $A\bar{B} \to AB$ with cycle paths shown in Fig. 14.53b. Here, the initiator is B, and the first and second invariants are y_3 and y_2, respectively. Note that y_1 remains inactive (logic 0). From this

14.17 ANALYSIS OF FUNDAMENTAL MODE STATE MACHINES

information and the cycle paths, it is clear that the ANDing race gate must be $y_2 A \bar{B}$ in Y_3. If the E_1-hazard is to form, a delay Δt_{E_1} of sufficient magnitude must occur on the B line to y_3 such that y_2 wins the race with B at the race gate. The indirect path (IP) must not be inconsistent with the initiating state a ($\bar{y}_3, \bar{y}_2, \bar{y}_1, y_0$) or with A (which is constant) and must contain the initiator as B or $\bar{B}$, all in Y_2. Thus, the indirect path must be via the p-term $y_0 AB$ in Y_2. Now, the minimum path delay requirement to form the E-hazard can be easily calculated to be $(\Delta t_{E_1} + \tau_{INV}) > 2\tau_P$. In this expression $2\tau_p$ derives from the $y_0 AB$ gate and the ORing of terms in Y_2; τ_{INV} on the left side results from the presence of a presumed inverter on the B line to the ANDing race gate, $y_2 A \bar{B}$ in Y_3. If now $\tau_{INV} = \frac{3}{5}\tau_P$ is introduced into this inequality, there results the minimum path delay requirement $\Delta t_{E_1} > \frac{7}{5}\tau_p$, where τ_p is propagation delay through any gate, as in previous examples. Once sufficiently activated, simulation results show that the E_1-hazard error transition path is $a \rightarrow 0101 \rightarrow 1100$, not the expected path $a \rightarrow 0101 \rightarrow c \rightarrow 1100 \rightarrow d$, and that the FSM remains stably in state 1100 (a cycle path state) — it never enters and stabilizes into the intended destination state 0100. Clearly, a serious malfunction of the FSM results.

2. The predicted E_2-hazard path is $b \rightarrow d \rightarrow c$ for input conditions $\bar{A}B \rightarrow AB$ with cycle paths indicated in Fig. 14.53c. In this case the initiator is A, the first and second invariants are y_2 and y_3, respectively, and y_0 stays inactive. Given this information and the cycle paths, the ANDing race gate is found to be $y_3 \bar{A} B$ in Y_2. To cause this E-hazard to form, a delay of at least Δt_{E_2} must exist on the A line to y_2, thereby allowing y_3 to win the race with A at the race gate. The indirect path must not be inconsistent with the initiating state b ($\bar{y}_3, \bar{y}_2, y_1, \bar{y}_0$) or with B (which is constant) and must contain the initiator as A or $\bar{A}$, all in Y_3. Therefore, it follows that the indirect path must be via the p-term $y_1 AB$ in Y_3. From this information, it is concluded that the minimum path delay requirement to form the E_2-hazard is $(\Delta t_{E_1} + \tau_{INV}) > 2\tau_P$ or $\Delta t_{E_1} > (2\tau_P - \tau_{INV})$, which is exactly the same as that calculated for the E_1-hazard formation. If sufficiently activated, simulation results indicate that the error transition path caused by E_2-hazard is $b \rightarrow 1010 \rightarrow 1100$, where again the FSM resides stably in the cycle state 1100, an obvious malfunction of the FSM.

The results of ideal simulations indicate that these two E-hazards begin to form under precisely the minimum path delay conditions predicted by the forgoing analyses. Comparing the minimum path delay requirements to activate E-hazards in the two designs, nested cell and one-hot, it is concluded that E-hazard activation is easier in the one-hot design than in the nested cell design by a gate delay τ_p. That is, to activate either of the E-hazards in the one-hot design requires a minimum path delay of $\Delta t_E > \frac{7}{5}\tau_p$, whereas for the nested cell design a delay of $\Delta t_E > \frac{12}{5}\tau_p$ is required. Here, it is assumed that $\tau_{INV} = \frac{3}{5}\tau_P$ and that τ_p is the path delay through any gate.

Remember, E-hazards are only potential timing defects that may never be activated under normal operation of a real FSM — even if the theoretical minimum path delay requirements are just exceeded. However, should these timing defects be sufficiently activated, malfunction of the FSM is guaranteed. By the expression "sufficiently activated" is meant that an asymmetric delay of sufficient magnitude must exist on a specific path (noted in the analysis) to cause the E-hazard to form. In a real circuit this may require exceeding the theoretical minimum path requirement to a significant extent before malfunction occurs. Nevertheless, the designer might be prudent to include some counteracting delay on specific feedback lines to further ensure that these timing defects will never occur. Modern logic circuits are now commonly constructed of very high-speed logic. If, for example, gate propagation delays exist in the subnanosecond range, it does not take much of a lead delay in a specific path to activate an E-hazard. Such delays may be caused by parasitic capacitance and resistance, by buffers, or by gates that have abnormally large path delays.

FURTHER READING

Unfortunately, significant reference material in the area of asynchronous state machines design and analysis is limited to a few text sources. Only the texts of Comer, Dietmeyer, Fletcher, Kohavi, Nelson *et al.*, Roth, Tinder, and Yarbrough devote one or more chapters to this subject. However, some of these texts provide only a superficial treatment. Though of an older vintage, the text of Unger is devoted entirely to this subject and still stands as one of the better sources of information on asynchronous FSMs. However, the reader will find this text, like those of Dietmeyer and Kohavi, somewhat difficult to grasp on first reading. Unger's text, for example, is theorem and lemma based.

[1] D. J. Comer, *Digital Logic and State Machine Design*, 3rd ed. Saunders College Publishing, Fort Worth, TX, 1995.
[2] D. L. Dietmeyer, *Logic Design of Digital Systems*, 2nd ed. Allyn and Bacon, Inc., Boston, Mass, 1978.
[3] W. I. Fletcher, *An Engineering Approach to Digital Design*. Prentice Hall, Englewood Cliffs, NJ, 1980.
[4] Z. Kohavi, *Switching and Finite Automata Theory*. McGraw-Hill, New York, 1978.
[5] V. P. Nelson, H. T. Nagle, B. D. Carroll, and J. D. Irwin, *Digital Logic Circuit Analysis and Design*. Prentice Hall, Englewood Cliffs, NJ, 1995.
[6] C. H. Roth, *Fundamentals of Logic Design*, 4th ed. West Publishing Co., St. Paul, MN, 1992.
[7] R. F. Tinder, *Digital Engineering Design: A Modern Approach*. Prentice Hall, Englewood Cliffs, NJ, 1991.
[8] S. H. Unger, *The Essence of Logic Circuits*. Prentice Hall, Englewood Cliffs, NJ, 1989.
[9] J. M. Yarbrough, *Digital Logic Applications and Design*. West Publishing Co., Minneapolis/St. Paul, MN, 1997.

Perhaps the most frequently cited reference on the STT approach to asynchronous FSM design is the article by Tracey. Other than that only the texts by Dietmeyer and Unger (previously cited) appear to be worthy of mention with regard to published work on this subject.

[10] J. H. Tracey, "Internal State Assignments for Asynchronous Sequential Machines," *IEEE Trans. on Electronic Computers*, Vol. EC-15, Aug. 1966, pp. 551–560.

The one-hot method in state machine design is apparently offered for significant further reading in only two texts, those by Hayes and by Nelson *et al.* (previously cited). Both contribute something different to the subject and are recommended. To a much lesser extent this subject is covered in the texts by Comer, Dietmeyer, and Unger (all previously cited).

[11] J. P. Hayes, *Introduction to Digital Design*, Addison-Wesley, Reading, MA, 1993.

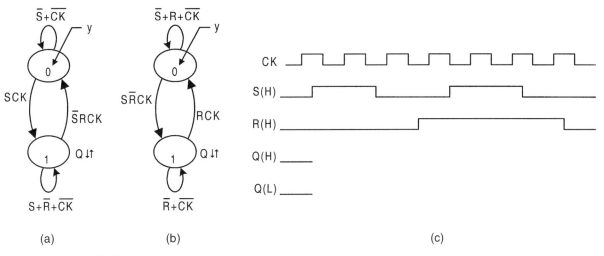

FIGURE P14.1

PROBLEMS

14.1 Problem 10.3 of Chapter 10 deals with the clocked set-dominant basic cell. There, questions are asked based on an expression that is provided without explanation of its origin. This exercise provides the basis for this expression together with that for the clocked reset-dominant basic cell.

Shown in Figs. P14.1a and p14.1b are the state diagrams for the clocked set- and reset-dominant basic cells, respectively. Notice the similarities with the state diagrams in Figs. 14.7a and 14.9a.

(1) Use the lumped path delay (LPD) model to obtain an optimum design for each of these asynchronous FSMs. (Hint: For the reset-dominant basic cell use maxterm code to extract minimum cover from the EV K-map.)

(2) Implement the *set-dominant basic cell* by using four NAND gates (nothing else), and let the inputs be active high. Implement the *reset-dominant basic cell* by using two NOR gates and two AND gates (nothing else), and let the inputs be active high.

(3) Complete the waveforms in Fig. P14.1c for each FSM by following the examples in Figs. 14.8 and 14.10. Keep in mind the action of clock and the nature of the set- and reset-dominant behavior of these clocked basic cells. Verify your results by using a simulator.

14.2 In Section 14.7 the two-input rendezvous module (RMOD) is designed by using the nested cell model.

(a) Given the state diagram for the two-input RMOD in Fig. 14.11, design this FSM by using the LPD model. End with an optimum logic circuit that will generate both $y(H)$ and $y(L)$, as is done in Fig. 14.11e.

(b) In what way does the LPD design of the two-input RMOD differ from that of the nested cell design? Use the waveforms for $y(H)$ and $y(L)$ in Fig. 14.12 to support your explanation.

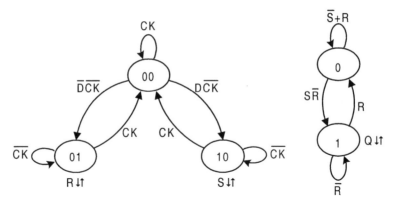

FIGURE P14.2

(c) Write a generalized expression for an n-input RMOD that is designed by the LPD model.

14.3 Presented in Fig. P14.2 are the state diagrams for the FET D flip-flop. Given the state code assignment indicated, design this flip-flop by using the LPD model. To do this, follow the example in Figs. 14.15 and 14.16 for the RET D flip-flop. End with an optimum logic circuit by using six NOR gates and a single inverter (nothing else). (Hint: Avoid using the don't care in the NS K-maps.)

14.4 In Fig. 12.12, a D flip-flop is used to design a toggle module (a divide-by-2 counter). Shown in Fig. P14.3 are the state diagrams for the toggle module.
(a) Design this flip-flop by using the LPD model and end with a logic circuit consisting of six NAND gates (nothing else). To do this follow the design of the RET D flip-flop in Section 14.8.
(b) Is this design the same as that in Fig. 12.12? Explain your answer.
(c) Demonstrate the operation of the toggle module by simulating the logic circuit of part (a).

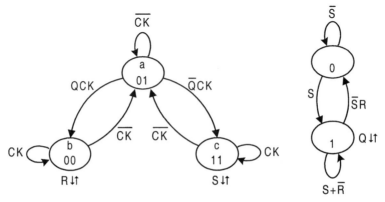

FIGURE P14.3

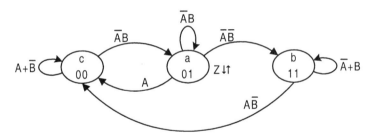

FIGURE P14.4

14.5 An asynchronous FSM that has two inputs, A and B, and one output, Z, operates as follows: Whenever $B = 1$, then $Z = 0$. But if $B = 0$ and $Z = 0$, a change in A causes Z to be $Z = 1$. The output Z cannot change to $Z = 0$ until B changes to $B = 1$.
  (a) Construct the state diagram for this FSM. Make certain that it is free of endless cycles, critical races, and ORGs. (Hint: The state diagram should be one of four states.)
  (b) Use the LPD model to obtain the NS and output logic that is free of static hazards. End with a gate-minimum NOR/INV logic circuit for this FSM. Assume that the inputs and output are all active high.

14.6 In Fig. P14.4 is a three-state FSM that is to be operated in the fundamental mode.
  (a) Analyze this FSM for possible endless cycles and critical races. If either of these timing defects exists, indicate how it can be eliminated.
  (b) Design this FSM by using the LPD model. To do this, find an optimum set of SOP NS and output logic expressions. Analyze the NS logic for possible static 1-hazards. If any exist, indicate their type (internally initiated or externally initiated) and add the necessary hazard cover to the original NS expressions. To do this, follow the examples in Subsection 14.10.3.
  (c) Repeat part (b) for an optimum set of POS NS logic expressions and analyze them for possible static 0-hazards.
  (d) Based on the results of part (b), construct an optimum NAND/INV logic circuit. Assume that the inputs arrive active high and that the output is active low.
  (e) Analyze this FSM for potential essential hazards (E-hazards) and d-trios. If any exist, give the direct and indirect paths, race gates, branching conditions, and the theoretical minimum path delay requirements for their formation. Use a timing diagram to illustrate their formation. On the logic circuit, show where the counteracting delay must be placed to reduce the probability of E-hazard formation. To accomplish all of this, follow the example in Subsection 14.10.4.

14.7 Repeat parts (b), (d) and (e) of Problem 14.6 for the design of the FSM in Fig. P14.4 by using the nested cell model. Thus, design this FSM by using two set-dominant basic cells as the memory. To do this, follow the example in Fig. 14.11, but now with two state variables y_1 and y_0. What conclusion do you come to with regard to the presence of static hazards in the NS functions? Explain. If your results indicate that E-hazards are possible in the nested cell design of this FSM, explain why this is so.

14.8 The state diagram for the resolver of an RET D flip-flop is shown in Fig. 14.15a, and the resulting NS and output expressions are given by Eqs. (14.9). Two potentially active d-trios exist in this FSM.

(a) Run a complete d-trio analysis of this FSM. To do this, give the direct and indirect paths, race gates, branching conditions, and theoretical minimum path delay requirements for their formation. Also, indicate what problems they could cause should they become active.

(b) On the logic circuit, show where the counteracting delays must be placed to reduce the probability of d-trio formation.

(c) Use timing diagrams to verify the results of part (a). Simulate if necessary.

14.9 The FSM represented by the state diagram in Fig. P14.4 is to be designed by using the STT array algebraic approach. To accomplish this, refer to Section 14.12 and do the following:

(a) Construct the state table (including the output) for this FSM. From the state table, obtain the state matrix **S** and the destination matrix **D**.

(b) Given the state code assignment indicated, use the array algebraic approach to obtain the next-state logic expressions for Y_1 and Y_0. Are static hazards possible in the NS expressions? If so, give the hazard cover for any hazard that may exist.

(c) Repeat the array algebraic approach to obtain the output function, Z. Prove that ORGs are not possible and that no static 1-hazard is associated with the output.

(d) Run a complete E-hazard analysis on this FSM (a logic circuit is not necessary). If an E-hazard or d-trio exists, give the direct and indirect paths, race gates, branching conditions, and the theoretical minimum path delay requirements for its formation. Also, if one of these potential defects exists, indicate the magnitude of the counteracting delay and the position where it must be placed to reduce the probability of E-hazard or d-trio formation to near zero.

14.10 In Fig. P14.5 is the state table for an FSM that has two inputs and three outputs.

(a) Design this FSM by using the *STT array algebraic approach*. To do this, follow the example in Section 14.12. Note that the FSM, as it stands, has two cycles that must be eliminated, while retaining the algorithm, before the STT approach can be applied. Plan to initialize the FSM into state $a = 000\cdots$. It is also a requirement that state d have a state code assignment of $11\cdots$. Remove any

$Y_{m-1}\cdots y_1 y_0$ \ AB	00	01	11	10	R	S	T
a	b	b	a	a	0	0	B
b	b	c	a	b	A	0	0
c	a	c	d	c	0	0	0
d	b	d	d	a	0	$\overline{A}$	0

FIGURE P14.5

static hazards that may be present in the NS and output logic. End with a valid optimized set of NS and output expressions for the array algebraic method. Do not implement the circuit. (Hint: The state diagram or state table for an FSM is useful in identifying and eliminating cycles.)

(b) Use EV K-maps and a logic minimizer (e.g., BOOZER) to obtain the NS and output functions directly from the state diagram for this FSM. To do this, use the LPD model together with the STT state code assignment, and include any static hazard cover that may be necessary. Use the gate/input tally (exclusive of inverters) to compare these results with those obtained in part (a). Are critical races and ORGs possible in either design when using the STT state code assignments? Explain.

(c) Run a complete E-hazard and d-trio analysis on the results of part (a). If any are present, give the direct and indirect paths, race gates, branching conditions, and theoretical minimum path delay requirements for their formation. To do this, follow the example in Subsection 14.10 but without a logic circuit. Also, if E-hazards or d-trios exist, indicate the location and magnitude of the counteracting delays that will reduce the probability of their formation to near zero.

14.11 Note: This problem should be undertaken only after completing Problem 14.10.

(a) Use the *one-hot approach* to design the FSM represented by the state table in Fig. P14.5, but only after removal of the cycles, as in Problem 14.10. End with a set of NS and output functions as read directly from the corrected state diagram or corrected state table obtained in Problem 14.10(a). Use the gate/input tally to compare the results of the one-hot design with the STT design in parts (a) and (b) of Problem 14.10.

(b) Analyze this FSM for possible static 1-hazards and ORGs. If any exist, indicate their origin and the means by which they can be eliminated. Are E-hazards possible in a one-hot design? Explain.

(c) Discuss the factors that affect the relative FSM speeds of the two designs (STT and one-hot). Which of the two designs is expected to be the faster, if either?

14.12 Presented in Fig. P14.6 is the state diagrams for an FSM with two inputs and four outputs.

(a) Construct the state table for this FSM and include the outputs.

(b) Use the *STT approach* to design this asynchronous FSM. To do this, find an STT state matrix **S** that satisfies the partial state assignment indicated and follow the example in Section 14.12. End with a complete set of NS and output logic expressions. Assuming that the transition $a \rightarrow b$ cannot occur, comment on its function in the design of this problem.

(c) Analyze the FSM for static 1-hazards and eliminate any that exist. If hazard cover is required, first check for redundant terms then eliminate the hazards. Assume that the inputs arrive active high. Also, prove that no ORGs exist.

(d) Construct a logic circuit for the results of parts (b) and (c), and initialize the FSM into the 111 state as required by the state diagram. For initialization, refer to Section 14.11.

(e) Verify the proper operation of this FSM by simulation.

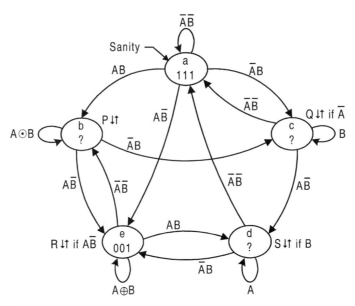

FIGURE P14.6

14.13 The state diagram for an FSM having two inputs and three outputs is shown in Fig. P14.7.
(a) Construct the state table for this FSM.
(b) Design this FSM by using the *STT approach*. To do this, find an STT state matrix **S** that satisfies the partial state codes indicated and follow the example

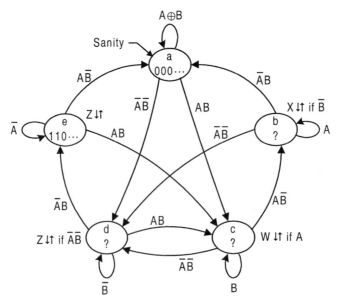

FIGURE P14.7

PROBLEMS

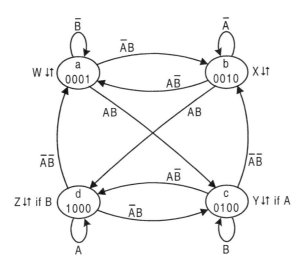

FIGURE P14.8

in Section 14.12. End with a complete set of NS and output logic expressions. (Hint: This FSM can be designed by using five state variables.)

(c) Analyze the FSM for static 1-hazards and eliminate any that exist. Also, prove that no ORGs exist. Assume that the external inputs arrive from positive logic sources.

(d) Construct a logic circuit for the results of part (b), and initialize the FSM into the all zero state as required by the state diagram. For initialization, refer to Section 14.11.

(e) Verify the proper operation of this FSM by simulation.

14.14 In Fig. P14.8 is the state diagram for an FSM that has two inputs and four outputs, and that is to be operated in the fundamental mode.

(a) Design this FSM by using the *one-hot method*. End with a valid set of NS and output expressions. The design must be free of critical races, ORGs, and static hazards. To do this, follow the example in Section 14.14.

(b) Run a complete E-hazard analysis on this FSM. If any are present, give the direct and indirect paths, race gates, branching conditions, and theoretical minimum path delay requirements for their formation. Also, if E-hazards or d-trios exist, indicate the location of the counteracting delays that will reduce the probability of their formation.

(c) What are the advantages and disadvantages to the one-hot method.

(d) Without constructing a logic circit, explain how this FSM can be initialized into the 00001 state (state a).

14.15 Shown in Fig. P14.9 are the state diagrams for two FSMs, each with two inputs and two outputs, that are to be operated in the fundamental mode.

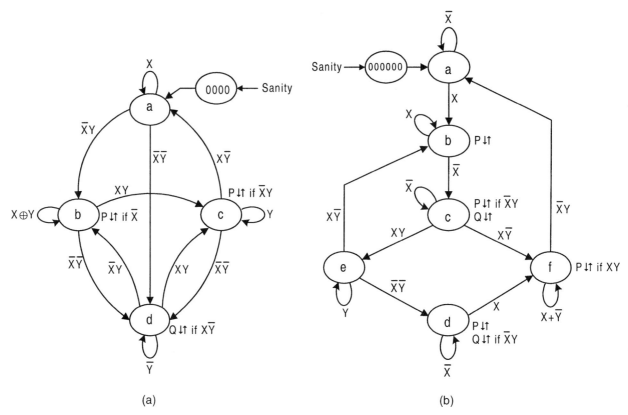

FIGURE P14.9

(1) Design each of these FSMs by using the *one-hot-plus-zero method* as in Section 14.14. Thus, obtain a complete set of NS and output expressions free of critical races, ORGs and static hazards.

(2) Construct the logic circuit for each FSM assuming that the inputs and outputs are all active high.

(3) Verify the proper operation of each design by simulation.

14.16 Presented in Fig. P14.10 is the state diagram of an asynchronous FSM that has two inputs and four outputs.

(a) Design this FSM by using the *one-hot code method*. End with a valid set of logic equations for the NS and output functions that are free of critical races, ORGs, and static hazards. Plan to use the one-hot-plus-zero approach to initialization such that the next transition is into state a.

(b) Construct the p-term table for this FSM that is suitable for a PLA implementation. Assume that the inputs are active high. Take the outputs as $P(H)$, $Q(H)$, $R(L)$, and $S(L)$. What are the minimum dimensions for the PLA? Can (or should) a ROM be used to implement this FSM? Explain.

14.17 (a) Construct the state table for the asynchronous FSM in Fig. P14.10.

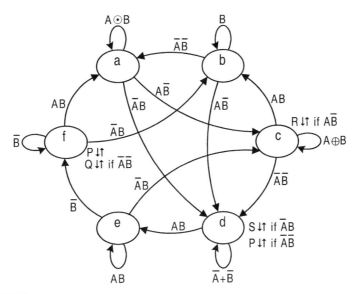

FIGURE P14.10

(b) Find the minimum number of state variables required to design the FSM in Fig. P14.10 by the *STT array algebraic approach*. (Hint: Fewer than eight state variables are required.) Given this result, which approach to design of this FSM (the one-hot or STT) would appear to be the most practical? Base your answer on the hardware commitment that is expected for each of these design methods.

(c) Obtain a suitable state matrix **S** and the corresponding destination matrix **D** from the results of parts (a) and (b).

(d) Obtain a complete set of NS and output functions from the results of parts (a), (b) and (c). What size PLA would be required to implement these results? (Hint: For the state assignment map use the format $y_6 y_5 / y_4 y_3 \| y_2 / y_1 y_0$, similar to that used in Fig. 5.7 except with third-order submaps.)

14.18 The following NS and output logic is read from an FSM that is designed to operate in the fundamental mode. Here, the inputs are A and B and the outputs are X and Z.

$$Y_2 = y_2 \bar{A} \bar{B} + A\bar{B}$$
$$Y_1 = y_1 \bar{A} \bar{B} + \bar{A} B + y_1 y_0 A B$$
$$Y_0 = y_2 \bar{A} \bar{B} + \bar{A} B + y_0 A B + A\bar{B}$$
$$X = y_2 \bar{A} \bar{B} + y_1 y_0 A B$$
$$Z = y_1 \bar{A} \bar{B} + \bar{y}_1 y_0 A B$$

(a) Construct the state diagram and state table for this FSM. Identify any don't-care states that are associated with it.

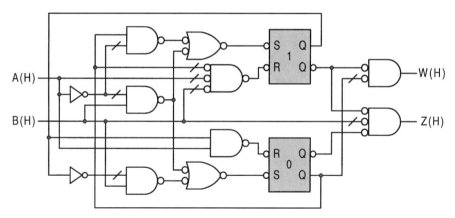

FIGURE P14.11

(b) From the state diagram, determine if this FSM has any obvious transition or output problems. Does it satisfy the basic rules for the proper operation of a fundamental mode FSM?

(c) Check the FSM for possible endless cycles, critical races, and static hazards in both the NS and output logic. If any exist, indicate their origin.

(d) What type of FSM (LPD, STT or one-hot) is this FSM?

14.19 Shown in Fig. P14.11 is the logic circuit for a fundamental mode FSM that has been designed by using the nested cell model.

(a) Analyze this FSM. To do this, first construct its state diagram by following the example in Figs. 14.47, 14.48, and 14.49. Analyze this FSM for critical races, ORGs, and static hazards. If any exist, indicate their type, origin, and the means to eliminate them.

(b) Run complete E-hazard and d-trio analyses on this FSM. If any are present, give the direct and indirect paths, race gates, branching conditions, and minimum path delay requirements for their formation, and indicate the location of the counteracting delays that will reduce the probability of their formation.

14.20 The following NS and output logic is read from an FSM that is designed to operate in the fundamental mode. Here, the inputs are A and B and the outputs are P and Q.

$$Y_3 = y_2 AB + y_0 \bar{A} B + y_3 \bar{y}_2 \bar{y}_1 + y_3 B$$

$$Y_2 = y_3 A \bar{B} + y_1 \bar{A} B + y_0 A \bar{B} + y_2 \bar{y}_3 \bar{y}_1 + y_2 \bar{A} B + y_2 A \bar{B}$$

$$Y_1 = y_3 \bar{A} \bar{B} + y_2 \bar{A} \bar{B} + y_0 \bar{A} \bar{B} + y_1 \bar{y}_3 \bar{y}_0 + y_1 \bar{B}$$

$$Y_0 = y_1 AB + y_0 \bar{y}_3 \bar{y}_1 + y_0 AB$$

$$P = y_2 \bar{A} B + y_1$$

$$Q = y_3 A + y_2 A \bar{B}$$

PROBLEMS   769

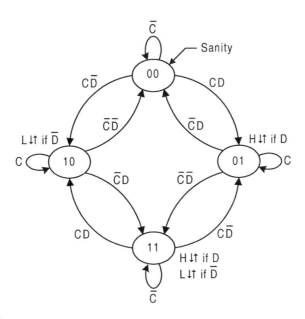

FIGURE P14.12

(a) Analyze this FSM by constructing its state diagram. What approach to FSM design is this? Check for any timing problems it may have. If any exist, indicate their type and origin.

(b) Run a complete E-hazard analysis following the example in Figs. 14.51 through 14.53.

14.21 The state diagram in Fig. P14.12 represents the *selector module*. It is the function of this module to steer input signals, C, to either the H (high) or L (low) output, depending on the activation level of input D.

(a) Use both the LPD model and nested cell model to obtain an optimal set of NS and output functions for the selector module that are free of static hazards. Is this also a valid STT design? Explain.

(b) Analyze this FSM for possible E-hazards and d-trios. What do you conclude from this analysis?

(c) Construct the logic circuit from the results of part (a). Assume that the inputs arrive from positive logic sources and that the outputs are issued active high. Initialize the FSM into the 00 state as indicated.

(d) Verify the proper operation of this FSM by simulation.

14.22 An asynchronous FSM is to be designed that will detect the direction of rotation of a circular shaft as indicated in Fig. P14.13. Two light beams are caused to fall incident on the end surface of the shaft half of which is reflecting and half nonreflecting. Two photocells, A and B, are located at the proper angle of reflection relative to the two beams so that whenever a beam strikes a reflecting surface the photocell receiving the reflected beam will generate a voltage signal. For the shaft position shown in Fig. P14.13, the logic input to the FSM is $AB = 01$. It is a requirement of this FSM

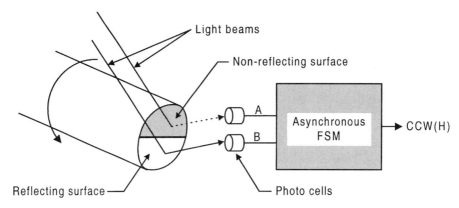

FIGURE P14.13

that the output *CCW* be active any time the shaft is rotating counterclockwise, and be inactive if rotating clockwise (*CW*). Note that the direction of rotation is taken with respect to the front face of the shaft.

(a) Use the LPD model to design an optimum logic circuit for this FSM that is free of endless cycles, critical races, ORGs, and static hazards. [Hint: Only four states are required. Also, the output *CCW* is best issued from each state as a function of both input variables so that XOR-type patterns can be extracted in maxterm code to yield a gate-minimum (three-gate) output that is free of hazards.]

(b) Discuss the limits of this design relative to shaft oscillation sensitivity. To do this, sketch the shaft face orientations that have the least and most light beam sensitivity to possible rotational oscillations.

(c) Analyze this FSM for possible E-hazards and d-trios. If any are present, give the direct and indirect paths, race gates, branching conditions, and minimum path delay requirements for their formation, and indicate the location of the counteracting delays that will reduce the probability of their formation. If they cannot exist, explain why.

14.23 The block diagram in Fig. P14.14 illustrates the handshake interface between a *call module* and a digital system. It is the function of the call module to issue a signal, r, to the system indicating that an access request signal has been made on one of two lines, *REQX* or *REQY*, but not on both. Then, if the system acknowledges receipt of the request by sending back a signal *ACK* to the call module while the request is

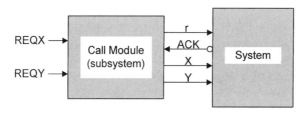

FIGURE P14.14

active, the call module will steer that access request (either *REQX* or *REQY*) to its respective output, *X* or *Y*. But this can happen only if that the "other" request line is inactive at the time *ACK* is received. Thus, *REQX* → *X* if *r* is sent to and *ACK* is received from the system when *REQY* is inactive. Similarly, *REQY* → *Y* if *r* is sent to and *ACK* is received from the system when *REQX* is inactive. A second request can be granted access if *ACK* is active following withdrawal of the first request.

(a) Construct the two state diagrams for the call module. (Hint: One version of the call module consists of two RMODs of the type shown in Fig. 14.11 together with the appropriate NS and output logic.)

(b) Construct the logic circuit for the call module by using two RMODs, an XOR gate, and two NOR gates (nothing else). Assume that the request signals, *REQX* and *REQY*, arrive active high, and that the *ACK* input is active low. Let the outputs be issued active high.

(c) Design the entire call module as a single two-state FSM. In this version of the call module, a repeating contender, *REQX* or *REQY*, can be granted access to the system without an active *ACK* signal prior to each grant of access. The call module version of parts (a) and (b) requires that an *ACK* signal be received from the protected system before access can be granted during a request. Consider both the LPD and nested cell design of this version of the call module. Choose the design that yields the simpler implementation of the module. Remember to eliminate any static hazards that might exist in either design.

14.24 (a) Repeat parts (b), (d) and (e) of Problem 14.6 for the design of the FSM in Fig. P14.4 by using two RMODs as the memory. (Hint: First design for the nested cell model and then convert to the RMOD design. Simple conversion logic can be obtained by comparing state transition tables.) What conclusion do you come to with regard the presence of static hazards in the NS functions? Explain. If your results indicate that E-hazards or d-trios are possible in this design, explain why this is so and give the information required by part (e) of Problem 14.6.

(b) Verify the proper operation of this design by simulation.

CHAPTER 15

The Pulse Mode Approach to Asynchronous FSM Design

15.1 INTRODUCTION

Asynchronous FSMs that are designed to operate with nonoverlapping pulsed inputs and that use "data-triggered" memory elements are called *pulse mode* sequential machines. The pulse mode approach offers a simple and reliable means of designing clock-independent FSMs, but at the price of greatly restricted input signal conditions. Chapter 14 dealt exclusively with asynchronous FSMs that are designed to operate in the fundamental mode. The fundamental mode, it will be recalled, is characterized, in part, by overlapping inputs signals and the potential to form certain types of timing defects such as endless cycles, critical races, and essential hazards, any of which, if present and active, is guaranteed to cause malfunction of the FSM. Furthermore, fundamental mode FSMs can also cause malfunction due to the presence of static hazards in the NS-forming logic. But like synchronous FSMs and unlike fundamental mode FSMs, properly designed pulse mode machines cannot have any of these timing defects — no endless cycles, no critical races, no essential hazards. Furthermore, pulse mode FSMs cannot malfunction because of static hazards in the NS logic. Thus, pulse mode FSMs would seem to have all the advantages of synchronous FSMs, but with none of the timing defects of fundamental mode machines. However, this apparent advantage is offset by the severe restrictions placed on the input signals. In fact, it is for this reason that treatment of the pulse mode approach to FSM design has been deferred until this time.

15.2 PULSE MODE MODELS AND SYSTEM REQUIREMENTS

The generalized (Mealy) model for pulse-mode FSM design is illustrated in Fig. 15.1. It is unique in the sense that its memory stage is composed of data-triggered *toggle modules* that include memory elements of the type featured in Fig. 12.12 or T flip-flops set to the toggle mode. Thus, data-triggered toggle modules are, in effect, *unclocked memory elements*. The degenerate forms of this model follow those shown for synchronous FSMs in Figs. 10.3 and 10.4.

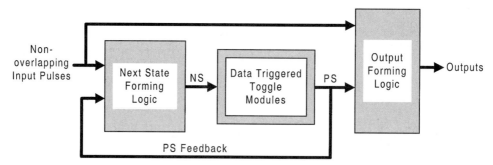

FIGURE 15.1
Mealy's (general) model for an asynchronous FSM that is designed to operate in the pulse mode.

The inputs to pulse mode FSMs must be nonoverlapping pulses that are at least *minimally separated* such that the leading edge of one pulse is sufficiently separated from the trailing edge of any previous pulse. Examples of such pulses are shown in Fig. 15.2. Here, the positive pulses are shown to have active durations (pulse widths) with no upper bound but with a lower limit sufficient to trigger the flip-flop memory elements and initiate a state change. Runt pulses must not be permitted since their effect on the flip-flops is unpredictable. It should be understood that the "at least minimally separated" restriction placed on these input pulses is governed by the stability criteria given by Eqs. (14.3) and (14.4). That is, it is equivalent to the requirement that a second input to a fundamental-mode circuit not be permitted to change until the stability criteria $y_j(t) = Y_j(t)$ (for all j) is satisfied following a previous input change. In fact, proper operation of any FSM (synchronous or asynchronous) can be ensured only if all memory elements of the FSM achieve stability prior to any successive change of an input logic level. The complement of the pulse trains shown in Fig. 15.2 are examples of negative pulses that have no upper bound on their inactive durations but, nevertheless, must be minimally separated.

15.2.1 Choice of Memory Elements

The choice of memory elements for pulse mode FSM design is quite limited. For reasons made clear in the subsequent discussions, it is best that triggering occurs on the trailing edge of the data input pulses. Thus, positive $(0 \to 1 \to 0)$ pulses of unrestricted active duration from positive logic sources require the use of FET toggle modules, while

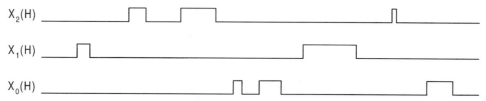

FIGURE 15.2
Examples of nonoverlapping and at least minimally separated positive pulses having active durations with no upper bound.

15.2 PULSE MODE MODELS AND SYSTEM REQUIREMENTS

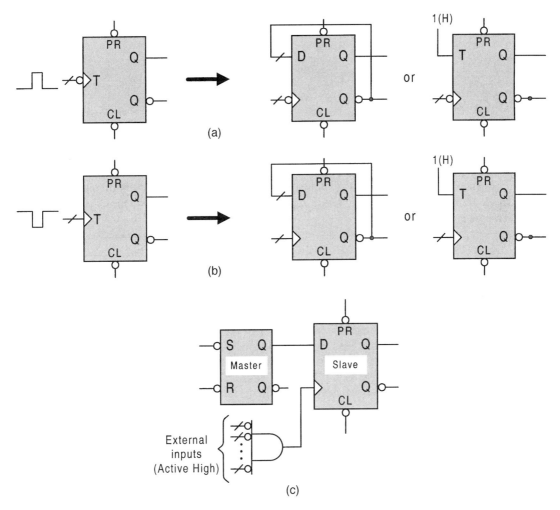

FIGURE 15.3
Data-triggered memory elements required for pulse mode FSMs that receive pulses having no upper bound on pulse width. (a) FET toggle modules required for positive pulses from active high sources. (b) RET toggle modules required for negative pulses from active low sources. (c) Master/slave memory element used for positive pulses from active high sources.

negative ($1 \rightarrow 0 \rightarrow 1$) pulses of unrestricted inactive duration from negative logic sources require RET toggle modules. The various memory elements recommended for use in pulse-mode designs are shown in Fig. 15.3. Use of the toggle modules shown in Figs. 15.3a and 15.3b are the simplest and most reliable memory elements that can be used for this purpose. These memory elements require no upper bound on pulse width. As a lower bound, the data pulses must be fully developed enough to trigger the flip-flops and initiate a state transition — also a requirement for synchronous state machines.

The master/slave configuration in Fig. 15.3c is an acceptable memory element for pulse-mode designs and, like the toggle modules of Figs. 15.3a and 15.3b, it triggers on trailing edge of the data pulse. However, use of the master–slave memory element requires a lower bound on the data pulse width determined by

$$\Delta t_{pulse} \geq (\text{NS logic} + \text{Master stage}). \tag{15.1}$$

Thus, pulses of active duration less than this lower bound may not be picked up. Because the master–slave configuration requires more hardware, is slower, and places a significant lower bound on pulse width, it is less desirable for use as a memory element than toggle modules. To use the master–slave configuration requires the use of the excitation table for the basic cell in Fig. 10.15c together with the mapping algorithm in Section 10.6 to obtain the NS logic in S and R form. Consequently, the data inputs will be present in both the S and R NS logic functions to the master stage and as inputs to slave stage D flip-flop via the multiple input NOR gate.

Under certain conditions the basic cell can be used solely as a memory element in pulse mode designs. However, it is not a good idea to use a basic cell for this purpose, since triggering must occur on the leading edge of the data pulse. This requires that delays be placed on the feedback lines and that an upper bound be placed on the active duration of the positive data pulses. Exceeding this upper bound risks the activation of more than one memory element in response to an input pulse. Sufficient overlap of active memory elements in a pulse mode FSM is tantamount to introducing overlapping pulses and, consequently, causes the malfunction of that FSM. To use basic cells solely as the memory in pulse-mode designs requires the application of the *nested cell model* as indicated in Fig. 15.4. Here, delays are required on all feedback lines if activation of more than one basic cell memory element is to be avoided. As a conservative limit, the pulse widths are limited to an upper bound Δt_{pulse} given by

$$\Delta t_{pulse} \leq (\Delta t_j + \text{best-case path delay through the system}), \tag{15.2}$$

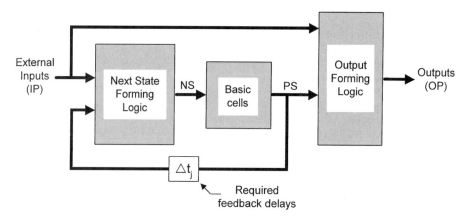

FIGURE 15.4
Generalized nested cell model of a pulse mode FSM for which delays in the feedback lines are required to avoid simultaneous activation of memory elements.

where Δt_j represents the feedback delays. The "best case path delay through the system" is a quantity that usually falls in the range of $2\tau_p$ to $3\tau_p$ for most systems, where τ_p is an average gate path delay. The lower limit is, as before, the requirement that the pulse be of sufficient strength to initiate a state change. This lower limit together with the upper limit expressed by Eq. (15.2) lead to what is called a *bounded pulse*. The bounded pulse requirement places a severe restriction on the pulse widths that a nested cell design can properly accept without malfunction. It is for this reason that the nested cell approach to pulse mode FSM design is of little or no practical importance. Should the nested cell model be used for this purpose, the NS functions must be generated by combining the excitation table for the basic cell given in Fig. 10.15c with the state diagram for the FSM by using the mapping algorithm in Section 10.6. Remember that for such designs, Eq. (15.2) must always be satisfied.

To summarize, all memory elements in a pulse mode design require nonoverlapping pulse waveforms of sufficient width to initiate a state transition and that are at least minimally separated. But it is only the memory elements in Fig. 15.3 that require no upper bounds to the pulse width. Both the toggle module and master/slave memory elements are triggered on the trailing edge of the data pulse and require no delays in the feedback lines. In contrast, the basic cells in Fig. 15.4 are triggered on the rising edges of the data pulses and require feedback delays and bounded data pulse widths. All pulse mode designs require that the data pulse widths be of sufficient duration as to cause a state change. Because of the bounded data pulse width requirement placed on the use of the nested cell model, this model is not recommended for use in the design of pulse mode FSMs. Furthermore, master–slave memory elements are significantly slower, require more hardware, and place a larger lower bound on the data pulse width than toggle modules. Consequently, the toggle modules of Figs. 15.3a and 15.3b are the memory elements of choice. The examples and discussions presented in this chapter will justify this fact. The examples will also utilize positive data pulse trains exclusively.

15.3 OTHER CHARACTERISTICS OF PULSE MODE FSMs

There are a number of interesting and advantageous pulse mode characteristics that result from the use of data-triggered toggle modules as memory elements:

1. Branching conditions in a pulse mode state diagram consist of *single variables* or *ORed single variables* that are *always* uncomplemented (for positive pulses) or *always* complemented (for negative pulses) — never mixed! *Unconditional branching in a state diagram is strictly forbidden* for obvious reasons.

2. Any state coding scheme will suffice, but, since toggle modules are used, a binary sequence is preferred where possible to minimize the NS logic. Recall the design of binary counters in Section 12.3.

3. The NS logic is obtained by combining the excitation table for the T flip-flop in Fig. 10.37c with the pulse mode state diagram by using the mapping algorithm in Section 10.6.

4. Since states in a toggle module design cannot toggle to themselves, only outgoing single variable or ORed single variable (e.g., $X + Y$) branching conditions need be

considered in mapping the NS logic. Thus, holding conditions should not be indicated in a state diagram or state table.

5. The sum rule in Eq. (10.3) is never observed — it has no meaning in the state diagram for a pulse-mode FSM. However, the mutually exclusive requirement is uniquely satisfied by the nonoverlapping inputs requirement (see Problem 10.24).

6. When it is appropriate to do so, outputs should be made conditional on the exciting branching variable. Use of conditional (Mealy) outputs results in two important benefits involving exclusively those outputs (explanations are given later) in which

   (a) Output race glitches (ORGs) are not possible.
   (b) Static hazards in the output forming logic are not possible.

   These benefits are not guaranteed if unconditional (Moore) outputs are used.

7. As stated earlier, pulse mode designs cannot have endless cycles, critical races, or essential hazards, and cannot have problems due to static hazards in the NS logic functions.

8. Initialization methods are exactly the same as those for synchronous FSMs discussed in Section 11.7.

9. Debouncing of inputs from switches is absolutely necessary since pulse-mode circuits are highly sensitive to transient signals of sufficient duration and strength.

10. The inputs to pulse mode FSMs need not be synchronized since the requirement of nonoverlapping data pulses, at least minimally separated, is a form of synchronization.

11. Properly designed and operated pulse mode FSMs cannot go metastable and hence have an infinite MTBF, assuming that the data pulses are of sufficient duration and strength (not runt pulses).

The 11 characteristics of pulse mode FSMs just given should seem impressive when compared to those of synchronous state machines and asynchronous FSMs that are operated in the fundamental mode. In fact, it appears that pulse mode FSMs have all the benefits of synchronous and asynchronous fundamental mode machines, but with none of their problems. This is true! However, the price to be paid for this "perfection" is the severe restrictions that are placed on the input signals — they must be nonoverlapping pulses that are at least minimally separated.

The reason why pulse mode FSMs with toggle modules and Mealy outputs cannot have either ORGs or static hazards in the output logic is because triggering occurs on the trailing edge of the data pulse. This means that the requirements for ORG and static hazard formation cannot be met, since all data inputs are inactive at the time the transitions occur, assuming positive data pulses. Remember that to initiate an externally or internally activated static hazard, the data variable must be active for positive pulses. But since the data variable is always inactive immediately following a transition, externally initiated s-hazards are unconditionally eliminated and internally initiated s-hazards cannot form if Mealy outputs are used. Using the same argument, ORGs are not possible for an output conditional on an active exciting input since, again, the transition occurs only after the input goes inactive (trailing-edge triggering).

If Moore outputs are used with toggle module memory elements, ORGs and internally initiated s-hazards in the output functions are possible. Because of the trailing-edge triggering of the toggle modules, such logic noise (if present) cannot be filtered with D flip-flops

15.4 DESIGN EXAMPLES

In this section three pulse mode FSMs of varying complexity will be designed by using toggle modules as memory elements and will feature different implementations of the NS and output forming logic. For this purpose use will be made of discrete logic, a ROM, and a PLA.

A SIMPLE PULSE MODE SEQUENCE RECOGNIZER. Consider the state diagram for a simple sequence recognizer in Fig. 15.5a that is suitably documented for a pulse mode design. In this case, toggle modules are to be used as the memory elements. Notice that the branching conditions are single uncomplemented variables, as required for nonoverlapping positive data pulses, and that no holding conditions are shown. Holding conditions have no relevance in a pulse mode design that uses toggle modules as memory elements, since a given present state variable cannot toggle to itself. A single output exists in state 10 and is conditional on the exciting condition Y. Shown in Figs. 15.5b and 15.5c are the excitation table for a T flip-flop and the resulting NS and output K-maps, from which the minimum functions are found to be

$$T_A = BX + AY, \quad T_B = \bar{A}X + BY, \quad \text{and} \quad Z = AY. \tag{15.3}$$

The entry in cell 1 for T_B is $X + Y$ because bit B must toggle in that state on the falling edge of either an X pulse or a Y pulse. Note that the term AY is a shared PI between T_A and Z, and that the FSM is initialized into the 00 state.

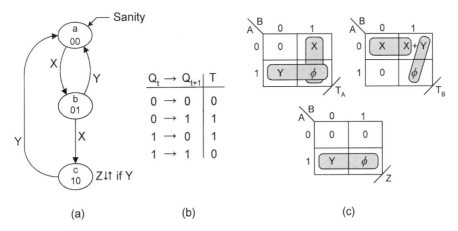

FIGURE 15.5
Design of a simple sequence recognizer by using the pulse mode approach. (a) State diagram applicable to a pulse mode design. (b) Excitation table for the T flip-flop. (c) NS and output K-maps and minimum cover.

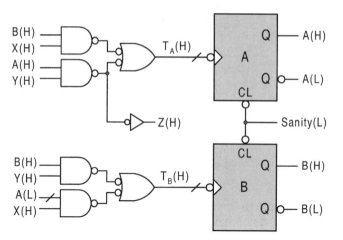

FIGURE 15.6
Discrete logic implementation of Eqs. (15.3) representing the pulse mode FSM in Fig. (15.5).

The logic circuit for the simple pulse mode sequence recognizer of Fig. 15.5 is shown in Fig. 15.6. Here, discrete logic is used to implement the NS and output functions represented by Eqs. (15.3). The FSM is initialized by a *Sanity(L)* input to the *CL* asynchronous overrides of the two toggle modules. Notice that the NS logic is presented active high to the FET toggle modules, a requirement for trailing-edge triggering by positive pulses.

The operation of this pulse mode FSM is illustrated by the timing diagram in Fig. 15.7, which is the result of a simulation. The vertical dashed lines are positioned so that the NS, PS, and output responses to the data pulses can be easily compared. The results show that the time elapsing between input and NS pulses is always $2\tau_p$, and that the rise and fall edges of the PS pulses lag the corresponding data input pulse edges by $4\tau_p$ and $5\tau_p$, respectively.

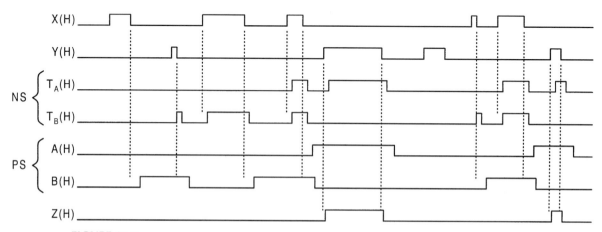

FIGURE 15.7
Simulation results for the pulse mode logic circuit in Fig. 15.6 showing the PS, NS, and output responses to input changes (compare by using vertical dashed lines).

15.4 DESIGN EXAMPLES

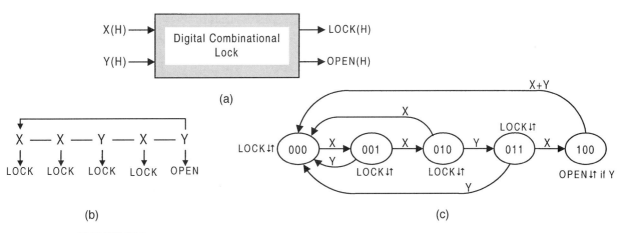

FIGURE 15.8
Pulse mode design of a digital combinational lock. (a) Block symbol circuit symbol. (b) input pulse sequence showing outputs. (c) Suitably documented state diagram.

The time between input pulse change and output response is always τ_p and the lower bound of input pulse width is $2\tau_p$. Notice that the output pulse width is the same as that of the Y pulse causing it. Here, as before, τ_p is the path delay through any gate, regardless of the type or number of inputs.

A PULSE MODE DIGITAL COMBINATIONAL LOCK. Shown in Fig. 15.8 are the block circuit symbol, the input pulse sequence required to open the lock, and a state diagram appropriately documented for the pulse-mode design of a digital combinational lock. The two inputs, X and Y, are assumed to be nonoverlapping pulses of sufficient strength and duration and to arrive from positive logic sources. Furthermore, it is assumed that these inputs are produced by debounced, interlocked mechanical switches that cannot be activated simultaneously. Simultaneous activation of the switches would violate the fundamental premise on which the pulse mode is based — that is, that the input pulses be nonoverlapping and at least minimally separated from each other. It is understood that the logic used in the implementation of the digital combination lock is very much faster than the mechanical switches delivering the input signal pulses. Another requirement is that the two outputs be free of all logic noise and be delivered active high to the next stage. Finally it is required that this FSM be designed by using toggle modules as the memory and by using a ROM to implement the NS- and output-forming logic.

The ROM program table for this pulse mode FSM is given in Fig. 15.9. Notice that the $X, Y = 1, 1$ conditions are absent in this table, since they are irrelevant in a pulse-mode design — the pulses are never permitted to overlap. Also, observe that don't-care states 101, 110 and 111 are presented as nonoutput states to avoid possible ORGs. The input conditions for these states are arbitrarily taken to be logic 0 although they are actually irrelevant as are the corresponding NS function values. The minimum ROM size required for this FSM is $2^5 \times 5$.

Implementation of the ROM program table is shown in Fig. 15.10a by using three toggle modules and a ROM of the minimum required dimensions. The outputs are delivered directly from the ROM free of logic noise. The operation of this digital combination lock is illustrated

ROM Inputs					ROM Outputs					ROM Inputs					ROM Outputs				
PS					NS					PS					NS				
I_4	I_3	I_2	I_1	I_0	Y_4	Y_3	Y_2	Y_1	Y_0	I_4	I_3	I_2	I_1	I_0	Y_4	Y_3	Y_2	Y_1	Y_0
A	B	C	X	Y	T_A	T_B	T_C	LOCK	OPEN	A	B	C	X	Y	T_A	T_B	T_C	LOCK	OPEN
			0	0	0	0	0	1	0				0	0	1	0	0	1	0
0	0	0	0	1	0	0	0	1	0	1	0	0	0	1	0	0	0	0	1
			1	0	0	0	1	1	0				1	0	0	0	0	1	0
			0	0	0	0	1	1	0				0	0	ϕ	ϕ	ϕ	0	0
0	0	1	0	1	0	0	0	1	0	1	0	1	0	0	ϕ	ϕ	ϕ	0	0
			1	0	0	1	0	1	0				0	0	ϕ	ϕ	ϕ	0	0
			0	0	0	1	0	1	0				0	0	ϕ	ϕ	ϕ	0	0
0	1	0	0	1	0	1	1	1	0	1	1	0	0	0	ϕ	ϕ	ϕ	0	0
			1	0	0	0	0	1	0				0	0	ϕ	ϕ	ϕ	0	0
			0	0	0	1	1	1	0				0	0	ϕ	ϕ	ϕ	0	0
0	1	1	0	1	0	0	0	1	0	1	1	1	0	0	ϕ	ϕ	ϕ	0	0
			1	0	1	0	0	1	0				0	0	ϕ	ϕ	ϕ	0	0

FIGURE 15.9
ROM program table obtained directly from the state diagram in Fig. 15.8c showing all input conditions except the $X, Y = 1, 1$ conditions, which are irrelevant in a pulse-mode design.

in Fig. 15.10b for a sequence of input pulses leading to the output *OPEN*. For simplicity, no logic delays are shown. Vertical dashed lines are placed on the trailing edges of the input pulses for the convenience in reading the timing diagram. Notice that the output *LOCK* is maintained active until *OPEN* is activated as required by the ROM program table. This is important only if it is assumed that the *LOCK/OPEN* mechanisms are such that one or the other of the two outputs must be active at all times, but never both inactive or both active.

DESIGN OF A CANDY-BAR VENDING MACHINE. As a third and final example, a candy-bar vending machine controller is designed by using the pulse mode approach. The candy bars each cost 40 cents (a bargain these days) and are dispensed automatically by the machine after correct change has been inserted. The vending machine accepts nickels (N), dimes (D), and quarters (Q) only. It consists of a controller (CONTROLLER), a coin receiver (CR), an electromechanically operated coin changer (CC) for nickel return (RN), a 4-bit accumulator (ACC), a 4-bit parallel loadable down counter (CNT), a comparator (COMP) to keep account of the coin exchange, an electromechanically operated candy bar drop mechanism (CBD), and a price strapping unit (PSU) to set the price of the candy. These components and their interconnections are illustrated in the block diagram of Fig. 15.11.

15.4 DESIGN EXAMPLES   783

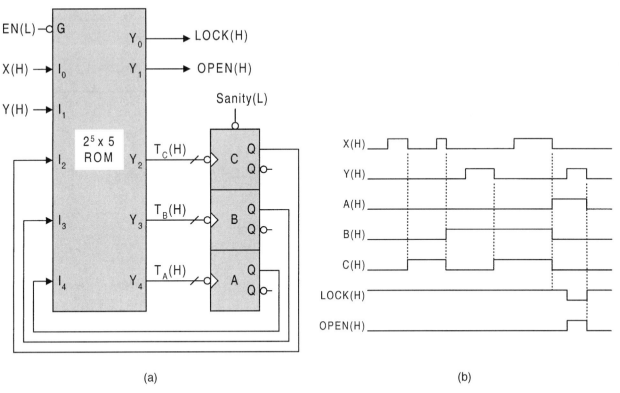

FIGURE 15.10
(a) ROM implementation of the pulse mode FSM represented by the program table in Fig. 15.9.
(b) Timing diagram leading to an output *OPEN* and excluding ROM and memory delays.

Only the controller is designed in this chapter. This will be accomplished by using FET toggle modules as the memory and by using a PLA to implement the NS and output logic. The controller must accept discrete nonoverlapping pulses generated by coin insertion and must generate well-developed output signals that are free of logic noise. The controller must be initialized into an origin state and must return to that state once the exact payment has been received by the vending machine and a candy bar has been dispensed to the customer.

The state diagram for the controller of the candy-bar vending machine is provided in Fig. 15.12 together with the meanings of the abbreviations used in the state diagram and in the block diagram of Fig. 15.11. Notice that the exiting condition from state b is *CFR* not $\overline{CIR}$. *CFR*, meaning coin free of receiver, is a positive pulse in keeping with the requirement of a positive pulse mode design. Except for the output *RN*, all outputs are conditional (Mealy) outputs that, with the state assignment given, ensure that no ORGs will be produced. The controller is to be initialized into the 000 state by using a sanity circuit of the type shown in Fig. 11.28.

Though the details of the data path devices are not needed at this time, it is important to have a general understanding of their function within the system so that the controller can be properly designed. With reference to Figs. 15.11 and 15.12, the following provides this

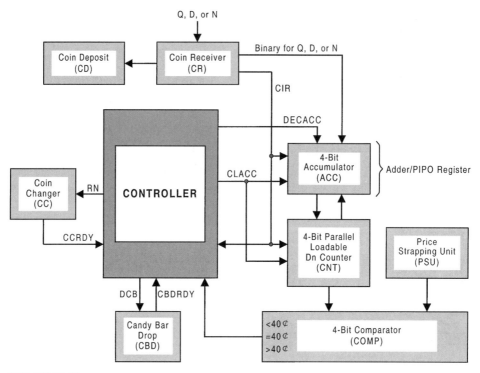

FIGURE 15.11
Block diagram for a candy-bar vending machine showing controller and data path devices.

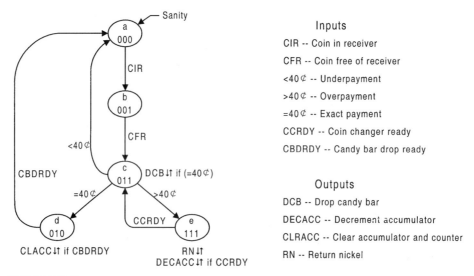

FIGURE 15.12
State diagram and definitions for the pulse mode design of the candy-bar vending machine controller shown in Fig. 15.11.

15.4 DESIGN EXAMPLES

general understanding:

- Coins are placed in the slots of the coin receiver (CR) and the 4-bit adder in the accumulator is automatically updated.
- Each quarter (Q), dime (D), or nickel (N) that is inserted into the coin receiver (CR) is encoded according to the number of nickels: N = 0001, D = 0010, and Q = 0101.
- The accumulator's PIPO register stores the current coinage count and the 4-bit counter is parallel loaded only after each coin has cleared the receiver (CFR). Thus, the register is triggered and the counter is parallel loaded on the trailing edge of the CFR pulse.
- The counter should be a data-triggered up/down binary counter with asynchronous parallel load as detailed in Fig. 13.46, but set for down count with $Up = 0(H)$.
- The output of the counter is compared in the comparator (COMP) with the value of a candy bar set by the price strapping unit (PSU), and the result ($<40¢$, $=40¢$ or $>40¢$) is sent to the controller and to the accumulator's adder. In this case the PSU is set at 40 cents = 1000 (eight nickels).
- If an underpayment ($<40¢$) signal is received by the controller, the system awaits the insertion another coin. If overpayment ($>40¢$) is received, a nickel is returned (RN) to the coin changer (CC) and the accumulator is decremented (DECACC) when the CC is ready (CCRDY); and this process is repeated until the exact amount is reached. When the exact amount ($= 40¢$) is received by the controller, a candy bar is dropped (DCB) and no nickel is returned.
- Immediately following the dispensing of a candy bar and after the candy bar drop is ready (CBDRDY), the accumulator and counter are cleared (CLRACC) and the controller is returned to the initialization state, 000. The controller is now ready to repeat the process.

In Fig. 15.13 are given the NS and output logic K-maps and minimum cover for the pulse-mode FSM represented by the state diagram in Fig. 15.12. It is the plan to implement the NS and output logic of this FSM by using a PLA so that a comparison can be made with the previous two examples where discrete logic and a ROM are used for the NS and output logic. Recall that it is strongly advisable, but not mandatory, to use minimum or reduced cover for a PLA implementation of the NS and output logic. ROMs must use canonical (minterm) data but not PLAs as discussed in Sections 7.2 and 7.3. From the K-maps in Fig. 15.13 the NS and output functions are easily read to be

$$\begin{cases} T_A = \bar{A}BC(>40¢) + A(CCRDY) \\ T_B = \bar{B}C(CFR) + \bar{A}BC(<40¢) + B\bar{C}(CBDRDY) \\ T_C = \bar{B}\bar{C}(CIR) + \bar{A}BC(<40¢) + \bar{A}BC(=40¢) \\ DCB = \bar{A}BC(=40¢) \\ CLACC = B\bar{C}(CBDRDY) \\ RN = A \\ DECACC = A(CCRDY) \end{cases} \qquad (15.4)$$

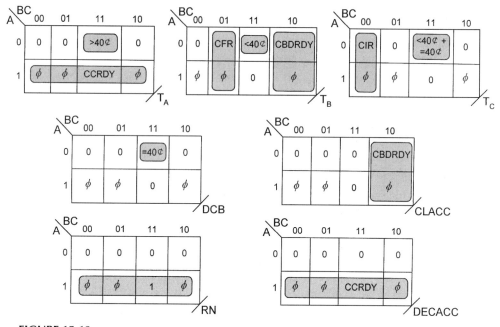

FIGURE 15.13
NS and output K-maps for the pulse mode FSM in Fig. 15.12 showing minimum NS and output logic cover.

Here, four shared PIs are indicated, $A(CCRDY)$, $\bar{A}BC(<40\ cents)$, $B\bar{C}(CBDRDY)$, and $\bar{A}BC(=40\ cents)$, which brings the total number of p-terms to eight for the combined NS and output functions. It is not uncommon for a relatively large number of shared PIs to exist in a pulse mode design that has several outputs, each of which is conditional on an exiting condition. Characteristic 6 in Section 15.3 makes the point that Mealy outputs should be used whenever possible so as to avoid ORGs and static hazards in the output functions. Obviously, another advantage in using Mealy outputs is that they tend to maximize the number of shared PIs, but add more input variables to the output functions. Notice that the single Moore output RN is state variable A and that no ORG results from it. Note also that in cell 3 of the T_C K-map, $(\overline{>40\ cents})$ must not be used in place of $(>40\ cents) + (=40\ cents)$. To do so would cause the FSM to malfunction, since a basic principle of this pulse mode design would have been violated — that is, all nonoverlapping pulses must be positive pulses, never a mixture of positive and negative pulses.

The p-term table for the PLA implementation of the candy-bar vending machine controller is constructed directly from Eqs. (15.4). It can be seen that there are 10 PLA inputs, 7 outputs, and 8 p-terms (including four shared PIs), which requires a PLA of minimum dimensions $10 \times 8 \times 7$. This p-term table is provided in Fig. 15.14 following the format given in Section 7.3. Recall that a dash (—) in the AND plane indicates the absence of an input variable in a p-term, and hence no connection for that input. Clearly, the use of a ROM to implement the NS and output logic for this FSM would be a gross overkill, since a ROM of 10 inputs requires 2^{10} minterms. Compared to only eight p-terms required by a PLA, it is obvious that a ROM would be a poor choice for such applications.

Shown in Fig. 15.15 is the block diagram for the PLA implementation of the candy-bar vending machine represented by the p-term table in Fig. 15.14. Observe that the NS functions

| P-terms | \multicolumn{10}{c}{PLA Inputs} | \multicolumn{7}{c}{PLA Outputs} |

	PS										NS						
	I_9	I_8	I_7	I_6	I_5	I_4	I_3	I_2	I_1	I_0	Y_6	Y_5	Y_4	Y_3	Y_2	Y_1	Y_0
P-terms	A	B	C	CIR	CFR	<40¢	=40¢	>40¢	CCRDY	CBDRDY	T_A	T_B	T_C	DCB	CLACC	RN	DECACC
$\bar{A}BC(>40¢)$	0	1	1	–	–	–	–	1	–	–	1	0	0	0	0	0	0
A(CCRDY)	1	–	–	–	–	–	–	–	1	–	1	0	0	0	0	0	1
$\bar{B}C(CFR)$	–	0	1	0	1	–	–	–	–	–	0	1	0	0	0	0	0
$\bar{A}BC(<40¢)$	0	1	1	–	–	1	–	–	–	–	0	1	1	0	0	0	0
$B\bar{C}(CBDRDY)$	–	1	0	–	–	–	–	–	–	1	0	1	0	0	1	0	0
$\bar{B}\bar{C}(CIR)$	–	0	0	1	–	–	–	–	–	–	0	0	1	0	0	0	0
$\bar{A}BC(=40¢)$	0	1	1	–	–	–	1	–	–	–	0	0	1	1	0	0	0
A	1	–	–	–	–	–	–	–	–	–	0	0	0	0	0	1	0

FIGURE 15.14
P-term table for implementation of the pulse mode candy-bar vending machine controller represented by the NS and output functions in Eqs. (15.4).

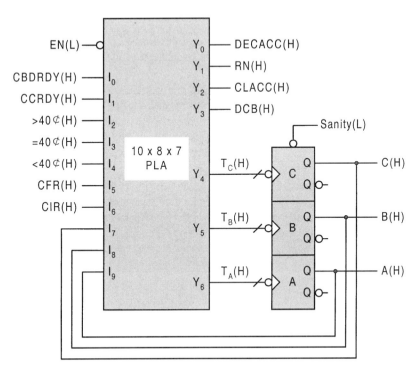

FIGURE 15.15
PLA implementation of the p-term table in Fig. 15.14 for the candy-bar vending machine designed to operate in the pulse mode.

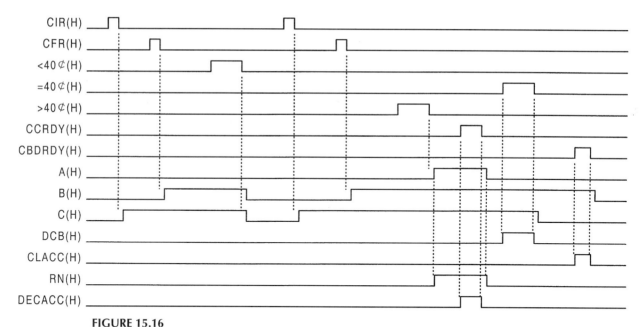

FIGURE 15.16
Simulation results of the candy-bar vending machine controller by using the discrete logic expressed by the NS and output functions in Eqs. (15.4).

T_A, T_B, and T_C are introduced active high to the FET toggle modules, a requirement for positive data pulses. Note also that no input or output conditioning circuits are necessary for this FSM.

The sequential behavior of the candy-bar vending machine is revealed by the simulation results in Fig. 15.16, which were produce by using discrete logic for the NS and output functions of Eqs. (15.4). Vertical dashed lines are provided to facilitate reading of the various responses to data pulses. The time elapsing between input and NS function pulses is always τ_p (not shown); the rising and falling edges of the present state pulses (A, B, and C) lag the corresponding falling edges of the data input pulses by $4\tau_p$, and $5\tau_p$, respectively. The output response time to input pulse change is τ_p, and the lower bound of input pulse width is $2\tau_p$. As always, τ_p is the propagation delay of a gate regardless of its type or number of inputs. The FET toggle modules are designed by using the D flip-flops as given in Fig. 12.12, but with inverters on the CK inputs.

15.5 ANALYSIS OF PULSE MODE FSMs

The procedure used to analyze pulse mode FSMs is basically the same as that used to analyze synchronous and asynchronous (fundamental mode) FSMs discussed in Sections 10.13 and 14.17. The NS and output functions are read from a logic circuit and the results are plotted in K-maps. The K-maps are converted to D form and the PS/NS table is constructed. It is at this point that the analyses of pulse mode FSMs differ from those of synchronous and

15.5 ANALYSIS OF PULSE MODE FSMS

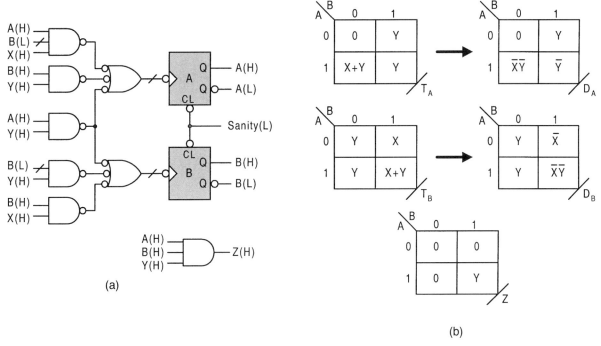

FIGURE 15.17
Analysis of a pulse mode FSM. (a) logic circuit. (b) NS and output K-maps as plotted from Eqs. (15.5).

fundamental mode FSMs. The following corrections to the PS/NS table are necessary for positive pulse mode FSMs having more than one external input:

> *Line out (disregard) all entries in the PS/NS table that are associated with either all inactive data inputs or that are associated with more than one active data input. Thus, only one active input is permitted for each entry.*

The state diagram for the pulse mode FSM is then constructed from the corrected PS/NS table and the result is analyzed for possible problems. For pulse mode FSMs that are designed to operate with negative pulses, valid entries in the PS/NS table must include only those having one inactive input.

A SIMPLE EXAMPLE. Shown in Fig. 15.17a is the logic circuit for a pulse mode FSM that is to be analyzed. This FSM is seen to have two external inputs, X and Y, two state variables, and a single output, Z. From the logic circuit the NS and output functions are easily read and found to be

$$\begin{cases} T_A = AY + BY + A\bar{B}X \\ T_B = AY + \bar{B}Y + BX \\ Z = ABY \end{cases}. \tag{15.5}$$

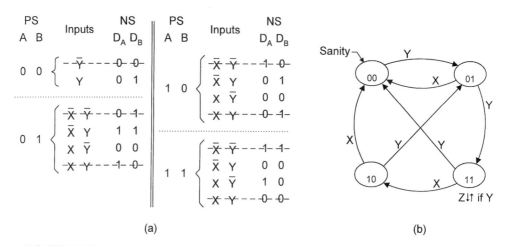

FIGURE 15.18
Analysis of the pulse mode FSM in Fig. 15.17 (contd.). (a) PS/NS table constructed from the D K-maps in Fig. 15.17b. (b) State diagram derived from the PS/NS table in (a) and from Eqs. (15.5).

Mapping Eqs. (15.5) yields the results shown in Fig. 15.17b, where map conversion is used to present the NS functions in D form.

The PS/NS table is now easily constructed from the D K-maps in Fig. 15.17b and is presented in Fig. 15.18a. Here, entries that are associated with either all inactive inputs or two active inputs are lined out and disregarded. Thus, entries with only one active input are considered. From the PS/NS table there results the state diagram given in Fig. 15.18b. The sanity input and output Z are not indicated in the PS/NS table but are known by inspection of Fig. 15.17 and Eqs. (15.5) and are shown in the state diagram.

The sequential behavior is easily deduced from the state diagram. Keeping in mind that the output Z is issued coincidentally with data pulse Y, it is clear that this FSM recognizes and issues an output only after three consecutive Y pulses. Interposition of one or more X pulses before three consecutive Y pulses occur requires the FSM to begin the Y sequence again. Notice that the Y sequence can be initiated from either the initiation state 00 or from state 10.

No ORG is possible from state 11 during the 10 → 01 transition, since the transition is executed on the trailing edge of the positive Y pulse. This means that Y is inactive at the time the transition occurs, making it impossible for output Z to be issued. This fact together with the proper operation of the FSM is verified by the simulation result provided in Fig. 15.19. Here, as in many examples given previously, dashed vertical lines are provided to facilitate the reading of the various transitional events. Also, as in Fig. 15.16, the rise and fall edges of the present state pulses (now A and B) lag the corresponding falling edges of the data pulses by $4\tau_p$ and $5\tau_p$, respectively; the output response time to input pulse change is τ_p. Although there is no upper bound of data pulse width, there still remains a lower bound at $2\tau_p$, where τ_p is the path delay through a gate regardless of type of number of inputs. Note that any number of X pulses while in state 00 retain the FSM in that state, since a Y pulse is required to initiate a transition from state 00 to state 01, as indicated in the state diagram of Fig. 15.18b.

15.5 ANALYSIS OF PULSE MODE FSMS

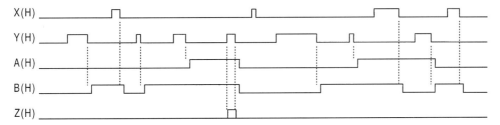

FIGURE 15.19
Simulation results of the logic circuit in Fig. 15.17a, verifying its proper operation including the absence of any ORGs.

A SIMPLE NESTED CELL EXAMPLE. Consider the logic circuit in Fig. 15.20 representing a pulse mode design by using the nested cell model of Fig. 15.4 but without delays in the feedback lines. Thus, all $\Delta t_j = 0$. This is done to test the validity of the inequality expressed in Eq. (15.2), as well as to reinforce the notion that the nested cell approach to pulse mode designs should be avoided except under very special circumstances to be discussed later.

Reading the logic circuit in Fig. 15.20 yields the following NS and output functions:

$$\begin{cases} S_A = BX & R_A = \bar{B}Y \\ S_B = \bar{A}\bar{B}X & R_B = BX + BY \\ \quad Z = BX & \end{cases}. \tag{15.6}$$

Also, an inspection of the initialization connections indicate that the FSM will initialize into the 10 state. That is, $S_A(L) = R_B(L) = Sanity(L)$ with $R_A(L) = S_B(L) = \overline{Sanity}(L)$ forces the FSM into state 10 when $Sanity(L) = 1(L)$ and preserves the mixed-rail output logic values.

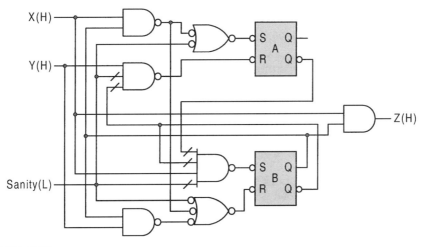

FIGURE 15.20
Logic circuit for the nested cell design of a pulse mode FSM to be analyzed showing initialization connections.

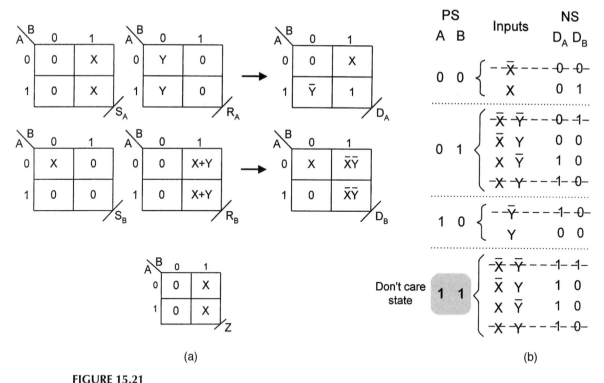

FIGURE 15.21
Analysis of the nested cell pulse mode circuit in Fig. 15.20. (a) NS and output K-maps and map conversion plotted from Eqs. (15.6). (b) PS/NS table constructed from the NS D K-maps in (a).

The NS functions in Eqs. (15.6) are plotted in the S, R K-maps shown in Fig. 15.21a, where they are converted to D form by using Eq. (14.37) with the appropriate change in the NS and output notation. The PS/NS table can now be easily constructed from the D K-maps as presented in Fig. 15.21b. Notice that the nonessential and invalid entries are lined out and will be disregarded in constructing the state diagram from this table. Also, observe that state 11 has no entry from any other state and is, therefore, a don't-care state. Thus, state 11 can be omitted in the state diagram.

The state diagram is constructed directly from the PS/NS table in Fig. 15.21b and is presented in Fig. 15.22a. Included are the sanity input and conditional output that are not shown in the PS/NS table but that are easily deduced from the logic circuit. The sequential behavior is easily discernible from the state diagram. Assuming valid bounded pulses, as discussed in Subsection 15.2.1, the output Z will be issued coincidentally (actually after a gate path delay) with the second X pulse in an uninterrupted sequence $Y \to X \to X$. The proper sequential behavior of this FSM can be verified by simulation of the logic circuit in Fig. 15.20. This is done with the result shown in Fig. 15.22b. Here, the pulse widths are set at $2\tau_p$ (two gate delays). Since no feedback delays are present ($\Delta t_j = 0$), the pulse widths must not fall significantly outside the range of $2\tau_p$ to $3\tau_p$ for this FSM, as expressed by the inequality of Eq. (15.2). The simulation result in Fig. 15.22c is an example of what can happen when the upper bound of permissible pulse width is exceeded. Here, the pulse

15.5 ANALYSIS OF PULSE MODE FSMS 793

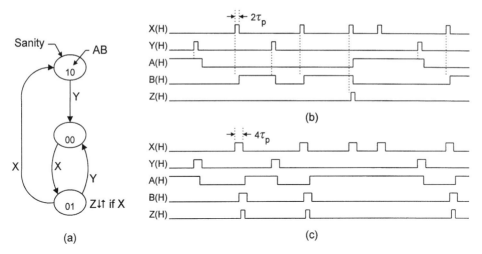

FIGURE 15.22
Analysis of the pulse mode FSM in Fig. 15.20 (contd.). (a) State diagram as deduced from the PS/NS table in Fig. 15.21b. (b) Simulation result for $2\tau_p$ pulse widths indicating correct operation. (c) Simulation result for $4\tau_p$ pulse widths showing malfunction of the FSM by exceeding the upper bound for pulse width.

widths are set at $4\tau_p$, causing the FSM to malfunction. Since the lower bound of permissible pulse width is about $2\tau_p$, it is obvious that pulse widths for nested cell designs of this type must be restricted to the narrow range of $2\tau_p$ to $3\tau_p$ if malfunction is to be avoided.

The upper bound problem for nested cell designs of pulse-mode FSMs can be solved in one or both of two ways. First, if no feedback delays are used, pulse narrowing circuits, as in Fig. 10.28a, can be used as input conditioning stages for all inputs. To do this requires that the pulses be set within that narrow range acceptable for proper operation of the FSM. But this may not achieve the desired result if output pulses of longer widths are required. Use of feedback delays Δt_F offer a partial solution to this problem. However, it may still be necessary to limit the width of incoming pulses, pulses that may vary greatly in pulse width. In this event, both pulse narrowing circuits and feedback delays are required, as indicated in Fig. 15.23. Now, incoming pulses are constrained to a width of Δt_P. That is, pulses introduced to the NS logic of the FSM can be no greater than Δt_P, assuming that $\Delta t_P < \Delta t_F$; and incoming pulses less than Δt_P will not reach the output of the pulse narrowing circuit. This scheme has the advantage that conditional outputs can have pulse widths significantly larger than that shown in Fig. 15.22b while allowing proper operation of the FSM. Remember that introducing feedback delays alone does not guarantee proper operation for incoming pulses of unrestricted pulse width. It is for this reason that pulse narrowing circuits are also necessary. If pulse narrowing circuits are used by themselves the delay element Δt_P must be set such that the pulse widths fall, say, within the range of $2\tau_p$ to $3\tau_p$ for most nested cell designs.

DELAY ELEMENTS. The delay elements, Δt_P and Δt_F, can be realized in any number of ways. Small delay elements are easily obtained by cascading gates, buffers, inverters,

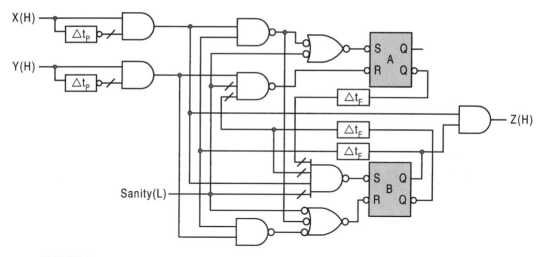

FIGURE 15.23
Logic circuit of Fig. 15.20 with input pulse narrowing circuits and feedback delays as required for incoming data pulses of unrestricted upper bound.

and/or Schmitt triggers in some combination to achieve the desired delay. For larger delays inertial elements are needed. Shown in Fig. 15.24a is an inertial delay element composed of diodes, resistors, and capacitors and a rendezvous module (RMOD). The two-input RMOD is designed in Fig. 14.11 by using the nested cell model.

The inertial delay element in Fig. 15.24a ensures the creation of the delay Δt indicated in the timing diagram provided in Fig. 15.24b. Although the analysis of this circuit is complex and beyond the scope of this text — it is a nonlinear second-order circuit — its operation can be understood qualitatively with little difficulty. On the rising edge of the X input pulse, the RC time constant at node A is smaller than that at node B because

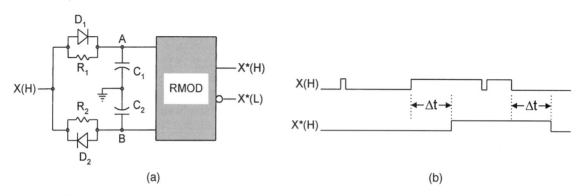

FIGURE 15.24
An inertial delay element for creation of large delays. (a) Circuit composed of resistors, R, diodes, D, capacitors, C, and a rendezvous module (RMOD). (b) Timing diagram showing formation of the delay, Δt, and the filtering action of the R-C circuit.

diode D_1 is turned ON (with low resistance in forward bias) while diode D_2 is OFF (with high resistance in reverse bias). As a result, capacitor C_1 charges up via the low resistance of diode D_1 allowing node A to reach the high-voltage threshold of the RMOD before node B. Assuming that R_2 has a greater resistance than diode D_1 in forward bias, capacitor C_2 charges up after a time Δt bringing node B to the threshold voltage of the RMOD. When both inputs to the RMOD reach the threshold (i.e., become active), the RMOD responds by issuing an active output, X^*. The reverse is true for the falling edge of the X input pulse. Now diode D_2 is ON and D_1 is OFF, resulting in a smaller time constant at node B than at node A. Thus, node B reaches the low threshold voltage of the RMOD before node A since the capacitor C_2 can discharge through the low resistance of diode D_2, now in forward bias. Since capacitor C_1 must discharge through high resistance R_1, node A reaches the low threshold voltage after a time Δt. Then when both inputs to the RMOD reach the low threshold (i.e., go inactive), the RMOD issues an inactive output. In this discussion it is assumed that $D_1 = D_2$, $R_1 = R_2$, and $C_1 = C_2$, which accounts for the ideal edge delay symmetry indicated in Fig. 15.24b. Actually, the falling edge of $X^*(H)$ is delayed by $\Delta t + \tau_p$, as can be deduced from the RMOD simulation in Fig. 14.12.

The magnitude of the delay Δt produced by the inertial delay element can be adjusted somewhat by altering the values of the R's and C's in the R-C circuit of Fig. 15.24a. The larger the time constant, the greater will be the Δt delay. Use of large time constants to generate large delays probably necessitates the use of Schmitt triggers on the outputs of the RMOD to minimize waveform distortion produced by the R-C components. Notice that the narrow input pulses have no effect on the delayed output response because of the low-pass filtering action of the R and C components. This, in effect, sets the lower bound on pulse width if the inertial delay element is used in the pulse narrowing circuits.

15.6 PERSPECTIVE ON THE PULSE MODE APPROACH TO FSM DESIGN

Clearly, pulse mode asynchronous FSMs have very limited practical application because of the stringent requirements placed on the input data signals. That is, the inputs must consist of nonoverlapping pulses at least minimally separated and with pulse widths of lower bound depending on the logic used. Also, an important distinction is made with regard to the memory elements that can or should be used in the design of asynchronous pulse-mode FSMs. The use of toggle modules as memory elements requires no upper bound on incoming pulse widths and has the advantage of eliminating ORGs and static hazards in the output logic by using outputs conditional on exiting pulses. This, of course, is made possible because it is a requirement that triggering occurs on the trailing edge of the data pulse. The toggle modules can be implemented by using D flip-flops, as in Fig. 15.3, or by using T or JK flip-flops operated in the toggle mode. The only down side to the use of toggle modules is that output logic noise, if it exists, cannot be filtered out by any of the conventional methods discussed so far in this text. This is so because the transitions occur on the trailing edges of the data pulses. However, such timing defects can occur only if Moore (unconditional) outputs are used.

Attempting to use the nested cell model in the design of pulse mode FSMs requires that special attention be paid to the bounds of pulse width that can be tolerated by the system. Pulse widths exceeding the upper bound limit will cause malfunction of the FSM. The only reliable means of dealing with unrestricted data pulse widths is to use feedback delays together with pulse narrowing circuitry as demonstrated in Fig. 15.23. Remember that use of inertial delay elements as in Fig. 15.24 may be necessary to generate the large delays required by some applications. It is true that without the use of these two types of delay the nested cell approach may enjoy a slight speed advantage over an equivalent design using toggle modules as the memory. However, safeguarding the nested cell design by adding feedback delays and pulse narrowing circuits negates any speed advantage the system may have had over a toggle module design. In fact, the logic circuit in Fig. 15.23 is likely to be considerably slower that its toggle module counterpart. With these facts in mind, one must conclude that there is little or no justification for using the nested cell approach to design pulse mode state machines. Therefore, if a pulse mode design is called for, it is recommended that toggle modules (or T flip-flops) be the memory elements of choice. The rather extensive discussion of this subject in this text is justified on the basis that most of the references in Further Reading at the end of this chapter deal with nested cell designs of pulse mode FSMs — often without discussing the critical pulse width problem.

If desirable, the nested cell design of a pulse mode FSM can be easily converted to the use of the MS memory elements of the type shown in Fig. 15.3c. Here, the S and R NS functions to the master stage remain the same as in the nested cell design, but now triggering occurs on the trailing edges of the data pulses similar to the toggle module approach. Also, as with toggle modules, the MS memory elements can be initialized via the asynchronous PR and CL overrides of the slave D flip-flops. However, the additional requirements on the lower bound of pulse width, expressed by Eq. (15.1), together with the additional hardware requirements and slower FSM performance, make this approach to pulse mode design less desirable than an equivalent toggle module design. Thus, toggle modules remain the memory elements of choice if given the option to use them in the design of pulse mode FSMs.

By their nature, pulse mode FSMs have limited applicability. But for those applications that are appropriate, pulse mode designs can offer the best approach. Sequence recognizers, digital combination locks, and vending machines, as described in Section 15.4, are good examples of appropriate applications of the pulse mode concept. Actually, the control or recognition of individual events of any kind often lend themselves quite naturally to the pulse mode design concept. For example, controlling, counting, or recognizing the passage or transport of individual people, fish, cans, coins, automobiles, boats, boxes, batteries, etc., is easily handled by the pulse mode method. Certain types of mechanical motion can also be recognized or controlled by pulse mode machines. Remember that the pulse mode approach to design of FSMs requires no clock oscillator circuitry, which can result in reduced hardware and power consumption.

The applications of the pulse mode also extends to counter design. Shown in Fig. 13.46 is the design of a 4-bit data-triggered up/down binary counter with asynchronous parallel load and asynchronous clear. This is a pulse mode design which requires that the *Up* and *Dn* input pulses never be active at the same time and that they always be at least minimally separated. For the counting of individual events, such data-triggered counters may be the best choice.

FURTHER READING

Unfortunately, few texts treat the subject of asynchronous pulse mode FSMs, and half of those give only passing mention to the design and analysis of these state machines. The texts of Kohavi, McCluskey, Nelson *et al.*, Tinder, Unger, and Yarbrough are the exceptions. All five of these texts cover the subject to one extent or another, but with different emphases. However, based on the subject as it is presented in this text, the texts of Nelson *et al.* and Tinder are the two recommended here for further reading. These two texts cover the use of both data-triggered T flip-flops and basic cells as memory elements in the design of pulse mode FSMs. Texts by Unger and Yarbrough tend to emphasize the use of basic cells. Unger provides a good discussion of the delay element requirements in the use of basic cell memory elements and is recommended for further reading. The text of McCluskey, on the other hand, provides a broadened definition of the pulse mode and covers different aspects of the pulse mode concept, those dealing with both synchronous and asynchronous FSMs. However, unless one is familiar with the ANSI/IEEE Standard for logic circuit symbols, McCluskey's text will be somewhat difficult to read. It should be mentioned that only McCluskey's text and the present text devote an entire chapter to the discussion of pulse mode machines.

[1] Z. Kohavi, *Switching and Finite Automata Theory*. McGraw-Hill, New York, 1978.
[2] E. J. McCluskey, *Logic Design Principles*. Prentice Hall, Englewood Cliffs, NJ, 1986.
[3] V. P. Nelson, H. T. Nagle, B. D. Carroll, and J. D. Irwin, *Digital Logic Circuit Analysis and Design*, Prentice Hall, Englewood Cliffs, NJ, 1995.
[4] R. F. Tinder, *Digital Engineering Design: A Modern Approach*. Prentice Hall, Englewood Cliffs, NJ, 1991.
[5] S. H. Unger, *The Essence of Logic Circuits*. Prentice Hall, Englewood Cliffs, NJ, 1989.
[6] J. M. Yarbrough, *Digital Logic Applications and Design*. West Publishing Co., Minneapolis/St. Paul, MN, 1997.

PROBLEMS

15.1 A simple digital combination lock (DCL) is to be designed for a vault that is to be operated in the pulse mode. It is the function of the DCL to issue a signal *OPNVLT* coincidentally with the last pulse in the pulse sequence $\cdots Y-X-Y-Y-X \cdots$, and then return immediately to the initialization state and reissue a *LOCK* signal. Note that the sequence cannot be overlapping.

(a) Construct the state diagram for the DCL by following the example in Fig. 15.8, keeping in mind that the sequence must be a nonoverlapping sequence. Give a state code assignment and output assignment that is free of ORGs.

(b) From the results of part (a), obtain an optimum set of NS and output functions for the DCL. To do this, use T flip-flops as the memory. Plan to use don't cares as permitted by the requirements of the design.

(c) Construct the logic circuit for the DCL. Use discrete logic for the NS and output logic and FET T flip-flops as the memory. Plan to initialize into the all-zero state.

(d) Verify the correct operation of the DCL by simulating the logic circuit of part (c).

15.2 Shown in Fig. P15.1 are the state diagrams for two FSMs that are to be operated in the pulse mode. These two FSMs are adaptations of those in Fig. P13.4 used earlier for synchronous FSM design.

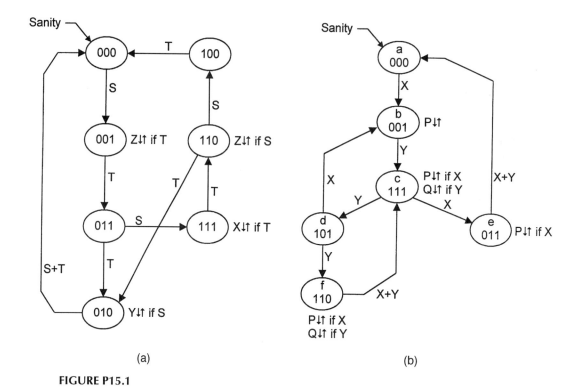

(a)   (b)

FIGURE P15.1

(1) Obtain an optimum set of NS and output functions for each of these pulse mode FSMs. Plan on using toggle modules as the memory. Indicate any problem these FSMs may have.

(2) Based on part (1), construct the logic circuit for each of these FSMs. To do this, use FET toggle modules as the memory and a PLA for the NS and output logic. Thus, construct the p-term table for the PLA. Initialize as indicated in the state diagrams, and assume that all inputs and outputs are active high.

15.3 The state diagram in Fig. P15.2 represents a pulse mode FSM.

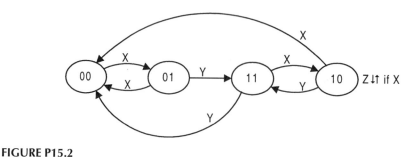

FIGURE P15.2

PROBLEMS

(a) Design this FSM by using the *nested cell model*. Thus, obtain an optimum set of NS and output equations appropriate for using basic cells as the memory. To do this, use $T \rightarrow S, R$, K-map conversion. (Hint: Refer to Algorithm 12.1, in Subsection 12.3.2, and Figs. 10.43 and 10.45. Thus, K-map conversions $T \leftrightarrow S, R$ and $T \leftrightarrow J, K$ are similar except in the way that don't cares are used.)

(b) Construct the logic circuit for the results of part (a). Assume that the inputs and output are all active high. Indicate the bounds of permissible pulse widths that can be used by this FSM. Plan to initialize into the 00 state. (Hint: To initialize a zero, force the basic cell into a reset condition.)

(c) Verify the proper operation of this FSM by simulating the results of part (b). Show the consequence of exceeding the upper bound in pulse width.

(d) State the algorithm (sequential function) for this FSM.

15.4 The following NS and output logic is read from an FSM that is designed to operate in the pulse mode. Here, the inputs are C and D, and the outputs are H and L. Initialization occurs via the active low PR and CL overrides to the two toggle modules, A and B.

$$T_A = \bar{A}\bar{B}C + AD + BD$$
$$T_B = C + D$$
$$H = D(A \oplus B)$$
$$L = AC$$
$$Sanity(L) = PR_A(L) = CL_B(L)$$

(a) Construct the state diagram for this FSM. Follow the example in Figs. 15.17 and 15.18.

(b) From the state diagram determine the sequential function of this FSM (its algorithm). Does this FSM satisfy all the requirements for operation in the pulse mode? Explain.

(c) Verify the results of parts (a) and (b) by simulating the circuit. (See Fig. 15.19 as an example.)

15.5 In Fig. P15.3 is the p-term table for a PLA implementation of an FSM that is designed to operate in the pulse mode. Here, X and Y are the inputs and P and Q are the outputs. The FSM is initialized via its PR and CL overrides according to the following: $Sanity(L) = CL_A(L) = CL_B(L) = CL_C(L)$.

(a) Construct the state diagram for this FSM by following the example in Figs. 15.17 and 15.18.

(b) Does this FSM satisfy all requirements for operation in the pulse mode? What limitations are placed on the pulse width limits? Are ORGs and static 1-hazards present in the output logic? Justify your answers to these questions. If ORGs and static hazards cannot be present, explain why that is so.

(c) Verify the proper operation of this FSM by simulating the circuit.

P-term	A	B	C	X	Y	T_A	T_B	T_C	P	Q
$\bar{B}CX$	–	1	0	1	–	1	0	0	1	0
BCY	–	1	1	–	1	1	0	0	1	0
$A\bar{B}Y$	1	0	–	–	1	1	0	0	0	0
ACX	1	–	1	1	–	1	0	0	1	0
$\bar{A}CY$	0	–	1	–	1	0	1	0	0	0
$\bar{A}BX$	0	1	–	1	–	0	1	0	0	0
$A\bar{C}X$	1	–	0	1	–	0	1	0	0	0
$\bar{C}X$	–	–	0	1	–	0	0	1	0	0
BY	–	1	–	1	–	0	0	1	0	0
CY	–	–	1	–	1	0	0	1	0	0
AX	1	–	–	1	–	0	0	1	0	0
AY	1	–	–	–	1	0	0	0	0	1

FIGURE P15.3

15.6 The following NS and output logic is read from a pulse mode FSM that has been designed by using the nested cell model. Here, X, Y, and Z are the inputs, and P and Q are the outputs. The logic circuit is initialized into the 00 state following the example in Fig. 14.47.

$$S_A = \bar{A}X + \bar{A}Z, \qquad S_B = A\bar{B}X + A\bar{B}Y + A\bar{B}Z$$
$$R_A = A\bar{B}X + AY, \qquad R_B = ABX + ABY + \bar{A}X + \bar{A}Z$$
$$P = X(A \oplus B)$$
$$Q = ABY + A\bar{B}Z$$

(a) Construct the state diagram for this FSM. To do this, follow the example in Figs. 15.21 and 15.22.

(b) Does this FSM satisfy all requirements for operation in the pulse mode? What limitations are placed on the pulse width limits? Are ORGs and static 1-hazards possible in the output logic? Justify your answers to these questions. If upper and lower bounds of pulse width exist for this FSM, quantify them.

(c) Verify the proper operation of this FSM by simulating the logic circuit. Follow the example in Fig. 15.22. Thus, show the consequence of exceeding the upper bound of pulse width.

(d) Indicate on the logic circuit how the circuit can be altered to accommodate pulses of greater widths than that specified in part (b).

15.7 A pulse mode asynchronous FSM is to be designed that functions as a controller for a security area shown in Fig. P15.4. It is required that when occupied the security area must be occupied by just two people, no more and no less. Access to the security area is through an outer door (D1) one person at time, along a narrow corridor and through

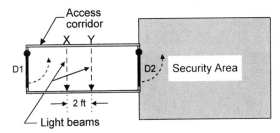

FIGURE P15.4

an inner door (D2), as shown in the figure. The corridor is equipped with two narrow (planar) light beams, X and Y, that fall incident on photodetector cells on the opposite wall. When the second person passes the Y check beam, the inner door (normally locked) is unlocked and the outer door (normally unlocked) is locked. The system permits one or both of the people in the access corridor to change their minds at any time and exit the corridor. However, any attempt by a third person to pass through the check beam X once two occupants have passed both check beams will set off an alarm (ALARM).

A red occupancy light (LT) is monitored on a remote control panel. It is initially OFF (LTOFF) and remains OFF until the second person passes the Y check beam on entering, at which time it is turned ON (LTON). Thereafter it remains ON until the second person passes the X check beam on exiting the corridor, at which time it is turned OFF.

(a) Construct a state diagram for the controller FSM that has no more than six states. Make certain it is free of ORGs and plan to initialize into the 000 state. (Hint: Use Gray code, and use conditional outputs only where necessary.)

(b) Construct the ROM program table directly from the state diagram in part (a). To do this, follow the example in Figs. 15.8 and 15.9. Assume that the inputs are all active high. Let all outputs be active high except those of *LTON* and *LTOFF*, which are issued active low to an LED display.

(c) Construct the logic circuit for the security area access controller. Use a block symbol for the ROM and assume the use of FET T flip-flops as the memory. Are static hazards and ORGs possible in the output logic from the ROM? If so, explain the consequences of their presence.

15.8 One severely limiting aspect to pulse mode FSM design is the requirement of nonoverlapping input pulses. Many applications of the pulse mode approach to design are prohibited because the inputs arrive as overlapping waveforms. Furthermore, if a nested cell design is to be used, further restrictions are placed on the upper and lower bounds of nonoverlapping pulses. Some of these problems can be solved by the use of a *bus arbiter*, which is the subject of this problem.

Shown in Fig. P15.5 are two basic two-input *bus arbiter modules*. Each consists of a mutual exclusion element (ME) and the external logic as shown. The ME is composed of a special basic cell and two line drivers. It is the function of the bus arbiter to arbitrate between two competing requests R_X and R_Y and grant access to a protected system based on a "first-in/first-out" principle. Thus, only one grant signal (G_X or G_Y) is active at any given time, even though the inputs may be overlapping. If the inputs are

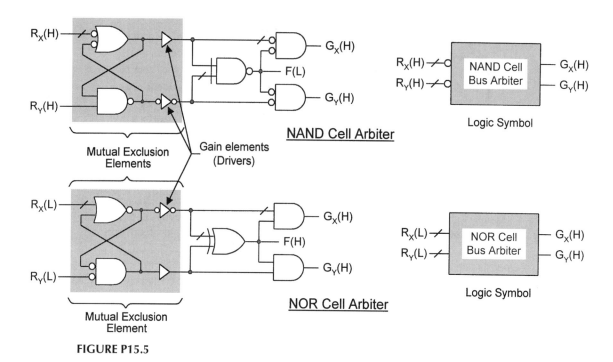

FIGURE P15.5

overlapping at some point in time, a second access is granted only after the first request goes inactive. Should both input requests go active at the same time, the ME must arbitrate a "winner" and grant access to that input. It is the specially built basic cell together with the gain elements and XOR gate that perform the arbitration function.

(a) Simulate each circuit in Fig. P15.5 as a logic circuit. To do this, omit the gain elements and treat the remainder as simple logic. Monitor all inputs and outputs shown, including the $F(L)$ and $F(H)$ outputs. From the simulation, determine the throughput response and the pulse width of the grant signals in each case.

(b) Multiple inputs can be handled by a multiple-input bus arbiter. Use a sufficient number of NOR cell bus arbiter modules (in Fig. P15.5) together with a sufficient number of RMODs to design a three-input bus arbiter. Let the three inputs be R_1, R_2, and R_3, and assume that all inputs are active low and outputs are active high. Consider that the number of bus arbiter modules required is determined by the number of combinations of n inputs taken q at a time given by

$$N_n = \left(\frac{n!}{q!(n-q)!} \right).$$

Plan to provide an asynchronous clear capability to the bus arbiter. To do this, use the LPD RMOD design of Problem 14.2 and follow Fig. 14.32a. End with a logic circuit by using logic (block) symbols for each bus arbiter module and each RMOD.

(c) Verify the proper operation of the three-input bus arbiter of part (b) by simulating the actual arbiter circuit. To do this, present the inputs to the arbiter as overlapping waveforms that do not change in close proximity to one another.

(d) Repeat part (b) for a four-input bus arbiter. (Hint: Use three-input RMODs designed by the LPD model.) How many bus arbiter modules and RMODs are required for five-input and six input bus arbiters, and how many inputs must each RMOD have?

15.9 Use the two-input NAND cell bus arbiter module in Fig. P15.5 to design the pulse mode FSM in Fig. P15.2 if it is assumed that the inputs, X and Y, arrive as overlapping waveforms.

(a) Obtain an optimum set of NS and output functions for this FSM by using toggle modules as the memory.

(b) Use the results of part (a) to construct the logic circuit for this FSM. Include the arbiter module and assume that the inputs and output are all active high. Plan to initialize into the 00 state.

(c) By using the logic circuit for the two-input bus arbiter module, show how the pulse widths from the arbiter module can be augmented for use in pulse mode FSMs. Recall that permissible pulse widths often fall in the range of $2\tau_p$ to $3\tau_p$ for most nested cell designs, but that they must be greater than $2\tau_p$ for pulse mode designs that use toggle modules as the memory.

(d) Simulate the logic circuit of part (b) by making use of part (c). To do this, use the logic equivalent of the arbiter module (exclusive of drivers) and apply the input waveforms given as follows:

XY 00–10–11–01–00–01–11–10–00–01–11–10–11–01–11–10–

00–10–00–01–00.

Make certain that the input waveform changes are sufficiently separated. An ideal logic simulator cannot arbitrate between two competing request signals that are changing in close proximity to one another.

CHAPTER 16

Externally Asynchronous/ Internally Clocked (Pausable) Systems and Programmable Asynchronous Sequencers

16.1 INTRODUCTION

Externally asynchronous/internally clocked (EAIC) systems represent a compromise between the synchronous and asynchronous design methodologies. While functioning asynchronously with respect to the external world, the EAIC system is controlled by a single internally generated clock signal that is produced when valid outputs exist from each memory element. In this scheme, input synchronizing registers and memory registers coordinate to generate the internal clock. The internal clocking of an EAIC system causes it to be free of critical races, essential hazards, and errors due to static hazards. In addition, the memory modules of an EAIC system are protected against errors due to metastability and, hence, are an integral part of a *pausable system* — one that is capable of an infinite MTBF. The speed of the internal clock is limited only by the actual logic delays within the system, rather than by the typical worst-case delay of synchronous systems, and can operate in excess of 400 MHz for state-of-the-art submicron CMOS designs. The EAIC memory elements are constructed of either static or dynamic domino logic, and each is protected by a unique metastable detection stage that prohibits any metastable condition from reaching the output. The internal clock generating circuitry is shown to be delay insensitive when operated within specified bounds.

This chapter concludes with the detailed development of two unique and important classes of asynchronous programmable sequencers that are designed to operate in the fundamental mode. These sequencers can be driven by discrete logic or by PLDs (e.g., PLAs or PALs) free of the numerous timing defects that can cause fundamental mode FSMs to fail. Furthermore, by multiplexing PLDs to drive a single sequencer, it is possible to instantly switch between radically different asynchronous FSMs. By this means multiple controllers can be operated asynchronously, on a time-shared basis, by the same sequencer. The PLDs that drive the sequencer are easily programmed directly from a state diagram or state table, or from K-maps plotted from the state diagram.

16.2 EXTERNALLY ASYNCHRONOUS/INTERNALLY CLOCKED SYSTEMS AND APPLICATIONS

The general (Mealy) model for an EAIC system is shown in Fig. 16.1. It consists of input (synchronization) and memory DFLOP registers of either the static logic (SL) or dynamic logic (DL) type, next-state-forming logic, and clock-generating circuitry. On the rising edge of each clock cycle, the inputs are stored in the input register and a new state is stored in the memory register as determined by the next-state logic during the previous clock cycle. As each DFLOP resolves, a data-ready (R) signal is issued to the majority gate (NOR gate) which, in turn, issues the falling edge of the clock when all DFLOPs have resolved. On the falling edge of the clock, the DFLOPs return to their unresolved state, causing the R signals to be deasserted and a new rising clock edge to be issued by the clock generating circuitry.

The memory element of an internally clocked system is called a DFLOP module. It functions in a manner similar to an edge-triggered D flip-flop, but with an added output that signals when the DFLOP is resolved and ready for a deactivating clock edge. This added output signal is required for proper operation of the DFLOP within the EAIC system. In addition, each DFLOP contains mutual-exclusion circuitry that protects the output stage of

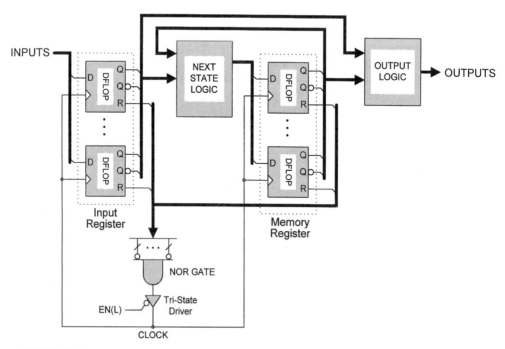

FIGURE 16.1
General architecture (Mealy model) for the EAIC system showing DFLOP input and memory registers and clock-generating circuitry with tri-state enable.

16.2 EXTERNALLY ASYNCHRONOUS/INTERNALLY CLOCKED SYSTEMS

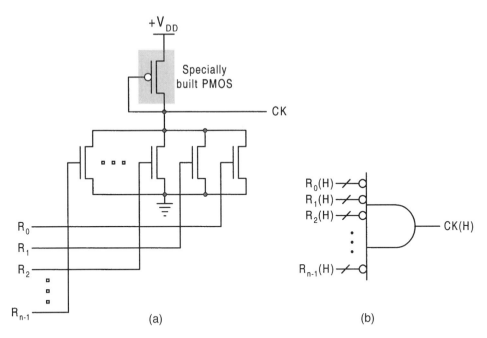

FIGURE 16.2
Multiple-input NOR gate specifically designed to minimize fan-in limitations and propagation delay. (a) Generalized CMOS circuit required for application in EAIC systems. (b) Generalized NOR gate symbol and input logic level requirements for EAIC system operation.

the DFLOP from errors caused by any metastable condition that may develop in the input stage. The design details of the DFLOPs are discussed later in Subsection 16.2.1.

The NOR gate shown in Fig. 16.1 is an important part of the clock generating system. As is pointed out in Section 8.9, the performance of a conventional CMOS NOR gate diminishes with increasing fan-in. Since the NOR gate in an EAIC system must be able to accommodate a large number of inputs, it is necessary to use a specially designed CMOS gate structure. Shown in Fig. 16.2 is the multiple input NOR gate featured in Fig. 8.46, but specifically labeled for use in an EAIC system. The number of permissible inputs is unlimited with negligible effect on the gate path delay. Thus, it retains essentially the same path delay of a two-input NOR gate regardless of the number of inputs. The output of this gate goes to high voltage (HV) only if *all* inputs are at low voltage (LV). If any one or more of the inputs go to HV, the output goes to ground level (LV). To work correctly, it is necessary that the PMOS be specially designed so that the drain-to-source resistance (actually impedance) remains sufficiently high so as to minimize drain current when one or more NMOS are turned ON. Note that the specially built PMOS can be replaced by a depletion-mode NMOS permitting the NOR gate of Fig. 16.2 to be replaced by the NMOS technology of Fig. A.1 in Appendix A.

16.2.1 Static Logic DFLOP Design

The general structure for a DFLOP is shown with block symbols in Fig. 16.3. Its structure is similar to that of a conventional D flip-flop except the DFLOP is equipped with a metastable

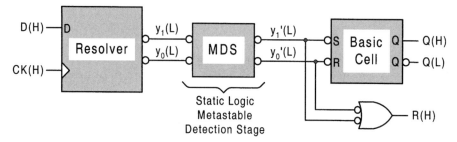

FIGURE 16.3
Block diagram for a DFLOP showing the static logic CMOS metastable detection stage and the NAND gate required to generate the data-valid signal (R).

detection stage (MDS) and data-ready (R) circuitry. It is the function of the MDS to detect any metastable condition in the resolver and block it from entering the output basic cell stage. The manner in which this is done is discussed later in this section.

The details required for the design of the static logic (SL) DFLOP are provided in Fig. 16.4. The resolver state diagram shown in Fig. 16.4a is similar to that for the RET D flip-flop given in Fig. 14.14a. The state code assignment differs from that of the D flip-flop because of the need for logic symmetry when connecting the resolver to the MDS logic shown in Fig.16.4b. The state diagram for the set-dominant basic cell in Fig. 16.4c is the same as that given in Fig. 14.14b, except for the branching condition labels that derive from the MDS outputs.

On the rising edge of clock (CK), the static-logic DFLOP (SL-DFLOP) resolver stores the value of the input data D and issues an output R via the MDS (see Fig. 16.3) indicating that it has resolved the data. Once the resolver has entered a resolved state (either 01 or 10), further changes in the input data D cannot affect the stored value until the next rising CK

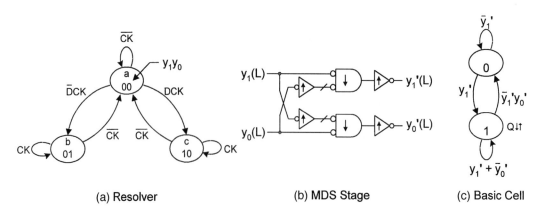

(a) Resolver   (b) MDS Stage   (c) Basic Cell

FIGURE 16.4
Design of the DFLOP for EAIC systems exclusive of preset and clear circuitry. (a) State diagram for the resolver FSM input stage. (b) Metastable detection stage (MDS) indicating raised ($\uparrow$) and lowered ($\downarrow$) thresholds for inverters and gates. (c) State diagram for the set-dominant basic cell.

16.2 EXTERNALLY ASYNCHRONOUS/INTERNALLY CLOCKED SYSTEMS

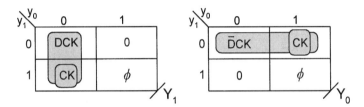

FIGURE 16.5
EV K-maps for the resolver of a DFLOP as plotted from the state diagram in Fig. 16.4a.

edge. This, of course, is the data lockout character of a D flip-flop. Only when CK goes inactive will the resolver return to the unresolved 00 state where it awaits another active CK signal.

The next-state (NS) K-maps are obtained directly from the resolver state diagram in Fig. 16.4a and are presented in Fig. 16.5. To obtain the activation levels and logic symmetry needed to interface with the MDS, an apparent nonoptimum cover is chosen as indicated by the shaded loops. Thus, the don't care in each K-map is ignored. From these K-maps the resulting NS logic expressions are found to be

$$\begin{cases} Y_1 = \bar{y}_0 DCK + y_1 \bar{y}_0 CK = (D + y_1) \cdot \bar{y}_0 CK \\ Y_0 = \bar{y}_1 \bar{D}CK + \bar{y}_1 y_0 CK = (\bar{D} + y_0) \cdot \bar{y}_1 CK \end{cases}, \quad (16.1)$$

where factoring is used to optimize the logic and for purposes of interfacing with the MDS stage.

The Set and Reset branching conditions of the set-dominant basic cell in Fig. 16.4c are easily defined in terms of the present state variables of the resolver. Recalling the connections shown in Fig. 16.3, these branching conditions are given by

$$\begin{cases} y'_1 = y_1 \bar{y}_0 & \text{Set condition} \\ \bar{y}'_1 y'_0 = \bar{y}_1 y_0 & \text{Reset condition} \end{cases}. \quad (16.2)$$

Clearly, the Set condition results from the resolver entering a resolved state 10, whereas the Reset condition is caused when the resolver enters the 01 state.

The complete logic circuit for the SL-DFLOP is constructed from Figs. 16.3, 16.4, and 16.5 and from Eqs. (16.1) and is presented in Fig. 16.6, where the MDS is highlighted for emphasis. Included are the preset (*PR*) and clear (*CL*) overrides, which are necessary for initialization and reset of the DFLOPs. The active low inputs to the MDS are provided by the outputs from the two four-input NAND gates of the resolver. This represents the logic level compatibility and logic symmetry mentioned earlier. In order to correctly implement the preset and clear functions, it is necessary to set the state of the resolver FSM as well as that of the output basic cell. Because extra time is necessary for the effect of the preset or clear signal to propagate through the MDS stage to the outputs, the duration of either the *PR*(L) or *CL*(L) signal to the DFLOPs must be long enough to assure that the correct mixed-rail outputs have time to propagate through the next state logic (indicated in Fig. 16.1) before the occurrence of the next clock event.

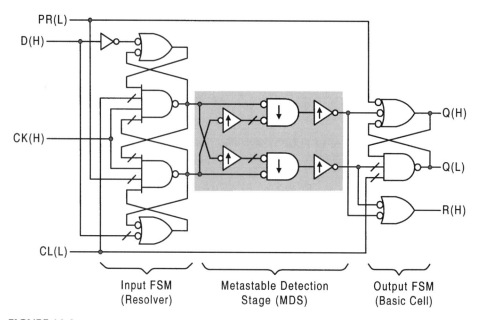

FIGURE 16.6
Logic circuit for the static logic SL-DFLOP as constructed from Figs. 16.3, 16.4, and 16.5 together with Eqs. (16.1) showing PR and CL override connections.

It is a basic cell output stage of the SL-DFLOP that stores the *set* or *reset* output from the resolver via its MDS circuit. The data-ready (R) signal issued by one of the resolved states of the resolver is formed by the logical OR of the *set* and *reset* conditions from the MDS. An active R signal indicates that the DFLOP outputs have been updated and signals a readiness of the resolver to receive a falling CK edge. Since the basic cell output stage is protected from any possible metastable conditions in the resolver, the Q(H) and Q(L) outputs will be error free and logically stable.

The Metastable Detection Stage Each SL-DFLOP used in the EAIC system employs a metastability detection stage (MDS) of the type shown in Fig. 16.6. The MDS operates as a mutual exclusion element to prevent a possible metastable state in the resolver from being passed on to the basic cell output FSM: If either y_1 or y_0 is active (not both active), the corresponding y_1' (set condition) or y_0' (reset condition) becomes active, signaling that the resolver has resolved into a logically definable state. Under any other set of input conditions, the outputs y_1' and y_0' are always deactivated — they drop low!

The simulated PSPICE response of the static logic MDS in Fig. 16.4b to a variety of input conditions is shown in Fig. 16.7. Correct operation of the resolver is simulated in the first input sequence (0–30 ns), where the MDS outputs y_1' and y_0' follow the inputs y_1 and y_0 as the resolver transits between resolved and unresolved states. Worst-case conditions exist in the next segment (30–100 ns) where the inputs are introduced as a damped sine-wave oscillation with a phase difference of 90 degrees causing the maximum difference between the inputs y_1 and y_0 to approach 2.5 volts. As can be seen, the straddling of the MDS switching threshold ($V_{th} = 1.1\ V$) by y_1 and y_0 causes the beginning of pulse formation

16.2 EXTERNALLY ASYNCHRONOUS/INTERNALLY CLOCKED SYSTEMS

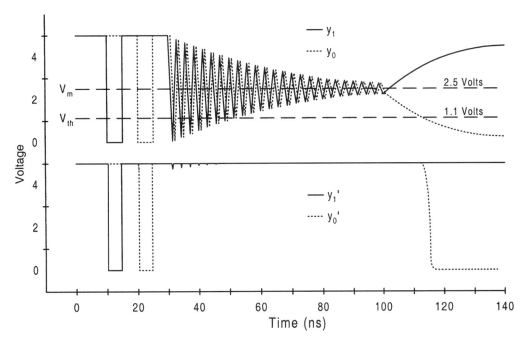

FIGURE 16.7
PSPICE simulation of the static logic MDS circuit in Fig. 16.4b.

on the MDS outputs, y_1' and y_0'. However, the formation of these erroneous output pulses is very small and directly dependent on the frequency of oscillation. An increase in oscillation frequency results in a decrease in straddling time and consequently permits the MDS to correctly filter the metastable condition. In addition, because a valid output pulse can only be generated if the inputs straddle the adjusted switching threshold, any input activity above the threshold of 1.1 volts cannot result in output pulses regardless of the frequency and phase difference of the input signals.

The metastable voltage V_m tends to lie in the range of mid-supply (see references in Further Reading), which in this case is taken to be 2.5 volts for a 5.0-volt supply, as indicated in Fig. 16.7. Consequently, shifting the switching threshold of the MDS away from the predicted voltage of V_m can greatly reduce the probability (possibly to zero) that a metastable state will occur and cause an erroneous output signal to be generated. In order to shift the threshold of the MDS circuit shown in Fig. 16.4b, the switching thresholds of the MDS gates (including inverters) are adjusted in the following way: A PMOS-to-NMOS width ratio of $wp/wn = 0.25$ is used in the MOSFETs of the low-threshold ($\downarrow$) gates, shifting their switching threshold to approximately 1.1 volts. For the high-threshold inverters ($\uparrow$), a PMOS-to-NMOS width ratio of $wp/wn = 8$ is used to raise their switching threshold to approximately 3.0 volts. By using these adjusted gates and inverters, the switching threshold of the entire MDS in Fig. 16.4b is lowered to approximately 1.1 volts. As a result, only cleanly asserted signals can pass through a given detection circuit, and the in-phase behavior of a metastable input condition will not produce an output signal. Thus, the outputs of the DFLOP cannot be updated until the resolver has cleanly resolved into the set or reset condition.

Should the resolver enter a metastable state, the logic state of the outputs would be undefined and a fatal error would result if the metastable state were permitted to propagate to the external system. Consequently, it is essential that the outputs of the resolver be stably resolved before the result is permitted to propagate to the output stage of the DFLOP. It is, of course, the MDS that performs this function within the DFLOP. Studies of the metastable condition relevant to this subject are cited in Further Reading at the end of this chapter. In a fully protected EAIC system, any pause in the issuance of the R signal due to metastability in the resolver of a DFLOP will result in a corresponding pause in the internal clock. Therefore, such an EAIC system can be categorized as a *pausable-clock system*.

A frequently reported study of the metastable state in the cross-coupled NAND gates is cited in Further Reading at the end of this chapter. With this study in mind, and since the signal rise/fall time tends to dominate the propagation delay of simple CMOS gates, it is reasonable to assume that a metastable condition in the DFLOP resolver would be characterized by an output voltage V_m and not by oscillatory behavior. However, an oscillatory metastable condition must be considered as possible. Therefore, the oscillation frequency and phase difference of any oscillatory metastable condition that is passed to the MDS circuit is important in evaluating the total performance of the MDS. Although most previous work supports the in-phase nature of metastable oscillation, little has been said about possible phase differences. The symmetrical nature of the cross-coupled NAND gates in the DFLOP resolver supports the assumption of minimal phase difference in oscillatory behavior. Thus, should a metastable oscillatory condition occur, any actual phase difference would be much less than the 90 degrees difference used in the simulation of Fig. 16.7, allowing the detection circuit to fully protect the outputs from any possible metastable input conditions.

16.2.2 Domino Logic DFLOP Design

Dynamic domino CMOS logic, or simply domino logic (DL), can be used advantageously in the design of DFLOP modules. Domino logic gates are noninverting and are fast, but require reasonably high clocking frequencies to control the precharge and evaluate phases of the dynamic operation. Low-frequency operation is excluded because of leakage current effects. Since an EAIC system provides a fast and regular clock signal, the DFLOP is ideally suited for implementation with dynamic domino logic — possibly the best usage of the DL technology. For this purpose, the resolver for the domino logic DFLOP (DL-DFLOP) must be designed to accommodate the requirements of domino logic. References on domino logic are cited in Further Reading at the end of this chapter. The following subsection provides an introduction to domino CMOS logic.

Represented in Fig. 16.8 are the essential components of the DL-DFLOP. Included are the state diagram for the DL-DFLOP resolver FSM, the DL MDS stage, and the familiar set-dominant basic cell as the output FSM. The dashed branching paths for CK and $\overline{CK}$ shown in the state diagram are used to indicate that the CK signal does not act directly to force a state-to-state transition but does so via the precharge and evaluate stages of dynamic domino logic operation. The asterisk (*), placed within a gate symbol, identifies a domino CMOS logic structure. Except for this notable difference, the resolver FSM for the DL-DFLOP operates the same as that for the SL-DFLOP in Fig. 16.6.

Presented in Fig. 16.9 are the NS K-maps for the DL-DFLOP resolver FSM. Again, the use of don't cares is avoided so to yield the proper logic level and symmetry characteristics

16.2 EXTERNALLY ASYNCHRONOUS/INTERNALLY CLOCKED SYSTEMS

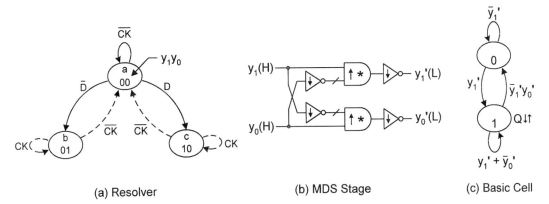

(a) Resolver   (b) MDS Stage   (c) Basic Cell

FIGURE 16.8
Design of the domino logic DL-DFLOP for EAIC systems exclusive of the preset and clear circuitry. (a) State diagram for the resolver input FSM. (b) Metastable detection stage (MDS) indicating lowered ($\downarrow$) and raised ($\uparrow$) thresholds for inverters and gates, and dynamic domino logic AND gates (*). (c) State diagram for the set-dominant basic cell output FSM.

needed to interface with the DL MDS stage. The K-maps are plotted as though the CK and $\overline{CK}$ branching paths in Fig. 16.8a were absent. The resulting NS functions for the DL-DFLOP, as read directly from the K-maps, are

$$\begin{cases} Y_1 = \bar{y}_0 D + y_1 \bar{y}_0 = (D + y_1) \cdot \bar{y}_0 \\ Y_0 = \bar{y}_1 \bar{D} + \bar{y}_1 y_0 = (\bar{D} + y_0) \cdot \bar{y}_1 \end{cases}, \qquad (16.3)$$

which are the same as those for the SL-DFLOP but with the CK input missing. The Set and Reset branching conditions for the basic cell output FSM are the same as those given by Eqs. (16.2).

As pointed out previously, domino logic is noninverting. This means that AND and OR gate forms are used in configuring the DL-DFLOP. The resulting logic circuit for the DL-DFLOP is easily constructed from Eqs. (16.3) and Figs. 16.8 and 16.9, and is shown in Fig. 16.10. As before, the gates (or inverters) with lowered switching thresholds are indicated with a down arrow ($\downarrow$) and those with raised switching thresholds are identified with an up-arrow ($\uparrow$). Because of the opposite oriented adjusted thresholds and the opposite activation levels of the inputs from the resolver, the switching threshold for the DL MDS

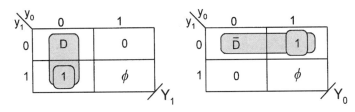

FIGURE 16.9
EV K-maps for the domino logic resolver of a DL-DFLOP as plotted from the state diagram in Fig. 16.8a.

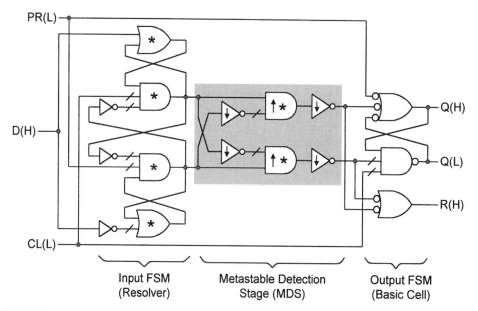

FIGURE 16.10
Logic circuit for the DL-DFLOP based on Figs. 16.8 and 16.9 and on Eqs. (16.3) showing PR and CL override connections.

circuit in Fig. 16.10 is raised to approximately 3.3 volts and not lowered as in the case of the SL MDS. Again, the asterisk symbol (*) indicates dynamic domino CMOS logic. Notice that DL logic us used for both resolver and MDS circuit. This helps to improve performance of the DL-DFLOP.

16.2.3 Introduction to CMOS Dynamic Domino Logic

Conventional CMOS gates of the general structure shown in Fig. 3.5 can be characterized as having a pull-up part (the PMOS) and a pull-down part (the NMOS) that are positioned to make the best use of the MOSFETs. The PMOS transistors are placed on the supply end (high side) because they pass HV well but not LV; the NMOS transistors are placed on the ground end (low side) because they pass LV well but not HV. Thus, conventional CMOS can be viewed as having to realize the same logic function twice in complementary fashion, once for the pull-up part and once for the pull-down part. The dynamic CMOS logic eliminates this redundancy by using one clocked pull-up PMOS (T_P) to precharge the output high, and one clocked pull-down NMOS (T_N) to evaluate low the intervening NMOS logic between these two transistors. Domino CMOS logic (DL) adds an inverter to the output of the dynamic structure. This can best be understood by viewing the generalized DL logic configuration shown in Fig. 16.11a. Here, the symbol Φ represents a single phase clock signal whose logic values have the following meaning:

$$\Phi = 0 \quad (LV) \quad \text{Precharge}$$
$$\Phi = 1 \quad (HV) \quad \text{Evaluate.}$$

16.2 EXTERNALLY ASYNCHRONOUS/INTERNALLY CLOCKED SYSTEMS

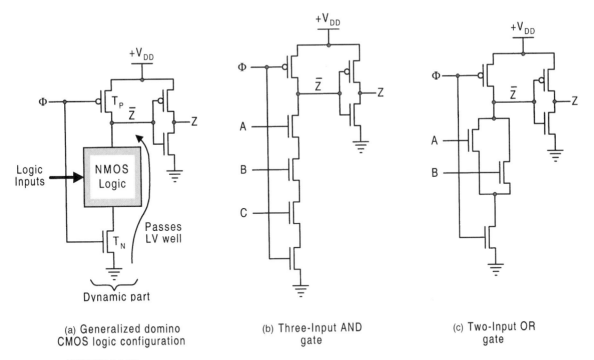

(a) Generalized domino CMOS logic configuration

(b) Three-Input AND gate

(c) Two-Input OR gate

FIGURE 16.11
Examples of domino CMOS logic (DL) structures suitable for use in an EAIC system. (a) Generalized DL configuration. (b) Three-input AND gate. (c) Two-input OR gate.

During the precharge stage, T_P is turned ON while T_N is turned OFF, bringing $\bar{Z}$ high and Z low. Thus, a single PMOS is required to pass HV during the precharge stage, which it does well. Then during the evaluate stage, the logic values of $\bar{Z}$ and Z depend on the intervening NMOS logic. The three-input DL AND gate example in Fig. 16.11b may help the reader better understand the evaluate stage. If all inputs to this gate are at HV during the evaluate stage, $\bar{Z}$ is forced to ground potential (low) while Z goes high. Thus, the node at $\bar{Z}$ is discharged to ground, a fast process. If, on the other hand, one or more of the three inputs are at LV, then $\bar{Z}$ remains at its previous precharge level (by stored charge in the capacitance of the system). So to prevent significant leakage current and static (quiescent) power dissipation during this precharge–hold state, the precharge/evaluate process must be driven at high frequency. The same arguments apply to the DL OR gate in Fig. 16.11c, and to all other DL gate structures.

The dynamic part of domino CMOS logic tends to be noisy because of switching transients. This problem is eliminated by adding the inverter (buffer) to the dynamic part as indicated in Fig. 16.11. It is because of the presence of the inverter buffer that DL logic is basically noninverting. Without the inverter, the gates in Figs. 16.11b and 16.11c would be a three-input NAND gate and a two-input NOR gate, respectively. But these gates should never be configured in that manner. If a three-input NAND gate is required, an inverter must be added to the DL AND gate in Fig. 16.11b. Similarly, an inverter must be added to the DL OR gate in Fig. 16.11c to form a NOR gate.

For relatively few inputs, domino logic requires more transistors than for static logic. However, as the number of inputs increases, a crossover point is reached beyond which domino logic has fewer transistors than static logic. With reference to Figs. 3.10 through 3.19, 16.11b, and 16.11c, the number of transistors (N) as a function of number of inputs (I) for dynamic logic (DL) and for static logic (SL) is given by the following equations:

$$\text{DL AND or OR} \quad N = (I + 4)$$
$$\text{DL NAND or NOR} \quad N = (I + 6)$$
$$\text{SL NAND or NOR} \quad N = 2I$$
$$\text{SL AND or OR} \quad N = 2I + 2$$

Clearly, beyond four inputs, the DL AND or OR gate requires fewer inputs than the SL AND or OR gate. However, for NAND or NOR gates the crossover point is at six inputs.

Domino CMOS logic is fast if operated correctly. However, it is difficult to make a valid comparison of the relative speeds of the DL and SL technologies. It is true that in DL gates, only a single PMOS needs to be precharged, which it does over a very short period of time. In SL gates, the complementary configuration of PMOS is required to pass HV (a charging process), which it does well but over period of time that depends on the complexity of the complementary logic. It is likely that precharging a single PMOS as in the DL case takes less time that does the charging process in SL gates. This difference may be especially significant for OR or NOR gates of the two technologies.

16.2.4 EAIC System Design

The general architecture for the EAIC system is illustrated in Fig. 16.1. The operation of this system centers mainly on the manner in which the internal clock is generated and the events that take place in triggering the DFLOPs of the input and memory registers. Otherwise the operation of the EAIC system is quite similar to the operation of a synchronous FSM that uses D flip-flops as the memory. In fact, the design and analysis of EAIC FSMs is exactly the same as the design and analysis of synchronous FSMs that use D flip-flops, as described in Sections 10.12 and 10.13. What must be done next is to discuss the details of the timing constraints and throughput characteristics for an EAIC system.

Next-State Logic and Input Pulse Constraints In order to guarantee the proper operation of an EAIC system, certain timing constraints must be observed. Within the bounds of these constraints, the EAIC system may be classified as completely delay-insensitive. To optimize throughput it is necessary that the updated Q outputs from the input register propagate through the next-state forming logic before the next rising-edge clock event. Consequently, it is required that

$$\delta_{NS} \leq (\delta_{DFLOP} + 2\delta_{NOR}), \tag{16.4}$$

where δ_{NS} is the propagation delay through next-state logic, $2\delta_{NOR}$ is the propagation delay through the external portion of the clock-generating circuitry (NOR gate plus driver), and

16.2 EXTERNALLY ASYNCHRONOUS/INTERNALLY CLOCKED SYSTEMS

δ_{DFLOP} is the propagation delay through DFLOP. There is a little more than about three or four (maximum) gate delays through a DFLOP, and one gate delay through the NOR gate. Consequently, the propagation delay through the next-state logic should not exceed five gate delays, if optimum and reliable results are to be achieved. This allows a good margin for error when two-level next-state logic is used. Exceeding the next-state logic constraint can cause error only in the outputs of some Mealy machines. Also, the minimum input pulse width must be greater than the period of the internal clock, guaranteeing that each input state will last long enough to be clocked into the input register.

Frequency and Throughput Characteristics of the EAIC System An inspection of the general EAIC architecture in Fig. 16.1 indicates that the internal clock generating path involves the propagation delay through the DFLOP registers combined with that of the NOR gate. Tracing this path beginning and ending at the NOR gate output provides the following expression for the internal clock frequency:

$$f_{CK} = (2\delta_{DFLOP} + 2\delta_{NOR})^{-1}. \tag{16.5}$$

With the gate delay equivalents given earlier, the internal clock period is estimated to lie in the range of 8 to 10 gate delays, but will depend on the fan-in of these gates. Given the propagation delays of modern state-of-the-art CMOS gates, frequencies in excess of 400 MHz can be expected for EAIC systems that employ this technology.

Throughput may be defined as the elapsed time between an external input change and a resulting output response from the DFLOP memory register. For an EAIC system, the throughput will normally be in the range

$$\left(f_{CK}^{-1} + 3\delta_{DFLOP} + 2\delta_{NOR}\right) \geq \delta_{Throughput} \geq (3\delta_{DFLOP} + 2\delta_{NOR}) \tag{16.6}$$

with a minimum exceeding the clock period by δ_{DFLOP}, or approximately three to four gate delays. The ranges expressed by Eq. (16.6) result from introducing Eqs. (16.4) and (16.5) into the minimum and maximum throughputs given by $(2\delta_{DFLOP}+\delta_{NS})$ and $(3\delta_{DFLOP}+2\delta_{NS})$, where the latter quantity is the minimum plus the feedback delay of $(\delta_{DFLOP} + \delta_{NS})$.

16.2.5 System Simulations and Real-Time Tests

Shown in Fig. 16.12 are the state diagram, and NS and output K-maps for a simple two-input/one-output sequence recognizer that is used to test the EAIC system. The minimum NS and output functions, as read from the K-maps, are

$$\begin{Bmatrix} D_A = A\bar{X}Y + BXY + AB \\ D_B = \bar{A}X\bar{Y} + BXY \\ Out = A\bar{B}\bar{X}Y \end{Bmatrix}. \tag{16.7}$$

The Mealy output is issued only in state 10 and then only under the input conditions $\bar{X}Y$. It is possible for this FSM to transit (cycle) with CK through states 10 and 00 under branching conditions $X\bar{Y}$ without issuing an output and without holding in either state. This is done deliberately to test throughput.

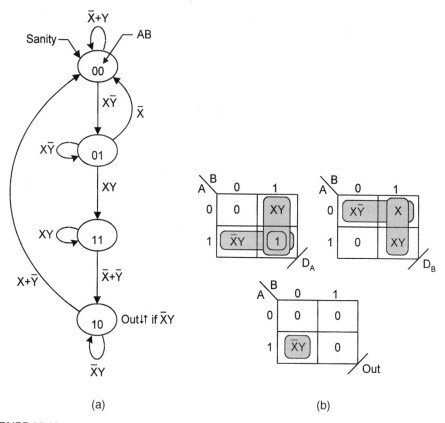

FIGURE 16.12

Design of a simple sequence recognizer for use in testing the EAIC system. (a) State diagram. (b) Next-state and output K-maps.

The EAIC circuit is constructed by using Eqs. (16.7) and is shown in Fig. 16.13. Notice that two DFLOPs are used for input register and two for the memory register, and that all four issue a data ready (R) signal that are part of the clock generating circuitry. All four R signals must be inactive and must rendezvous at the NOR gate before an active $CK(H)$ signal is issued. Then, on the rising edge of $CK(H)$, all four DFLOPs are triggered simultaneously. At this time the inputs are stored in the input register and delivered to the memory register via the NS logic and, at the same time, the new state is stored in the memory register as determined by the NS logic during the previous clock cycle. When the four DFLOPs are triggered, the four R signals become active, which deactivates $CK(H)$ at the NOR gate. When the DFLOPs receive the falling edge of $CK(H)$, the MDS outputs go low, which deactivates the R signals while retaining the current input values in the basic cell output stages. The inactive R signals cause CK to go active again and the process just described is repeated.

The EAIC system in Fig. 16.13 was simulated by using PSPICE (Level 3), with 1.0μ n-well MOSFET transistor models obtained from MOSIS fabrication runs. The 1μ model provides a suitable reference point between old and new industrial standards. All gates were designed by using CMOS technology, with the goal of optimizing for size while setting an

16.2 EXTERNALLY ASYNCHRONOUS/INTERNALLY CLOCKED SYSTEMS

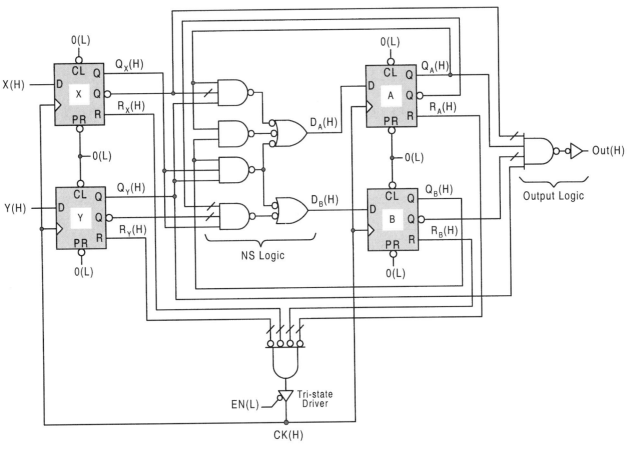

FIGURE 16.13
Logic circuit for the EAIC system applied to a simple two-input/one-output sequence recognizer.

arbitrary rise/fall time ceiling at approximately 0.5 ns for both 1.0μ static and domino logic gates. In the simulations, the parasitic effects of line resistance, capacitance, and inductance were assumed to be negligible, a reasonable assumption with the possible exception of line capacitive effects. (See Further Reading for reference on this work.)

PSPICE simulations were performed on both SL-DFLOP and DL-DFLOP designs of the sequence recognizer in Fig. 16.13. A typical result is shown in Fig. 16.14, which is that for a DL-DFLOP design with a conventional CMOS NOR gate in the clock-generating circuitry. The measured internal frequency for this design is 280 MHz with a minimum and maximum throughput of 4.9 and 8.5 ns, respectively. The maximum allowable NS logic delay for this design is found to be 2.3 ns. Simulations performed on SL-DFLOP design of the sequence recognizer yield an internal frequency of 220 MHz and minimum and maximum throughputs of 7.6 and 12.1 ns, respectively, with a maximum allowable NS logic delay of 3.2 ns.

In order to test the functionality of a static logic EAIC system in real time and make comparisons with the simulations results, the sequence recognizer of Fig. 16.13 was fabricated

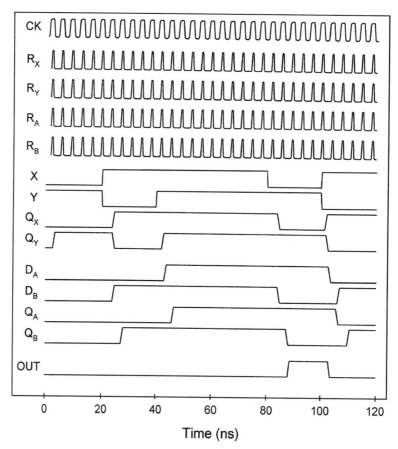

FIGURE 16.14
PSPICE simulation tracing of the EAIC circuit in Figure 16.13 using DL-DFLOPs and a conventional CMOS NOR gate in the clock generating circuitry.

by using a 2μ n-well CMOS process. The chips were tested by using a laboratory test rig and, more extensively, by using the HP 82000 test station with a 0.5-ns resolution, both monitoring the internal frequency directly. The chips were found to operate correctly and revealed variable internal clock frequencies in the range of 25–35 MHz. The lower frequencies, which were observed by the test station, fell well below the predicted PSPICE value of 38 MHz for a 2μ design. However, frequency measurements on a laboratory test rig accounted for frequencies up to about 35 MHz, very close to the predicted PSPICE value. The frequency differences are believed to be due mainly to capacitive loading effects by the measurement leads.

16.2.6 Variations on the Theme

A very interesting aspect of the EAIC approach to FSM design is that nearly all D flip-flop synchronous design considerations, methods, and associated alternative architectures are applicable to EAIC systems. This should not be surprising since the only difference between the two design methods is the way in which the clock is generated — internally in the EAIC

16.2 EXTERNALLY ASYNCHRONOUS/INTERNALLY CLOCKED SYSTEMS

system, externally for the conventional approach. For reference purposes, the following are examples of the overlap between the two approaches:

1. DFLOPs can be converter to either TFLOPs or JKFLOPs, as is done for D flip-flops in Section 10.8.
2. The design and analysis of FSMs by using the EAIC system follows the discussion for synchronous FSMs in Sections 10.12, 10.13, and 11.9.
3. Logic noise (including ORGs and static hazards) in the output functions of EAIC FSMs can be filtered by using conventional edge triggered D flip-flops as discussed in Subsection 11.2.2. In contrast to synchronous FSMs, the filtering D flip-flops should be triggered in phase to the internal clock.
4. Sanity circuits and debouncing circuits can and should be used in EAIC systems following Sections 11.7 and 11.8.
5. The array algebraic approach to FSM design and the one-hot design method, as discussed in Sections 11.11 and 13.5, also apply to EAIC system designs.
6. Any shift register or counter discussed in Chapter 12 can be designed by using DFLOPs or TFLOPs in place of D flip-flops or T flip-flops.
7. All alternative architectures and system-level design methods discussed in Chapter 13 are also applicable to EAIC system design. Thus, ROMs, PLAs, nonregistered PALs, etc., are all applicable to EAIC system design. Data path FSMs in a given system can be controlled by the internal clock of the controller.
8. As in synchronous FSM design, endless cycles, critical races and essential hazards cannot exist in EAIC systems — an advantage that both synchronous and EAIC FSMs have over fundamental mode FSMs.

Although there are several features of the EAIC system that are in common with conventional synchronous systems, sharply distinct differences exist as discussed in the following subsection.

16.2.7 How EAIC FSMs Differ from Conventional Synchronous FSMs

- Perhaps the most important difference between the EAIC approach and the synchronous approach is the fact that EAIC systems are inherently protected from metastability and require no other synchronizing scheme. The reason why this is so rests with the nature of DFLOP and the clock generating circuitry in the EAIC system. A properly designed EAIC system is *pausable* in the sense that if any one or more of the DFLOPs should go metastable, the system is held up (paused) until those DFLOPs exit from the metastable state and issue a clean set or reset. Thus, a properly designed EAIC system cannot fail because of metastability and its MTBF becomes infinite. In contrast, to achieve a large (but not infinite) MTBF for a synchronous FSM, synchronizing schemes of the type discussed in Section 11.4 must be applied. Application of such schemes to synchronous systems would, in many cases, lower the performance well below that of a comparable EAIC system.

- A second important distinction between the two approaches is that clock skew is not possible within a properly designed EAIC system consisting of both controller and data path FSMs that coordinate to produce the internal clock. Clock skew is always a potential problem in synchronous system-level designs.
- A third important difference is that EAIC systems are delay insensitive when operated within the bounds given in Subsection 16.2.4. To this extent, unexpected delays in the NS logic have no effect on the operation of the EAIC system. Even asymmetric delays in the one or more of the DFLOPs or in any part of the clock-generating circuitry, including clock skew, will not cause malfunction — the system simply performs more slowly if such delays exist. The same claims cannot be made with regard to synchronous FSMs. Asymmetric delays in fundamental mode FSMs are likely to cause malfunction of the FSM as discussed in Chapter 14.
- Other differences exist that are also advantages of the EAIC system. The internal clock frequency can be easily lowered by simply adding a delay (or counter) to the output of the external clock generating circuitry. Furthermore, the EAIC FSM's internal clock can be paused at any time by use of a tri-state enable/driver in the clock-generating circuit as indicated in Fig. 16.1. This can result in a savings of power during periods when the EAIC system must remain idle.

16.2.8 Perspective on EAIC Systems as an Alternative Approach to FSM Design

The EAIC system offers the designer an innovative alternative to synchronous and asynchronous (fundamental mode) approaches to FSM design. The EAIC approach has the advantages of high speed, operational reliability, low power consumption, and relatively low real estate commitment, all in the absence of an external clock oscillator circuit, as required for a comparable synchronous design. The EAIC system appears to have most all the benefits of the synchronous system and none of the disadvantages of asynchronous fundamental mode machines — the best of both worlds. Also, it may offer one of the most effective and appropriate applications of domino CMOS technology. The high-frequency internal clock seems ideally suited to the precharge/evaluate rates required by domino logic. Because of its pausable nature, the EAIC is essentially immune to clock distribution problems (clock skew) within a closed system, that is, within one controlled by a single internal clock. So why is the EAIC system not the approach of choice of designers for most modern applications? The answer to this question is explored in the following paragraphs.

The EAIC system designs are not without their drawbacks. One potential drawback to the EAIC approach to large system-level design is the fact that multiple controllers within a large system must communicate by means of handshake signals. Interfacing two independent clocked systems is never a simple task. But it is necessary since each controller establishes its own clock frequency independent of the others. In contrast, a fully synchronous system, consisting of multiple controllers, can operate on a single system clock. However, such a synchronous system is definitely subject to clock skew problems which may require handshake interfacing as well.

Another disadvantage to the EAIC system is due to the fact that the internal clock is not precise. That is, its frequency may vary slightly depending on a variety of factors including temperature effects. Also, duty cycle cannot generally be altered. Crystal-controlled oscillator circuits, of the type used in high-quality synchronous designs, are precise and have a number of desirable characteristics not found in the internal clock-generating circuits of the EAIC system. These desirable characteristics are discussed in Section 11.6.

16.3 ASYNCHRONOUS PROGRAMMABLE SEQUENCERS

In Chapter 14, it is made clear that any FSM that is designed to operate in the fundamental mode must be free of certain timing defects that would otherwise cause the FSM to fail. Such timing defects include endless cycles, critical races, static hazards in the NS logic, and essential hazards. Normally, it is not difficult to eliminate these defects, but the task can be tedious and does require a fair understanding of asynchronous FSM design methods. The EAIC system, presented in the first portions of this chapter, offers one means of avoiding these problems, and does so by operating from an internally generated clock, somewhat similar to a synchronous FSM. But the EAIC system cannot be used as a programmable sequencer owing to the mechanism required to generate the internal clock. In this section a distinctive, versatile, and highly reliable class of asynchronous programmable sequencers is considered in detail.

16.3.1 Microprogrammable Asynchronous Controller Modules and System Architecture

A unique family of high-speed asynchronous programmable sequencers is now described that combine fundamental mode operation with the programmability power of PLDs. These sequencers have been dubbed *microprogrammable asynchronous controller (MAC) modules*. Shown in Fig. 16.15 is the generalized architecture for a fully programmable system capable of operating as any one of 2^k asynchronous controllers that operate by means of a single n-bit (2^n-state) MAC module. The basic components are a 2^k bank of PLDs (ROMs, PLAs, nonregistered PALs, or any combination thereof), a k-to-2^k decoder for PLD selection, an interfacing and deactivate inputs (DI) stage, and the n-input MAC module with initialization and enable inputs. If several PLDs are used to drive the MAC module, the interfacing and DI stage should be a bank of n 2^k-input MUXs, one MUX for each input to the MAC module. If only two PLDs are used to drive the MAC module, the decoder is reduced to a simple inverter. Also, if one PLD is used, the interfacing and DI stage is simply composed of discrete logic. These and other related subjects will be explored more fully in later sections.

The DI signal, which is introduced into the interfacing logic from the MAC module, as shown in Fig. 16.15, plays an essential role in the operation of the MAC module. Following each successful transition of the FSM, all inputs to the MAC module must be deactivated for a short time by the DI signal so as to make ready for the next transition as determined by the PLD program driving the system. A handshake mechanism involving two fundamental-mode state machines within the MAC module coordinates this process so that each transition is guaranteed to occur in an orderly and reliable fashion, even if cycles exist under conditional or unconditional branching. The handshake process guarantees that endless cycles, critical races, static hazards in the NS functions, and essential hazards cannot cause the MAC module operation to fail. Furthermore, ORGs are not possible, since every state-to-state transition must be logically adjacent. In short, the operation of the MAC

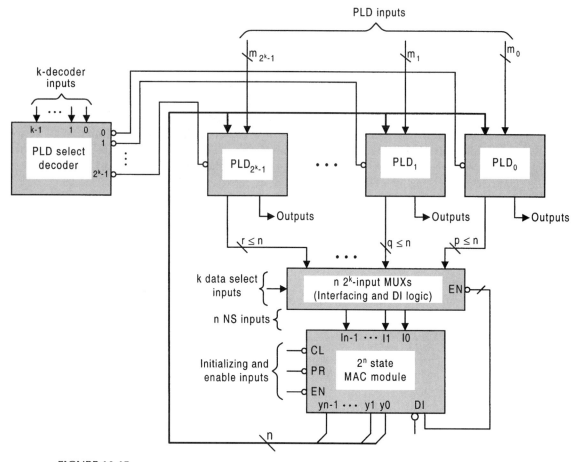

FIGURE 16.15
Generalized architecture for programming an n-bit MAC module to operate with any one of 2^k PLDs (ROMs, PLAs, or PALs) by using k inputs to a PLD select decoder and to n 2^k-input interfacing MUXs.

module cannot fail by any timing defect common to fundamental mode FSMs, and clean outputs are guaranteed to be issued.

16.3.2 Architecture and Operation of the MAC Module

Shown in Fig. 16.16 are the two fundamental mode FSMs of which an n-bit MAC module is composed. One FSM represents a 2^n state array machine (SAM for short) with n-way branching capability. The other is a timing control machine or TCM. The two machines coordinate the handshake process that permits the MAC module to operate correctly. Notice that five outputs of the SAM are the inputs to the TCM. These are the select parameters, *Se* and *So*, parity parameters *EP* and *OP*, and *Reset*. Completing the handshake, the inputs to the SAM received from the TCM are the transition enable parameters, *Te* and *To*. The SAM issues n present state signals ($yn - 1, \ldots, y2, y1, y0$) to the PLDs and back to itself

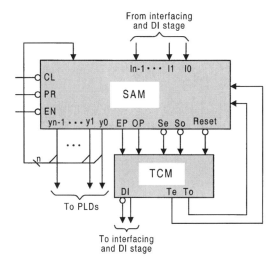

FIGURE 16.16
Components of an *n*-input MAC module consisting of a 2^n state array machine (SAM) and a timing control machine (TCM) and their interconnections.

as feedback. In return, the SAM receives from the selected PLD (via the interface and DI stage) one of *n* programmed NS instructions, $In-1, \ldots, I2, I1$, or $I0$. This is the one bit that must change to produce the required SAM transition, but only when the transition is enabled by a transition enable parameter (*Te* or *To*) from the TCM. After the transition is complete, the TCM issues the DI signal to the interfacing and DI stage, thereby deactivating the NS instruction input so that the process can begin all over again.

The SAM consists of an array of states such that any given reference state in the array has transitions paths to states that are logically adjacent and, hence, of opposite parity to the reference state. Thus, a transition from any state to another (adjacent) state involves a change of only one state variable and a change in parity [odd parity (OP) to even parity (EP) or vice versa]. The structure of the SAM is best illustrated by example. Shown in Fig. 16.17a is the 2×4 state SAM required by a 3-input (2^3-state) MAC module. This SAM can be used to operate any 2-, 4-, 6-, or 8-state controller FSM — an odd number of states is strictly forbidden in any MAC module controller design. Notice that the states are coded in 3-bit Gray code and that there is three-way branching from each state, permitting it to transition to any one of three logically adjacent states. The branching condition for each transition path is the Boolean product of a transition enable parameter (*Te* or *To*) and a single NS instruction input (via the interfacing and DI stage) given by $I2, I1$, or $I0$. The specific NS instruction input represents the positional weight $2^2, 2^1$, or 2^0 of the bit programmed to change during a given transition. For example, a transition from even parity (EP) state 101 to odd parity (OP) state 001 will occur only if $Te \cdot I2$ is valid (active). The holding conditions in the SAM state diagram are those required to maintain the SAM in a given even or odd parity state during the time that the inputs are deactivated.

The outputs *Se*, *So*, and *Reset* from the 2^3-state SAM in Fig. 16.17a are the inputs to the TCM. These outputs are issued conditionally on the functions to which they are equated. For example, *Se* is issued in an even-parity state (EP active) when *To* is inactive and when

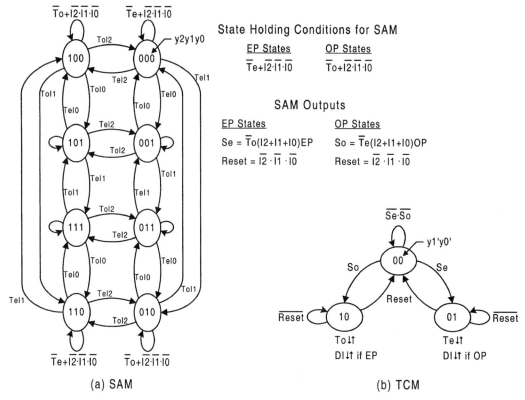

FIGURE 16.17
State diagrams for a 3-input (2^3-state) MAC module. (a) The 2×4 state array machine (SAM) showing the holding conditions and outputs separately. (b) Timing control machine (TCM).

one of the NS instruction inputs ($I2$, $I1$ or $I0$) is active. The parity parameters EP and OP are defined by the relations

$$OP = y2 \oplus y1 \oplus y0 = \text{Odd parity}$$
$$EP = \overline{OP} = \text{Even parity}. \qquad (16.8)$$

The output *Reset* is issued to the TCM from any state of the SAM provided that the NS instruction inputs are all in the deactivated condition such that $\overline{I2 + I1 + I0} = \overline{I2} \cdot \overline{I1} \cdot \overline{I0} = 1$. It is important that all inputs be in the deactivated state after each successful transition and before the NS instruction is received by the MAC module. Thus, the intermediate interfacing and DI stage is ideally suited for this purpose. However, the inputs can be deactivated either by the *DI* signal or by the PLD, whichever action occurs first.

The TCM, shown in Fig. 16.17b, is a resolver FSM. When *Reset* is active, meaning that $\overline{I2} \cdot \overline{I1} \cdot \overline{I0} = 1$, the TCM must reside in the unresolved 00 state for as long as both select inputs from the SAM are inactive ($\overline{Se} \cdot \overline{So}$). When one of the select inputs becomes active, the TCM must transit to a resolved state (01 or 10) and must issue a transition enable command (*Te* or *To*) to the SAM. Then, when the SAM successfully transits and parity is changed, a DI signal is issued to the interfacing and DI stage thereby deactivating the NS instruction input that caused the transition. This, in turn, causes the SAM to issue an active

16.3 ASYNCHRONOUS PROGRAMMABLE SEQUENCERS

Reset signal which forces the TCM back to the unresolved 00 state ready to receive the next active select input, *Se* or *So*.

The operation of the MAC module can best be understood by following the sequence of steps leading to a state-to-state transition of a controller FSM that is implemented with the 2^3-state MAC module shown in Figs. 16.16 and 16.17. To begin with, assume that the MAC module (hence the controller also) is initialized into the 000 ($EP = 1$) state, that all NS instruction inputs to the SAM are inactive, and that the TCM is in the unresolved 00 state where both *Te* and *To* are inactive. Now, assume that the PLD issues one of three NS instructions ($I2$, $I1$, or $I0$) to the SAM, which in turn issues the conditional output *Se* to the TCM. The TCM receives the *Se* signal and transits to the 01 state where the output *Te* is issued to the SAM ($Te = 1$ and $To = 0$). After receiving the *Te* signal, the SAM transits from the 000 state to an OP state (100, 010, or 001) under one of the branching conditions $TeI2$, $TeI1$, or $TeI0$, respectively. When the SAM successfully completes the transition, parity changes requiring that $OP = y2 \oplus y1 \oplus y0 = 1$, and the TCM issues the conditional output *DI* to the interfacing and DI stage, which deactivates the single NS instruction input that caused the SAM transition (now all instructions are deactivated). This causes the SAM to issue the output *Reset*, which forces the TCM to transit from state 01 to the unresolved 00 state where $Te = To = 0$. Then when one of the NS instruction inputs from the PLD goes active, the SAM issues an *So* signal to the TCM. This forces the TCM to transit to the 10 state where *To* is issued to the SAM ($To = 1$ and $Te = 0$). The *To* signal is received by the SAM, causing it to transit from an *OP* state to an *EP* state, changing the parity back to $EP = 1$. The *DI* signal is again issued by the TCM which, in turn, causes the SAM to issue a *Reset* command that forces the TCM back to the 00 state ($Te = To = 0$), allowing the process to be repeated.

It is important to understand that the *DI* signal maintains strict control over all state-to-state transitions. This is especially important when cycles occur in the controller's state diagram, or when buffer states must be added to the controller's state diagram to satisfy the state logic adjacency requirement of the MAC module. Clearly any alteration of a state diagram or state table must be done prior to programming the PLD. Because of the handshake between the TCM and SAM, no SAM transition (and hence no controller transition) can take place until a sequence of events occurs leading to the deactivated state of all inputs. This fact alone eliminates any possibility of essential hazard or d-trio formation. Thus, a transition via a cycle or buffer state is treated no differently from any other state-to-state transition in the MAC module — oscillatory endless cycles, for example, are not possible.

16.3.3 Design of the MAC Module

Presented in Fig. 16.18 are the NS K-maps for the 3-input SAM as plotted from Fig. 16.17a. Optimum two-level results for the NS-forming logic of the SAM are easily read directly from these K-maps and are given by the following equations:

$$\begin{cases} Y2 = \overline{y2}\,\overline{y1}\,\overline{y0}\,Te\,I2 + \overline{y2}\,\overline{y1}\,y0\,To\,I2 + \overline{y2}y1\overline{y0}\,Te\,I2 + \overline{y2}y1\,\overline{y0}\,To\,I2 \\ \quad + y2\overline{y1}\,\overline{y0}\,\overline{To} + y2\overline{y1}\,y0\,\overline{Te} + y2y1\overline{y0}\,\overline{To} + y2\,y1\overline{y0}\,\overline{Te} + y2\overline{I2} \\ Y1 = \overline{y2}\,\overline{y1}\,\overline{y0}\,Te\,I1 + \overline{y2}\,\overline{y1}y0\,To\,I1 + y2\overline{y1}\,\overline{y0}\,To\,I1 + y2\overline{y1}y0\,Te\,I1 \\ \quad + \overline{y2}y1y0\,\overline{Te} + \overline{y2}y1\overline{y0}\,\overline{To} + y2y1y0\,\overline{To} + y2y1\overline{y0}\,\overline{Te} + y1\overline{I1} \\ Y0 = \overline{y2}\,\overline{y1}\,\overline{y0}\,Te\,I0 + \overline{y2}y1\,\overline{y0}\,To\,I0 + y2\overline{y1}\,\overline{y0}\,To\,I0 + y2y1\overline{y0}\,Te\,I0 \\ \quad + \overline{y2}\,\overline{y1}y0\,\overline{To} + \overline{y2}y1y0\,\overline{Te} + y2\overline{y1}y0\,\overline{Te} + y2y1y0\,\overline{To} + y0\overline{I0} \end{cases}. \quad (16.9)$$

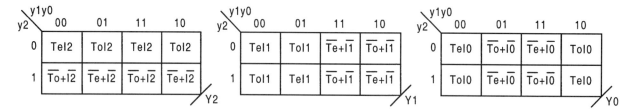

FIGURE 16.18
NS K-maps plotted from the state diagram for the 3-input SAM in Fig. 16.17a.

Notice that there are a total of 22 p-terms for these three expressions, including four shared PIs. Thus, the PLA required by Eqs. (16.9) must have minimum dimensions of $8 \times 22 \times 3$. It is important to note that because of the nature of the state code assignments together with the NS instructions, static hazards are not possible in the optimized expressions for $Y2$, $Y1$, and $Y0$ functions given by Eqs. (16.9).

Alternative approaches to the implementation of the NS-forming logic of the SAM are possible. A nonregistered $8 \times 27 \times 3$ PAL can be used for this purpose, but the shared PIs must be treated as separate p-terms. Use of an 8-input ROM would not be a good choice since it would be an "overkill" when compared to the more efficient use of a PLA or PAL. A more interesting alternative design of the SAM can be obtained by making use of the XOR patterns that exist in the NS K-maps (see Section 5.2). Again, reading the K-maps directly leads to the following multilevel expressions:

$$\left\{ \begin{array}{l} Y2 = \overline{y2}\, Te\, I2(\overline{y1 \oplus y0}) + \overline{y2}\, To\, I2(y1 \oplus y0) + y2\, \overline{To}(\overline{y1 \oplus y0}) \\ \quad + y2\overline{Te}(y1 \oplus y0) + y2\overline{I2} \\ Y1 = \overline{y1}Te\, I1(\overline{y2 \oplus y0}) + \overline{y1}\, To\, I1(y2 \oplus y0) + y1\overline{To}(\overline{y2 \oplus y0}) \\ \quad + y1\overline{Te}(y2 \oplus y0) + y1\overline{I1} \\ Y0 = \overline{y0}\, Te\, I0(\overline{y2 \oplus y1}) + \overline{y0}\, To\, I0(y2 \oplus y1) + y0\overline{To}(\overline{y2 \oplus y1}) \\ \quad + y0\overline{Te}(y2 \oplus y1) + y0\overline{I0} \end{array} \right\}. \quad (16.10)$$

Use of these expressions has two advantages over the two-level function of Eqs. (16.9): Reduced fan-in for discrete logic implementation, and use of the parity expression in Eq. (16.8) to generate one of the three required XOR terms. As is true for Eqs. (16.9), static hazards are not possible in the NS functions of Eqs. (16.10).

The TCM is best designed by using the nested cell model. The NS- and output-forming logic for the TCM can be deduced directly from the state diagram in Fig. 16.17 without the need for K-maps. When this is done the results become

$$\left\{ \begin{array}{ll} S_1 = So & S_0 = Se \\ R_1 = Reset & R_0 = Reset \\ \multicolumn{2}{c}{To = y1'} \\ \multicolumn{2}{c}{Te = y0'} \\ \multicolumn{2}{c}{DI = y1'EP + y0'OP} \\ \multicolumn{2}{c}{\quad = To\, EP + Te\, OP} \end{array} \right\}, \quad (16.11)$$

16.3 ASYNCHRONOUS PROGRAMMABLE SEQUENCERS

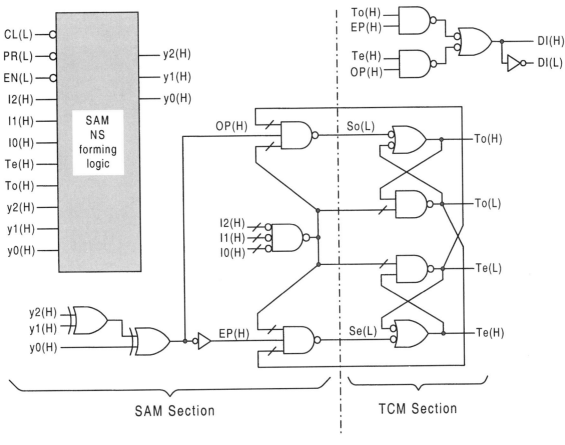

FIGURE 16.19
Logic diagrams for the SAM and TCM sections of a 3-input (8-state) MAC module.

where the subscripts 1 and 0 refer to basic cells 1 and 0. The optimized NS logic results for S_1 and S_0 that are given in Eqs. (16.11) are a result of the internal handshake configuration used by the MAC module — all cells of the K-maps (not shown) for these functions are don't cares except for cell 00, which contains So and Se, respectively.

Putting it all together, there results the logic circuits for the 3-input MAC module shown in Fig. 16.19. Here, the optimal results for the TCM given by Eqs. (16.11) are implemented, and the two XOR gates represent the parity logic expressed by Eqs. (16.8). The SAM is presented as a block diagram since its implementation is a matter of choice by the designer.

The TCM logic shown in Fig. 16.19 remains the same regardless of the SAM dimensions, which can be of any 2^n-state size. A generalization of the 2^n-state SAM is shown in Fig. 16.20, as required by an n-input MAC module. Here, each transition path into an EP state is from an OP state, and each transition path out of an EP state must go to an OP state. Similarly, each transition path into and out of an OP state is from and to an EP state, respectively. Notice that a 2^n-state a SAM has up to n-way out-branching capability of any one of its states to a logically adjacent state of opposite parity. Thus, a 2^2-state SAM is a 2×2 array with up to 2-way out-branching capability, a 2^3-state SAM is a 2×4 array with up to 3-way

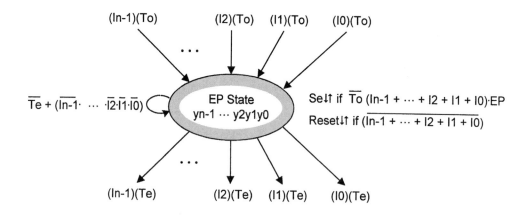

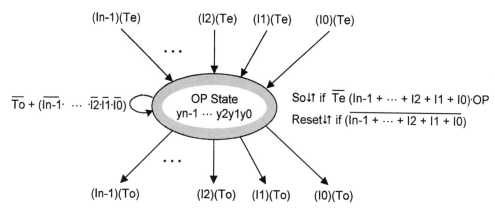

FIGURE 16.20
Generalized transition conditions and outputs for the EP and OP states of a 2^n state SAM with n-way out-branching.

out-branching capability, a 2^4-state SAM is either a 4×4 array or a 2×8 array, both with up to 4-way out-branching capability, and so on. Clearly, it is not necessary that all n-way branching possibilities be used in any state-to-state transition. But it is required that for any FSM design, the SAM transition paths include an even number of states — never an odd number. Thus, a 4×4 array SAM can be used to design any controller FSM with states numbering 2, 4, 6, ..., etc., up to 16 states. If, for example, a 5-state FSM is to be designed, one or an odd number of buffer states must be added so that either a 2×4 or 4×4 SAM array (meaning a 3- or 4-input MAC module) can be utilized in the design. Initialization of any SAM array requires that the procedures discussed in Section 14.11 be followed.

16.3.4 MAC Module Design of a Simple FSM

Perhaps the most stringent test of the MAC module's capabilities is its use in the design of an unrestrained Gray code counter. This is done for the 3-bit Gray code counter shown in

16.3 ASYNCHRONOUS PROGRAMMABLE SEQUENCERS

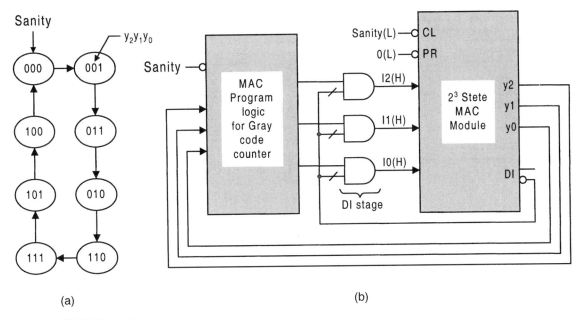

(a) (b)

FIGURE 16.21
MAC module design of a simple 3-bit Gray code up-counter. (a) State diagram for the counter. (b) Logic diagram showing the block symbol for the MAC module program logic chosen from Eqs. (16.12), the DI stage, and the block symbol for the 2^3-state MAC module.

Fig. 16.21. To determine the program logic required to drive the 2^3-state MAC module in Fig. 16.21b, it is only necessary to read the state diagram in Fig. 16.21a for the changing y-variable bits. As an example, the y_2 bit changes only in transitions from states 010 and 100, hence $\bar{y}_2 y_1 \bar{y}_0 + y_2 \bar{y}_1 \bar{y}_0$. Continuing with bits y_1 and y_0 there results the following NS instruction equations in both two-level and multilevel form:

$$\left\{\begin{array}{l} I2 = \bar{y}_2 y_1 \bar{y}_0 + y_2 \bar{y}_1 \bar{y}_0 \\ \quad = \bar{y}_0 (y_2 \oplus y_1) \\ I1 = \bar{y}_2 \bar{y}_1 y_0 + y_2 y_1 y_0 \\ \quad = y_0 (y_2 \odot y_1) \\ I0 = \bar{y}_2 \bar{y}_1 \bar{y}_0 + \bar{y}_2 y_1 y_0 + y_2 \bar{y}_1 y_0 + y_2 y_1 \bar{y}_0 \\ \quad = \bar{y}_1 (y_2 \odot y_0) + y_1 (y_2 \oplus y_0) \\ \quad = y_2 \oplus y_1 \odot y_0 \end{array}\right\} . \quad (16.12)$$

Shown in Fig. 16.22 is the mixed logic simulation for the MAC module design of the 3-bit Gray code counter in Fig. 16.21b. This is done by using the two-level NS instructions functions in Eqs. (16.12) to drive the 3-input MAC module presented in Fig. 16.19. The results show that the y-variable response to an I-instruction change occurs following a time period that varies between $4\tau_p$ and $5\tau_p$, where τ_p is the average gate propagation delay. Also, the results indicate that the average time period between y-variable changes is about

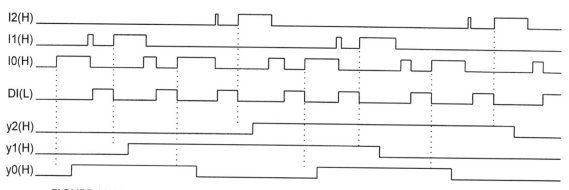

FIGURE 16.22
Mixed logic simulation results for the MAC module design of the 3-bit Gray code counter in Fig. 16.21a when the two-level forms of Eqs. (16.12) are used as transition instructions to a 2^3-state MAC module.

$12\tau_p$. These time periods are predictable, given the logic circuits for the SAM and TCM of the MAC module.

The additional pulses observed in the $I2$, $I1$, and $I0$ waveforms of Fig. 16.22 are a consequence of the unrestrained nature of the state-to-state transitions in the Gray code counter together with the action of the DI input. However, as can be seen from the simulation results, these additional pulses have no effect on the state variables. It is the strict control maintained by the MAC's internal handshake mechanism and the action of the DI signal that ensures reliable transitions of the FSM even under these severe operating conditions.

16.3.5 Cascading the MAC Module

Cascading two 2^3-state MAC modules of the type indicated in Fig. 16.19 increases the state capacity of the system to $2^3 \times 2^3 = 64$ states with six state variables and up to 6-way branching capability. Shown in Fig. 16.23 is such a cascading arrangement where the DI stage is properly placed on the outputs of the PLD. With this arrangement, the NS instruction logic (the PLD) must provide a separate set of three NS instruction inputs to the SAM of each MAC module for a total of six, as indicated in Fig. 16.23. Also, two DI inputs are needed, but these must *not* be ORed to the DI stage. Thus, immediately following the completion of a transition involving a change in a state variable $y5$, $y4$, or $y3$, the DI output of MAC module 1 goes active while the DI output of MAC module 0 remains inactive. The reverse is true for a change in a state variable $y2$, $y1$, or $y0$ of MAC module 0. It is important to understand that since these modules are cascaded in parallel, speed and reliability are not compromised.

A second approach to cascading MAC modules makes use of the $EN(L)$ inputs to multiplex the MAC modules so that only one is enabled at any given time. This cascading method requires that the proper multiplexing instructions be programmed into the NS instruction logic of the PLD. Of the two cascading methods, the one illustrated in Fig. 16.23 is likely to be the simpler for most applications.

Multiple MAC modules can be cascaded for greatly enhanced capability. For example, cascading three 2^3-state MAC modules increases the state capacity to $(2^3)^3 = 2^9 = 512$ states

16.3 ASYNCHRONOUS PROGRAMMABLE SEQUENCERS

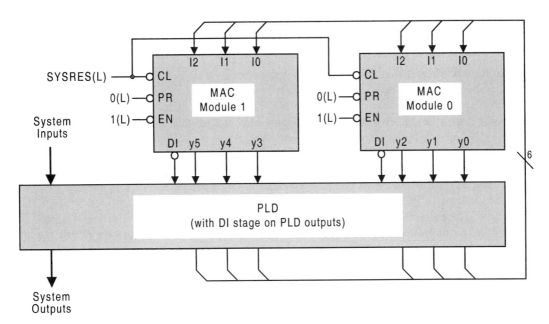

FIGURE 16.23
Cascading configuration for two 3-bit MAC modules that is independent of the EN inputs.

with nine state variables and up to 9-way branching capability. Cascading three 2^4-state MAC modules results in a system having $(2^4)^3 = 2^{12} = 4096$ states with 12 state variables and up to 12-way branching capability. Generally, MAC modules having SAMs with state variables numbering $l, m, n, \ldots$ can be cascaded to produce a system state capacity of $2^l \times 2^m \times 2^n \times \cdots$ with up to $(l+m+n+\cdots)$-way branching capability. Since the modules are cascaded in parallel, speed and reliability are not compromised. Cascading n 2-bit MAC modules to produce larger systems having state capacities of $(2^2)^n$ is an attractive alternative for improved speed capability. This is so because use can be made of a single XOR gate (e.g., $y1 \oplus y0$) for the parity circuit of each 2-bit module. Remember, that the internal handshake mechanism depends on the parity parameters EP and OP, where for an n-input MAC module $OP = yn - 1 \oplus \cdots \oplus y2 \oplus y1 \oplus y0$. Therefore, the larger the state capacity of a MAC module, the slower will be its response time.

16.3.6 Programming the MAC Module

To begin with, it is necessary that each state-to-state transition of the controller FSM be logically adjacent. If this is not the case initially, then either buffer states must be added or the number of state variables must be increased to accomplish this. State code assignments of Hamming distances greater than 1 cannot be used in any MAC module controller design. After the state logic adjacency requirement is satisfied and before programming a MAC module, there still remains one important requisite that must be satisfied. The conditions for an endless cycle, as defined in Subsection 14.10.1, must never exist in the state diagram for the controller. Such conditions lead to the formation of static hazards in the NS instruction logic — defects that can cause MAC module failure.

Once the requirements just mentioned have been met, programming the NS instruction and output logic (hence, the PLD) for the MAC module design of a controller is easily accomplished with an I/O table obtained directly from either the state diagram or state table for the controller. To do this, the state variables ($yn-1, \ldots, y2, y1, y0$) are placed on the input side, and the NS instruction inputs ($In-1, \ldots, I2, I1, I0$) and the controller outputs are placed on the output side of the I/O table. Then, one simply places the branching conditions for any given transition in the appropriate output column. Thus, programming the MAC module does not require reference to the state array of the SAM since the position of the changing bit in the state diagram or state table for the controller indicates the required PS-to-NS transition in the SAM.

The NS instruction inputs can be mapped from the I/O table and then minimized for discrete logic or PLA (or PAL) implementation. Alternatively, the I/O table can be read directly in canonical form for ROM implementation. In any case, the NS instruction logic will be free of static hazards. Static hazards produced in the output-forming logic of a PLA device can be eliminated by either adding static hazard cover or by using an output holding register (D flip-flops) triggered by the *DI* signal. If the PLD is a ROM, an output holding register must be used since redundant cover for hazard elimination is not possible.

16.3.7 Metastability and the MAC Module: The Final Issue

The MAC module is a programmable asynchronous sequencer that is designed to operate in the fundamental mode. When used in the design of one or more asynchronous controllers, the MAC module will not fail by any of the timing defects common to fundamental mode machines. However, there is the possibility that a MAC module-designed controller could go metastable because of a runt pulse, or be forced into irregular behavior if the setup and hold times requirements are not met by the external inputs — problems that can occur in any fundamental mode FSM. To avoid possible metastable behavior, the MAC module controller should be protected by using a bus arbiter on its inputs. Such an arbiter is described logically in Problem 15.8 and is available commercially as the Signetics fast (7 ns) 74F786 asynchronous bus arbiter. This commercial arbiter is designed with a metastable detection stage and should be used on the output of the interfacing and DI stage shown in Fig. 16.15, or in place of the DI state located on the output of the PLD shown in Fig. 16.23. When this is done, a highly reliable and robust system results that will not fail under any set of input conditions.

16.3.8 Perspective on MAC Module FSM Design

Interchanging PLDs or reprogramming an existing PLD permits the PLD/MAC module system to be easily converted from one asynchronous FSM to another radically different one without the need to run timing defect analyses on either FSM. This is a very attractive feature, given all that is required to ensure the proper and reliable operation of each asynchronous FSM. Endless cycles, critical races, static hazards in the NS logic, and potentially active essential hazards are all automatically eliminated in a MAC module design. By multiplexing PLDs, any number of asynchronous controllers can be operated reliably on a time-shared basis and without clock skew problems within any given controller. Also, MAC modules can be cascaded in parallel to greatly enhance state machine capacity without compromising either speed or reliability, another important feature of the MAC module approach.

16.4 ONE-HOT PROGRAMMABLE ASYNCHRONOUS SEQUENCERS   835

Of course, the one major drawback to MAC module FSM design is the fact that each state-to-state transition must involve one, and only one, state variable change. Thus, the controller state machine must be composed of an even number of states that are unit-distance coded. When this is not the case, the logical adjacency requirement must be met by either adding buffer states or by increasing the number of state variables. Adding a buffer state may or may not be acceptable, since it does introduce a delay in executing a given transition. Adding state variables may require increasing the size (capacity) of the MAC module. As an example, the vending machine controller in Fig. P13.2 would require three state variables and two properly positioned buffer states to satisfy the logical adjacency requirement. In this case the code for state g would be changed to 010. But there are some FSMs whose branching paths may not be amenable to MAC module design. The FSMs in Figs. P14.6 and P14.10 would appear to fall into this category.

16.4 ONE-HOT PROGRAMMABLE ASYNCHRONOUS SEQUENCERS

One-hot asynchronous programmable sequencers can be designed by applying Eqs. (14.40) for all possible branching conditions in an n-state FSM. This type of asynchronous programmable sequencer enjoys some attractive advantages over the MAC module approach discussed in the previous section. The one-hot approach requires only a state array machine and can support any state-to-state transition in an FSM that is *void of cycle conditions*, including, in particular, endless cycles. Because of the one-hot coding, a timing control machine is not needed. That is, no parity detection or deactivation of inputs is required. Furthermore, the programming of a one-hot sequencer is exceedingly simple since it is only necessary to provide the sequencer with the branching condition for each 1-hot state-to-state transition as read from a state table or state diagram of the FSM to be designed.

16.4.1 Architecture for One-Hot Asynchronous Programmable Sequencers

Shown in Fig. 16.24 is the generalized architecture for an n-state asynchronous programmable one-hot sequencer. Here, it is seen that 2^k PLDs representing 2^k different asynchronous FSMs can be selected by a decoder to drive the sequencer on a time-shared basis. This, of course, is no different than in the case of the MAC module architecture in Fig. 16.15. What is different is the interfacing logic which, in the case of the one-hot approach, is nothing more than an array of OR gates.

An inspection of Fig. 16.24 indicates that an n-state one-hot sequencer requires specification of n^2 inputs, one for each branching condition in an $n \times n$ state array. As indicated in Fig. 16.25, each kth state of a completely specified n-state one-hot sequencer requires n input branching paths, including the required holding condition — hence, n-way branching capability to and from each state. Thus, for n states, n^2 branching conditions must be specified in a one-hot sequencer that contains all possible branching paths. This can be viewed as a significant down side to one-hot sequencer design and application. For example, consider that a 10-state sequencer requires that 100 branching conditions be specified for a given FSM design, though many of these branching conditions are set to logic 0 if their corresponding branching paths do not exist in the FSM. However, this seemingly impractical requirement of dealing with n^2 outputs from PLDs can be handled by using the output augmentation scheme shown in Fig. 7.16. For a large number of ORing operations in the interfacing logic of Fig. 16.24, it is recommended that the CMOS NOR gate form in Fig. 8.46

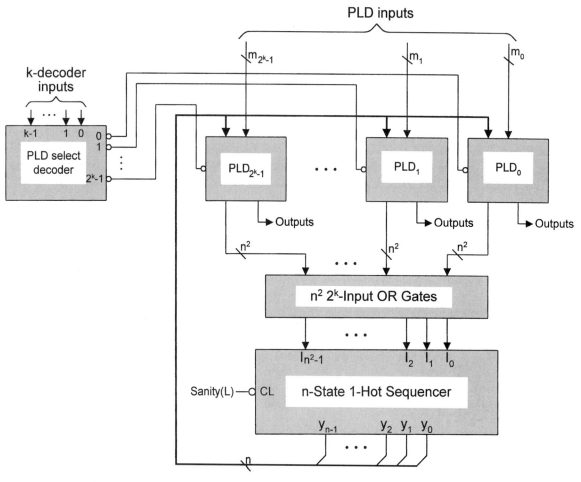

FIGURE 16.24
Generalized architecture for programming an asynchronous n-state one-hot sequencer with n^2 inputs.

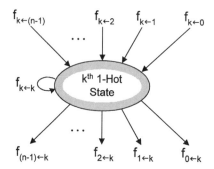

FIGURE 16.25
Generalized transition conditions for the kth state of an n-state one-hot sequencer.

16.4 ONE-HOT PROGRAMMABLE ASYNCHRONOUS SEQUENCERS

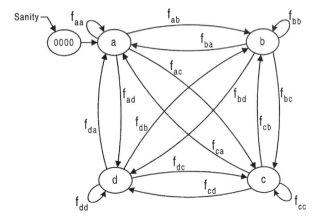

FIGURE 16.26
State diagram for a 4-state 1-hot sequencer that will initialize into the 0001 state (state a).

be used with the appropriate changes in activation levels of the inputs. This can avoid costly fan-in delays or delays due to possible OR tree forms. Finally, for a large n-state one-hot sequencer, the NS-forming logic is best implemented by using a PLD, preferably a PLA. Further discussion on this and other related subjects is included in Subsection 16.4.4.

16.4.2 Design of a Four-State Asynchronous One-Hot Sequencer

To illustrate the design of a one-hot sequencer, consider the state array for a fully specified, 4-state, one-hot sequencer shown in Fig. 16.26. Notice that each state requires specification of four branching conditions. Once this sequencer is designed, its programming for use in the design of an asynchronous FSM requires specification of 16 branching conditions many of which may be zero if those branching paths are nonexistent. The design of this sequencer is straightforward. By using Eqs. (14.40) together with the one-hot-plus-zero approach, there results the following set of four NS logic equations given in array form:

$$\begin{bmatrix} Y_a \\ Y_b \\ Y_c \\ Y_d \end{bmatrix} = \left\{ \begin{bmatrix} f_{aa} & f_{ba} & f_{ca} & f_{da} \\ f_{ab} & f_{bb} & f_{cb} & f_{db} \\ f_{ac} & f_{bc} & f_{cc} & f_{dc} \\ f_{ad} & f_{bd} & f_{cd} & f_{dd} \end{bmatrix} \begin{bmatrix} y_a \\ y_b \\ y_c \\ y_d \end{bmatrix} + \begin{bmatrix} \bar{y}_b \bar{y}_c \bar{y}_d \\ y_b \bar{y}_a \bar{y}_c \bar{y}_d \\ y_c \bar{y}_a \bar{y}_b \bar{y}_d \\ y_d \bar{y}_a \bar{y}_b \bar{y}_c \end{bmatrix} \right\} \overline{Sanity}. \quad (16.13)$$

Here, each f_{ij} term in the 4 × 4 state array represents the branching condition from the ith state "into" the jth state. Also, the four leading diagonal (holding conditon) terms f_{ii}, when ANDed with y_i to give $f_{ii}y_i$, provide the cover required to eliminate the static 1-hazards that develop between an "into" term and the single "out of" term in each NS function. Notice that initialization into state a follows the one-hot-plus-zero method discussed in Section 14.14. In this case, however, the "out of" term combines with the "one-hot-plus-zero" term to give $y_a \bar{y}_b \bar{y}_c \bar{y}_d + \bar{y}_a \bar{y}_b \bar{y}_c \bar{y}_d = \bar{y}_b \bar{y}_c \bar{y}_d$ for initialization into state a via the 0000 state. Any outputs associated with the sequencer design of an FSM are generated from the PLD or discrete logic by applying Eqs. (14.41).

Another important advantage of the one-hot sequencer approach to FSM design is the fact that the NS logic equations for any size fully specified sequencer can be written down directly without the aid of a state table or state diagram. The generalized NS equations are put into tensor notation form as

$$Y_i = \left\{ \sum_{j=0}^{n-1} \sum_{i=0}^{n-1} f_{ij} y_j + y_i \cdot \overline{\sum_{\substack{k=0 \\ k \neq i}}^{n-1} y_k} \right\} \quad (16.14)$$

where an additional term $\bar{y}_0 \bar{y}_1 \bar{y}_2 \bar{y}_3 \cdots \bar{y}_{n-1}$ must be added to a specific Y_i for use with the one-hot-plus-zero approach. Eqs. (16.14) are of importance for CAD purposes in dealing with relatively large sequencers, particularly if a PLD is to be used to implement the NS functions. An inspection of Eqs. (16.14) reveals how this can be easily accomplished. Any n-state sequencer would require an $n \times n$ non-symmetrical matrix of f_{ij} branching condition terms, n state variables, and n^2 inputs. The "out of" term for each NS function is a p-term consisting of the uncomplemented state variable for that function ANDed with the complement of each of the remaining state variables. For initialization purposes, the one-hot-plus-zero approach requires that a term of ANDed complements of all state variables (exclusive of that for the initialization state) be added to the specific Y_i variable into which the FSM is to be initialized. When the initialization term is combined with the "out of" term for that initialization state, the result is a reduced p-term consisting of the complement of all state variables exclusive of that for the initialization state, as in Eqs. (16.13). Once the one-hot-plus-zero implementation is complete, the sanity circuit can be used to drive all state variables initially and momentarily into the inactive state.

16.4.3 Design and Operation of a Simple FSM by Using a Four-State One-Hot Sequencer

To demonstrate the application of a one-hot sequencer, consider the state diagram for the FSM of Figure P14.8, which is reproduced in Fig. 16.27 for the convenience of the reader.

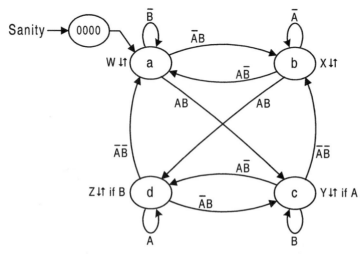

FIGURE 16.27
State diagram for the FSM to be designed by using the four-state one-hot sequencer of Fig. 16.26.

16.4 ONE-HOT PROGRAMMABLE ASYNCHRONOUS SEQUENCERS

Notice that there are 12 in-branching paths including the four holding conditions. Thus, 4 of the 16 branching conditions are set to zero. From the state diagram, the information required for programming the state array in Fig. 16.26 is easily deduced by inspection to be the following:

$$\begin{cases} \bar{A}\bar{B} = f_{da} = f_{cb} & \bar{A} = f_{bb} & f_{ad} = 0 \\ \bar{A}B = f_{ab} = f_{dc} & A = f_{dd} & f_{bc} = 0 \\ A\bar{B} = f_{ba} = f_{cd} & \bar{B} = f_{aa} & f_{ca} = 0 \\ AB = f_{ac} = f_{bd} & B = f_{cc} & f_{db} = 0 \\ & W = y_a & \\ & X = y_b & \\ & Y = y_c \cdot A & \\ & Z = y_d \cdot B & \end{cases} \quad (16.15)$$

From this information it is clear that only six two-input gates and two inverters are needed to program the sequencer. Notice that any given Moore output is a one-hot state variable and that any given Mealy output is a one-hot state variable ANDed with the input condition on which it depends.

The logic circuit for the one-hot sequencer design of the FSM in Fig. 16.27 is shown in Fig. 16.28 where all 16 branching conditions and outputs are implemented by using discrete logic. When $Sanity(L) = 1(L)$, the sequencer is forced into the 0000 state. Then, when $Sanity(L) = 0(L)$, the sequencer initializes into state a, after which normal operation of the FSM can occur.

The operation of the one-hot sequencer design of the FSM in Figs. 16.27 and 16.28 is best represented by the simulation results provided in Fig. 16.29. For simplicity, only the external inputs, outputs, and state variables are represented. An analysis of the simulation results indicate that the time elapsing between an input change and an output response (response time) varies from τ_p to $5\tau_p + \tau_{INV}$, where τ_p is the average path delay through a gate and τ_{INV} is the path delay of an inverter, both in keeping with the usage in this text. The time spent in a state having two 1's during the transition between one-hot states is found to be $2\tau_p + \tau_{INV}$, or about half the maximum response time. Comparing the response times of the one-hot sequencer and the MAC module indicates that the two sequencers operate at approximately the same speed, but which will vary for some applications because of their inherently different design features.

16.4.4 Perspective on Programmable Sequencer Design and Application

Obviously, the most serious problem in using a fully specified n-state one-hot sequencer is in dealing with n^2 inputs. If the FSM to be implemented by the sequencer has relative few inputs and its branching conditions are relative simple, the problem is easily manageable with discrete logic. The example shown in Fig. 16.28 is a case in point. For FSMs having a large number of states with fairly complex branching conditions, the n^2 requirement for one-hot sequencer design most likely will require a CAD approach. A simple program will suffice for programming a PLD. As a reminder, ROMs should not be used as a PLD for designing the sequencer. ROMs are noisy and can create serious problems in one-hot sequencer operation.

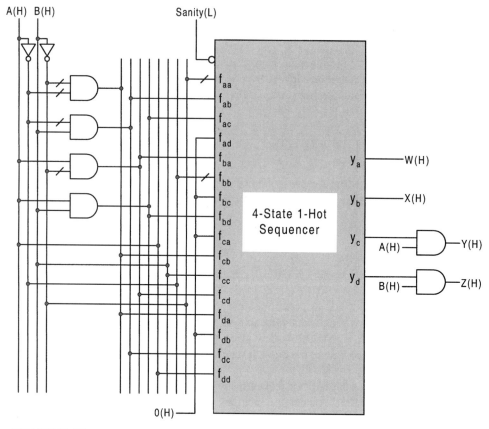

FIGURE 16.28
Implementation of the FSM in Fig. 16.27 by using the 4-state one-hot sequencer in Fig. 16.26 with discrete transition and output-forming logic.

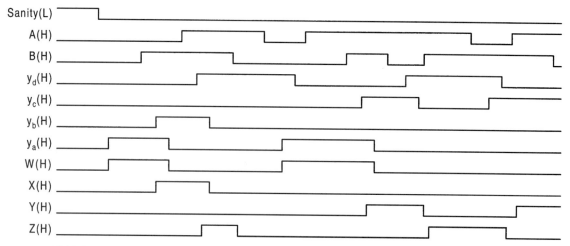

FIGURE 16.29
Simulation results for the one-hot design of the FSM in Figs. 16.27 and 16.28 showing state transition and output response to input change following initialization into the 0001 state by using the one-hot-plus-zero method.

16.4 ONE-HOT PROGRAMMABLE ASYNCHRONOUS SEQUENCERS

The main advantage in using a programmable one-hot sequencer is the ease with which it can be designed and the relative simple means by which it can be programmed. True, the hardware commitment is substantial. But if the objective is to use a single machine to reliably operate as any number of asynchronous controller FSMs on a time-shared basis, then the one-hot sequencer approach should be considered as a viable option. With the large PLDs that are available and with the software provided by the vendors to program them, the design and programming of a one-hot sequencer can be carried out with little difficulty. For large sequencers, the use of an FPGA is recommended, provided that care is taken in selecting the routing paths, a precaution that is essential in dealing with any asynchronous FSM. Obviously, none of the flip-flops in an FPGA can be used in the design of an asynchronous one-hot sequencer, but they can and should be used in the design of a one-hot synchronous sequencer. The genalized NS equations for a synchronous one-hot sequencer are obtained from Eqs. (16.14) exclusive of the "out of" terms expressed by the second portion of Eqs. (16.14). Remember that it is the action of the flip-flop that maintains the state variable of the origin state active until the transition to the destination state is complete.

Thus, a synchronous programmable one-hot sequencer can be designed by applying Eqs. (13.9) in much the same way that Eqs. (16.14) are applied to the design of an asynchronous one-hot sequencer. Actually, if the y-variables are fed back externally in the design of a one-hot sequencer, there is the option of converting from an asynchronous design to one that is synchronous. With feedback and output-forming logic taken from the flip-flop outputs, a synchronous one-hot sequencer results. To do this, however, requires that the "out of" terms be disabled. With this information in mind, Fig. 16.24 is applicable to the design of a generalize one-hot sequencer.

At this point, it is interesting to compare the two asynchronous sequencers that have been considered in this chapter. Both the MAC module sequencer and the one-hot sequencer operate in the fundamental mode and can be programmed by a bank of PLDs or by discrete logic. Both types make use of a fixed-state array machine, both are immune to most of the asynchronous timing problems, and both may need arbiter protection on the relatively few external inputs. However, the similarity stops here. The MAC module can be cascaded to produce modules of much greater capacity without compromising speed and reliability. In contrast, a one-hot sequencer cannot be cascaded because of the branching character between one-hot states. Whereas the MAC module approach requires that all state-to-state transitions be logically adjacent, which, in turn, requires that an even number of states be used in the controller design, the one-hot approach has no such requirement. The MAC module sequencer requires n inputs and is limited to n-way branching capability for a 2^n-state array, but the one-hot sequencer requires n^2 inputs and has n-way branching for an n-state array. Also, recall that the MAC module sequencer requires the use of a timing control machine to complete the required handshake configuration with the state array machine. A one-hot sequencer, on the other hand, requires no timing control machine for its proper operation.

Continuing this comparison, it is known that a one-hot sequencer should not be used to implement an FSM with cycles. An endless cycle condition between two states causes the one-hot sequencer to stick in the intervening state with two 1's. In contrast, the MAC module permits, and sometimes requires, cycle conditions, but never between adjacent states (see subsection 16.3.6). For example, a one-hot sequencer could not be used to design a ring counter as in Fig. 12.30, but the MAC module can easily be used to design the Gray code counter in Fig. 16.21. Both of these counters are examples of FSMs with continuous cycles.

The two types of asynchronous programmable sequencers that are discussed in this chapter are each, by their nature, unique approaches to multiple controller design. Beyond these two approaches there are no other prospects. The one feature that accounts for the success of an asynchronous programmable sequencer is the fact that each state-to-state transition is predictable in some unique way. For the MAC module, it is a parity shift between logically adjacent states, whereas for the one-hot approach it is a logic 1 shift between one-hot coded states.

The applications of asynchronous programmable sequencers to multiple controller use on a time-shared basis may appear to be highly specialized and somewhat limited. And to some extent this is so. However, with greater need for high-speed processing void of internal clock skew, there are some applications for which asynchronous programmable sequencers are better suited than conventional synchronous controller designs. The response times characteristic of asynchronous sequencer controllers will be considerably less than those associated with conventional clock-driven (synchronous) controllers of the same technology. Also, it is asserted here that as modern synchronous systems become more complex and are operated at increasingly higher speeds, failure due to clock distribution problems (clock skew) becomes more probable. Use of asynchronous programmable sequencers offers a practical means of avoiding such problems while meeting the demands for greater speeds. Communication between multiple asynchronous sequencer controllers operated simultaneously within a given system can be accomplished reliably by using appropriate handshake interfaces, again avoiding clock skew problems.

16.5 EPILOGUE TO CHAPTER 16

It is hoped that the subject matter presented in this chapter will serve to stimulate new ideas in both teaching and research. It is the author's position that teaching and research are closely interrelated and that innovation often arises from a spirit of inquiry. Upon completing a second-level course in digital design, students should be left with the notion that it is proper to challange "old" ideas and to seek new and innovative approaches to logic design. If this text can engender these concepts and instill in the reader the spirit of inquiry, then it has accomplished an important feat. To accept without question work of the past and present is to surrender to a future of stagnant technology and lackluster innovation.

FURTHER READING

A variety of systems have been studied that utilize both internally fixed and pausable clocks, but for various reasons are inherently more complex and slower than the EAIC system described in this text. In further contrast, these systems offer little or no protection from metastable effects — an important feature of the EAIC approach. The six selected references that follow are typical of these studies.

[1] W. Lim, "Design Methodology for Stoppable Clock Systems," *Proc. IEE* **133E**, 65–69 (1986).
[2] M. Afghahi and C. Svensson, "Performance of Synchronous and Asynchronous Schemes for VLSI Systems," *IEEE Trans. Comput.* **41**(7), 858–872 (1992).

FURTHER READING

[3] W. Lim and J. R. Cox, "Clocks and the Performance of Synchronizers," *Proc. IEE* **130E**, 57–64 (1983).

[4] A. B. Hayes, "Stored State Asynchronous Sequential Circuits," *IEEE Trans. Comput.* **C-30**(8) 596–600 (1981).

[5] H. Y. H. Chuang and S. Das, "Synthesis of Multiple-Input Change Asynchronous Machines using Controlled Excitation and Flip-Flops," *IEEE Trans. Comput.* **C-22**(12) (1973).

[6] S. M. Nowick and D. L. Dill, "Automatic Synthesis of Locally-Clocked Asynchronous State Machines," Proc. ICCAD-1991, pp. 318–321.

Closely related to this chapter is the work of Rosenberger *et al.*, who describe the design and analysis of Q-flops in an internally clocked configuration. The Q-flops are designed with an internal handshaking mechanism that ensures that the inputs are not stored until the input stage is ready to accept them and the outputs are not updated until the input stage has fully resolved and is stable in its new state. This allows the design of sequential delay-insensitive modules that require fewer delay constraints than other functionally equivalent design methodologies. Tinder provides a logic interpretation and discussion of Q-flops in EAIC systems.

[7] F. U. Rosenberger, C. E. Molnar, T. J. Chaney, and T. Fang, "Q-Modules: Internally Clocked Delay-Insensitive Modules," *IEEE Trans. Comput.* **37**(9), 1005–1018 (1988).

[8] R. F. Tinder, *Digital Engineering Design: A Modern Approach.* Prentice Hall, Englewood Cliffs, NJ, 1991.

Extensive studies have been conducted on the effects of the metastable state in D latches and synchronizers. Typical among these are the works of Jackson and Albicki and those of Pechoucek. Other studies relevant to the subject of metastability and to this chapter are those of Kacprzak and Albicki and of Chaney and Molner, the latter notable for work on metastability in cross-coupled NAND gates (the set-dominant basic cell).

[9] T. A. Jackson and A. Albicki, "Analysis of Metastable Operation in D Latches," *IEEE Trans. on Circuits and Systems* **36**(11), 1392 (1989).

[10] M. Pechoucek, "Anomalous Response Times of Input Synchronizers," *IEEE Trans. Comput.* **C-25**(2), 133–139 (1976).

[11] T. Kacprzak and A. Albicki, "Analysis of Metastable Operation in RS CMOS Flip-Flops," *IEEE J. Solid-State Circuits* **SC-22**(1), 57–64 (1987).

[12] T. J. Chaney and C. E. Molnar, "Anomalous Behavior of Synchronizer and Arbiter Circuits," *IEEE Trans. Comput.* **C-22**, 421–422 (1973).

A number of texts cover the subject of CMOS domino logic. Among these the texts of Fabricius, Mavor *et al.*, and Weste and Eshraghian are recommended.

[13] E. D. Fabricius, *Introduction to VLSI Design.* McGraw-Hill, New York, 1990.

[14] J. Mavor, M. A. Jack, and P. B. Denyer, *Introduction to MOS LSI Design.* Addison-Wesley, Reading, MA, 1983.

[15] N. Weste and K. Eshraghian, *Principles of CMOS VLSI Design.* Addison-Wesley, Reading, MA, 1985.

The portions of this chapter dealing with the EAIC system are based in part on the work of VanScheik and Tinder, which includes additional studies not mentioned in this chapter. In this reference, a comparison is made between the EAIC system featured in this chapter and the Q-Flops described in the article by Rosenberger *et al.* (cited previously). The part of this chapter, describing a unique class of asynchronous sequencers (MAC modules), is based in part on the work of Tinder, Klaus, and Snodderley, cited below. There is no known previous work on the subject of one-hot programmable asynchronous sequencers. For information on one-hot asynchronous FSM design, refer to Further Reading at the end of Chapter 14.

[16] W. S. VanScheik and R. F. Tinder, "High Speed Externally Asynchronous/Internally Clocked Systems," *IEEE Trans. Computers* **46**(7), 824–829 (1997).

[17] R. F. Tinder, R. I. Klaus, and J. A. Snodderley, "High Speed Microprogrammable Asynchronous Controller Modules," *IEEE Trans. Computers* **43**(10), 1226–1232 (1994).

PROBLEMS

16.1 (a) Convert the SL-DFLOP in Fig. 16.6 to a static logic JKFLOP by using the flip-flop conversion given by Eq. (14.10) and illustrated in Fig. 14.17. Use conventional gates and inverters for the MDS — that is, keep the conversion at the gate level only.

(b) Test the JKFLOP by simulation. To do this plan to use the logic input waveforms, including CK, similar to those of Fig. 10.42c. Include $Q(L)$ in the simulation.

16.2 (a) Construct the logic circuit for the three-bit binary up/down counter of Fig. 10.55a by using the general architecture for EAIC system given in Fig. 16.1. To do this, use TFLOPs for the input and memory registers, and use Eqs. (10.17) for the NS and output logic. Convert the DFLOPs to TFLOPs as in Fig. 10.39.

(b) Predict how the D-to-TFLOP conversion logic will affect the NS logic constraints, and the frequency and throughput characteristics, given by Eqs. (16.4), (16.5), and (16.6), for the up/down counter of part (a).

(c) Test by simulation at the gate level the up/down counter in part (a) by using both an up-count and a down-count. Keep in mind that this is an ideal simulation. This means that input conditions that might lead to metastability in a real-time test of the EAIC system cannot be resolved by the simulator program — the MDS in a real EAIC system is part analog and part digital. As a result, the simulation will likely show momentary oscillation at certain nodes in the circuit.

16.3 (a) Repeat all parts of Problem 16.2 by using Eqs. (10.16) for the NS and output functions. Replace part (b) in Problem 16.2 with the following:

(b) Predict how the multilevel logic of Eqs. (10.16) will affect the NS logic constraints, and the frequency and throughput characteristics, given by Eqs. (16.4), (16.5), and (16.6), for the up/down counter.

16.4 (a) Construct the logic circuit for the two-input/two-output FSM in Fig. 11.43b by using the general architecture for EAIC system given in Fig. 16.1. Plan to use Eqs. (11.11) for the NS and output logic and to initialize into the 000 state.

(b) Test the logic circuit of part (a) by simulation. To do this, set the input conditions necessary to traverse the state diagram in the following way:

$$a \to b \to c \to b \to d \to c \to e \to a$$

PROBLEMS

Keep in mind that this is an ideal simulation. This means that input conditions that might lead to metastability in a real-time test of the EAIC system cannot be resolved by the simulator program — the MDS in a real EAIC system is part analog and part digital. As a result, the simulation will likely show momentary oscillation at certain nodes in the circuit.

16.5 (a) Design the rotation detector in Problem 14.22 by using the EAIC system. Implement the NS and output function with discrete logic. Plan to use the two-level SOP expression for *CCW* output and then use an edge-triggered D flip-flop to filter out any logic noise that is generated by static hazards.

(b) Address the issues of endless cycles, critical races, ORGs, static hazards, and E-hazards as they relate to the EAIC design of the rotation detector.

(c) Simulate the results of part (a) to verify the proper operation of the rotation detector EAIC design.

16.6 Design and test the 2-input (2^2-state) MAC module as follows:

(a) Construct the fully documented state diagrams for the SAM and TCM by following the example in Subsections 16.3.2 and 16.3.3. Include all branching conditions and outputs.

(b) Obtain an optimum set of NS and output functions for the SAM and TCM, and end with a complete logic circuit by using these results.

(c) Use the 2-input MAC module to design a 2-input Gray code counter by following the example in Subsection 16.3.4.

(d) Test the design in part (c) by simulation. Thus, initialize the system into the 00 state and then cycle the counter through all four states in a manner similar to that shown in Fig. 16.22 for the 3-bit Gray code counter.

16.7 The FSM in Fig. P14.4 (see Problems at the end of Chapter 14) is to be designed by using the 2-input MAC module of Problem 16.6.

(a) Construct the state diagram for the FSM that is appropriate for a MAC module design.

(b) Obtain the NS instructions from the state diagram. (Hint: First construct the K-maps for the instruction inputs.)

(c) Construct a complete logic diagram for this MAC module design. Include all branching conditions and outputs. Note: Block symbols may be used for the 2-input MAC module.

(d) Test the design of part (c) by simulating the logic circuit. To do this, initialize the FSM into the 00 state and then cause it to transit through all state-to-state paths.

16.8 Repeat Problem 16.7 for the design of the FSM in Fig. 16.12a. Note: When simulating the logic circuit, include the cycle from state 11 to state 01 under input conditions $X\bar{Y}$ as one of the tests.

16.9 Obtain the NS instruction functions for the FSM in Fig. P14.12 assuming the use of the 3-input MAC module, presented in Fig. 16.17, as the sequencer. (Hint: First assign a 3-bit state code to the FSM in Fig. P14.12 by adding a 0 (zero) to each state code in the MSB position.

16.10 Obtain the NS instruction functions for the FSM in Fig. P13.4a required to drive the 3-input MAC module of Fig. 16.17 as the sequencer.

16.11 Design and test by simulation the one-hot asynchronous sequencer design of the FSM in Fig. P14.9a by using the programmable sequencer discussed in Subsection 16.4.2. To do this, use discrete logic for both the sequencer design and the external logic required to program it.

16.12 By following the example in Subsection 16.4.2, write the NS functions for a fully specified, six-state, one-hot asynchronous sequencer by using the one-hot-plus-zero approach to initialization. To do this, use the form of Eqs. (16.13) and assume that initialization is to occur into state a.

16.13 (a) Write the program logic required to operate the one-hot sequencer of Problem 16.12 as the FSM represented by the state diagram in Fig. P14.10. To do this, follow the format of Eqs. (16.14) and (16.15).

(b) Construct the logic circuit for this design in a manner similar to that used to represent the sequencer design shown in Fig. 16.28.

EAIC System Design Projects at the Advanced Level

For projects at the advanced level, more complex EAIC device and system designs can be carried out together with simulations of those designs. The projects that can be used for this purpose are extensive in number and really limited only by one's imagination. A few examples are as follows:

Device Category Examples

Four-bit parallel loadable shift register
Four-bit parallel loadable up/down binary counter
Four-bit autonomous linear feedback shift register (ALFSR) counter
Four-bit ripple counter by using toggle modules as memory elements
One-hot EAIC design of a serial 2's complementer (see Subsection 13.5.2)

System Category Example   Four-bit parallel-to-serial adder/subtractor system. Follow the design in Subsection 13.6.1, but use the parallel loadable right shift registers designed in Subsection 12.2.2.

Notes

1. Remember that it is not necessary to add external synchronizing circuitry to an EAIC system since the input register serves to synchronize the inputs.
2. To correctly design an EAIC system containing both controller and data path devices, it may be necessary to generate the internal clock with data ready (R) signals from controller and data path devices of the same triggering edge, e.g., RET.
3. For simulation purposes, it will be necessary to use a conventional NOR gate in the clock generating circuit. For large number of inputs, a NOR tree configuration similar to the OR tree in Fig. 4.49 may be necessary.

PROBLEMS

Asynchronous Programmable Sequencer Design Projects at the Advanced Level

Advanced designs with MAC module and one-hot sequencers are, of course, required to operate in the fundamental mode. Keeping this in mind, the following devices and system designs are examples that can be undertaken at the advanced level with simulations included.

Device Category Examples

A 4-bit MAC module with preset, clear, and enable inputs.
A 4-bit Gray code up/down counter by using two 2-input MAC modules.
The rotation detector of Problem 14.22 designed by using the four-state one-hot programmable asynchronous sequencer.
A 6-state one-hot programmable sequencer and its use to implement the 5-state FSM in Fig. P14.7.

System Category Example A four-bit parallel-to-serial adder/subtractor system similar to the eight-bit parallel-to-serial adder/subtractor in Subsection 13.6.1. To do this, use a 2-input (2^2-state) MAC module for the controller, and trigger the appropriate data path devices with the *DI* signal.

An asynchronous one-hot design of the voter booth system of Problem 13.20 that will tabulate continuously the individual count of the contestants and give, on command, the difference in the count at any point in the voting period. It is required that a parallel-to-serial adder/subtractor be used, one similar to that in Subsection 13.6.1 but enlarged to handle the number of voters that may vote. (Hint: To generate a periodic triggering signal for use with the adder/subtractor, consider activating a very simple one-hot "cycle" FSM properly initialized.)

Notes

1. In most cases, it will be necessary to use minimized functions in these designs. Therefore, except for the one-hot design, use of a logic minimizer is highly recommended.
2. Since fundamental mode FSMs are involved, avoid external input changes in near proximity to one another when simulating the designs.
3. In the system-level design problem that uses the 2-input MAC module for the controller, pay particular attention to the timing of events when triggering the peripherals with the *DI* signal.

APPENDIX A

Other Transistor Logic Families

A.1 INTRODUCTION TO THE STANDARD NMOS LOGIC FAMILY

Though CMOS is currently the most important member of the MOS family, NMOS still occupies a significant position is modern technology. Shown in Fig. A.1 is the generalized NMOS logic configuration similar to that for CMOS given by Fig. 3.5. The depletion-mode NMOS serves a similar purpose as the PMOS in CMOS logic — it produces a high resistance (impedance) when the enhancement-mode NMOS logic (NL) is evaluated (shorted to ground), but becomes a low resistance otherwise. Thus, the CMOS NOR gate in Fig. 8.46 could be replaced by a NOR gate built with the NMOS technology of Fig. A.1.

Figure A.1 implies that NL can represent a variety of logic devices. Four simple examples of NMOS logic gates are presented in Fig. A.2. The AND and OR gates are implemented by adding inverters to the NAND and NOR gates, as is done in CMOS (see Figs. 3.16 and 3.17).

The relative simplicity of NMOS logic compared to CMOS is illustrated by the examples in Fig. A.3. Here, the comparison is made between AND-OR-invert (AOI) gates of the two MOS families. These circuits and those of Fig. A.2 are classified as static logic. Dynamic domino CMOS logic is discussed in Subsection 16.2.3. Dynamic domino NMOS logic is similar to dynamic domino CMOS logic except that in dynamic domino NMOS logic the depletion mode NMOS is replaced by an enhancement mode NMOS transistor for the precharge stage.

The main advantage of CMOS logic over NMOS logic is in power dissipation. For example, when the input to an NMOS inverter is at low voltage (LV) no DC power is dissipated. However, when the input goes to HV, the depletion mode NMOS draws a saturation current which causes "quiescent" power dissipation. When the packing density of NMOS gates reaches into the hundred of thousands (small by modern standards), Joule heat dissipation becomes a problem. This heat must be sinked; otherwise it could accelerate chip failure due to impurity and dopant diffusion. Remember that diffusion processes are exponentially temperature dependent.

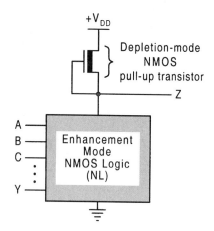

FIGURE A.1

A.2 INTRODUCTION TO THE TTL LOGIC FAMILY

The standard transistor–transistor logic (TTL) family is composed mainly of bipolar junction transistors (BJTs) and resistors. Shown in Fig. A.4 are three examples of standard TTL gates. In these figures B is the base, E is the emitter, and C is the collector. The phenomenological operation of these gates can be easily understood by first considering the inverter. Qualitatively, the TTL inverter operates as follows: When X_{in} is at LV (E = LV), transistor T_1 forces transistor T_2 to be turned OFF, thereby bringing X_{out} to V_{CC} level, hence X_{out} = HV. But when X_{in} goes to HV (E = HV), T_1 causes T_2 to be turned ON, which brings X_{out} to ground level, hence, X_{out} = LV. Thus, this behavior obeys the physical truth table in Fig. 3.6b. In a sense, a BJT is turned ON when E = HV and is turned OFF when E = LV, which is similar to the behavior of an NMOS transistor when these voltages are

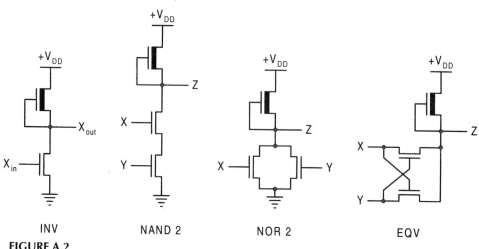

FIGURE A.2

A.2 INTRODUCTION TO THE TTL LOGIC FAMILY

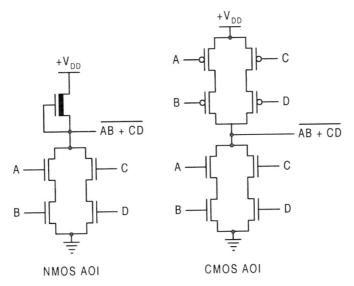

FIGURE A.3

applied to the gate, G. The resistors in the inverter of Fig. A.4 are called "pull-up" resistors and serve basically the same function as the depletion mode NMOS in Fig. A.2 — they are current-limiting elements. The symbol V_{CC} is internationally accepted to represent the supply voltage to the bipolar logic families.

The operation of the two-input BJT NAND gate in Fig. A.4 follows directly from the description of the inverter. Here, T_1 is a dual-emitter BJT, the operation of which is not unlike that of T_1 for the inverter. Thus, any time either input X or Y (or both) is at LV, T_2 is turned OFF, causing Z to go to HV. Only when both X and Y are at HV is T_2 turned ON, bringing Z to LV. This behavior is, of course, expressed by the truth table in Fig. 3.10b. The number of inputs to a TTL NAND gate can be increased by increasing the number of emitter connections. However, this usually limited to eight or fewer for technological reasons. Standard TTL AND gates are produced by attaching TTL inverters to the NAND gate outputs.

The operation of the two-input NOR gate can be explained in a similar manner. If either input X or Y (or both) is at HV, T_3 or T_4 (or both) is turned ON and the output Z goes to

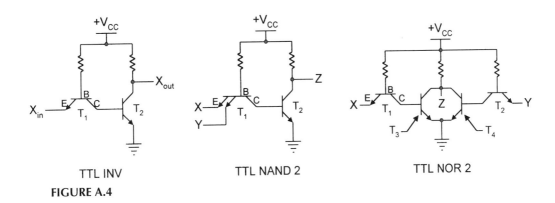

FIGURE A.4

ground ($Z = LV$). Only when both X and Y are at LV will the output Z be at HV. This is the same physical behavior expressed by the truth table in Fig. 3.12b. Standard TTL OR gates are produced by attaching TTL inverters to the NOR gate outputs.

The presentation here does not do justice to the field of TTL devices, which is extensive and requires much more time and space than is permitted here. The TTL family is actually divided into several subfamilies. These include Schottky TTL (S), low-power Schottky (LS), and advanced low-power Schottky (ALS). Also belonging to the bipolar group of families is the emitter-coupled logic (ECL) family. Further Reading at the end of this appendix cites several sources that will carry the reader well beyond the present treatment.

A.3 PERFORMANCE CHARACTERISTICS OF IMPORTANT IC LOGIC FAMILIES

All members of the TTL family suffer from high power consumption in comparison to the MOS family, and especially in comparison to CMOS. The one advantage that bipolar logic families may have over MOS is speed. Generally, all members of the TTL family of gates are faster than either NMOS or CMOS. However, modern high-speed CMOS (HC) has closed the gap in speed somewhat. Of the bipolar logic families, ECL is the fastest but suffers from high power consumption. It is because of its speed that the ECL family is currently the fourth important logic family next to CMOS, NMOS, and TTL, CMOS being the most important. Members of the MOS family have the lowest power consumption of any of the logic families, with CMOS logic having the lowest.

Any summary of logic family performance characteristics is risky because the relative assessments change, sometimes rapidly, with technological developments and because these assessments often have to be qualified to be valid. Nevertheless, an attempt is made in Table A.1 to present a qualitative assessment of these characteristics for some of the more important logic families.

Not included in Table A.1 are several other logic families that are important for certain specialized applications. The families include integrated injection logic (I^2L), low voltage injection logic (LVIL), gallium arsenide logic (GAL), and silicon-on-sapphire CMOS (CMOS/SOS) logic. Also, there are the rather esoteric optical and superconducting families that appear to have relatively little use.

FURTHER READING

To one extent or another, every text in digital design or digital electronics contributes something to the subject of logic families and their characteristics. Some are technology

Table A.1 Characteristics of commonly used IC logic families

Parameter	TTL	ECL	NMOS	CMOS
Switching speed	Very good	Excellent	Fair	Good
Power dissipation	Medium	High	Low	Very low
Noise immunity	Very good	Fair	Good	Very good
Fan-out	Fair	Fair	Very good	Excellent
Packing density	Medium	Low	High	High

dependent and some are not. The text of McCluskey is recommended for its technology-dependent coverage of TTL, diode–transistor logic (DTL), and integrated injection logic (I^2L) logic. Another technology-oriented coverage of the TTL logic family is found in the text by Wakerly. Other texts that cover TTL and DTL to a lesser extent include those by Katz and Tinder. Good electronics-oriented coverage of the integrated logic families are provided in the texts of Jones and Tocci. For CMOS logic, the VLSI text of Fabricius and that of Weste and Eshraghian are recommended. Another source of information on the characteristics of the most commonly used logic families is *The Electrical Engineering Handbook*, R. C. Dorf, Editor-in-Chief.

[1] R. C. Dorf, Editor-in-Chief, *Electrical Engineering Handbook*, 2nd ed. CRC Press, Boca Raton, FL, 1997, pp. 1769–1790.
[2] E. D. Fabricius, *Introduction to VLSI Design*. McGraw-Hill, New York, 1990.
[3] L. D. Jones, *Principles and Applications of Digital Electronics*. Macmillan, New York, 1986.
[4] R. H. Katz, *Contemporary Logic Design*. Benjamin/Cummings Publishing, Redwood City, CA, 1994.
[5] E. J. McCluskey, *Logic Design Principles*. Prentice-Hall, Englewood Cliffs, NJ, 1986.
[6] R. F. Tinder, *Digital Engineering Design: A Modern Approach*. Prentice Hall, Englewood Cliffs, NJ, 1991.
[7] R. J. Tocci, *Digital Systems, Principles and Applications*, 4th ed. Prentice Hall, Englewood Cliffs, NJ, 1988.
[8] J. F. Wakerly, *Digital Design Principles and Applications*. Prentice Hall, Englewood Cliffs, NJ, 1994.
[9] N. H. E. Weste, and K. Eshraghian, *Principles of CMOS VLSI Design*. Addison-Wesley, Reading, MA, 1985.

APPENDIX B

Computer-Aided Engineering Tools

B.1 FUNCTION MINIMIZATION TOOLS

An excellent software minimization tool, called *BOOZER* (for BOOlean ZEro-one Reduction), is bundled with this text and is recommended. It can accept entered variables (EVs) or canonical data from multioutput functions and can return an optimal or near-optimal multioutput SOP solution. Incompletely specified functions are no a problem for BOOZER. These features are especially important in synchronous and asynchronous FSM design of relatively large systems. No other minimization algorithm appears to have the capability of accepting EVs without patching them in. Additionally, BOOZER can handle up to 16 functions of 16 variables when used on modern computers. The program operates in the DOS mode on conventional PCs and requires only the use of the text editor. The computer memory requirements are minimal. Complete instructions and additional information regarding BOOZER are included with the software, together with the author's name and address. See Further Reading at the end of this appendix for more information about BOOZER.

A well-known software tool used in industry for minimization of large Boolean functions, called Espresso, is available through the University of California, Berkeley, 1986 VLSI tools distribution. It supports advanced algorithms for minimization of two-level, multioutput Boolean functions but does not accept EVs unless patched in. The algorithms are described in an article by Rudell cited in Further Reading at the end of this appendix.

B.2 SCHEMATIC CAPTURE, SIMULATION, AND TIMING ANALYSIS TOOLS

An outstanding gate-level, interactive, schematic-capture and simulation program for logic design, called *EXL-Sim2000*, is bundled with this text. It is the student version of a more powerful professional-level program. EXL-Sim2000 is unique for its intuitive approach, yet it is powerful enough to handle most any problem associated with this text. Its features include a drag-and-drop capability, rubber banding, mixed logic capability, primitive (gate) libraries, macro generation capability, library development, project management, individual or global delay assignments, unrestricted timing intervals, multiple zoom levels, simple editing and labeling capability, multiple windows, waveform scrolling, a variety of printout capabilities, and a host of other features. EXL-Sim2000 operates in the Windows environment and requires relatively little computer memory. In developing this program, an attempt was made to incorporate the best features of other available gate-level simulators, but also to add some unique capabilities that set it apart from the others. EXL-Sim2000 is designed for both beginning and advanced students, and even instructors. Complete instructions (including examples) for its use and information regarding the professional version of EXL-Sim2000 are included with the software. See also EXL-Sim2000 home page at http://www.tbdgroup.com.

There are other schematic capture and simulation software available for logic design at the student level. Examples include *Beige Bag* V3.0 for Windows (http://www.beigebag.com), *LogicWorks* from Capilano Computing for either PCs or Macs (http://www.logicworks.com), and the student edition of *Workview Office* by VIEWlogic for Windows (http://www.prenhall.com/workview). These three offer similar features, which are given at their respective Web sites.

At the professional level, *Workview Office* (V 7.31A or higher) is one of the most powerful tools available. It includes such features as front-end and project-management tools, project navigation and library maintenance, design entry and schematic capture, PLD and FPGA design entry, digital/analog simulation, timing analysis, synthesis/FPGA design, netlisting, graphical analysis and editing, and PCB layout. Workview Office operates in the Windows environment, and its memory requirements are substantial. Information about this program can be obtained on Viewlogic's home page: http://www.viewlogic.com.

A quality CAD tool, called *Cedes* (C++ Engine for Discrete Event Simulator), is available to students and professionals. Cedes is an affordable, efficient, object-oriented design tool for modeling and simulation of digital systems. The program permits use of the MS graphical user interface (GUI) to describe a design in schematic format and can automatically generate a simulation program based on the schematic drawing. The user can click icons to run the simulation, make changes to the diagram as needed, and then rerun the simulation. Cedes is VHDL and Verilog compatible in netlist formats. Custom libraries can be generated that support local design environments. Information regarding this program is available on its Web site: http://www.fbeedle.com.

The three best-known professional tools for schematic capture, simulation, VLSI chip design, and circuit board layout are Mentor Graphics, Cadence, and OrCad. Since these three CAD tools provide essentially the same features and since more detailed information can be obtained from their Web sites, only a brief description is given of Mentor Graphics' features. Mentor Graphics permits a variety of inputs into *ModelSim*, perhaps the world's most powerful simulation engine. These inputs include schematic capture, VHDL or Verilog description, state table representation, and Quicksim. Cell and FPGA libraries can be generated from the VHDL description, as can circuit board layout files. ModelSim (Elite Edition) is used for ASIC systems on UNIX, while ModelSim (Personal Edition) is targeted toward FPGA designs on PCs operating in the Windows environment. To obtain further information on these powerful professional tools, see their Web sites: http://www.mentor.com; http://www.cadence.com; and http:/www.orcad.com.

A powerful tool for modeling, analyzing, and documenting digital systems is called *TimingDesigner Professional*, by Chronology. This program is unique in the sense that it uses a built-in static timing analysis engine and worst-case simulator to help the designer create and analyze the timing and sequence protocol at any stage in the design process.

Design requirements such as timing constraints, delays, and sequence protocols can be modeled within the graphical diagram window. The program's primary function is to identify potential problems, evaluate design alternatives, and explore alternative design solutions, all at an early stage in the design process. Use of either VHDL or Verilog expression syntax in simulation equations is permitted since the choice of language does not affect simulation output. The program operates in either the Windows or UNIX environments. For further information visit Chronology's Web site: http://www.chronology.com.

FURTHER READING

The algorithm for the original BOOZER program is discussed in Fletcher. The program has since been revised by its author to include entered variables. The basic algorithms of Espresso are described in an article by Rudell.

[1] W. I. Fletcher, *An Engineering Approach to Digital Design*. Prentice Hall, Englewood Cliffs, NJ, 1980.
[2] R. Rudell, "Espresso-MV: Algorithms for Multiple-Valued Logic Minimization," *Proc. Int. Circ. Conf.*, May 1985.

APPENDIX C

IEEE Standard Symbols

C.1 GATES

The standard ANSI/IEEE Std. 91-1984 is extensive and will not be covered in its entirety in this appendix. Shown in Fig. C.1a is the basic logic gate rectangle used in the standard gate library, and in Fig. C.1b are the most commonly used standard and nonstandard qualifying logic operation symbols that are placed in the open box at the top of the standard rectangle. In Fig. C.1c are the qualifying input/output (I/O) symbols that are used as logic level indicators on the inputs and outputs of the standard rectangle.

Presented in Fig. C.2 are three circuit representations of the function $F = A\bar{B} + \bar{C}D + \bar{E}$. Here, the distinctive shape symbols are used in Fig. C.2a and is compared with two IEEE standard representations of the same circuit in Figs. C.2b and c. Clearly, the identity of the gates (NAND or AND/OR) is lost in the compact IEEE format of Fig. C.2c.

C.2 COMBINATIONAL LOGIC DEVICES

Several types of dependency are used in the IEEE standard symbols of combinational and sequential MSI devices. A truncated list is provided as follows:

G — AND	V — OR	N — Negate (Exclusive-OR)
Z — Interconnection	C — Control	S and R — Set and Reset
EN — Enable	M — Mode	A — Address

Shown in Fig. C.3 are MSI combinational devices that are represented as IEEE standard symbols. The meaning of the symbol $G\frac{x}{y}$ is somewhat self-explanatory. For example, the 8-to-1 MUX shows a grouping of three data select inputs that steer one of eight (0-to-7) data inputs to the output. Hence, $G\frac{0}{7}$ indicates AND dependency with signals given in the range 0 to 7. The symbol G4 EN in the dual 1-to-4 DMUX is an enable input affecting four outputs such that when G4 is inactive-low the outputs are inactive-low and vice versa.

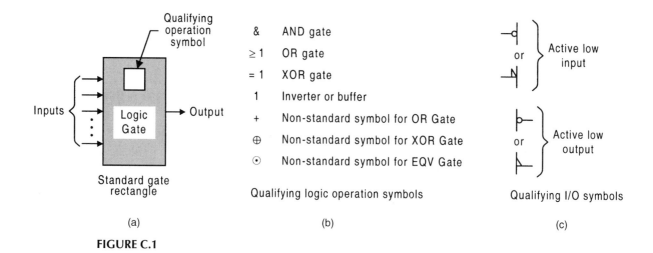

FIGURE C.1

C.3 FLIP-FLOPS, REGISTERS, AND COUNTERS

An explanation of the IEEE standard symbology for sequential machines is somewhat more involved. Shown in Fig. C.4 are five examples. The symbology for the two flip-flops is straightforward. Single data and clock inputs are indicated by 1D and C1, respectively. The active low asynchronous preset and clear inputs are given the symbols S and R for set and reset.

The symbology used for the MSI devices in Fig. C.4 is more complex, although that for the storage register appears to be self-evident. The four-bit universal shift register (USR) uses $M\frac{0}{3}$ to indicate the mode control inputs that set the FSM for true hold, shift right, shift left, or parallel load: hence, four mode dependencies (0-to-3). Inputs D0–D3 are subject to simultaneous dynamic control by the clock input. The 3 in the symbol 3,4D indicates that the input is enabled for parallel loading ($M = 3$). The dynamic (clock) symbol C4, $1 \rightarrow /2 \leftarrow$ on the clock input simply means that when $M = 1$ the USR shifts right ($\rightarrow$) and when $M = 2$ it shifts left ($\leftarrow$), and that the 4D inputs are controlled by clock at input C4. The label SRG4 indicates a four-input shift register.

For the 4-bit up-counter, the symbol CT means "content" and has a somewhat different meaning when applied to an input vs an output. For example, in this counter $5CT = 0$

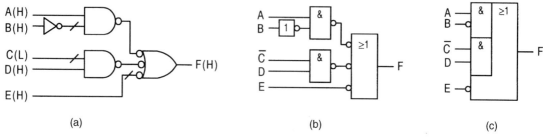

FIGURE C.2

C.3 FLIP-FLOPS, REGISTERS, AND COUNTERS

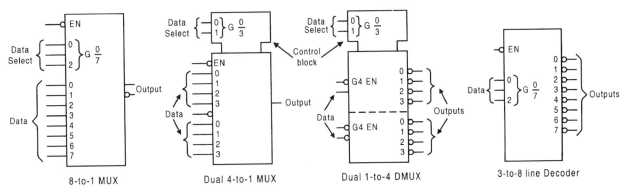

FIGURE C.3

represents a synchronous clear, meaning that a clear command must be clocked into the FSM. The output symbol 3CT = 15 means that an output signal becomes active only at the count of 15 when G3 is active. In this counter, symbols G3 and G4 are dual count enable inputs, such that both must be active before the counter will count up. Thus, the dynamic input symbol C5/2, 3, 4+ signifies that clock input C5 controls inputs 5D and that the counter will count up only if M2, G3, and G4 are active. An up/down counter would have two dynamic input strings, one ending in plus (+) and the other in minus (−). The label CTRDIV16 simply indicates a 4-bit (÷16) counter.

The IEEE standard symbol forms are not for everyone and certainly not for the beginning student in the subject area. As can be seen from these examples, the standard language is complicated and should not be used by anyone but the most experienced user. Thus, for pedagogical reasons, this text has avoided the use of the standard forms in favor of the more traditional symbols. The main advantage of the IEEE standard symbology seems to

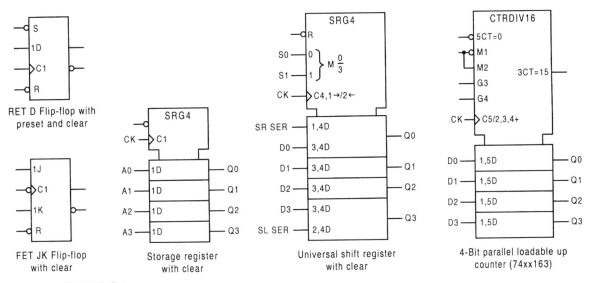

FIGURE C.4

be that it is, in fact, a standard that professionals can adhere to. The problem is that not all of industry uses the standard, which leads to a mix of notation and symbology. The new ANSI/IEEE Std 91-1984, *IEEE Standard Graphic Symbols for Logic Functions*, is based on the International Electrotechnical Commission (IEC) standard 617, and its use is required by the U.S. Department of Defense (DOD). For those wishing more information on the IEEE standard, the references cited in Further Reading should be helpful.

FURTHER READING

Perhaps the best way to begin to learn the IEEE standard symbology is to read those texts that either emphasize its use or have included a detailed summary of it. Such texts include those of McCluskey, Nelson *et al.*, Wakerely, and Yarbrough and are recommended. For a more complete treatment of this symbology the reader should visit the original IEEE documentation cited here.

[1] *Standard Graphic Symbols for Logic Functions, IEEE/ANSI Standard 91-1984*. Institute of Electrical and Electronics Engineers, Inc., IEEE Standards Office, 345 East 47th St. New York, NY 10017, 1984.
[2] E. J. McCluskey, *Logic Design Principles*. Prentice Hall, Englewood Cliffs, NJ, 1986.
[3] V. P. Nelson, H. T. Nagle, B. D. Carroll, and J. D. Irwin, *Digital Logic Circuit Analysis and Design*. Prentice Hall, Englewood Cliffs, NJ, 1995.
[4] J. F. Wakerly, *Digital Design Principles and Practices*, 2nd ed. Prentice-Hall, Englewood Cliffs, NJ, 1994.
[5] J. M. Yarbrough, *Digital Logic Applications and Design*. West Publishing Co., Minneapolis/St. Paul, MN, 1997.

Index

A

ABEL, 329
Absolute minimum expressions, 198
Absorptive Laws
    AND/OR forms, 108
    EQV/XOR forms, 112
Accumulator, parallel design, 607
Actel FPGAs, 319–321
    Act-1 family, 320–321
Activation level indicators
    Active high, 80
    Active low, 80
Active state, 79
Active transition point, 465
Adders
    Binary, 335–340
    Binary coded decimal (BCD), 386–387
    Carry look-ahead, 345–349
    Carry-save, 349–350
    Excess 3 (XS3), 387
    Full, 337–338
    Half, 336
    Ripple-carry, 338–340
Adder/subtractors
    Binary, 342–388
    Binary coded decimal (BCD), 387
    Excess 3 (XS3), 387–388
    Parallel-to-serial, 645–651, 651–655
    Ripple-carry, 342–345
Addition
    Binary, 52–53
    Binary coded decimal (BCD), 62–63
    Excess 3 (XS3), 75
    Floating point, 64–65
    Hexadecimal, 75
Adjacent XOR patterns, 198–206
Algorithmic state machine (ASM) charts
    In one-hot FSM designs, 640–645
    Symbology, 537–538
    Vs state diagrams, 538, 642–644, 659–660
Algorithms
    BCD addition, 63
    BCD subtraction (10's complement), 63–64
    BCD-to-binary conversion, 260
    Binary addition, 52

Binary division, restoring, 59
Binary multiplication, 56
Binary-to-BCD conversion, 261
Booth's, 57–58
Carry-save addition of multiple integers, 349
D-to-JK K-map conversion, 475
D-to-T K-map conversion, 471
Diminished radix complement, 48
Direct quadratic convergence, nonrestoring, 62
Floating point multiplication/division, 67
Fraction conversion, 41
Positive integer conversion, 38, 39
Radix complement, 46
Round off for fraction conversion, 42
T-to-JK K-map conversion, 576
Two's complement, 46
Two's complement multiplication, 57–58
Two's complement subtraction, 54
Alternative race paths
    In analysis of ORGs, 492
    In analysis of races and critical races, 703
Alternative synchronous FSM architecture
    Choice of components, 613–614
    One-hot method, 636–649
    Parallel loadable up/down counters as the memory, 632–637
    Universal shift registers as the memory, 626–632
    Use of ROMs, PLAs, and PALs, 614–626
Analog noise, 499
Analysis of finite state machines (FSMs)
    Asynchronous, 741–758, 788–795
    Synchronous, 476–479
Analysis of synchronous FSMs
    Examples, 476–479
    Procedure, 476
    PS/NS table, use of, 476
AND
    Definition, 87
    Logic circuit symbols, 88, 92–94
    Operator symbols, 87
AND array, 301 (*see also* AND stage)
AND function, 94
AND gate
    Conjugate gate symbols, 92–94
    CMOS, 92

863

AND gate (*cont.*)
    Domino logic configuration, 815
    Mixed logic interpretations, 92
    Multiple inputs, 93
    Physical truth table, 92
AND laws, 106
AND operator, 87
AND-OR-Invert (AOI) gate
    CMOS, 317–318, 851
    Logic equivalent circuits, 318, 319
    NMOS, 851
    Truth tables, 318
    Use of in ALU design, 364
AND stage (plane or section)
    In PLDs, 297–298, 301–303, 307–309
ANSI/IEEE Standard gate symbology, 859–860
ANSI/IEEE Std91-1984 Standard, 859–862
Antiphase triggering
    Of output holding registers, 497, 498
    Of synchronizers, 512, 513, 515
Apolar inputs, 615
Arbiters
    Bus, 801–803
Arithmetic and logic units (ALUs), 357–380
    Carry look-ahead configurations, 361–363, 378–380
    Dedicated and with CLA capability, 358–363
    Dual-rail systems with completion signals, 369–380
    MUX approach for VLSI application, 363–369
Arithmetic codes
    Binary coded decimal (BCD), 34–36
    Excess 3 (XS3), 35–36, 49
    Nine's complement, 48
    One's complement, 47, 48
    Signed-magnitude, 44–45
    Ten's complement, 45–46, 48
    Two's complement, 45–47
    Unsigned binary, 33
    Vs degree of difficulty of arithmetic operations, 68
Arithmetic combinational devices, 335–380
    Adders, 335–340, 345–349, 349–350, 386–387
    Adder/subtractors, 342–345, 387–388
    Arithmetic and logic units, 357–380
    Comparators, 265–272
    Dividers, 353–357
    Multipliers, 350–353, 389–390
    Subtractors, 340–342
    VHDL description of a full adder, 381–382
Array algebraic approach to logic design
    Asynchronous FSMs, 720–730
    Partitioning methods for state code assignments, 721–724

Single transition time (STT) FSMs, 720–730, 734, 738–740
    Synchronous FSMs, 542–547
ASCII character code, table, 71
ASICs, 238
ASMs, 536–538, 640–644, 659 (*see also* Algorithmic state machine charts)
Associative laws
    AND/OR forms, 108
    EQV/XOR forms, 111
Associative XOR patterns, 198–204
Asynchronous binary counters
    Data triggered, 664–665
    Ripple counters, 600–605
Asynchronous FSM analysis
    Critical races, 703–705
    Endless cycles, 701–793
    Essential hazards, 711–719, 746, 750, 752, 756–757
    Examples, 741–758
    LPD model FSMs, 743–747
    Nested cell model FSMs, 747–752
    One-hot FSMs, 752–757
    Procedure, 741–742
    Static hazards in the NS and output logic, 705–711
Asynchronous FSM design
    Array algebraic approach (STT FSMs), 720–711
    Call module, 770–771
    Flip-flops, 438–461, 698–701
    Fundamental mode, defined, 686
    Hazard-free FSMs by using the nested cell model, 730–734
    Initialization, 719–720
    Latches, 441–444, 460–464
    LPD model, 686
    Lumped path delay (LPD) approach, 692–695, 698–700, 705–710, 714–715
    One-hot approach, 734–740, 835–842
    Nested cell approach, 441–448, 460–461, 695–700, 730–734
    Perspective on state code assignments, 738–740
    Rendezvous module (RMOD), 695–698
    Rotation detector, 769–770
    Rules for use of PLDs, 740–741
    Selector module, 769
    Use of PLDs, 740–742
Asynchronous inputs, 510–517
    Branching dependency rule, 510
    Conditional output rule, 510
    Definition, 510
    Mean time between failures (MTBF), 516–517
    Metastability and the synchronizer, 514–517
    Multiple stage synchronizers, 515–517
    Stretching and synchronizing the inputs, 512–514
    Synchronizing the inputs, 511–517

INDEX

Asynchronous preset and clear overrides, 463–464
Asynchronous parallel loading
    Counters, 579–581, 588–589
    Shift registers, 568–570, 588–589
Asynchronous programmable sequencers
    Microprgrammable asynchronous controller modules, 823–835
    One-hot programmable modules, 835–839
Asynchronous state machines
    Analysis, 741–758
    Array algebraic approach to the design, 720–730
    Design examples, 695–698, 698–701, 720–733, 734–738, 740–741
    Detection and elimination of timing defects, 701–719
    Excitation table for the LPD model, 688–689
    Features, 684
    Fully documented state diagrams, 689–690
    Fundamental mode, defined, 686
    Hazard-free design, 730–734, 734–740, 837–839
    Initialization and reset, 719–720
    Lumped path delay (LPD) model, 685–689
    Models, 439, 685–687, 773–774
    Nested cell model, 439, 776
    Need for clock-independent FSMs, 685
    Pausable system approach to the design, 806–823
    Perspective on state code assignments of fundamental mode FSMs, 738–740
    Programmable sequencer systems, 823–835
    Pulse mode approach, 773–796
    One-hot approach to design, 734–738, 835–842
    Single transition time (STT) machines, 720–730
    Stability criteria, 688
    State tables, use of, 691–692
    Timing defects, 701–719
AutoLogic VHDL, 649
Autonomous linear feedback shift register (ALFSR) counters, 594–600
    Correction for all zero state, 596–600
    Decade, 598
    Maximum length, 596–600
    Near maximum length, 596, 599
    Table of near maximum length, 599
A word of warning, 5

B

Barrel shifter, 275
Base (radix) of a number, 32
Basic (memory) cells
    Circuit symbols, 436
    Combined excitation table, 433
    EV K-maps, 429, 431
    Excitation tables, 430, 432, 433
    Logic circuits, 434, 435, 436
    Mixed-rail output response, 436
    Mixed-rail outputs, 434–435
    Operation tables, 429, 431
    Reset-dominant, 431–433, 435, 436, 694–695
    Set-dominant, 428–431, 434, 436, 692–694
    Timing diagrams (examples), 430, 432, 436, 694, 695
Basic model, 422
BCD addition, algorithm, 63
BCD multiplier, 390
BCD representation, 34–35
    Negative, 47–48
BCD subtraction
    Algorithm, 63–64
    Ten's complement, 63
BCD-to-creeping code converter, 625
BCD-to-decimal conversion
    Polynomial representation, 35
    Positional weight representation, 35
    Table, 35
BCD-to-seven-segment display converter, 261–265
Biased-weighted codes
    Excess, 127, 49, 51
    Offset, 49
    XS3, 35, 49, 68
Bi-directional counters, 466–469, 579–588
Binary adders
    Carry look-ahead, 345–349
    Carry save, 349–350
    Full, 337–338
    Half, 336–337
    Ripple-carry, 338–340
Binary addition, 52–53
    Algorithm, 52
Binary arithmetic, 52–67
    Addition, 52–53
    Algorithms, 52, 54, 56, 57–58, 59, 62, 63–64, 67
    BCD, 62–64
    Division, 58–62
    Division by direct quadratic convergence, 59–62
    Floating point, 64–67
    Multiplication, 55–58
    Subtraction, 53–55
    Two's complement, 53–54
Binary coded decimal (BCD) code, 34–35
Binary codes
    BCD, 34
    Biased-weighted, 68
    Decimal codes, 68
    Error detection, 69
    Gray, 70, 140
    One-hot, 70
    Reflective, 70
    Unit distance, 70, 71
    Unweighted, 69
    Weighted, 69
    XS3, 68, 69

Binary coded hexadecimal (BCH), 36–37
Binary coded octal (BCO), 36–37
Binary decision diagrams (BDDs), 405, 407–408, 410–411
Binary derived radices, 37
Binary digit (bit), 33
Binary division, 58–62
   Algorithms, 59, 62
   Direct quadratic convergence, 59–62
   Restoring type, 58
Binary number system, 33–34
Binary multiplication, 55–58
   Algorithms, 56, 57–58
   Two's complement, 56–58
Binary state terminology, 79–81
   Activation level indicators, 80
   Logic domain Vs the physical domain, 80–81
   Mixed logic notation, 80–81
   Negative logic, 81
   Positive logic, 81
Binary subtraction, 53–55
   Algorithm, 54
   Direct, 53
   One's complement, 54–55
   Two's complement, 53–54
Binary subtractors, 340–342
Binary-to-2's complement conversion, 45–47
   Algorithm, 46
   Negation, 47
Binary-to-BCD conversion, 260–261
   Algorithm, 261
   Converter, 292
Binary-to-decimal conversion, 32–34
   Method of positional weights, 33
   Polynomial, 32–33
   Table, 34
Bipolar junction transistors (BJTs), 850–852
Biquinary code, 70
BIST, 599
Bit, 33
Bit slice, 8
Bond set, 210
Boolean algebra, 105–116
   Absorptive laws, 108–112
   AND laws, 106
   Associative laws, 108, 111
   Commutative laws, 108, 111
   Consensus laws, 108, 112
   Corollaries, 114
   DeMorgan's laws, 110, 112
   Distributive laws, 108, 112
   Duality, 107
   EQV laws, 111
   Factoring laws, 108, 112
   OR laws, 107
   Useful identities, 115
   Worked examples, 118–120
   XOR algebra
Boolean product, 87
Boolean sum, 87
BOOZER logic minimizer, 855
Borrow-in, 340, 341
Borrow-out, 340, 341
Bounded pulse, 777
Branching action of registers and counters, 570, 589
Branching conditions (BCs), 425, 690
Branching dependency rule, 510
Branching paths, 425, 690
Buffer, 87, 94
Buffering and gating the clock, 522
Buffer (fly) state, 496, 498, 835
Built-in-self-test (BIST), 599
Bus arbiters, 801–803
Bus arbiter modules, 801–802
Bus lines, 274, 313

C

CAD, 552, 554, 838, 839, 856
Cadence, 329, 856
CAD help in programming PLDs, 328–329
   ABEL, 329
   Mentor Graphics design architecture, 329
   ORCAD's SDT, 329
   PALASM (PAL assembler), 329
   XACT (Xilinx automated CAE tools), 329
   X-BLOX, 329
   Xilinx-ABEL, 329
CAE, 329 (*see also* computer aided engineering design)
Call module, 770–771
Candy bar vending machine design
   Pulse mode approach, 782–788
Canonical forms
   Produce-of-sums, 135, 136
   Sum-of-products, 132, 133
Canonical truth tables, 133, 136
Capacitors
   In debouncing circuits, 526–527
   In inertial delay elements, 794–795
   In sanity circuits, 523–524
Cardinality of a function, 151–152
Carry generate/propagate (CGP) networks, 346
Carry-in, 337, 338, 339
Carry look-ahead (CLA) adders, 345–349
   Carry generate/propagate network, 346
   CLA module, 346
   Logic circuits, 346–348
Carry-out, 337, 338, 339
Carry overflow, 53, 54
Carry propagate, 345

INDEX

867

Carry-save (CS) adders, 349–350
    Algorithm, 349
    Logic circuit, 350
Cascadable binary counters, 575–588
Cedes, 856
Cell coordinates, 138, 140, 143
Character codes, 70–72
    ASCII, 71
    EBCDIC, 72
Chips classification, 238
Clear asynchronous overrides
    In flip-flops, 463–464
Clock buffering and gating, 522
Clock frequency, 437, 521
Clock generating circuitry, 520–521
Clock logic waveforms, 437
Clock Period, 437, 521
Clock signal specifications, 521–522
    Factor of safety, 521–522
Clock skew, 517–520, 685, 822
    Examples, 518, 519
Clock sources
    Clock oscillator circuits, 520–521
    Duty cycle, 521
C-module, 696 (*see also* rendezvous module)
CMOS, definition, 82
CMOS domino logic, 814–816
    DFLOP design, 812–814
    Gates, 815
    Generalized configuration, 815
CMOS gate configurations, generalized, 82–83
CMOS terminology and symbology, 82–83
    Ideal equivalent circuits, 82
    NMOS, 82
    PMOS, 82
Code conversion between number systems, 37–43
    Fractions, 40–43
    Integers, 38–40
Code converters, 257–265
    Algorithms for binary/BCD conversion, 260, 261
    BCD-to-binary, 261–263
    BCD-to-creeping code, 292
    BCD-to-seven-segment display, 261–265
    BCD-to-XS3, 258–260
    Binary-to-BCD, 260–261, 292
    Binary-to-2's complement, 291–292, 643–645
    Gray-BCD, 291
    Gray-to-binary, 257–258
    Procedure, 257
Codes. *See* binary codes
Combinational logic devices, non-arithmetic
    Building blocks, 237–238
    Classification of chips, 238
    Code converters, 257–265
    Decoders/Demultiplexers, 248–254
    Design procedure, 241–242

Encoders, 254–256
Magnitude comparators, 265–272
Multiplexers (MUXs), 242–248
Parity generators and detectors, 273–275
Part numbering systems, 241
Performance characteristics, 238–241
Shifters, 275–278
Steering logic, 278–279
VHDL description, 279–287
Combinational shifters, 275–278
Common anode LED configuration, 263, 265
Common cathode LED configuration, 263, 265
Commutative laws
    AND/OR forms, 108
    EQV/XOR forms, 111
Comparators, 265–272
Complementary MOSFET (CMOS) switching
    circuits, 82–83
Complementation, 95
Composite output maps, 629, 635
Compressed entered variable (EV) truth table, 242, 244
Computer aided engineering (CAE) design
    Logic minimization tools, 329, 855
    Schematic capture, simulation and timing
        analysis tools, 329, 855–857
Conditional branching, 492, 496
Conditional outputs, 424, 425
Conditional output rule, 510–511
Conjoint terms, 114, 209
Conjugate mixed logic gate symbols
    AND, 92, 93, 94
    Buffer, 94
    EQV, 101, 103
    Inverter, 84, 94
    NAND, 88, 89, 94
    NOR, 90, 91, 94
    OR, 93, 94
    Summary, 94, 103
    Tri-state driver, 86
    XOR, 100, 103
Connectives
    AND, 87
    EQV, 98
    OR, 87
    XOR, 98
Consensus laws
    AND/OR, 108
    EQV/XOR, 112
Construction of mixed-logic circuits, 97–98
Contracted Reed-Muller transformation (CRMT)
    minimization, 209–229
    Heuristics, 217–218
    Incompletely specified functions, 218–228
    Multiple output functions with don't cares,
        222–228

Contracted Reed-Muller transformation (*cont.*)
  Perspective, 229
  Subfunction partitioning, 225–228
Controlled inverters. *See* controlled logic level conversion
Controlled inversion, 103
Controlled logic level conversion, 103–104
  Adder/subtractor designs, 342–343
  ALU designs, 374–375
  BCD adder/subtractor designs, 387
  Binary counter designs, 586–587
  Mixed logic interpretation, 103–104
  Overflow error detection circuits, 344
  Positive logic interpretation, 104
  XS3 adder/subtractor designs, 387–388
Controlled system, 349
  Data path unit (DPU), 650
Controller
  In system-level design, 649–650
Conventional K-maps, 137–158, 167
Conversion between flip-flops, 450–461 (*see also* flip-flop conversion)
Conversion between number systems
  Algorithms, 38, 39, 41
  Fractions, 40–43
  Integers, 38–40
  Rounding off, 42–43
  Summaries, 39, 41
Conversion of fractions, 40–43
  Algorithms, 41, 42
  Rounding off, 42
  Table, 41
Conversion of integers, 38–40
  Algorithms, 38, 39
  Table, 39
Corollaries in XOR algebra, 114, 204
Counteracting delay
  Elimination of essential hazards, 712, 715
Counters, 572–605
  Asynchronously parallel loadable, 579–581, 587–589, 664–665
  Bi-directional, 579–588, 664–665
  Binary, 572–605
  Binary up/down. *See* bi-directional
  Branching action of a parallel loadable up/down, 587–590
  Cascadable BCD, 575–579
Cascadable binary, 579–588, 664–665
  Cascadable up/down, 579–584
  Data triggered, 664–665
  Johnson (twisted ring), 593–594
  Linear feedback shift register (LFSR), 594–600
  Multimode, 607–608
  One-bit modular design, 584–588
  Parallel loading, a perspective, 588–589
  Ripple (asynchronous), 600–605
  Ring, 590–593
  Shift register, 590–600
  Synchronous parallel loadable, 581–587, 588–589
  True hold capability, 581, 582, 584, 589
  Twisted ring (Johnson), 593–594
Coupled term, definition, 392
Coupled variable, definition, 392
Cover, definition, 10
  Minimum POS, examples, 146–148
  Minimum SOP, examples, 146–148
Creeping code, 69
Critical races, 703–705
CRMT coefficients, 210–212
CRMT forms, 210–216
CRMT minimization, heuristics for, 217–218
Cross branching, definition, 10
  Relative to STT designs, 740
Cube representation, 173
Cycles (In asynchronous FSMs), 702

D

Data path (In system-level design), 649, 650
Data bus, 274
Data lockout character of flip-flops, 445
Data selector, 242 (*see also* multiplexer)
Data-triggered counters, 664–665
Data-triggered memory elements, 773–775
  Toggle modules, 573, 774–775
Debouncing circuits, 526–530
Decade counters, 575–579, 598, 603–604
Decimal codes, 68
  BCD, 34–35, 69
  Table, 69
  XS3, 36, 49, 69
  Unweighted, 68–69
  Weighted, 68–69
Decimal-to-BCD conversion. *See* BCD-to-decimal
Decoders/demultiplexers, 248–254
  Design, 248–251
  Mixed logic inputs, 252
  Stacked configurations, 251
  Steering logic implementation, 279–280
  Use in Combinational logic design, 251–253
Decomposition (Shannon's expansion theorem), 177–180
D flip-flops, 440–450
  Analysis of FSMs with D flip-flops, 476–480
  Conversion from JK flip-flops, 456–458
  Conversion to SR flip-flops, 485
  Conversion to T flip-flops, 452–453
  Data lockout, 461
  Edge triggered, 444–448, 698–700
  Excitation table, 441, 458
  Logic circuit symbols, 443, 444, 448, 449
  Master/slave, 448–450

INDEX **869**

Operation table, 441
State diagrams, 441, 442, 444, 445–446, 449, 698
Timing diagrams, 443, 448, 449
Use as a filter for logic noise, 497–499
Use as a synchronizer, 511–517
VHDL behavioral description, 480
Delay elements, 794–795
DeMorgan relations, 95
DeMorgan's laws
    AND/OR forms, 110–111
    EQV/XOR forms, 112
Demultiplexers/decoders, 248–254 (*see also* Decoders/demultiplexers)
Depletion mode NMOS, 378, 807, 849–850
Design area Vs performance, 180–181
Design of synchronous FSMs with edge triggered flip-flops
    Design procedure, 530–532
    One-to-three pulse generator, 532–536, 615–622
    More complex FSM, 622–626
    Sequence recognizer, 471–476
    Three-bit binary up/down counter, 466–471
Design procedure (general)
    Combinational logic, 241–242
    Finite state machines (FSMs), 530–532
Destination matrix, 542–543, 724
Destination and origin states (in ORG analysis), 492–493
DFLOPs
    Dynamic logic design, 812–814
    State diagrams, 808, 813
    Static logic design, 807–810
    Use of Metastable detection stages, 807–808, 810–812
    Use in EAIC systems, 806–807, 816–820
Diagonal XOR patterns, 198–199, 201–203
Digital combination lock (DCL), 781–782, 797
Digital machines, an overview, 684
Diminished radix complement representation, 48
Diodes
    In inertial delay elements, 794–795
    In Sanity circuits, 523–524
Diode-transistor logic (DTL), 853
Distributed path delay model, 688
Distributive laws
    AND/OR forms, 108
    EQV/XOR forms, 112
Divide-by-N binary counters
    Divide-by-2, 573
    Divide-by-3, 573–574
    Divide-by-4, 574
    Perspective on divide-by-N, 574–575
Dividers
    Parallel (fast), 353–357
Division (binary)
    Algorithms, 59, 62

    By direct quadratic convergence, 59–62
    Restoring, 58–59
D-latch
    Design, 441–444, 705–707
    Logic circuit, 443, 444
    Logic circuit symbol, 443, 444
    Next state functions, 443, 706
    State diagram, 443, 706
    Timing diagrams, 443, 706
    Transparency character, 443–444
DL-DFLOPs
    Design, 812–814
    State diagrams, 813
    Use in EAIC systems, 819–820
DMUX (demultiplexer/decoder), 248–254
Domain boundary, 814–816
Domino logic
    CMOS, 815
    Gate examples, 815
    Generalized gate configuration, 815
    Precharge and evaluate stages, 814–815
    Use of in DFLOP design, 812–814
Don't cares, 150–158 (*see also* incompletely specified functions)
    As entered variables, 164
    As nonessential minterms or maxterms, 150, 151
    In canonical forms, 150, 151
    Rules in multiple-output minimization, 153
DPU (data path unit), 650
Drivers
    Buffers, 87, 94
    Tri-state, 85–87
D-trios (*see also* essential hazards)
    Analysis and elimination, 716–718
    Requirements for formation, 711–714
Dual-emitter BJT, 851
Dual-rail systems
    ALUs with completion signals, 369–380
Duality, definition of, 107
    Dual forms of Boolean laws, 107–116
Duty cycle, 521
    In counters, 574, 578
Dyad groups of logic adjacencies, 145
Dynamic hazards
    In multilevel XOR-type functions, 409–411
    Use of binary decision diagrams (BDDs), 411
    Use of lumped path delay diagrams (LPDDs), 410
Dynamic power dissipation, 239–240

E
EAIC systems, 806–823 (*see also* externally asynchronous/internally clocked systems)
Edge-triggered flip-flops
    Conversion between, 450–459
    D, 444–448

Edge-triggered flip-flops (*cont.*)
    JK, 454–456
    T, 452–453
    Used in the design of FSMs, 466–476, 530–536, 562–605, 617–666
    Unusual types, 459–460, 461
EEPROMS, 298
E-hazards, 711–719 (*see also* Essential hazards)
Electronically erasable PROMs (EEPROMs), 298–299
    Floating gate NMOS transistors, 299
Emitter-coupled logic (ECL), 852
Encoders
    Priority, 254–256, 291
    Stacked, 256
Endless cycles, 702–703
Enhancement mode NMOS logic, 849
Entered variable (EV) K-map minimization, 158–169, 198–207
    Don't cares as EVs, 164
    Five or more variables, 165–169
    Map compressions, 158–169
    Map Key, 160
    Subfunction rules, 164, 165
    Use of submaps, 159, 163, 164, 182, 184, 187
    Worked examples, 181–188
    XOR patterns, 198–207
Entered variables (EVs)
    In K-maps, 158–169, 198–207
    In truth tables, 183, 244, 246, 268, 269, 358, 361, 372, 374
    In XOR-type patterns, 198–207
Epilogue to Chapter 16, 842
EPI, 149
EPROMs, 298–299
EQPOS functions, 208–209
EQPOS/POS functions, 225, 228
Equivalence. *See* EQV
EQV function
    Defining relations, 101
    Definition, 99
    Logic circuit symbols, 101, 103
    Multiple gate realizations, 101–102
    Operator symbol, 98
EQV gate
    CMOS, 101
    Conjugate logic circuit symbols, 101, 103
    Effect of active low inputs, 102
    In controlled logic level conversion, 103–104
    Mixed-logic interpretations, 101
    Physical truth table, 101
    Tree forms for multiple inputs, 99
EQV laws, 111
Erasable programmable read-only memory (EPROM), 298–299
Error catching in MS JK flip-flops, 462–463

Error checking systems, 274–275
Error checking circuits
    Parity circuits, 273–274
Error detection codes, 69–70
    Even and odd parity, table, 70
Espresso, 173, 855
Espresso algorithm (reference), 855
    Qualitative description, 173–174
Essential hazards in fundamental mode FSMs, 711–718
    Analysis examples, 714–718, 743–758
    Counteracting delays, 712, 715
    D-trios, 712–713, 714, 716–717
    General requirements for formation, 711
    Indirect path requirements, 714
    In LPD FSMs, 743–747
    In nested cell FSMs, 747–752
    In one-hot FSMs, 752–758
    Minimum requirements for formation, 712–713
    Perspective, 718–719
    Timing diagrams, 718
Essential prime implicants (EPIs), 149–150
EV K-maps, 158–169, 198–207
    Worked examples, 181–188
Exact minimum expression, 198
Excess representations, 49
Excitation table for the LPD model, 688–689
    Comparison with the D flip-flop, 689
Excitation tables for basic (memory) cells
    Combined form, 433
    Reset-dominant, 432, 433
    Set-dominant, 430, 433
Excitation tables for flip-flops and latches
    D, 441
    JK, 454
    SR, 433
    Summary of, 457–458
    T, 452
EXL-Sim2000, 855–856
EXSOP functions, 207–208
EXSOP/SOP, 226–227
Externally asynchronous/internally clocked (EAIC) systems
    DFLOP conversion, 821
    Domino logic DFLOP design, 812–814
    EAIC system architecture, 806
    Features, 805–807, 816–817
    Memory elements, 806–814
    Metastable detection stage, 810–812
    Models, 806, 808
    MTBF, 805, 821
    Pausable systems, 805
    Perspective, 822–823
    Real time tests, 819–820
    Simple sequence recognizer example, 817–820

INDEX   **871**

System simulations, 819–820
Static logic DFLOP design, 807–812
Timing constraints, 816–817
Variations on the theme, 820–821
Vs conventional synchronous FSMs, 821–823

F
Factoring law
   AND/OR, 108
   EQV/XOR, 112
Factorization, 175–176
Factor of safety for clock signals, 521–522
Falling edge-triggering (FET)
   Defined, 437–438
   In flip-flops, 432
False data rejection (FDR)
   In ALU design, 358–359, 366
   In code converter design, 257, 259–260
Fan-in, 240–241
Fan-out, 240–241
FDR. *See* false data rejection
Feedback delays
   Counteraction E-hazard formation, 712, 715
   Nested cell designs of pulse mode FSMs, 776–777, 793–794
Feedback paths
   In models for FSMs, 423–424, 439, 686, 774
   In PLDs, 309–310
FET. *See* falling edge triggering
Field programmable gate arrays (FPGAs), 319–329
   Actel, 319–321
   Configurable logic blocks (CLBs), 321–327
   I/O blocks (IOBs), 321–326
   Logic cell arrays (LCAs), 328
   Xilinx, 321–328
Field programmable logic arrays, 301–306 (*see also* Programmable logic arrays)
Fill bit, in combinational shifters, 275–278
Finite state machines (FSMs), 421
Fixed-point numbers, 32
Flip-flop conversions, 450–460
   D-to-JK, 454–456
   D-to-T, 452–453
   D-to-unusual flip-flops, 459–460
   JK-to-D, 456–458
   JK-to-T, 455–457
   Model for conversion from D, 451
Flip-flop design, general, 438–440
   Mapping algorithm, 440
   Models, 439, 451
   Procedure, 440
Flip-flops (FFs)
   D, 440–450, 698–700
   Data lockout, MS, 461
   Edge triggered, 437–438
   Hierarchical flow chart, 439

JK, 454–456, 700–701
Master slave, 448–450, 462–463
Models, 439, 451
SR, 485
T, 451–453
Unusual, 459–460
Floating gate NMOS transistors, 298–299
Floating point addition, 64–65
Floating point arithmetic, 64–67
Floating point division
   Algorithm, 67
   Quadratic convergence, 66–67
Floating point multiplication
   Algorithm, 67
   Signed-magnitude, 65–66
Floating point number (FPN) systems, 49–52
   IEEE standard, 50–51
   Normalized, 50
Floating point subtraction
   Two's complement, 65
Flow charts, 533–534
FPGAs
   Actel, 319–321
   Xilinx, 321–328
FPLAs, 302 (*see also* Programmable logic arrays)
Fraction conversion, 40–43
   Algorithms, 41, 42
   Rounding off and error bounds, 41–43
   Summary of methods, 41
Free set
   In CRMT minimization method, 210
Frequency division
   In binary counters, 572–575
   In ripple counters, 600–601
Frequency synthesizers, 521
"From rule", 540
FSM. *See* finite state machine
Full adders (FAs)
   Design of, 337–338
   In adder/subtractor design, 342–345
   In carry-save adder design, 349–350
   In multiplier design, 350–353
   In parallel-to-serial adder/subtractor design, 651–652
   In ripple-carry adder design, 338–340
Full subtractors (FS)
   Design, 340–342
   Use of in parallel divider design, 354–355
Fully documented state (FDS) diagrams, 425
   Features, 424–425, 689–690
   Mutually exclusion requirement, 46–428, 490, 686–690
   Sum rule, 426, 689–690
Functional partition
   In system-level design, 650, 652, 658
Function generators, 245

Function hazards
  Combinational, 412
  Internally initiated in FSMs, 491 (*see also* output race glitches)
Function matrix, 543–546, 724–727
Function minimization
  Cube notation, 173
  CRMT method, 210–229
  Decomposition (Shannon's expansion theorem), 177–180
  Factorization, 175–176
  K-map, 144–169
  Perspective on, 181
  Reed-Muller transformation, 207–209
  Re-substitution, 176–177
  Tabular (Quine-McCluskey algorithm), 169–172
  Worked EV K-map examples, 181–188
  XOR-type patterns, 198–204
Fundamental Mode FSMs
  LPD model, 685–687
  Nested cell model, 687, 696, 730, 776
  Requirements for operation, 686
  Stability criteria, 688
Fuse map, 329
Fusible links
  In FPLAs, 302–303
  In PROMs, 297
  On transistors, 297–298
  On diodes (bipolar form), 297, 298–299

G

Gain element, 100
Gates (CMOS) and symbols
  AND, 92, 93, 94
  AND-OR-invert (AOI), 317–319
  EQV, 100–101, 103
  CMOS configuration, generalized, 83
  IEEE standard symbols, 859–860
  Inverter, 83–84, 94
  NAND, 88–89, 94
  NOR, 89–90, 91, 94
  OR, 93, 94
  OR-AND-invert (OAI), 317–319
  XOR, 100, 103
Gate/input tally, 151–152
  Minimum, 198
  Vs cardinality, 151–152
Gated basic cell, 483–485
Gate-minimum cover, 198
Gate propagation time delay, defined, 239
General-purpose PLDs
  Erasable programmable logic devices (EPLDs), 328
  Field programmable gate arrays (FPGAs), 317–328
  Field programmable logic sequencers (FPLSs), 328
  Generic array logic (GAL) devices, 328
  Programmable array logic (PAL) devices, 307–310
  Programmable logic arrays (PLAs), 301–306
  Read-only memories (ROMs), 295–301
Glitches, types
  Negative, 391, 492
  Output race glitches (ORGs), 491–492
  Positive, 391, 492
  Static hazards, 391–392
Glossary of terms, expressions and abbreviations, 5–29
GO/NO-GO configuration, 647
Gray code, 140

H

Half adder (HA), 336–337
Half-adder/half-subtractor counter design, 584–588
Half subtractor, 341–342
  Use of in a 1-bit modular counter design, 585
Hamming distance
  In state code assignments of fundamental mode FSMs, 739–740
Handshake interface
  In system-level design, 650
Hardware description languages (HDLs)
  Verilog, 856
  VHDL, 279, 288, 380, 480, 856
Hazard cover, 392
  Effect on stuck-at faults, 412–413
  Static 1-hazards (SOP hazards), 393–394, 397–398, 401–402, 501–502, 505–509
  Static 0-hazards (POS hazards), 394–396, 398, 402–403, 499–501, 506
Hazard-free design of asynchronous FSMs, 730–734
Hazards
  Dynamic, 392, 409–411
  Essential, 711–718
  Function, 412, 491
  Static 0 (POS hazard), 391
  Static 1 (SOP hazard), 391
HDLs, 279
Hexadecimal addition, 75
  Table, 77
Hexadecimal multiplication, 75
  Table, 77
Hexadecimal number system, 36–37
  Fraction conversion to/from radix r, 40
  Integer conversion to/from radix r, 38
Holding condition, 425
Holding (storage) register
  Applications, 499, 614–615, 625, 630–631, 636–637, 664
  Design, 562–563

INDEX 873

Hold time, 495
Hybrid forms
    AND/OR, 175–180
    XOR/SOP/EQV/POS, 225–228

I
IEEE standard graphic symbols for logic functions, 859–862
    Combinational logic devices, 859–860
    Flip-flops, registers and counters, 860–862
    Gates, 860
Inactive state, 80
Inactive transition point, 465
Incompatibility and complementation, 95–96
Incompatibility slash, 95
Incompletely specified functions, 150–152 (*see also* don't cares)
    Rules for use in EV K-maps, 164
    Rules in multiple output minimization, 153
    Use in canonical forms, 150, 151
Inertial delay elements, 794–795
Initialization and reset of the FSM
    Asynchronous FSMs, 719–720
    Sanity circuits, 523–526
    Synchronous FSMs, 523
Initiator input in E-hazard analysis, 711
In-phase triggering
    In EAIC systems, 821
    In filtering out logic noise, 499
Input matrix, 544, 725
Input/state map, 425, 426
Internally pausable clocked systems, 806–823 (*see also* externally asynchronous/internally clocked systems)
Intersection, 87 (*see also* Boolean product)
"Into rule", 540
Introductory remarks and glossary, 1–29
    Automatic control systems, 2
    Communications, 2
    Computing, 1
    Entertainment, 2
    Glossary, 5–29
    Information retrieval, 1–2
    Instrumentation, 2–3
    What is so special about digital systems?, 1–3
    Word of warning, 5
    Year 2000 and beyond?, 3–4
Invariant state variable in E-hazard analysis, 711
Inverters
    CMOS, 83–84
    Circuits, 84, 850, 851
    Conjugate logic circuit symbols, 84, 94
    Mixed logic interpretations, 84
    NMOS, 850
    Physical truth table, 84
    TTL, 851

Involution, 106
Irredundant cover, 173–174
Irrelevant input, 249

J
JEDEC, 17, 329
JK flip-flops
    Analysis of FSMs with JK flip-flops, 476–479
    Conversion from edge triggered D flip-flops, 454–455, 700–701
    Conversion to D flip-flops, 456–458
    Conversion to T flip-flops, 456–457
    Design of FSMs with JK flip-flops, 471–475, 562–564
    Excitation table, 454, 458
    Master-slave, 462–463
    Operation table, 454
    PR and CL overrides, 701
    State diagram, 454
    Timing diagram for edge triggered, 456
Jump state, 603

K
Karnaugh maps (K-maps)
    Domain boundaries, 145
    Entered variable (EV), 158–169
    First-order, 138
    Forbidden groups of minterms or maxterms, 145–146
    Fourth-order, 143
    Loop-out protocol, 145
    Map key, 160
    POS extraction procedure, 145
    Reduction rule, 145
    Second-order, 138
    Third-order, 140
K-map conversion
    Algorithms, 471, 474, 576
    D-to-JK, 473–474
    D-to-T, 470–471
    JK-to-D, 477, 478
    JK-to-T, 576, 577
K-map minimization
    Conventional (1's and 0's), 138–158
    Entered variable (EV), 158–169
    XOR patterns, 198–207
K-maps. *See* Karnaugh maps
K-map subfunction partitioning, 225–228
Keywords in VHDL, 281

L
Large-scale integrated circuits (LSI), 238
Latch
    D, 441–444, 464, 705–707
    JK, 461–462

Latch (*cont.*)
    SR, 460–461, 483–484
    T, 461–462
Laws of Boolean algebra, 105–116 (*see also* Boolean algebra)
LED
    In seven-segment display designs, 265
Least significant bit (LSB), 33
Linear Feedback shift register (LFSR) counters, 594–600
Line drivers
    Buffers, 87
    Tri-state, 84–87
Linear state machine, 627, 632
Logic adjacency
    In cube notation, 173
    In Espresso algorithm, 173
    In K-maps, 145
    In Quine-McCluskey algorithm, 170
    Requirement for in the MAC module, 825–826
Logic cell
    Configurable logic block (CLB), 321
Logic circuit symbols
    Summary of conjugate mixed logic symbols, 94, 103
Logic compatibility, 96
Logic domain, 80
Logic function graphics. *See* Karnaugh maps
Logic instability
    By E-hazard formation, 719
    By s-hazard formation, 705–706, 710–711
    Due to endless cycles in asynchronous FSMs, 702–703
    Due to metastability, 514–515
    In basic cells, 694, 695
Logic level conversion, 83–84
    Controlled inverter, 103–104
    Inverter, 83
    Logic circuit symbols, 84
    NAND gate, 90–92
    NOR gate, 90–92
Logic level incompatibility, 95–96
    Complementation, 95
    Examples, 96
    Incompatibility indicator slash, 95, 96
Logic minimization tools
    BOOZER, 855, 857
    Espresso, 173–174, 329, 855, 857
Logic noise
    Filtering, 497–499
    Output race glitches (ORGs), 491–499
    Static hazards, 499–510, 705–711
Logic simulators. *See* simulators, logic
Logic state, definition, 421
Logic waveforms, 105 (*see also* timing diagrams)

Look-ahead-carry (LAC) adder
    Same as carry look-ahead adder, 345–349
Loop-out protocol, 145
LPD model. *See* Lumped path delay model
LPD-to-SR conversion, 730–732
Lumped path delay (LPD) model, 685–687
    Excitation table, 688–689
    Functional relationships, 687–688
    Stability criteria, 688

M

Magnitude comparators, 265–267
    Cascadable, 265–272
    Non-cascadable, 388
Majority functions, 115 [SOP form of Eq. (3.33)], 293
Majority gate, 696
Map key
    Use in EV K-map minimization, 160
Mapping algorithm for FSM design, 440
Master/Slave D flip-flop
    Circuit symbol, 449
    CMOS implementation, 450
    Conversion to MS JK flip-flops, 463
    Logic circuit, 449
    State diagrams, 449
    Timing diagram, 449
Maxterm, 134
Maxterm code
    Defined, 134
    Table, 135
Mealy machine, 422
Mealy output, 424, 426
Mealy's (general) model
    For fundamental mode (LPD) FSMs, 686
    For nested cell designs of pulse mode FSMs, 776
    For pulse mode FSMs with toggle modules, 774
    For synchronous FSMs, 424
Mean time between failures (MTBF), 516–517, 805, 821
Medium-scale integrated (MSI) circuits, 238
Memory cells, 428–436
    Set-dominant basic cell, 428–431
    Reset-dominant basic cell, 431–433
Memory elements
    In EAIC system design, 806
    In fundamental mode (LPD) FSM design, 686
    In nested cell designs, 687
    In pulse mode FSM design, 774
    In synchronous FSM design, 438
Mentor Graphics, 329, 856
Merging of states
    State minimization, 547–549
Metastability
    And the synchronizer, 514–517
    Mean time between failures (MTBF), 516–517

INDEX   875

Metastable exit time, 514
Practical solutions to the synchronizer problem, 515–517
Metastable detection stage in EAIC systems, 808
Domino logic design, 812–813
Simulation, 810–811
Static logic design, 810–812
Metastable exit time, 514
Microprogrammable asynchronous controller (MAC) modules
Application to a Gray code counter design, 830–832
Architecture, generalized, 823–825
Cascading, 832–833
Components of an n-input, 824–825, 829–830
Design of a 3-input MAC module, 827–829
Features, 823–824, 834–835
Metastability considerations, 834
Perspective, 834–835
Programming, 833–834
Simulation results, 831–832
State array machine (SAM), 824–827, 829–830
Timing control machine (TCM), 824–827
Minimization algorithms
Espresso, 173–174
Quine-McCluskey, 169–172
Minimization, degrees of, 198
Minimization, logic function
Contracted Reed-Muller transformation (CRMT), 209–229
CRMT, 210–218
EV K-map, 158–169, 198–207
Decomposition, 177–180
Factorization, 175–176
Multiple output, 152–158, 222–229
Reed-Muller transformation, 207–209
Tabular (Quine-McCluskey), 169–172
XOR pattern, 198–204
Minterm, defined, 132
Minterm code
Defined, 132
Table, 133
Missing state analysis, 475–476
Mixed-logic inputs and outputs
ROMs, PLAs and PALs, 310–311
Mixed logic notation, 81
Mixed-mode design entry, 329
Mixed-rail outputs
Basic cells, 434–435
Combinational logic circuits, 105
Flip-flops, 451
Mobius counter, 573
Models for sequential machines
Asynchronous FSMs, 686, 774, 776
Basic model, 422, 439
EAIC systems, 806

Mealy's model, 424, 686, 774
Moore's model, 423
Pulse mode FSMs, 773–774, 776
Synchronous FSMs, 421–424
ModelSim, 856
Modular and bit slice devices
Registers, 561–572
Counters, 572–605
Modular approach to design, 561, 562
Modulo-N counters, 572
Moore machine, 422
Moore's model, 423
Moore output, 422
MOS
CMOS, 82–83, 814–816, 849–851, 852
NMOS, 82, 849–851, 852
PMOS, 82, 849
MOSFET, 82
MTBF, 516–517, 805, 821
Muller C module, 696 (see also Rendezvous module)
Multilevel logic minimization forms
Due to CRMT methods, 210–229
Due to factorization, resubstitution, or decomposition, 174–179
Due to Reed-Muller transformations, 207–209
Due to use of XOR K-map patterns, 198–207
Multiple number addition
Carry-save adder, 349–350
Multiple output functions, 152–158, 222–227
Multiple output minimization
CRMT approach, 222–225
Examples, two-level, 154–158
Maxterm ORing rules, 153
Minterm ANDing rules 153
Multiple PLD schemes, 312–316
Input augmentation, 312–315
Output augmentation, 313–316
Partitioned program tables, 315
Use of tri-state enables, 312
Multiple pulse generator system, 679–680
Multiple stage synchronizers, 515–517
Multiplexers (MUXs), 242–248
As function generators, 245
Design, 242–245
Mixed logic inputs, 247
Steering logic implementation, 278–279
Use in combinational logic design, 245–248, 363–365
Use in FSM design, 563–564, 567, 587–588
Multiplicand and multiplier, 55–57, 351
Multiplication
Algorithms, 56, 57–58, 67
Binary, 55–57
Floating point number (FPN), 65–66
Two's complement, 56–58

Multipliers
  BCD, 389–390
  Binary, 350–353
  Four-by-four bit, 350–353
  Iterative carry-save with CLA, 352–353
  XS3, 390
Mutual exclusion elements
  Bus arbiters, 801–802
Mutually conjoint terms, 114, 207
Mutually disjoint terms, 114, 207
Mutually exclusive requirement, 426–428, 490
  Defined, 427
  Exceptions, 428
MUX approach
  ALU design, 363–365

N

NAND gate
  CMOS, 88
  Conjugate gate symbols, 88, 94
  Logic level converter, 90–91
  Mixed logic interpretations, 88, 89
  Multiple inputs, 89
  NMOS, 850
  Physical truth table, 88
  TTL, 851
Natural binary, 33
Nested cell designs
  Conversion from LPD designs, 730
  Flip-flops, 444–448
  Hazard-free design of fundamental mode FSMs, 730–734
  Latches, 441–444, 460–461
  Pulse mode FSMs, 776–777, 791–794
  Rendezvous modules (RMODs), 695–698
  STT FSMs, 730–734
  Vs the LPD approach for STT FSM design, 734
Nested cell model, 439, 776
Nested inverse radix, 40
Nested radix form, 38
Next state, 421, 422
Next state function matrix, 543, 724
Next state (NS) function, 423
Next state table and the state assignment rules, 539–542
NMOS
  Ideal equivalent circuits, 82
  Simplified circuit symbol, 82
NMOS logic family, 849–851
  Gate examples, 850
  Generalized configuration, 850
  NMOS AOI gate Vs CMOS AOI gate, 851
Noise immunity, 525
Noise margins, 81, 240
Non-restoring logic
  Steering logic, 278–279
Non-arithmetic combinational logic, 237–279
  Building blocks, 237–238
  Code converters, 257–265
  Combinational shifters, 275–278
  Decoders/demultiplexers, 248–253
  Design procedure, 241–242
  Encoders, 254–256
  Multiplexers, 242–248
  Magnitude comparators, 265–272
  Part numbering systems, 241
  Parity generators and error checking systems, 273–275
  Steering logic and tri-state applications, 278–279
  VHDL description, 279–287
Nonessential minterms and maxterms, 150–151 (*see also* don't cares)
Nonoverlapping sequences, 472
NOR gates
  CMOS, 90, 91
  Configurations that eliminate fan-in problems, 377–378, 849
  Conjugate gate symbols, 90, 94
  In EAIC system design, 807
  In one-hot programmable sequencer design, 835–837
  Logic level converter, 90–91
  Mixed logic interpretations, 90, 91
  Multiple inputs, 94, 377–378, 807, 849
  NMOS, 850
  Physical truth table, 90
  TTL, 851
NOT function, 106
Number systems
  BCD, 34–36
  Biased weighted representation, 35
  Binary, 33–34
  Binary coded hexadecimal (BCH), 37
  Binary coded octal (BCO), 37
  Conversion of fractions, 40–43
  Conversion of integers, 38–40
  Diminished radix complement, 48
  Excess (offset) representation, 49
  Fixed-point, 32
  Floating point, 49–52
  Important characteristics, 31
  Positional and polynomial representations, 32
  Radix complement, 45–48
  Signed binary, 43–48
  Signed magnitude, 44–45
  Ten's complement, 45–46
  Two's complement, 46–47
  Unsigned binary, 33–43
  XS3, 35–36

O

Octal system, 36–37
Odd parity BCD code, 70
Offset patterns, 198–203

INDEX   877

One-bit modular counter design, 584–588
One-hot code, 70
One-hot design and analysis of asynchronous FSMs
    Analysis, 752–758
    Guidelines for design, 735
One-hot-plus-zero approach, 735–738
    Perspective, 738–740
    Programmable sequencers, 835–840
One-hot design of FSMs
    Asynchronous, 734–740, 835–840
    Synchronous, 636–649, 841
One-hot design of synchronous FSMs
    One-hot-plus-zero initialization, 639–640
    Parallel-to-serial adder/subtractor controller, 645–648
    Perspective, logic noise and use of PLDs, 647–649
    Serial 2's complementer, 643–645
    Use of ASM charts Vs state diagrams, 640–644
One-hot programmable asynchronous sequencers, 835–840
    Application to a 4-state FSM, 838–840
    Architecture, generalized, 835–837
    Design of a four-state sequencer, 837–838
    NS equations, generalized, 387–838
    Perspective on programmable sequencer design and applications, 839–842
    Simulation results, 839–840
One-hot programmable synchronous sequencers, 841
One's complement subtraction, 54–55
One-to-three pulse generator designs, 532–536, 615–622
Operation tables
    Flip-flops, 441, 452, 454, 459
    For counters, 582, 584, 588
    For shift registers, 564, 566, 569
OR
    Definition, 87
    Logic circuit symbols, 88, 93, 94
    Operator symbols, 87
OR-AND-Invert (OAI) gate
    CMOS, 319
    Logic equivalent circuit, 319
    Truth tables, 319
OrCad, 839, 856
ORGs. See Output race glitches
OR gate
    CMOS, 93
    Conjugate gate symbols, 93, 94
    Domino logic configuration, 815
    Logic interpretations, 93
    Multiple inputs, 94
    Physical truth table, 93
Origin and destination states, 493–494, 496
Output discontinuity, 493
Output forming logic, 423, 424, 686, 774

Output holding register
    Filtering of logic noise, 625, 630–631, 636–637
Output K-maps
    In asynchronous FSM design, 690–691
    In static hazard analysis, 500–502, 621–622
    In synchronous FSM design, 472–473, 535
Output race glitches (ORGs)
    Analysis procedure, 496
    As an internally initiated function hazard, 491
    Elimination, 496–499
    Examples, 492–495
    In asynchronous FSMs, 705, 721, 727, 739–740
    In synchronous FSMs, 491–499
Overflow error detection circuits, 343–345
Overlapping sequence, 472

P

P-term tables
    In programming PLAs and PALs, 304–305, 317
Packing density, 239
PAL, 307–310 (*see also* Programmable array logic devices)
PALUs, 363–380 (*see also* Programmable arithmetic and logic units)
Parallel accumulator, 607
Parallel adders, 338–340
Parallel dividers, 353–357
    Subtractor modules, 354–356
Parallel loadable up/down counters
    Asynchronous parallel loading, 579–581, 587–588, 664–665
    Branching action, 589
    Cascadable, 575–588
    Data triggered, 664–665
    One-bit modular design, 584–588
    Operation tables, 582, 584, 588
    Perspective on parallel loading, 588–589
    State diagrams, 579, 582, 584
    Synchronous parallel loading, 581–584, 584–587
    With true hold, 581–588
Parallel loadable shift registers
    Operation tables, 564, 566, 569
    Right shift register with synchronous parallel loading, 562–565
    State diagrams, 564, 566, 569
    Timing diagrams, 565
    Universal shift register with asynchronous parallel loading, 568–570
    Universal shift register with synchronous parallel loading, 565–568
Parallel loading of counters and registers, perspective, 588–589
Parallel-to-serial adder/subtractor controller
    Conventional design with JK flip-flops, 652–655
    One-hot design, 645–647

Parallel-to-serial adder/subtractor system
    Controller, 645–648, 652–655
    Design, 651–655
    Functional partition, 651–652
    Timing diagram, 653
Parity bit, 273
Parity generators and detectors
    Design, 273–274
    Use of in error checking systems, 274–275
Partitioning method for state code assignments, 721–723
    Procedure, 721–722
    π-partitions, 721–722
    Seed sets, 722–723
    τ-partitions, 722
Part numbering systems
    CMOS and TTL logic families, 241
    ECL logic family, 241
Parasitic capacitance effects, 517
Passive switching devices, 84, 278
Pass transistor switches, 84–85 (*see also* transmission gates)
Pausable clock systems
    Externally asynchronous/internally clocked (EAIC) systems, 806–823
PDP, 240 (*see also* Power-delay product)
Performance characteristics of IC logic families
    Table, qualitative assessments, 852
Performance characteristics of switching devices, 238–241
    Cost, 241
    Fan-in and fan-out, 240–241
    Noise margins, 240
    Packing density, 239
    Power-delay product, 240
    Power dissipation, 239–240
    Propagation delay (switching speed), 239
Phase-locked loops, 521
PLAs. *See* Programmable logic arrays
PLDs. *See* Programmable logic devices
PMOS
    Ideal equivalent circuit, 82
    Simplified circuit symbol, 82
Polarized mnemonics, 80, 81
Polynomial representations
    Binary numbers, 33
    Number of radix r, 32
POS hazards (*see also* static hazards)
    In asynchronous FSMs, 710–711
    In combinational logic circuits, 394–396, 398
    In synchronous FSMs, 501, 506–507
POS representation, 134
    Canonical form, 135–137
    Expansion of reduced forms, 135–136
    Use of maxterm code, 134–137
Positional representation of a number, 32

Positional weight, 33, 35
Power-delay product (PDP), 240
Power dissipation
    Dynamic, 240
    Static (quiescent), 815, 849
Powers of 2, table, 76
Present state/next state (PS/NS) table
    Use of in analysis of FSMs, 476, 741–742
Preset asynchronous overrides
    In flip-flops, 463–464
Prime implicants, 148–150
    Essential, 149
    Optional, 149
    Redundant, 149
Priority encoders
    Cascadable, 254–256
    Collapsed truth tables, 255
    Logic circuits, 255–256
    Noncascadable, 256
Product-of-sums (POS) representation, 134
    Canonical forms, 135–137
    Use of maxterm code, 134–137
Programmable array logic (PAL) devices, 307–310
    Applications, 617–619
    Basic I/O type, 309
    L-type, 309
    Mixed logic inputs and outputs, 310–311
    R-type, 309
    Symbolic representation, 308
    V-type, 310
Programmable logic arrays (PLAs)
    Applications, 302–306
    Architecture, 301, 303
    Dimensions, 301
    FPLAs, 302
    Fusible links, 304
    Mixed logic inputs and outputs, 310–311
    NMOS connections (switches), 302–303
    Programming, 302–304
    P-term tables, 304–305
    Symbolic representation, 306
    Types, 302
Programmable logic devices (PLDs)
    FPGAs, 319–328
    EPLDs, 328
    FPLSs, 328
    GALs, 328
    PALs, 307–310
    PLAs, 302–306
    ROMs, 295–301
Programmable read-only memories (PROMs)
    Application, 299–301
    Bipolar, 298
    Dimensions, 296
    Fusible links, 297

INDEX

MOS architecture, 297
Symbolic representation, 300
Programmable sequencers, 823–842
    Microprogrammable asynchronous controller (MAC) modules, 823–835
    One-hot, 835–839
    Perspectives, 834–835, 839–842
PROMs. *See* Programmable read-only memories
Propagation delay, 239
    Levels of, 197–198
Pulse mode FSMs
    Analysis, 788–794
    Candy bar vending machine system, 782–788
    Characteristics, 777–778
    Design examples, 779–788
    Digital combinational lock, 781–783
    Feedback delays for nested cell designs, 793–794
    Input requirements, 774–775
    Memory elements, 774–777
    Models, 774, 776
    Nested cell approach, 776, 791–794
    Perspective, 795–796
    Pulse narrowing circuits for nested cell designs, 793
    Security area controller, 800–801
    Sequence recognizer, 779–781
    Simulations, 780, 788, 791, 793
Pulse narrowing circuits, 444
Pulse synchronizer module, 743–747
Pulse width adjuster, 547–549

Q

Quad, groupings of logic adjacencies, 145
Quadratic convergence
    Algorithm, 62
    In fast binary division, 59–62
Quiescent power dissipation, 815, 849
Quine-McCluskey algorithm, 169–173
    Applications, 170–172
    Notation, 170

R

Race conditions
    In critical race analysis, 703–704
    In E-hazard analysis, 711–713
    In ORG analysis, 492–494
    Race gate, 711
Radix complement representation
    Algorithms, 46
    Radix, r, 45
    Table for 2's complement, 47
    Ten's complement, 45–47
    Two's complement, 46–47
Radix divide method, 38
Radix multiply method, 41
Reading mixed-logic circuits, 97–98

Read only memories (ROMs), 295–301
    Applications, 299–301, 618–626
    Architecture, 296, 297
    Dimensions, 296
    EEPROMs, 298
    EPROMs, 298–299
    Fusible links, 297
    Mixed logic inputs and outputs, 310, 311
    NMOS connections, 297
    Programming, 297–299
    PROMs, 297
    Symbolic representation, 300
    UVEPROMs, 298
Redundant cover (*see also* static hazard cover)
    As used to eliminate s-hazards, 392
Redundant prime implicant, 149
Reed-Muller coefficients, 207
Reed-Muller transformations, 207–209
    Minimum function extraction, 209–217
    POS-to-EQPOS, 208–209
    SOP-to-EXSOP, 207–208
Reflective codes
    Gray, 70, 71
    XS3 gray, 71
Registered PLDs
    FPGAs, 321–328
    General purpose, 328
    R-type PALs, 307, 309
    V-type PALs, 307, 309–310
Registers, 561–572 (*see also* Shift registers)
    Shift, 562–572
    Storage, 561–563
Rendezvous modules (RMODs)
    As memory elements in asynchronous FSM design, 771
    In bus arbiters, 802–803
    Logic circuits, 696, 697
    Logic circuit symbol, 794
    LPD design, 759–760
    Nested cell design, 695–698
    Timing diagram, 697
    Use in delay circuit design, 794–795
Reset-dominant basic cell
    EV K-maps for, 431, 433
    Excitation table, 432, 433
    Logic circuit, 431, 435, 436, 695
    Mixed-rail output response, 436
    Mixed-rail outputs, 435
    Next state function, 432, 694
    Operation table, 431
    State diagram, 432, 695
    Timing diagrams, 432, 436, 695
Residue
    In finding static hazard cover, 392
Restoring (active) switching devices, 84
Re-substitution method, 176–177

880　　INDEX

RET, 437–438 (*see also* Rising edge triggering)
RET D flip-flop, 444–448, 698–700
RET JK flip-flop, 454–456, 700–701
Reverse bias
　In diodes of inertial delay elements, 794–795
　In diodes of sanity circuits, 524
Ring counters, 590–593
Ripple counters, 600–605
　Bi-directional, 604–605
　Decade, 603–604
　Design, 600–605
　Choice of memory elements, 600, 601–602
　Logic circuits, 601, 604
　Propagation delay, 602
　State diagram, 603
　Timing diagram, 601
Rising edge triggered D flip-flop
　Design, 444–448, 698–700
　Logic circuits, 448, 700
　Logic circuit symbol, 448
　Next state and output functions, 447, 698
　NS and output K-maps, 446, 699
　State diagrams, 445, 446, 698, 699
　Timing diagram, 449
Rising edge triggering (RET), 437–438
　Flip-flop, 438
　Latches, 438
ROMs. *See* Read-only memories
Rotation direction detector, 469–470
Round-off error, 42–43
Runt pulse, 465, 515–516

S
Sampling interval, 465
Sampling variable, 465, 510
Sanity circuits, 523–526
Schematic capture, simulation and timing analysis tools
　Beige Bag, 856
　Cadence, 856
　Cedes, 856
　EXL-Sim2000, 855
　Logic works, 856
　Mentor Graphics, 856
　ModelSim, 856
　OrCad, 856
　Timing Designer, 856–857
　Workview Office, 856
Schmitt triggers
　CMOS circuit, 525–526
　In debouncing circuits, 527
　In inertial delay elements, 793–794
　In sanity circuits, 523
　In synchronizing circuits, 516
　Logic symbols, 526
　V-t characteristics, 525
Security area controller, 800–801

Seed sets, 722
Selector module, 769
Self-complementing codes, 69
Self-correcting counters, 591–592
Sequence of states, 420, 422
Sequence recognizers, 471–475, 488, 674–675, 675–676, 791–793
Sequential machines, overview, 684
Serial one-bit adder, 554–555
Serial 2's complementer, 488, 643
　One-hot design, 643–645
Serial BCD-to-XS3 converter, 555–556
Serial odd parity detector, 556
Set, 430
Set-dominant basic cell
　EV K-maps, 429, 693
　Excitation table, 430, 433
　Logic circuit, 429, 434, 436, 693
　Mixed-rail output response, 435–436
　Mixed-rail outputs, 434
　Next state function, 430, 693
　Operation table, 429
　State diagram, 430, 693
　Timing diagrams, 430, 436, 694
Setup and hold time requirements of flip-flops, 465–466
　Hold time, 465
　Improper sampling, 466
　Proper sampling, 466
　Sampling interval, 465, 466
　Setup time, 465
Seven-segment display counter, 671
Shannon's expansion theorem
　Single variable decomposition, 177–178
　Multiple variable decomposition, 178
Shift-left/add 3 algorithm, 261
Shift register counters, 590–600
　Autonomous linear feedback shift register (ALFSR), 594–600
　Linear feedback shift register (LFSR), 594
　Ring, 590–593
　Twisted ring (Johnson), 593–595
Shift registers, 562–572
　Asynchronous parallel load, 568–571
　Branching action of a universal shift register (USR), 570–571
　Modes of operation, 561
　Operation tables, 564, 566, 569
　Right shift with synchronous parallel load, 562–565
　State diagrams, 564, 566, 569
　Synchronous parallel load, 562–568
　Universal with asynchronous parallel load, 568–571
　Universal with synchronous parallel load, 565–568
Shift-right/subtract 3 algorithm, 260

INDEX

Shifters, combinational, 275–278
    Barrel, 275
    Examples of operations, 276
    General n-bit, 275
    MUX design, 276–278
Sign bit, 44, 50, 53, 54, 66, 343–344
Sign-bit error detection circuits, 343–345
Sign-complement arithmetic, 53–55, 56–58, 63–64
Signed binary numbers, 43–48
Signed-magnitude representation, 44–45
Simulators, logic
    Beige Bag, 856
    Cedes, 856
    EXL-Sim2000, 855
    Logicworks, 856
    ModelSim, 856
    Workview Office, 856
Single transition time (STT) machines, 720–730
    Analysis, 747–752
    Array algebraic approach, 724–727
    Design example, 722–730
    Hazard-free design, 730–734
    Partitioning methods for state code assignment, 721–724
SOP hazards (*see also* Static hazards)
    In asynchronous FSMs, 707–709, 727–729, 731–732, 750–751
    In combinational logic circuits, 392–394, 397–398, 400–401
    In synchronous FSMs, 500–502, 505–507
SOP representation, 131–133
    Canonical forms, 132–133, 136–137
    Expansion of reduced forms, 133
    Use of minterm code, 132–137
SPDT switch, 528–529
    Use of basic cells, 528–529
Special-purpose flip-flops, 459–460
Spill bit, 276
SPST switch, 526–528
    Use of R-C circuits, 527
Stability criteria
    For fundamental mode FSMs, 688
State adjacency sets, 543–544, 546, 724–725
State code assignment rules and application
    For one-hot approach to design, 636, 735
    For STT FSM design, partitioning methods, 721–722
    For synchronous FSMs, 539–541
State matrix, 542, 543, 723–724
State minimization, 547–549
    Merging of states, 547–548
State tables
    Use in obtaining state code assignments in STT FSM design, 721–724
    Use with state assignment rules, 540–541
    Vs state diagrams, 539
State variable, 421

Static hazard cover, 392
Static hazards in asynchronous FSMs
    Detection and elimination in NS logic, 705–711
Static hazards in two-level combinational logic circuits, 392–398
    Procedure for elimination, 392, 397
    Static 1-hazards, 391
    Static 0-hazards, 391
    Terms associated with, 392
    Timing diagrams, 393–396
    Use of K-maps, 393–397
Static hazards in multilevel XOR-type functions
    Alternative path configuration, 399
    Detection and elimination procedure in complex functions, 408–409
    Dynamic hazards, 409–411
    Timing diagrams, 402, 403, 405, 407
    Use of binary decision diagrams (BDDs), 405, 408, 411
    Use of lumped path delay diagrams (LPDDs), 404, 406, 410
    XOP and EOS functions, 400–403
Static hazards in synchronous FSMs
    Detection and elimination, 499–510
    Examples, 500–502, 505–509
    Externally initiated, 500–502
    Internally initiated, 502–505
    Prospective, 509–510
Static logic DFLOP design
    In EAIC systems, 807–812
Static (quiescent) power dissipation, 815
Steering logic, 84
    Buffering rule, 279
    Decoder design, 279–280
    MUX design, 278–279
Stepping motor control system, 655–665
    ASM chart for controller, 659
    Functional partition, 658
    Logic circuit for controller, 665
    NS and output K-maps and minimum cover, 662
    Operational characteristics, 656
    P-term table for PLA implementation, 663
    State diagram, 660
    Timing requirements, 657
Stretching and synchronizing the input, 512–514
Stuck-at-faults, 412–413
    Effect of hazard cover, 413
Subfunction partitioning
    Applied to the CRMT approach, 225–228
Subtraction
    Algorithms, 54, 63–64
    BCD, 63–64
    Binary, 52–55
    Direct, 53
    FPN, 64–65
    Hexadecimal, 75

Subtraction (*cont.*)
   One's complement, 54–55
   Two's complement, 53–54
Subtractors
   Full, 340–342
   Ripple-borrow, 341
   Use in comparators, 388
   Use in dividers, 355–356
Sum-of-products (SOP) representation, 131–133
   Canonical forms, 132–133
   Use of minterm code, 132–133
Sum rule
   Defined, 426
   Exceptions, 428
Switch bounce periods, 526, 527
Switch debouncing circuits
   Rotary selector switch, 529–530
   Single-pole/double throw (SPDT) switch, 528–529
   Single-pole/single throw (SPST) switch, 526–528
Switching speed, 852
Switching threshold, 465, 516
Synchronizer circuits
   Mean time between failures (MTBF), 516–517
   Metastability, 514–517
   Multiple stage, 517
   Single stage, 512
   Two stage with counters, 515
Use of Schmitt triggers, 517
Synchronous binary counters, 572–589
   Branching action of a parallel loadable up/down counter, 589
   Cascadable BCD up counters, 575–579
   Cascadable up/down with asynchronous parallel load, 579–581
   Cascadable up/down with synchronous parallel load and true hold, 581–584
   Divide-by-N, 573–575
   One-bit modular up/down design with parallel load and true hold, 584–588
   Perspective on parallel loading, 588–589
   Timing diagrams, 578
   Types, 572–573
Synchronous FSM architectures, 426–424
Synchronous state machine analysis
   Analysis procedure, 476
   Output race glitches (ORGs), 491–499
   Static hazards, 499–510
   Use of the PS/NS table, 476, 476–479
Synchronous state machine design
   Array algebraic method, 542–547
   Choice of components, 613–614
   Counters, 572–589
   Design considerations, 491–530, 536–549
   Design procedure, 440, 530–532
   Mapping algorithm, 440

   Memory elements, 438
   One-hot method, 636–649
   Shift register counters, 590–600
   Shift registers, 562–572
   Simple state machines, examples, 466–476, 532–536
Synchronous state machines, general
   Analysis procedure, 476
   Design procedure, 440, 530–532
   Fully documented state diagram, 424–428
   Models, 423–424
   Mutually exclusion requirement, 426–428, 490
   Sequence of states, 420–421
   Sum rule, 426, 490
Synchronous Vs asynchronous parallel loading, a perspective
   In counters, 588–589
   In shift registers, 588–589
System clock, 420, 657
System level design
   Architecture, 650
   Candy bar vending machine system, 782–788
   Controllers and data paths, 649–650, 652, 658, 784
   Dealing with unusually large controllers, 666–668
   Functional partitions, 652, 658, 784
   Parallel-to-serial adder/subtractor control system, 651–655
   Stepping motor control system, 655–665
   Use of ASM charts, 659

T

Tabular minimization, 169–172
Ten's complement arithmetic
   Algorithm, 63–64
   BCD subtraction, 63–64
T flip-flops
   Analysis of FSMs with T flip-flops, 488, 489, 490
   Conversion from D flip-flops, 452–453
   Conversion from JK flip-flops, 455–457
   Design of FSMs with T flip-flops, 466–470, 474–475, 579–581
   Excitation table, 452
   Master/slave, 453
   Operation table, 452
   PR and CL overrides, 464
   State diagram, 452
   Timing diagram, 453
Time constant, RC, 524, 527
Timing defect analyses in fundamental mode FSMs
   Critical races, 703–705
   Endless cycles, 702–703
   Essential hazards, 711–718, 746, 750–752, 756–757
   Static hazards, 705–711, 750–751

INDEX

Timing defects in combinational logic
   Dynamic hazards, 392, 409–411
   Functional hazards, 392, 412
   Static hazards in two-level logic circuits, 392–398
   Static hazards in multilevel XOR-type circuits, 399–409
Timing Designer Professional, 586–587
Timing diagrams (examples), 410, 443, 448, 464, 548, 657, 694, 718, 751, 780, 793, 840
Timing problems
   In latches, 461–462
   In master-slave JK flip-flops, 462–463
Toggle modules
   Design from D flip-flops, 573
   Use in counter design, 600–605, 664–665
   Use in pulse mode FSMs, 773–775
Traffic light control system, 681–682
Transistor-transistor logic (TTL), 850–852
Transmission gates
   CMOS, 84–85
   Circuit symbols, 84–85
   Ideal equivalent circuits, 84–85
   NMOS, 84–85
   PMOS, 84–85
Transparent D latch. *See* D-latch
Tree structures, 90, 180
Triggering mechanisms
   Falling edge triggering (FET), 437–438
   Master-slave, 438
   Rising edge triggering (RET), 437–438
Triggering threshold. *See* switching threshold
Tri-state bus, 313
Tri-state drivers, 85–87
   CMOS logic circuits, 86
   Idealized equivalent circuits, 86
   Logic circuit symbols, 86
   Use in augmentation schemes for PLDs, 312
   Use in PLD implementation, 297, 300, 303, 306, 307, 309–310
   Use in steering logic designs, 278–280
True hold
   In counters, 582, 584, 588
   In shift registers, 561, 566, 569
Truth tables (non-conventional)
   Compressed (collapsed), 249, 255, 256, 264, 280
   Compressed entered variable, 244, 278
   Entered variable (EV), 246, 268, 269, 277, 358, 361, 372, 374
TTL logic family, 850–852
   Gate examples, 851
Twisted ring counters
   Universal shift register design, 593–594
   Self correcting, 593, 595
Two-phase clocking, 450
Two's complement arithmetic
   Algorithms, 54, 57–58

   Multiplication, 56–58
   Subtraction, 53–54
Two's complement representation, 45–47
   Algorithm, 46
   Negation, 47
   Table, 47
   True value, 47

U

Unary operator, 106
Unclocked memory elements
   Pulse mode approach to FSM design, 773, 774–777
Unconditional branching, 496, 497
Unconditional output, 425
Union, 87 (*see also* OR)
Unit distant codes
   Decimal code, 71
   Gray code, 71
   Table, 71
   XS3 Gray code, 71
Universal flip-flops
   JK flip-flops, 438
Universal shift register (USR), 562
   Asynchronous parallel loaded, 568–570
   Branching actions, 570–571
   Cascaded, 568
   Circuit symbols, 567, 571
   Logic circuits, 567, 568, 571
   NS functions, 566, 569
   Operation tables, 566, 569
   State diagrams, 566, 569
   Synchronous parallel loaded, 565–568
Unsigned binary coded decimal (BCD), 34–35
Unsigned binary coded hexadecimal (BCH), 36–37
Unsigned binary coded octal (BCO), 36–37
Unsigned binary number systems, 33–34
Unstable state, 688, 702
Unusually large controllers, 666–668
Unweighted codes
   Creeping, 68, 69
Up/down counters
   Cascadable with asynchronous parallel load, 579–581, 587–588
   Cascadable with synchronous parallel loading, 581–584, 584–587
   Data triggered with parallel load, 664–665
   With true hold, 581–588
USR. *See* Universal shift register
USR/Up-Down counter combination, 609–610
UVEPROMs, 298

V

Very-large-scale integrated circuits (VLSI), 238
VHDL description of combinational primitives
   Cascadable bit-comparator, 283–285

VHDL description of combinational (*cont.*)
   Four-to-one MUX, 282
   Full adder, 381–382, 383
   NOR gate, 281
VHDL description of sequential machines
   Simple FSM, 481–482
VHDL, an introduction to, 279–287
   Behavior level, 280
   Dataflow level, 280
   IEEE standard package *std-logic-1164*, 286
   Key words, 281
   Logic data types, 287
   Operator overloading, 285
   Structural level, 280
   VHDL operator list, 286
VHSIC, 279
Viewlogic, 856
Voltage waveform, 80, 522, 526
Voter booth tabulation system, 680–681

W

Waveform analysis
   Combinational logic circuits, 97, 105
Weighted codes, 68, 69
Wired OR technology
   Multiple PLD schemes, 312, 313, 316
WorkView Office
   Professional version, 856
   Student version, 856
WSI circuits, 238

X

Xilinx FPGAs, 321–329
   Configurable logic blocks (CLBs), 321, 322, 325, 327
   I/O blocks (IOBs), 321, 322, 324, 326
   Range of Xilinx FPGAs, 324
   XACT software, 329
XNOR, 99 (*see also* EQV)
XOR algebra
   Absorptive laws, 112
   Associative laws, 111
   Commutative laws, 111
   Consensus laws, 112
   Corollaries, 114–115
   DeMorgan's laws, 112
   Distributive laws, 112
   EQV laws, 111
   Summary of useful identities, 115
   Worked examples, 119–120
   XOR laws, 111
XOR function
   Defining relations, 101
   Definition, 99
   Logic circuit symbols, 100, 103
   Multiple gate realizations, 101–102
   Operator symbol, 98
XOR gate
   CMOS, 100, 117
   Conjugate gate symbols, 100, 103
   Effect of active low inputs, 102
   In controlled logic level conversion, 103–104
   Mixed-logic interpretations, 100
   Physical truth table, 100
   Tree forms for multiple inputs, 99
XOR patterns, 197–207
   Adjacent, 198–199
   Associative, 198–199
   Compound, 198–199
   Diagonal, 200
   Extraction procedure, 200
   K-map plotting, 205–207
   Offset, 198–199
   Vs two-level logic minimization, table, 204

Y

Year 200 and beyond, 3–4

Z

Zero blanking
   BCD-to-seven-segment displays, 265, 267